Lecture Notes in Computer Science 14062

The series Lecture Notes in Computer Science (LNCS), including its subseries Lecture Notes in Artificial Intelligence (LNAI) and Lecture Notes in Bioinformatics (LNBI), has established itself as a medium for the publication of new developments in computer science and information technology research, teaching, and education.

LNCS enjoys close cooperation with the computer science R & D community, the series counts many renowned academics among its volume editors and paper authors, and collaborates with prestigious societies. Its mission is to serve this international community by providing an invaluable service, mainly focused on the publication of conference and workshop proceedings and postproceedings. LNCS commenced publication in 1973.

Antonio Pertusa · Antonio Javier Gallego ·
Joan Andreu Sánchez · Inês Domingues
Editors

Pattern Recognition and Image Analysis

11th Iberian Conference, IbPRIA 2023
Alicante, Spain, June 27–30, 2023
Proceedings

Editors
Antonio Pertusa
University of Alicante
Alicante, Spain

Antonio Javier Gallego
University of Alicante
Alicante, Spain

Joan Andreu Sánchez
Universitat Politècnica de València
Valencia, Spain

Inês Domingues
IPO Porto
Coimbra, Portugal

ISSN 0302-9743 ISSN 1611-3349 (electronic)
Lecture Notes in Computer Science
ISBN 978-3-031-36615-4 ISBN 978-3-031-36616-1 (eBook)
https://doi.org/10.1007/978-3-031-36616-1

This Springer imprint is published by the registered company Springer Nature Switzerland AG
The registered company address is: Gewerbestrasse 11, 6330 Cham, Switzerland

Preface

We are pleased to present the proceedings of the 11th Iberian Conference on Pattern Recognition and Image Analysis, IbPRIA 2023, a biennial conference organized by the national IAPR associations for pattern recognition in Spain (AERFAI) and Portugal (APRP) since 2003.

This year's IbPRIA was held during June 27–30, 2023, in Alicante, Spain, and was hosted by the University of Alicante with the support of the University Institute for Computing Research (IUII).

After 20 years, in its 11th edition, IbPRIA has become a key research event in pattern recognition and image analysis on the Iberian Peninsula. Therefore, most of the research in this edition comes from Spanish and Portuguese authors. Of the 198 authors whose works were accepted, 57% are from Spain, and 15% are from Portugal. The rest are from another 16 countries: Belgium, Sweden, the USA, Norway, Mexico, Ukraine, France, Germany, New Zealand, Czech Republic, Italy, India, Bangladesh, Cuba, Switzerland, and Estonia.

IbPRIA 2023 received 86 submissions. The review process was diligent and required careful consideration by more than 100 reviewers who spent significant time and effort reviewing the papers, with an average of 2.9 single-blind reviews per paper and an average number of papers per reviewer of 2.2. In the end, 56 papers were accepted, an acceptance rate of 65%, lowering the rate from previous editions. For the final program, 26 papers were selected for oral presentations (30% acceptance rate) and 30 as poster presentations.

We hope these proceedings will be a valuable resource for the pattern recognition and machine learning research community, and we want to thank all who made this possible, including authors, reviewers, chairs, and members of the conference committees. Special thanks to the invited speakers, Timothy Hospedales, Nuria Oliver, and Gustau Camps-Valls, and tutorial presenters Sergio Orts-Escolano, Mikel Artetxe, and Karteek Alahari. And, of course, a final word to the outstanding local committee members.

Thanks!

May 2023

Antonio Pertusa
Antonio Javier Gallego
Joan Andreu Sánchez
Inês Domingues

Organization

General Co-chair AERFAI

Joan Andreu Sánchez	Polytechnic University of Valencia, Spain

General Co-chair APRP

Inês Domingues	Instituto Superior de Engenharia de Coimbra, Portugal

Local Chair

Antonio Pertusa	University of Alicante, Spain

Program Chairs

Antonio Javier Gallego	University of Alicante, Spain
Raquel Justo	Universidad del País Vasco, Spain
Manuel J. Marín	Universidad de Córdoba, Spain
Hélder Oliveira	INESC TEC, University of Porto, Portugal

Tutorial Chairs

Jose Javier Valero	University of Alicante, Spain
Verónica Vasconcelos	Instituto Superior de Engenharia de Coimbra, Portugal

Local Committee

María Alfaro	University of Alicante, Spain
Francisco Castellanos	University of Alicante, Spain
Luisa Micó	University of Alicante, Spain
Antonio Ríos	University of Alicante, Spain
Marcelo Saval	University of Alicante, Spain

Program Committee

Daniel Acevedo	Universidad de Buenos Aires, Argentina
Roberto Alejo	Tecnológico Nacional de México, Campus Toluca, México
Luís A. Alexandre	UBI and NOVA LINCS, Portugal
Francisco Antunes	University of Coimbra, Portugal
Antonio Bandera	University of Malaga, Spain
Antonio Jesús Banegas Luna	Universidad Católica de Murcia (UCAM), Spain
Zuria Bauer	ETH Zürich, Switzerland
Jose Miguel Benedi	Universitat Politècnica de València, Spain
Emmanouil Benetos	Queen Mary University of London, UK
Larbi Boubchir	University of Paris 8, France
Susana Brás	Universidade de Aveiro, Portugal
Jorge Calvo-Zaragoza	University of Alicante, Spain
Pedro J. S. Cardoso	Universidade do Algarve, Portugal
Jaime Cardoso	University of Porto, Portugal
Gonçalo Carnaz	Politécnico de Coimbra, ISEC, DEIS, Portugal
Jesus Ariel Carrasco-Ochoa	Instituto Nacional de Astrofísica, Óptica y Electrónica, INAOE, México
Francisco Casacuberta	Universitat Politècnica de València, Spain
Francisco J. Castellanos	University of Alicante, Spain
Francisco Manuel Castro	University of Malaga, Spain
Miguel Coimbra	Universidade do Porto, Portugal
Diego Sebastián Comas	Universidad Nacional de Mar del Plata, Argentina
Paulo Correia	Universidade de Lisboa, Portugal
Paulo Costa	Politécnico de Lisboa, Portugal
Hanz Cuevas-Velasquez	University of Edinburgh, UK
António Cunha	University of Trás-os-Montes and Alto Douro, Portugal
Manuel Curado Navarro	Universidad Católica de Murcia (UCAM), Spain
Inês Domingues	Politécnico de Coimbra, ISEC, DEIS, Portugal
Hugo Jair Escalante	Instituto Nacional de Astrofísica, Óptica y Electrónica, INAOE, México
Félix Escalona	Universidad de Alicante, Spain
Jacques Facon	Universidade Federal do Epirito Santo, Sao Mateus, Brazil
Francesc J. Ferri	University of Valencia, Spain
Vítor Filipe	University of Trás-os-Montes and Alto Douro, Portugal
Alicia Fornés	Universitat Autònoma de Barcelona, Spain
Giorgio Fumera	University of Cagliari, Italy

Vicente Garcia	Universidad Autónoma de Ciudad Juárez, Mexico
Petia Georgieva	University of Aveiro, Portugal
Francisco Gomez-Donoso	University of Alicante, Spain
Teresa Goncalves	University of Evora, Portugal
Sónia Gouveia	University of Aveiro, Portugal
Miguel Angel Guevara Lopez	Polytechnic Institute of Setubal, Portugal
Jose M. Iñesta	Universidad de Alicante, Spain
Alfons Juan	Universitat Politècnica de València, Spain
Martin Kampel	Vienna University of Technology, Austria
Vitaly Kober	Centro de Investigación Científica y de Educación Superior de Ensenada, CICESE, México
Pedro Latorre Carmona	Universidad de Burgos, Spain
Marcos A. Levano	Universidad Católica de Temuco, Chile
Fernando Lopes	ISEC, Portugal
Damián López	Universitat Politècnica de València, Spain
Juan Valentín Lorenzo-Ginori	Universidad Central "Marta Abreu" de Las Villas, Cuba
Miguel Angel Lozano	University of Alicante, Spain
F.J. Madrid-Cuevas	Cordoba University, Spain
Andre R. S. Marcal	University of Porto, Portugal
Carlos David Martinez Hinarejos	Universitat Politècnica de València, Spain
Rafael Medina-Carnicer	Cordoba University, Spain
Ana Mendonça	University of Porto, Portugal
Ramón A. Mollineda Cárdenas	University Jaume I, Spain
Fernando Monteiro	Polytechnic Institute of Bragança, Portugal
Henning Müller	University of Applied Sciences and Arts of Western Switzerland, Switzerland
António J. R. Neves	University of Aveiro, Portugal
João Carlos Neves	NOVA-LINCS, Universidade da Beira Interior, Portugal
Lawrence O'Gorman	Nokia Bell Labs, USA
Joel P. Arrais	University of Coimbra, Portugal
Kalman Palagyi	University of Szeged, Hungary
Cristina Palmero	Universitat de Barcelona, Spain
Roberto Paredes	Universidad Politécnica de Valencia, Spain
Billy Peralta	Universidad Andrés Bello, Chile
Carlos Pereira	ISEC, Portugal
Alicia Pérez	University of the Basque Country, Spain
Petra Perner	Institute of Computer Vision and Applied Computer Sciences, Germany
Armando Pinho	University of Aveiro, Portugal
Filiberto Pla	University Jaume I, Spain

Nicolaie Popescu-Bodorin	Lumina University of South-East Europe, Romania
Diogo Pratas	University of Aveiro, Portugal
Luis Rato	University of Évora, Portugal
Pedro Real Jurado	Seville University, Spain
Arsénio Reis	University of Trás-os-Montes and Alto Douro, Portugal
Francesco Renna	Universidade do Porto, Portugal
Bernardete Ribeiro	University of Coimbra, Portugal
João M.F. Rodrigues	Universidade do Algarve, Portugal
Luis Javier Rodriguez-Fuentes	University of the Basque Country, Spain
Antonio Rodriguez-Sanchez	University of Innsbruck, Austria
Francisco J Romero-Ramirez	University of Cordoba, Spain
Rubén San Segundo	Universidad Politécnica de Madrid, Spain
João Sanches	Universidade de Lisboa, Portugal
Joan Andreu Sanchez	Universitat Politècnica de València, Spain
Jose Salvador Sanchez	Universitat Jaume I, Spain
Antonio-José Sánchez-Salmerón	Universitat Politècnica de València, Spain
Alberto Sanchis	Universitat Politècnica de València, Spain
Emilio Sanchis	Universitat Politècnica de València, Spain
Carlo Sansone	University of Naples Federico II, Italy
Marcelo Saval-Calvo	University of Alicante, Spain
Raquel Sebastião	University of Aveiro, Portugal
Ana F. Sequeira	InescTEC, Portugal
Samuel Silva	Universidade de Aveiro, Portugal
Catarina Silva	University of Coimbra, Portugal
Jose Silvestre Silva	Portuguese Military Academy, Portugal
Luis Teixeira	University of Porto, Portugal
César Teixeira	University of Coimbra, Portugal
Ruben Tolosana	Universidad Autónoma de Madrid, Spain
M. Inés Torres	Universidad del País Vasco, Spain
Alejandro Toselli	Universitat Politècnica de València, Spain
V. Javier Traver	Universitat Jaume I, Spain
Jose J. Valero-Mas	University of Alicante, Spain
Verónica Vasconcelos	ISEC-IPC, Portugal
Enrique Vidal	Universitat Politècnica de València, Spain
Jordi Vitria	Autonomous University of Barcelona, Spain

Additional Reviewers

Sophie Noiret
Jose Vicent
Ricardo Veiga
Julian Stroymayer
Nicolás Cubero
Pau Torras
Irene Ballester
Paula Ruiz-Barroso

Sponsoring Institutions

AERFAI - Asociación Española de Reconocimiento de Formas y Análisis de Imágenes
APRP - Associação Portuguesa de Reconhecimento de Padrões
UA - University of Alicante
DLSI - Department of Software and Computing Systems
IUII - University Institute for Computing Research
PRAIG - Pattern Recognition and Artificial Intelligence Group
GVA - Conselleria de Innovación, Universidades, Ciencia y Sociedad Digital, Generalitat Valenciana

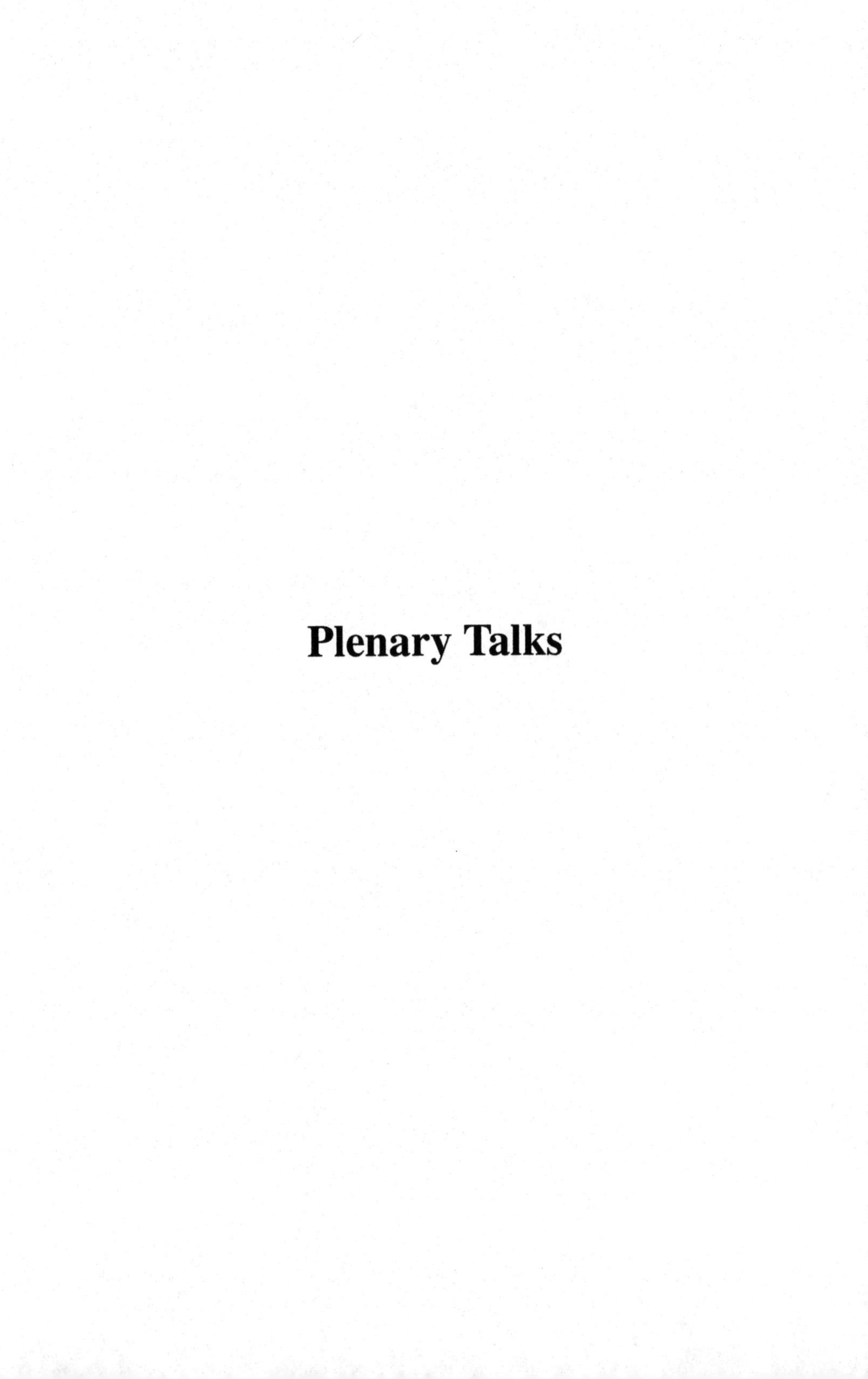

Plenary Talks

Distribution Shift: The Key Bottleneck for Pattern Recognition in Practice?

Timothy Hospedales

School of Informatics, University of Edinburgh, UK

Abstract. AI has made rapid progress as measured by benchmark dataset performance. In this talk, I will argue that distribution shift between training and deployment is very difficult to avoid in practice, and that its detrimental impact is often an underlying cause of practical system failures. As such, distribution shift is one of the key reasons why real-world pattern recognition impact has lagged benchmark performance. I will introduce a variety of paradigms and tools that the community is developing to tackle this challenge and discuss where they are having success, where they are failing, where they are insufficiently imaginative, and what we can do better as a community to facilitate progress towards more robust AI.

Data Science Against COVID-19

Nuria Oliver

ELLIS Alicante, Spain

Abstract. Data Science against COVID-19 is the work of a multidisciplinary team of 20+ volunteer scientists between March of 2020 and April of 2022, working very closely with the Presidency of the Valencian Government to support their decision-making during the COVID-19 pandemic in Spain. In my talk, I will describe our work and share the lessons learned in this very special initiative of collaboration between the civil society at large (through a citizen survey), the scientific community (through this taskforce) and a public administration (through our collaboration with the Presidency of the Valencian Government).

AI for Sustainable Earth Sciences

Gustau Camps-Valls

Image Processing Laboratory (IPL), University of València, Spain

Abstract. AI has tremendous potential to achieve sustainable development goals in the Earth sciences. In this talk, I will discuss how AI can improve environmental data collection and analysis, reduce waste and pollution, improve weather and climate forecasting, and develop more efficient renewable energy sources. Additionally, I will show how AI can positively impact climate science and natural resource management, from monitoring crops, forests, and oceans to detecting, characterizing, and interpreting extreme events, such as droughts and heatwaves, in massive Earth data. A full AI agenda will be introduced based on hybrid machine learning modeling, explainability, and causal inference. Finally, I will discuss the challenges and ethical considerations associated with AI for sustainability, where interdisciplinary education and diversity will be key. This talk will provide an overview of the potential of AI for achieving a more sustainable future and provide insight into the innovative possibilities of AI for sustainability in the Earth sciences.

Invited Tutorials

Machine Learning for Computational Photography

Sergio Orts

Google Research, San Francisco, CA, USA

Abstract. In this tutorial, we will explore the use of deep learning techniques in the field of computational photography. In recent years, we have seen how deep learning techniques have enabled the creation of high-quality pictures and videos captured using smartphones that look like they were taken with a professional DSLR camera. In particular, deep learning techniques have shown great potential to improve image quality and to perform post-capture image edits, e.g. creative retouching, improving low-light photography, creating depth of field effect, etc. Throughout the talk, we will present multiple research works and show how deep learning techniques can be applied to real-world computational photography applications: rendering a natural camera bokeh effect, relighting human portraits, realistic background replacement, etc. We will also discuss the challenges and limitations of using deep learning in this field, as well as future directions for research and development.

A Brief History of Unsupervised Machine Translation: from a Crazy idea to the Future of MT?

Mikel Artetxe

Reka AI Inc, USA

Abstract. Machine Translation (MT) has traditionally relied on millions of examples of existing translations. In 2011, Ravi and Knight attempted the impossible—training MT systems without parallel data—but their statistical decipherment approach was only shown to work in very limited settings. Barely a decade later, we have seen the first serious claims of state-of-the-art MT results without using any explicit parallel data. Interestingly, this progress has come from increasingly simpler ideas combined with scale, an illustrative example of the broader trend in AI. In this talk, I will present the journey that has led to this progress, and reflect on what it means to be a researcher in the era of large language models.

Continual Visual Learning: Where are we?

Karteek Alahari

Univ. Grenoble Alpes, Inria, CNRS, Grenoble INP, LJK,

Abstract. Several methods are being developed to tackle the problem of incremental learning in the context of deep learning-based models, i.e., adapting a model, originally trained on a set of classes, to additionally handle new classes, in the absence of training data of the original classes. They aim to mitigate "catastrophic forgetting"—an abrupt degradation of performance on the original set of classes, when the training objective is adapted to the new classes. In this tutorial, we plan to provide a comprehensive description of the main categories of incremental learning methods, e.g., based on distillation loss, growing the capacity of the network, introducing regularization constraints, or using autoencoders to capture knowledge from the initial training set, and analyze the state of affairs. We will then study the new challenges of learning incrementally in frameworks that are not fully supervised, such as semi-or self-supervised learning.

Contents

Machine Learning

CCLM: Class-Conditional Label Noise Modelling 3
Albert Tatjer, Bhalaji Nagarajan, Ricardo Marques, and Petia Radeva

Addressing Class Imbalance in Multilabel Prototype Generation for *k*-Nearest Neighbor Classification 15
Carlos Penarrubia, Jose J. Valero-Mas, Antonio Javier Gallego, and Jorge Calvo-Zaragoza

Time Series Imputation in Faulty Systems 28
Ana Almeida, Susana Brás, Susana Sargento, and Filipe Cabral Pinto

DARTS with Degeneracy Correction 40
Guillaume Lacharme, Hubert Cardot, Christophe Lenté, and Nicolas Monmarché

A Fuzzy Logic Inference System for Display Characterization 54
Khleef Almutairi, Samuel Morillas, Pedro Latorre-Carmona, and Makan Dansoko

Learning Semantic-Visual Embeddings with a Priority Queue 67
Rodrigo Valério and João Magalhães

Optimizing Object Detection Models via Active Learning 82
Dinis Costa, Catarina Silva, Joana Costa, and Bernardete Ribeiro

Continual Vocabularies to Tackle the Catastrophic Forgetting Problem in Machine Translation 94
Salvador Carrión and Francisco Casacuberta

Evaluating Domain Generalization in Kitchen Utensils Classification 108
Carlos Garrido-Munoz, María Alfaro-Contreras, and Jorge Calvo-Zaragoza

Document Analysis

Segmentation of Large Historical Manuscript Bundles into Multi-page Deeds 121
Jose Ramón Prieto, David Becerra, Alejandro Hector Toselli, Carlos Alonso, and Enrique Vidal

A Study of Augmentation Methods for Handwritten Stenography Recognition 134
Raphaela Heil and Eva Breznik

Lifelong Learning for Document Image Binarization: An Experimental Study 146
Pedro González-Barrachina, María Alfaro-Contreras, Mario Nieto-Hidalgo, and Jorge Calvo-Zaragoza

Test-Time Augmentation for Document Image Binarization 158
Adrian Rosello, Francisco J. Castellanos, Juan P. Martinez-Esteso, Antonio Javier Gallego, and Jorge Calvo-Zaragoza

A Weakly-Supervised Approach for Layout Analysis in Music Score Images 170
Eric Ayllon, Francisco J. Castellanos, and Jorge Calvo-Zaragoza

ResPho(SC)Net: A Zero-Shot Learning Framework for Norwegian Handwritten Word Image Recognition 182
Aniket Gurav, Joakim Jensen, Narayanan C. Krishnan, and Sukalpa Chanda

Computer Vision

DeepArUco: Marker Detection and Classification in Challenging Lighting Conditions 199
Rafael Berral-Soler, Rafael Muñoz-Salinas, Rafael Medina-Carnicer, and Manuel J. Marín-Jiménez

Automated Detection and Identification of Olive Fruit Fly Using YOLOv7 Algorithm 211
Margarida Victoriano, Lino Oliveira, and Hélder P. Oliveira

Learning to Search for and Detect Objects in Foveal Images Using Deep Learning 223
Beatriz Paula and Plinio Moreno

Relation Networks for Few-Shot Video Object Detection 238
Daniel Cores, Lorenzo Seidenari, Alberto Del Bimbo, Víctor M. Brea, and Manuel Mucientes

Optimal Wavelength Selection for Deep Learning from Hyperspectral Images 249
S. Dehaeck, R. Van Belleghem, N. Wouters, B. De Ketelaere, and W. Liao

Can Representation Learning for Multimodal Image Registration be Improved by Supervision of Intermediate Layers? 261
Elisabeth Wetzer, Joakim Lindblad, and Nataša Sladoje

Interpretability-Guided Human Feedback During Neural Network Training 276
Pedro Serrano e Silva, Ricardo Cruz, A. S. M. Shihavuddin, and Tiago Gonçalves

Calibration of Non-Central Conical Catadioptric Systems from Parallel Lines 288
James Bermudez-Vargas, Jesus Bermudez-Cameo, and Jose J. Guerrero

S^2-LOR: Supervised Stream Learning for Object Recognition 300
César D. Parga, Gabriel Vilariño, Xosé M. Pardo, and Carlos V. Regueiro

Evaluation of Regularization Techniques for Transformers-Based Models 312
Hugo S. Oliveira, Pedro P. Ribeiro, and Helder P. Oliveira

3D Computer Vision

Guided Depth Completion Using Active Infrared Images in Time of Flight Systems 323
Amina Achaibou, Nofre Sanmartín-Vich, Filiberto Pla, and Javier Calpe

StOCaMo: Online Calibration Monitoring for Stereo Cameras 336
Jaroslav Moravec and Radim Šára

Smart-Tree: Neural Medial Axis Approximation of Point Clouds for 3D Tree Skeletonization 351
Harry Dobbs, Oliver Batchelor, Richard Green, and James Atlas

A Measure of Tortuosity for 3D Curves: Identifying 3D Beating Patterns of Sperm Flagella 363
Andrés Bribiesca-Sánchez, Adolfo Guzmán, Alberto Darszon, Gabriel Corkidi, and Ernesto Bribiesca

The ETS2 Dataset, Synthetic Data from Video Games for Monocular Depth Estimation 375
David María-Arribas, Alfredo Cuesta-Infante, and Juan J. Pantrigo

Computer Vision Applications

Multimodal Human Pose Feature Fusion for Gait Recognition 389
Nicolás Cubero, Francisco M. Castro, Julián R. Cózar, Nicolás Guil, and Manuel J. Marín-Jiménez

Proxemics-Net: Automatic Proxemics Recognition in Images 402
Isabel Jiménez-Velasco, Rafael Muñoz-Salinas, and Manuel J. Marín-Jiménez

Lightweight Vision Transformers for Face Verification in the Wild 414
Daniel Parres and Roberto Paredes

Py4MER: A CTC-Based Mathematical Expression Recognition System 426
Dan Anitei, Joan Andreu Sánchez, and José Miguel Benedí

Hierarchical Line Extremity Segmentation U-Net for the SoccerNet 2022 Calibration Challenge - Pitch Localization 442
Miguel Santos Marques, Ricardo Gomes Faria, and José Henrique Brito

Object Localization with Multiplanar Fiducial Markers: Accurate Pose Estimation 454
Pablo García-Ruiz, Rafael Muñoz-Salinas, Rafael Medina-Carnicer, and Manuel J. Marín-Jiménez

Real-Time Unsupervised Object Localization on the Edge for Airport Video Surveillance 466
Paula Ruiz-Barroso, Francisco M. Castro, and Nicolás Guil

Identifying Thermokarst Lakes Using Discrete Wavelet Transform–Based Deep Learning Framework 479
Andrew Li, Jiahe Liu, Olivia Liu, and Xiaodi Wang

Object Detection for Rescue Operations by High-Altitude Infrared Thermal Imaging Collected by Unmanned Aerial Vehicles 490
Andrii Polukhin, Yuri Gordienko, Gert Jervan, and Sergii Stirenko

Medical Imaging and Applications

Inter vs. Intra Domain Study of COVID Chest X-Ray Classification with Imbalanced Datasets ... 507
Alejandro Galán-Cuenca, Miguel Mirón, Antonio Javier Gallego, Marcelo Saval-Calvo, and Antonio Pertusa

Automatic Eye-Tracking-Assisted Chest Radiography Pathology Screening ... 520
Rui Santos, João Pedrosa, Ana Maria Mendonça, and Aurélio Campilho

Deep Neural Networks to Distinguish Between Crohn's Disease and Ulcerative Colitis ... 533
José Maurício and Inês Domingues

Few-Shot Image Classification for Automatic COVID-19 Diagnosis ... 545
Daniel Cores, Nicolás Vila-Blanco, Manuel Mucientes, and María J. Carreira

An Ensemble-Based Phenotype Classifier to Diagnose Crohn's Disease from 16s rRNA Gene Sequences ... 557
Lara Vázquez-González, Carlos Peña-Reyes, Carlos Balsa-Castro, Inmaculada Tomás, and María J. Carreira

Synthetic Spermatozoa Video Sequences Generation Using Adversarial Imitation Learning ... 569
Sergio Hernández-García, Alfredo Cuesta-Infante, and Antonio S. Montemayor

A Deep Approach for Volumetric Tractography Segmentation ... 581
Pablo Rocamora-García, Marcelo Saval-Calvo, Victor Villena-Martinez, and Antonio Javier Gallego

MicrogliaJ: An Automatic Tool for Microglial Cell Detection and Segmentation ... 593
Ángela Casado-García, Estefanía Carlos, César Domínguez, Jónathan Heras, María Izco, Eloy Mata, Vico Pascual, and Lydia Álvarez-Erviti

Automated Orientation Detection of 3D Head Reconstructions from sMRI Using Multiview Orthographic Projections: An Image Classification-Based Approach 603
Álvaro Heredia-Lidón, Alejandro González, Carlos Guerrero-Mosquera, Rubèn Gonzàlez-Colom, Luis M. Echeverry, Noemí Hostalet, Raymond Salvador, Edith Pomarol-Clotet, Juan Fortea, Neus Martínez-Abadías, Mar Fatjó-Vilas, and Xavier Sevillano

Machine Learning Applications

Enhancing Transferability of Adversarial Audio in Speaker Recognition Systems 617
Umang Patel, Shruti Bhilare, and Avik Hati

Fishing Gear Classification from Vessel Trajectories and Velocity Profiles: Database and Benchmark 629
Pietro Melzi, Juan Manuel Rodriguez-Albala, Aythami Morales, Ruben Tolosana, Julian Fierrez, and Ruben Vera-Rodriguez

Multi-view Infant Cry Classification 639
Yadisbel Martinez-Cañete, Hichem Sahli, and Abel Díaz Berenguer

Study and Automatic Translation of Toki Pona 654
Pablo Baggetto, Damián López, and Antonio M. Larriba

Detecting Loose Wheel Bolts of a Vehicle Using Accelerometers in the Chassis 665
Jonas Schmidt, Kai-Uwe Kühnberger, Dennis Pape, and Tobias Pobandt

Clustering ECG Time Series for the Quantification of Physiological Reactions to Emotional Stimuli 680
Beatriz Henriques, Susana Brás, and Sónia Gouveia

Predicting the Subjective Responses' Emotion in Dialogues with Multi-Task Learning 693
Hassan Hayat, Carles Ventura, and Agata Lapedriza

Few-Shot Learning for Prediction of Electricity Consumption Patterns 705
Javier García-Sigüenza, José F. Vicent, Faraón Llorens-Largo, and José-Vicente Berná-Martínez

Author Index 717

Machine Learning

CCLM: Class-Conditional Label Noise Modelling

Albert Tatjer[1], Bhalaji Nagarajan[1(✉)], Ricardo Marques[1], and Petia Radeva[1,2]

[1] Dept. de Matemàtiques i Informàtica, Universitat de Barcelona, Barcelona, Spain
{acataltal1@alumnes.ub.edu, bhalaji.nagarajan,ricardo.marques,petia.ivanova}@ub.edu
[2] Computer Vision Center, Cerdanyola, Barcelona, Spain

Abstract. The performance of deep neural networks highly depends on the quality and volume of the training data. However, cost-effective labelling processes such as crowdsourcing and web crawling often lead to data with noisy (i.e., wrong) labels. Making models robust to this label noise is thus of prime importance. A common approach is using loss distributions to model the label noise. However, the robustness of these methods highly depends on the accuracy of the division of training set into clean and noisy samples. In this work, we dive in this research direction highlighting the existing problem of treating this distribution globally and propose a class-conditional approach to split the clean and noisy samples. We apply our approach to the popular DivideMix algorithm and show how the local treatment fares better with respect to the global treatment of loss distribution. We validate our hypothesis on two popular benchmark datasets and show substantial improvements over the baseline experiments. We further analyze the effectiveness of the proposal using two different metrics - Noise Division Accuracy and Classiness.

Keywords: Learning with Noisy Labels · Label Noise Modelling · Class-conditional data splitting

1 Introduction

Deep Neural Networks (DNNs) have gained immense attention in the research community due to their success in various challenging domains. High-performing models in all verticals require good quality data of huge volume. However, it is difficult to create such large datasets with high precision as it is labour-intensive, both in collecting and labelling the samples [16]. Crowdsourcing and using semi-automatic labelling pipelines on web-scrapped data reduce the cost of creating datasets. However, they lead to the introduction of noise in the assigned labels [23,24,30]. As a result, real-world datasets tend to have significant label noise,

A. Tatjer and B. Nagarajan—Joint First Authors.
P. Radeva—IAPR Fellow.
R. Marques—Serra Húnter Fellow.

A. Pertusa et al. (Eds.): IbPRIA 2023, LNCS 14062, pp. 3–14, 2023.
https://doi.org/10.1007/978-3-031-36616-1_1

estimated to be in the range of 8.0% to 38.5% [26]. Modern DNNs have a high number of learnable parameters compared to the size of the dataset and often result in overfitting to the label noise [32].

Learning with Noisy Labels (LNL) was introduced in the late 1980s s [1] and has been a long-studied problem with a focus on making the models robust to the label noise. Loss modification-based methods use noise distribution to create more robust loss functions [19], replacing the cross entropy loss or using a noise transition matrix [25]. Correctly estimating the noise and creating methods that are robust to high noise levels are very challenging tasks. Other methods that use sample selection [18] or reweighting [33] use *ad-hoc* criteria to select noisy samples and reduce the impact on the learning process. The challenge here is the selection of a good criterion to split the clean and noise samples.

Semi-supervised learning (SSL) is a helpful strategy for reducing the cost of annotating datasets. SSL-based frameworks use a limited subset of the original dataset being carefully labelled and learn the unlabelled data, which is multi-fold larger. Recent LNL methods aimed at combining SSL methods and robust loss methods. DivideMix [15] is a popular benchmark that uses two networks to learn the noisy data. At each iteration, one of the networks divides the data into a clean and a noisy datasets, which are then used to train the other network. DivideMix formed the basis for a new family of algorithms [13,27,34]. Several methods based on DivideMix have improved various stages of the pipeline. One of the important questions in most of these methods is *How to effectively demarcate the clean and the noisy samples?* As can be seen with all the works, the selection of clean samples plays an important role in the subsequent sub-tasks. The usage of loss distribution to split the samples has been a well-documented technique. However, not all samples in the training set are learned in the same rate, which hinders noise detection using this technique [21]. Here, we explore this global loss conundrum. We study the division in detail and propose a per-class label noise modelling. We adapt the noise model for each class and create a local threshold used to separate the clean and noisy samples. In order to further study this effect, we propose new metrics to identify the class behaviour of the LNL algorithm. The key contributions of this paper are as follows:

- First, we highlight and study the noise modelling obstacle present in the existing LNL algorithms.
- Second, we propose a class-conditional label noise modelling approach. By replacing the existing global-noise modelling in the popular DivideMix algorithm, we show how the class-conditional approach benefits the algorithm.
- Finally, we introduce new metrics to understand the importance of learning with class importance and show the improvement using two public datasets.

2 Related Works

Typical approaches in LNL algorithms include the use of advanced techniques such as sample selection, loss correction, label correction, sample reweighting or using robust loss methods [8]. Sample selection methods deal with finding clean samples using different strategies such as small loss [9] and topological information [31]. Loss correction is achieved by reweighting the loss to avoid overfitting on noisy samples, which helps mitigate the errors introduced by wrongly labelled samples. Similarly, in terms of sample reweighting, the samples that are noisiest are identified and weighted less compared to the clean samples [18]. Another family of LNL algorithms attempt to correct the noise (i.e., errors) in the labels, typically resorting to the notion of *transition matrix*. [7,11,25]. Most of these approaches are based on the assumption that there is a single transition probability between the noisy label and the ground-truth label [10]. Several noise tolerance loss functions have also been proposed to replace the conventional cross-entropy loss [19,20,35]. Recent studies on LNL have shown that hybrid approaches can further boost the algorithms and make them more robust towards label noise [6]. A comprehensive survey on the various LNL algorithms is presented in [26].

Co-teaching [9] of two networks by iteratively selecting clean samples (training samples having small loss) from the other network proved to be very successful. DivideMix [15] used two independent networks for sample selection and then adopted MixMatch [3] to further boost the label correction. DivideMix is a popular benchmarking algorithm in LNL problems. Augmentation plays an important role in LNL problems and was studied over DivideMix [22]. C2D further explored the warmup stage of DivideMix [34]. Probabilistic Noise Prediction [27] used two networks, one to predict the category label and the other to predict the noise type. ProMix [29] used an iterative selection process to select the clean samples using a high-confidence selection technique. SplitNet [13] used an additional learnable module to assist in splitting the clean and noisy samples.

Importance-weighted risk minimization [4] has been studied and employed in different DNNs over the years. A common scheme of importance weighting is weighting the loss terms [12] between different learning tasks. Class-conditional approaches were adapted to LNL algorithms. Generalized Data Weighting via Class-level Gradient Manipulation [5] converted the loss gradient into class-level gradients by applying the chain rule and reweighting the gradients separately. Noise transition matrix [33] was used to capture the class conditional label corruption and used total variation regularization to create more distinguishable predictions. Class-dependent softmax classifier [17] was used to entangle the classification of multi-class features. Our proposed approach is motivated by the LNL methods that deal with the separation of clean and noisy data using different means. As we observe during our study, using a global approach to all the samples is suboptimal and using a class-conditional split at the early stages of the algorithm makes the models more robust towards label noise.

3 The Label Noise Modelling Obstacle

Algorithms which allow efficient learning in the presence of noisy labels often require the division of the training dataset between clean and noisy samples [15]. The samples are then treated differently by the learning algorithm, to extract

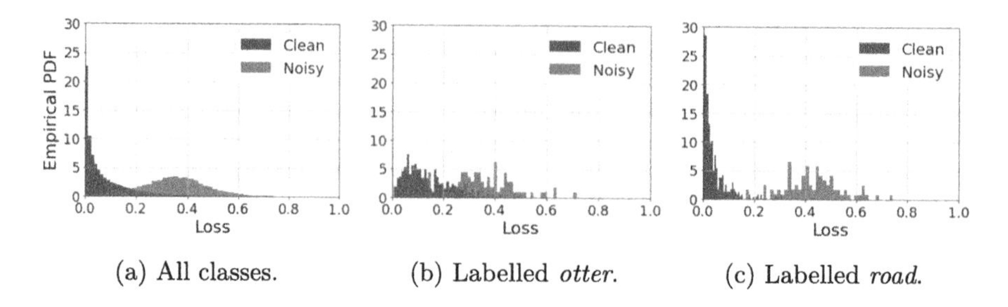

(a) All classes. (b) Labelled *otter*. (c) Labelled *road*.

Fig. 1. Empirical per-sample loss distribution between the clean samples and the noisy samples after the warm-up phase (CIFAR-100 20% sym. noise).

the maximum amount of information possible from each of the partitions of the data via semi-supervised learning [3]. The correct division between clean and noisy data is thus crucial for effective learning in the presence of noisy labels. Indeed, if too many noisy labels are classified as clean, then the algorithm risks trusting wrongly labelled data, eventually introducing important biases in the training process and leading to poor results. Conversely, if too many clean samples are deemed noisy, the algorithm would risk disposing of important information, hence potentially impairing the learning outcome. A common approach to splitting the dataset into clean and noisy samples is to create a noise model of the *whole* training data, allowing to characterize the probability of a given sample being clean (i.e., correctly labelled). Given the obvious limitations, it is common to rely on some proxy measure such as loss, which can be used to estimate to which extent a sample is likely to be clean or noisy [2]. We provide a more detailed analysis of this approach and show how the global noise model is suboptimal for some particular classes.

Global Label Noise Modelling. The typical approach to the label noise detection problem relies on the samples' loss to determine whether a given sample is deemed noisy or clean. The sample loss is obtained using a NN model, and the underlying principle is that the NN will tend to yield larger loss values to noisy samples because wrongly labelled data are a more complex pattern and therefore are more difficult to learn [28]. Given a sample x in dataset D, its corresponding one-hot encoded label y and a model p_θ parameterized with the set of parameters θ, the individual loss l of sample x is defined using the cross-entropy loss $l\left(p_\theta(x), y\right)$. The distribution of the cross-entropy loss is shown in Fig. 1a. The distribution appears to follow two modes, with the clean samples (in blue) exhibiting a characteristic loss generally lower than the noisy samples (in red). To turn the cross-entropy loss into a probability of a given sample being correctly (or wrongly) labelled, Li et al. [15] fit a two-component 1-D Gaussian Mixture model (GMM) to the loss distribution l. They select the GMM component g with lower mean and use it to model the probability ω_i of a given sample x_i with loss l_i to be clean. Finally, they use a threshold τ on ω_i to partition the dataset into a clean set $\mathcal{C}_D$ and a noisy set $\mathcal{N}_D$, given by:

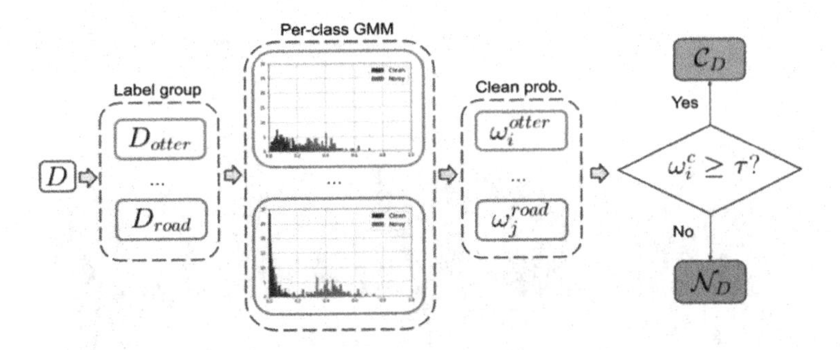

Fig. 2. CCLM splits the dataset according to labels, fitting a GMM to each independent set and yielding the Clean-Noise partition from a decision boundary on each label group's probability distribution.

$$\begin{aligned}\mathcal{C}_D &= \{(x_i, y_i)|\, \omega_i \geq \tau\}_{(x_i,y_i)\in D} \\ \mathcal{N}_D &= \{(x_i, y_i)\,|\, \omega_i < \tau\}_{(x_i,y_i)\in D}\,,\end{aligned} \tag{1}$$

where $\omega_i = p(g|l_i)$ and $l_i = l(p_\theta(x_i), y_i)$.

Limitations of Global Label Noise Modelling: The global noise modelling approach described above assumes that all samples are i.i.d and, therefore, that their per-sample loss tends to behave consistently. However, with the presence of noisy labels, it is obvious that some classes might be harder to learn than others, leading to potentially different characteristic losses across different classes. This situation can be appreciated in Fig. 1, where the per-sample loss distribution for the 'otter' class (Fig. 1b) differs from that of the 'road' class (Fig. 1c) and, more importantly, from the global per-sample loss distribution of the whole dataset (Fig. 1a). We identify this difference between the loss distribution of a particular class and the global loss distribution as an obstacle towards the correct partition of the data into a clean set, $\mathcal{C}_D$ and a noisy set, $\mathcal{N}_D$ which, in its turn, might hinder the training of NN models in the presence of noisy labels. Moreover, we conjecture that taking a class-conditional approach to the problem of label noise detection can lead to better dataset partitions, and thus to superior model accuracy after training in the presence of noisy labels. In the following, we propose a class-conditional approach to the problem of noise label detection, aimed at overcoming the aforementioned limitations of global label noise modelling.

4 Proposed Methodology

Typical approaches in LNL assume all samples to be i.i.d. Consequentially, label noise modelling is performed leveraging solely per-sample loss information, which can impair the accuracy of the trained model. We propose an alternative label-noise model we call Class-Conditional Local Noise Model (CCLM) that allows

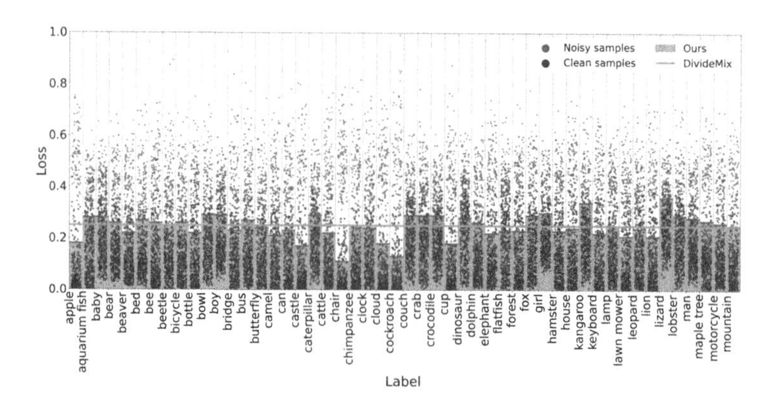

Fig. 3. Per-sample loss distribution of 50 of the CIFAR-100 label groups (20% sym.). Label groups are formed according to the potential noisy labels.

to locally adapt the noise model to the specificity of each considered class, hence improving the overall accuracy of the noisy label detection task. To achieve this, we explicitly include the label class information in the noise modelling process. Our proposed approach is depicted in Fig. 2. We take as input the training set D composed of pairs $(x_i, y_i)_{i=1}^{N}$, N being the total number of samples in D. Then, the dataset is split by sample label y, yielding C disjoint subsets of D such that $D = \cup_{c=1}^{C} D_c$, with C being the total number of possible classes. Then, for each subset D_c, we fit a two-component GMM to the per-sample loss and use the GMM component with a lower mean to model the probability of the samples being clean. We use the threshold τ over the probability of a sample being clean to split the data into noisy and clean within each class. In other words, the probability of a sample being clean is conditional on its loss and its label, $\omega_i^{y_i} = p(g|l_i, y_i)$. Finally, the Clean-Noisy partition is given by:

$$\begin{aligned} \mathcal{C}_D &= \cup_{c=1}^{C} \mathcal{C}_{D_c} = \cup_{c=1}^{C} \{(x_i, y_i) \,|\, \omega_i^c \geq \tau\}_{(x_i, y_i) \in D_c} \\ \mathcal{N}_D &= \cup_{c=1}^{C} \mathcal{N}_{D_c} = \cup_{c=1}^{C} \{(x_i, y_i) \,|\, \omega_i^c < \tau\}_{(x_i, y_i) \in D_c}, \end{aligned} \tag{2}$$

Figure 3 illustrates the result of our noise detection model on 20% noise of the CIFAR-100 dataset. The y-axis represents the per-sample loss associated with each sample, with noisy samples being represented in red, and clean samples in blue. The orange horizontal line depicts the split between noisy and clean labels using the typical global noise modelling approach. The green area depicts the per-class split between noisy and clean labels using our CCLM. We clearly see that the shape of the per-sample loss distribution varies across classes (e.g., 'apple' vs. 'aquarium fish'). Moreover, it can be seen that the global noise approach does not distinguish between classes, yielding a single decision boundary for all classes independently of their particular loss features. In contrast, our CCLM provides a locally adapted division between clean and noisy data which allows improving both the split accuracy and the NN accuracy when trained in the presence of noisy labels.

Table 1. Best and Last accuracy on CIFAR-10 and CIFAR-100.

Method		CIFAR-10					CIFAR-100				
		20%	50%	80%	90%	Asym. 40%	20%	50%	80%	90%	Asym. 40%
DivideMix [15]	Best	96.1	94.6	93.2	76	**93.4**	77.3	74.6	60.2	31.5	72.2**
	Last	95.7	94.4	92.9	75.4	92.1	76.9	74.2	59.6	31.0	72.4**
CCLM (ours)	Best	**96.5**	**95.6**	**93.7**	**83.6**	92.6	**79.5**	**76.4**	**61.1**	**33.5**	**75.4**
	Last	96.3	95.3	93.6	82.4	91.5	79.1	75.9	60.9	33.0	75.1

** Results reported from Contrast2Divide [34]. DivideMix does not report this setting.

5 Experiments and Results

In this section, we present the datasets and the implementation details for evaluating our proposal. We show the performance with two types of label noise - symmetric and asymmetric and compare them against the baseline DivideMix.

5.1 Datasets and Implementation Details

We conduct our evaluations using two benchmark datasets - CIFAR-10 and CIFAR-100 [14]. Both CIFAR datasets contain images of 32×32 RGB pixels, with 50k training samples and 10k test samples. To maintain consistency across the benchmark methods, we follow the same train/test split [15,34]. We conduct our experiments on two different noise types. *Symmetric* (sym.) noise is generated by replacing the labels for a percentage of the training data with a uniform distribution over all the possible labels. *Asymmetric* (asym.) noise is injected by replacing labels with similar classes (e.g. deer → horse, dog ↔ cat). For both datasets, we use 18-layer PreAct Resnet following the settings of DivideMix [15]. We keep all the hyperparameters the same as in DivideMix. We use a batch size of 128 and train the models for 300 epochs using SGD optimizer with a momentum of 0.9 and weight decay of 0.0005. We set an initial learning rate of 0.02 and reduce it by a factor of 10 after 150 epochs. For CIFAR-10, we use a warmup of 10 epochs and for CIFAR-100, we use a warmup of 30 epochs. We perform all experiments using PyTorch on NVIDIA RTX2080Ti GPU. Following previous works [15,34], we report the best test accuracy across all epochs and the average test accuracy over the last 10 epochs (identified as Last).

5.2 Results

First, we show the overall performance of the CCLM against DivideMix, which follows a global noise model approach. We compare it against DivideMix as it is considered a benchmark in LNL. The improvements brought by our proposed method in terms of label noise detection are likely to have a positive impact on algorithms which rely on the partition of the data between clean and noisy.

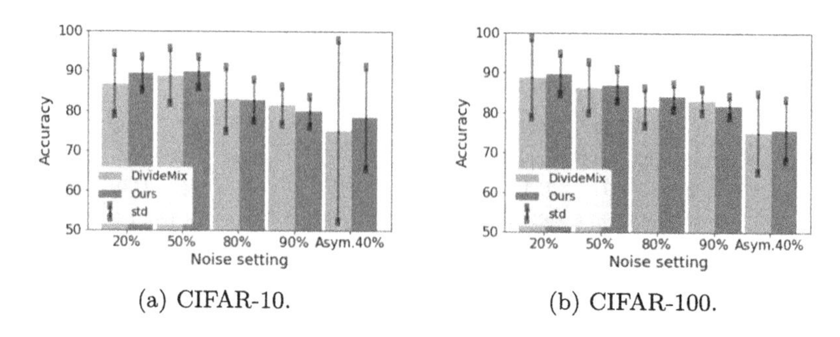

(a) CIFAR-10. (b) CIFAR-100.

Fig. 4. Clean-Noise split accuracy and std. dev. at the first epoch after warm-up.

Performance. Table 1 shows the test accuracy on CIFAR-10 and CIFAR-100 with different levels of label noise. Our CCLM outperforms the global noise label modelling approach of DivideMix in all the noise settings on CIFAR-100. On CIFAR-10, our approach fares better in all the noise settings except for the asym. noise setting. This superior performance is explained by the local adjustment (i.e., per class) of the label noise model, yielding a more accurate split between clean and noisy samples, hence improving the final model accuracy. Regarding CIFAR-10 asym. noise, where label groups are characterized by a limited number of classes, adopting a class conditional strategy may lead to overly confident discrimination between clean and noisy samples. This phenomenon could be attributed to a lack of related classes, resulting in insignificant "help" on the label group. The results in Table 1 show that our CCLM consistently outperforms the original global noise modelling used in DivideMix, being also more stable at low-noise settings. For both approaches, the accuracy of the models decreases considerably as the noise ratio increases. One of the reasons for it is the overfitting of the models in large noise settings which might make loss-based noise detection models less efficient. Hence, we further analyse the behaviour of CCLM using two different metrics - Noise Division Accuracy and Classiness.

Noise Division Accuracy. We evaluate the noise detection accuracy at the first epoch after the warmup. To this end, we first collect the per-sample loss provided by the pre-trained DL model. Then, we feed the collected loss values to both the global (DivideMix, using the threshold τ proposed for each noise level) and our CCLM noise detection approaches, yielding two distinct clean-noisy splits. Finally, using the reference solution, we evaluate the accuracy of the splits. The result is shown in Fig. 4, where the average split accuracy and the corresponding standard deviation for CIFAR-10 (Fig. 4a) and CIFAR-100 (Fig. 4b) are depicted, averaged over two runs. Our proposed method achieves a better average accuracy in all noise levels except for 90% sym. noise. It should be noted that a more accurate split after the warm-up does not guarantee better overall model accuracy. In the case of 90% noise, predicting all the samples as noisy would be very bad. The results also show that our method consistently yields a smaller standard deviation than the original one, which indicates that

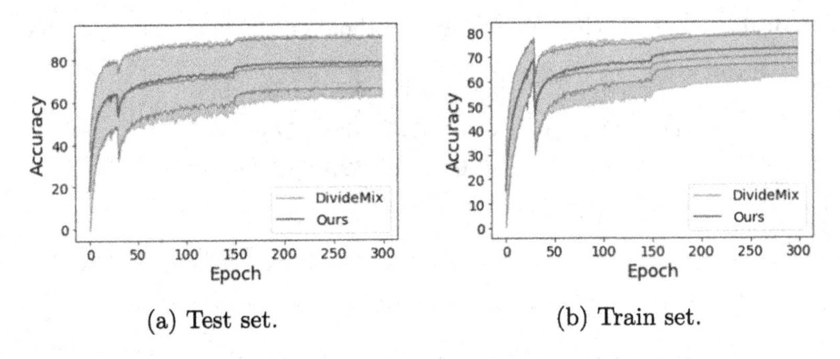

(a) Test set.

(b) Train set.

Fig. 5. Model accuracy and *Classiness* on CIFAR-100 with 20% sym. noise. The accuracy drop is a result of the ending of the warm-up phase (at epoch 30).

Table 2. Training and Testing Classiness of CIFAR-100.

		Train				Test			
Noise Rate		20%	50%	80%	90%	20%	50%	80%	90%
DivideMix [15]	Average	11.0	14.3	18.2	18.1	16.0	16.7	20.0	19.8
	Best	9	12.7	17.6	18.8	14	15	19	**21**
Ours	Average	7.7	10.9	17.5	17.1	13.6	14.9	19.4	20.2
	Best	**6**	**8.8**	**17.1**	**17.1**	**12**	**13**	**19**	22

our proposal yields a clean-noisy split that performs more evenly over the different classes. This results in a reduced bias towards the easier classes.

Classiness. In the classification task, it is clear that some classes can be more robust to label noise than others. A desired property of any classification model is to achieve similar accuracy in all the classes. In order to capture this accuracy distribution over classes, we introduce the notion of *classiness*, measured as the standard deviation of the model accuracy over the different latent classes. A lower classiness value thus represents a more evenly accurate network. We study classiness during training and testing. Training classiness allows us to inspect whether the training process is biased towards some of the classes. With class-agnostic label-noise modelling, *difficult* classes would be treated as mostly noise. Testing classiness is the desired model property informing about the achieved evenness. We show the training and testing model accuracy and classiness evolution during the training process using 20% sym. noise of CIFAR-100 in Fig. 5, where the first 30 epochs correspond to the warm-up phase. The training classiness (Fig. 5b (painted area)) is smaller on average for our proposed CCLM. This indicates that the model is less biased during training, allowing for better learning in the posterior epochs. The average accuracy of the baseline model and our proposal are similar, but the smaller classiness tends towards better learning seen in the later epochs. Table 2 shows the average training classiness

Table 3. Impact of the clean noise split threshold on the final model accuracy.

τ	0.5	0.6	0.7	0.8	0.9
DivideMix [15]	30.3	**31.5**	31.6	27.6	27.6
Ours	**30.6**	31.1	**32.6**	**33.5**	**30.8**

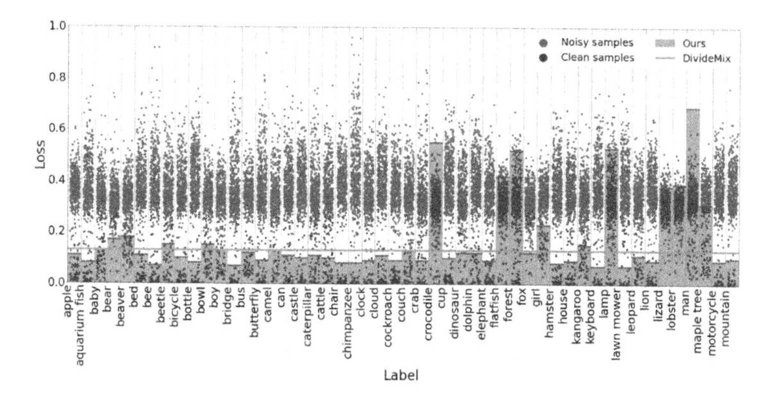

Fig. 6. Per-sample loss distribution of 50 classes of CIFAR-100 (90% sym.).

on CIFAR-100 with all different noise levels and the test classiness of the best epoch. The best epoch is chosen as the one with the better test accuracy. We see that our method has reasonably better test and train classiness on low noise settings and comparable test and train classiness on higher noise settings.

Effect of Clean-Noise Splitting Threshold. In current LNL practice, the label noise modelling step is most commonly used to perform the Clean-Noise split. However, the resulting split also depends on the threshold used as a decision boundary. A critical aspect of the analysis entails examining the impact of the selected threshold on the split outcome. To this end, we present Fig. 6 and Fig. 3, which illustrate the feature space we aim to split. The label-noise modelling procedure yields the probability of a sample belonging to the clean set, $\omega_i = p(g|l_i)$. Following Eq. (1) and Eq. (2), the Clean-Noisy partition is obtained in a sample level by $\omega_i > \tau$. In Table 3, we compare the model accuracy between the global LNL model, and our CCLM, across different split thresholds. Upon inspecting Fig. 6, we observe that a class-agnostic label noise modelling procedure, with a high threshold, disregards harder classes entirely. In contrast, our CCLM is not susceptible to this particular pitfall and is therefore capable of managing higher thresholds.

6 Conclusions

Learning with Noisy Labels is a very important data-centric Machine learning problem. Modern algorithms rely on the division of clean and noisy samples to make further learning decisions. In this paper, we study this division of training samples and show how the global division is sub-optimal. On this front, we propose a class-conditional label noise model, which uses a local division of clean and noisy samples. We validate our approach comparing it with the baseline on two benchmarking datasets. We further introduce two metrics - noise division accuracy and classiness to show the improvements. Our future work focuses on studying the impact of global vs. local threshold in other parts of the algorithm, using other metrics to understand this division and validating the method on other large LNL datasets.

Acknowledgements. This work was partially funded by the Horizon EU project MUSAE (No. 01070421), 2021-SGR-01094 (AGAUR), Icrea Academia'2022 (Generalitat de Catalunya), Robo STEAM (2022-1-BG01-KA220-VET-000089434, Erasmus+ EU), DeepSense (ACE053/22/000029, ACCIÓ), DeepFoodVol (AEI-MICINN, PDC-2022-133642-I00) and CERCA Programme/Generalitat de Catalunya. B. Nagarajan acknowledges the support of FPI Becas, MICINN, Spain. We acknowledge the support of NVIDIA Corporation with the donation of the Titan Xp GPUs. As Serra Húnter Fellow, Ricardo Marques acknowledges the support of the Serra Húnter Programme.

References

1. Angluin, D., Laird, P.: Learning from noisy examples. Mach. Learn. **2**, 343–370 (1988)
2. Arazo, E., Ortego, D., Albert, P., O'Connor, N., McGuinness, K.: Unsupervised label noise modeling and loss correction. In: International Conference on Machine Learning, pp. 312–321. PMLR (2019)
3. Berthelot, D., Carlini, N., Goodfellow, I., Papernot, N., Oliver, A., Raffel, C.A.: Mixmatch: a holistic approach to semi-supervised learning. In: NIPS, vol. 32 (2019)
4. Byrd, J., Lipton, Z.: What is the effect of importance weighting in deep learning? In: International Conference on Machine Learning, pp. 872–881. PMLR (2019)
5. Chen, C., et al.: Generalized data weighting via class-level gradient manipulation. In: NIPS, vol. 34, pp. 14097–14109 (2021)
6. Chen, Z., Song, A., Wang, Y., Huang, X., Kong, Y.: A noise rate estimation method for image classification with label noise. In: Journal of Physics: Conference Series, vol. 2433, p. 012039. IOP Publishing (2023)
7. Cheng, D., et al.: Instance-dependent label-noise learning with manifold-regularized transition matrix estimation. In: CVPR, pp. 16630–16639 (2022)
8. Ding, K., Shu, J., Meng, D., Xu, Z.: Improve noise tolerance of robust loss via noise-awareness. arXiv preprint arXiv:2301.07306 (2023)
9. Han, B., et al.: Co-teaching: robust training of deep neural networks with extremely noisy labels. In: NIPS, vol. 31 (2018)
10. Han, J., Luo, P., Wang, X.: Deep self-learning from noisy labels. In: ICCV, pp. 5138–5147 (2019)
11. Hendrycks, D., Mazeika, M., Wilson, D., Gimpel, K.: Using trusted data to train deep networks on labels corrupted by severe noise. In: NIPS, vol. 31 (2018)
12. Khetan, A., Lipton, Z.C., Anandkumar, A.: Learning from noisy singly-labeled data. arXiv preprint arXiv:1712.04577 (2017)
13. Kim, D., Ryoo, K., Cho, H., Kim, S.: SplitNet: learnable clean-noisy label splitting for learning with noisy labels. arXiv preprint arXiv:2211.11753 (2022)

14. Krizhevsky, A., Hinton, G., et al.: Learning multiple layers of features from tiny images (2009)
15. Li, J., Socher, R., Hoi, S.C.: DivideMix: learning with noisy labels as semi-supervised learning. arXiv preprint arXiv:2002.07394 (2020)
16. Liao, Y.H., Kar, A., Fidler, S.: Towards good practices for efficiently annotating large-scale image classification datasets. In: CVPR, pp. 4350–4359 (2021)
17. Liu, S., Zhu, Z., Qu, Q., You, C.: Robust training under label noise by over-parameterization. In: ICML, pp. 14153–14172. PMLR (2022)
18. Liu, X., Luo, S., Pan, L.: Robust boosting via self-sampling. Knowl.-Based Syst. **193**, 105424 (2020)
19. Ma, X., Huang, H., Wang, Y., Romano, S., Erfani, S., Bailey, J.: Normalized loss functions for deep learning with noisy labels. In: ICML, pp. 6543–6553 (2020)
20. Miyamoto, H.K., Meneghetti, F.C., Costa, S.I.: The Fisher-Rao loss for learning under label noise. Inf. Geometry 1–20 (2022)
21. Nagarajan, B., Marques, R., Mejia, M., Radeva, P.: Class-conditional importance weighting for deep learning with noisy labels. In: VISIGRAPP (5: VISAPP), pp. 679–686 (2022)
22. Nishi, K., Ding, Y., Rich, A., Hollerer, T.: Augmentation strategies for learning with noisy labels. In: CVPR, pp. 8022–8031 (2021)
23. Northcutt, C., Jiang, L., Chuang, I.: Confident learning: estimating uncertainty in dataset labels. J. Artif. Intell. Res. **70**, 1373–1411 (2021)
24. Oyen, D., Kucer, M., Hengartner, N., Singh, H.S.: Robustness to label noise depends on the shape of the noise distribution in feature space. arXiv preprint arXiv:2206.01106 (2022)
25. Patrini, G., Rozza, A., Krishna Menon, A., Nock, R., Qu, L.: Making deep neural networks robust to label noise: a loss correction approach. In: CVPR, pp. 1944–1952 (2017)
26. Song, H., Kim, M., Park, D., Shin, Y., Lee, J.G.: Learning from noisy labels with deep neural networks: a survey. IEEE Tran. NNLS (2022)
27. Sun, Z., et al.: PNP: robust learning from noisy labels by probabilistic noise prediction. In: CVPR, pp. 5311–5320 (2022)
28. Valle-Pérez, G., Camargo, C.Q., Louis, A.A.: Deep learning generalizes because the parameter-function map is biased towards simple functions. arXiv e-prints arXiv:1805.08522 (2018)
29. Wang, H., Xiao, R., Dong, Y., Feng, L., Zhao, J.: ProMix: combating label noise via maximizing clean sample utility. arXiv preprint arXiv:2207.10276 (2022)
30. Wei, J., Zhu, Z., Cheng, H., Liu, T., Niu, G., Liu, Y.: Learning with noisy labels revisited: a study using real-world human annotations. arXiv preprint arXiv:2110.12088 (2021)
31. Wu, P., Zheng, S., Goswami, M., Metaxas, D., Chen, C.: A topological filter for learning with label noise. In: NIPS, vol. 33, pp. 21382–21393 (2020)
32. Zhang, C., Bengio, S., Hardt, M., Recht, B., Vinyals, O.: Understanding deep learning (still) requires rethinking generalization. ACM **64**(3), 107–115 (2021)
33. Zhang, Y., Niu, G., Sugiyama, M.: Learning noise transition matrix from only noisy labels via total variation regularization. In: ICML, pp. 12501–12512 (2021)
34. Zheltonozhskii, E., Baskin, C., Mendelson, A., Bronstein, A.M., Litany, O.: Contrast to divide: self-supervised pre-training for learning with noisy labels. In: WACV, pp. 1657–1667 (2022)
35. Zhou, X., Liu, X., Zhai, D., Jiang, J., Ji, X.: Asymmetric loss functions for noise-tolerant learning: theory and applications. IEEE Trans. PAMI (2023)

Addressing Class Imbalance in Multilabel Prototype Generation for *k*-Nearest Neighbor Classification

Carlos Penarrubia[1], Jose J. Valero-Mas[1,2](✉), Antonio Javier Gallego[1], and Jorge Calvo-Zaragoza[1]

[1] U.I. for Computer Research, University of Alicante, Alicante, Spain
carlos.penarrubia@ua.es, {jjvalero,jgallego,jcalvo}@dlsi.ua.es
[2] Music Technology Group, Universitat Pompeu Fabra, Barcelona, Spain

Abstract. Prototype Generation (PG) methods seek to improve the efficiency of the k-Nearest Neighbor (kNN) classifier by obtaining a reduced version of a given reference dataset following certain heuristics. Despite being largely addressed topic in multiclass scenarios, few works deal with PG in multilabel environments. Hence, the existing proposals exhibit a number of limitations, being label imbalance one of paramount relevance as it constitutes a typical challenge of multilabel datasets. This work proposes two novel merging policies for multilabel PG schemes specifically devised for label imbalance, as well as a mechanism to prevent inappropriate samples from undergoing a reduction process. These proposals are applied to three existing multilabel PG methods—Multilabel Reduction through Homogeneous Clustering, Multilabel Chen, and Multilabel Reduction through Space Partitioning—and evaluated on 12 different data assortments with different degrees of label imbalance. The results prove that the proposals overcome—in some cases in a significant manner—those obtained with the original methods, hence validating the presented approaches and enabling further research lines on this topic.

Keywords: Multilabel Learning · Imbalanced classification · Prototype Generation · Efficient k-Nearest Neighbor

1 Introduction

Due to its conceptual simplicity and good statistical properties, the k-Nearest Neighbor (kNN) algorithm represents one of the most well-known non-parametric methods for supervised classification [6]. This method labels a given query element with the most common class among its k nearest neighbors from a reference data collection, based on a specified similarity measure [1].

This work was supported by the I+D+i project TED2021-132103A-I00 (DOREMI), funded by MCIN/AEI/10.13039/501100011033.

A. Pertusa et al. (Eds.): IbPRIA 2023, LNCS 14062, pp. 15–27, 2023.
https://doi.org/10.1007/978-3-031-36616-1_2

As a *lazy* learning method, kNN does not derive a classification model out of the training data, but performs an exhaustive search across the reference corpus for every single query [5]. In this sense, kNN generally entails low-efficiency figures in both classification time and memory usage, thereby becoming a bottleneck when addressing large-scale collections.

One of the most considered solutions to tackle this limitation consists in reducing the size of the reference set for improving the overall efficiency while not significantly degrading the performance, namely Data Reduction (DR) [8]. This process is typically carried out by resorting to one of the following approaches [11]: *(i)* Prototype Selection (PS) methods, which retain a representative subset of elements out of the entire reference collection; and *(ii)* Prototype Generation (PG) techniques, which derive an alternative set by merging the elements of the initial dataset following certain heuristics. The focus of this work is that of the latter family as it generally yields the sharpest reduction rates with the least performance decrease [7].

While PG has been largely explored in *multiclass* cases—each instance is assigned one category from a set of mutually excluding labels—, its research in *multilabel* scenarios—each query is assigned an undetermined number of categories—has been scarcely addressed, with the sole exception of a few recent proposals [12,16]. Hence, as reported in those works, there exist a plethora of open questions that need to be addressed as, for instance, studying possible correlations among the label spaces for devising adequate reduction strategies, proposing manners of tackling noisy datasets, or addressing the issue of *class imbalance*—*i.e.*, the uneven distribution of labels in the data collection at hand. Note that this latter point has been reported to hinder the performance of existing reduction schemes remarkably, being hence of paramount relevance to research on it due to the inherent imbalanced nature of multilabel data [10].

Due to its relevance, a number of methods have been proposed to deal with data imbalance, which can be grouped into four categories [9]: (i) *algorithm adaptation*, which modify the basis of the algorithm to account for the possible bias; (ii) *resampling*, which artificially balance the data distributions; (iii) *cost-sensitive training*, which bias the learning method towards minority classes; and (iv) *ensembles*, which combine some of the previous techniques. The reader is referred to the work of Tarekegn et al. [13] for a thorough review of the field.

Furthermore, to our best knowledge, there barely exist works that address the issue of class imbalance in DR tasks, especially in multilabel learning. One of them is that by Valero-Mas et al. [15], which studied the combination of data resampling techniques with PS methods to tackle class-imbalance scenarios. However, they only addressed multiclass cases. Another example is the so-called MultiLabel edited Nearest Neighbor (MLeNN) [3] that proposes a pipeline to prevent samples with a severe imbalance level from undergoing a multilabel PS strategy that would, eventually, remove them. Nevertheless, no existing work has addressed the issue of data imbalance when considering multilabel PG processes.

This work proposes a set of novel mechanisms and extensions to existing multilabel PG methods to cope with their limitations when addressing scenarios with class imbalance. More precisely, we select three representative multilabel

PG strategies—namely, the *Multilabel Reduction through Homogeneous Clustering* (MRHC) strategy [12] as well as the *Multilabel Chen* (MChen) [16] and the *Multilabel Reduction through Space Partitioning* (MRSP3) [16] methods—and introduce several generation mechanisms that contemplate the imbalance levels to perform an adequate reduction process. The results obtained in 12 datasets with various levels of class imbalance prove that the proposals overcome—in some cases in a significant manner—the results obtained with the original methods. Such an outcome proves the validity of the proposals and enables further research lines on this topic of multilabel PG in scenarios with class imbalance.

The rest of the paper is structured as follows: Sect. 2 introduces the proposed methodology; after that, Sect. 3 presents the experimental set-up considered; then, Sect. 4 shows and discusses the results; and finally, Sect. 5 concludes the work and poses future research lines to address.

2 Methodology

Let $\mathcal{X} \in \mathbb{R}^f$ denote an input f-dimensional feature space, $\mathcal{L} = \{\lambda_1, \dots, \lambda_L\}$ an L-sized set of labels, and $\mathcal{Y} = \{0, 1\}^L$ an L-dimensional label space. Set $\mathcal{T} = \{(\boldsymbol{x}_i, \boldsymbol{y}_i)\}_{i=1}^{|\mathcal{T}|} \subset \mathcal{X} \times \mathcal{Y}$ represents a data collection that relates the i-th sample $\boldsymbol{x}_i \in \mathcal{X}$ to label vector $\boldsymbol{y}_i \in \mathcal{Y}$, where $y_{ij} \in \{0, 1\}$ with $1 \leq j \leq L$ denotes that label λ_j is present—or not—in the label space. Note that this $\mathcal{T}$ set is the one to be reduced with the PG proposals studied in this work.

The three considered multilabel PG strategies—MRHC, MChen, and MRSP3—perform the reduction process following a *space partitioning* policy [2]. These techniques subdivide the feature space into different non-overlapping regions that gather one or more prototypes following certain criteria—namely, *Space splitting* stage—for then deriving a single representative prototype per region—process referred to as *Prototype merging*. In this work we focus on the latter stage of the process, assuming that the former retrieves a set C comprising clusters of prototypes that accomplish conditions $\cup_{m=1}^{|C|} C(m) = \mathcal{T}$ and $\cap_{m=1}^{|C|} C(m) = \emptyset$.

Figure 1 presents the devised methodology to adequate these PG techniques to scenarios with label imbalance. Note that the stages marked with dashed borders denote those where the proposed mechanisms are performed.

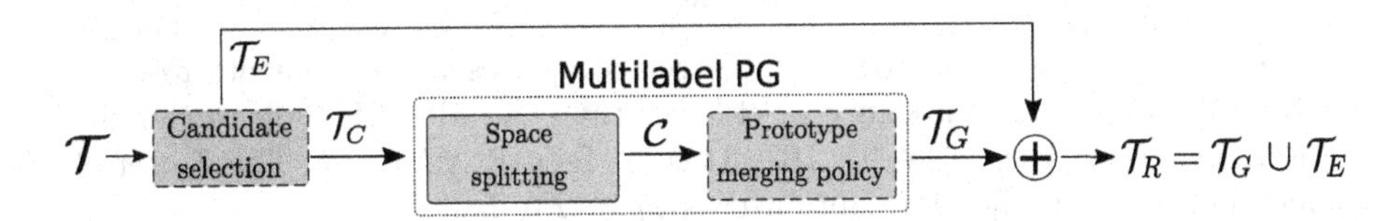

Fig. 1. Graphical description of the methodology proposed in the work.

As it may be observed, the *Candidate selection* stage splits the initial set $\mathcal{T}$ into a collection of severely imbalanced samples—subset $\mathcal{T}_E$—that must remain unaltered throughout the reduction process and another collection—set $\mathcal{T}_C$—comprising the elements that are suitable for being processed by the PG method using a specific merging policy. The set of generated prototypes $\mathcal{T}_G$—with $|\mathcal{T}_G| << |\mathcal{T}_C|$—is then joined with the former $\mathcal{T}_E$ collection of excluded instances, retrieving the target $\mathcal{T}_R$ reduced set. The rest of the sections thoroughly describe these individual processes.

2.1 Candidate Selection Stage

The candidate selection approach is based on that performed by the aforementioned MLeNN method [3]. Based on the imbalance level of the labels in the instances, the process divides the set $\mathcal{T}$ into two disjoint subsets $\mathcal{T}_C$ and $\mathcal{T}_E$ that respectively comprise the elements to be further processed by the PG methods and the ones that are prevented from undergoing it. The rationale behind this approach is that the subsequent PG method, if applied to a set of instances depicting severely imbalanced labels, may consider them as noise, hence obviating the contributions of such particular labels. Thus, only samples with a low imbalance level are further considered by the PG method.

To formally describe this division process, we introduce the *imbalance ratio per label* (IRLbl) metric that estimates the imbalance level of each label in a given data collection. For a given label $\lambda \in \mathcal{Y}$, this indicator is defined as:

$$\mathrm{IRLbl}\left(\lambda\right) = \frac{\displaystyle\max_{\forall\lambda' \in \mathcal{Y}} \left(\sum_{i=1}^{|\mathcal{T}|} [\![\lambda' \in \boldsymbol{y}_i]\!] \right)}{\displaystyle\sum_{i=1}^{|\mathcal{T}|} [\![\lambda \in \boldsymbol{y}_i]\!]} \tag{1}$$

where $[\![\cdot]\!] \rightarrow \{0, 1\}$ represents the Iverson bracket, which outputs the unit value when the condition in the argument is met and zero otherwise. Note that this descriptor ranges in $\mathrm{IRLbl}\left(\lambda\right) \in [1, \infty)$ and denotes higher imbalance as the value increases. Based on this, the *mean imbalance ratio* (MeanIR) index summarizes, in a single value, the entire imbalance level of a collection as:

$$\mathrm{MeanIR} = \frac{1}{|\mathcal{Y}|} \sum_{\lambda \in \mathcal{Y}} \mathrm{IRLbl}\left(\lambda\right) \tag{2}$$

Considering these indicators, the $\mathcal{T}_C$ set comprises those elements from $\mathcal{T}$ whose labels depict an imbalance level lower than the MeanIR indicator of the dataset, *i.e.*, $\mathcal{T}_C = \{(\boldsymbol{x}_i, \boldsymbol{y}_i) \in \mathcal{T} : \mathrm{IRLbl}\left(\lambda\right) < \mathrm{MeanIR} \ \forall \lambda \in \boldsymbol{y}_i\}$. The $\mathcal{T}_E$ set of excluded elements is therefore obtained as $\mathcal{T}_E = \mathcal{T} \setminus \mathcal{T}_C$.

2.2 Prototype Merging Policies

Once the $\mathcal{T}_C \subseteq \mathcal{T}$ set has been obtained, the PG method must perform the reduction process to retrieve $\mathcal{T}_G$. As aforementioned, we assume that the *Space splitting* stage has obtained the C set of clusters (process done resorting to the original partitioning proposals of each method that may be checked in [12] for the MRHC approach and [16] for the MChen and MRSP3 cases), and we exclusively focus on the merging process. Hence, since the merging stage is expected to derive a single prototype per cluster, this process outputs a set $\mathcal{T}_G = \{(\boldsymbol{x}_m, \boldsymbol{y}_m)\}_{m=1}^{|C|}$.

We now detail the three prototype merging policies for obtaining $\boldsymbol{y}_m$ contemplated in this work: the base case used by the original PG methods as well as the two novel proposals specifically devised to deal with label imbalance.

i) **Base case.** The base case assigns the labels with a certain representation level to the new sample. Considering that $|C(m)|_\lambda$ denotes the number of prototypes with label λ in cluster $C(m)$, this is done as:

$$\boldsymbol{y}_m = \left\{\lambda \in C(m) : |C(m)|_\lambda \geq \frac{|C(m)|}{2}\right\} \tag{3}$$

ii) **Policy 1.** The first proposal biases the label space generation resorting to the imbalance level depicted by the IRLbl indicator. This is done as:

$$\boldsymbol{y}_m = \left\{\lambda \in C(m) : |C(m)|_\lambda \geq \left\lfloor \frac{|C(m)|}{2 \cdot \text{IRLbl}(\lambda)} \right\rfloor\right\} \tag{4}$$

iii) **Policy 2.** The second proposal maintains the labels with certain representation level as well as those that, if underrepresented in the cluster, are severely imbalanced—*i.e.*, $\text{IRLbl}(\lambda) > \text{MeanIR}$. This is done as:

$$\begin{aligned}\boldsymbol{y}_m = \Bigg\{\lambda \in C(m) : \left(|C(m)|_\lambda \geq \frac{|C(m)|}{2}\right) \vee \\ \left([\![\text{IRLbl}(\lambda) > \text{MeanIR}]\!] \wedge |C(m)|_\lambda < \frac{|C(m)|}{2}\right)\Bigg\}\end{aligned} \tag{5}$$

Finally, while the merging stage must perform such a generation process for both the feature and the label spaces, the issue of label imbalance only depends on the latter. Thus, based on the contemplated methods, prototype $\boldsymbol{x}_m$ related to $\boldsymbol{y}_m$ is obtained by averaging the features of the instances in cluster $C(m)$.

3 Experimental Set-Up

This section presents the datasets used for assessing the proposals as well as the classification method and the evaluation metrics considered.

3.1 Datasets

We have considered 12 multilabel datasets from the Mulan repository [14] comprising different domains, sizes, initial space dimensionalities, target label spaces, and imbalance ratios. Table 1 summarizes these details for each collection.

Table 1. Summary of the datasets considered for the experimentation. Each collection is described in terms of its data domain, partition sizes, dimensionality of input data (features) and output space (labels), and imbalance ratio (MeanIR).

Name	Domain	Set size		Dimensionality		MeanIR
		Train	Test	Features (f)	Labels (L)	
Bibtex	Text	4,880	2,515	1,836	159	12.78
Birds	Audio	322	323	260	19	6.10
Corel5k	Image	4,500	500	499	374	183.29
Emotions	Music	391	202	72	6	1.49
Genbase	Biology	463	199	1,186	27	31.60
Medical	Text	333	645	1,449	45	48.59
rcv1subset1	Text	3,000	3,000	47,236	101	191.42
rcv1subset2	Text	3,000	3,000	47,236	101	177.89
rcv1subset3	Text	3,000	3,000	47,236	101	192.48
rcv1subset4	Text	3,000	3,000	47,229	101	170.84
Scene	Image	1,211	1,196	294	6	1.33
Yeast	Biology	1,500	917	103	14	7.27

For the sake of analysis, we group the presented assortments attending to their label imbalance based on the MeanIR indicator: **Low imbalance** when MeanIR ≤ 15, hence grouping the *Bibtex*, *Birds*, *Emotions*, *Scene*, and *Yeast* collections; and **High imbalance** when MeanIR > 15, which gathers the *Corel5k*, *Genbase*, *Medical*, *rcv1subset1*, *rcv1subset2*, *rcv1subset3*, and *rcv1subset4* sets.

3.2 Classification Method and PG Parameterization

Regarding the classification schemes, we resort to the Multilabel-kNN (ML-kNN) [17] method as it constitutes a reference adaptation of the multiclass kNN classifier to multilabel frameworks. Besides, we fix $k = 1$ as the number of neighbors to be consulted since increasing this parameter could affect the selection process of the minority samples. Also, $k > 1$ is supposed to help with noisy datasets, which is not the study objective of the present paper.

Finally, in terms of the PG methods, note that the MRHC and MRSP3 strategies depict an autonomous and parameter-free space splitting stage, whereas the MChen method requires the use of a reduction rate value. This parameter, denoted in this work as ϕ, specifies the expected size of the resulting set as a percentage of the initial one, *i.e.*, $\phi = |\mathcal{T}_G|/|\mathcal{T}_C|$. In this regard, we contemplate the values of $\phi \in \{10, 50, 90\}$ for this PG method as representative cases of sharp, middle, and low reduction rates, being denoted as MChen$_\phi$.

3.3 Evaluation Metrics

As done in the reference works [12,16], we consider two different criteria to assess the goodness of our proposals: classification performance and efficiency figures.

Regarding the performance assessment, we resort to the macro F_1 score as it constitutes a reference metric for imbalance data:

$$F_1 = \frac{1}{L} \sum_{\forall \lambda \in \mathcal{Y}} \frac{2 \cdot TP_\lambda}{2 \cdot TP_\lambda + FP_\lambda + FN_\lambda} \quad (6)$$

where TP_λ, FP_λ, and FN_λ respectively denote the True Positive, False Positive, and False Negative estimations for label λ.

In terms of efficiency, we assess the results based on the size of the reduced set $\mathcal{T}_R$, normalized by that of the original dataset $\mathcal{T}$. Computation time is discarded due to its variability depending on the load of the computing system.

Besides, since PG methods seek to optimize two contradictory goals—efficiency and efficacy—, we frame this problem within a Multi-objective Optimization task. The solutions under this framework are retrieved by considering the concept of non-dominance: one solution is said to dominate another if it is better or equal in each goal function and, at least, strictly better in one of them. Those elements are known as non-dominated and constitute the Pareto frontier, being all of them optimal solutions without any order.

4 Results

This section introduces and discusses the results obtained with the presented extensions to the existing PG proposals considering the aforementioned evaluation set-up. More specifically, we first analyze the efficiency and efficacy of each method to then evaluate the insights observed when addressing the task as a Multi-objective Optimization framework; eventually, we perform a set of significance tests to further validate the remarks observed in the initial analyses.[1]

Table 2 shows the performance—F_1—and reduction—resulting set size—figures obtained by each PG method and imbalance scenario. The case disregarding reduction—denoted as ALL—is included for comparison purposes.

Attending to the results, the imbalance level of the datasets—given by the MeanIR indicator—severely affects the overall performance. Focusing on the exhaustive search—ALL case—, the scheme achieves a value of $F_1 = 42.79\%$ in the low-imbalance case—MeanIR ≤ 15—, decreasing this figure to $F_1 = 12.15\%$ when addressing datasets with high label imbalance—MeanIR > 15.

The introduction of the reference PG methods—the case of *No candidate selection* with the *Base* merging policy—shows a general degradation in terms of performance, most noticeably in the MChen_{10} and MChen_{50} cases. While this effect is somehow expected due to the sharp reduction, it also denotes that these techniques are not adequately devised for label imbalance cases.

[1] An implementation of this experimental procedure together with the assessed methods are available in: https://github.com/jose-jvmas/imbalance-MPG_IbPRIA23.

Table 2. Results in terms of F_1 (%) and resulting set size for the different PG methods and extensions as well as the case of exhaustive search. Bold values highlight the best performance per PG method and imbalance scenario whereas underlined elements depict performance improvement over the ALL case.

	No candidate selection ($\mathcal{T}_C = \mathcal{T}$)				Using candidate selection ($\mathcal{T}_C \subseteq \mathcal{T}$)			
	Size (%)	Merging policy			Size (%)	Merging policy		
		Base	Policy 1	Policy 2		Base	Policy 1	Policy 2
Low imbalance								
ALL ■	100	42.79	–	–	100	42.79	–	–
MRHC ▼	57.83	42.44	**43.70**	42.87	70.65	42.34	43.36	42.34
MChen_{10} ◆	9.98	30.11	36.87	27.81	40.98	36.36	**40.33**	36.36
MChen_{50} ◆	49.97	37.15	41.64	38.26	67.17	40.46	**41.66**	40.46
MChen_{90} ◆	89.89	42.20	42.29	42.26	93.23	42.99	**43.00**	42.99
MRSP3 ▲	66.84	40.73	**43.58**	41.43	78.12	41.76	43.04	41.76
High imbalance								
ALL ■	100	12.15	–	–	100	12.15	–	–
MRHC ▼	47.55	12.03	**12.48**	12.03	46.89	11.92	**12.48**	11.92
MChen_{10} ◆	9.98	7.23	9.80	7.36	12.64	7.48	**9.94**	7.48
MChen_{50} ◆	49.96	9.96	**11.93**	10.04	51.45	9.97	11.78	9.97
MChen_{90} ◆	89.58	11.91	**12.08**	11.91	89.74	11.83	12.04	11.83
MRSP3 ▲	60.94	11.65	**13.96**	12.11	61.08	8.80	10.59	8.80

Focusing on the merging proposals, it may be observed that *Policy 1* systematically improves the results compared to the base strategy. Oppositely, while *Policy 2* also shows some improvements with respect to the same baseline, the results are not that consistent. Regarding the *Candidate selection* stage, the figures obtained show that this process generally boosts the recognition rates by increasing the resulting set size. Nevertheless, the improvement observed in these cases is less prominent than that achieved by the assessed merging policies.

Finally, note that some of cases—*e.g.*, the MRHC and MRSP3 with *Policy 1* for all imbalance scenarios and disregarding the selection stage—do improve the performance achieved by the exhaustive search (ALL case). This fact suggests that some of the presented extensions are capable of removing noise inherent to the datasets, hence showing higher robustness than the reference PG strategies.

4.1 Multi-objective Optimization

Having individually analyzed the performance and reduction rates for the different PG methods and the proposed extensions, we now resort to the Multi-objective Optimization framework for a joint analysis of these criteria. Figure 2 shows the results of this assessment for the two considered imbalance levels.

The Pareto frontiers for both the low and high imbalance levels comprise solutions solely based on the proposed extensions. The reference PG

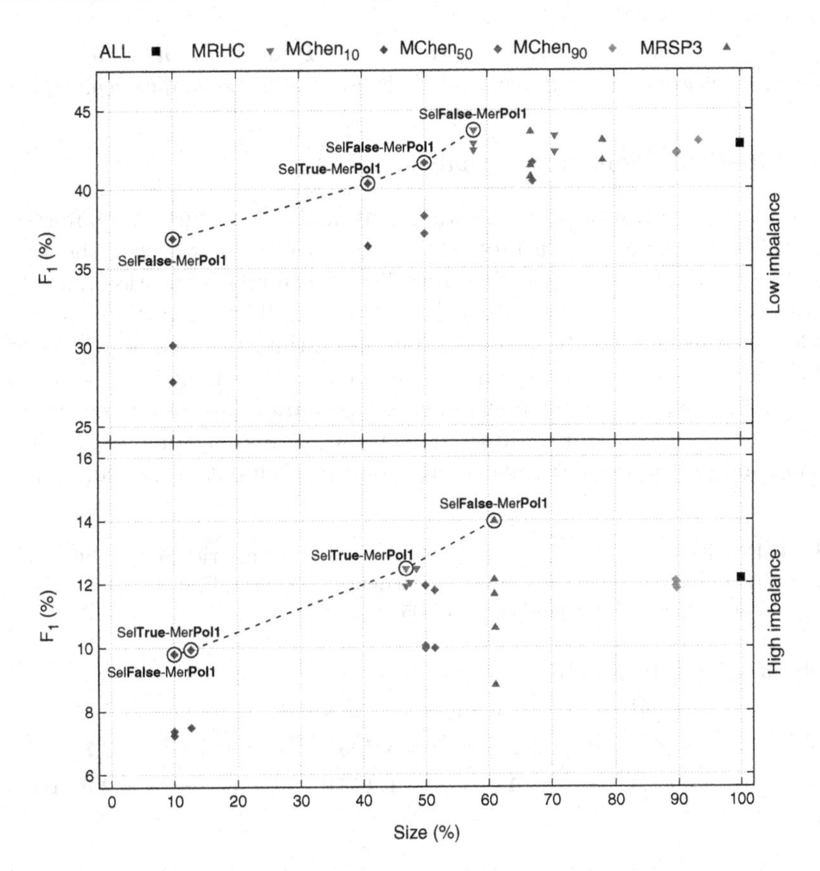

Fig. 2. Results in terms of the F_1 and resulting size for the base PG methods, their extensions, and the exhaustive search case (ALL) for the imbalance levels contemplated. Circled cases and dashed lines represent the non-dominated elements and the Pareto frontiers in each scenario, respectively. The *Sel* and *Mer* labels of the optimal solutions specify the selection and merging policies used.

implementations—those disregarding mechanisms to deal with data imbalance—result in dominated cases or non-optimal solutions.

Focusing on the low-imbalance case, all non-dominated cases—the MChen_{10}, MChen_{50}, and MRHC methods—depict the novel *Policy 1* merging approach introduced in this work (Eq. 4). In addition, it may be checked that only one of these cases considers the *Candidate selection* mechanism—labeled as MChen_{10} with *Sel**True**-Mer**Pol1***—, hence suggesting that the merging policy is capable enough of dealing with the low levels of label imbalance.

Similarly, all non-dominated solutions in the high-imbalance scenario—the MChen_{10}, MRHC, and MRPS3 methods—also exclusively contemplate the *Policy 1* merging approach. However, it must be highlighted that two of these schemes consider the *Candidate selection* mechanism introduced in the work, being them $MChen_{10}$ and MRHC, both with *SelTrue-MerPol1*. This suggests that such a selection process may be beneficial in high-imbalance scenarios.

4.2 Statistical Significance Analysis

This last section presents a set of statistical analyses to further reinforce the aforementioned claims. More precisely, we aim to assess whether the performance improvements observed with the different candidate selection and prototype merging processes may be deemed as significant. The exhaustive search case is no longer considered as this analysis focuses on the PG strategies themselves.

Table 3 shows the results of the non-parametric Friedman test [4] for each of the PG families in the different imbalance scenarios. Note that, for each PG family, the data distributions compared are those given by the different candidate selection and prototype merging combinations for each data collection.

Table 3. Results of the Friedman test in terms of the F_1 metric for the different PG methods and imbalance scenarios. Bold values denote the cases that depict a significant difference considering a threshold of $p < 0.05$.

Imbalance level	Reduction strategy				
	MRHC ▼	$MChen_{10}$ ◆	$MChen_{50}$ ◆	$MChen_{90}$ ◆	MRSP3 ▲
Low	$5.4 \cdot 10^{-1}$	$\mathbf{1.2 \cdot 10^{-3}}$	$\mathbf{6.7 \cdot 10^{-3}}$	$3.8 \cdot 10^{-1}$	$2.7 \cdot 10^{-1}$
High	$1.3 \cdot 10^{-1}$	$\mathbf{2.4 \cdot 10^{-4}}$	$\mathbf{1.4 \cdot 10^{-3}}$	$6.3 \cdot 10^{-1}$	$\mathbf{2.9 \cdot 10^{-3}}$

Attending to Table 3 and considering a significance threshold of $\rho < 0.05$, only the $MChen_{10}$, $MChen_{50}$—in the low and high imbalance scenarios—and MRSP3—only for the high imbalance case—methods depict significant differences. Hence, we now perform a post-hoc analysis focusing on these cases to further gain insights about possible statistical differences among the proposed extensions. For that, we resort to the Bonferroni-Dunn test whose results are provided in Fig. 3 considering as references the base PG cases—*i.e.*, the PG methods that use the base merging strategy (Eq. 3) with no candidate selection.

As it may be observed, the merging *Policy 1* is the only one that depicts a statistical improvement with respect to the base PG methods in all contemplated cases. Such a fact validates the initial premise that using imbalance indicators—in this case, the IRLbl one—for biasing the merging stage is beneficial in label imbalance scenarios. On the contrary, since *Policy 2* does not show any statistical improvement, this strategy may not be deemed as adequate for the case at hand.

Regarding the *Candidate selection* stage, this analysis proves this mechanism as particularly relevant in scenarios with a high label imbalance. This is especially noticeable in the MRSP3 methods since this selection mechanism coupled with the *Policy 1* merging approach is the only configuration that achieves a significant improvement with respect to the baseline strategy considered.

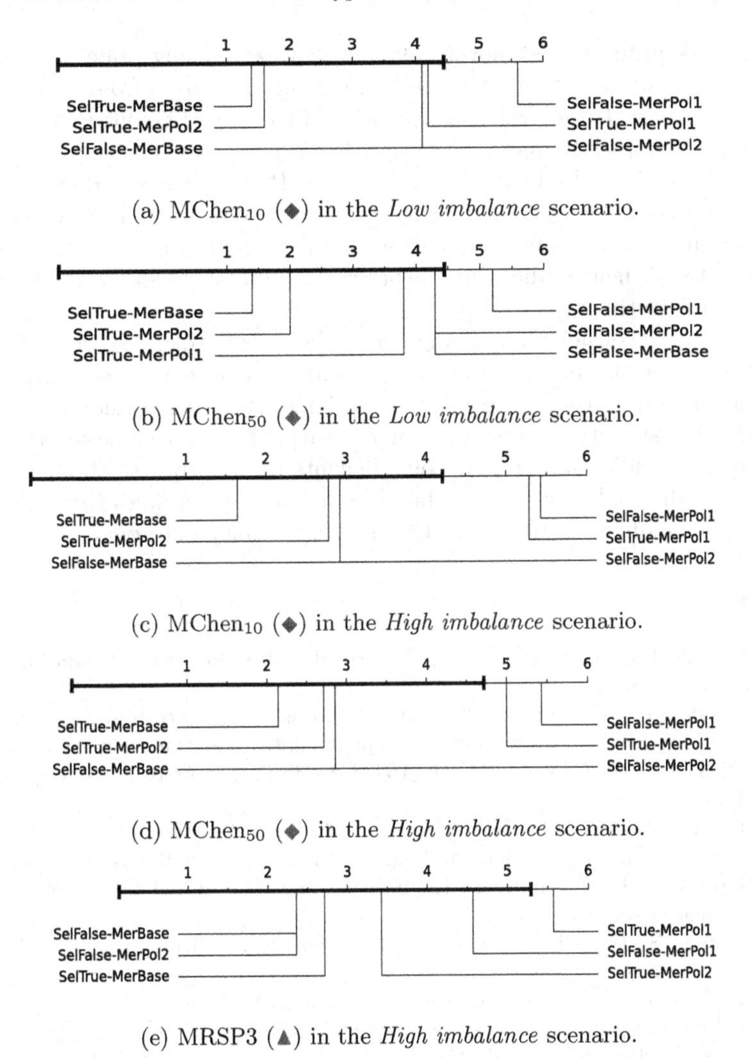

Fig. 3. Post-hoc Bonferroni-Dunn tests considering the F_1 metric with a significance threshold of $\rho < 0.05$. Solutions connected by the bold line depict no statistical differences among them.

5 Conclusions

Prototype Generation (PG), the family of methods that obtain a reduced version of a given reference data collection following certain merging heuristics, represents one of the main paradigms for improving the efficiency of the k-Nearest Neighbor (kNN) classifier. Nevertheless, since research on PG in multilabel environments has been scarcely addressed, the few existing proposals exhibit a number of limitations, being label imbalance one of paramount relevance.

This work proposes two novel imbalance-aware merging policies for PG as well as a mechanism to prevent severely imbalanced samples from undergoing a reduction process. These proposals are applied to three different multilabel PG methods—Multilabel Reduction through Homogeneous Clustering, Multilabel Chen, and Multilabel Reduction through Space Partitioning—and evaluated on 12 datasets with different imbalance degrees. The results prove that the proposals overcome—in some cases in a significant manner—the results obtained with the original methods, hence validating the presented approaches and enabling further research lines on this topic.

Future works consider developing similar policies for other stages within the PG pipelines to obtain fully imbalance-aware reduction strategies as well as extending these proposals to other PG methods. Also, we consider that incorporating other statistical indicators for measuring the imbalance level different from the ones studied may report some benefits in the process. Finally, we consider that additional insights may be obtained by considering other multilabel kNN-based classifiers or other scenarios such as a noisy-data one.

References

1. Bishop, C.M., Nasrabadi, N.M.: Pattern Recognition and Machine Learning. Springer, Cham (2006)
2. Castellanos, F.J., Valero-Mas, J.J., Calvo-Zaragoza, J.: Prototype generation in the string space via approximate median for data reduction in nearest neighbor classification. Soft. Comput. **25**(24), 15403–15415 (2021). https://doi.org/10.1007/s00500-021-06178-2
3. Charte, F., Rivera, A.J., del Jesus, M.J., Herrera, F.: MLeNN: a first approach to heuristic multilabel undersampling. In: Proceedings of the 15th International Conference on Intelligent Data Engineering and Automated Learning (IDEAL), Salamanca, Spain, pp. 1–9 (2014)
4. Demšar, J.: Statistical comparisons of classifiers over multiple data sets. J. Mach. Learn. Res. **7**, 1–30 (2006)
5. Deng, Z., Zhu, X., Cheng, D., Zong, M., Zhang, S.: Efficient kNN classification algorithm for big data. Neurocomputing **195**, 143–148 (2016)
6. Duda, R.O., Hart, P.E., et al.: Pattern Classification. Wiley, Hoboken (2006)
7. Escalante, H.J., Graff, M., Morales-Reyes, A.: PGGP: prototype generation via genetic programming. Appl. Soft Comput. **40**, 569–580 (2016)
8. García, S., Luengo, J., Herrera, F.: Data Preprocessing in Data Mining, vol. 72. Springer, Cham (2015)
9. Liu, B., Tsoumakas, G.: Dealing with class imbalance in classifier chains via random undersampling. Knowl.-Based Syst. **192**, 105292 (2020)
10. Mishra, N.K., Singh, P.K.: Feature construction and smote-based imbalance handling for multi-label learning. Inf. Sci. **563**, 342–357 (2021)
11. Nanni, L., Lumini, A.: Prototype reduction techniques: a comparison among different approaches. Expert Syst. Appl. **38**(9), 11820–11828 (2011)
12. Ougiaroglou, S., Filippakis, P., Evangelidis, G.: Prototype generation for multilabel nearest neighbours classification. In: Sanjurjo González, H., Pastor López, I., García Bringas, P., Quintián, H., Corchado, E. (eds.) HAIS 2021. LNCS (LNAI), vol. 12886, pp. 172–183. Springer, Cham (2021). https://doi.org/10.1007/978-3-030-86271-8_15

13. Tarekegn, A.N., Giacobini, M., Michalak, K.: A review of methods for imbalanced multi-label classification. Pattern Recogn. **118**, 107965 (2021)
14. Tsoumakas, G., Spyromitros-Xioufis, E., Vilcek, J., Vlahavas, I.: MULAN: a Java library for multi-label learning. J. Mach. Learn. Res. **12**, 2411–2414 (2011)
15. Valero-Mas, J.J., Calvo-Zaragoza, J., Rico-Juan, J.R., Iñesta, J.M.: A study of prototype selection algorithms for nearest neighbour in class-imbalanced problems. In: Proceedings of the 8th Iberian Conference Pattern Recognition and Image Analysis (IbPRIA), Faro, Portugal, pp. 335–343 (2017)
16. Valero-Mas, J.J., Gallego, A.J., Alonso-Jiménez, P., Serra, X.: Multilabel prototype generation for data reduction in k-nearest neighbour classification. Pattern Recogn. **135**, 109190 (2023)
17. Zhang, M.L., Zhou, Z.H.: ML-KNN: a lazy learning approach to multi-label learning. Pattern Recogn. **40**(7), 2038–2048 (2007)

Time Series Imputation in Faulty Systems

Ana Almeida[1,2(✉)], Susana Brás[1,3], Susana Sargento[1,2], and Filipe Cabral Pinto[4]

[1] Departamento de Eletrónica, Telecomunicações e Informática, Universidade de Aveiro, 3810-193 Aveiro, Portugal
{anaa,susana.bras,susana}@ua.pt
[2] Instituto de Telecomunicações de Aveiro, 3810-193 Aveiro, Portugal
[3] IEETA, DETI, LASI, Universidade de Aveiro, 3810-193 Aveiro, Portugal
[4] Altice Labs, Aveiro, Portugal
filipe-c-pinto@alticelabs.com

Abstract. Time series data has a crucial role in business. It reveals temporal trends and patterns, making it possible for decision-makers to make informed decisions and mitigate problems even before they happen. The existence of missing values in time series can bring difficulties in the analysis and lead to inaccurate conclusions. Thus, there is the need to solve this issue by performing missing data imputation on time series.

In this work, we propose a Focalize KNN that takes advantage of time series properties to perform missing data imputation. The approach is tested with different methods, combinations of parameters and features. The results of the proposed approach, with overlap and disjoint missing patterns, show Focalize KNN is very beneficial in scenarios with disjoint missing patterns.

Keywords: Missing data imputation · K-Nearest Neighbors · Time series · Overlap missing data · Disjoint missing data

1 Introduction

Missing data in time series can hide existing patterns and trends, which can influence the analysis and conclusions drawn by decision-makers. A wrong analysis can have high costs and negative impacts on business and people's lives, contributing to a lack of confidence in data and methods. This can prevent decision-makers from taking valuable insights from data and from performing a coherent analysis.

This work is supported by FEDER, through POR LISBOA 2020 and COMPETE 2020 of the Portugal 2020 Project CityCatalyst POCI-01-0247-FEDER-046119. Ana Almeida acknowledges the Doctoral Grant from Fundação para a Ciência e Tecnologia (2021.06222.BD). Susana Brás is funded by national funds, European Regional Development Fund, FSE, through COMPETE2020 and FCT, in the scope of the framework contract foreseen in the numbers 4, 5 and 6 of the article 23, of the Decree-Law 57/2016, of August 29, changed by Law 57/2017, of July 19. We thank OpenWeather for providing the datasets.

A. Pertusa et al. (Eds.): IbPRIA 2023, LNCS 14062, pp. 28–39, 2023.
https://doi.org/10.1007/978-3-031-36616-1_3

In smart cities, the missingness of data is mainly associated with faults and failures in the infrastructure. For instance, they can happen on the network or in the equipments and services from the edge and core. Moreover, they can have distinct causes, such as humans (e.g., maintenance in the infrastructure or constructions), natural (e.g., bad weather conditions) and problems in the software or the hardware (e.g., bugs in the software). Besides, these faults and failures can have different durations, leading to missing data lasting a few minutes, hours, days, weeks, months, etc. This type of missing data is called block missing data [8].

These faults and failures can severely impact the application of analytical, statistical and machine learning methods, such as the time series decomposition in components, the detection of the most relevant lags, the clustering of time series, and forecasting tasks. Besides, they can lead to service degradation, negatively impacting the services provided by smart cities [11].

There are several methods that allow to perform missing data imputation. The most used method is replacing the missing values with a specific value, such as the mean, median, mode, zero, or another specific value. However, these are usually not suitable for time series data. Methods such as forward and backwards fill and using the nearest, or even based on interpolation (e.g., linear, quadratic, cubic, polynomial, spline, etc.), are usually better choices when working with time series; however, these methods are not expecting long sequences of missing data.

With this work, we aim to decrease the impact that faults and failures can have on data analysis in smart cities. More specifically, we aim to decrease the impact of missing data by proposing a method to perform missing data imputation on time series. Most works in the literature focus on generic missing data, not specific patterns in missing data, such as consecutive missing data. The main contributions of this work are:

- A mechanism to generate synthetic missing data.
- The study of the impact of using different types of feature engineering in time series to solve missing data.
- A method to perform imputation in time-series, the Focalize K-Nearest Neighbors (FKNN).

The results of the proposed mechanism and study show that FKNN presents small benefits in the overlap missing patterns, but is able to capture the missing data, and perform missing data imputation in a disjoint missing patterns.

The organization of the paper is as follows. Section 2 presents related work on missing data imputation, and Sect. 3 describes the proposed approach. Section 4 depicts the experiments and results, and Sect. 5 contains the conclusions and directions for future work.

2 Related Work

K-Nearest Neighbors (K-NN) is a predictive model that can be used for data imputation by using similar points to guess the missing data. This model is vastly used; however, it suffers from the curse of high dimensionality [5]. Besides, KNN stores the complete dataset in memory [13]. Several improvements have been proposed over the years. For instance, Wettschereck et al. [13] presented a locally adaptive nearest neighbor for classification. The proposed method adapts the value of k locally within a subset with specific characteristics instead of using the global value of k computed during cross-validation. They presented four different variations of the method and concluded that their method was very beneficial in certain datasets. Regarding the imputation of missing values in time series, Oehmcke et al. [10] used a weighted version of K-NN ensembles with penalized Dynamic Time Warping (DTW) as a distance measure to perform multivariate time series imputation. They created windows for DTW using linear interpolation and the correlation coefficient for the global weights. The model penalizes consecutive missing data, expects highly correlated time series, and has proven to be more beneficial at higher missing ratios. Sun et al. [12] used a temporal gap-sensitive windowed K-NN, and they focused their study on traffic data. The model has proven to be robust even with high missing ratios. This technique can be useful in several cases; however, since we expect blocks of missing data, we can have windows with only missing data. Besides, the dataset is v times bigger that the original dataset, being v the window size. This can be problematic, especially when the dataset by itself is bigger enough and, therefore, computationally expensive.

Some methods use Artificial Neural Networks (ANNs) for missing data imputation. In [2,3,8] the authors used Recurrent Neural Networks (RNNs). Cao et al. [2] used a bidirectional Long Short-Term Memory (LSTM) graph. According to the authors, their model can handle multiple correlated missing values and is able to generalize. Che et al. [3] proposed a model based on Gated Recurrent Unit (GRU) with trainable decays to capture previous properties. Their goal is to make predictions, even though the dataset may contain missing data. The model depends on the missing data patterns and the correlation between those patterns and the prediction phase. Therefore, the model can perform poorly if the two tasks are not correlated. Li et al. [8] presented a multi-view learning method that combines LSTM with Support Vector Regression (SVR) and collaborative filtering. They consider spatial-temporal dependencies and focus their study on block missing patterns, with different missing patterns and with a high missing ratio.

Other approaches include Generative Adversarial Networks (GANs) and AutoEncoders. For example, Luo et al. [9] proposed an End-to-End Generative Adversarial Network (E^2GAN) that uses discriminative and squared error loss to perform imputations using the closest generated complete time series. Kuppannagari et al. [7] proposed a spatio-temporal model to deal with missing data based on Graph Neural Network (GNN) and Denoising Autoencoder (DAE). Their method was able to adapt to different configurations of missing blocks.

Khayati et al. [6] evaluated imputation methods for time series. They compared twelve methods and improved some of them. Some of the methods are based on Principal Component Analysis (PCA), Singular Value Decomposition (SVD), Expectation-Maximization (EM), Autoregressive (AR), among others. They used several datasets with different properties and different missing data patterns. They concluded that no single method is the best for all cases.

Deep learning techniques require a huge amount of data for efficiency and not biased results. In this context, traditional machine learning models may be the solution. However, methods such as K-NN can be problematic with high-dimensional datasets. In this work, a modification of the K-NN method is proposed and evaluated for data imputation in time series.

3 Methods

A multivariate time series can be represented by a matrix $\mathcal{M}$ of dimensions $m \times n$, as is shown in Fig. 1. m is the number of rows or the number of time steps, and n is the number of columns or the number of features. $\mathcal{M}_{ij}$ is the value of the feature j (f_j) at the time i (t_i).

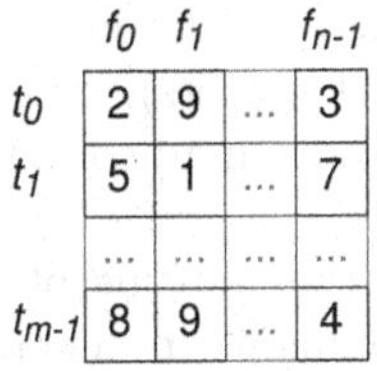

	f_0	f_1		f_{n-1}
t_0	2	9	...	3
t_1	5	1	...	7
	...	...	...	...
t_{m-1}	8	9	...	4

Fig. 1. Matrix $\mathcal{M}$.

3.1 Dataset and Feature Engineering

To start our study, we used one dataset obtained from OpenWeather that contains weather data. It has one year of hourly measurements, of the year 2022, from twenty cities taken in 2022. Regarding the weather information, we have data from temperature, pressure, humidity, wind speed, wind direction, wind gust, and cloudiness. Table 1 contains information regarding the dataset. Since we have features from distinct types of sensors, it was essential to ensure they all had the same range of values. Therefore, we applied the Min-Max scaler.

Feature engineering is essential in data science, especially when we have time series related problems. A common approach with time series is the selection of the most relevant lags, since we can have cyclic and seasonal patterns [1]. For instance, in the context of weather, we know that depending on the season, we can have warmer or colder weather, some months are more prone to rain, while others are not, and nights are colder than days. Therefore, we decided to use, for instance, the previous 24 h (1 day) and 168 h (7 days) as lags. However, since our problem is missing data and not a forecasting task, we also decide to use, as features, the next 24 h and 168 h.

Table 1. Dataset properties.

Dataset	Weather Dataset
Number of cities	20
Number of sensors (per city)	7
Number of features	140
Number of samples	8 760
Total data points	1 226 400
Time interval	January 1st–December 31th, 2022
Periodicity	Hourly

3.2 Generation of Synthetic Missing Data

The impact of each fault or failure in the system can leave sequences of missing data with different durations. According to the reason that generated the fault or failure, we can have missing data in only one time series (e.g., a sensor is damaged), some time series (e.g., we lost communication with a collecting unit that aggregates data from several sensors), or in all time series (e.g., we have network issues and lost connection with all communication entities, and we have a blackout situation).

Since we are expecting a particular type of pattern regarding our missing data, we decided to create mechanisms to generate similar patterns. Figure 2a contains a type of failure called overlap [6]. In this type of failure, k sensors fail at the same time during a period p. In a more extreme case, if all sensors fail at the same time, we will have a blackout. If sensors fail at different times, we have a disjoint situation [6], as is represented in Fig. 2b.

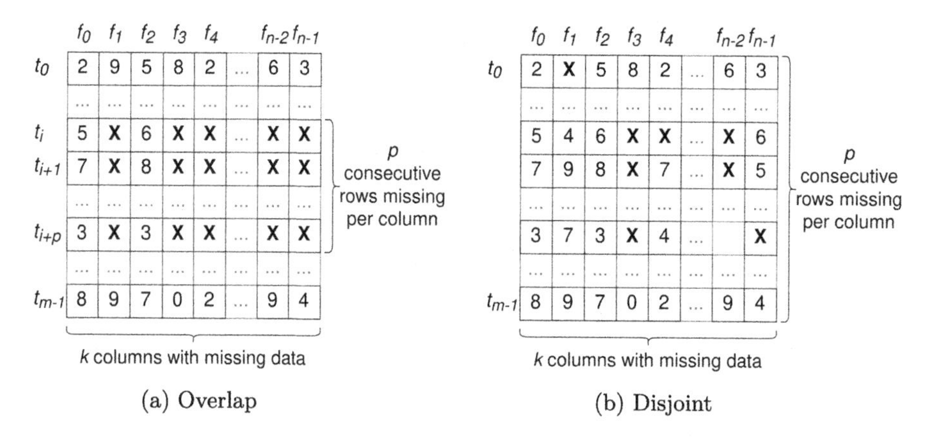

(a) Overlap (b) Disjoint

Fig. 2. Missing block patterns.

Figure 3 represents how the duration of the fault or failure, and the number of sensors and stations affected, influence the percentage of missing data. For

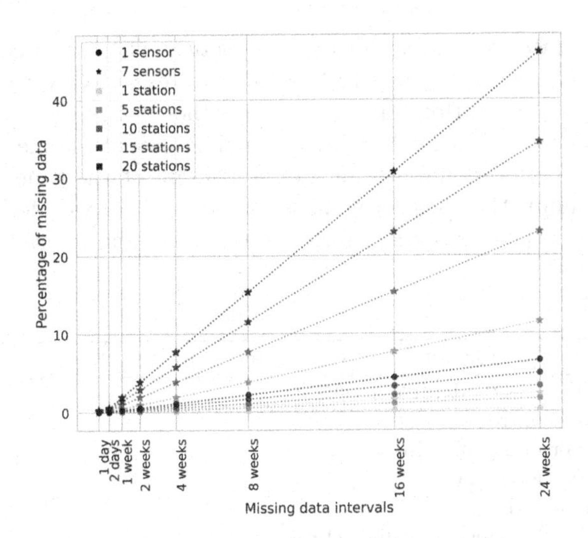

Fig. 3. Percentage of missing data.

instance, if we have a block of 24 weeks of missing data affecting 20 stations and 7 sensors per station, we will have 46.03% of missing data.

Note that, when we have, for instance, one sensor affected in 10 different stations, we force that sensor to be the same in all stations (e.g., temperature sensor). The same happens if we have, for instance, seven sensors affected in one station. This brings additional challenges, since we can have missing data in very correlated sensors. Besides, we are testing each missing data pattern with 80 different configurations of missing data (8 time intervals $\times$ 2 sensor combinations $\times$ 5 station combinations).

3.3 Algorithms Implementation and Evaluation

We selected 20 statistical baseline methods to evaluate our algorithm. Some of the chosen methods replaced missing data using a specific value such as the mean, median, last value, previous value, the nearest value, and zero. Others used interpolation techniques, such as the *barycentric*, *pchip*, *splines* with different orders, polynomial functions, piecewise polynomial, and *akima*[1]. From these 20 methods, we decided to select the best 3 to compare them with K-NN imputation and our proposed method.

Regarding K-NN imputation, we experimented with the regular K-NN in which the weights are uniforms and the weighted version. We only present the results for the best K-NN.

We created the Algorithm 1 based on K-NN. However, K-NN is an algorithm based on distances; therefore, if we have a high number of features or irrelevant

[1] https://pandas.pydata.org/docs/reference/api/pandas.DataFrame.interpolate.html.

features on our dataset, K-NN will perform poorly. Given the context of our problem, we have time series in which some features are very correlated with others. We decided to perform imputation on the matrix by applying K-NN imputation focalized on each column with missing data of the matrix. The proposed algorithm performs imputation in one column at the time. We start by selecting the column that has fewer missing data. Then, we identify the best features and select the relevant lags for the selected column to use them as input features of the K-NN.

Algorithm 1. FKNN for Time Series Missing Data Imputation

```
1: procedure TimeSeriesImputation(M)
2:     m ← number of rows in M
3:     n ← number of columns in M
4:     f ← empty dictionary
5:     for i ← 1 to n do
6:         if M[:, i] has missing values then
7:             Add most correlated features of M[:, i] to f[i]
8:             Add temporal lags of M[:, i] to f[i]
9:         end if
10:    end for
11:    while M has missing data do
12:        i ← index of column with minimum missing values (number of missing
13:            values bigger than 0)
14:        c ← column with index i
15:        X ← matrix with column c, relevant features, and lags
16:        X_imputed ← apply KNN to matrix X
17:        M[:, i] ← X_imputed[:, i]
18:    end while
19:    return M
20: end procedure
```

For the identification of the best features, we tried different approaches. Since we need to make this identification even when there is a high percentage of missing data, we decided to use *Spearman*'s correlation to help to choose the best features. We considered using a constant value for the number of features selected, for instance, by selecting the 10 most correlated features, and by using a threshold and electing all the features with a correlation value bigger than the threshold. Figure 4 contains an example with four cities, three sensors, and using a threshold of 0.5. For this case, we considered features very correlated, with values bigger than 0.5 or lower than -0.5. We excluded from the analysis the diagonal of the matrix, since it compares the features with themselves. Besides, we also used lag selection, as explained in Sect. 3.1.

Figure 5 presents the workflow used in this work. We started with the dataset composed of a multivariate time series without missing data; then, we generated synthetic missing data to obtain a multivariate time series with missing data.

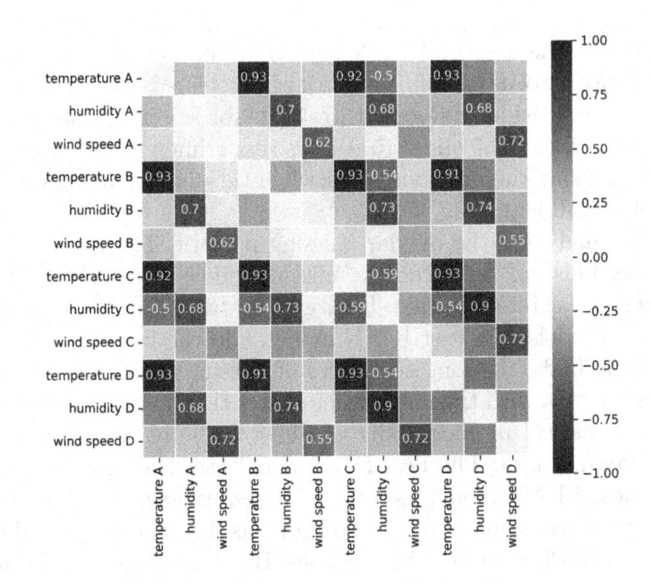

Fig. 4. Correlation matrix using data from 4 cities and 3 sensors.

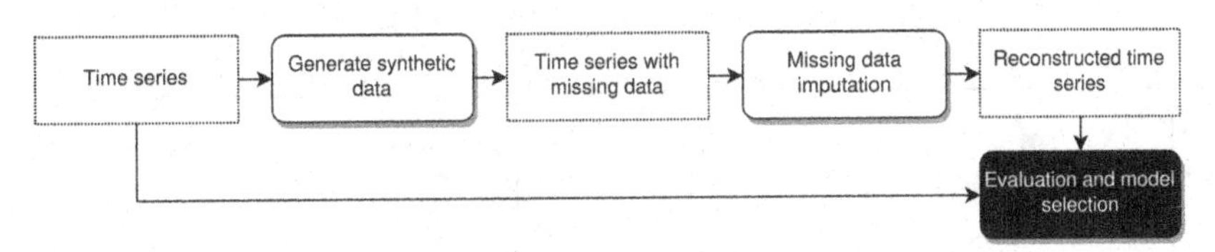

Fig. 5. Workflow.

We applied the imputation methods to the time series with missing data and got a reconstructed time series. Finally, we evaluated the methods by comparing the original time series with the reconstructed one. We selected two evaluation metrics, Mean Squared Error (MSE) and R^2-Score, to evaluate our models, and we decided to use R^2-Score to choose the best model [1,4]. The MSE can have values from 0 to $+\infty$, and it penalizes big errors when compared to similar metrics such as Mean Absolute Error (MAE). A good model has a low MSE. R^2-Score provides valuable information regarding how well the model can fit the data. Its values range from $-\infty$ to 1. A good model has an R^2-Score close to 1.

4 Results and Discussion

The correction of each missing pattern is based on statistical baseline methods, K-NN and the proposed FKNN. In addition, we test with different parameters to improve the model's suitability. Two versions of the K-NN imputation are tested: the uniform and the weighted version. Considering the number of neighbors, different hypotheses are explored: 2, 3, 5, 7, 9, 11, 13, 15, and 19. The lags combination may influence the FKNN performance, so under this work scope, one hour, one day, two days, and seven days are tested. We also consider the use of no lags. We consider different thresholds for selecting the most relevant features, such as values of 0.5 and 0.75. We also experiment with selecting at most 10 features when using the previously described thresholds, or even selecting the best 10, 20, 30, or 40 features.

We start by analyzing the overlap missing pattern. The three best statistical baseline methods are replacing the missing values with the mean, median, and performing linear interpolation. Figure 6a contains the R^2-Score of the best three statistical models, the best K-NN model, and the best FKNN. Regarding the statistical methods, the long sequences of missing data make strategies like replacing missing data with the mean value better than interpolation techniques. Note that all statistical methods only consider the univariate time series they are trying to impute data. The best KNN model uses 9 neighbors and uniform weights. The best FKNN model uses 9 neighbors, uniform weights, a threshold of 0.5, and the lags corresponding to the previous day, two days, and one week. Figure 6b compares the obtained MSE values. By comparing Figs. 6a and 6b, we can conclude that our method's improvement is low compared to the standard version of K-NN.

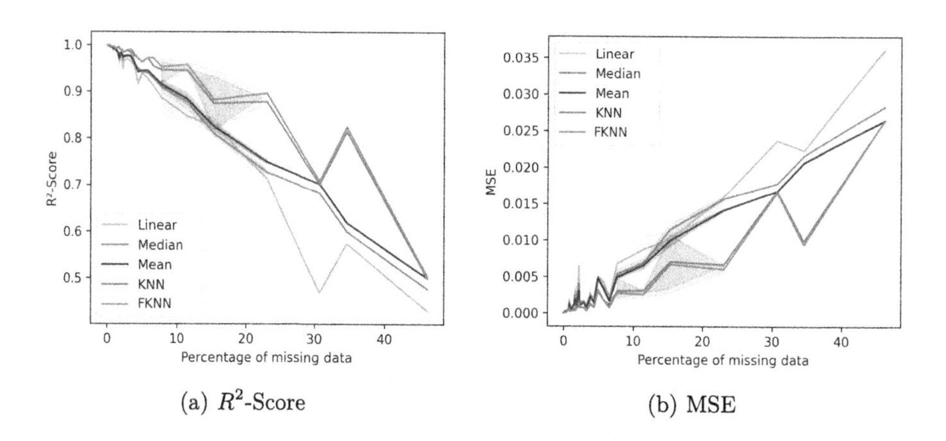

(a) R^2-Score (b) MSE

Fig. 6. Overlap missing pattern.

Choosing the best k (number of neighbors) in K-NN is a challenging task. One of the strategies used is to compute the square root of the number of samples. However, in our case, we would get a value of 94, which is a very high value of k. Larger values of k can contribute to high bias and lower variance; besides, they are computationally expensive. On the other side, lower values present high variance and low bias. Therefore, we decided to evaluate the model's performance when we changed the values of k, as can be visualized in Fig. 7. We decided to focus on odd numbers for k; however, since we expect more than one neighbor,

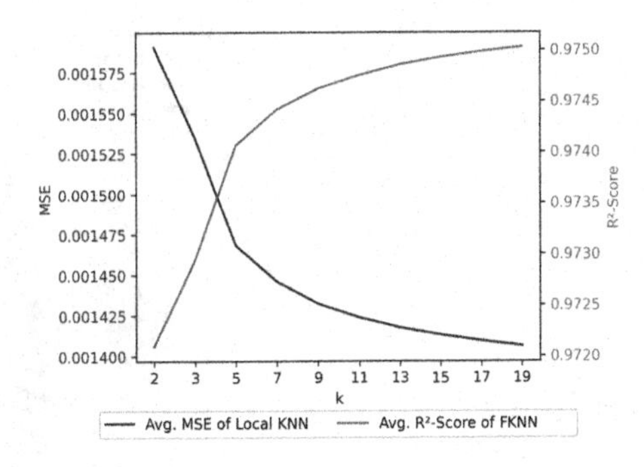

Fig. 7. Performance of FKNN versus k - overlap missing pattern.

we started k with a value of 2 instead of 1. As we can observe, the curves start to flatten around 9 neighbors; therefore, we decided to use k with the value of 9.

For the disjoint pattern, the best models are the linear interpolation, replacement of missing data with the median and the mean, and the K-NN with 9 neighbors and uniform weights. The best version of our algorithm also uses 9 neighbors and uniform weights. Besides, for the best version, we use a threshold of 0.5, and the lags correspond to the previous hour and the previous day. Our algorithm achieves a good performance for the disjoint pattern, as shown in Fig. 8a, especially when we have a more significant percentage of missing data. In Fig. 8b, the MSE presents a similar evidence.

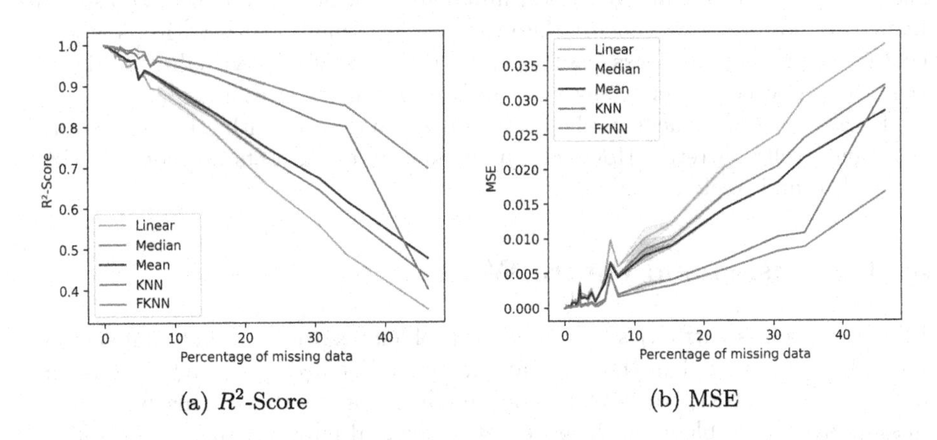

(a) R^2-Score (b) MSE

Fig. 8. Disjoint missing pattern.

Regarding the performance of the FKNN versus k, we make similar conclusions to the ones presented for Fig. 7. The results can be observed in Fig. 9.

We usually obtain the best results using the uniform version of K-NN rather than the weighted version of K-NN. Furthermore, we notice that using the next

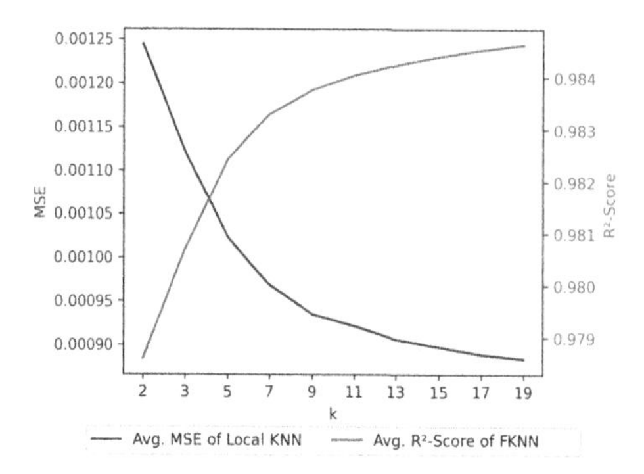

Fig. 9. Performance of FKNN versus k - disjoint missing pattern.

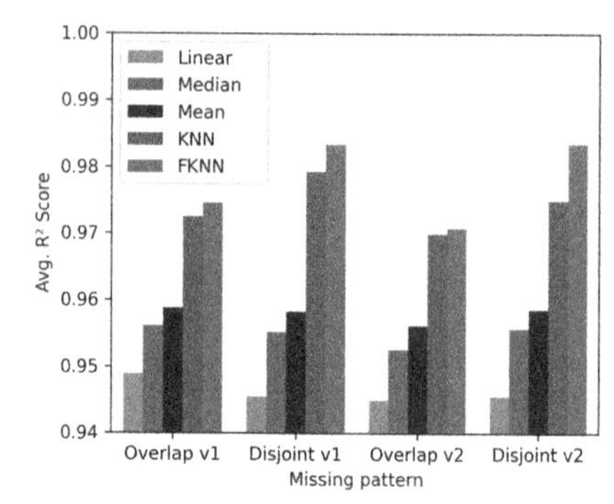

Fig. 10. Avg. R^2-Score per method, overlap and disjoint patterns.

temporal periods as features does not improve the performance compared to using only past lags. We also experimented using only the best positive correlation versus the best absolute correlations, and noticed that the performance of our model slightly improved when considering the best absolute correlations.

We also decided to compare the average of the R^2-Score for all missing data values. Also, we repeated the experiment and generated more missing data patterns. The overlap v1 and disjoint v1 are the first datasets in which we tested the proposed method, and overlap v2 and disjoint v2 are the datasets we used to finally evaluate the experiment. Note that we have more values with a low percentage of missing data, less than 10%. Nevertheless, we can visualize in Fig. 10 that the proposed model achieves the best performance overall.

Finally, we aim to achieve the best results in a dataset with stronger seasonality and cyclic patterns. However, these time-series datasets are not straightforward to find freely available.

5 Conclusion and Future Work

Missing data in smart cities is a common problem that can severely impact the normal operation of algorithms and applications. Usually, when we have missing data in this context, we have a block missing data pattern. This pattern of missing data is problematic since we lose sequential information.

This work proposed a method to perform block missing data imputation on multivariate time series. Our method, FKNN, proved to be more beneficial when there is a disjoint missing pattern rather than an overlapping missing pattern. This is a great improvement, since it makes possible the use of data accurately even in the presence of data losses.

In the future, we aim to develop other methods to perform imputation in time series.

References

1. Almeida, A., Brás, S., Oliveira, I., Sargento, S.: Vehicular traffic flow prediction using deployed traffic counters in a city. Futur. Gener. Comput. Syst. **128**, 429–442 (2022). https://doi.org/10.1016/j.future.2021.10.022
2. Cao, W., Wang, D., Li, J., Zhou, H., Li, L., Li, Y.: Brits: bidirectional recurrent imputation for time series. In: Bengio, S., Wallach, H., Larochelle, H., Grauman, K., Cesa-Bianchi, N., Garnett, R. (eds.) Advances in Neural Information Processing Systems, vol. 31. Curran Associates, Inc. (2018)
3. Che, Z., Purushotham, S., Cho, K., Sontag, D., Liu, Y.: Recurrent neural networks for multivariate time series with missing values. Sci. Rep. **8** (2018). https://doi.org/10.1038/s41598-018-24271-9
4. Chicco, D., Warrens, M.J., Jurman, G.: The coefficient of determination R-squared is more informative than SMAPE, MAE, MAPE, MSE and RMSE in regression analysis evaluation. PeerJ Comput. Sci. **7** (2021). https://doi.org/10.7717/peerj-cs.623
5. Grus, J.: Data Science from Scratch: First Principles with Python. O'Reilly Media, Inc. (2019)
6. Khayati, M., Lerner, A., Tymchenko, Z., Cudre-Mauroux, P.: Mind the gap: an experimental evaluation of imputation of missing values techniques in time series. Proc. VLDB Endow. **13**, 768–782 (2020). https://doi.org/10.14778/3377369.3377383
7. Kuppannagari, S.R., Fu, Y., Chueng, C.M., Prasanna, V.K.: Spatio-temporal missing data imputation for smart power grids. In: Proceedings of the Twelfth ACM International Conference on Future Energy Systems, e-Energy 2021, pp. 458–465. Association for Computing Machinery, New York (2021). https://doi.org/10.1145/3447555.3466586
8. Li, L., Zhang, J., Wang, Y., Ran, B.: Missing value imputation for traffic-related time series data based on a multi-view learning method. IEEE Trans. Intell. Transp. Syst. **20**(8), 2933–2943 (2019). https://doi.org/10.1109/TITS.2018.2869768
9. Luo, Y., Zhang, Y., Cai, X., Yuan, X.: E^2GAN: end-to-end generative adversarial network for multivariate time series imputation. In: Proceedings of the Twenty-Eighth International Joint Conference on Artificial Intelligence, IJCAI 2019, pp. 3094–3100. International Joint Conferences on Artificial Intelligence Organization (2019). https://doi.org/10.24963/ijcai.2019/429
10. Oehmcke, S., Zielinski, O., Kramer, O.: KNN ensembles with penalized DTW for multivariate time series imputation, pp. 2774–2781 (2016). https://doi.org/10.1109/IJCNN.2016.7727549
11. Shamsi, J.A.: Resilience in smart city applications: faults, failures, and solutions. IT Prof. **22**(6), 74–81 (2020). https://doi.org/10.1109/MITP.2020.3016728
12. Sun, B., Ma, L., Cheng, W., Wen, W., Goswami, P., Bai, G.: An improved k-nearest neighbours method for traffic time series imputation. In: 2017 Chinese Automation Congress (CAC), pp. 7346–7351 (2017). https://doi.org/10.1109/CAC.2017.8244105
13. Wettschereck, D., Dietterich, T.: Locally adaptive nearest neighbor algorithms. In: Cowan, J., Tesauro, G., Alspector, J. (eds.) Advances in Neural Information Processing Systems, vol. 6. Morgan-Kaufmann (1993)

DARTS with Degeneracy Correction

Guillaume Lacharme(✉), Hubert Cardot, Christophe Lenté, and Nicolas Monmarché

Université de Tours, LIFAT EA 6300, Tours, France
{guillaume.lacharme,hubert.cardot,christophe.lente, nicolas.monmarche}@univ-tours.fr

Abstract. The neural architecture search (NAS) is characterized by a wide search space and a time consuming objective function. Many papers have dealt with the reduction of the cost of the objective function assessment. Among them, there is DARTS paper [1] that proposes to transform the original discrete problem into a continuous one. This paper builds an overparameterized network called hypernetwork and weights its edges by continuous coefficients. This approach allows to considerably reduce the computational cost, but the quality of the obtained architectures is highly variable. We propose to reduce this variability by introducing a convex depth regularization. We also add a heuristic that controls the number of unweighted operations. The goal is to correct the short term bias, introduced by the hypergradient approximation. Finally, we will show the efficiency of these proposals by starting again the work developed in Dots paper [2].

1 Introduction

The neural architecture search (NAS) is characterized by several hyperparameters such as the number of layers, the size/type of these layers and finally the arrangement between the different layers. The choices concerning these architectural hyperparameters have a direct impact on the model degrees of freedom and thus on its ability to overfit. We know from numerous experiments that deep architectures tend to generalize better. Therefore, a good search space must contain a lot of deep architecture to be relevant. Classical black-box optimization methods are not easily applicable for this kind of search space. This comes from the following fact: the computational cost of training an architecture increases with its depth. In order to reduce this training cost, many approaches have been developed such as Network mophism [15], hyperband [18] and zero shot NAS [17]. However, at the moment, the most impactful approach in NAS field comes from Differentiable Architecture Search (DARTS) paper [1]. The transformation of the discrete optimization problem into a continuous one has considerably reduced the computational cost of NAS. Using this approach, it is possible to train both the network weights and the architectural hyperparameters.

In this paper, we will present an approach that aims to correct the degeneracy problem observed with the DARTS algorithm. We characterize degeneracy with

A. Pertusa et al. (Eds.): IbPRIA 2023, LNCS 14062, pp. 40–53, 2023.
https://doi.org/10.1007/978-3-031-36616-1_4

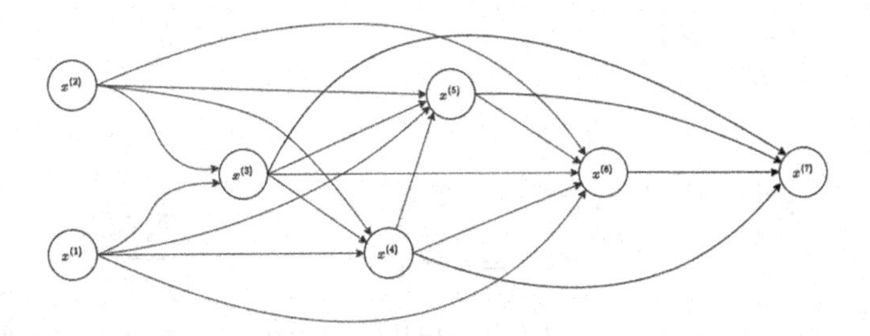

Fig. 1. Complete DAG up to the node $x^{(6)}$

the following observation: The algorithm defined in DARTS paper mainly selects architectures that converge quickly. This last observation can be explained using the work of PR-DARTS [4] and short horizon bias [9]. To correct this bias, we will introduce a depth metric and use it to penalize shallow architectures more heavily. Moreover, we will apply a modification of the discretization operator. This modification allows a better selection when the continuous coefficients are not totally sparse. Finally, we will apply a heuristic that prevents the selection of too many weightless operations. We list the contributions of this paper as follows:

- Convex depth metric
- Modification of the discretization operator
- Heuristic that controls the number of weightless operations
- Generalization of Dots approach [2] for high number of incomings edges.

2 Search Space

The DARTS paper [1] characterizes the architecture of a convolutional neural network (CNN) using a normal cell and a reduction cell. Therefore, it reduces the search space by focusing only on the architecture of the normal and reduction cells. In order to further reduce the computational cost, this same paper proposes to search for the architecture of a so-called "proxy" neural network that contains 8 cells and then transfer the resulting architecture to a target network with 20 cells. The DARTS algorithm is divided into two phases:

- Search phase: we look for the architecture of the normal/reduction cell of the proxy network
- Evaluation phase: the obtained architectures are transferred to the target network then we train it to know the final generalization error

The cell of a convolutional neural network can be characterized using a direct acyclic graph (DAG). DARTS paper [1] defines the nodes and the edges of the DAG as follows:

- edges $o^{(i,j)}$: Operation
- nodes $x^{(j)}$: Sum

$$o^{(i,j)}: \mathbb{R}^{h_i \times w_i \times d_i} \longrightarrow \mathbb{R}^{h_j \times w_j \times d_j} \qquad x^{(i)} \mapsto y^{(i,j)} = o^{(i,j)}(x^{(i)}) \qquad x^{(j)} = \sum_i y^{(i,j)} \tag{1}$$

An architecture that belongs to the DARTS space corresponds to a subgraph of the complete DAG illustrated in Fig. 1. Input nodes $x^{(1)}$ and $x^{(2)}$ match the output of the two previous cells. The internal nodes $x^{(3)}, x^{(4)}, x^{(5)}, x^{(6)}$ are attached to only two of its predecessors. There are 7 choices of operations for blue edges. The black edges are fixed and correspond to skip connections. For the final node $x^{(7)}$, a concatenation operator is applied instead of the sum. There are:
$(7^{4\times 2} \times C_2^2 \times C_2^3 \times C_2^4 \times C_2^5)^2 = 7^{16} \times 32400 \approx 10^{18}$ possible architectures.

3 Continuous Transformation

For this section, we will introduce the continuous transformation taken from the Dots DARTS paper [2]. We will use this transformation for the rest of our work. This paper proposes to build a hypernetwork whose nodes and edges are expressed as follows:

$$o^{(i,j)}: [0,1]^7 \times \mathbb{R}^{h_i \times w_i \times d_i} \longrightarrow \mathbb{R}^{h_j \times w_j \times d_j} \qquad (\beta^{(i,j)}, x^{(i)}) \mapsto y^{(i,j)} = \sum_{o \in O} \beta_o^{(i,j)} \times o(x^{(i)}) \qquad x^{(j)} = \sum_{i=1}^{j-1} \alpha_i^{(j)} y^{(i,j)}, \quad \alpha_i^{(j)} \in [0,1] \tag{2}$$

Instead of considering a single operation per edge we will consider a mixture of operations. We will also apply a mixture of edges for the internal nodes. From now on, each internal node is linked to all its predecessors. Therefore, the hypernetwork is characterized by a complete DAG weighted by coefficients α and β. The mathematical expression of the coefficients is defined below:

$$\beta^{(i,j)} \doteq \sum_{o \in O} \beta_o^{(i,j)} e_o = \sum_{o \in O} \frac{\exp\left(\frac{\lambda_o^{(i,j)}}{T}\right)}{\sum_{o' \in O} \exp\left(\frac{\lambda_{o'}^{(i,j)}}{T}\right)} e_o$$

$$\alpha^{(j)} = \sum_{v \in V^{(j)}} \gamma_v^{(j)} v = \sum_{v \in V^{(j)}} \frac{\exp\left(\frac{\lambda_v^{(j)}}{T}\right)}{\sum_{v' \in V^{(j)}} \exp\left(\frac{\lambda_{v'}^{(j)}}{T}\right)} v \tag{3}$$

with

- $O = \{(Skip, 1), (AvgP, 2), (MaxP, 3), (Sep3 \times 3, 4), (Sep5 \times 5, 5), (Dil3 \times 3, 6), (Dil5 \times 5, 7)\}$
  A number is associated to each operation
- $e_o \in \{E : e \in \{0,1\}^7, ||e||_1 = 1\}$ the standard basis

- $v \in \{V^{(j)} : v^{(j)} \in \{0,1\}^{n_j}, ||v^{(j)}||_1 = 2\}$
  with $n_j = j - 1$ the number of incoming edges to node j
- $Card(V^{(j)}) = C^2_{n_j}$
- T corresponds to the temperature

The final goal is to keep only one operation per edge and two incoming edges per node, therefore weighting must check $||\beta^{(i,j)}||_1 = 1$ and $||\alpha^{(j)}||_1 = 2$. It is important to note that hypernetwork construction adds a significant memory cost.

To understand the benefit of a discrete transformation into a continuous one, we need to deal with the optimization problem characterizing the neural architecture search. This is a bilevel optimization problem defined as follows:

$$\text{argmin}_{\lambda} \quad \mathcal{L}_{valid}(w^*(\lambda), \lambda) \tag{4}$$
$$s.t. \;\; w^*(\lambda) = \text{argmin}_w \mathcal{L}_{train}(w, \lambda), \tag{5}$$

with

- $\boldsymbol{\lambda} \in \mathbb{R}^m$: architectural hyperparameters of the neural network
- $\boldsymbol{w} \in \mathbb{R}^n$: weights of the neural network
- $\boldsymbol{\mathcal{L}_{valid}|\mathcal{L}_{train}}$: the validation and training error respectively

(4) corresponds to the "outer" optimization problem and (5) corresponds to the "inner" optimization problem. Instead of alternately solving the inner optimization problem and then the outer optimization problem, DARTS paper proposes to solve both problems simultaneously. To do so, it alternates between a gradient descent on the architectural hyperparameters and a gradient descent on the network weights. When the algorithm has finished "converging" we apply an argmax operator to discretize the continuous coefficients α and β.

$$i^* \in \text{argmax}^{(1,2)}_{i \in \{1,\ldots,j-1\}} \; \alpha^{(j)}_i \tag{6}$$
$$o^{(i^*,j)}_* \in \text{argmax}_{o \in O} \beta^{(i^*,j)}_o \tag{7}$$

$\text{argmax}^{(1,2)}$ returns the two indices associated with the two largest elements of the vector $\alpha^{(j)}$.

3.1 DARTS Algorithm Limits

This method significantly reduces the time spent for architectural search. However, the DARTS algorithm yields degenerate architectures that mainly contain operations without weight (Skip-Connection, Avg-Pooling, Max-Pooling). There are several explanations for this behavior:

1. Hypernetwork training avoids having to train each architecture independently. However, DropNAS paper [16] notice a coadaptation phenomenon between the different operations. We are faced with a correlated variable selection problem,

2. Correlation between discretized network performance and the continuous one is not necessarily established,
3. Hyperparameter gradient approximation (hypergradient) leads to a short horizon bias [9].

4 DARTS Improvement

4.1 Batch Norm

Unlike DARTS paper we use an α coefficient that allows to select the desired number of edges per node. This approach enables direct application of the batch norm without causing the unreliability of β as mentioned in Variance-stationary paper [14]. It is possible to have a stationary variance without using the β-continuous relaxation. As a reminder, we apply batch norms before weighting by coefficients α and β so that they correctly reflect the operation strengths.

4.2 Discretization Error

Concerning the discretization error, several papers ([2,4], ...) propose to add a temperature term T in the softmax. We have chosen to initialize the temperature to $T\_max = 10$ and then we decrease it towards $T\_min = 0.01$ using an exponential decay:

$$T = T\_max \times \left(\frac{T\_min}{T\_max}\right)^{\frac{epoch}{num\_epoch}}$$

4.3 Hypergradient Approximation

The DARTS paper [1] proposes two hypergradient approximations. The first order DARTS approximation is based on the hyperparameter direct gradient and the second order DARTS approximation is based on the one-step lookahead approximation [10]. This last approximation [10] is a truncated version of the unrolling gradient [12]. Due to storage limitations, we cannot apply a weaker truncation. This approximation creates the short horizon problem detailed in the paper [9].

In our case, the main alternatives to approximate the hypergradient relies on Evolution Strategies (ES) [5,6], the Implicit Function Theorem (IFT) [13] and hypergradient distillation [7,11]. The Truncated Evolution Strategies can approximate hypergradient by using stochastic finite differences. This approach removes the memory cost of unrolling gradient however it increases considerably the computation time. In an other hand, the use of the IFT requires the convergence of the inner optimization problem. It is possible to apply it in online fashion with the warm start approach [21]. However, this approach depends on a hyperparameter ϵ that must be carefully defined. Moreover, with the IFT approach we need to approximate inverse hessian vector product. Concerning paper [11], numerous approximations are made and access to the code is not yet available.

In order to avoid approximations on the hypergradient, several papers [2,8,16] have decided to simplify the bilevel optimization problem by incorporating hyperparameters into the inner optimization problem. As a result, we end up with a single level problem. This approach greatly improves the architectures found by DARTS algorithm. However, the use of single level approach increases the risk of overfitting by selecting architectures that contain too many parameters. In DARTS space this risk is negligeable because we look for architectures that contain only two incoming edges per node. However for larger search space it could be a real problem.

In this paper, we have chosen to use the hyperparameter direct gradient. Paper SGAS [22] empirically showed with kendal tau correlation measure that second order approach was worse than first order approach. Thus, the 1st order approach allows to improve both the results and the memory cost (divided by 2 compared to the 2nd order approach).

4.4 Reduction of the Number of Hyperparameters

In Sect. 3 it can be noted that $\alpha^{(j)}$ corresponds to a convex combination of vertices belonging to $V^{(j)}$ ($\{V^{(j)} : v \in \{0,1\}^{n_j}, ||v||_1 = 2\}$). The expression of the coefficient $\alpha^{(j)}$ requires $Card(V^{(j)}) = C^2_{n_j}$ hyperparameters. Thus, when the number of incoming edges to the node j ($n_j = j-1$) is large, there is a memory limit problem due to the number of hyperparameters to be stored. Furthermore, the computational cost of $\alpha^{(j)}$ increases. In order to reduce the number of hyperparameters used in DOTS DARTS paper [2], we express $\alpha^{(j)}$ as follows:

$$\alpha^{(j)} = \sum_{i=1}^{j-1} S\left(\frac{\lambda_i^{(j)}}{T} + b\right) e_i \tag{8}$$

With $S(.)$ a Logistic function, e_i belongs to the standard basis and $b = \log\left(\frac{2/n_j}{1-2/n_j}\right)$. As we can see, we used only n_j hyperparameters instead of $C^2_{n_j}$. In Sect. 6 there is a detailed explanation of how to respect the constraint $||\alpha^{(j)}||_1 = 2$.

4.5 Convex Depth Regularization

In our case, the short horizon bias is symptomatic of the selection of architectures that converge quickly. We propose to correct this bias by regularizing more shallow architectures. To do so, we add a term in the cost function to penalise shallow architectures. The following new cost function is defined:

$$F(\lambda) = \delta \times \mathcal{L}_{val}(w(\lambda), \lambda) - (1-\delta) \times \left(\sum_{j=4}^{6}\sum_{i=1}^{j-1} d^{(i,j)} \times \lambda_i^{(j)}\right) \tag{9}$$

with $\delta \in [0,1]$ the degree of constraint imposed on the depth and $d^{(i,j)} \in \mathbb{R}$ the penalty assigned to $\lambda_i^{(j)}$. The penalty is linear and therefore convex. We define the depth of an edge with the following formulation:

$$d^{(i,j)} = \mathbb{1}_{\{i \neq 1\}} \times (j - i - 1) + \mathbb{1}_{\{i=1\}} \times (j - i + 3) \tag{10}$$

Thus the depth of an edge corresponds to the number of nodes skipped by this one. For the gradient calculation we use the same trick used in the AdamW paper [24]. The following update is applied to $\lambda_i^{(j)}$:

$$\lambda_{i,t+1}^{(j)} = \lambda_{i,t}^{(j)} - \eta_\lambda \left(\theta_t(\lambda_{i,t}^{(j)}) \times \delta - \frac{d^{(i,j)}}{||d||_1} \times 14 \times (1 - \delta) \right) \tag{11}$$

With $\theta_t(\lambda_{i,t}^{(j)}) = \frac{m_{i,t}^{(j)}}{\sqrt{v_{i,t}^{(j)}+\epsilon}}$ the normalized gradient using Adam optimizer [23] and the vector d contains all coefficients $d^{(i,j)}$, $\forall j \in \{1, ..., 3\}$, $\forall i \in \{1, ..., j-1\}$. We have $dim(d) = 14$. We note $\lambda^{(\alpha)}$ the vector that contains all hyperparameters $\lambda_i^{(j)}$, $\forall j \in \{1, ..., 3\}$, $\forall i \in \{1, ..., j-1\}$. If we assume that $m_t^{(\alpha)}$ and $v_t^{(\alpha)}$ are good approximations of the first and second momentum of the stochastic gradient $g_t(\lambda_t^{(\alpha)}) = \nabla_{\lambda^{(\alpha)}} \mathcal{L}_{val}(w(\lambda_t^{(\alpha)}), \lambda_t^{(\alpha)})$. We have:

$$\theta_t(\lambda^{(\alpha)}) = \frac{\mathbb{E}[g_t(\lambda^{(\alpha)})]}{\sqrt{\mathbb{E}[g_t(\lambda^{(\alpha)})^2]} + \epsilon} = \frac{\mathbb{E}[g_t(\lambda^{(\alpha)})]}{\sqrt{\mathbb{E}[g_t(\lambda^{(\alpha)})]^2 + \mathbb{V}[g_t(\lambda^{(\alpha)})]} + \epsilon} \tag{12}$$

θ_t reaches its maximum amplitude when $\mathbb{V}[g_t(\lambda^{(\alpha)})] = 0$. When this is the case we have:

$$||\theta_t(\lambda^{(\alpha)})||_1 = ||sign(\mathbb{E}[g_t(\lambda^{(\alpha)})])||_1 = 14$$

If $g_t(\lambda^{(\alpha)})$ points in the constant direction d, we would have $\mathbb{V}[g_t(\lambda^{(\alpha)})] = 0$. Therefore, we scale the term $\frac{d^{(i,j)}}{||d||_1}$ by 14. The search interval for δ can be reduced to $[0.5, 1]$.

4.6 Customized Discretization Operator

Weighted operations extract information from the database, while unweighted operations apply processing independent from the dataset. It seems natural to distinguish these two categories.

We define the following two sets:

- $O_1 = \{(Skip, 1), (AvgP, 2), (MaxP, 3)\}$
- $O_2 = \{(Sep3 \times 3, 4), (Sep5 \times 5, 5), (Dil3 \times 3, 6), (Dil5 \times 5, 7)\}$

The new discretization operator is defined hereafter:

If $\sum_{o \in O_1} \beta_o^{(i,j)} > \sum_{o \in O_2} \beta_o^{(i,j)}$: $\text{argmax}\left(\beta_{skip}^{(i,j)}, \beta_{maxp}^{(i,j)}, \beta_{avgp}^{(i,j)}\right)$

If $\sum_{o \in O_1} \beta_o^{(i,j)} < \sum_{o \in O_2} \beta_o^{(i,j)}$: $\text{argmax}\left(\beta_{sep3\times3}^{(i,j)}, \beta_{sep5\times5}^{(i,j)}, \beta_{dil3\times3}^{(i,j)}, \beta_{dil5\times5}^{(i,j)}\right)$

Let's consider the example below:

$$\left[\beta_{skip}^{(i,j)}, \beta_{maxp}^{(i,j)}, \beta_{avgp}^{(i,j)}, \beta_{sep3\times3}^{(i,j)}, \beta_{sep5\times5}^{(i,j)}, \beta_{dil3\times3}^{(i,j)}, \beta_{dil5\times5}^{(i,j)}\right] = [0.26, 0, 0, 0.21, 0.13, 0.23, 0.17]$$

A classic argmax will return "skip" while the modified operator will return "dil3 × 3". When we initialize the elements of $\beta^{(i,j)}$, we obtain the following vector:

$$\beta^{(i,j)} = \left[\frac{1}{7}, \frac{1}{7}, \frac{1}{7}, \frac{1}{7}, \frac{1}{7}, \frac{1}{7}, \frac{1}{7}\right]$$

The operator previously defined will necessarily choose weighted operations because:

$$\sum_{o \in O_1} \beta_o^{(i,j)} = \frac{3}{7} < \sum_{o \in O_2} \beta_o^{(i,j)} = \frac{4}{7} \tag{13}$$

In order to avoid a bias at initialization, we reexpress the vector $\beta^{(i,j)}$ as follows:

$$\beta^{(i,j)} = S\left(\frac{\lambda_{\overline{w}}^{(i,j)}}{T}\right) \times \sum_{o \in O_1} \frac{\exp\left(\frac{\lambda_o^{(i,j)}}{T}\right)}{\sum_{o' \in O_1} \exp\left(\frac{\lambda_{o'}^{(i,j)}}{T}\right)} e_o + \left(1 - S\left(\frac{\lambda_{\overline{w}}^{(i,j)}}{T}\right)\right) \times \sum_{o \in O_2} \frac{\exp\left(\frac{\lambda_o^{(i,j)}}{T}\right)}{\sum_{o' \in O_2} \exp\left(\frac{\lambda_{o'}^{(i,j)}}{T}\right)} e_o \tag{14}$$

$S(.)$ corresponds to a logistic function and $\overline{w}$ means without weight. For the edge mixture (i, j), we get at initialization:

$$\beta^{(i,j)} = [0.1666, 0.1666, 0.1666, 0.125, 0.125, 0.125, 0.125]$$

About the new discretization operator it returns the same value as the classical argmax when the solution is sufficiently sparse.

4.7 Degeneracy Correction

Degeneracy is also characterised by a predominance of weightless operations. To control this aspect of degeneracy we propose to define a competition between weightless operations when the algorithm starts to degenerate. To do so, we define a maximum number of weightless operations (noted N) the algorithm can select. During the learning process, we follow algorithme 1. ξ corresponds to the blue edges set of the DAG described in Fig. 1 and $max^{(N)}$ refers to the Nth largest value of $\lambda_{\overline{w}}^{(i,j)}, \forall (i,j) \in \xi$. Thus, the algorithm will not be able to select more than N weightless operations (Fig. 2).

Algorithm 1. Maximum number of weightless operations

1: **for** $t = 0, ..., T_{train}$ **do**
2:     **for** $(i,j) \in \xi$ **do**
3:         $\lambda_{\overline{w},t+1}^{(i,j)} = \lambda_{\overline{w},t}^{(i,j)} - \eta \nabla_{\lambda_{\overline{w},t}^{(i,j)}} \mathcal{L}_{val}(w(\lambda_{\overline{w},t}^{(i,j)}), \lambda_{\overline{w},t}^{(i,j)})$
4:     **end for**
5:     $M = \frac{1}{2}\left(\max_{(i,j)\in\xi}^{(N)} \lambda_{\overline{w},t+1}^{(i,j)} + \max_{(i,j)\in\xi}^{(N+1)} \lambda_{\overline{w},t+1}^{(i,j)}\right)$
6:     **if** $M > 0$ **then**
7:         **for** $(i,j) \in \xi$ **do**
8:             $\lambda_{\overline{w},t+1}^{(i,j)} = \lambda_{\overline{w},t+1}^{(i,j)} - M$
9:         **end for**
10:     **end if**
11: **end for**

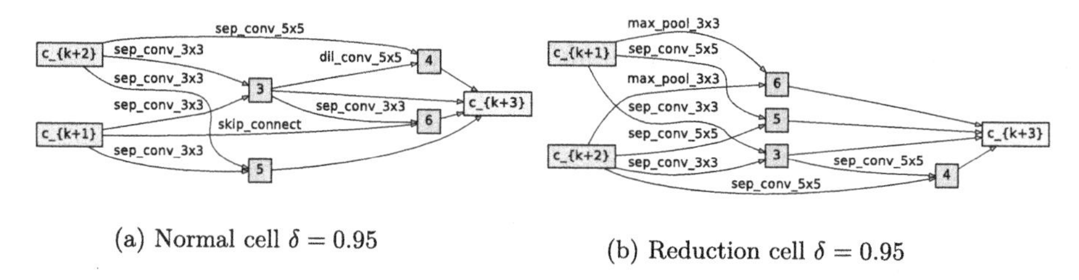

(a) Normal cell $\delta = 0.95$

(b) Reduction cell $\delta = 0.95$

Fig. 2. Our best architectures

5 Experiments

We follow the procedure defined in DARTS paper [1]. The search phase is run 4 times on CIFAR10 dataset. It takes approximately 10×4 hours with a NVIDIA Quadro RTX 8000. For each run, we get a normal cell architecture $\mathcal{A}_n$ and a reduction cell architecture $\mathcal{A}_r$. Then, we use $\mathcal{A}_n$ and $\mathcal{A}_r$ to build a target network with 20-cells. We train it on CIFAR10 then we compute its test error. This last stage takes 40×4 hours on our machine. Finally, we have 4 architectures with 4 test errors. We keep the architecture with the best test error.

5.1 Settings

We mainly use the same settings as in the DARTS paper [1]. For the search phase we add cutout. This helps to regularize the hypernetwork. For adam optimizer we use the following momentum $(0.9, 0.999)$ instead of $(0.5, 0.999)$. The hyperparameter δ is very expensive to optimize. We skip this step and we start the 4 runs of the research phase with different values δ. We define the 4 values taken by δ as follows: $\{1, 0.95, 0.9, 0.8\}$. Thus, we will obtain 4 different depth architectures. On average, the results will be slightly worse because architectures with a non-optimal depth have been selected.

Concerning the maximum number of weightless operations N, we have chosen to consider it as a priori information. We fix N to 2 for the normal cell and $N = 8$ for the reduction cell. For the evaluation phase, the settings are unchanged.

Table 1. Results on CIFAR10

Architecture	Test Error	Params (M)	Cost GPU days
DARTS (1st order)	3.00 ± 0.14	3.3	0.5
DARTS (2nd order)	2.76 ± 0.09	3.3	1
Dots DARTS	2.49 ± 0.06 (best: 2.37)	3.5	0.26
DropNAS	2.58 ± 0.14 (best: 2.26)	4.1	0.6
MiLeNAS	2.51 ± 0.11 (best: 2.34)	3.87	0.3
Ours	2.75 ± 0.15 (best: 2.53)	4.62	0.42

5.2 Results

Table 2.

δ	Average length path
1	17.6100
0.95	24.3603
0.9	32.4375
0.8	35.2419

Results from the different runs are summarized in Table 1. We could have hoped for better results compared to Dots [2]. However, we used a sub-optimal δ value during the search phase. On average our approach will be less good. But if we look at our best run, we can see that we are not far from the other papers results. On the other hand, it is important to note that Dots [2], DropNAS [16], MileNAS [3] papers rely on the one-level optimization approach. As mentioned earlier, one-level optimization is more stable but there is a risk of overfitting. In DARTS space, this risk is negligible because we force the algorithm to select only 2 edges per node. On the other hand, Dots paper [2] divides the search phase into two stages. It selects at first the best operations for each edge and then selects the 2 best edges per node. This approach creates a bias because operation selection depends on edge selection and vice versa. In contrast to Dots paper [2], our approach optimizes operation and edge selection at the same time. Therefore, our algorithm can reach better solutions.

In Table 2 we compute for each run the average length path of the target network DAG. When we analyse Table 2, we can see a clear correlation between the architecture depth and the δ value. This confirms the effectiveness of our penalisation.

It is important to remember that the quality of an approach is difficult to quantify with only 4 runs. Due to the computational cost we did not set up any additional run.

6 High Number of Incomings Edges

In this section we propose a low-cost alternative to move in C, the convex hull of $\{V : v \in \{0,1\}^n, ||v||_1 = 2\}$. For the sake of clarity, we have removed the indices (j). We define α as follows:

$$\alpha = \sum_{i=1}^{n} S\left(\frac{\lambda_i}{T} + b\right) e_i \tag{15}$$

With $S(.)$ a Logistic function and e_i belongs to the standard basis.
We initialize b and λ_i in the following manner:

$$b = \log\left(\frac{2/n}{1-2/n}\right), \qquad \lambda_i = 0 \tag{16}$$

At the initialization α belongs to the convex hull of V since $\forall i \in \{1, ..., n\}$, $\alpha_i = \frac{2}{n} \in [0,1]$ and $||\alpha||_1 = 2$. During the learning process, we follow Algorithm 2.

Algorithm 2. Projection onto C

1: **for** $t = 0, ..., T_{train}$ **do**
2: $\quad \lambda_{t+1} = \lambda_t - \eta \nabla_{\lambda_t} \mathcal{L}_{val}(w(\lambda_t), \lambda_t)$
3: $\quad \alpha_{t+1} = \sum_{i=1}^{n} S\left(\frac{\lambda_{i,t+1}}{T} + b\right) e_i$
4: $\quad$ Solve the following optimization problem:

$$\lambda_{t+1}^* = \min_{\{\lambda' | \alpha(\lambda') \in C\}} ||\alpha_{t+1} - \alpha(\lambda')||_2 \tag{17}$$

5: $\quad \lambda_{t+1} = \lambda_{t+1}^*$
6: **end for**

Algorithm 3.

1: **for** $t = 0, ..., T_{train}$ **do**
2: $\quad \lambda_{t+1} = \lambda_t - \eta \nabla_{\lambda_t} \mathcal{L}_{val}(w(\lambda_t), \lambda_t)$
3: $\quad \alpha_{t+1} = \sum_{i=1}^{n} S\left(\frac{\lambda_{i,t+1}}{T} + b\right) e_i$
4: $\quad$ Solve the following optimization problem:

$$k^* = \min_{k \in \mathbb{R}} \frac{1}{2}\left(2 - \sum_{i=1}^{n} S\left(\frac{\lambda_{i,t+1} + k}{T} + b\right)\right)^2 \tag{18}$$

5: $\quad \lambda_{t+1} = \lambda_{t+1} + k^* \mathbf{1}$ $\quad \triangleright$ **1** vector of ones in $\mathbb{R}^n$
6: **end for**

It can be noted that after each update of λ (with respect to the validation error) we modify λ in order to reposition α on C. Before solving (17), we are interested in the following problem:

$$\alpha_t^* = \min_{\alpha \in C} ||\alpha_t - \alpha||_2 \tag{19}$$

α_t^* is the projection of α_t on C. Paper [25] has shown that α_t^* can be expressed as follows:

$$\alpha_t^* = \sum_{i=1}^{n} HS\left(\alpha_{i,t} + k^*\right) e_i \quad \text{with } HS(x) := \begin{cases} 0 & \text{if } x < 0\,, \\ x & \text{if } x \in [0,1]\,, \\ 1 & \text{if } x > 1\,. \end{cases} \tag{20}$$

with $k^* \in \mathbb{R}$ solution of the following equation:

$$\sum_{i=1}^{n} HS\left(\alpha_{i,t} + k\right) = 2 \tag{21}$$

Paper [25] proposes an efficient algorithm to solve (21). Once we have found α_t^* we can access to λ_t^* by solving:

$$\sum_{i=1}^{n} S\left(\frac{\lambda_{i,t}}{T}+b\right) e_i = \alpha_t^* \quad \Rightarrow \lambda_{i,t}^* = T \times \log\left(\frac{\alpha_{i,t}^*}{1-\alpha_{i,t}^*}\right) - T \times b \tag{22}$$

$\lambda_{i,t}^*$ exists if $\alpha_{i,t}^* \in]0,1[$ but we have $\alpha_{i,t}^* \in [0,1]$. Therefore, instead of expressing $\lambda_{i,t}^*$ with respect to $\alpha_{i,t}^*$, we will search directly $\lambda_{i,t}^*$ by solving the following equation:

$$\sum_{i=1}^{n} S\left(\frac{\lambda_{i,t}+k}{T}+b\right) = 2 \tag{23}$$

With $k \in \mathbb{R}$. This is a sum of strictly increasing functions, there is a unique solution to equation (23). At initialization $k = 0$. Knowing that we are repositioning α after each update it can be deduced that $k = 0$ is not far from the optimum k^*. We used a dichotomy to solve equation (23) instead of newton descent as mentionned in paper [25]. When temperature T is low we notice that newton's descent diverged which is not the case for dichotomy. Once k^* has been found we replace $\lambda_{i,t}$ by $\lambda_{i,t} + k^*$. To sum up, we follow the procedure described in Algorithm 3.

Thus, in this section, we have proposed a way to move in the convex hull of V with only n hyperparameters instead of C_n^2 hyperparameters. And the computational cost is reduced to solving (18). Our approach can be generalized to m-capped simplex $\{x \in [0,1]^n, ||x||_1 = m\}$ with $m \in [0,n]$.

7 Conclusion

In this paper, we propose a new way to incorporate prior information about the size and the depth of our neural network. It has helped to correct the degeneracy phenomenon observed with DARTS algorithm. From now on, the algorithm is able to explore new and deeper architectures. Our approach can be applied to all DARTS algorithms that are subject to short term bias. On the other hand, we propose a projection algorithm in order to reduce the number of hyperparameters used in dots approach [2].

References

1. Liu, H., Simonyan, K., Yang, Y.: Darts: Differentiable architecture search. In: International Conference on Learning Representations (ICLR) (2019)
2. Gu, Y.C., et al.: DOTS: decoupling operation and topology in differentiable architecture search. In: Conference on Computer Vision and Pattern Recognition (CVPR) (2021)
3. He, C., Ye, H., Shen, L., Zhang, T.: MiLeNAS: efficient neural architecture search via mixed-level reformulation. In: Conference on Computer Vision and Pattern Recognition (CVPR) (2020)

4. Zhou, P., Xiong, C., Socher, R., Hoi, S.C.H.: Theory-inspired path-regularized differential network architecture search. In: Neural Information Processing Systems (NeurIPS) (2020)
5. Sohl-Dickstein, J.: Persistent unbiased gradient estimation in unrolled computation graphs with persistent evolution strategies. In: International Conference on Machine Learning (2021)
6. Metz, L., Maheswaranathan, N., Sun, R., Daniel Freeman, C., Poole, B., Sohl-Dickstein, J.: Using a thousand optimization tasks to learn hyperparameter search strategies. In: Neural Information Processing Systems (NeurIPS) (2020)
7. Fu, J., Luo, H., Feng, J., Low, K.H., Chua, T.S.: DrMAD: distilling reverse-mode automatic differentiation for optimizing hyperparameters of deep neural networks. In: International Joint Conference on Artificial Intelligence (IJCAI) (2016)
8. Hou, P., Jin, Y., Chen, Y.: Single-DARTS: towards stable architecture search. In: IEEE/CVF International Conference on Computer Vision Workshops (ICCVW) (2021)
9. Wu, Y., Ren, M., Liao, R., Grosse, R.: Understanding short-horizon bias in stochastic meta-optimizations. In: International Conference on Learning Representations (ICLR) (2018)
10. Luketina, J., Berglund, M., Greff, K., Raiko, T.: Scalable gradient-based tuning of continuous regularization hyperparameters. In: International conference on machine learning (ICML) (2016)
11. Lee, H.B., Lee, H., Shin, J., Yang, E., Hospedales, T.M., Hwang, S.J.: Online hyperparameter meta-learning with hypergradient distillation. In: International Conference on Learning Representations (ICLR) (2022)
12. Franceschi, L., Donini, M., Frasconi, P., Pontil, M.: Forward and reverse gradient-based hyperparameter optimization. In: International Conference on Machine Learning (ICML) (2017)
13. Lorraine, J., Vicol, P., Duvenaud, D.: Optimizing millions of hyperparameters by implicit differentiation. In: International Conference on Artificial Intelligence and Statistics (2020)
14. Choe, H., Na, B., Mok, J., Yoon, S.: Variance-stationary differentiable NAS. In: British Machine Vision Conference (BMVC) (2021)
15. Wei, T., Wang, C., Rui, Y., Chen, C.W.: Network morphism. In: Proceedings of Machine Learning Research (PMLR) (2016)
16. Hong, W., et al.: DropNAS: grouped operation dropout for differentiable architecture search. In: International Joint Conference on Artificial Intelligence (IJCAI) (2020)
17. Lin, M., et al.: Zen-NAS: a zero-shot NAS for high-performance image recognition. In: International Conference on Computer Vision (ICCV) (2021)
18. Li, L., Jamieson, K., DeSalvo, G., Rostamizadeh, A., Talwalkar, A.: Hyperband: a novel bandit-based approach to hyperparameter optimization. J. Mach. Learn. Res. **18**, 6765–6816 (2018)
19. Zoph, B., Vasudevan, V., Shlens, J., Le, Q.V.: Learning transferable architectures for scalable image recognition. In: Conference on Computer Vision and Pattern Recognition (CVPR) (2019)
20. Real, E., Aggarwal, A., Huang, Y., Le, Q.V.: Regularized evolution for image classifier architecture search. In: Conference on Artificial Intelligence (AAAI) (2019)
21. Vicol, P., Lorraine, J.P., Pedregosa, F., Duvenaud, D., Grosse, R.B.: On implicit bias in overparameterized bilevel optimization. In: International Conference on Machine Learning (ICML) (2022)

22. Li, G., Qian, G., Delgadillo, I.C., Muller, M., Thabet, A., Ghanem, B.: SGAS: sequential greedy architecture search. In: Conference on Computer Vision and Pattern Recognition (CVPR) (2020)
23. Kingma, D.P., Ba, J.L.: ADAM: a method for stochastic optimization. In: International Conference on Learning Representations (ICLR) (2015)
24. Loshchilov, I., Hutter, F.: Decoupled weight decay regularization. In: International Conference on Learning Representations (ICLR) (2019)
25. Ang, A., Ma, J., Liu, N., Huang, K., Wang, Y.: Fast projection onto the capped simplex with applications to sparse regression in bioinformatics. In: Neural Information Processing Systems (NeurIPS) (2021)
26. Wang, W., Lu, C.: Projection onto the capped simplex. arXiv preprint arXiv:1503.01002 (2015)

A Fuzzy Logic Inference System for Display Characterization

Khleef Almutairi[1,2], Samuel Morillas[1], Pedro Latorre-Carmona[3(✉)], and Makan Dansoko[4]

[1] Instituto de Matemática Pura y Aplicada, Universitat Politècnica de València, Camino de Vera sn, 46022 Valencia, Spain
kkalmuta@doctor.upv.es, smorillas@mat.upv.es

[2] Mathematics Department, Faculty of Science, Albaha University, Al Baha, Saudi Arabia
khalmotiri@bu.edu.sa

[3] Departamento de Ingeniería Informática, Universidad de Burgos, Avda. Cantabria s/n, 09006 Burgos, Spain
plcarmona@ubu.es

[4] Technopôle du Madrillet, Avenue Galilée - BP 10024, 76801 Saint-Etienne du Rouvray Cedex, France
makan.dansoko@groupe-esigelec.org

Abstract. We present in this paper the application of a fuzzy logic inference system to characterize liquid-crystal displays. We use the so-called fuzzy modelling and identification toolbox (FMID, Mathworks) to build a fuzzy logic inference system from a set of input and output data. The advantage of building a model like this, aside from its good performance, relies on its interpretability. Once trained, we obtain a physical interpretation of the model. We use training and testing datasets relating device dependent RGB data with device independent XYZ or xyY coordinates, measured with a colorimeter. We study different configurations for the model and compare them with three state-of-the-art methods in terms of $\Delta E00$ visual error. This study is restricted to a single display and therefore we also point out what features of the learned model we think are more display dependent and might possibly change for a different display.

Keywords: Fuzzy logic · Fuzzy inference system · Fuzzy modeling · Display characterization

1 Introduction

Display characterization is becoming an increasingly important field of research. The new variety of displays that appear in the market have ever-increasing features and capabilities. Consumers and manufacturers are continuously looking for new experiences and more accurate visualization features [3,12]. On the other hand, these devices can only achieve accurate image reproduction through an accurate display characterization process [12].

Supported by Generalitat Valenciana under AICO 2023 program.

A. Pertusa et al. (Eds.): IbPRIA 2023, LNCS 14062, pp. 54–66, 2023.
https://doi.org/10.1007/978-3-031-36616-1_5

Display characterization aims to create a model to connect a digital-to-analogue converter (DAC) RGB [4] input feature vector with the outputs of the display expressed in a device-independent colour space (usually in terms of XYZ tristimulus values or xyY colour coordinates). These relationships are generally complex: They are non-linear and there are interactions between all RGB inputs and all device independent colour coordinates. The procedure to develop a display model, from a general point-of-view, would include the following steps:

1. Measure a colour dataset (DAC RGB and corresponding XYZ and/or xyY) for model fitting.
2. Apply a model fitting procedure to the dataset measured in **step 1**.
3. Measure a colour test dataset of different RGB and XYZ/xyY coordinates.
4. Use the fitted model to compute the RGB values that should generate the corresponding aimed XYZ/xyY values of the dataset measured in **step 3**.
5. Measure the obtained XYZ/xyY from the RGB dataset in **step 4**.
6. Compute the visual error between desired outputs measured in **step 3**, and obtained output values measured in **step 5**.

In order to carry out the measurement of the XYZ output values, related to the DAC RGB vectors, we usually use either a colorimeter or more sophisticated devices, such as spectrophotometers. In the past, these measurements have traditionally been manually acquired, and several physical models were developed. For instance, the so-called RIT models proposed in [3,6], rely on relating the input and output values via a simple linear transformation, after compensating certain nonlinear behaviour of the inputs with a power function or a look-up table. A different example of a display characterization model can be found in [5,10], which used the measurements in the xyY colour space, and processed the Y component separately with a power function of the input and the xy chromaticity using linear interpolation between the measurements.

On the other hand, some mathematical models have also been used in display characterization. These models are highly dependent on the amount of data. An example of a mathematical model for a display is the Pseudo-Inverse model, which is based on finding the best linear application able to find the RGB-XYZ relationships. The measurements used to find this linear application are used to build an over-determined system of equations, and the optimal solution is obtained via squared error minimization using the so-called Penrose pseudo-inverse [11] method. Other mathematical models have been proposed for the task. Among these, the model proposed in [8] is able to assess the colourimetric information of the red, green, and blue *subpixels* on an independent basis. Another example of building a trained mathematical model is the one proposed in [12], which proposes to use a neural network (NN) that computes what RGB inputs need to be used to obtain a desired colour expressed either in XYZ or xyY coordinates. However, these types of methods are usually considered as *black box* methods, thus restricting the model explainability.

In this work, we approach the display characterization problem by building a mathematical model that is interpretable as well. In order to do this, we have considered the application of the Fuzzy modelling and identification toolbox

(FMID) [2] that can be used to build a fuzzy logic rule system able to compute what RGB coordinates need to be used to generate a desired colour expressed in device independent colour coordinates.

The advantage of such a system is that the rules found by the system can be extracted and, so, the corresponding knowledge that has been learned/inferred from the training data that explains how the system performs. A fuzzy logic system [15] is a relatively simple method that can be seen logically close to human thinking, which eases its interpretation. It uses **IF-THEN** structures to establish the qualitative relationships between the input and output variables. Then, the so-called inference and defuzzyfication processes are employed to determine output variables values from the input ones [16].

The rest of the paper is divided as follows: Sect. 2 presents the FMID method, and the trained system characterization. Section 3 presents the simulation and the experimental results in a comparative framework. Section 4 presents the conclusions.

2 Fuzzy Logic Modelling Approach

2.1 Fuzzy Modelling and Identification Toolbox

A fuzzy inference system (FIS) has been shown to be an effective and universal fitting function. If the relevant implication rules, membership functions, inference operators, and defuzzification method are appropriately chosen, any relationship (mapping) between inputs and outputs may be described using a FIS up to any degree of accuracy. A FIS is often created using professional expertise. In such a case, it is necessary to have a human solution for the particular problem being addressed. In many instances, however, ground truth data that connects the inputs and outputs of the intended system exists instead of this expert knowledge [1]. So it is also interesting to build FIS from datasets. Processing the collected data into a model that can be related to a human way of thinking can be achieved by building a fuzzy system [13] that provides a transparent representation of a non-linear system, giving a linguistic interpretation in the form of implication rules.

The rules in the FIS follow two types of modelling approaches: (a) The so-called Mamdani system [13], which is more straightforward and intuitive. This form usually has linguistic terms for both antecedents and consequents, such as: **IF** X is *low* **AND** Y is *medium*, **AND** Z is *high*, **THEN** G is *medium*, and (b) The Takagi-Sugeno system, which have linguistic terms in the antecedents but functions in the consequents so that it can be seen as a soft combination of different functions, that are activated to a higher or lesser degree depending on antecedent certainty and which provide the final outputs. In these systems we would find rules such as: **IF** X is *low* **AND** Y is *medium*, **AND** Z is *high*, **THEN** $G = f(X, Y, Z)$.

In our case, we are interested in finding a set of rules that associate which RGB inputs should be used for the display to show a particular colour expressed in the so-called device-independent coordinates (such as XYZ or xyY). And what we have available is data measured with a colorimeter. So we aim to build a FIS from a color dataset. For this, we use the Fuzzy modelling and identification toolbox (FIMD) [2] that generates FIS of Takagi-Sugeno (TS) type.

The FMID toolbox allows for different parameter settings to use depending on the problem [2]. The experimental (training and validation) data is critical for any identification process since it specifies the content that will be used during the generation and validation of the model. After feeding the data into the system, FMID uses fuzzy clustering, with a certain degree of overlapping, to divide the dataset into a specific number of clusters. Since the FIS rules are created using the clustering results, the designer decides in advance how many clusters to obtain and, so, how many local sub-models of the fuzzy model there should be. The Gustafson-Kessel (GK) algorithm [7] (extensively used in clustering) is applied to determine the division and so define the membership functions of the antecedents. Then, the activation functions of the consequents need to be fitted by least squares minimization. There are different options of function types that can be tested and the optimal option depends very much on the particular problem and data.

Once all of the FMID parameters have been determined and set to sub-optimal performance, the next step is to test the outcome using a new, independent color dataset. The trained system receives the validation dataset of desired colors expressed in device indepent color coordinates and predicts what RGB values need to be used to generate each of them. Here, we can use the mean squared error between RGB computed and corresponding RGB of the validation dataset (MSE_{RGB}) as performance metric used to describe the model's level of accuracy which is useful for FMID parameter fitting. However, final visual error should be measured between colors displayed with the computed RGB and desired colors in the validation dataset. Once the FIS is trained with acceptable results, we can proceed to extract the information that the FIS has learned and so interprete the knowledge codified in it.

2.2 Training a Fuzzy System with FMID for Display Characterization

We used a Dell Ultra-sharp UP2516D25 LCD display, and an X-rite eye-one display colorimeter, in order to create the datasets for model fitting and validation. To build them, a mesh of equally spaced points in the RGB color space (cube) was considered, where the RGB data take integer values in the $[0, 255]$ interval. In particular, for the training dataset, we divided each $DACRGB$ input into 22 levels, starting from 0 to 252, with a step size of 12 units. We considered all the potential combinations in the RGB cube, accounting for $22^3 = 10648$ different samples. The corresponding XYZ colour are obtained with the colorimeter. So we have a dataset of 10648 samples, with corresponding $RGB - XYZ$ data. Regarding the validation dataset, we defined another mesh with points in between those of the training dataset for which we chose 21 equally spaced RGB levels starting from 6 to 246 with the same step size as before and considering all combinations to have a total of $21^3 = 9261$ samples with corresponding $RGB - XYZ$ data.

Table 1. MSE_{RGB} performance varying the number of rules between 2–15 and the overlapping parameter m between 1.05–2.25 for the FIS system trained with XYZ inputs. The best two results are highlighted in red.

	Clusters													
m	2	3	4	5	6	7	8	9	10	11	12	13	14	15
1.05	60.8	13.5	6.2	4.1	2.8	2.5	1.9	2.7	1.8	1.6	1.9	2.3	2.1	1.8
1.25	71.7	21.2	6.0	4.5	4.3	3.1	2.4	2.0	1.9	2.2	1.5	1.7	2.3	1.7
1.50	74.3	26.6	9.6	4.5	3.0	2.7	2.4	2.3	1.7	1.9	2.0	1.5	2.2	2.0
1.75	78.2	28.1	13.3	6.4	4.0	3.1	2.7	2.2	2.0	1.9	1.9	2.2	2.4	2.5
2.00	84.7	32.6	16.3	9.4	5.7	4.4	3.8	2.7	2.9	2.9	2.7	2.2	3.4	3.4
2.25	93.0	38.0	20.0	11.9	8.2	6.2	5.6	3.9	3.3	3.5	3.6	3.3	4.3	3.1

This FMID toolbox allows to test different options for the parameter setting. Nevertheless, its number is huge (i.e. t-norm and s-norm, membership function shape, defuzzification method, to cite a few). We study the most important and critical parameters that have a higher impact on MSE_{RGB} performance: (1) The type of response functions in the rules consequents, (2) the number of clusters (rules) in the system, and (3) the overlapping parameter between the clusters membership functions (fuzziness in the system). Also we consider using XYZ and xyY to train two different systems with these inputs and using RGB as an output.

First, we analyse performance depending on the type of activation function in the consequents. We considered linear, parabolic and square root functions. In the case of XYZ inputs the linear option is the best in terms of MSE_{RGB} independently of the number of clusters. On the other hand, using xyY coordinates the parabolic response function is the best option.

Second, we analyze the performance as a function of the number of rules (clusters) that will be used in the model and the overlapping degree between the clusters (m value in FMID). For this we varied the number of rules between 2–15 and the m parameter between 1.05–2.25. Performance in terms of MSE_{RGB} is shown in Tables 1 and 2. For XYZ inputs we can see that $m = 1.5$ is a good choice as it provides a good performance for many numbers of rules. A decision about the number of rules is more complex. Optimal performance is achieved for 13 rules but having so many rules makes the system more difficult to interprete. However, performance does not drop much when reducing the number of rules a bit and it would be a system easier to interprete. So, we conclude that it could also be worth to build an FIS with only 6 rules and compare the visual performance with the 13 rules one. For xyY inputs performance is much worse. However, for a more complete study, we will also include the visual evaluation of the best system found (15 rules and $m = 1.25$).

We can conclude that, for the XYZ case, the response function will be a linear function and with different clustering result, for each output. We will study the performance in terms of the ΔE_{00} visual error, for the 13 clusters (rules)

Table 2. MSE_{RGB} performance varying the number of rules between 2–15 and the overlapping parameter m between 1.05–2.25 for the FIS system trained with xyY inputs. The best result is highlighted in red.

	Clusters													
m	2	3	4	5	6	7	8	9	10	11	12	13	14	15
1.05	151.4	89.1	54.0	38.4	28.5	23.2	18.3	13.9	12.2	11.2	10.1	8.7	8.2	8.0
1.25	135.1	76.7	48.4	33.0	26.0	20.3	15.4	12.4	10.7	9.3	8.3	7.6	7.0	6.3
1.50	131.7	69.8	46.9	36.8	28.4	23.9	18.8	16.1	13.7	11.8	10.8	9.6	8.9	8.3
1.75	145.4	94.7	89.9	80.8	66.2	70.3	53.4	45.9	42.4	45.2	46.6	42.2	35.1	27.7
2.00	159.0	109.1	177.5	195.0	204.5	227.4	233.4	240.3	240.7	249.4	264.3	240.9	193.1	154.5
2.25	166.0	184.2	188.4	266.9	223.1	116.5	290.0	323.2	309.5	333.1	376.1	344.7	367.4	315.8

case, which yields the best MSE_{RGB} value, as well as 6 clusters (rules) as we consider this to be the best trade-off between performance and interpretability with overlapping parameter $m = 1.5$. Regarding the FMID model using xyY coordinates, we are using the parabolic function with a different clustering for each output, and 15 clusters as this option has the lowest MSE_{RGB} with overlapping parameter $m = 1.25$.

2.3 Trained System Interpretation

In this section we focus on providing an interpretation of the FIS built using XYZ inputs, 6 rules only, overlapping parameter $m = 1.5$, and linear activation functions in the consequents. The option with 13 clusters basically follows the same principles that we will use to describe the 6 clusters case but adding more rules for marginal cases.

First, for the system interpretation we need to provide a linguistic meaning to the fuzzy inference rules. For that we can have a look first at Table 3 which shows the centers of each one of the 6 clusters for each of the RGB outputs. We can see that clusters for R output are mainly distributed covering the range of the X input with less variability for the rest of inputs. Similar behaviour is observed for G and Y, and B and Z. This makes sense as these are the strongest relationship between inputs and outputs. However, the cluster centers are only the peak of the membership functions in the antecedent of the rules. So, for a better understanding we must have a look at these functions and analyze what they are modelling. Figure 1 shows the membership function of each output. To complete the information about the rules, the function consequent used in each of them is provided in Table 3.

By analyzing all this information we can identify an overall behaviour consisting on the 6 rules considering for each R, G and B, 6 levels of the correspondingly related variables X, Y, and Z, in each case. This behaviour is expected, so possibly happens for other displays as well. In addition, each output ruleset accounts for the correlation with the less related variables in different ways. This latter point is more display dependent. In more detail, the implication rules can be interpreted as follows:

For the red component we have the rules:

1. IF X is extremely low AND Y is medium-low, AND Z is medium-low THEN use f_1^R.
2. IF X is very low AND Y is medium-low, AND Z is medium-low THEN use f_2^R.
3. IF X is low AND Y is medium-low, AND Z is low OR medium-low THEN use f_3^R.
4. IF X is medium-low AND Y is medium-low OR low, AND Z is low OR medium-low THEN use f_4^R.
5. IF X is medium AND Y is low OR medium-low OR medium, AND Z not-high THEN uses f_5^R.
6. IF X is high AND Y not-high THEN uses f_6^R.

For the green component we have the rules:

1. IF Y is extremely low AND X is medium-low THEN use f_1^G.
2. IF Y is very low AND X medium-low THEN use f_2^G.
3. IF Y is low AND X medium-low THEN use f_3^G.
4. IF Y is medium-low low AND X is medium-low OR medium THEN use f_4^G.
5. IF Y is medium AND X is medium-low OR medium OR medium-high THEN use f_5^G.
6. IF Y is high AND X is not-high THEN use f_6^G.

It can be seen how the different rules account for the correlation between RG and XY by considering different combinations of their possible values in the antecedents. For both R and G, Z is less important in general, but specially for G, Z is just not used in the antecedents of the rules. Notice that this happens for this display but other displays for which RGB spectral components are closer or more separated can behave quite differently.

For the blue component we have the rules:

1. IF Z is extremely low AND Y is medium-low, AND X is not high, THEN use f_1^B.
2. IF Z is very low AND Y is not high THEN use f_2^B.
3. IF Z is low THEN use f_3^B.
4. IF Z is medium-low THEN use f_4^B.
5. IF Z is medium THEN use f_5^B.
6. IF Z is high THEN use f_6^B.

It can be seen how the different rules account for the correlation between GB and YZ by considering different combinations of their possible values in the antecedents. Also, X is not used in the antecedents of the rules excepting for rule 1. This behaviour is also display dependent.

Now we can extract more information of the model by looking at the functions $f_i^K, K = R, G, B, i = 1, ..., 6$, used as a consequence of the rules: In all cases, the most important value to determine R (G or B) is X (Y or Z), correspondingly. In particular, they show a proportional relationship. The change in the coefficients that relate them in the different rules accounts for the nonlinear relationship between them. This can be seen as piecewise linear functions approximating a nonlinear ones. All this makes sense to happen for other displays as well.

More in detail, when determining R, Z is almost negligible. The same happens for G with respect to Z and B with respect to X, which makes sense as these are the weakest relationships between RGB and XYZ coordinates. On the other hand, we observe some inverse relations: R is inversely related to Y, G is inversely related to X and B is inversely related to Y. We interpret this as the way the model has found to compensate for the correlations and overlappings that exist between the RGB primaries in terms of XYZ coordinates of the generated colours. For instance, in the case of B, we see that B is inversely proportional to Y. We see that this is accounting for the existing overlapping in terms of the Z coordinates with respect to the G primary. That is: increasing the G will also produce some increment in Z. And the higher the Y, the higher the G, so some amount of Z will be produced from G when Y is high. So, the model accounts for this by establishing this inverse relationship between B and Y. The same explanation holds for R and Y, and G and X. These considerations are more display dependent as for other primaries these compensations can be quite different.

Table 3. Cluster centres of the trained model $FMID_{XYZ}$ and fitted output regression TS functions for each rule and RGB output.

Rules for R	Cluster (X, Y, Z) centroid	output function
1	$(11.83, 23.77, 34.51)$	$R = f_1^R(X,Y,Z) = 51.56X - 16.83Y - 5.72Z + 4.69$
2	$(14.50, 24.86, 34.74)$	$R = f_2^R(X,Y,Z) = 13.47X - 4.43Y - 1.50Z + 30.18$
3	$(19.52, 27.37, 10.11)$	$R = f_3^R(X,Y,Z) = 5.83X - 1.92Y - 0.67Z + 64.74$
4	$(26.58, 30.08, 66.06)$	$R = f_4^R(X,Y,Z) = 5.93X - 1.95Y - 0.67Z + 64.50$
5	$(36.49, 35.03, 35.12)$	$R = f_5^R(X,Y,Z) = 3.75X - 1.26Y - 0.41Z + 95.31$
6	$(55.99, 43.86, 34.47)$	$R = f_6^R(X,Y,Z) = 2.86X - 0.96Y - 0.32Z + 120.93$
Rules for G	Cluster (X, Y, Z) centroid	output function
1	$(21.92, 10.21, 33.22)$	$G = f_1^G(X,Y,Z) = -18.66X + 40.77Y + 0.13Z + 5.21$
2	$(23.04, 13.64, 33.60)$	$G = f_2^G(X,Y,Z) = -4.72X + 10.21Y + 0.03Z + 31.63$
3	$(23.79, 23.99, 9.86)$	$G = f_3^G(X,Y,Z) = -2.06X + 4.48Y - 0.02Z + 67.04$
4	$(30.75, 27.10, 65.78)$	$G = f_4^G(X,Y,Z) = -2.08X + 4.53Y - 0.00093Z + 66.88$
5	$(33.66, 45.28, 35.80)$	$G = f_5^G(X,Y,Z) = -1.41X + 2.98Y + 0.01Z + 96.42$
6	$(40.99, 67.44, 36.83)$	$G = f_6^G(X,Y,Z) = -1.08X + 2.31Y + 0.0051Z + 121.65$
Rules for B	Cluster (X, Y, Z) centroid	output function
1	$(24.88, 29.83, 1.91)$	$B = f_1^B(X,Y,Z) = 0.90X - 1.92Y + 25.10Z + 4.31$
2	$(25.44, 30.10, 5.96)$	$B = f_2^B(X,Y,Z) = 0.21X - 0.53Y + 6.74Z + 28.70$
3	$(26.48, 30.56, 14.09)$	$B = f_3^B(X,Y,Z) = 0.09X - 0.27Y + 3.48Z + 51.58$
4	$(28.47, 31.40, 29.64)$	$B = f_4^B(X,Y,Z) = 0.05X - 0.17Y + 2.25Z + 74.63$
5	$(31.76, 32.84, 55.10)$	$B = f_5^B(X,Y,Z) = 0.04X - 0.13Y + 1.66Z + 96.87$
6	$(36.21, 34.83, 89.13)$	$B = f_6^B(X,Y,Z) = 0.03X - 0.10Y + 1.26Z + 123.07$

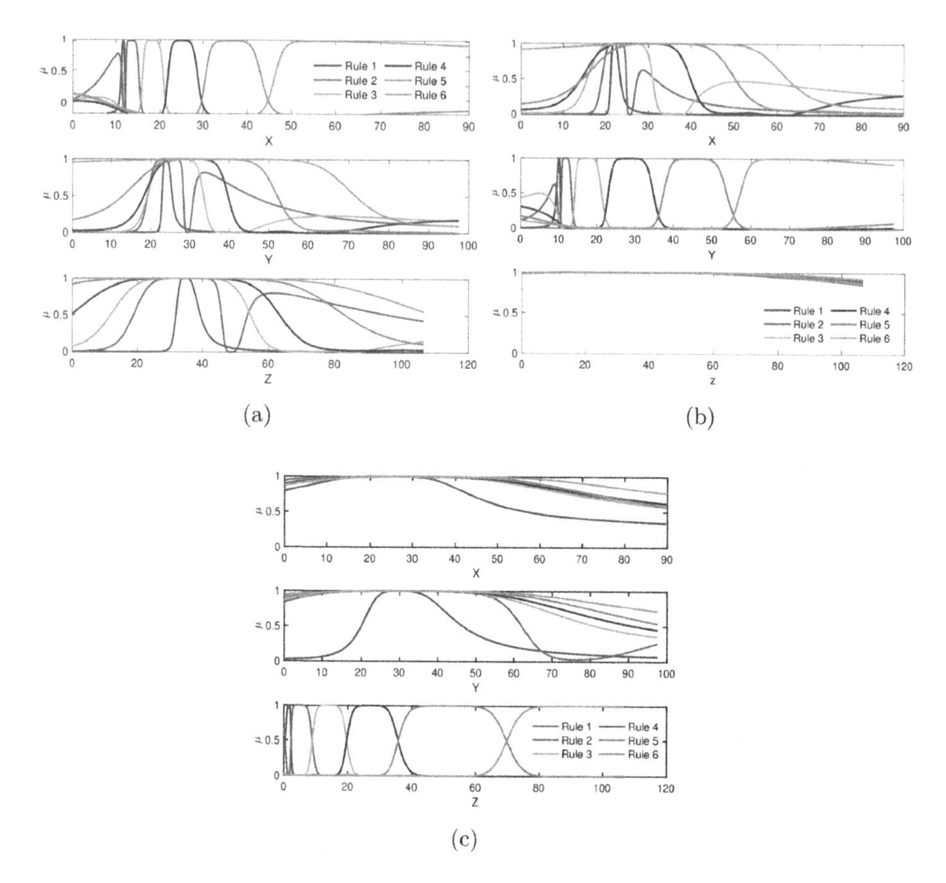

Fig. 1. Membership functions (μ) for $FMID_{XYZ}$, using different clustering results for each output: (a) The first output (R), (b) The second output (G), (c) The third output (B).

3 Experimental Comparison and Discussion

In order to assess the performance of our models, we computed the visual error ΔE_{00} [9,14] between the desired colors of the validation dataset and displayed colors using RGB inputs computed by each method. These results are summarized in Table 4. We compare the three models created with FMID detailed in Sect. 2.2 against three state-of-art models: (1) A neural network (NN) using XYZ tristimulus values and xyY colour coordinates [12], and the so-called PI model [11].

Table 4 shows, for each model: (a) The mean square error (MSE_{RGB}), (b) The mean of the visual error (ΔE_{00}), (c) The standard deviation (STD) of ΔE_{00}; (d)-(f) The last three columns show the number of colours for which the model produces a correct display value (with a $\Delta E_{00} \leq 1$), it makes a small mistake ($1 < \Delta E_{00} \leq 2$), and the number of colours for which the mistake would be clearly appreciated by a *normal* observer ($\Delta E_{00} > 2$). We can see that the FMID models for XYZ inputs show the best performance for the 13 clusters one closely followed by the 6 clusters one. For them, average of ΔE_{00} is less than one which is below the threshold and means that on average the colors are satisfactorily displayed. This also happens for NN_{XYZ} model. However, the difference between them is related to the number of colors for which the model makes a clear error, which is 5 times bigger for FMID with 6 clusters and NN_{XYZ}. The FMID model using xyY has an average ΔE_{00} of one unit, which is in the human discrimination threshold level, but the number of wrongly displayed colors is much larger. On the other hand, PI has the highest error, over 2 in terms of ΔE_{00}, which means the colour error can be visible at a quick glance by humans.

Figure 2 shows the ΔE_{00} in terms of chrominance Y and luminance xy for the best models (i.e., $FMID_{XYZ}$ using 13 clusters, $FMID_{XYZ}$ using 6 clusters, and NN_{XYZ}). The $FMID_{XYZ}$ using 6 clusters performs better when Y is getting bigger, but when Y is smaller, the model performs in low and high error. The addition of more clusters clearly improves performance for dark colors. On the other hand, NN_{XYZ} performs poorly when Y increases, being worse for the bright greenish and yellowish colors. In terms of xy, we can see that the $FMID_{XYZ}$ with 13 clusters has an non negligible error in a small number of colours, showing a better performance than the other two models in dark greenish and blueish colors, where the other two models perform similarly.

Table 4. MSE_{RGB}, mean of the ΔE_{00} visual error, visual error standard deviation (STD), and number of points bounded by error equal to 1 (non perceptually distinguishable) and 2.

Method	MSE_{RGB}	Mean ΔE_{00}	STD	$\Delta E_{00} \leq 1$	$1 < \Delta E_{00} \leq 2$	$\Delta E_{00} > 2$
$FMID_{XYZ}$ (6 clust.)	3.00	0.61	0.47	8172	933	156
$FMID_{XYZ}$(13 clust.)	1.5	0.50	0.32	8697	528	36
$FMID_{xyY}$	6.30	1.00	1.12	6178	2467	616
NN_{XYZ}	2.77	0.62	0.46	8055	1052	154
NN_{xyY}	4.82	0.73	0.41	7650	1496	115
PI	48.93	2.19	1.08	1162	3246	4853

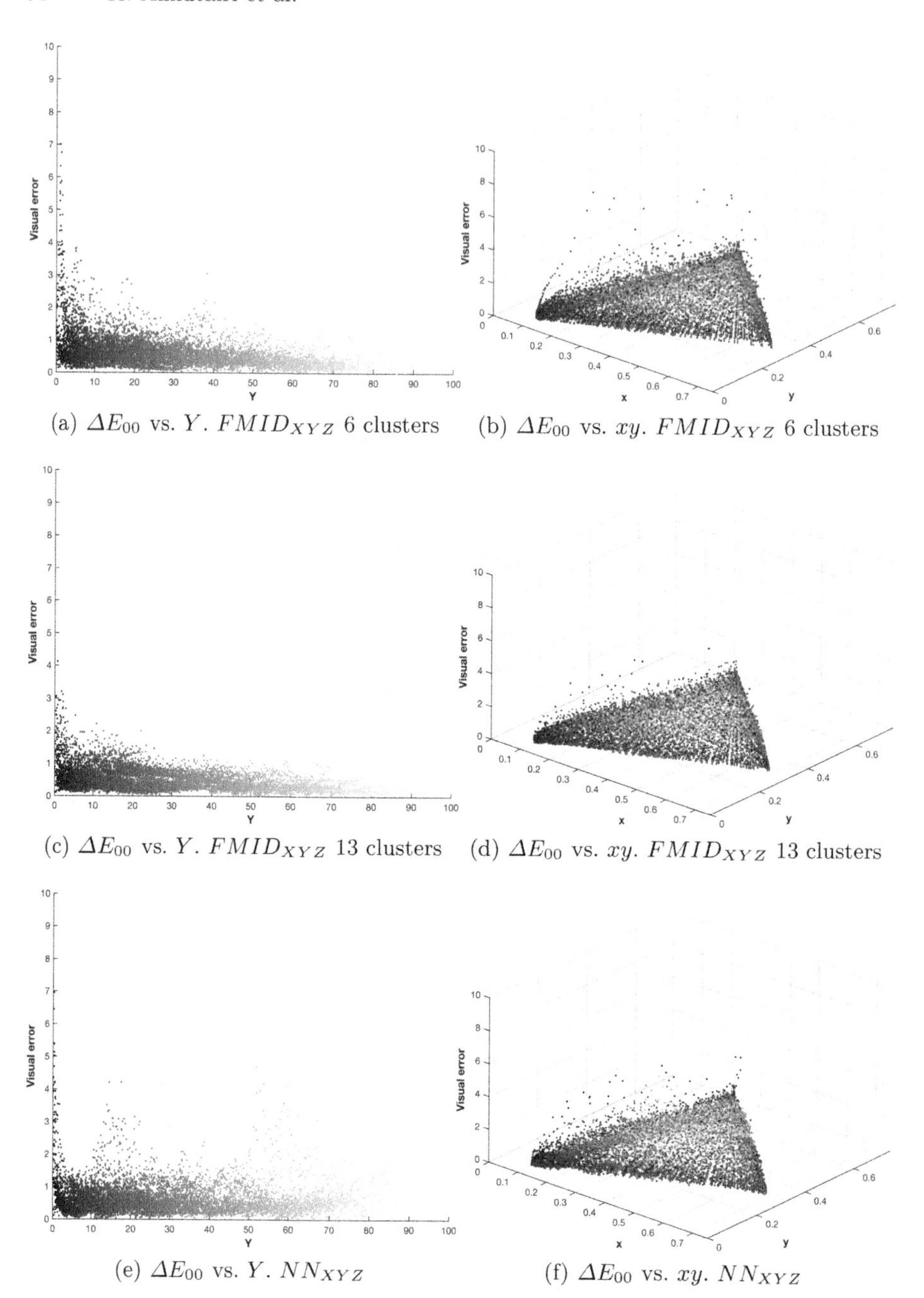

(a) ΔE_{00} vs. Y. $FMID_{XYZ}$ 6 clusters (b) ΔE_{00} vs. xy. $FMID_{XYZ}$ 6 clusters

(c) ΔE_{00} vs. Y. $FMID_{XYZ}$ 13 clusters (d) ΔE_{00} vs. xy. $FMID_{XYZ}$ 13 clusters

(e) ΔE_{00} vs. Y. NN_{XYZ} (f) ΔE_{00} vs. xy. NN_{XYZ}

Fig. 2. ΔE_{00} visual error analysis for three trained models ($FMID_{XYZ}$ with 6 clusters, $FMID_{XYZ}$ with 13 clusters, and NN_{XYZ}). First column: ΔE_{00} versus Y luminance values of desired colors; Second column: ΔE_{00} versus xy chrominance values of desired colors.

4 Conclusions

In this paper, we built three Takagi-Sugeno fuzzy inference systems for display characterization using the FMID toolbox. A thorough study of toolbox parameter concluded with three models being trained and tested in terms of visual error against other three, which they clearly outperform. The models relating $XYZ - RGB$ data show better performance than when $xyY - RGB$ was used. Also, a model with 13 clusters showed the best performance closely followed by a model with only 6 clusters. We analysed in detail the information encoded in this latter model which was easier to interpret. We found that 6 different levels of X, Y, and Z where considered to determine R, G, and B, respectively. Also, activation output functions of the FIS account with positive relations between X and R, Y and G, and Z and B and either negligible or inverse relations among the rest. Inverse relations are interpreted as the way the model has found to compensate for the correlations and overlappings that exist between the RGB primaries in terms of XYZ coordinates of the generated colours.

References

1. Maali Amiri, M.: Spectral reflectance reconstruction using fuzzy logic system training: color science application. Sensors **20**(17), 4726 (2020)
2. Babuška, R.: Fuzzy Modeling for Control, vol. 12. Springer, Cham (2012)
3. Berns, R.S.: Methods for characterizing CRT displays. Displays **16**(4), 173–182 (1996)
4. Bryant, J., Kester, W.: Data converter architectures. In: Analog-Digital Conversion, pp. 1–14. Analog Devices (2005)
5. Capilla, P., Dıez-Ajenjo, M.A., Luque, M.J., Malo, J.: Corresponding-pair procedure: a new approach to simulation of dichromatic color perception. JOSA A **21**(2), 176–186 (2004)
6. Fairchild, M., Wyble, D.: Colorimetric characterization of the apple studio display (flat panel LCD) (1998)
7. Gustafson, D.E., Kessel, W.C.: Fuzzy clustering with a fuzzy covariance matrix. In: 1978 IEEE Conference on Decision and Control Including the 17th Symposium on Adaptive Processes. IEEE (1978)
8. Kim, J.M., Lee, S.W.: Universal color characterization model for all types of displays. Opt. Eng. **54**(10), 103103 (2015)
9. Luo, M.R., Cui, G., Rigg, B.: The development of the CIE 2000 colour-difference formula: CIEDE 2000. Color Res. Appl. **26**(5), 340–350 (2001)
10. Malo, J., Luque, M.: Colorlab: a matlab toolbox for color science and calibrated color image processing. Servei de Publicacions de la Universitat de Valencia (2002)
11. Penrose, R.: On best approximate solutions of linear matrix equations. In: Mathematical Proceedings of the Cambridge Philosophical Society, vol. 52, pp. 17–19. Cambridge University Press (1956)
12. Prats-Climent, J., et al.: A study of neural network-based LCD display characterization. In: London Imaging Meeting, vol. 2021, pp. 97–100. Society for Imaging Science and Technology (2021)
13. Ross, T.J.: Fuzzy Logic with Engineering Applications. Wiley, Hoboken (2009)

14. Sharma, G., Wu, W., Dalal, E.N.: The CIEDE2000 color-difference formula: implementation notes, supplementary test data, and mathematical observations. Color Res. Appl. **30**(1), 21–30 (2005)
15. Terano, T., Asai, K., Sugeno, M.: Fuzzy Systems Theory and Its Applications. Academic Press Professional, Inc. (1992)
16. Zadeh, L.A.: Fuzzy sets. Inf. Control **8**(3), 338–353 (1965)

Learning Semantic-Visual Embeddings with a Priority Queue

Rodrigo Valério(✉) and João Magalhães

NOVA LINCS, School of Science and Technology, Universidade NOVA de Lisboa, Caparica, Portugal
r.valerio@campus.fct.unl.pt, jm.magalhaes@fct.unl.pt

Abstract. The Stochastic Gradient Descent (SGD) algorithm and margin-based loss functions have been the learning workhorse of choice to train deep metric learning networks. Often, the random nature of SGD will lead to the selection of sub-optimal mini-batches, several orders of magnitude smaller than the larger dataset. In this paper, we propose to augment SGD mini-batch with a priority learning queue, i.e., SGD+PQ. While the mini-batch SGD replaces all learning samples in the mini-batch at each iteration, the proposed priority queue replaces samples by removing the *less informative ones*. This novel idea introduces a sample update strategy that balances two sample removal criterion: (i) removal of stale samples from the PQ that are likely outdated, and (ii) removal of samples that are not contributing to the error, i.e. their sample error is not changing during training. Experimental results demonstrate the success of the proposed approach across three datasets.

Keywords: Embeddings learning · batch size · priority batch priority queue

1 Introduction

Learning visual-semantic representations is a core task in AI with applications to problems that need to compute the semantic similarity between images, language and other types of data. To compute the visual-semantic similarity between two samples [22], projection functions are learned to project the samples from their original format into a high-dimensional embedding representation where a metric function is used to compute the semantic similarity. Typically, these projection functions follow deep neural network architectures to better capture both linear and non-linear patterns of the data. Content-based retrieval [31,37], image categorisation [4], siamese networks [3], face verification and recognition [12,30], and visual question-answering [5] are some of the examples where learning semantic representations play a critical role[1].

The goal of deep metric learning is to project samples of one category into the same neighborhood and samples that share no semantics to neighborhoods that

[1] https://github.com/rcvalerio/Priority-Queue.

A. Pertusa et al. (Eds.): IbPRIA 2023, LNCS 14062, pp. 67–81, 2023.
https://doi.org/10.1007/978-3-031-36616-1_6

are far apart [22]. SGD is the tool of choice to find the model that minimizes the errors in projecting such samples. SGD iteratively computes the error gradients from mini-batch, to update the projection function parameters. Unfortunately, the learning process is severely constrained by the mini-batch size [2]: its limited number of samples will hinder the accurate and robust estimation of the loss and its gradients. The constraints on the mini-batch size will increase the stochastic nature of the learning process. Hence, to reduce the variance and improve the robustness of the loss and gradient estimates it is critical to move beyond the mini-batch size barrier.

Recently, there has been works that aim to mitigate the limited number of samples in a single mini-batch [2] and employ triplet mining [38], gradient accumulation [6], external memory [36] or specialized loss functions [14]. In the triplet mining arena, the seminal work of [38] has carefully examined the impact of mining soft and hard negatives to create more informative triplets. Gradient accumulation has observed a renewed interest thanks to the field of distributed learning [18]. Other works have also explored specialized loss functions [14,21] that summarize a larger number of samples into a proxy sample. Moreover, memory-based extensions [36] of the mini-batch, have been proposed to enrich the mini-batch samples with negative samples from previous iterations.

In this work, we propose to move beyond the standard limitations of mini-batch SGD [32] and propose to extend the mini-batch with a priority queue. The rationale is to improve deep metric learning with a priority queue that updates the queue by adding fresher samples and removing less informative samples. The priority queue update strategy needs to have a minimal computational cost, preserving the low computational footprint of the SGD algorithm. We argue that being able to keep a priority queue many times larger than the mini-batch with the most informative learning samples leads to a better estimation of the loss calculation and reduces the variance of the gradients, delivering significant improvements in terms of model accuracy and computational efficiency. The increase in memory usage is below 1% of the total memory required by the gradients of the back-propagation algorithm. While the mini-batch iteratively replaces all the learning samples, the priority queue replaces samples by removing the less *useful* samples. The PQ is aware of the contribution of each sample to the learning effectiveness by keeping track of each sample error. Hence, leveraging this fact, the proposed update policy balances two strategies: a sample can be removed either because it has been in the priority queue for too long, or because the model is no longer learning from it.

2 Related Work

Deep metric learning benefits from having more samples for the margin-based loss function, however, increasing the size of the mini-batch has several computational costs – sampling grows cubically as the dataset gets larger, and the gradients become to large-scale to be stored efficiently [8,32]. There are several principled approaches to overcome this bottleneck. Sample mining builds on the rationale that most of the triplets do not contribute to the loss since they are already learned, so, sample selection plays a crucial part in training [38]. There are two approaches to create a mini-batch of triplets: Offline triplet mining which generates triplets offline every n steps, and online triplets mining where the triplets are created during training by being sampled directly from mini-batches [30]. The sampling is usually done via online triplet mining, which is more efficient. Wu et al. [38] showed that sampling only hard negatives is problematic. This is because the gradients with respect to the negative examples are small, and given enough training noise introduced by the training algorithm, the direction of the gradient becomes dominated by the noise. This leads to noisy gradients that cannot effectively push examples apart, and consequently, a collapsed model. Wu et al. [38] also proposed a new distance weighted sampling, which selects more informative and stable examples than traditional approaches. However, sample mining is computationally intensive and the number of informative triplets depends on the mini-batch size [12]. Which brings us to other family of principled approaches that shows how the size of the mini-batch or memory augmentations of the batch are of extreme importance [8–10, 36].

Recently, [2] proposed SimCLR which is a framework for self-supervised learning and contrastive learning of visual representations. In self-supervised learning a model is trained using labels that are naturally part of the data, rather than requiring separate external labels. In contrastive learning the objective is to find similar and dissimilar items. SimCLR learns representations by maximizing agreement between differently augmented views of the same data example via a contrastive loss in the latent space. This framework samples a minibatch of N examples and defines the contrastive prediction task on pairs of augmented examples derived from the minibatch with the data augmentation module, resulting in $2N$ data points. A different technique called proxyNCA loss was proposed by [21] in which the learning is done over a different space of points, called proxies. Proxies are learned as part of the model parameters such that for each data point in the original space there is a proxy point close to it (usually there is one proxy per class). Since this space is usually much smaller than the original space, it becomes possible to write the loss over all the possible triplets, without the need of sampling. As a result, this re-defined loss is easier to optimise, and trains faster. However it does not consider the rich data-to-data relations.

In order to take advantage of the fast and reliable convergence of the proxy-based losses and the fine-grained semantic relations between data points of pair-based losses, [14] proposed a new proxy based loss called Proxy-Anchor loss. This loss utilises each proxy as an anchor and associates it with all data in a batch. It takes data-to-data relations into account by associating all data in a batch with each proxy so that the gradients with respect to a data point are weighted by its relative proximity to the proxy (i.e., relative hardness), affected by the other data in the batch. The key difference and advantage of Proxy-Anchor over Proxy-NCA is the active consideration of relative hardness based on data-to-data relations.

3 Learning with SGD + Priority Queue

Given a set $X = (x_1, y_1), ..., (x_K, y_K)$ of x_i image samples and its corresponding labels y_i, the function $f_\Theta(x_i) = \mathbf{v}_i$ is an embedding function that projects samples x_i onto a high-dimensional space where neighbouring samples $\mathbf{v}_*$ share the same label y_i with a probability $< \epsilon$. The parameters Θ can be learned with gradient based optimization algorithm, that minimizes the model loss, computed across the entire set of K data samples. Using all K data samples, leads to a robust, but expensive, estimation of the loss and its gradient. The SGD algorithm can use a mini-batch with $b << K$ samples, which provides a frequent, unbiased but highly invariant, update of the model loss and loss gradient. This paper hypothesis, is that *leveraging SGD with a priority queue, instead of the traditional mini-batch, leads to low variance and efficient estimations of the model loss and loss gradient, thanks to the larger sample size.*

3.1 Low Variance Loss Estimate

In this section, we formalize how a priority queue can provide the SGD algorithm with low-variance estimates of the model loss for the current set of model parameters.

Mini-Batch Based Estimation. In mini-batch gradient descent, only a small set of samples is accessible in each iteration, and the loss is estimated from that single mini-batch, which is limited by the number of informative pairs/triplets. Increasing the mini-batch size [2,8] drastically increases the number of pairs/triplets but is restricted by GPU memory.

PQ-Based Estimation. Some recent methods [16,36,41] memorize the features of old batches in a memory bank to improve the loss estimation. In our implementation, the PQ stores r mini-batches of size b, thus,

$$\mathrm{PQ}_{\mathrm{size}} = b \cdot r. \tag{1}$$

During the forward pass, the stored features are reused to better estimate the loss function. The model loss with a PQ, compares each instance of the current mini-batch to all high-priority embeddings stored in the PQ using a metric loss. The total model loss is given by the loss among the samples of the new batch and the loss across all samples in the PQ,

$$\mathcal{L} = \frac{1}{b}\sum_{i=1}^{b} \mathcal{L}_B + \frac{1}{b \cdot r}\sum_{i=1}^{b \cdot r} \mathcal{L}_{\mathrm{PQ}}, \tag{2}$$

where the loss $\mathcal{L}_B$ in the mini-batch is straightforward and the loss in the priority queue, $\mathcal{L}_{\mathrm{PQ}}$, is given by

$$\mathcal{L}_{\mathrm{PQ}} = \sum_{i=1}^{b} \Big[\sum_{\widetilde{y}_j \neq y_i}^{b \cdot r} w_{ij} \widetilde{\mathbf{s}}_{ij} - \sum_{\widetilde{y}_j y_i}^{b \cdot r} w_{ij} \widetilde{\mathbf{s}}_{ij} \Big], \tag{3}$$

where the similarity s_{ij} between the batch embeddings $\mathbf{v}_i$ and the PQ embeddings $\widetilde{\mathbf{v}_j}$ corresponds to $\widetilde{s}_{ij} = \mathbf{v}_i^T \widetilde{\mathbf{v}}_j$, and w_{ij} is the loss specific weight

$$w_{ij} = criteria(\mathbf{v}_i, y_i, \widetilde{\mathbf{v}}_j, \widetilde{y}_j) \tag{4}$$

of each negative-anchor pair (x_i, n_i) or positive-anchor pair (x_i, p_i). Note that most metric learning losses [22] can be used, e.g. contrastive loss [9] and triplet loss [13].

In the backward pass, the estimation of the model error is less invariant, which improves the convergence of the gradient descent algorithm:

$$\Theta^{(t+1)} = \Theta^{(t)} - \eta \cdot \frac{\partial \mathcal{L}(\Theta^{(t)}, x_i)}{\partial \Theta}. \tag{5}$$

Considering the decomposition result of Eq. 3 we highlight how the extra embeddings in the PQ improve the loss estimation. Each sample $\mathbf{v}_i$ in the newest mini-batch is compared to all other samples $\mathbf{v}_j$ in the PQ. This will reduce SGD variance estimates because of the increased sample size and the prioritisation strategy. The gradient of the $\mathcal{L}_{\text{PQ}}$, Eq. 3, w.r.t. v_i is:

$$\frac{\partial \mathcal{L}_{\text{PQ}}}{\partial v_i} = \sum_{\widetilde{y}_j \in y_i}^{b \cdot r} w_{ij} \widetilde{\mathbf{v}}_j - \sum_{\widetilde{y}_j \in y_i}^{b \cdot r} w_{ij} \widetilde{\mathbf{v}}_j, \tag{6}$$

Let us now depart from the general case, and consider a deep convolutional network with L layers and activations $a^{(l)}$. The error gradient on layer (l) is decomposed into the gradient of the current layer and the gradient of previous layers, i.e.,

$$\frac{\partial \mathcal{L}}{\partial \Theta} = \delta^{(l)} \cdot o^{(l)} = \frac{\partial \mathcal{L}}{\partial a^{(l)}} \cdot \frac{\partial a^{(l)}}{\partial \Theta}. \tag{7}$$

Expanding the first part, the loss gradient with respect to the model output embedding $\mathbf{v}_i = a^{(L, x_i)}$, we obtain

$$\delta^{(l, x_i)} = \begin{cases} w_{i*} \cdot s_{\mathbf{v}_i, \mathbf{v}_*} & , l = L \\ a^{(l, x_i)} \cdot \sum_k \delta_k^{(l+1)} \theta_k^{(l+1)} & , l \neq L, \end{cases} \tag{8}$$

which highlights the key factor $w_{i*} \cdot s_{\mathbf{v}_i, \mathbf{v}_*}$, i.e. the similarity $s_{\mathbf{v}_i, \mathbf{v}_*}$ between between embedding $\mathbf{v}_i$ and other positive/negative embedding vectors $\mathbf{v}_*$, and a loss specific weight w_{i*}. This key factor is the exact spot where the PQ increases the support to compute the loss and gradient, $K > \text{PQ}_{\text{size}} >> b$. This approach improves on memory-based methods [36,39,41], and augments the mini-batch with a PQ that prioritises the embeddings of previous iterations. The embeddings prioritisation implemented by the PQ, allows the model to reuse the high-priority mini-batches and to obtain more hard-negative pairs with minimal computational cost. When using this result in Eq. 5 we end up with the final expression

$$\theta_j^{(l)} \leftarrow \theta_j^{(l)} - \eta \cdot \delta^{(l)} \cdot o^{(l)}. \tag{9}$$

The importance of the batch size among the set of hyperparameters is becoming clearer [2,3,17] and the PQ is explicitly leveraged by the gains obtained by computing the loss over a set of $b \cdot r$ samples, instead of $b << b \cdot r$ samples. More importantly, the key novelty of the PQ is the update strategy of the memory. While previous works used a FIFO strategy, we introduced a priority strategy that keeps the most informative samples in the learning queue.

Priority-Memory Based Sampling. A FIFO queue is generally used in memory-based sampling [36,39,41] to store the embeddings of past iterations, and, when full, the oldest instances are removed. In the FIFO based approaches [36,39,41], authors have reported only marginal differences between embeddings computed at different training iterations, and they found that after a few iterations of warmup period [20], mining across mini-batches can provide negative pairs with valid information. Recently, image retrieval transformers [4] achieved state-of-the-art results for category-level image retrieval. It consists of a transformer architecture, fine-tuned with a contrastive loss and augmented with a FIFO module.

Given past evidence regarding FIFO based memory learning [36,39,41], the priority-based memory PQ consists of a priority queue that functions similarly to a FIFO queue but has a priority oriented removal criterion. The criterion is chosen in a way that it maximizes the relevance of the samples present in the PQ memory. The priority of each batch in the PQ is given by

$$priority(j) = \alpha \cdot staleness(j) + \beta \cdot \mathcal{L}_{\mathrm{PQ}}^{(j)} \tag{10}$$

where $staleness(j)$ is the number of iterations of the batch j in the PQ, and α and β control the tradeoff between the two removal criteria.

It is important to elaborate on this aspect. The PQ indirectly performs hard sample mining through its prioritisation strategy and therefore maintains more relevant samples in memory than the FIFO queue. Thus, it has more informative samples for training, which improves the loss estimation. Moreover, note that it is crucial that stale embeddings are removed from the PQ or periodically updated. Since the embeddings are not back-propagated, they will get stale due to the change in the model parameters, and the training will stagnate.

Experimentally, we observed that updating all the embeddings in memory at once should be avoided since it is computationally expensive and results in abrupt changes in the loss function puts SGD off the previous optimization path. Instead, it is best to update only the oldest embeddings or update using a moving average update [39,41]:

$$\widetilde{\mathbf{v}}_i = m\widetilde{\mathbf{v}}_i + (1-m)v_i \tag{11}$$

3.2 Robust Gradient Estimation

Gradient Accumulation. In mini-batch gradient descent, the gradients are calculated using only a single mini-batch, which might not accurately represent the gradients of the dataset. To address this limitation, gradient accumulation, originally used in distributed and multi-GPU settings [18], simulates a bigger batch size without extra computational costs by accumulating the gradients over multiple mini-batches. Instead of having a single forward pass with $b \cdot r$ samples through the network, acc forward passes are made with $n = b \cdot r/acc$ examples each (Eq. 12). The loss is not accumulated across multiple mini-batches. Instead, each mini-batch has an individual loss

$$\Theta^{(t+1)} = \Theta^{(t)} - \eta \cdot \sum_{i=1}^{acc} \sum_{j=1}^{n} \frac{\partial err(\Theta^{(t)}, x_{j \cdot n+i})}{\partial \Theta}. \tag{12}$$

However, gradient accumulation is not equivalent to having a mini-batch of size $b \cdot r$ since some losses are highly dependent on the actual batch size. Note that in our implementation we let $r \neq acc$, so that gradient accumulations can occur more frequently.

PQ + Gradient Accumulation. Together, gradient accumulation and the priority queue complement each other. This framework can be easily applied to many different tasks since the implementation is generic and independent from the problem. Priority queue prioritises the embeddings of past iterations, which allows the model to collect more informative pairs across multiple mini-batches and have a better estimate of the loss. Moreover, gradient accumulation enriches the model with more accurate gradients calculated from multiple prioritised mini-batches.

4 Experiments

4.1 Datasets

We trained and evaluated our experiments on three large-scale datasets for zero-shot image retrieval:

Stanford Online Products (SOP). [24] contains 120,053 online product images and 22,634 categories. There are only 2 to 10 images per category. Following [24], we use 59551 images (11,318 classes) for training, and 60,502 images (11,316 classes) for testing.

In-shop Clothes Retrieval (In-shop). [19] contains 72,712 clothing images of 7,986 classes. Following [19], we use 3,997 classes with 25,882 images as the training set. The test set is partitioned to a query set with 14,218 images of 3,985 classes, and a gallery set having 3,985 classes with 12,612 images

CUB-200-2011 (CUB) contains 11,788 birds images of 200 classes. There are about 60 images/class. Following [34], we use 5,864 images of 100 classes for training and the remaining 5,924 images for testing.

4.2 Training Details

Unless otherwise stated, training was performed on a single GPU with a ResNet-50 [11] pretrained on ImageNet [28]. The model's last layer is a 128-d fully

Table 1. Recall@K(%) for K=1,10 on SOP. The CNN backbone is **G**oogleNet, Inception**B**N or **R**esNet50. The superscript is the embedding size.

Recall@k(%)		1	10
Contrastive loss	R^{128}	71.9	85.9
Triplet loss	R^{128}	72.8	86.4
HDC	G^{384}	69.5	84.4
A-BIER [25]	G^{512}	74.2	86.9
ABE [15]	G^{512}	76.3	88.4
SM [33]	G^{512}	75.2	87.5
Clustering [23]	B^{64}	67.0	83.7
ProxyNCA	B^{64}	73.7	–
HTL [7]	B^{512}	74.8	88.3
MS [35]	B^{512}	78.2	90.5
SoftTriple [26]	B^{512}	78.6	86.6
Margin [38]	R^{128}	72.7	86.2
Divide [29]	R^{128}	75.9	88.4
FastAP [1]	R^{128}	73.8	88.0
MIC	R^{128}	77.2	89.4
XBM [36]	R^{128}	80.6	91.6
SGD+PQ-50k w/ Cont.	G^{512}	78.6	90.4
SGD+PQ-50k w/ Cont.	B^{512}	79.4	90.5
SGD+PQ-8k w/ Cont.	R^{128}	81.1	91.4
SGD+PQ-50k w/ Cont.	R^{128}	**81.9**	**91.9**

connected layer with l_2 normalization. For all datasets, the input images are first resized to 256×256, and then cropped to 224×224. Random crops and random flips are used as data augmentation during training to compute the embedding of each instance. For testing, only the single center crop is used. In all experiments we use the contrastive loss and the Adam optimizer with $5e^{-4}$ weight decay and a batch of size 64 together with a sampler that guarantees at least four samples of the same category for every sample in the mini-batch.

4.3 Baselines and Metrics

To assess the success of the proposed method, we compared it to other state-of-the-art approaches: HDC [40], MIC [27], A-BIER [25], ABE [15], SM [33], Clustering [23], ProxyNCA [14], HTL [7], MS [35], SoftTriple [26], Margin [38], Divide [29], FastAP [1], XBM [36]. Also, since the PQ memory provides additional informative positives and negatives embeddings for training, we evaluated the proposed PQ update strategies:

- **SGD+PQ dequeue by staleness:** When the PQ is full, the lowest priority batch is the one that has been in PQ for more iterations, this is the FIFO strategy (same as [36] but without the gradient accumulation).

Table 2. Recall@K(%) for K=1,8 on CUB.

Recall@k(%)		1	8
HDC	G^{384}	53.6	85.6
A-BIER [25]	G^{512}	57.5	86.2
ABE [15]	G^{512}	60.6	87.4
SM [33]	G^{512}	49.8	83.3
Clustering [23]	B^{64}	48.2	81.9
ProxyNCA	B^{64}	49.2	72.4
HTL [7]	B^{512}	57.1	86.5
MS [35]	B^{512}	65.7	91.2
SoftTriple [26]	B^{512}	65.4	90.4
XBM [36]	B^{512}	65.8	89.9
SGD+PQ-1k w/ Cont.	B^{512}	**68.9**	**91.7**

Table 3. Recall@K(%) for K=1,10 on InShop.

Recall@k(%)		1	10
HDC	G^{384}	62.1	84.9
A-BIER [25]	G^{512}	83.1	95.1
ABE [15]	G^{512}	87.3	96.7
HTL [7]	B^{512}	80.9	94.3
MS [35]	B^{512}	89.7	97.9
Divide [29]	R^{128}	85.7	95.5
MIC	R^{128}	88.2	97.0
FastAP [1]	R^{512}	90.9	97.7
XBM [36]	R^{128}	91.3	**97.8**
SGD+PQ-8k w/ Cont.	B^{512}	90.0	97.1
SGD+PQ-8k w/ Cont.	R^{128}	**92.0**	**97.8**

- **SGD+PQ dequeue by loss:** Similar to the previous one, by the lowest priority batch is the one with the lowest loss.
- **SGD+PQ dequeue by staleness and loss:** Implements the full model and the lowest priority batch is a combination of staleness and loss.

The retrieval performance was evaluated with Recall@K (R@K) metric.

4.4 Results and Discussion

General Results. The general results on the SOP dataset, Table 1, the CUB dataset Table 2 and the InShop dataset Table 3 confirm the validity of augmenting the mini-batch with the PQ and demonstrates the achieved improvements. Improvements were achieved in the three datasets. These results use the hybrid removal criterion where the oldest batch is removed, followed by the four batches with the lowest loss. The results show that the contrastive loss paired with the PQ can outperform state-of-the-art losses on all tested datasets. Our method with a PQ-50k improved the R@1 of the state-of-the-art method XBM [36] on all datasets: SOP 80.6%→81.9%, CUB 65.8%→68.9% and InShop 91.3%→92.0%.

4.5 Ablation Study

Batch Size and Gradient Accumulation. Our implementation of contrastive loss with a batch of size 16 where the gradients are accumulated over four iterations has a slight decrease in R@1 performance compared to the baseline contrastive loss of batch size 64. However, our implementation requires 13% less memory. Figure 1 (left) illustrates these results. For bigger batch sizes, the difference is more prominent. A batch of size 256 requires 3.71x more memory than accumulating four times a batch of size 64. This indicates that the batch size can be compensated by accumulating gradients for a tradeoff of training time.

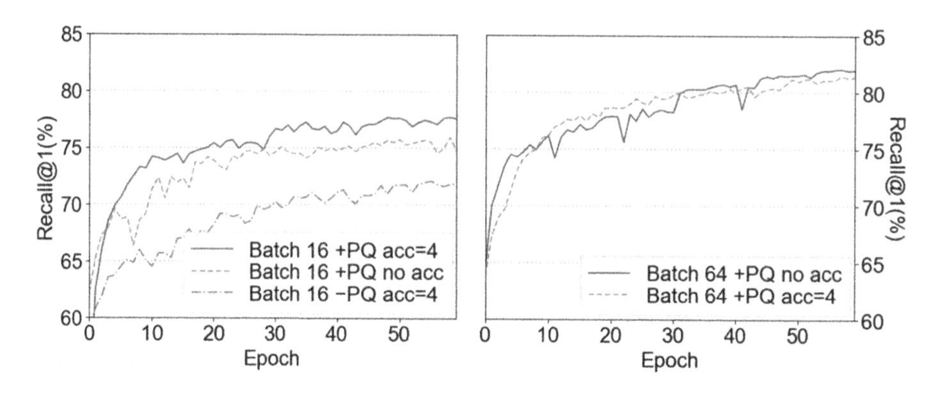

Fig. 1. Recall@1 on SOP with different batch sizes.

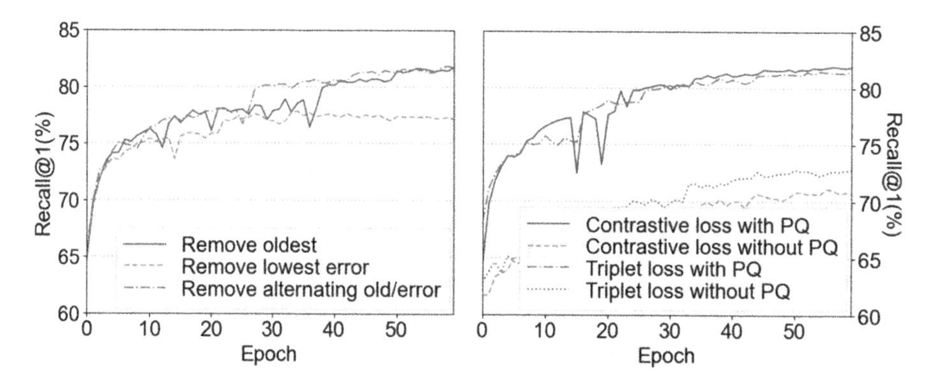

Fig. 2. Recall@1 for different training strategies.

However, it is not feasible for huge batches since it takes too many iterations for a single step. Augmenting the PQ with gradient accumulation improved the performance on batches of size 16, but we found no benefit on batches of size 64, Fig. 1 (right).

Loss Function and PQ Size. The PQ memory provides thousands of examples without increasing the batch size and does not require gradients. It drastically enhances the performance of pair-based losses such as the contrastive (+10%) and triplet loss (+8.8%), Table 4, with less than 1% memory overhead. The size of the PQ and the removal criterion have the most significant impact on model performance, Fig. 2 (left). R@1 can increase by up to 6.15% by increasing the PQ size from 256 entries to 50k entries.

Priority Queue Removal Criterion. We experimented many different removal criteria for the PQ. In most datasets, removing individual samples by staleness or loss decreased the R@1 after a few iterations. Removing the batch with the lowest loss stagnated the training. Removing the oldest batch achieved

Table 4. Comparison of the contrastive loss and triplet loss functions with and without the PQ on SOP.

Recall@k(%)		1	10	1000	10000
SGD	Cont.	71.9	85.9	94.0	98.1
SGD+PQ-50k	Cont.	**81.9**	**91.9**	**96.4**	**98.7**
SGD	Triplet	72.8	86.4	94.2	98.2
SGD+PQ-50k	Triplet	81.6	**91.9**	**96.4**	**98.7**

Table 5. Recall@1 of priority queue with multiple memory sizes and different removal criteria.

SGD+PQ update strategy					
Loss	–	× 1	× 1	× 1	× 4
Staleness	× 1	–	× 1	× 4	× 1
Methods					
SGD+PQ-256	**77.10**	74.73	75.76	75.69	77.05
SGD+PQ-1k	78.21	74.77	79.07	**79.24**	79.11
SGD+PQ-8k	80.76	75.21	81.05	**81.13**	80.18
SGD+PQ-50k	81.72	77.18	81.70	**81.84**	81.75

the best performance on small memories, but the hybrid approach has a slight performance gain and converges faster on bigger memories, Table 5.

We also experimented with removing individual samples instead of full batches. Under this setting, we observed that removing individual samples from the PQ creates a significant class imbalance, thus leading to poor results. As the training progresses, the number of samples of each class in memory decreases since removing individually does not respect the sampler property. The result is a lack of positive samples that degrades training. Removing full batches fixes this issue.

Figure 2, shows that removing old batches guarantees that the model will train on the most updated samples, whereas removing the batches with the lowest losses will focus more on the hardest samples. Since the sample embeddings that are in the queue are never recomputed with the new model parameters, misclassified images will never leave the queue and it will become filled with stale samples that provide noisy gradients. We found that alternating between the removal of the oldest batch and the batch with the lowest loss achieves the best result, Table 5. This hybrid approach guarantees that the PQ stays updated and is simultaneously more focused on the harder samples.

Priority Queue Update. In order to prevent the queue from being filled up with stale entries, we experimented different update strategies. We experimented with periodic updates of the PQ individual entries but it was not successful. Updating the whole memory at once, produced spikes in the loss function due to

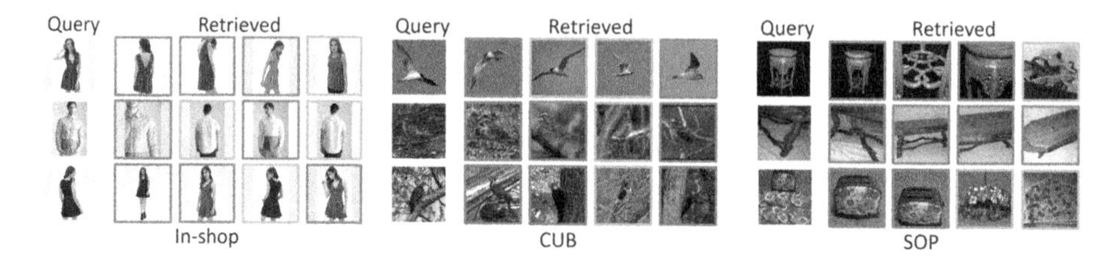

Fig. 3. Top 4 retrieved images with correct results highlighted green and incorrect results highlighted orange. (Color figure online)

the abrupt change of the stored embeddings. Another strategy we tried was to backpropagate the error to the stored embeddings, however it turned out to be prohibitively expensive. We also tried employing a moving average update and removing only the most outdated entries, but we could not surpass the results obtained with our hybrid approach. This suggests that it is enough to periodically remove the oldest batch. The results of these experiments are reported in annex.

5 Conclusion

In this paper we proposed to replace the mini-batch with a priority queue as a way to provide SGD with a better pool of learning samples. The PQ can effectively improve the learning of visual semantic embeddings. Figure 3 illustrate the results of the proposed method in a retrieval task across the three datasets. We can observe that the images that are most similar to the query image are in fact semantically related.

The learning framework described in this paper brings a series of advantages with key takeaways:

- **SGD + Priority queue** is a memory-efficient method with less than 1% memory overhead. The PQ stores the pre-computed embeddings of past mini-batches and reuses them for enhancing the SGD loss.
- **SGD + PQ prioritisation strategies** can drastically alter the performance of the model. Prioritisation strategies implements a sampling mining strategy at the batch level. Our experiments demonstrated that alternating between removing the oldest batch and the lowest loss batch possesses the best balance between keeping the PQ updated and maintaining the PQ with hard samples, hence steering the sets of triplets used for learning.
- **Performance gains:** The size of the PQ is positively correlated with performance gains. By using only the contrastive loss augmented with the PQ-50k, we achieved state-of-the-art results on three metric learning datasets.

Overall, the key challenge of the proposed method is rooted in the PQ update strategy. While there are clear gains in using the PQ for learning, the strategy to select the data to be removed from the PQ is still subject of research.

Acknowledgements. This work was partially funded by the FCT project NOVA LINCS (UIDP/04516/2020), and the CMU Portugal project iFetch (LISBOA-01-0247-FEDER-045920).

References

1. Cakir, F., He, K., Xia, X., Kulis, B., Sclaroff, S.: Deep metric learning to rank. In: Proceedings of the IEEE/CVF Conference on Computer Vision and Pattern Recognition, pp. 1861–1870 (2019)
2. Chen, T., Kornblith, S., Norouzi, M., Hinton, G.: A simple framework for contrastive learning of visual representations. In: International Conference on Machine Learning, pp. 1597–1607. PMLR (2020)
3. Chen, X., He, K.: Exploring simple siamese representation learning. In: Proceedings of the IEEE/CVF Conference on Computer Vision and Pattern Recognition, pp. 15750–15758 (2021)
4. El-Nouby, A., Neverova, N., Laptev, I., Jégou, H.: Training vision transformers for image retrieval. arXiv preprint arXiv:2102.05644 (2021)
5. Gao, L., Dai, Z., Fan, Z., Callan, J.: Complementing lexical retrieval with semantic residual embedding. arXiv preprint arXiv:2004.13969 (2020)
6. Gao, L., Zhang, Y., Han, J., Callan, J.: Scaling deep contrastive learning batch size under memory limited setup. In: Proceedings of the 6th Workshop on Representation Learning for NLP (RepL4NLP-2021), pp. 316–321 (2021)
7. Ge, W.: Deep metric learning with hierarchical triplet loss. In: Proceedings of the European Conference on Computer Vision (ECCV), pp. 269–285 (2018)
8. Goyal, P., et al.: Accurate, large minibatch sgd: Training imagenet in 1 hour. arXiv preprint arXiv:1706.02677 (2017)
9. Hadsell, R., Chopra, S., LeCun, Y.: Dimensionality reduction by learning an invariant mapping. In: 2006 IEEE Computer Society Conference on Computer Vision and Pattern Recognition (CVPR 2006), vol. 2, pp. 1735–1742 (2006). https://doi.org/10.1109/CVPR.2006.100. https://ieeexplore.ieee.org/abstract/document/1640964
10. He, K., Fan, H., Wu, Y., Xie, S., Girshick, R.: Momentum contrast for unsupervised visual representation learning. In: Proceedings of the IEEE/CVF Conference on Computer Vision and Pattern Recognition, pp. 9729–9738 (2020)
11. He, K., Zhang, X., Ren, S., Sun, J.: Deep residual learning for image recognition. In: Proceedings of the IEEE Conference on Computer Vision and Pattern Recognition, pp. 770–778 (2016)
12. Hermans, A., Beyer, L., Leibe, B.: In defense of the triplet loss for person re-identification (2017)
13. Hoffer, E., Ailon, N.: Deep metric learning using triplet network. In: Feragen, A., Pelillo, M., Loog, M. (eds.) SIMBAD 2015. LNCS, vol. 9370, pp. 84–92. Springer, Cham (2015). https://doi.org/10.1007/978-3-319-24261-3_7
14. Kim, S., Kim, D., Cho, M., Kwak, S.: Proxy anchor loss for deep metric learning. In: Proceedings of the IEEE/CVF Conference on Computer Vision and Pattern Recognition, pp. 3238–3247 (2020)
15. Kim, W., Goyal, B., Chawla, K., Lee, J., Kwon, K.: Attention-based ensemble for deep metric learning. In: Proceedings of the European Conference on Computer Vision (ECCV), pp. 736–751 (2018)
16. Li, S., Chen, D., Liu, B., Yu, N., Zhao, R.: Memory-based neighbourhood embedding for visual recognition. In: The IEEE International Conference on Computer Vision (ICCV), pp. 6102–6111 (2019)

17. Lin, T., Kong, L., Stich, S., Jaggi, M.: Extrapolation for large-batch training in deep learning. In: International Conference on Machine Learning, pp. 6094–6104. PMLR (2020)
18. Lin, Y., Han, S., Mao, H., Wang, Y., Dally, W.: Deep gradient compression: Reducing the communication bandwidth for distributed training (2018). https://openreview.net/pdf?id=SkhQHMW0W
19. Liu, Z., Luo, P., Qiu, S., Wang, X., Tang, X.: Deepfashion: powering robust clothes recognition and retrieval with rich annotations. In: Proceedings of IEEE Conference on Computer Vision and Pattern Recognition (CVPR) (2016)
20. Loshchilov, I., Hutter, F.: Sgdr: Stochastic gradient descent with warm restarts. arXiv preprint arXiv:1608.03983 (2016)
21. Movshovitz-Attias, Y., Toshev, A., Leung, T.K., Ioffe, S., Singh, S.: No fuss distance metric learning using proxies. In: Proceedings of the IEEE International Conference on Computer Vision, pp. 360–368 (2017)
22. Musgrave, K., Belongie, S., Lim, S.-N.: A metric learning reality check. In: Vedaldi, A., Bischof, H., Brox, T., Frahm, J.-M. (eds.) ECCV 2020. LNCS, vol. 12370, pp. 681–699. Springer, Cham (2020). https://doi.org/10.1007/978-3-030-58595-2_41
23. Oh Song, H., Jegelka, S., Rathod, V., Murphy, K.: Deep metric learning via facility location. In: Proceedings of the IEEE Conference on Computer Vision and Pattern Recognition, pp. 5382–5390 (2017)
24. Oh Song, H., Xiang, Y., Jegelka, S., Savarese, S.: Deep metric learning via lifted structured feature embedding. In: Proceedings of the IEEE Conference on Computer Vision and Pattern Recognition, pp. 4004–4012 (2016)
25. Opitz, M., Waltner, G., Possegger, H., Bischof, H.: Deep metric learning with bier: boosting independent embeddings robustly. IEEE Trans. Pattern Anal. Mach. Intell. **42**(2), 276–290 (2018)
26. Qian, Q., Shang, L., Sun, B., Hu, J., Li, H., Jin, R.: Softtriple loss: deep metric learning without triplet sampling. In: Proceedings of the IEEE/CVF International Conference on Computer Vision, pp. 6450–6458 (2019)
27. Roth, K., Brattoli, B., Ommer, B.: Mic: mining interclass characteristics for improved metric learning. In: Proceedings of the IEEE/CVF International Conference on Computer Vision, pp. 8000–8009 (2019)
28. Russakovsky, O., et al.: Imagenet large scale visual recognition challenge. Int. J. Comput. Vis. **115**(3), 211–252 (2015)
29. Sanakoyeu, A., Tschernezki, V., Buchler, U., Ommer, B.: Divide and conquer the embedding space for metric learning. In: Proceedings of the IEEE/CVF Conference on Computer Vision and Pattern Recognition, pp. 471–480 (2019)
30. Schroff, F., Kalenichenko, D., Philbin, J.: Facenet: A unified embedding for face recognition and clustering. In: CVPR, pp. 815–823. IEEE Computer Society (2015). http://dblp.uni-trier.de/db/conf/cvpr/cvpr2015.html#SchroffKP15
31. Semedo, D., Magalhães, J.: Cross-Modal Subspace Learning with Scheduled Adaptive Margin Constraints (2019). https://doi.org/10.1145/3343031.3351030
32. Smith, S.L., Kindermans, P.J., Ying, C., Le, Q.V.: Don't decay the learning rate, increase the batch size. In: International Conference on Learning Representations (2018)
33. Suh, Y., Han, B., Kim, W., Lee, K.M.: Stochastic class-based hard example mining for deep metric learning. In: Proceedings of the IEEE/CVF Conference on Computer Vision and Pattern Recognition, pp. 7251–7259 (2019)
34. Wah, C., Branson, S., Welinder, P., Perona, P., Belongie, S.: The caltech-ucsd birds-200-2011 dataset (2011)

35. Wang, X., Han, X., Huang, W., Dong, D., Scott, M.R.: Multi-similarity loss with general pair weighting for deep metric learning. In: Proceedings of the IEEE/CVF Conference on Computer Vision and Pattern Recognition, pp. 5022–5030 (2019)
36. Wang, X., Zhang, H., Huang, W., Scott, M.R.: Cross-batch memory for embedding learning. In: Proceedings of the IEEE/CVF Conference on Computer Vision and Pattern Recognition, pp. 6388–6397 (2020)
37. Wohlhart, P., Lepetit, V.: Learning descriptors for object recognition and 3d pose estimation. In: Proceedings of the IEEE Conference on Computer Vision and Pattern Recognition, pp. 3109–3118 (2015)
38. Wu, C.Y., Manmatha, R., Smola, A.J., Krahenbuhl, P.: Sampling matters in deep embedding learning. In: Proceedings of the IEEE International Conference on Computer Vision, pp. 2840–2848 (2017)
39. Wu, Z., Efros, A.A., Yu, S.X.: Improving generalization via scalable neighborhood component analysis. In: Proceedings of the European Conference on Computer Vision (ECCV), pp. 685–701 (2018)
40. Yuan, Y., Yang, K., Zhang, C.: Hard-aware deeply cascaded embedding. In: Proceedings of the IEEE International Conference on Computer Vision, pp. 814–823 (2017)
41. Zhong, Z., Zheng, L., Luo, Z., Li, S., Yang, Y.: Invariance matters: exemplar memory for domain adaptive person re-identification. In: Proceedings of the IEEE/CVF Conference on Computer Vision and Pattern Recognition, pp. 598–607 (2019)

Optimizing Object Detection Models via Active Learning

Dinis Costa[1(✉)], Catarina Silva[1], Joana Costa[1,2], and Bernardete Ribeiro[1]

[1] Department of Informatics Engineering, CISUC, Coimbra, Portugal
ddcosta@student.uc.pt, {catarina,joanamc,bribeiro}@dei.uc.pt
[2] School of Technology and Management, Polytechnic Institute of Leiria, Leiria, Portugal

Abstract. Object detection models have made significant progress and achieved state-of-the-art performance, which can now be comparable to human experts in various domains. However, training such models often requires a large amount of labeled data, which can be challenging to obtain. To address this issue, Active Learning (AL) has emerged as a technique to enhance the efficiency of deep learning models by reducing the amount of data and time required to train models to a satisfactory level.

In this paper, in the context of smart farming, we propose to study the impact of AL in object detection models trained with a small dataset of labelled images of whitefly-infested tomato leaves. We use YOLOv5 and fit the bounding box with confidence as a score function to select the most active relevant examples. The results show a trade-off between performance and cost suggesting AL outweigh the associated costs when labelled training data is limited.

Keywords: Machine Learning · Deep Learning · Active Learning · Object Detection

1 Introduction

Deep Learning (DL) algorithms are broadly classified into two categories: supervised and unsupervised learning. Supervised learning algorithms necessitate labeled data for training, whereas unsupervised learning algorithms do not. However, supervised learning algorithms need all data to be labeled in advance, which can be a laborious and demanding task that requires expertise. Active Learning (AL) techniques are often employed to address this challenge [7].

Object detection models fall within the supervised learning category. Active Learning techniques aim to reduce the overall amount of data required to train a model while achieving satisfactory results. This is achieved by iteratively selecting the most relevant data to be used in the training process. Such approach is often mandatory in different applications where there are inherent costs or difficulties in acquiring and annotating examples, notably when people (e.g. health)

A. Pertusa et al. (Eds.): IbPRIA 2023, LNCS 14062, pp. 82–93, 2023.
https://doi.org/10.1007/978-3-031-36616-1_7

or hardware (e.g. sensors) are involved. Hence, AL can improve the efficiency of object detection models when they are trained on large datasets [2,4].

In this study, we investigate the efficacy of AL techniques in object detection models when dealing with small datasets. This scenario is challenging as acquiring and labeling examples can be costly and difficult. We employ state-of-the-art object detection models such as YOLOv5 and use the bounding box with confidence as a score function to select the most relevant examples as active examples in each iteration.

The evaluation of the proposed approach is carried out in a real application of smart farming. We trained on a dataset of images captured in real plantations featuring whitefly-infested tomato leaves. Whiteflies are considered one of the top pests that cause significant greenhouse crop losses [8], making their early detection crucial for preventing their spread in plantations.

The article is organized as follows. Section 2 presents background on the object detection architecture and on AL, and related work involving AL and large datasets. In Sect. 3, we introduce the proposed approach, including the dataset used for the experiment, the experimental design, and the metrics used to evaluate the models. Section 5 describes the experimental results, and Sect. 6 presents the conclusions and future work.

2 Background and Related Work

2.1 Object Detection Models

Object detection and classification involves predicting the location of objects in images or videos by using bounding boxes, and classifying them into corresponding categories. There are various models with different architectures that process images in distinct ways.

YOLOv5 is a real-time object detection system developed by Joseph Redmon and Ali Farhadi. It is the fifth version in the You Only Look Once (YOLO) series of object detection models [6]. This model is referred to as a "one-phase" model because it processes the image only once. For this reason, this model is known to be fast at detecting objects. Since our aim is to provide farmers with a tool they can rely on to make decisions about the action to take when pests are present, the YOLO family is the right architecture for this purpose. YOLOv5 was used in [3] for a similar task, making its use highly relevant due to the possibility of using transfer learning for its models. The model can be subdivided into three parts (see Fig. 1):

- **Backbone:** Composed of a Cross Stage Partial Network (CSPNet), the backbone is based on a CNN architecture where each layer is responsible for extracting features from the input. This network was pre-trained on the ImageNet dataset. The feature map of the base layer is divided into two parts by the CSPNet and then the two parts are merged through a cross-stage hierarchy, starting with low-level features such as edges and corners and gradually building up to higher-level features such as patterns and shapes. The backbone also includes a Spatial Pyramid Pooling (SPP) which is based on the idea of dividing the input data into a grid of cells and then applying pooling operations within each cell.

- **Neck:** uses a Path Aggregation Network (PANet) that processes the feature maps from the backbone network by passing them through multiple paths, including a series of convolutional layers. The PANet is responsible for aggregating features from different scales and providing context for the detection head.
- **Head:** the model Head uses as input the features map outputted by the previous layers. It is composed from three convolution layers that predicts the location of the bounding boxes, the scores and the objects classes. YOLOv5 uses the SiLU function in the hidden layers of the convolution operations, while the sigmoid function is used in the output layers.

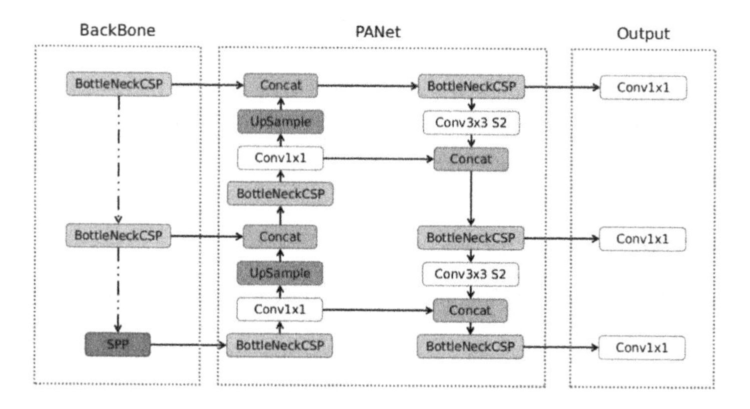

Fig. 1. Overview of YOLOv5 [6]

2.2 Active Learning

Active Learning (AL) is a learning technique that aims to train models with a small amount of labeled data by actively selecting the next set of data to be labeled [4]. The selection of appropriate data is critical to the success of this process.

Figure 2 summarizes the AL cycle. The first batch of data that the model uses to acquire initial knowledge is selected randomly since the model has no prior knowledge and cannot determine which data is most relevant for learning. Once trained, the model classifies the unlabeled data, providing information such as the probability of belonging to a certain class based on the model's classification. This information is fed into a score function, which is the key factor for classifying the unlabeled data from most relevant to least relevant, with the most relevant being the data that the model is most uncertain about. A batch of the most relevant data is then selected and added to the training set, where the model will be trained again and gain more knowledge. This process is repeated until a desirable performance is achieved.

AL can improve the efficiency of ML by reducing the amount of data needed for training, resulting in a model with satisfactory performance. This process can save time on labelling and training tasks and also require less effort from experts who may be needed to label the data.

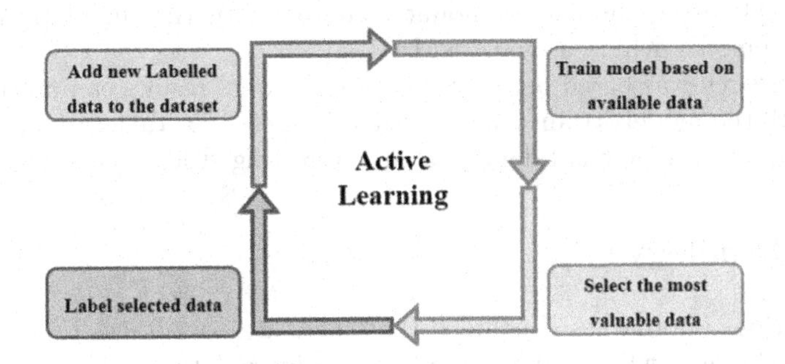

Fig. 2. Active Learning cycle

2.3 Active Learning Approaches

A recent study by [5] aimed to compare the effectiveness of different scoring functions for selecting relevant data to be added to the training set of a model. The study compared four scoring functions against random selection, using six different models. The dataset used in the study comprised over 800k images, each annotated with up to five classes: car, pedestrian, bicycle, traffic sign, and traffic light. The scoring functions used were:

- **Entropy:** which involves computing the probability map of a specific class.
- **Mutual Information (MI):** which makes use of the ensemble of the 6 models to measure disagreement.
- **Gradient of the output layer (Grad):** which measures the uncertainty of the model based on the magnitude of "hallucinated" gradients [1].
- **Bounding boxes with confidence (Det-Ent):** assuming that the final predicted bounding boxes the model detectors have an associated probability. This allows to compute the uncertainty in the form of entropy for each bounding box.

In the study, the researchers set a score function (MI) and trained the set of models with an initial random selected set of 100k images. They then iterated the process three times, adding 200k images to the training set in each iteration. During each iteration, they compared three different methods for selecting the data to be added: 200k randomly selected images from the unlabelled set (X_u),

200k images selected based on a score function from the X_u set, and 200k images selected based on a score function from the X_u set combined with the set of images already used for training (X_I). In this last method there can be repeated images, but can lead to better results since they are being selected for being considered the most relevant for the model to learn.

The achieved results by [5] appoints that the model can benefit from selecting the most relevant data for training, since the model achieved better results (73.2% weight mean Average Precision (wMAP)) using 700k sample images selected using AL techniques over random selection where the model achieved 69.2% wMAP using the same amount of data. With the full data set (850k images) the model achieved 69.0% wMAP.

This research was based on a large dataset, and its results can not be generalized to all the models trained for all sized datasets. For this reason, exploring AL techniques using a small dataset is an interesting study to conduct.

3 Methodology

While previous approaches concluded that there are advantages to using active learning techniques when training models with large datasets [5], in the present paper, we propose to study the impact of AL techniques in building object detection models trained with small datasets. With this in mind, we propose the workflow shown in Fig. 3.

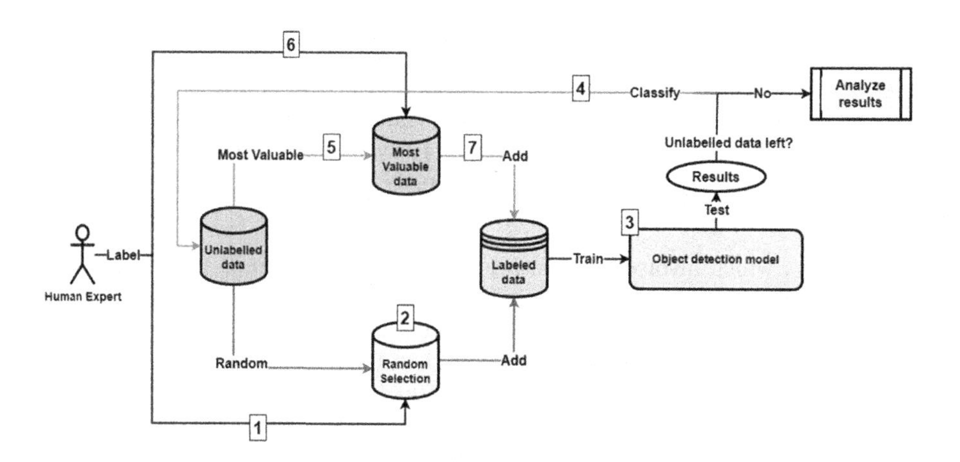

Fig. 3. Detailed Active Learning (AL) Workflow

First, a small batch of images from the unlabeled set is randomly selected and annotated by a human expert (1). After this, the set is randomly split into training and validation sets (2). Using that small set, a model is trained, and the time taken to train it is recorded (3). The trained model is used to classify all the data from the unlabeled set (4). The obtained results feed a score function, which is used to sort the unlabeled data from the most to the least valuable data (5). A small batch of the most valuable data is selected, labeled by a human expert (6), and added to the labeled set (7). A new model is trained from scratch using the data from the labeled set. The process is repeated until the unlabeled set is empty. The numbers in parentheses represent the steps shown in the Fig. 3.

4 Experimental Setup

In this section, we outline the experimental setup utilized to derive insights into the effectiveness of AL techniques when training object detection models with small datasets. We provide a detailed description of the dataset used for the experiment, outline the AL methodology implemented, and present the performance metrics used for evaluating the models.

(a) Dataset image before labelling

(b) Dataset image after labelling

Fig. 4. Labelling Process

4.1 Case Study

This study was initially motivated by the smart farming industry, where the objective was to develop a model capable of detecting pests in non-controlled environments, such as identifying pests directly on leaves. To accomplish this goal, we endeavored to construct a dataset to address the current shortage of data in this field.

We built a dataset of images containing whitefly-infested tomato leaves. The images were recorded in a tomato greenhouse located in Coimbra, Portugal. The greenhouse was in the late stages of cultivation and was approximately 200×100 meters in size. The tomatoes were planted in rows, and we collected images from three randomly selected rows of plants. In total, we recorded five hundred images, which were resized to 3000 × 3000 resolution. It was then that the need to label all the whiteflies in each image arose, and motivated the use of AL techniques. The first 200 images were selected and the contained whiteflies were annotated using *labelImg*[1] Fig. 4 shows the same image before and after the annotation process.

A total of 2479 whiteflies were included in the testing subset, which consisted of 20% of the images randomly selected from the dataset. The dataset specifications are outlined in Table 1.

Table 1. Dataset specifications

200 Images (10747 Whiteflies)	
Train	Test
160 Images (8268 Whiteflies)	40 Images (2479 Whiteflies)

4.2 Experimental Design

Since we are working with a small dataset, with only 200 images, we want to start with a very low sized batch of images and iteratively increment the training set with small batches. Table 2 summarizes the experiment to be conducted.

Due to limited time and computational resources, we decided to use YOLOv5 small architecture to conduct this experiment, which is one of the lightest model of its family and therefore, faster to train. The weight initialization method was set to random and in each AL setup defined, the model was trained from scratch. The score function used to select the most valuable data to add to the train set was *bounding box with confidence*, using the images with the lowest average confidence, which can be computed as follows:

$$\mathcal{H}(p_{obj}) = p_{obj} \log p_{obj} + (1 - p_{obj}) \log(1 - p_{obj}) \tag{1}$$

where p_{obj} is the confidence of the object detected belonging to the class of whitefly.

4.3 Performance Metrics

To evaluate the performance of the model, we use three metrics:

- **Mean Average Precision (mAP):** In object detection and classification models, it is assumed that an object in an image has a predicted bounding box (the bounding box predicted by the model) and a ground truth bounding box (the real location of the object as given by its annotation in the dataset).

[1] Open-source online tool: https://github.com/heartexlabs/labelImg.

Table 2. Summary of Experimental Setups, detailing the total number of training images and the percentage of the dataset allocated to each setup

Setup name	Total number of Images	Dataset Proportions (%)		
		Train	Validation	Test
0	60	5	5	20
1	75	10	7.5	
2	90	15	10	
3	105	20	12.5	
4	120	25	15	
5	135	30	17.5	
6	150	35	20	
7	160	40	20	
8	170	45	20	
9	180	50	20	
10	190	55	20	
full	200	60	20	

Intersection over Union (IoU) measures the overlap between the predicted bounding box and the ground truth bounding box. In Fig. 5 a visual representation of this metric is presented.
In mAP, an object is considered a true positive (TP) if $IoU \geq 0.5$, and a false positive (FP) if $IoU < 0.5$. This, if a threshold of $IoU = 0.5$ is considered. The average precision (AP) is calculated by finding the Area under the Curve (AuC) of the precision-recall curve, which plots the precision of the model at different levels of recall at different thresholds of confidence assigned to a bounding box by the model. These thresholds indicate how confident the model is that a certain bounding box contains an object. The mAP averages the AP of all classes.

- **Ratio of the predicted counted objects:** When the goal is to control pests by counting them, a metric that involves the total number of objects found in an image and compares it to the real number of objects in the image would be a good choice. For this purpose, the following calculation can be used:

$$Ratio_{Pred_{obj}} = \frac{Total\ Object\ counted}{Actual\ Count} \tag{2}$$

where the *Total Object counted* is the total objects detected by the model in a set of images, assuming that it only has one class, while *Actual Count* is the total object annotated in the same set of images. This gives the ratio of the predicted to the counted objects. If $Ratio_{Pred_{obj}} > 1$, it means that the model is detecting more objects than the ones annotated in the dataset, otherwise, it means that the model is detecting fewer objects than the ones annotated in the dataset.

- **Time taken to train:** Since we are conducting an AL experiment aimed at reducing the total amount of data required to train a model with satisfactory results, the time taken to achieve those satisfactory results is also a factor to take into account.

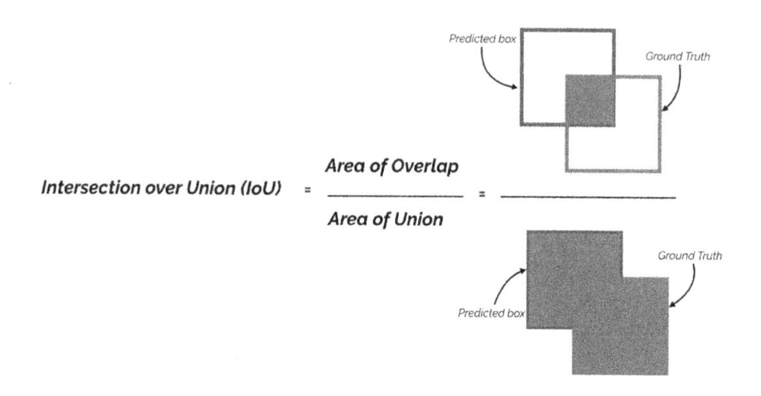

Fig. 5. Intersection over Union calculation [9]

5 Experimental Results and Analysis

Table 3 outlines the results obtained on the test set. The first split was achieved by randomly selecting 5% of the data for training and 5% for validation. As expected, it was the split with the poorest performance.

Comparing the mAP results on the second split, we observe that random selection outperforms AL techniques by 2 mAP percentage points. This could be explained by the weak learning that the model experienced in the previous split, using only 10% of the data for training and validation. Therefore, the images selected for the second split were chosen based on a poor model.

From the third split to the second last, AL outperformed random selection of the data. This highlights that the AL experiment achieved optimal performance when 70% (140 images) of the data was used in training and validation, while the Random experiment achieved optimal performance in the last setup using 80% (160 images) of the data for training and validation. Even though the optimal performance of the Random experiment outweighed AL by 0.3 mAP percentage points, we consider that the trade-off of the data used and the time required to gain that 0.3% of mAP is not worth it. To achieve 92.6% of mAP in the AL experiment, 1.51 images per mAP point were needed, while 1.72 images were required in the Random experiment.

Given the significant amount of time and manpower required to annotate all of the whiteflies in each image, the marginal improvement in performance achieved by adding 20 additional images may not be worth the effort. Instead, alternative approaches such as improving annotation quality or leveraging other sources of data could be explored to further enhance the performance of the model.

Regarding the *Ratio. Pred. counted Objects*, with the exception of the first setup, all the values were greater than one, meaning that the model is detecting more objects than those annotated. This can lead us to conclude one of two options: either the model is detecting more whiteflies than were annotated in the dataset, or the model is more accurate at detecting whiteflies than the annotator. We cannot exclude the second option due to the nature of the dataset, which is composed of low-resolution images captured under different light exposures, making pest identification a difficult task. Additionally, the model could be detecting pests in earlier stages of their lives, where they are quite small and even harder to detect.

Table 3. Results Table: Comparison of mAP (%), training time (using an NVIDIA GeForce RTX 4080), and ratio of predicted counted objects when the model is trained on various quantities of images selected randomly or with Active Learning (AL)

Training and Validation Images	mAP (%)		Training time (s)		Ratio. Pred. counted Objects	
	AL	Random	AL	Random	AL	Random
60	5.9	5.9	289.2	310.2	0.35	0.35
75	49.5	51.5	386.4	375.5	1.58	1.40
90	60.4	59.4	463.3	462.9	1.74	1.47
105	79.4	76.5	472.2	461.1	1.32	1.37
120	85.8	85.5	548.2	534.5	1.53	1.34
135	86.6	86.1	612.6	602.4	1.42	1.36
150	89.5	87.6	672.9	658.9	1.39	1.43
160	89.4	87	731.6	731.5	1.43	1.43
170	90.7	90.1	764.9	769.2	1.33	**1.24**
180	**92.6**	91.7	835.3	828.8	**1.27**	1.34
190	92	91	890.2	890.7	1.44	1.30
200	92.6	**92.9**	932.1	936.5	1.30	1.26

Analyzing Fig. 6, we observe an expected trend: a significant increase in mAP in the first setups, followed by a stabilization of the curve. When we superimpose the curve of training time onto the mAP curve, we see that the two curves converge, suggesting that further improvements in performance come at the cost of increased training time.

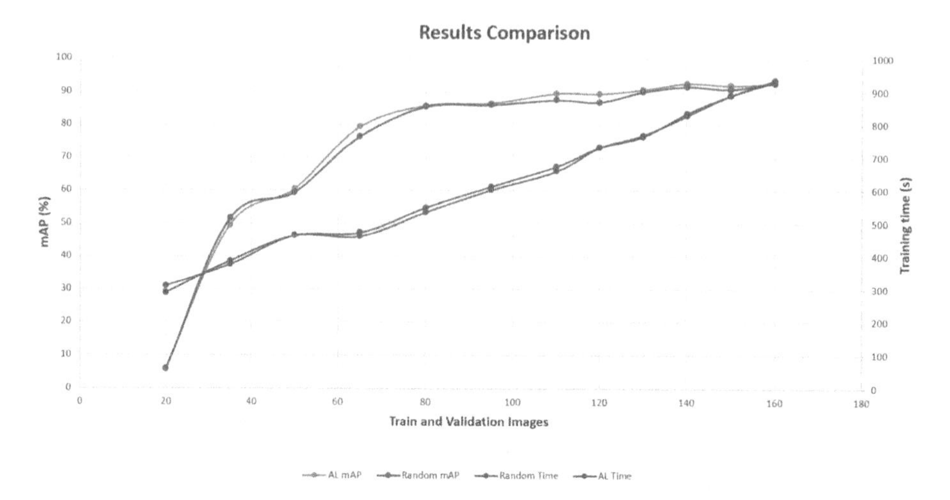

Fig. 6. Comparison Chart: mAP Achieved with Incremental Number of Images Used in Training and Validation, Selected via Active Learning (AL) vs Random Selection. Training Time Comparison Included

6 Conclusions and Future Work

In this paper we examined the effectiveness of active learning techniques in object detection models when working with small datasets. The findings suggest that object detection models can achieve optimal performance without accessing all available data when using active learning techniques, but require all the data when using random selection. Nevertheless, if there is a trade-off between the time required to train a model and the level of performance achieved, the additional performance gains may not justify the added training time.

Further work will involve conducting more comprehensive experiments to further substantiate the findings. As the first setup was randomly chosen, and the train-validation split was also random, we intend to conduct multiple experiments to introduce variability into the dataset. Furthermore, while we explored one scoring function in this study, future work may explore other functions to identify the most relevant data for training.

Acknowledgments. This work was supported by project PEGADA 4.0 (PRR-C05-i03-000099), financed by the PPR - Plano de Recuperação e Resiliência and by national funds through FCT, within the scope of the project CISUC (UID/CEC/00326/2020).

References

1. Ash, J.T., Zhang, C., Krishnamurthy, A., Langford, J., Agarwal, A.: Deep batch active learning by diverse, uncertain gradient lower bounds (2019). https://doi.org/10.48550/ARXIV.1906.03671

2. Brust, C.A., Käding, C., Denzler, J.: Active and incremental learning with weak supervision. KI - Künstliche Intelligenz **34**(2), 165–180 (2020)
3. Cardoso, B., Silva, C., Costa, J., Ribeiro, B.: Internet of things meets computer vision to make an intelligent pest monitoring network. Appl. Sci. **12**(18) (2022). https://doi.org/10.3390/app12189397. https://www.mdpi.com/2076-3417/12/18/9397
4. Haussmann, E., et al.: Active learning techniques and impacts. In: 2020 IEEE Intelligent Vehicles Symposium, pp. 1430–1435 (2017). https://doi.org/10.1109/IV47402.2020.9304793
5. Haussmann, E., et al.: Scalable active learning for object detection. In: 2020 IEEE Intelligent Vehicles Symposium (IV), pp. 1430–1435 (2020). https://doi.org/10.1109/IV47402.2020.9304793
6. Jocher, G., et al.: Ultralytics/YOLOv5: v7.0 - YOLOv5 SOTA Realtime Instance Segmentation (2022). https://doi.org/10.5281/zenodo.7347926
7. Kanteti, D., Srikar, D., Ramesh, T.: Active learning techniques and impacts. In: 2017 5th IEEE International Conference on MOOCs, Innovation and Technology in Education (MITE), pp. 131–134 (2017). https://doi.org/10.1109/MITE.2017.00029
8. Nieuwenhuizen, A., Hemming, J., Suh, H.: Detection and classification of insects on stick-traps in a tomato crop using faster R-CNN (2018)
9. Yohanandan, S.: Map (mean average precision) might confuse you! (2020). https://towardsdatascience.com/map-mean-average-precision-might-confuse-you-5956f1bfa9e2

Continual Vocabularies to Tackle the Catastrophic Forgetting Problem in Machine Translation

Salvador Carrión(✉) and Francisco Casacuberta

Universitat Politècnica de València, Camí de Vera, 46022 València, Spain
{salcarpo,fcn}@prhlt.upv.es

Abstract. Neural Machine Translation (NMT) models are rarely decoupled from their vocabularies, as both are often trained together in an end-to-end fashion. However, the effects of the catastrophic forgetting problem are highly dependent on the vocabulary used. In this work, we explore the effects of the catastrophic forgetting problem from the vocabulary point of view and present a novel method based on a continual vocabulary approach that decouples vocabularies from their NMT models to improve the cross-domain performance and mitigate the effects of the catastrophic forgetting problem. Our work shows that the vocabulary domain plays a critical role in the cross-domain performance of a model. Therefore, by using a continual vocabulary capable of exploiting subword information to construct new word embeddings we can mitigate the effects of the catastrophic forgetting problem and improve the performance consistency across domains.

1 Introduction

Language, words, and meanings are constantly evolving. Therefore, we must design strategies to incorporate continual learning mechanisms into the existing machine translation systems. However, trained models have to deal with the catastrophic forgetting problem, whereby a model may forget previously learned information when learning new one. Similarly, this phenomenon is closely related to the domain adaptation problem, whereby a model may perform well in a given domain and very poorly in a similar one.

Over the years, researchers have proposed multiple approaches to mitigate this problem, given that the original training data is not always available, nor are the resources necessary to train the model from scratch. These strategies are usually based on regularization techniques, which penalize changes in weights that deviate too much from another model in order to minimize the forgetting [16,21]; dynamic architectures, which grow linearly with the number of tasks [10,31]; or Complementary Learning Systems (CLS), which are inspired by the human learning to generate synthetic data to avoid forgetting [15] previous tasks. However, these strategies are often very computationally intensive (e.g., they require

Supported by Pattern Recognition and Human Language Technology Center (PRHLT).

A. Pertusa et al. (Eds.): IbPRIA 2023, LNCS 14062, pp. 94–107, 2023.
https://doi.org/10.1007/978-3-031-36616-1_8

two models in memory simultaneously). Besides, these strategies are usually not specifically designed for natural language tasks.

For these reasons, we decided to tackle this problem from a much simpler approach, based on: i) decoupling vocabularies from translation models; ii) exploiting subword information to deal with the open vocabulary problem and construct new word embeddings compositionally; and iii) sharing a universal feature space between embeddings to easily extend the vocabulary of a pretrained model (no alignment needed). This approach allows us to extend the vocabulary of a pre-trained NMT model in a plug&play manner to cope with unknown words, typos, or spelling variations, which ultimately help to mitigate the domain-shift effects or the catastrophic forgetting problem in our case. Furthermore, this method also allowed us to perform zero-shot translation at the vocabulary level, which led to slight performance improvements upon expanding the vocabulary, regardless of the domain.

The contributions of this work are the following:

- First, we show that the effects of the catastrophic forgetting problem are highly dependent on the vocabulary domain.
- Next, we discuss the problematic of continual vocabularies and show that embedding alignment is a strong requirement for a lifelong approach.
- Finally, we propose a novel method based on a continual vocabulary approach to improve the cross-domain consistency after fine-tuning and mitigate the effects of the catastrophic forgetting problem.

2 Related Work

McCloskey and Cohen [26] introduced the term *Catastrophic Interference.* Since then, many researchers have tried to delve into the causes that produce it and have developed strategies to mitigate its effects. For example, Carpenter and Grossberg [5] was one of the first pioneers in studying the stability-plasticity dilemma, whereby there is a trade-off between the ability of a model to preserve past knowledge (stability) and the ability to learn new information effectively (plasticity).

To deal with this problem, some researchers tried to regularize or adjust the network weights. For example, Li and Hoiem [21] introduced a model with shared parameters across tasks and task-specific parameters; Kirkpatrick et al. [16] identified which weights are important for the old task in order to penalize the updates on those weights; Jung et al. [14] penalized changes in the final hidden layer; Zenke et al. [41], introduced intelligent synapses that accumulate task-relevant information; Hu et al. [13] introduced a model with a set of parameters that was shared by all tasks and the second set of parameters that were dynamically generated to adapt the model to each new task.

Similarly, other researchers have tackled this issue by using data from past tasks when training for the new tasks. Lopez-Paz and Ranzato [23] proposed a model that alleviates the catastrophic forgetting problem by storing a subset of the observed examples from an old task (episodic memory), and Shin et al. [36], instead of storing actual training data from past tasks, trained a deep generative model that replayed past data (synthetically) during training to prevent forgetting.

In addition to these works, there are many more works on the subject such as iCaRL [30], PathNet [11], FearNet [15], Incremental Moment Matching [20], or based on dynamic architectures [10,31], distillation [33], Memory Aware Synapses [1] or Dynamically Expandable Networks [19], amongst many others, that deal with this problem using similar techniques as the ones previously mentioned.

However, despite the recent breakthroughs in the field of machine translation [34,37,38,42], the catastrophic forgetting problem (or, more generally, the lifelong learning problem) has not been so widely studied. Along these lines, Xu et al. [40] proposed a meta-learning method that exploits knowledge from past domains to generate improved embeddings for a new domain; Qi et al. [29] showed that pre-trained embeddings could be effective in low-resource scenarios; Liu et al. [22] learned corpus-dependent features by sequentially updating sentence encoders (previously initialized with the help of corpus-independent features) using Boolean operations of conceptor matrices; and more recently, Sato et al. [32] presented a method to adapt the embeddings between domains by projecting the target embeddings into the source space, and then fine-tuning them on the target domain.

3 Models

3.1 Compositional Embeddings

Word embeddings allow models to operate on textual data by mapping words or phrases to numerical values. Nowadays, there are many techniques for learning good embeddings representations. However, all of them share the same core idea: "*You shall know a word by the company it keeps*"[1]. In practice, this means that words keep linear relationships amongst them, so by performing basic algebraic operations, we can transform one word into another (e.g., king+woman = queen).

Even though many approaches exist, we needed one that allowed us to deal with: i) the open vocabulary problem (e.g., bytes, chars, subwords); ii) the embedding alignment problem (seamlessly), and iii) generate embeddings for unknown words through composition.

As a result, we decided to use FastText [4] as our initial embeddings, given that it can learn character n-grams representations to construct words as the sum of the n-gram vectors, and, therefore, exploit the subword information to construct word embeddings compositionally. It is essential to point out that the core idea here are the requirements specified for our approach (open vocabulary strategy, universal feature space, and compositionality), not the use of FastText itself (more on this later).

[1] Firth, J. R. (1957:11).

3.2 Transformer Architecture

Neural encoder-decoder architectures such as the Transformer [38] are the current standard in Machine Translation [3] and most Natural Language Tasks [9].

This state-of-the-art architecture is based entirely on the concept of *attention* [2,24] to draw global dependencies between the input and output. Because of this, it can process all its sequences in parallel and achieve significant performance improvements compared to previous architectures [8,37,39]. Furthermore, this architecture does not use any recurrent layer to deal with temporal sequences. Instead, it uses a mask-based approach along with positional embeddings to encode the temporal information of its sequences.

4 Experimental Setup

4.1 Datasets

All the datasets used for this work can be found in Table 1.

Table 1. Datasets partitions

Dataset	Langs	Train size
SciELO (Health)	es-en, pt-en	120K/100K
SciELO (Biological)	es-en, pt-en	120K/100K
SciELO (H+B)	es-en, pt-en	240K/100K
Europarl	de-en	2M/100K
Multi30K	de-en	30K

The SciELO datasets contain parallel sentences extracted from scientific publications of the SciELO database used in the WMT'17 biomedical translation task. Specifically, there are three versions: one for the health domain, another for the biological domain, and a last one (created by us) that combines both domains. Similarly, the Europarl and Multi30K datasets contain parallel sentences from the European Parliament website and a multimodal WMT16 task.

4.2 Training Details

All datasets were preprocessed using Moses [17]. Then, SentencePiece [18] was used to build domain-specific vocabularies using the health, biological, and merged corpora, with the following vocabulary limits: 4000, 8000, 16000, and 32000. This was done to study the effects of the catastrophic forgetting problem as a function of the vocabulary size and domain.

Our NMT models have been trained using AutoNMT [6], a tool to streamline the research of seq2seq models, by automating the preprocessing, training, and evaluation of NMT models. After performing some tests with the standard Transformer (with around 92M parameters), we found out that the size of our model was not a bottleneck for us, so to speed up our research, we used a simplified version of the standard Transformer with 4.1M to 25M parameters depending on the vocabulary size.

This small Transformer consisted of 3 layers, 8 heads, 256 for the embedding dimension, and 512 for the feedforward layer. Similarly, the training hyperparameters were quite standard for all models: CrossEntropy (without label smoothing), Adam as the optimizer, 4096 tokens/batch or a batch of 128 sentences, max token length of 150, clip-norm of 1.0, a maximum epoch of 200 epochs with early stopping (patience = 15).

All training was done using two NVIDIA TITAN XP, with 12 GB each.

4.3 Evaluation Metrics

Automatic metrics compute the quality of a model by comparing its output with a reference translation written by a human.

Given that BLEU [27] is the most popular metric for machine translation, but it is pretty sensitive to chosen parameters and implementation, we used SacreBLEU [28], the reference BLEU implementation for the WMT conference. Additionally, we contrasted our results using BERTScore [42].

- **BiLingual Evaluation Understudy (BLEU)**: Computes a similarity score between the machine translation and one or several reference translations, based on the n-gram precision and a penalty for short translations.

5 Experimentation

5.1 Effects of the Vocabulary on the Catastrophic Forgetting Problem

In this section, we study the effects of the catastrophic forgetting problem and cross-domain performance, as a function of the vocabulary size and domain.

In Fig. 1, we have the evaluations of 36 translation models trained on the SciELO dataset (es-en), under different vocabularies[2]. These models have been trained on three different domains (Health, Biological, and Merged) and evaluated on the same three domains (each model); each training set had the same amount of sentences to account for potential training biases. Moreover, this experimentation was repeated with vocabularies from the Health, Biological, and Merged domains and three vocabulary sizes (8000, 16000, and 32000). Finally, we repeated part of this experiment using other datasets (i.e., Europarl), language pairs (i.e., en-es, en-pt, en-de, en-fr), and domains, obtaining similar and consistent results.

[2] The "Domain" in the title refers to the domain of the vocabulary used, while the name of the column groups refers to the training dataset, and the legend refers to the evaluation domain.

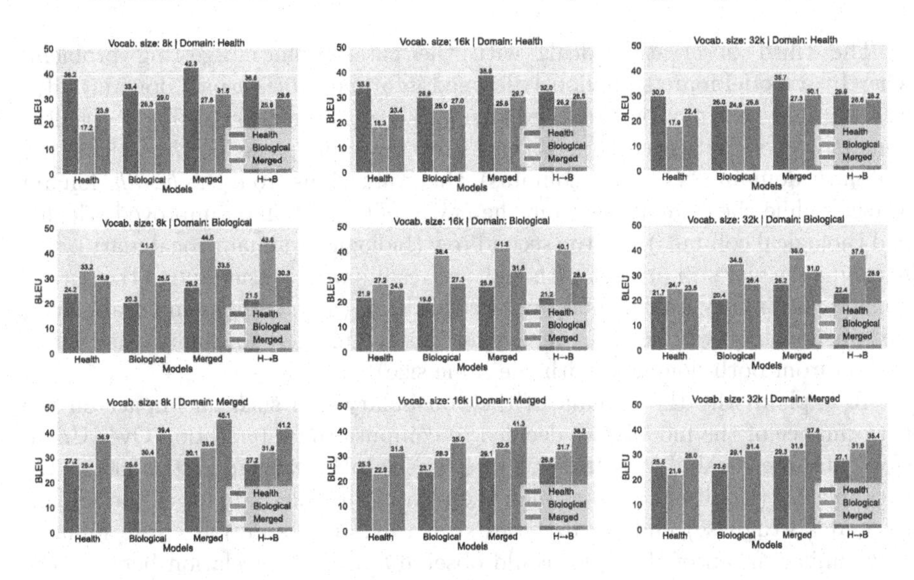

Fig. 1. The domain of a vocabulary has a significant impact on the performance of the model (see columns, from top to bottom). The best results were always obtained when the vocabulary and evaluation domain matched, regardless of the training domain. Furthermore, larger vocabularies appear to improve the cross-domain consistency (see rows, from left to right). Finally, a more generic vocabulary (merged domain) helped to significantly mitigate the effects of the catastrophic forgetting problem.

The first observation we can make is to point out the significant role that the vocabulary domain plays in the performance of a model since, regardless of the training domain (and vocabulary size), the best performance was always obtained when the test-set domain matched the domain of the vocabulary (see Figure Fig. 1 columns, from top to bottom), which means that the performance of a model is highly dependent on the chosen vocabulary.

The second observation relates to the vocabulary size since, by increasing the vocabulary size from 8K to 32K entries, the cross-domain differences were significantly reduced, regardless of the training domain, evaluation domain, or vocabulary domain. That is, models seem to become more consistent across domains as the number of entries in their vocabulary increases (see Fig. 1 rows, from left to right), probably due to the better coverage. However, we noticed that the performance of the models with 32K vocabularies was, on average, worse than the models with 8K vocabularies. We speculate that this could be due to three reasons. First, it is known that small or character-level vocabularies are usually beneficial in medium-to-low datasets [7]. Second, each dataset has an optimal vocabulary size with the shape of an inverted “U” [12]. And third, large vocabularies usually require more training data [35].

The third observation deals with the catastrophic forgetting problem, whereby a model forgets previously learned information (i.e., past domain) after learning new information (i.e., new domain). In the first row of Fig. 1 (health-domain vocabulary), it can be seen that after fine-tuning the *health* model in the *biological* domain ("$H \rightarrow B$" columns), the evaluations from the *health* domain worsen, while the evaluations from the new *biological* domain improved (Health and Biological columns). On the second row (biological-domain vocabulary), this phenomenon was still present, although the results were noisier due to the out-of-domain vocabulary used. Finally, on the third row (merged-domain vocabulary), this phenomenon was not present anymore as the merged vocabulary had information from both domains (with the same size).

To explain why the domain of the vocabulary had such an impact on the performance of the models, we decided to compute the Intersection Over Union (IoU) of those vocabularies. In Fig. 2 we have the IoU Heatmap for the 8000-entry vocabularies (English). Even though the results for other languages (i.e., Spanish and Portuguese) and vocabularies sizes (4K, 16K, and 32K) were remarkably similar, in all of them, we could observe a strong correlation between the domain-overlapping of vocabulary and the performance of the model in that domain. Therefore, we hypothesized that by closing the gap between domains, we could get performance improvements across domains, along with an increased robustness against the effects of the catastrophic forgetting phenomenon.

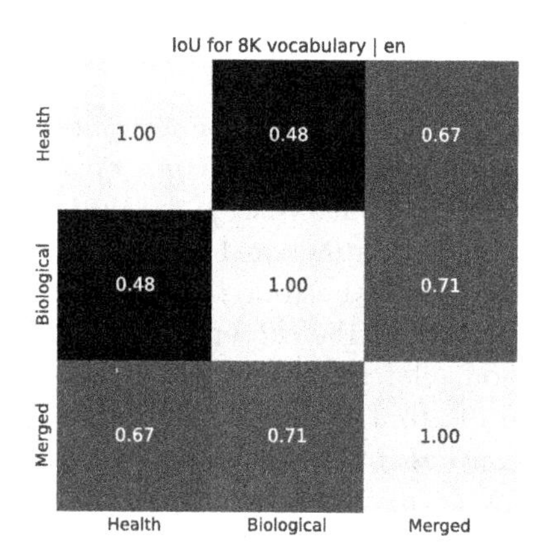

Fig. 2. Domain overlapping: The domain overlapping between vocabularies is strongly correlated with the model performance on that domain.

These results reflect the importance of choosing good vocabularies from the beginning if one wants to be able to generalize to other tasks or domains, since a seemingly suboptimal vocabulary for a training set, can lead to improvements in the generalization of the model for other domains and, help to mitigate the effects of catastrophic forgetting phenomenon.

5.2 On the Continual Vocabulary Problem

From the above experiments, we hypothesized that if we increased the overlap between the domain vocabularies, we could improve the cross-domain consistency and mitigate the effects o catastrophic forgetting. However, the most straightforward way to do this would be to use character- or byte-based vocabularies, but this would lead to other problems due to the use of longer sequences, more memory consumption, longer inference times, or the additional model complexity needed to deal with the generalization required by these vocabularies.

Consequently, we decided to use a continual vocabulary approach. That is, we would start with an initial vocabulary that would be expanded on demand with each new task or domain. However, a continual vocabulary needs to deal with two problems: i) the open vocabulary problem, which can be solved (practically) by using bytes, chars, or subwords; and ii) The alignment problem, which is needed in order to add new pretrained embeddings into an existing vocabulary (also pretrained) so that all embeddings can share the same feature space consistently.

In order to extend a pretrained vocabulary (embedding), the new embeddings must have the same dimensionality as the previous ones (projection). However, this requirement is not sufficient by itself because even though that as we increase the amount of training data, the quality of these embeddings starts to improve and form consistent clusters of related words, the position at which these clusters are located in the latent space depends on the initial values of these embeddings and the training process. That is, related words might converge to similar clusters, but their position in the feature space does not have to (alignment problem).

This idea can be seen more clearly illustrated in Fig. 3, where the red and orange points form two clusters corresponding to the embedding set #1, and the blue and green points form another two clusters corresponding to a different embedding set #2.

To get a better intuition on the alignment problem, we projected three different high-quality embeddings into a 2-dimensional space using t-SNE [25] to visualize the clusters of words that had emerged (See Fig. 4). These embeddings correspond to the Europarl dataset[3] and the GloVe and FastText pre-trained embeddings. In all of them, we can see that similar words are grouped together (i.e., *man, woman, children*), but their clusters are entirely different, in addition to the distance between non-related words (i.e., distance between *bird* and *man*). It is worth pointing out that, to account for projection variations that could bias our results, we ran t-SNE multiple times, but we obtained very similar projections in each of them.

As expected, these embeddings were not aligned despite their high quality and sharing the same dimension. Therefore, it is clear that a continual vocabulary needs a mechanism to project new words into a common feature space and align them with the previous ones.

[3] Custom embeddings trained in Europarl-2M (de-en).

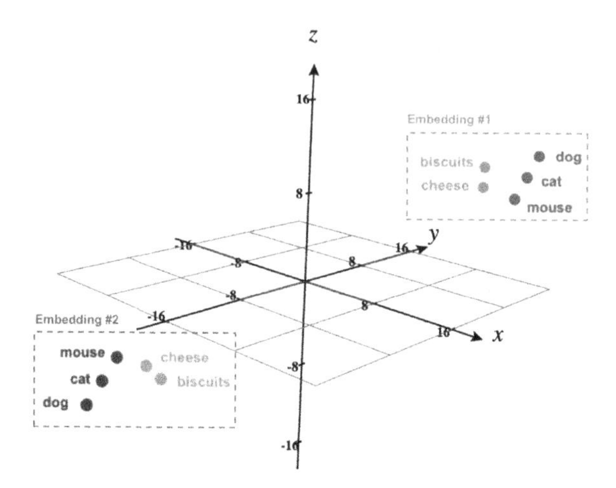

Fig. 3. Two different embeddings projected in 3D dimensions. Similar embeddings might end up in different regions of the same latent space

5.3 Tackling the Catastrophic Forgetting Problem

In this section, we study if the catastrophic forgetting problem can be tackled using the continual vocabulary approach theorized in the previous section (see Sect. 5.2).

The idea behind this continual vocabulary approach is to create an initial vocabulary V_0, and expand it as the number of tasks or domains increases:

$$V_n = \bigcup_{i=1}^{n} V_i \tag{1}$$

For example, suppose we have a model with a vocabulary of 16K words built for a task A, and we want to learn a new task B. In that case, we will add to this initial vocabulary only the words of the new task that are not in the previous vocabulary.

Similarly, if we want to make use of new pretrained embeddings, we will need to align them w.r.t the previous pretrained embeddings by minimizing the alignment error between the shared words of the new and the old embeddings, and predicting the aligned location for the unknown ones:

$$\min_{V_{i-1} \cap V_i} \sum_{j=1}^{m} (v_j^{(i-1)} - v_j^{(i)})^2 \tag{2}$$

Interestingly, we can also exploit the subword information to construct new word embeddings compositionally that will be automatically aligned with the previous ones. Furthermore, this approach allowed us to perform zero-shot translation at the vocabulary level because if the words *cardio* and *radiologist* are

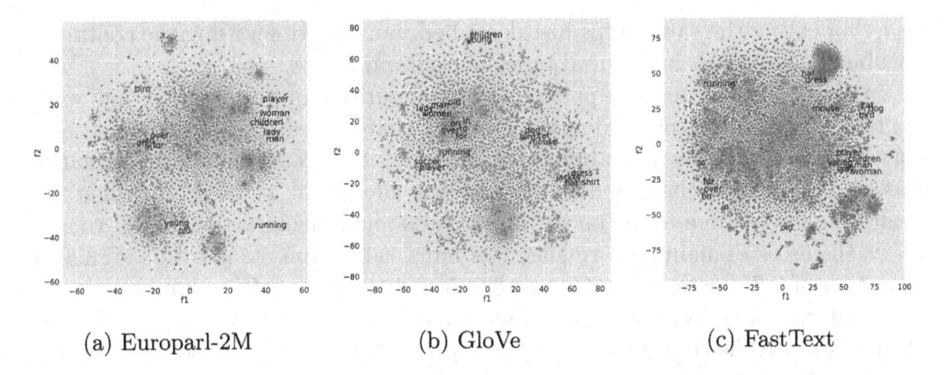

(a) Europarl-2M (b) GloVe (c) FastText

Fig. 4. t-SNE projections for the Europarl, GloVe and FastText embeddings. High-quality embeddings cluster similar words together but locate them in different regions of the feature space.

known but the word *cardiologist* is it not, then we will be able to get the embeddings for this unknown word compositionally (i.e., *cardio* + *radiologist* = *cardiologist*).

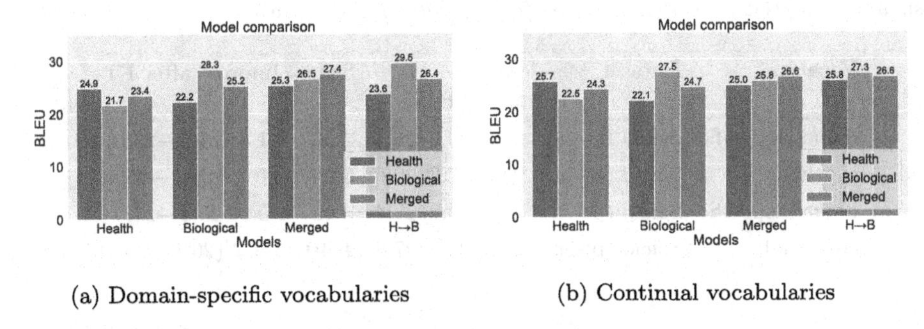

(a) Domain-specific vocabularies (b) Continual vocabularies

Fig. 5. For the models trained from scratch on a single domain (Health, Biological and Merged), both approaches performed remarkably similar, as the pre-trained embeddings could not provide enough information during training to make the models able to generalize to other domains. In contrast, after fine-tuning, the models with continual vocabularies achieved better cross-domain consistencies and helped to mitigate the catastrophic forgetting effects.

To test our hypothesis, we first started by training four Transformer models. Three models were trained on the Health, Biological, and Health+Biological (Merged) domains (separately), and another one was pre-trained on the Health domain but then fine-tuned on the Biological domain. All these models had domain-specific vocabularies, and their weights were randomly initialized (embeddings included). This part was done in order to establish the baselines needed to test our hypothesis. In our second experiment, we tested our hypothesis by repeating the same steps just described, but this time, using the continual vocabulary approach introduced in the paragraph above.

In Fig. 5 we can see the performance comparison of the models that were trained from scratch, using domain-specific vocabularies (baseline models) and continual vocabularies (our approach). Although the cross-domain performance was quite similar in the raw domains (Health, Biological, or Merged), when the health model was fine-tuned on the biological domain using the continual vocabulary, the cross-domain performance was remarkably consistent with a standard deviation of 0.75 instead of 2.95 (see Fig. 5 and the "$H \rightarrow B$" columns).

Similarly, the effects of catastrophic forgetting were significantly smaller for the model with the continual vocabulary approach than with the domain-specific vocabularies. This can be seen comparing the "$H \rightarrow B$" columns from Figs. 5a and 5b against the health models. While the health model with the domain-specific vocabulary lost −1.3pts of BLEU (24.9 vs. 23.6pts), the model with the continual vocabulary not only did not lose performance but was slightly increased by +0.1pts of BLEU (25.7 vs. 25.8pts).

Table 2. Models fine-tuned with continual vocabularies led to more consistent results across domains than those trained with domain-specific vocabularies (see *Std.* column: the lower, the better). Similarly, the effects of the catastrophic forgetting were much smaller using these vocabularies (see *Improv. after FT* column).

Vocabs	Dataset	tr.: H → B ts.: H/B/M	Std.	Improv. after FT
domain-specific	Scielo (es-en)	23.6/29.5/26.4	2.95	−1.3 (24.9→23.6)
continual	Scielo (es-en)	25.8/27.3/26.6	**0.75**	**+0.1** (25.7→25.8)
domain-specific	Scielo (pt-en)	24.5/30.8/27.4	3.15	−0.8 (25.3→24.5)
continual	Scielo (pt-en)	26.3/29.1/27.8	**1.40**	**+0.2** (26.1→26.3)

Finally, we repeated the same experiment for the pt-en pair available in the SciELO dataset, where we obtained similar results (See Table 2). That is, models fine-tuned with continual vocabularies presented better consistency across domains and more robustness against the effects of the catastrophic forgetting phenomenon.

6 Conclusions

In this work, we have studied the effects of the catastrophic forgetting problem in NMT from the vocabulary point of view. In addition to this, we have introduced a novel method to mitigate the effects of the catastrophic forgetting phenomenon and improve the cross-domain consistency after fine-tuning.

From studying the effects of the catastrophic forgetting problem in NMT as a function of the vocabulary domain and size, we have shown that the vocabulary domain plays a critical role in the cross-domain performance of a model. Consequently, we have discussed the requirements to build a continual vocabulary, showing that embedding alignment is needed to expand a pre-trained vocabulary with new embeddings. Finally, we have introduced a novel method to mitigate the effects of the catastrophic forgetting problem and improved the cross-domain consistency after fine-tuning, based on using a decoupled continual vocabulary capable of exploiting the subword information to construct word embeddings compositionally to deal with the open vocabulary problem and so that their embeddings are automatically aligned with previous ones.

Acknowledgment. Work supported by the Horizon 2020 - European Commission (H2020) under the SELENE project (grant agreement no 871467) and the project Deep learning for adaptive and multimodal interaction in pattern recognition (DeepPattern) (grant agreement PROMETEO/2019/121). We gratefully acknowledge the support of NVIDIA Corporation with the donation of a GPU used for part of this research.

References

1. Aljundi, R., Babiloni, F., Elhoseiny, M., Rohrbach, M., Tuytelaars, T.: Memory aware synapses: learning what (not) to forget. CoRR abs/1711.09601 (2017). http://arxiv.org/abs/1711.09601
2. Bahdanau, D., Cho, K., Bengio, Y.: Neural machine translation by jointly learning to align and translate. CoRR abs/1409.0473 (2015)
3. Barrault, L., et al.: Findings of the 2020 conference on machine translation (WMT20). In: Proceedings of the Fifth Conference on Machine Translation, pp. 1–55 (2020)
4. Bojanowski, P., Grave, E., Joulin, A., Mikolov, T.: Enriching word vectors with subword information. CoRR abs/1607.04606 (2016)
5. Carpenter, G.A., Grossberg, S.: A massively parallel architecture for a self-organizing neural pattern recognition machine. Comput. Vis. Graph. Image Process. **37**(1), 54–115 (1987). https://doi.org/10.1016/S0734-189X(87)80014-2. https://www.sciencedirect.com/science/article/pii/S0734189X87800142
6. Carrión, S., Casacuberta, F.: AutoNMT: a framework to streamline the research of Seq2Seq models (2022). https://github.com/salvacarrion/autonmt/
7. Cherry, C., Foster, G.F., Bapna, A., Firat, O., Macherey, W.: Revisiting character-based neural machine translation with capacity and compression. CoRR abs/1808.09943 (2018). http://arxiv.org/abs/1808.09943
8. Cho, K., et al.: Learning phrase representations using RNN encoder-decoder for statistical machine translation. In: Proceedings of the 2014 Conference on EMNLP, pp. 1724–1734 (2014)
9. Devlin, J., Chang, M., Lee, K., Toutanova, K.: BERT: pre-training of deep bidirectional transformers for language understanding. CoRR abs/1810.04805 (2018)
10. Draelos, T.J., et al.: Neurogenesis deep learning. CoRR abs/1612.03770 (2016)
11. Fernando, C., et al.: PathNet: evolution channels gradient descent in super neural networks. CoRR abs/1701.08734 (2017). http://arxiv.org/abs/1701.08734
12. Gowda, T., May, J.: Finding the optimal vocabulary size for neural machine translation. In: Findings of the ACL: EMNLP 2020, pp. 3955–3964 (2020)
13. Hu, W., et al.: Overcoming catastrophic forgetting for continual learning via model adaptation. In: ICLR (2019)

14. Jung, H., Ju, J., Jung, M., Kim, J.: Less-forgetting learning in deep neural networks. CoRR abs/1607.00122 (2016). http://arxiv.org/abs/1607.00122
15. Kemker, R., Kanan, C.: FearNet: brain-inspired model for incremental learning. CoRR abs/1711.10563 (2017)
16. Kirkpatrick, J., et al.: Overcoming catastrophic forgetting in neural networks. CoRR abs/1612.00796 (2016)
17. Koehn, P., et al.: Moses: open source toolkit for statistical machine translation. In: Proceedings of the 45th Annual Meeting of the ACL on Interactive Poster and Demonstration Sessions, pp. 177–180. ACL 2007 (2007)
18. Kudo, T., Richardson, J.: SentencePiece: a simple and language independent subword tokenizer and detokenizer for neural text processing. In: Blanco, E., Lu, W. (eds.) Proceedings of the 2018 Conference on Empirical Methods in Natural Language Processing, EMNLP 2018: System Demonstrations, Brussels, Belgium, October 31 - November 4, 2018, pp. 66–71 (2018). https://doi.org/10.18653/v1/d18-2012
19. Lee, J., Yoon, J., Yang, E., Hwang, S.J.: Lifelong learning with dynamically expandable networks. CoRR abs/1708.01547 (2017). http://arxiv.org/abs/1708.01547
20. Lee, S., Kim, J., Ha, J., Zhang, B.: Overcoming catastrophic forgetting by incremental moment matching. CoRR abs/1703.08475 (2017). http://arxiv.org/abs/1703.08475
21. Li, Z., Hoiem, D.: Learning without forgetting. CoRR abs/1606.09282 (2016)
22. Liu, T., Ungar, L., Sedoc, J.: Continual learning for sentence representations using conceptors. ArXiv abs/1904.09187 (2019)
23. Lopez-Paz, D., Ranzato, M.: Gradient episodic memory for continuum learning. CoRR abs/1706.08840 (2017). http://arxiv.org/abs/1706.08840
24. Luong, T., Pham, H., Manning, C.D.: Effective approaches to attention-based neural machine translation. In: Proceedings of the 2015 Conference on EMNLP, pp. 1412–1421 (2015)
25. van der Maaten, L., Hinton, G.: Visualizing data using t-SNE. J. Mach. Learn. Res. **9**, 2579–2605 (2008)
26. McCloskey, M., Cohen, N.J.: Catastrophic interference in connectionist networks: the sequential learning problem. In: Psychology of Learning and Motivation, vol. 24, pp. 109–165. Academic Press (1989). https://doi.org/10.1016/S0079-7421(08)60536-8, https://www.sciencedirect.com/science/article/pii/S0079742108605368
27. Papineni, K., Roukos, S., Ward, T., Zhu, W.J.: BLEU: a method for automatic evaluation of machine translation. In: Proceedings of the 40th Annual Meeting on ACL, p. 311–318. ACL 2002 (2002)
28. Post, M.: A call for clarity in reporting BLEU scores. In: Proceedings of the Third Conference on Machine Translation: Research Papers, pp. 186–191 (2018)
29. Qi, Y., Sachan, D., Felix, M., Padmanabhan, S., Neubig, G.: When and why are pretrained word embeddings useful for neural machine translation? In: Proceedings of the 2018 Conference of the North American Chapter of the Association for Computational Linguistics: Human Language Technologies, vol. 2 (Short Papers) (2018)
30. Rebuffi, S., Kolesnikov, A., Lampert, C.H.: icarl: Incremental classifier and representation learning. CoRR abs/1611.07725 (2016). http://arxiv.org/abs/1611.07725
31. Rusu, A.A., et al.: Progressive neural networks. CoRR abs/1606.04671 (2016)

32. Sato, S., Sakuma, J., Yoshinaga, N., Toyoda, M., Kitsuregawa, M.: Vocabulary adaptation for domain adaptation in neural machine translation. In: Findings of the Association for Computational Linguistics: EMNLP 2020, pp. 4269–4279. Association for Computational Linguistics, Online (2020). https://doi.org/10.18653/v1/2020.findings-emnlp.381.https://aclanthology.org/2020.findings-emnlp.381
33. Schwarz, J., et al.: Progress & compress: a scalable framework for continual learning (2018). https://doi.org/10.48550/ARXIV.1805.06370. https://arxiv.org/abs/1805.06370
34. Sennrich, R., Haddow, B., Birch, A.: Neural machine translation of rare words with subword units. In: Proceedings of the 54th Annual Meeting of the ACL (Long Papers), vol. 1, pp. 1715–1725 (2016)
35. Sennrich, R., Zhang, B.: Revisiting low-resource neural machine translation: a case study. In: Proceedings of the 57th Annual Meeting of the Association for Computational Linguistics, pp. 211–221. Association for Computational Linguistics, Florence, Italy (2019). https://doi.org/10.18653/v1/P19-1021. https://aclanthology.org/P19-1021
36. Shin, H., Lee, J.K., Kim, J., Kim, J.: Continual learning with deep generative replay. CoRR abs/1705.08690 (2017). http://arxiv.org/abs/1705.08690
37. Sutskever, I., Vinyals, O., Le, Q.V.: Sequence to sequence learning with neural networks. In: Ghahramani, Z., Welling, M., Cortes, C., Lawrence, N., Weinberger, K.Q. (eds.) NIPS, vol. 27 (2014)
38. Vaswani, A., et al.: Attention is all you need. In: Proceedings of the 31st NeurIPS, pp. 6000–6010. NIPS 2017 (2017)
39. Wu, Y., et al.: Google's neural machine translation system: bridging the gap between human and machine translation. CoRR abs/1609.08144 (2016)
40. Xu, H., Liu, B., Shu, L., Yu, P.S.: Lifelong domain word embedding via meta-learning. In: Proceedings of the Twenty-Seventh International Joint Conference on Artificial Intelligence, IJCAI-2018, pp. 4510–4516 (2018)
41. Zenke, F., Poole, B., Ganguli, S.: Continual learning through synaptic intelligence. In: Precup, D., Teh, Y.W. (eds.) Proceedings of the 34th International Conference on Machine Learning. Proceedings of Machine Learning Research, vol. 70, pp. 3987–3995. PMLR (2017). https://proceedings.mlr.press/v70/zenke17a.html
42. Zhang, T., Kishore, V., Wu, F., Weinberger, K.Q., Artzi, Y.: BERTscore: evaluating text generation with BERT. CoRR abs/1904.09675 (2019). http://arxiv.org/abs/1904.09675

Evaluating Domain Generalization in Kitchen Utensils Classification

Carlos Garrido-Munoz(✉), María Alfaro-Contreras, and Jorge Calvo-Zaragoza

University Institute for Computing Research (IUII), University of Alicante, Alicante, Spain

carlos.garrido@ua.es, {malfaro,jcalvo}@dlsi.ua.es

Abstract. The remarkable performance of deep learning models is heavily dependent on the availability of large and diverse amounts of training data and its correlation with the target application scenario. This is especially crucial in robotics, where the deployment environments often differ from the training ones. Domain generalization (DG) techniques investigate this problem by leveraging data from multiple source domains so that a trained model can generalize to unseen domains. In this work, we thoroughly evaluate the performance in the classification of kitchen utensils of several state-of-the-art DG methods. Extensive experiments on the seven domains that compose the Kurcuma (Kitchen Utensil Recognition Collection for Unsupervised doMain Adaptation) dataset show that the effectiveness of some of the DG methods varies across domains, with none performing well across all of them. Specifically, most methods achieved high accuracy rates in four of the seven datasets, and while there was a reasonable improvement in the most difficult domains, there is still ample room for further research.

Keywords: Deep Learning · Domain Generalization · Robotics · Computer Vision

1 Introduction

In computer vision, image classification is the task of automatically assigning a label, out of a set of possible categories, to a given image [29]. State-of-the-art approaches rely on neural models based on deep learning (DL), which require large amounts of labeled data—consisting of problem images together with their corresponding classes—to be adequately trained. This constraint often requires performing expensive and error-prone labeling campaigns to gather and label high-quality data. Moreover, these models assume that the training (source) and testing (target) data are independent and identically distributed (i.i.d.), meaning they come from the same domain/dataset and thus follow the same distribution. However, in practice, this assumption is often violated, leading to steep drops in recognition performance [38].

When labeled, partially labeled, or unlabeled target data are available during training time, several mechanisms—including semi-supervised learning, weakly

A. Pertusa et al. (Eds.): IbPRIA 2023, LNCS 14062, pp. 108–118, 2023.
https://doi.org/10.1007/978-3-031-36616-1_9

labeled learning, transfer learning, and domain adaptation—can be used to adapt the model to it [33]. But what happens when the target data are fed to the system only during deployment? How can we leverage the knowledge gathered from multiple labeled source domains so that it can generalize to unseen target domains without any adaptation? Domain generalization (DG) is the research field that studies how to overcome the domain shift problem in the absence of target data [32].

Towards this goal, most of the existing domain generalization methods can be roughly divided into three main groups [38]: (i) domain alignment, in which domain-invariant representations are learned by minimizing the difference among source domains; (ii) meta-learning, in which the domain shift is simulated by dividing the source domains into meta-train and meta-validation sets and learning a model on the former to reduce the error on the latter; and (iii) data augmentation, in which new samples are generated to increase the source diversity.

Although DG provides a great alternative to mitigate the significant discrepancies that often exist between controlled training scenarios and real-world applications in the robotics field, most existing literature in that area tackles the domain shift problem by leveraging target data [6,25,35]. As this may not always be a realistic scenario, we explore the use of state-of-the-art DG strategies in the context of autonomous home assistants, with a focus on household tasks.

Specifically, we thoroughly review ten state-of-the-art DG approaches proposed for image classification. We conduct extensive experiments on the Kurcuma (Kitchen Utensil Recognition Collection for Unsupervised doMain Adaptation) dataset, a collection of seven domains for the classification of kitchen utensils [25]—a task of relevance in home-assistance robotics. The results show that the effectiveness of DG methods depends on the target domain, with none showing consistent performance across all domains.

The remainder of the paper is organized as follows: Sect. 2 provides the background to this work, while Sect. 3 thoroughly develops the domain generalization framework considered. Section 4 describes the experimental setup and Sect. 5 presents and analyses the results. Finally, Sect. 6 concludes the work and discusses possible ideas for future research.

2 Background

DG is a challenging problem in machine learning that refers to the ability of a model to generalize across multiple domains that have not been seen during training [38]. Currently, the literature groups the different proposed approaches into three categories: domain alignment [2,17,34], meta-learning [3,20,37], and data augmentation [27,36,39].

Domain alignment methods, which are adapted from the unsupervised domain adaptation (DA) literature, aim to align representations between domains to obtain invariant features [32]. This is achieved by applying techniques that align the representations of the training domains, as we do not have access to the target domain. These methods assume that obtaining invariant features will lead to better generalization.

Meta-learning, also known as "learning to learn", is a paradigm that tries to improve the model's ability to generalize by learning through training episodes. It has been rapidly growing in the machine learning and computer vision literature [14]. Currently, state-of-the-art approaches are based on Model-Agnostic Meta-Learning (MAML) method [10], which can be seen as a bi-level optimization problem that aims to learn the best update of weights in the training set to improve performance on the validation set. In the case of DG, test adaptation is simulated by dividing the training set into meta-train and meta-test.

Data augmentation methods attempt to augment the training data or the obtained features through classical augmentations [36] or (adversarial) transformation networks [39] or to map the target images to the source domains via style-transfer methods [5].

Several other approaches have emerged that support generalization across domains. One such approach is self-supervised learning, which involves learning from pseudo-labeled data—no human annotation is involved—to converge in one or more downstream tasks such as classification [24]. This can be achieved by predicting the transformations applied to the image data, such as the shuffling order [7,8]. Another approach is regularization strategies, which aim to suppress over-dominant features based on some heuristics [15,31]. Lastly, learning disentangled representations can also support domain generalization by separating the underlying factors of variation in the data either by decomposing the model into domain-specific and domain-agnostic parts [21] or learning independent latent subspaces for class, domain, and object, respectively [16].

3 Methodology

In this work, we focus on the multi-source Domain Generalization setting, where we have a set $S = \{S_1, S_2, \ldots, S_n\}$ representing the n available data sources with pairs $\{x_i^{(j)}, y_i^{(j)}\}_{i=1}^{N_j}$ for each domain $S_j \in S$, where y_i denotes the label of the corresponding x_i image. The objective is that a classifier $f : X \rightarrow Y$, trained only on source data S_{train}, generalizes to an unknown target domain $\{x_i^{(t)}, y_i^{(t)}\}_{i=1}^{N_t} \in T$, with T and S sharing the same set of categories but coming from different marginal distributions (the i.i.d. assumption is violated). Thus, we aim to overcome the covariate shift problem [38], formalized as $P_{train}(y \mid X) = P_{test}(y \mid X)$, $P_{train}(X) \neq P_{test}(X)$.

We evaluate ten different state-of-the-art DG algorithms proposed for image classification and compare their performance in the classification of kitchen utensils. To highlight the effectiveness of the DG algorithms in mitigating the domain shift problem, we compare them with the standard learning approach, known as **Empirical Risk Minimization (ERM)** [30]. ERM minimizes the average loss across all the training examples from all the domains.

3.1 Domain Generalization Methods

The goal of this study is to conduct a thorough evaluation of state-of-the-art DG methods on the task of kitchen utensils classification. To ensure a diverse and representative selection, we have chosen ten widely used DG algorithms. Our selection encompasses a variety of techniques, including domain alignment methods, meta-learning methods, data augmentation methods, and regularization strategies. The considered algorithms are described below.

Invariant Risk Minimization (IRM) [2]. This is a domain alignment algorithm that learns a feature representation such that the optimal linear classifier on top of that representation space matches for all domains.

Group Distributionally Robust Optimization (GroupDRO) [26]. It performs ERM while minimizing the worst-domain loss—in other words, it increases the importance of domains with larger errors.

Mixup [36]. A data augmentation method that generates new virtual data by linearly interpolating pairs of samples and their labels.

Meta-Learning for Domain Generalization (MLDG) [20]. This method is based on the MAML algorithm [10], the model is trained to adapt to new domains by splitting the source domains into meta-train and meta-test domains.

Correlation Alignment (CORAL) [28]. A domain alignment method—originally proposed for DA—that aligns the second-order statistics (covariance) between the source domains.

Minimizing Maximum Mean Discrepancy (MMD) [22]. This work extends that of Makhzani et al. [23] by aligning the domains using the MMD distance and matching the aligned distributions to an arbitrary prior distribution (Laplacian, in this case) via adversarial learning.

Domain-Adversarial Neural Network (DANN) [12]. This is a domain alignment method—originally proposed for DA—that directly aligns inter-domain distributions by adversarial training the class classifier together with a domain classifier. It maximizes the domain loss to extract features that are invariant/common to all domains.

Marginal Transfer Learning (MTL) [4]. It builds a classifier $f(x^d, \mu^d)$, where $d \in S \cup T$ and μ^d is a kernel mean embedding that summarizes the domain d associated to the example x^d. Since the distributional identity of the target domain T is unknown, μ^T is estimated using single target examples at test time.

Variance Risk Extrapolation (VREx) [19]. This method performs ERM while minimizing the variance between the domain predictions so as to obtain a domain-invariant classifier.

Representation Self-Challenging (RSC) [15]. A regularization strategy that masks the dominant features to force the model to predict using only the remaining information.

4 Experimental Setup

This section presents the corpus and the implementation details considered.

4.1 The Kurcuma Dataset

The Kurcuma (Kitchen Utensil Recognition Collection for Unsupervised doMain Adaptation) dataset includes seven domains/subsets of kitchen utensils with a varying number of 300×300 pixels of color images distributed in the same nine categories [25]. These domains differ in whether the objects and background in each of them are considered real or synthetic. Below is a description of the seven subsets:

- *Edinburgh Kitchen Utensil Database (EKUD).* This subset comprises real-world pictures of utensils with uniform backgrounds.
- *EKUD Real Color (EKUD-M1).* This subset combines the images of EKUD with patches from the Berkeley Segmentation Data Set and Benchmarks 500 (BSDS500) [1], following a similar approach as the one used for the creation of the MNIST-M dataset [11]. The background of the EKUD images was modified while keeping the original color of the objects.
- *EKUD Not Real Color (EKUD-M2).* This subset is an extension of EKUD-M1 in which the color of the objects was altered by being mixed with the color of the patches used for the background.
- *EKUD Not Real Color with Real Background (EKUD-M3).* This is a variation of EKUD-M2, in which the distortion process of EKUD-M2 is directly applied to the original EKUD subset.
- *Alicante Kitchen Utensil Database (AKUD).* This collection depicts real objects in different real-world backgrounds (photographs), covering a wide range of lighting conditions and perspectives.
- *RENDER.* This subset comprises synthetic images rendered using different base public models of utensils and backgrounds from the Internet.
- *CLIPART.* This subset consists of draw-like images gathered from the Internet with a plain white background.

A summary of the main features of the Kurcuma dataset is depicted in Table 1 together with some image examples in Fig. 1.

Table 1. Summary of the characteristics of the seven corpora comprising the Kurcuma collection in terms of the nature of the objects and background, the number of samples per class, and their total size.

Corpus	Description		Total Size
	Objects	Background	
EKUD	Real	Uniform	618
EKUD-M1	Real	Synthetic	
EKUD-M2	Real	Synthetic	
EKUD-M3	Real	Uniform	
AKUD	Real	Real	1,681
RENDER	Synthetic	Synthetic	1,647
CLIPART	Synthetic	None	1,069

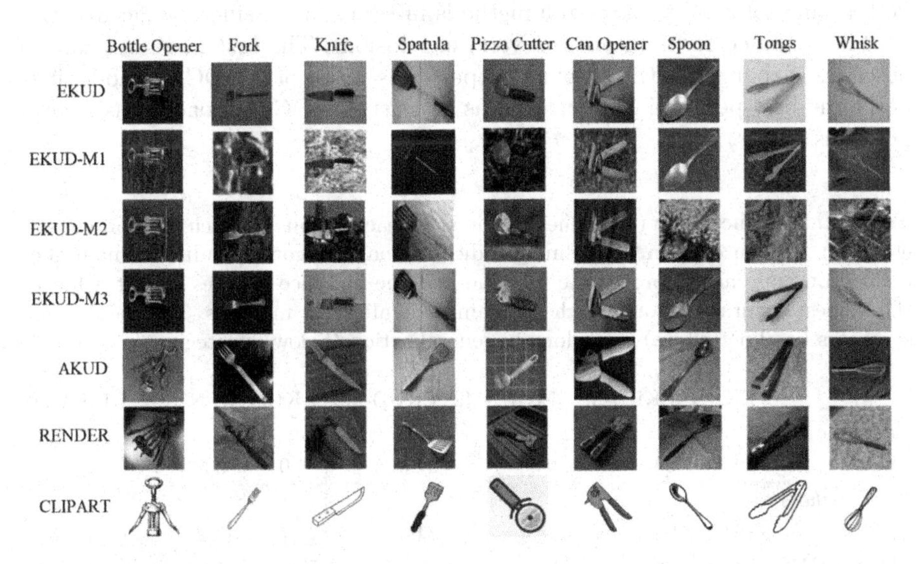

Fig. 1. Examples of the nine different classes for the different image domains comprising the Kurcuma collection. Image taken from [25].

4.2 Implementation Details

We use the DomainBed suite presented in [13] to train the different models described in Sect. 3.1. In this sense, all models follow a ResNet-50 [9] configuration and employ ADAM [18] as the optimizer. We keep their hyper-parameter grids and modify the training steps to 2 500.

Selecting the best-performing model in DG is a rather complicated process due to the lack of access to the target data. Gulrajani and Lopez-Paz [13] address this issue by arguing that a DG algorithm is incomplete without a model selection strategy and that any oracle-selection results should be discarded since access to the target domain is prohibited in DG.

Following the aforementioned recommendations, we consider the most commonly used training-domain validation set method for model selection. The source domains are split into training and validation subsets. The validation subsets are aggregated to create an overall validation set that is used for model selection. In other words, we keep the model with the highest accuracy on the overall validation set.

It must be noted that the experiments are repeated 3 times to obtain a better estimate of the average performance of each DG method.

5 Results

This section discusses the results obtained on the Kurcuma dataset for the proposed experimental scheme. In this regard, Table 2 and Fig. 2 report the average performance of all the considered methods in terms of classification accuracy.

The obtained results lead to several conclusions. The first idea that can be observed is that while the baseline outperforms many of the DG methods, it is never the best-performing algorithm, as at least one DG approach consistently performs better.

Table 2. Mean accuracy (%) values obtained on each corpus for each method considered. For a given domain, underlining indicates that the corresponding domain generalization approach improves the baseline and the boldface denotes that it achieves the highest accuracy among all the domain generalization methods. The dashed line separates baseline (above) from domain generalization (below) strategies.

	EKUD	EKUD-M1	EKUD-M2	EKUD-M3	AKUD	RENDER	CLIPART
Baseline							
ERM	99.7	99.4	94.5	97.6	69.0	66.5	61.4
DG methods							
IRM	80.3	94.5	94.7	74.4	74.7	43.2	66.4
GroupDRO	98.8	98.9	91.6	97.1	77.5	71.8	67.9
Mixup	99.3	72.9	**96.4**	98.1	57.8	68.0	66.1
MLDG	**99.8**	99.1	95.6	**98.9**	77.2	73.6	**71.1**
CORAL	99.6	98.6	**96.4**	**98.9**	**78.9**	72.8	70.0
MMD	72.4	99.1	96.2	97.8	73.3	**77.2**	48.1
DANN	98.7	70.5	90.7	71.9	67.0	76.9	47.2
MTL	99.6	**99.5**	95.1	98.2	77.8	68.5	66.5
VREx	77.1	98.4	92.0	73.9	50.3	63.2	67.2
RSC	96.4	98.6	96.1	97.4	65.4	56.0	64.0

We also notice that the extent of improvement varies significantly across domains. On one hand, in the EKUD dataset and its variants (EKUD-M1, EKUD-M2, and EKUD-M3), there is relatively little room for improvement since the baseline already achieves accuracy rates of over 90%. Although the baseline's performance is almost perfect in EKUD and EKUD-M1 (above 99%), there is always one method that performs better (MLDG and MTL, respectively). In EKUD-M2 and EKUD-M3, the number of DG methods outperforming the baseline increases to at least half of the ten selected. On the other hand, the margin of improvement is much higher for the remaining domains (AKUD, RENDER, and CLIPART), with the baseline being one of the least accurate methods in these cases. There is still ample room for further improvement since the accuracy rates for these domains are around 20 points lower than those of the previously mentioned EKUD group.

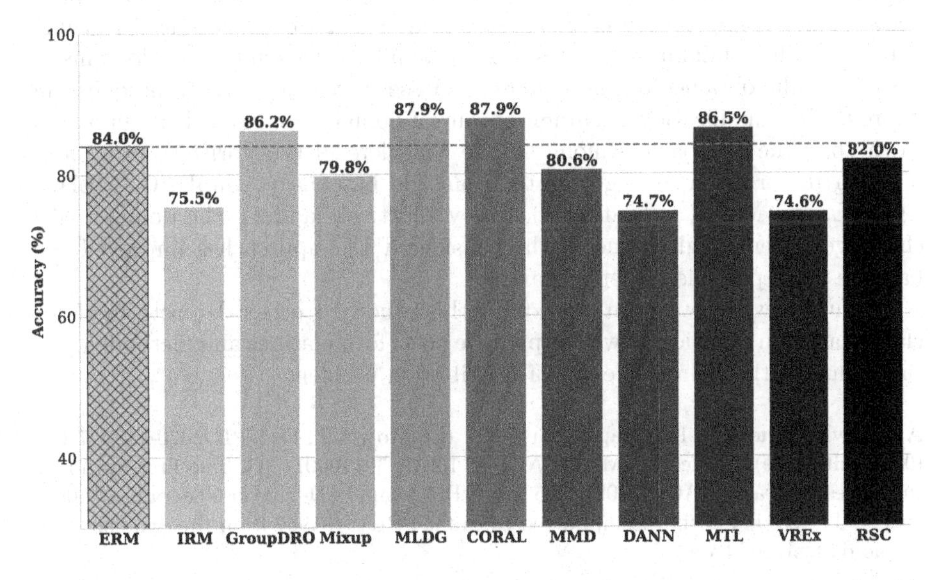

Fig. 2. Mean accuracy values (%) on all datasets for each method considered. The dashed line marks the baseline result to allow for simple comparison with respect to each domain generalization method.

Among the different DG methods, we observe that some consistently outperform others across all corpora. For instance, MLDG achieves the highest accuracy on EKUD and CLIPART, while CORAL does so on EKUD-M2 and AKUD. Both of them achieve the best performance rate on EKUD-M3. Looking at Fig. 2, we see that both methods have a similar average performance and position themselves as the best-performing algorithms in this study. We also notice that MTL follows closely since it attains higher accuracy rates than the baseline in six out of the seven domains, being the only algorithm to do so in the EKUD-M1 domain.

It is also important to remark that IRM, VREx, and DANN obtain lower results than the baseline on average—they only outperform the baseline in one domain, at most. In this regard, while DANN and CORAL are both designed for DA problems, DANN fails to adapt to the DG setting, becoming the worst-performing algorithm.

Overall, the results demonstrate that the effectiveness of DG methods varies across domains, and no single method is consistently superior. However, as seen in Fig. 2, MLDG and CORAL stand out from the rest, closely followed by MTL and GroupDRO, since the four of them outperform the baseline on average.

6 Conclusions

This work presents an experimental study of different DG methods for the kitchen utensils classification task. Ten state-of-the-art DG methods have been evaluated and compared among them and against the standard learning paradigm, which minimizes the loss for all samples across all source domains.

The results obtained on the Kurcuma dataset reveal that DG techniques can improve the classification accuracy on out-of-domain samples when compared to the baseline approach. Among all the considered DG algorithms, only four of them perform, on average, better than the baseline (GroupDRO, MLDG, CORAL, and MTL), indicating that they effectively address the domain shift challenge. Interestingly, none of the considered DG approaches improved the baseline in all possible target scenarios.

In future work, we plan to extend this study to other datasets used for classification in robotics, as well as propose novel formulations and methodologies that consistently address the out-of-distribution problem.

Acknowledgments. This paper is part of the project I+D+i PID2020-118447RA-I00 (MultiScore), funded by MCIN/AEI/10.13039/501100011033. The first author is supported by grant CIACIF/2021/465 from "Programa I+D+i de la Generalitat Valenciana". The second author is supported by grant FPU19/04957 from the Spanish Ministerio de Universidades.

References

1. Arbelaez, P., Maire, M., Fowlkes, C., Malik, J.: Contour detection and hierarchical image segmentation. IEEE Trans. Pattern Anal. Mach. Intell. **33**(5), 898–916 (2010)
2. Arjovsky, M., Bottou, L., Gulrajani, I., Lopez-Paz, D.: Invariant risk minimization. arXiv preprint arXiv:1907.02893 (2019)
3. Balaji, Y., Sankaranarayanan, S., Chellappa, R.: Metareg: towards domain generalization using meta-regularization. Adv. Neural Inf. Process. Syst. **31** (2018)
4. Blanchard, G., Deshmukh, A.A., Dogan, Ü., Lee, G., Scott, C.: Domain generalization by marginal transfer learning. J. Mach. Learn. Res. **22**(1), 46–100 (2021)
5. Borlino, F.C., D'Innocente, A., Tommasi, T.: Rethinking domain generalization baselines. In: International Conference on Pattern Recognition, pp. 9227–9233. IEEE (2021)
6. Bousmalis, K., et al.: Using simulation and domain adaptation to improve efficiency of deep robotic grasping. In: IEEE International Conference on Robotics and Automation, pp. 4243–4250. IEEE (2018)

7. Bucci, S., D'Innocente, A., Liao, Y., Carlucci, F.M., Caputo, B., Tommasi, T.: Self-supervised learning across domains. IEEE Trans. Pattern Anal. Mach. Intell. **44**(9), 5516–5528 (2021)
8. Carlucci, F.M., D'Innocente, A., Bucci, S., Caputo, B., Tommasi, T.: Domain generalization by solving jigsaw puzzles. In: IEEE/CVF Conference on Computer Vision and Pattern Recognition, pp. 2229–2238 (2019)
9. Deng, J., Dong, W., Socher, R., Li, L.J., Li, K., Fei-Fei, L.: ImageNet: a large-scale hierarchical image database. In: IEEE/CVF Conference on Computer Vision and Pattern Recognition, pp. 248–255. IEEE (2009)
10. Finn, C., Abbeel, P., Levine, S.: Model-agnostic meta-learning for fast adaptation of deep networks. In: International Conference on Machine Learning, pp. 1126–1135. PMLR (2017)
11. Ganin, Y., Lempitsky, V.: Unsupervised domain adaptation by backpropagation. In: International Conference on Machine Learning, pp. 1180–1189. PMLR (2015)
12. Ganin, Y., et al.: Domain-adversarial training of neural networks. J. Mach. Learn. Res. **17**(59), 1–35 (2016)
13. Gulrajani, I., Lopez-Paz, D.: In search of lost domain generalization. In: International Conference on Learning Representations (2021)
14. Hospedales, T., Antoniou, A., Micaelli, P., Storkey, A.: Meta-learning in neural networks: a survey. IEEE Trans. Pattern Anal. Mach. Intell. **44**(9), 5149–5169 (2021)
15. Huang, Z., Wang, H., Xing, E.P., Huang, D.: Self-challenging improves cross-domain generalization. In: Vedaldi, A., Bischof, H., Brox, T., Frahm, J.-M. (eds.) ECCV 2020. LNCS, vol. 12347, pp. 124–140. Springer, Cham (2020). https://doi.org/10.1007/978-3-030-58536-5_8
16. Ilse, M., Tomczak, J.M., Louizos, C., Welling, M.: Diva: domain invariant variational autoencoders. In: Medical Imaging with Deep Learning, pp. 322–348. PMLR (2020)
17. Jia, Y., Zhang, J., Shan, S., Chen, X.: Single-side domain generalization for face anti-spoofing. In: IEEE/CVF Conference on Computer Vision and Pattern Recognition, pp. 8484–8493 (2020)
18. Kingma, D.P., Ba, J.: Adam: a method for stochastic optimization. In: Bengio, Y., LeCun, Y. (eds.) International Conference on Learning Representations. San Diego, USA (2015)
19. Krueger, D., et al.: Out-of-distribution generalization via risk extrapolation (REx). In: International Conference on Machine Learning, pp. 5815–5826. PMLR (2021)
20. Li, D., Yang, Y., Song, Y.Z., Hospedales, T.: Learning to generalize: meta-learning for domain generalization. In: AAAI Conference on Artificial Intelligence, vol. 32 (2018)
21. Li, D., Yang, Y., Song, Y.Z., Hospedales, T.M.: Deeper, broader and artier domain generalization. In: IEEE International Conference on Computer Vision, pp. 5542–5550 (2017)
22. Li, H., Pan, S.J., Wang, S., Kot, A.C.: Domain generalization with adversarial feature learning. In: IEEE/CVF Conference on Computer Vision and Pattern Recognition, pp. 5400–5409 (2018)
23. Makhzani, A., Shlens, J., Jaitly, N., Goodfellow, I., Frey, B.: Adversarial autoencoders. In: International Conference on Learning Representations (2016)
24. Ohri, K., Kumar, M.: Review on self-supervised image recognition using deep neural networks. Knowl.-Based Syst. **224**, 107090 (2021)

25. Rosello, A., Valero-Mas, J.J., Gallego, A.J., Sáez-Pérez, J., Calvo-Zaragoza, J.: Kurcuma: a kitchen utensil recognition collection for unsupervised domain adaptation. Pattern Anal. Appl. 1–13 (2023)
26. Sagawa, S., Koh, P.W., Hashimoto, T.B., Liang, P.: Distributionally robust neural networks for group shifts: on the importance of regularization for worst-case generalization. In: International Conference on Learning Representations (2020)
27. Shi, Y., Yu, X., Sohn, K., Chandraker, M., Jain, A.K.: Towards universal representation learning for deep face recognition. In: IEEE/CVF Conference on Computer Vision and Pattern Recognition, pp. 6817–6826 (2020)
28. Sun, B., Saenko, K.: Deep CORAL: correlation alignment for deep domain adaptation. In: Hua, G., Jégou, H. (eds.) ECCV 2016. LNCS, vol. 9915, pp. 443–450. Springer, Cham (2016). https://doi.org/10.1007/978-3-319-49409-8_35
29. Szeliski, R.: Computer Vision: Algorithms and Applications. Springer, London (2022). https://doi.org/10.1007/978-1-84882-935-0
30. Vapnik, V.N.: An overview of statistical learning theory. IEEE Trans. Neural Netw. **10**(5), 988–999 (1999)
31. Wang, H., He, Z., Lipton, Z.C., Xing, E.P.: Learning robust representations by projecting superficial statistics out. In: International Conference on Learning Representations (2019)
32. Wang, J., et al.: Generalizing to unseen domains: a survey on domain generalization. IEEE Trans. Knowl. Data Eng. (2022)
33. Wang, M., Deng, W.: Deep visual domain adaptation: a survey. Neurocomputing **312**, 135–153 (2018)
34. Wang, Z., Loog, M., Van Gemert, J.: Respecting domain relations: hypothesis invariance for domain generalization. In: International Conference on Pattern Recognition, pp. 9756–9763. IEEE (2021)
35. Wulfmeier, M., Bewley, A., Posner, I.: Incremental adversarial domain adaptation for continually changing environments. In: IEEE International Conference on Robotics and Automation, pp. 4489–4495. IEEE (2018)
36. Zhang, H., Cisse, M., Dauphin, Y.N., Lopez-Paz, D.: mixup: beyond empirical risk minimization. In: International Conference on Learning Representations (2018)
37. Zhao, Y., et al.: Learning to generalize unseen domains via memory-based multi-source meta-learning for person re-identification. In: IEEE/CVF Conference on Computer Vision and Pattern Recognition, pp. 6277–6286 (2021)
38. Zhou, K., Liu, Z., Qiao, Y., Xiang, T., Loy, C.C.: Domain generalization: a survey. IEEE Trans. Pattern Anal. Mach. Intell. (2022)
39. Zhou, K., Yang, Y., Hospedales, T., Xiang, T.: Learning to generate novel domains for domain generalization. In: Vedaldi, A., Bischof, H., Brox, T., Frahm, J.-M. (eds.) ECCV 2020. LNCS, vol. 12361, pp. 561–578. Springer, Cham (2020). https://doi.org/10.1007/978-3-030-58517-4_33

Document Analysis

Segmentation of Large Historical Manuscript Bundles into Multi-page Deeds

Jose Ramón Prieto[1(✉)], David Becerra[4], Alejandro Hector Toselli[1], Carlos Alonso[3], and Enrique Vidal[1,2,3]

[1] PRHLT Research Center, Universitat Politècnica de València, Valencia, Spain
{joprfon,ahector,evidal}@prhlt.upv.es

[2] Valencian Graduate School and Research Network of Artificial Intelligence, Valencia, Spain

[3] tranSkriptorium IA, Valencia, Spain

[4] Universidad de Sevilla, Seville, Spain

Abstract. Archives around the world have vast uncatalogued series of image bundles of digitized historical manuscripts containing, among others, notarial records also known as "deeds" or "acts". One of the first steps to provide metadata which describe the contents of those bundles is to segment these bundles into their individual deeds. Even if deeds are page-aligned, as in the bundles considered in the present work, this is a time-consuming task, often prohibitive given the huge scale of the manuscript series involved. Unlike traditional Layout Analysis methods for page-level segmentation, our approach goes beyond the realm of a single-page image, providing consistent deed detection results on full bundles. This is achieved in two tightly integrated steps: first, the probabilities that each bundle image is an "initial", "middle" or "final" page of a deed are estimated, and then an optimal sequence of page labels is computed at the whole bundle level. Empirical results are reported which show that this approach achieves almost perfect segmentation of bundles of a massive Spanish series of historical notarial records.

Keywords: Handwritten document image processing · multi-page layout analysis · bundle segmentation · historical manuscripts

1 Introduction

Large series of historical manuscripts which contain important notarial records are stored in kilometers of shelves in archives around the world. Many of these series are digitized; i.e., converted into high-resolution digital images.

In general, these manuscript images are sequentially piled into folders, books or boxes, often called "image bundles", each containing thousands of page images and hundreds records. Hereafter, bundles, boxes, books, or folders of manuscript images are just called "*bundles*". A bundle may contain several, often many

A. Pertusa et al. (Eds.): IbPRIA 2023, LNCS 14062, pp. 121–133, 2023.
https://doi.org/10.1007/978-3-031-36616-1_10

"image documents", also called "files", "acts" – or *"deeds"* in the case of notarial image documents considered in this work.

For series of documents so massive, it is generally difficult or impossible for archives to provide detailed metadata to adequately describe the contents of each bundle. Specifically, no information is generally available about where in each bundle lays the sequence of page images of each of its deeds. Clearly, automatic methods are needed to assist the archive experts in the arduous task of cataloging these vast series. One of the first steps in this task is the segmentation of bundles into their individual deeds, which is the problem considered in the present work.

Previous approaches to this or similar problems were made in the field of Layout Analysis (LA), which considers a multitude of document analysis tasks, such as line detection, page layout segmentation, document understanding, etc. Many approaches for line detection and extraction, currently one of the basic steps for Handwritten Text Recognition (HTR) systems, rely on text baseline detection [6,23], using a variety of approaches. Other recent works apply convolutional networks [4], or encoder-decoder architectures to obtain, simultaneously, both layout segmentation and line detection [3,18,25]. Others add a spatial attention model at different resolutions to existing architectures [10].

Region Proposal Networks (RPN) are also used for single-page LA. MaskR-CNN [11] can be used for complex-layout segmentation as shown in [2,23]. In [1] the same system is used to deal with many lines close to each other in order to do information extraction in tables.

Also, substantial progress has been made by using transformer-based architectures for LA, also including HTR in some cases. In [14], the authors present DONUT, an approach to perform document understanding on the whole page in an end-to-end fashion. Another end-to-end model is DAN [7], where the authors show an end-to-end segmentation-free model to obtain the logical layout (without the need of geometrical information from LA) and recognize the text at the same time.

The interest of these kind of holistic approaches notwithstanding, none of these works consider text-image processing beyond the realm of a single-page image, which is the problem we face here.

Methods for automatic classification of untranscribed image documents (deeds) into their typological classes (such as "Letter of Payment", "Will", etc.) have been developed and successfully tested in [8,9,19,21]. However, in these works, the successive page images of each deed are assumedly given. But, as discussed above, in real applications, deeds are typically embedded into bundles, without any explicit separation of the specific page images encompassed by each deed.

Therefore, in order to provide really practical solutions towards automated documentary management of these important series of manuscripts, a pending problem is how to segment a large bundle into its constituent deeds. This is the task considered in the present work. Previous approaches to this problem exists, but they should be considered only rather tentative. In [5,6] and [20], the authors propose text-line-oriented, page-level segmentation using Hidden Markov

Models and recurrent CTC-based systems, respectively, in order to detect initial, medium and final slices of deeds on each page image. In [27], Tarride et al. present a complete workflow designed for extracting information from registers. However, none of these works provide results on full-bundle segmentation or any method to make the detection consistent at the full-bundle level.

2 The JMDB Series of Notarial Record Manuscripts

The notarial record manuscripts considered in the present work belong to a massive collection of 16 849 "notarial protocol books" (here called bundles), each composed of 250 notarial deeds and about 800 pages on average. They are preserved in the Spanish Provincial Historical Archive of Cádiz (AHPC), which was created in 1931, with the main objective of collecting and guarding the said notarial documentation more than one hundred years old.

The JMBD Series of this collection encompasses the notarial records produced by Juan Manuel Briones Delgado, notary of the city of Cádiz between 1712 and 1726. Specifically, each AHPC bundle is organized into sequential sections, each corresponding to a deed with exception of a first section of about 50 pages containing a kind of table of contents of the bundle. This initial section was identified but not used in the present experiments.

For this series, the deeds are *page-aligned*, where each deed always begins on a new recto page and can contain from one to dozens of pages, some of which may be (almost) blank. The first and last pages of each deed are often easily identifiable visually because of slight layout differences from other pages, However, separating the deeds of each bundle is not straightforward because many regular pages may also look very similar to initial and/or final pages, which often leads to confusion. Figure 1 shows a typical deed which encompasses initial, final and two regular mid-page images.

Fig. 1. A four-page deed from the JMBB-4949 bundle of the AHPC.

Although all this may seem very specific to these manuscripts, tote that the massive size of the series does deserve development of ad-hoc methods to deal specifically with it. Moreover, there are innumerable notarial record series with similar organization of their bundles – in particular those where every deed starts on a new page. Nevertheless, it is also worth noting that there are other notarial record series, for which the deeds are *not* page-level aligned. Among others, we can mention *Chancery* (HIMANIS) [20], *Oficio de Hipotecas de Girona* (OHG) [24,26] and *The Cartulary of the Seigneury of Nesle* (Nesle) [13].[1] These types of manuscripts are not considered in the present work, even though some of the concepts and methods here introduced may be useful also to deal with these kinds of bundles.

3 Proposed Approach

The bundle segmentation problem is formalized in this section, along with the approaches we propose based on page-image classification and whole-bundle consistence consolidation by Dynamic Programming.

3.1 Problem Statement and Proposed Approach

As discussed in Sect. 2, all the deeds within a bundle start on a new page and also end with a full page, even if part of the page is unused. Initial and final deed pages will be referred to as "Initial" (I) and "Final" (F), respectively. All the pages within I and F will be referred to as "Mid" (M). In addition, a deed may also contain blank or otherwise useless pages, which will be referred to as *"junk"*. These pages may appear either at the end of a deed, or within it. Junk pages are almost trivial to detect and, for the sake of simplicity, we assume that these useless pages are eliminated from the bundle in a trivial previous step.

Let $B = D^1, \ldots, D^K$ be a bundle which sequentially encompasses K deeds.[2] Each deed D^k, in turn, is a sequence of $N(k) \geq 2$ non-junk page images, denoted as $D^k = G_1^k, \ldots, G_{N(k)}^k$. In the deed segmentation task, a bundle B is given just as a plain, unsegmented sequence of M page images $B = G_1, \ldots, G_M$ and the problem is to find $K+1$ boundaries b_k, $0 \leq k \leq K$, such that $b_0 = 0$, $b_{k-1} < b_k$, $b_K = M$, and B becomes described as a sequence of deeds $D^1, \ldots, D^K$, where $D^k = G_{b_{k-1}+1}, \ldots, G_{b_k}$ and $N(k) = b_k - b_{k-1} + 1$, $1 \leq k \leq K$.

Note that the ultimately goal of segmentation is to preserve the textual information of each bundle deed so that this information can be retrieved when/if needed. To this end, each deed segment, $D^k, 1 \leq k \leq K$, must fulfil the following *Consistency Constraints* (CC): $G_{b_{k-1}+1}$ is an I-page, G_{b_k} is an F-page and, if $N(K) > 2$, $G_{b_{k-1}+2}, \ldots, G_{b_k-1}$ are all M-pages. A Markov Chain representing these CCs is shown in Fig. 2.

---

[1] https://deeds.library.utoronto.ca/cartularies/0249.

[2] To avoid cumbersome equations, we will abuse the notation and index the elements of a sequence of sequences with a plain, rather than parenthesized superindex. That is, we will write D^k, rather than $D^{(k)}$.

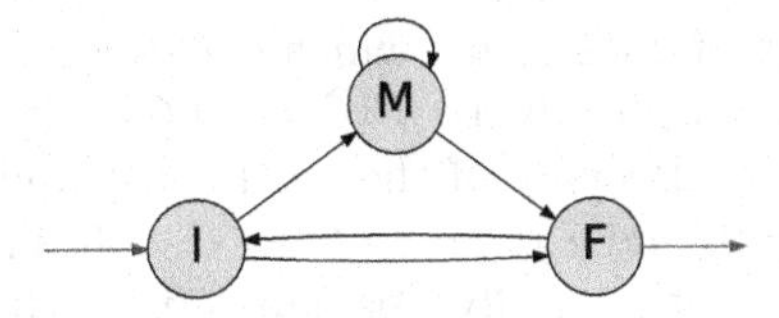

Fig. 2. Consistency Constraints Markov chain, where I, M, and F correspond to the initial, middle, and final page of a deed, respectively.

The following approach is proposed to solve this problem. First, using one or more segmented training bundles, a visual image classifier is trained to label each individual image as I, M or F. Then, for each image G_j of a test bundle $B = G_1, \ldots, G_M$, use the trained classifier to estimate the class-posteriors $P(c \mid G_j)$, $c \in \{\mathsf{I}, \mathsf{M}, \mathsf{F}\}$. We will refer to the sequence of these posteriors for $1 \leq j \leq M$ as the *posteriorgram*[3] of B, $\tilde{B} \stackrel{\text{def}}{=} \boldsymbol{g}_1, \ldots, \boldsymbol{g}_M$, where $g_{jc} = P(c \mid G_j)$, $c \in \{\mathsf{I}, \mathsf{M}, \mathsf{F}\}$.

Clearly $\tilde{B}$ is *not* a segmentation, but from $\tilde{B}$ a sequence of M class labels can be straightforwardly obtained as:

$$\hat{c}_j = \operatorname*{arg\,max}_{c \in \{\mathsf{I}, \mathsf{M}, \mathsf{F}\}} g_{jc}, \quad 1 \leq j \leq M \tag{1}$$

And, assuming this label sequence satisfies the CCs, a segmentation of B is readily obtained by setting the boundaries b_k to the posteriorgram positions labelled with F, that is: $b_0 = 0$; and for each $j : b_{k-1} < j \leq M$, $b_k = j$ iff $\hat{c}_j = \mathsf{F}$.

However, CCs are by no means guaranteed by Eq. (1). Therefore, the posteriorgram needs to be *"decoded"* into a *consistent sequence of class labels* with maximum probability, where, assuming the Markov property, the probability of a label sequence is the product of the probability of all its labels.

A simple *Greedy decoder* can be implemented by locally applying the CCs while computing the maximisation of Eq. (1):

$$\hat{c}_1 = \mathsf{I}; \quad \hat{c}_j = \operatorname*{arg\,max}_{c \in \rho(\hat{c}_{j-1})} g_{jc}, \; 2 \leq j \leq M-2; \quad \hat{c}_{M-1} = \pi(\hat{c}_{M-2}); \quad \hat{c}_M = \mathsf{F} \tag{2}$$

where the CCs (Fig. 2) are denoted by the function ρ, defined as: $\rho(\mathsf{I}) = \rho(\mathsf{M}) = \{\mathsf{M}, \mathsf{F}\}$; $\rho(\mathsf{F}) = \{\mathsf{I}\}$, and π is defined as $\pi(\mathsf{I}) = \pi(\mathsf{M}) = \mathsf{M}$; $\pi(\mathsf{F}) = \mathsf{I}$. By construction, this greedy decoder yields a consistent label sequence. But it does not guarantee that the probability of this sequence is maximum.

To achieve global optimization, a *Dynamic Programming decoder* is needed. This is in fact akin to a Viterbi procedure to obtain the best path through the Markov Chain of Fig. 2. Moreover, it allows to easily take into account the prior

[3] Following time-honored tradition in signal processing and automatic speech recognition, the term *posteriorgram* is used for this type of (variable-length) sequences of posterior probability vectors.

probabilities of the different constraints which correspond to probabilities in the edges of graph in Fig. 2. Specifically, let $P(C = c \mid C' = c')$ be the probability of the label c coming after c'. Because of the hard constraints, $P(C = \mathsf{I} \mid C' = \mathsf{M}) = P(C = \mathsf{M} \mid C' = \mathsf{F}) = 0$, and $P(C = c \mid C' = c) = 0$, iff $c \neq \mathsf{M}$. All the other probabilities can be straightforwardly estimated from Ground Truth reference segmented bundles.

Let $V(j, c)$ denote the probability of a consistent, max-probability labelling of the first j page images of B, with $V(1, \mathsf{I}) = 1$, $V(1, \mathsf{M}) = V(1, F) = 0$. Then the following recurrence relation holds for $2 \leq j \leq M$:

$$V(j, c) = \max_{c' \in \{\mathsf{I},\mathsf{M},\mathsf{F}\}} g_{jc} \, P(c \mid c') \, V(j-1, c'), \quad c \in \{\mathsf{I}, \mathsf{M}, \mathsf{F}\} \tag{3}$$

Once $V(M, \mathsf{F})$ is computed, backtracking yields a globally optimal consistent sequence of I,M,F labels.

In the segmentation of a bundle into deeds which is provided by this decoding process, each dead boundary is conditioned by the predictions from other parts of the bundle. As a result, the process yields a consistent, globally optimal sequence of $\hat{K}$ deeds $\hat{B} = \hat{D}^1, \ldots, \hat{D}_k, \ldots, \hat{D}^{\hat{K}}$, each deed $\hat{D}^k$ containing the corresponding $N(k)$ page images.

3.2 Individual Page Image Type Classification

To compute the posteriorgram $\boldsymbol{B} = \boldsymbol{g}_1, \ldots, \boldsymbol{g}_M$ introduced above, a page-image classifier is needed. This classifier should obtain the class posterior $P(c \mid G_j)$ for each page image $G_j, 1 \leq j \leq M$ and each $c \in \{\mathsf{I}, \mathsf{M}, \mathsf{F}\}$.

In the present work, we tried several classifiers, such as ResNet-{18,50,101} [12], ConvNeXt [16], and other transformer-based image classifiers, such as Swin [15]. All of them have been pre-trained on ImageNet from [29].

ResNet and ConvNeXt are both convolutional neural networks with a last, linear layer, which is used to classify into the $c \in \{\mathsf{I}, \mathsf{M}, \mathsf{F}\}$ classes. ResNet is a commonly used architecture for image classification tasks. It consists of multiple convolutional layers with residual connections between them. Residual connections enable the network to train much deeper architectures by mitigating the vanishing gradient problem. The specific architecture is determined by the number of ResNet blocks. As more blocks we set, the larger and deeper the network, and the greater number of parameters to be trained.

ConvNeXt, as the authors explain [16], is a "modernized" ResNet with the latest advances from training ConvNets and Vision Transformers. In the present work, we used ConvNeXt base. We also tried with ConvNeXt tiny but the results were quite worse.

We also tried to train Vision Transformers, such as Swin [15], but the model had convergence difficulties and results were rather poor. A possible cause is the difference in the resolution of images from the pre-training stage, where the original images were of 224×224 pixels, while we need to use an image size of 1024×1024. A larger size is required because, in our classification task, the clues are usually words-sized visual features or relatively small boxes in specific parts of the image and a higher resolution is needed to avoid losing this information.

4 Evaluation Measures

Raw classification error rates achieved by the different page-image classifiers using Eq. (1), without applying any consistency enforcing decoding, will be measured with the conventional classification error rate. In addition, to cope with the imbalanced number of samples per class in the datasets (as we show in Table 1) we will also provide a class-weighted classification error rate, which is essentially a macro-average of the error rate per class over the three classes I, M, F.

Most importantly, to assess the ultimate goal of the proposed approaches, we need to measure how well a whole bundle is segmented into its deeds. To this end, we propose a metric called Book Alignment Error Rate (BAER). It measures the amount of textual information lost beacouse of deed segmentation. Therefore, a page with little text should have less weight than a dense page. On the other hand, poorly recognized text should entail a penalty, even though misrecognition of a junk page should have a negligible impact.

Let $\hat{B} = \hat{D}^1, \dots, \hat{D}^{\hat{K}}$ be a sequence of $\hat{K}$ deeds, obtained as a hypothesis for the whole bundle as explained in Sect. 3, and let $B = D^1, \dots, D^K$ be the corresponding reference Ground Truth (GT). Following [9,20], we rely on Probabilistic Indexing [22] and Information Gain (IG) to compute a feature vector $\hat{\boldsymbol{D}} \in \mathbb{R}^n$ for each deed in $\hat{B}$, and similarly $\boldsymbol{D} \in \mathbb{R}^n$ for the reference B. The n components of these vectors correspond to the n words with higher IG, as determined in [20], and the value of each component is the number of occurrences of the corresponding word. Such a document representation provides compact information of the most relevant textual contents of the set of images that constitute a deed.

So, a system hypothesis deed-segmented bundle becomes represented as a sequence $\hat{\boldsymbol{D}} = \hat{\boldsymbol{D}}^1, \dots, \hat{\boldsymbol{D}}^{\hat{K}}$ and, similarly, the corresponding reference GT as $\boldsymbol{D} = \boldsymbol{D}^1, \dots, \boldsymbol{D}^K$. Using these vector sequences, we compute by dynamic programming the minimum number of edit operations to transform D into $\hat{D}$, according the following recurrence relation:

$$
\begin{aligned}
\mathrm{E(i,j)} \;=\; & \min(\mathrm{E}(i, j-1) + \mathrm{L}(\lambda, \boldsymbol{D}^j), \\
& \mathrm{E}(i-1, j-1) + \mathrm{L}(\hat{\boldsymbol{D}}^i, \boldsymbol{D}^j), \\
& \mathrm{E}(i-1, j) + \mathrm{L}(\hat{\boldsymbol{D}}^i, \lambda))
\end{aligned} \tag{4}
$$

where λ is an "empty vector" and $L(X, Y)$ is the bWER distance between X and Y as defined in [28].

The edit operations obtained by solving Eq. 4 are deed insertions, substitutions and deletions, corresponding in order to the three terms of the min function. An insertion indicates that a deed from the hypothesis does not appear in the reference. The cost of such an insertion, $L(\lambda, \boldsymbol{D}^j)$, is thus the number of running words that were not in D_j and were inserted. Similarly for deletion, where, for example, a deed $\hat{D}^i$ is misrecognized, with a cost $L(\hat{\boldsymbol{D}}^i, \lambda)$ equal to the number of running words in $\hat{D}^i$. In the case of substitution, the bWER distance $L(\hat{\boldsymbol{D}}^i, \boldsymbol{D}^j)$ is calculated between the two aligned deeds. Finally, the BAER is defined as:

$$\mathrm{BAER}(\hat{D}, D) = \frac{1}{\mathcal{W}} \mathrm{E}(\hat{K}, K) \tag{5}$$

where $\mathcal{W} = \sum_{j=0}^{K} ||\boldsymbol{D}^j||_1$ is the total number of running words in the bundle.

5 Data Set and Experimental Setup

The data set statistics are shown in this section, along with the empirical settings details needed to make the experiments reproducible.

5.1 Data Set

Among the JMDB Series of Notarial Record Manuscripts described in Sect. 2, 50 bundles were included in the collection compiled in the Carabela project [8].[4] From these bundles, in the present work we selected four: JMBD4946, JMBD4949, JMBD4950 and JMBD4952, dated between years 1722 and 1726, to be manually tagged with GT annotations.

In Table 1 the statistics for these bundles are shown. As we can see, each bundle has more than one thousand pages, except the JMBD4952, with 980 pages. The number of deeds is also shown, with more than two hundred per bundle, and almost three hundred in JMBD4950. An interesting fact that complicates the segmentation is the variation in the number of pages between deeds. In JMBD4946, for instance, we observe a variance of more than 14 pages per deed. In fact, one of the deeds spans up to 200 pages. This great page length variability poses a challenge for automated segmentation, since methods which may rely solely on page count or structure are not sufficiently effective. Additional strategies, such as the methods presented in the present work, need to be employed to accurately identify and separate individual deeds within a bundle.

For this study, we selected JMBD4949 and JMBD4950 for training and the other two bundles for testing. In fact, initially the only GT available was for JMBD4949 and JMBD4950. So, after training all models with these two bundles, we segmented the other two, JMBD4946 and JMBD4952, and requested experts to correct the few segmentation errors in order cost-effectively produce the GT needed to evaluate the results on the new bundles.

[4] In http://prhlt-carabela.prhlt.upv.es/carabela the images of this collection and a search interface based on Probabilistic Indexing are available.

Table 1. Number of page images and deeds for the bundles JMBD4946, JMBD4949, JMBD4950 and JMBD4952.

	JMBD4946	JMBD4949	JMBD4950	JMBD4952
N. Pages	1399	1615	1481	980
N. Deeds	248	295	260	236
Avg Pages per Deed	5.79	5.47	5.69	4.20
Min-max pages per Deed	2–200	2–122	2–62	2–38
St-dev pages per Deed	14.27	9.93	8.19	4.08

5.2 Empirical Settings

All of the image-classifiers explained in Sect. 3.2 are trained, at least, for 15 epochs and a maximum of 30, with an early stopping of 5 epochs concerning the evaluation loss. The evaluation was set equally for all the models, and randomly selected by using a 15% of the training set. The AdamW optimiser [17] was used with a learning rate of 0.001. Decay to the learning rate of 0.5 was applied after every 10 epochs. All of the images have been resized to 1024×1024.

6 Results

Table 2 shows the classification error rate and class-weighted classification error rate at the bundle level without consistence-enforcing decoding. ConvNeXt works better for the JMBD4946 bundle but ResNet50 provides better results for the JMBD4952. Even so, the classification error differences are small, and as mentioned, no decoding process is applied. So, only the maximum probability from the posteriorgram for a class c is used, and this does not reflect well the possible uncertainty that the model may have on an image sample. This is why now we will see that when consistency-enforcing decoding is applied, a model with lower raw classification error rate is not always better.

Table 2. Training with bundles JMBD4949 and JMBD4950 and testing on JMBD4945 (1399 pages) and JMBD4952 (980 pages), the table reports the classification errors without consistency-enforcing decoding.

Classifier	JMBD4946			JMBD4952		
	Err.	(%)	W.Err. (%)	Err.	(%)	W.Err. (%)
ResNet18	58	4.15	4.77	47	4.80	5.81
ResNet50	46	3.29	3.84	16	1.63	1.73
ResNet101	58	4.15	5.46	25	2.55	2.55
ConvNeXt	34	2.43	2.33	29	2.96	3.42

In Table 3 the alignment error at bundle level, BAER, is shown, along the total number of detected deeds (TDD columns), and the insertions (I), Deletions (D), and Substitutions (S) of deeds between the detected deeds and the GT. For each classifier we tried both decoders, previously explained in Sect. 3. As we might expect, the Viterbi decoder is always better than the greedy one, which is only used as a baseline result to compare with. In terms of the classifier, ResNet50 works better than others using the Viterbi decoder. In the bundle JMBD4946 a 0% BAER is achieved, which means that the system obtains a perfect segmentation. In the other bundle, JMBD4952, ResNet50 achieves a 1.11% of BAER, with only 2 insertions and 2 substitutions. Usually, an insertion (or a deletion) is accompanied by a substitution, since when we insert a deed we should cut the part of this deed from another deed that is misrecognized. We can observe that a classifier with a higher classification error, such as ResNet50, provides better segmentation results than another classifier, like ConvNeXt, that performs better in classification. This may be due to various reasons, but it is most likely that a model with enough trainable parameters but not as deep as ResNet101 or ConvNeXt, yields better-calibrated (i.e., less "extreme") probabilities which can then be more effectively used in the decoding process.

Finally, in Table 4 we report the results of the best classifier with both decodings, using the ResNet50 classifier, jointly for both test bundles. These results simulate an even larger bundle with a total of 484 deeds, distributed among 2379 pages. The resultant BAER is 0.44%, which is almost perfect, with only 2 insertions and 2 substitutions, coming from the bundle JMBD4952, since we know that the model can perfectly segment the other bundle.

Table 3. Training with bundles JMBD4949 and JMBD4950 and testing on JMBD4945 (248 deeds) and JMBD4952 (236 deeds). BAER is given in percentage and I, D and S are counts of deed Insertions, Deletions and Substitutions, respectively. Total detected deeds for each method are shown in the TDD columns.

Classifier	Decoder	JMBD4946					JMBD4952				
		BAER↓	TDD	I	D	S	BAER↓	TDD	I	D	S
ResNet18	Greedy	28.97	250	19	17	51	24.67	208	1	29	40
	Viterbi	7.91	252	5	1	6	3.75	238	4	2	8
ResNet50	Greedy	22.42	245	12	15	44	10.17	232	3	7	17
	Viterbi	0	248	0	0	0	1.11	238	2	0	2
ResNet101	Greedy	31.45	233	14	29	54	11.76	236	7	7	26
	Viterbi	10.65	246	3	5	10	4.35	234	2	4	11
ConvNeXt	Greedy	16.26	267	21	2	32	10.28	228	2	10	30
	Viterbi	3.20	249	3	2	12	5.60	229	0	7	14

Table 4. Training with bundles JMBD4949 and JMBD4950 and testing on JMBD4945 and JMBD4952, simultaneously. BAER is given in percentage and I, D and S are counts of deed Insertions, Deletions and Substitutions, respectively. The total of detected deeds is shown in the TDD (Total Detected Deeds) column. As reference, there are a total of 484 deeds between both bundles.

Classifier	Decoder	JMBD4946-4952				
		BAER↓	TDD	I	D	S
ResNet50	Greedy	17.57	477	15	22	61
	Viterbi	0.44	486	2	0	2

7 Conclusion

We have proposed a system that, for the first time, can segment a large historical manuscript bundle into several multi-page deeds. It is based on training an image classifier to compute a posteriorgram of page-classes along a full test bundle. This sequence of page-image class posteriors is decoded to attempt obtaining class-label sequence with maximum probability which fulfills the full-bundle consistency constraints required for a proper deed segmentation.

Different classifiers and decoders have been tried to demonstrate the robustness of the system. We also show that using the bundle's global context leads to significantly better segmentation decisions. In addition, we have introduced a new metric to measure the segmentation accuracy at the bundle level. It measures the loss of textual information caused by segmentation errors.

In future works, we intent to integrate deed segmentation and typology classification [9,19,21], taking into account both textual and visual information in a multimodal way [20].

Acknowledgements. Work partially supported by the research grants: the SimancasSearch project as Grant PID2020-116813RB-I00a funded by MCIN/AEI/10.13039/501100011033 and ValgrAI - Valencian Graduate School and Research Network of Artificial Intelligence and the Generalitat Valenciana, co-funded by the European Union. The second author's work was partially supported by the Universitat Politècnica de València under grant FPI-I/SP20190010. The third author's work is supported by a María Zambrano grant from the Spanish Ministerio de Universidades and the European Union NextGenerationEU/PRTR.

References

1. Andrés, J., Prieto, J.R., Granell, E., Romero, V., Sánchez, J.A., Vidal, E.: Information extraction from handwritten tables in historical documents. In: Uchida, S., Barney, E., Eglin, V. (eds.) DAS 2022. LNCS, vol. 13237, pp. 184–198. Springer, Cham (2022). https://doi.org/10.1007/978-3-031-06555-2_13
2. Biswas, S., Riba, P., Lladós, J., Pal, U.: Beyond document object detection: instance-level segmentation of complex layouts. Int. J. Doc. Anal. Recogn. (IJDAR) **24**(3), 269–281 (2021). https://doi.org/10.1007/s10032-021-00380-6

3. Boillet, M., Kermorvant, C., Paquet, T.: Multiple document datasets pre-training improves text line detection with deep neural networks. In: International Conference on Pattern Recognition (ICPR), pp. 2134–2141 (2020)
4. Boillet, M., Kermorvant, C., Paquet, T.: Robust text line detection in historical documents: learning and evaluation methods. Int. J. Doc. Anal. Recogn. **25**, 95–114 (2022)
5. Bosch, V., Toselli, A.H., Vidal, E.: Statistical text line analysis in handwritten documents. In: 2012 International Conference on Frontiers in Handwriting Recognition, pp. 201–206. IEEE (2012)
6. Campos, V.B.: Advances in document layout analysis. Ph.D. thesis, Universitat Politècnica de València (2020)
7. Coquenet, D., Chatelain, C., Paquet, T.: DAN: a segmentation-free document attention network for handwritten document recognition. IEEE Trans. Pattern Anal. Mach. Intell. 1–17 (2023)
8. Vidal, E., et al.: The Carabela project and manuscript collection: large-scale probabilistic indexing and content-based classification. In: 16th ICFHR (2020)
9. Flores, J.J., Prieto, J.R., Garrido, D., Alonso, C., Vidal, E.: Classification of untranscribed handwritten notarial documents by textual contents. In: Pinho, A.J., Georgieva, P., Teixeira, L.F., Sánchez, J.A. (eds.) IbPRIA 2022. LNCS, vol. 13256, pp. 14–26. Springer, Cham (2022). https://doi.org/10.1007/978-3-031-04881-4_2
10. Grüning, T., Leifert, G., Strauß, T., Michael, J., Labahn, R.: A two-stage method for text line detection in historical documents. Int. J. Doc. Anal. Recogn. (IJDAR) **22**(3), 285–302 (2019). https://doi.org/10.1007/s10032-019-00332-1
11. He, K., Gkioxari, G., Dollar, P., Girshick, R.: Mask R-CNN, vol. 2017-October, pp. 2980–2988. Institute of Electrical and Electronics Engineers Inc. (2017)
12. He, K., Zhang, X., Ren, S., Sun, J.: Deep residual learning for image recognition (2015)
13. Hélary, X.: Le cartulaire de la seigneurie de nesle [chantilly, 14 f 22] (2006)
14. Kim, G., et al.: OCR-free document understanding transformer (2022)
15. Liu, Z., et al.: Swin transformer: hierarchical vision transformer using shifted windows. In: Proceedings of the IEEE/CVF ICCV (2021)
16. Liu, Z., Mao, H., Wu, C.Y., Feichtenhofer, C., Darrell, T., Xie, S.: A convnet for the 2020s. In: Proceedings of the IEEE/CVF CVPR (2022)
17. Loshchilov, I., Hutter, F.: Decoupled weight decay regularization (2017)
18. Oliveira, S.A., Seguin, B., Kaplan, F.: dhSegment: a generic deep-learning approach for document segmentation, vol. 2018-August, pp. 7–12. IEEE (2018)
19. Prieto, J.R., Bosch, V., Vidal, E., Alonso, C., Orcero, M.C., Marquez, L.: Textual-content-based classification of bundles of untranscribed manuscript images. In: 2020 25th International Conference on Pattern Recognition (ICPR), pp. 3162–3169. IEEE (2021)
20. Prieto, J.R., Bosch, V., Vidal, E., Stutzmann, D., Hamel, S.: Text content based layout analysis. In: 2020 17th International Conference on Frontiers in Handwriting Recognition (ICFHR), pp. 258–263. IEEE (2020)
21. Prieto, J.R., Flores, J.J., Vidal, E., Toselli, A.H., Garrido, D., Alonso, C.: Open set classification of untranscribed handwritten documents. arXiv preprint arXiv:2206.13342 (2022)
22. Puigcerver, J.: A probabilistic formulation of keyword spotting. Ph.D. thesis, Univ. Politècnica de València (2018)
23. Quirós, L.: Layout analysis for handwritten documents. A probabilistic machine learning approach. Ph.D. thesis, Universitat Politècnica de València (2022)

24. Quirós, L., Bosch, V., Serrano, L., Toselli, A.H., Vidal, E.: From HMMs to RNNs: computer-assisted transcription of a handwritten notarial records collection. In: 2018 16th International Conference on Frontiers in Handwriting Recognition (ICFHR), pp. 116–121. IEEE (2018)
25. Quirós, L., Toselli, A.H., Vidal, E.: Multi-task layout analysis of handwritten musical scores. In: Morales, A., Fierrez, J., Sánchez, J.S., Ribeiro, B. (eds.) IbPRIA 2019. LNCS, vol. 11868, pp. 123–134. Springer, Cham (2019). https://doi.org/10.1007/978-3-030-31321-0_11
26. Quirós, L., et al.: Oficio de Hipotecas de Girona. A dataset of Spanish notarial deeds (18th Century) for Handwritten Text Recognition and Layout Analysis of Historical Documents (2018)
27. Tarride, S., Maarand, M., Boillet, M., et al.: Large-scale genealogical information extraction from handwritten Quebec parish records. Int. J. Doc. Anal. Recogn. (2023)
28. Toselli, A.H., Vidal, E.: Revisiting bag-of-word metrics to assess end-to-end text image recognition results. Preprint (2023)
29. Wolf, T., et al.: Transformers: state-of-the-art natural language processing. In: Proceedings of the 2020 Conference on Empirical Methods in Natural Language Processing: System Demonstrations, pp. 38–45. Association for Computational Linguistics (2020)

A Study of Augmentation Methods for Handwritten Stenography Recognition

Raphaela Heil(✉) and Eva Breznik

Centre for Image Analysis, Department of Information Technology, Uppsala University, Uppsala, Sweden
{raphaela.heil,eva.breznik}@it.uu.se

Abstract. One of the factors limiting the performance of handwritten text recognition (HTR) for stenography is the small amount of annotated training data. To alleviate the problem of data scarcity, modern HTR methods often employ data augmentation. However, due to specifics of the stenographic script, such settings may not be directly applicable for stenography recognition. In this work, we study 22 classical augmentation techniques, most of which are commonly used for HTR of other scripts, such as Latin handwriting. Through extensive experiments, we identify a group of augmentations, including for example contained ranges of random rotation, shifts and scaling, that are beneficial to the use case of stenography recognition. Furthermore, a number of augmentation approaches, leading to a decrease in recognition performance, are identified. Our results are supported by statistical hypothesis testing. A link to the source code is provided in the paper.

Keywords: Stenography · Handwritten text recognition · CNNs · Augmentation study

1 Introduction

Deep learning-based approaches form the majority of the state-of-the-art methods for handwritten text recognition (HTR) at the time of writing. While these approaches have been shown to reach high levels of recognition performance, they typically require large amounts of annotated data during training. This can for example pose a challenge to HTR for historic manuscripts. Here the acquisition of annotated data can be costly and time-consuming, as it often requires trained professionals, such as historians or palaeographers. Data availability may further be limited by the use of a rarely-used script or language, as is for example the case for the Khmer language [26].

When the acquisition of additional training data is not feasible, or possible, a commonly used approach for artificially increasing the dataset size is to use data

Supplementary Information The online version contains supplementary material available at https://doi.org/10.1007/978-3-031-36616-1_11.

A. Pertusa et al. (Eds.): IbPRIA 2023, LNCS 14062, pp. 134–145, 2023.
https://doi.org/10.1007/978-3-031-36616-1_11

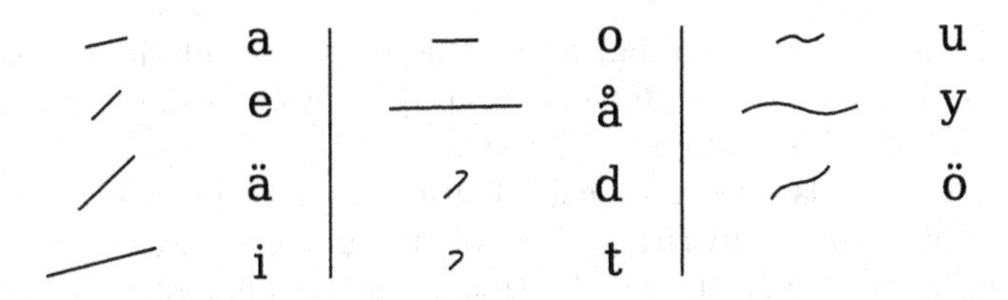

Fig. 1. Selected examples of symbols and respective character transliterations from the Swedish "Melin" stenography system.

augmentation [13,23]. Here, slight alterations, for example rotation and scaling, are applied to the images in order to increase the visual variety.

In the particular case of stenography, the acquisition of more data is limited both by the special skill required to transliterate the writing, as well as the limited use of the script, making data augmentation options especially interesting. In contrast to scripts like Latin, stenography (also called shorthand) typically uses short strokes to represent characters, n-grams or even whole words. As can be seen in the excerpt of the Swedish stenography alphabet in Fig. 1, features like rotation and scale play a considerable role in differentiating between certain symbols. This raises the question whether typical HTR augmentation techniques, which often include changes in rotation and scale, can also be applied to stenography or whether they may cause certain symbols to be interpreted as others (e.g. "a" as "e" and vice-versa).

In this work, we aim to address this question by studying the applicability of a selection of commonly used HTR augmentation techniques for the case of handwritten stenography recognition. We experiment with a deep learning architecture that has performed best in prior investigations for Swedish stenography, and observe how different augmentation techniques affect the text recognition performance on the public LION dataset (cf. [17]), consisting of stenographic manuscripts, written by the Swedish children's book author Astrid Lindgren.

2 Related Work

2.1 Handwritten Stenography Recognition

At the time of writing, very little research has been conducted on deep learning-based handwritten stenography recognition (HSR). In [16,18], convolutional neural networks (CNNs) are used to classify a selection of words written in Pitman's, respectively Gregg's, stenography system. Both types of stenography are primarily used in English-language contexts.

Zhai et al. [32] propose to use a CNN-based feature extractor, followed by a recurrent neural network to generate transliterations of individual words, written in Gregg's stenography.

2.2 Data Augmentation Methods for Handwritten Text Recognition

A variety of augmentation approaches for HTR has been proposed in the literature. Some of the most commonly used augmentations include rotations, translations, shearing and scaling, as well as greyscale dilations and erosions [20,24,30]. Similarly to these, Retsinas et al. [21] propose the use of rotations, skewing and additionally applying Gaussian noise.

Wick et al. [27] also use greyscale dilation and erosion and further simulate the variability in handwriting by applying grid-like distortions, originally proposed by Wigington et al. [28]. Krishnan and Jawahar [12] follow a similar approach, applying translation, scaling, rotation and shearing and combine these with the elastic transformations, proposed by Simard et al. [23].

All of the aforementioned works apply each augmentation with an independent probability (often 0.5) to each image in the training set. Furthermore, the augmentation parameters, for example the angle of rotation, are sampled per image from a user-defined range, yielding a wide variety of altered datasets.

Wilkinson et al. [31] use the same augmentation approach and parameters as [30] but arrange the perturbed images into lines and whole pages, thereby creating artificial data for segmentation-free word spotting.

Zhai et al. [32] apply similar transforms to all of the above but do not sample the respective parameters. Instead, they use fixed, pre-defined values and generate a total of eight images (with fixed choice of augmentation parameters) for every input.

Instead of perturbing the available data, a different line of augmentation techniques is centred around creating entirely synthetic datasets. Circumventing the use of the data at hand entirely, Krishnan et al. [11] generate words by using publicly available handwriting fonts. This approach is currently not applicable in the case of stenography, as such a library of handwritten symbols and fonts is not available for these scripts, to the best of our knowledge.

While Alonso et al. [1], Kang et al. [10] and Mattick at al. [14] also propose to generate synthetic data, they employ generative adversarial networks to synthesise words in the style of a given dataset or sample image. To the best of our knowledge, approaches in this line of research currently employ word-level annotations (segmentation and transliteration), which are not available for the dataset in our study.

3 Study Design

In order to investigate the effect of different augmentations on the text recognition performance, we evaluate a variety of image transformations that are applied during training. We compare the results with those obtained from a baseline model, trained on the original data, without any augmentations.

3.1 Examined Augmentations

Below, the examined augmentations and the respective parameter configurations are briefly presented. Each experiment configuration is denoted with a name by which it will be referred to in the remainder of this paper. A summary of names, augmentation types and parameters is shown in Table 1. Additionally, the

Table 1. Summary of augmentation configurations and the names by which they are referred to in this paper.

Name	Augmentation Type	Parameters
baseline	none	N/A
rot1.5	Random Rotation	$[-1.5, 1.5]$ degrees
rot5	Random Rotation	$[5, 5]$ degrees
rot10	Random Rotation	$[10, 10]$ degrees
positive	Random Rotation	$[0, 1.5]$ degrees
negative	Random Rotation	$[-1.5, 0]$ degrees
rot+2	Fixed Rotation	$2°$
rot−2	Fixed Rotation	$-2°$
square-dilation	Random Dilation	square SE, [1..4] px
disk-dilation	Random Dilation	disk SE, [1..4] px
square-erosion	Random Erosion	square SE, [1..3] px
disk-erosion	Random Erosion	disk SE, [1..3] px
shift	Random Shift	$horiz. = [0, 15]$, $vert. = [-3.5, 3.5]$
elastic	Random Elastic Transform. [23]	$\alpha = [16, 20]$, $\sigma = [5, 7]$
shear	Random Horizontal Shearing	$[-5, 30]$ degrees
shear30	Random Horizontal Shearing	$[-30, 30]$ degrees
scale75	Random Scaling	$[0.75, 1]$
scale95	Random Scaling	$[0.95, 1]$
mask10	Random Column Masking	10% of columns
mask40	Random Column Masking	40% of columns
noise	Gaussian Noise	$\sigma = \{0.08, 0.12, 0.18\}$
dropout	Pixel Dropout	[0, 20]% of pixels
blur	Gaussian Blur	kernel $= 5$, $\sigma = [0.1, 2]$

connected supplementary [8] contains a visualisation of the impact of selected augmentation parameters on the original line image.

The implementations for most of the examined augmentations were provided by [9,15,25]. All others are available in our project repository (cf. Sect. 5 - Source Code).

Baseline. In order to establish a baseline performance, the model is trained on the original data, not using any augmentations.

Rotations. We examine a variety of rotations, largely based on configurations used in related text recognition works. Three models are trained with random rotations in the ranges $[-1.5, 1.5]$ ("rot1.5") [24], $[-5, 5]$ ("rot5") [30] and $[-10, 10]$ ("rot10"). Following [32], one model each is trained with a fixed rotation of $+2$, respectively -2, degrees ("rot+2", "rot-2"). Lastly, one model each is trained with only positive, respectively negative rotations, in the ranges $[0, 1.5]$ ("positive") and $[-1.5, 0]$ ("negative").

Morphological Operations. With respect to morphological operations, we consider greyscale dilations and erosions with a square structuring element (SE) [20,24,30]. Additionally, we evaluate the use of a disk SE, because the primary writing implement in the LION dataset is a pencil, which typically features a round footprint. We refer to these augmentations as "square-dilation", "disk-dilation", "square-erosion" and "disk-erosion", respectively. For each application of these augmentations (i.e. per image) the size of the SE is sampled from [1..3] for erosions and [1..4] for dilations. It should be noted that in the case of a square SE, the size refers to the width, whereas it indicates the radius for disks.

Geometric Augmentations. In addition to the aforementioned rotations, we consider a number of other geometric augmentations.

Concretely, we evaluate shearing with angles in the range of $[-5, 30]$ ("shear", following [30]) and $[-30, 30]$ ("shear30"). Furthermore, we examine downscaling with factors in the range of $[0.95, 1]$ ("scale95", [24]) and $[0.75, 1]$ ("scale75"). The rescaled image is zero-padded at the top and bottom to the original size for further processing.

Following Krishnan and Jawahar [12] we investigate the applicability of the elastic transformation ("elastic"), proposed by [23]. In this work, we sample α from the range $[16, 20]$ and σ from $[5, 7]$.

Lastly, we consider random shifts ("shift") in horizontal and vertical direction, within the pixel ranges of $[0, 15]$, respectively $[-3.5, 3.5]$. For both directions, non-integer shifts may occur.

Intensity Augmentations. Finally, we consider a selection of augmentations that affect an image's pixel intensities. We examine both pixel dropout ("dropout"), i.e. zero-masking of random pixels, with rates in the range of [0, 20]% of the image, and random column masking with fixed rates of 10% ("mask10") and 40% ("mask40") of columns. Additionally, we consider the application of Gaussian noise ("noise"), with $\sigma = \{0.08, 0.12, 0.18\}$, and Gaussian blurring ("blur") with a kernel size of 5 and $\sigma = [0.1, 2]$.

3.2 Dataset

This study is centred around the LION dataset, which is based on stenographed manuscripts by Swedish children's book author Astrid Lindgren (1907–2002) (cf. [17]). It consists of 198 pages, written in Swedish, using the *Melin* shorthand system. For each page, line-level bounding boxes and transliterations are available, the latter of which were provided by stenographers through expert crowdsourcing [2]. Overall, the dataset has an alphabet size of 51, entailing the 26 lower-case

Latin characters (a–z), the Swedish vowels "äöå", digits 0–9, and a selection of punctuation marks and quotes appearing in the text.

We employ a five-fold cross-validation setup, resulting in 306 lines per fold. This portion of the dataset contains texts from *The Brothers Lionheart* (Chaps. 1–3, 5, 6; original title: Bröderna Lejonhjärta). The disjoint test split consists of 474 in-domain lines, covering chapter four of *The Brothers Lionheart*, and 191 out-of-domain lines, containing portions from other literary works by Lindgren. All lines in the dataset are reduced to the value channel of the HSV space and the contrast is stretched. For concrete implementation details, the interested reader is referred to our implementation (cf. Sect. 5). Figure 2 shows an example line from the dataset in its original form (top) and the preprocessed version (bottom).

3.3 Model Architecture

We focus our study on the deep neural network architecture proposed by Neto et al., who demonstrate that this model performs well in limited-resource line-based HTR settings [24]. This architecture outperformed other connectionist temporal classification (CTC) [6,7] architectures on the LION dataset in preliminary experiments. The model consists of a convolutional block, mixing regular and gated convolutions [5], for feature extraction, followed by a recurrent block, employing bi-directional gated recurrent units [4]. The output sequence is obtained via best-path decoding.

Further details regarding the model architecture can be found in our PyTorch [19] implementation in the accompanying code (cf. Sect. 5 - Source Code).

3.4 Experimental Settings

All of the experiments below follow the same general procedure and only differ in regard to the augmentation that is applied to the data during training. In each epoch, each image is augmented at a rate of 50%. Afterwards, all images are scaled to a fixed height of 64 pixels and padded with the background colour (black) to a width of 1362 pixels. Target sequences are padded with the token indicating blanks in the CTC-loss (here zero), to a length of 271.

The detailed training protocol is included for reference in the supplementary material [8].

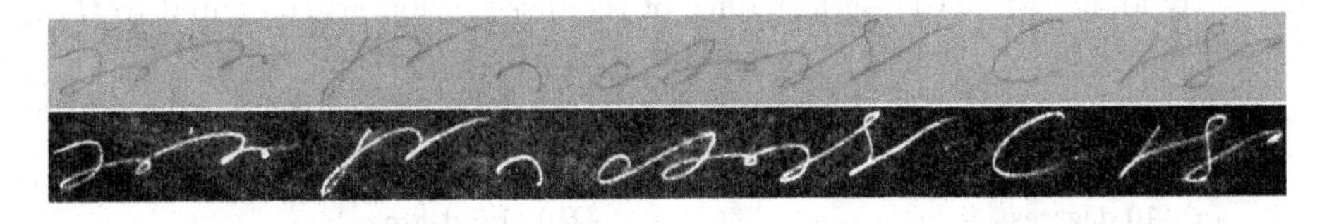

Fig. 2. Example image from the LION dataset. Top: original line image, bottom: preprocessed line image. The transliteration reads: "jonatan hette inte lejonhjärta från början" (English: *jonatan was not called lionheart from the beginning*).

Each augmentation configuration is trained 30 separate times for each of the five cross-validation folds, yielding a total of 150 model checkpoints (i.e. sets of weights). All checkpoints are evaluated on the separate test set by measuring the character error rate (CER) and word error rate (WER) for the transliterations, obtained via best path decoding [6,7] on the test set. We do not consider advanced decoding techniques, such as word beam search [22] or language models, in order to rule out any effects that these may have on the obtained transliterations. CER and WER are defined as:

$$\frac{S+D+I}{N} \tag{1}$$

where S is the number of character (word) substitutions, D the number of character (word) deletions and I the number of character (word) insertions, that are required to convert a given sequence to a reference sequence. N denotes the number of characters (words) in the reference string.

Final statistical testing for significant differences is performed by a Wilcoxon paired signed-rank test [29] on the test set results averaged over repetitions, with a Bonferroni correction [3] for multiplicity.

4 Results and Discussion

The mean error rates for the 22 different augmentations, as well as for the augmentation-free baseline, are shown in Table 2. As can be seen, the augmentations *rot10*, *square-erosion*, *disk-erosion*, *square-dilation*, *disk-dilation*, *mask40* and *blur* all result in significant decreases in performance.

For *rot10*, this result is likely attributable to the extreme rescaling of the line contents that occurs when height-normalising the rotated image. In order to avoid cutting off parts of a rotated line, the image size is expanded accordingly and padded with black. Larger rotations require a larger target image to accommodate the whole line, resulting in a smaller text size, as compared to smaller rotations, when scaled to the same height. Figure 3 shows an example of the impact on the image content, when rotating the original line image to by 10, respectively 1.5°.

Regarding the increase in error rates for both kinds of erosions, it can be observed that these augmentations thin the original strokes considerably. We argue that the selected erosions thin the strokes too much, leaving too little information for the recognition behind. This argument is supported by the significantly lower ($p < 0.01$) performance of the disk-erosion, as compared to the

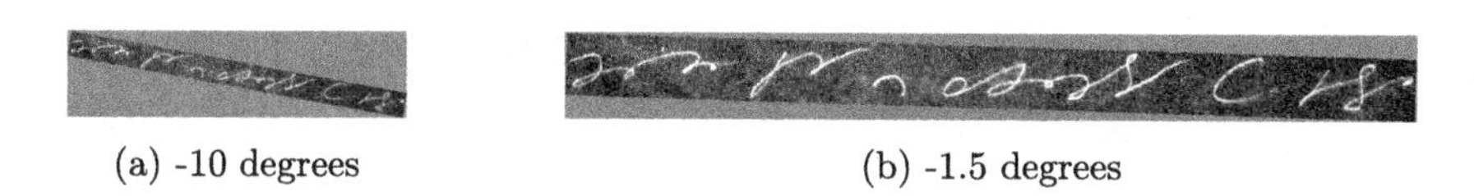

(a) -10 degrees (b) -1.5 degrees

Fig. 3. Examples for the impact of different rotations on the original image content size. Padding value set to grey to emphasise the required padding.

square one. As indicated earlier (cf. Sect. 3.1), the disk SEs are larger than their square counterparts, thus thinning the strokes more strongly and leaving even less information behind. It should be noted here, that this is not an issue that can be directly attributed to stenography but rather to the use of a relatively thin pencil, which results in thin strokes that are more susceptible to erosions.

Considering the two dilation-based augmentations, it can firstly be noticed that, in contrast to the WER, no significant difference for the CER can be determined in the case of the *square-dilation.* This may be attributable to large dilations closing the gap between words. Such word boundary errors will only marginally affect the CER as they require a single character substitution or insertion to establish the gap. They will however lead to considerable increases in the WER, as one word substitution and one word insertion will be required to recover the recognition mistake. This may also be a contributing factor for the *disk-dilation.* In this regard it can however also be observed that the disk SE, whose footprint is larger than the square's, leads to the loss of details, regarding small, and long and thin, loops, which are frequently used in Melin's stenography system. This loss of definition, an example of which is shown in Fig. 4, may lead to incorrect transliterations.

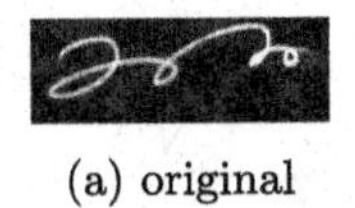

(a) original

(b) disk-dilation, radius = 4px

Fig. 4. Example demonstrating the filling of several loops due to a dilation with a large structuring element.

Since the performance of *mask10* is on par with the baseline, a probable explanation for the decrease in performance when using *mask40* is that the level of masking is simply too high, again removing too much of the information and thus affecting the recognition performance. The decrease in performance for the *blur* augmentation may be attributable to the loss of definition between smaller symbols, like "a" and "o", and "u" and "ö".

For six of the augmentations, namely rot5, rot+2, rot-2, mask10, noise and dropout, no differences can be observed with respect to the baseline. While including these may not directly harm the performance, they also do not appear to contribute to the learning and are therefore not of immediate interest to the concrete recognition task.

Finally, ten of the examined augmentations yield significant decreases in error rates, as compared to the baseline. Considering the improvements obtained when applying conservative rotations, in the overall range of $[-1.5, 1.5]$ (*rot1.5*, *positive*, *negative*), we cannot find any indications supporting the initial concern that rotations in general may negatively affect the transliteration of certain characters. Obviously, the angles between the similar symbols, for example "a" and "e", tend to be much larger than the ones examined in these experiments. However, as discussed for the case of rot10, even rotations closer to the actual angle between similar characters are likely to lead to scaling related issues, outweighing any potential recognition errors caused directly by rotated symbols.

Regarding the decrease in error rates for the remaining six augmentations, *shift*, *shear*, *shear30*, *elastic*, *scale75* and *scale95* it can be noted that all of these correspond to variations that, to some degree, naturally occur in the dataset. Applying these augmentations therefore constitutes plausible transformations within the dataset domain and increases the visual variety of the training data.

Given the reduction in error rates when applying several of the augmentations individually, the follow-up question arose whether a combination of some of these would have a similar positive effect. We therefore selected the three best performing augmentations, *rot1.5*, *shift* and *scale75*, and repeated the experiment with a combination of these. Here, each augmentation in the sequence

Table 2. Mean (and std. dev.) CER and WER per augmentation type, in percent. For both metrics, lower values are better. Comparison columns indicate if the mean for the inspected augmentation type is significantly higher (>, red), lower (<, blue), or no different (–) than the baseline mean, according to the Wilcoxon paired signed-rank test ($p < 0.01$, after applying a Bonferroni correction with $n = 22$).

Augmentation	CER (Std.dev.)	CER Comparison	WER (std.dev.)	WER Comparison
baseline	31.74 (1.23)	N/A	57.15 (1.45)	N/A
rot1.5	30.48 (2.98)	<	55.61 (3.34)	<
rot5	31.67 (3.51)	–	56.74 (3.83)	–
rot+2	31.14 (2.03)	–	56.16 (2.45)	<
rot-2	31.34 (3.14)	–	56.40 (3.67)	–
positive	30.66 (2.36)	<	55.84 (2.78)	<
negative	30.62 (2.25)	<	55.84 (2.73)	<
rot10	35.85 (7.53)	>	61.13 (6.59)	>
square-erosion	31.89 (1.56)	>	57.40 (1.87)	>
disk-erosion	33.52 (5.68)	>	58.99 (3.90)	>
square-dilation	31.82 (1.35)	–	57.39 (1.73)	>
disk-dilation	32.03 (1.64)	>	57.64 (2.09)	>
shift	30.65 (1.26)	<	55.83 (1.62)	<
shear	31.03 (1.31)	<	56.49 (1.73)	<
shear30	31.02 (1.28)	<	56.41 (1.57)	<
elastic	31.39 (1.28)	<	56.75 (1.65)	<
scale75	30.40 (1.90)	<	55.57 (2.35)	<
scale95	30.77 (1.19)	<	56.14 (1.56)	<
mask10	31.82 (1.47)	–	57.32 (1.92)	–
mask40	33.70 (7.97)	>	58.75 (5.40)	>
noise	31.72 (1.56)	–	57.15 (1.87)	–
dropout	31.72 (1.27)	–	57.27 (1.59)	–
blur	31.95 (1.53)	>	57.58 (1.96)	>

is applied with an independent probability of 0.5. This configuration achieves significantly lower error rates than the baseline, with a mean CER of **30.90%** (2.08) and WER of **56.03%** (2.53).

5 Conclusions

In this work we have examined 22 different augmentation configurations and their impact on handwritten stenography recognition. Based on the obtained results, we conclude that small rotations, shifting, shearing, elastic transformations and scaling are suitable augmentations for the recognition task and dataset at hand. Furthermore, an increase in performance could be observed for a combination of the top three augmentations, i.e. rotations in the range of $[-1.5, 1.5]$ degrees, horizontal and vertical shifting by $[0, 15]$ px, respectively $[-3.5, 3.5]$ px, and scaling by a factor in the range of $[0.75, 1]$.

A decrease in performance can be observed for larger rotations (± 10 degrees), which is attributable to the extreme rescaling resulting from the padding of augmented images. Besides this, the results obtained for erosions and dilations indicate that the examined configurations have adverse effects on the recognition performance. This can largely be attributed to the writing implement that was used in the dataset, highlighting that aspects like the stroke width should be taken into account when choosing augmentations for any kind of text recognition system.

Overall, initial concerns that commonly-used HTR augmentations may lead to confusions between certain symbols in the stenographic alphabet could not be confirmed.

Interesting avenues for future research include the creation of text lines by combining segmented words, for example as proposed by Wilkinson and Brun [31], as well as the investigation of text generation approaches similar to the ones proposed by Alonso et al. [1], Kang et al. [10] and Mattick at al. [14].

Source Code. The source code is available via Zenodo: https://doi.org/10.5281/zenodo.7905299.

Acknowledgements. This work is partially supported by Riksbankens Jubileumsfond (RJ) (Dnr P19-0103:1). The computations were enabled by resources provided by the National Academic Infrastructure for Supercomputing in Sweden (NAISS) at Chalmers Centre for Computational Science and Engineering (C3SE) partially funded by the Swedish Research Council through grant agreement no. 2022-06725. Author E.B. is partially funded by the Centre for Interdisciplinary Mathematics, Uppsala University, Sweden.

References

1. Alonso, E., Moysset, B., Messina, R.: Adversarial generation of handwritten text images conditioned on sequences. In: 2019 International Conference on Document Analysis and Recognition (ICDAR), pp. 481–486 (2019)

2. Andersdotter, K., Nauwerck, M.: Secretaries at work: accessing Astrid Lindgren's stenographed manuscripts through expert crowdsourcing. In: Berglund, K., Mela, M.L., Zwart, I. (eds.) Proceedings of the 6th Digital Humanities in the Nordic and Baltic Countries Conference (DHNB 2022), Uppsala, Sweden, 15–18 March 2022, vol. 3232, pp. 9–22 (2022)
3. Bonferroni, C.: Teoria statistica delle classi e calcolo delle probabilita. Pubblicazioni del R Istituto Superiore di Scienze Economiche e Commericiali di Firenze **8**, 3–62 (1936)
4. Cho, K., van Merriënboer, B., Bahdanau, D., Bengio, Y.: On the properties of neural machine translation: encoder-decoder approaches. In: Proceedings of SSST-8, Eighth Workshop on Syntax, Semantics and Structure in Statistical Translation, pp. 103–111 (2014)
5. Dauphin, Y.N., Fan, A., Auli, M., Grangier, D.: Language modeling with gated convolutional networks. In: Precup, D., Teh, Y.W. (eds.) Proceedings of the 34th International Conference on Machine Learning. Proceedings of Machine Learning Research, vol. 70, pp. 933–941. PMLR (2017)
6. Graves, A., Fernández, S., Gomez, F., Schmidhuber, J.: Connectionist temporal classification: labelling unsegmented sequence data with recurrent neural networks. In: Proceedings of the 23rd International Conference on Machine Learning (ICML 2006), pp. 369–376. ACM, New York, NY, USA (2006)
7. Graves, A., Liwicki, M., Fernández, S., Bertolami, R., Bunke, H., Schmidhuber, J.: A novel connectionist system for unconstrained handwriting recognition. IEEE Trans. Pattern Anal. Mach. Intell. **31**(5), 855–868 (2009)
8. Heil, R., Breznik, E.: Supplementary material (2023). https://arxiv.org/abs/2303.02761
9. Jung, A.B., et al.: imgaug (2020). https://github.com/aleju/imgaug. Accessed 01 Feb 2020
10. Kang, L., Riba, P., Wang, Y., Rusiñol, M., Fornés, A., Villegas, M.: GANwriting: content-conditioned generation of styled handwritten word images. In: Vedaldi, A., Bischof, H., Brox, T., Frahm, J.-M. (eds.) ECCV 2020. LNCS, vol. 12368, pp. 273–289. Springer, Cham (2020). https://doi.org/10.1007/978-3-030-58592-1_17
11. Krishnan, P., Jawahar, C.V.: Matching handwritten document images. In: Leibe, B., Matas, J., Sebe, N., Welling, M. (eds.) ECCV 2016. LNCS, vol. 9905, pp. 766–782. Springer, Cham (2016). https://doi.org/10.1007/978-3-319-46448-0_46
12. Krishnan, P., Jawahar, C.V.: HWNet v2: an efficient word image representation for handwritten documents. Int. J. Doc. Anal. Recognit. **22**(4), 387–405 (2019)
13. Krizhevsky, A., Sutskever, I., Hinton, G.E.: ImageNet classification with deep convolutional neural networks. In: Pereira, F., Burges, C., Bottou, L., Weinberger, K. (eds.) Advances in Neural Information Processing Systems, vol. 25. Curran Associates, Inc. (2012)
14. Mattick, A., Mayr, M., Seuret, M., Maier, A., Christlein, V.: SmartPatch: improving handwritten word imitation with patch discriminators. In: Lladós, J., Lopresti, D., Uchida, S. (eds.) ICDAR 2021. LNCS, vol. 12821, pp. 268–283. Springer, Cham (2021). https://doi.org/10.1007/978-3-030-86549-8_18
15. Michaelis, C., et al.: Benchmarking robustness in object detection: autonomous driving when winter is coming. arXiv preprint arXiv:1907.07484 (2019)
16. Montalbo, F.J.P., Barfeh, D.P.Y.: Classification of stenography using convolutional neural networks and canny edge detection algorithm. In: 2019 International Conference on Computational Intelligence and Knowledge Economy (ICCIKE), pp. 305–310 (2019)

17. Nauwerck, M.: About the Astrid Lindgren code (2020). https://www.barnboksinstitutet.se/en/forskning/astrid-lindgren-koden/. Accessed 05 May 2023
18. Padilla, D.A., Vitug, N.K.U., Marquez, J.B.S.: Deep learning approach in Gregg shorthand word to English-word conversion. In: 2020 IEEE 5th International Conference on Image, Vision and Computing (ICIVC), pp. 204–210 (2020)
19. Paszke, A., et al.: PyTorch: an imperative style, high-performance deep learning library. In: Wallach, H., Larochelle, H., Beygelzimer, A., d'Alché Buc, F., Fox, E., Garnett, R. (eds.) Advances in Neural Information Processing Systems, vol. 32, pp. 8024–8035. Curran Associates, Inc. (2019)
20. Puigcerver, J.: Are multidimensional recurrent layers really necessary for handwritten text recognition? In: 2017 14th IAPR International Conference on Document Analysis and Recognition (ICDAR), vol. 01, pp. 67–72 (2017)
21. Retsinas, G., Sfikas, G., Gatos, B., Nikou, C.: Best practices for a handwritten text recognition system. In: Uchida, S., Barney, E., Eglin, V. (eds.) Document Analysis Systems, pp. 247–259. Springer, Cham (2022). https://doi.org/10.1007/978-3-031-06555-2_17
22. Scheidl, H., Fiel, S., Sablatnig, R.: Word beam search: a connectionist temporal classification decoding algorithm. In: 2018 16th International Conference on Frontiers in Handwriting Recognition (ICFHR), pp. 253–258 (2018)
23. Simard, P.Y., Steinkraus, D., Platt, J.C.: Best practices for convolutional neural networks applied to visual document analysis. In: Proceedings of the Seventh International Conference on Document Analysis and Recognition (ICDAR 2003), vol. 2, p. 958. IEEE Computer Society, USA (2003)
24. de Sousa Neto, A.F., Leite Dantas Bezerra, B., Hector Toselli, A., Baptista Lima, E.: A robust handwritten recognition system for learning on different data restriction scenarios. Pattern Recognit. Lett. **159**, 232–238 (2022)
25. TorchVision Maintainers and Contributors: TorchVision: PyTorch's computer vision library (2016)
26. Valy, D., Verleysen, M., Chhun, S.: Data augmentation and text recognition on Khmer historical manuscripts. In: 2020 17th International Conference on Frontiers in Handwriting Recognition (ICFHR), pp. 73–78 (2020)
27. Wick, C., Zöllner, J., Grüning, T.: Rescoring sequence-to-sequence models for text line recognition with CTC-prefixes. In: Uchida, S., Barney, E., Eglin, V. (eds.) Document Analysis Systems, pp. 260–274. Springer, Cham (2022). https://doi.org/10.1007/978-3-031-06555-2_18
28. Wigington, C., Stewart, S., Davis, B., Barrett, B., Price, B., Cohen, S.: Data augmentation for recognition of handwritten words and lines using a CNN-LSTM network. In: 2017 14th IAPR International Conference on Document Analysis and Recognition (ICDAR), vol. 01, pp. 639–645 (2017)
29. Wilcoxon, F.: Individual comparisons by ranking methods. Biometrics Bull. **1**(6), 80–83 (1945)
30. Wilkinson, T., Brun, A.: Semantic and verbatim word spotting using deep neural networks. In: 2016 15th International Conference on Frontiers in Handwriting Recognition (ICFHR), pp. 307–312 (2016)
31. Wilkinson, T., Lindstrom, J., Brun, A.: Neural Ctrl-F: segmentation-free query-by-string word spotting in handwritten manuscript collections. In: Proceedings of the IEEE International Conference on Computer Vision (ICCV) (2017)
32. Zhai, F., Fan, Y., Verma, T., Sinha, R., Klakow, D.: A dataset and a novel neural approach for optical Gregg shorthand recognition. In: Sojka, P., Horák, A., Kopeček, I., Pala, K. (eds.) TSD 2018. LNCS (LNAI), vol. 11107, pp. 222–230. Springer, Cham (2018). https://doi.org/10.1007/978-3-030-00794-2_24

Lifelong Learning for Document Image Binarization: An Experimental Study

Pedro González-Barrachina(✉), María Alfaro-Contreras, Mario Nieto-Hidalgo, and Jorge Calvo-Zaragoza

University Institute for Computing Research (IUII), University of Alicante, Alicante, Spain
pcg71@alu.ua.es, {malfaro,jcalvo}@dlsi.ua.es, mnieto@dtic.ua.es

Abstract. Binarization is one of the ubiquitous processes in the field of document image analysis. In recent years, there has been an increasing number of approaches based on deep learning, which in turn require labeled data of each document collection. In many real cases, however, the possible domains of document collections are not known in advance but appear incrementally. Adapting a model to accommodate new domains is known as lifelong (or continual) learning. In this scenario, neural networks suffer from catastrophic forgetting; i.e., when learning about a new domain, the knowledge from the previous ones is lost. This paper presents an experimental study of alternatives to prevent catastrophic forgetting in a continuous learning scenario for document binarization. We present results with different evaluation protocols considering four different document collections of heterogeneous features. Our results show that while there is no single approach that consistently outperforms the others, all of the proposed mechanisms help prevent catastrophic forgetting to some extent. Interestingly, we found that relatively simple mechanisms such as transfer learning tend to perform better on average than specifically designed strategies for continual learning.

Keywords: Binarization · Continual Learning · Deep Learning · Transfer Learning · Discrete Representations

1 Introduction

Historically, information has been stored and transmitted by physical engraving onto documents, either by hand or using printing machines. Several initiatives exist to ensure the preservation of these historical documents through optical scanning. Within this context, the field of Document Image Analysis (DIA) studies the automatic retrieval of (total or partial) information from document images.

Binarization is one of the most well-known, extensively used processes in the context of DIA. It involves detecting pixels containing relevant information—for instance, those depicting ink—from those of the background. Such a process,

A. Pertusa et al. (Eds.): IbPRIA 2023, LNCS 14062, pp. 146–157, 2023.
https://doi.org/10.1007/978-3-031-36616-1_12

usually performed in the preprocessing stage of DIA workflows, enables better performance of ensuing processes by reducing the complexity of the image because irrelevant information is discarded.

In recent years, this task has shifted from methods based on handcrafted algorithms to learning methods that leverage Deep Learning (DL) techniques [3,12]. Auto-Encoders (AE) based on Fully Convolutional Networks are considered the current state of the art for document image binarization [3,5,12]. AE are feed-forward neural networks with input and output shapes that are exactly the same [11]. The network is typically divided into two stages: (i) the *encoder*, which maps the high-dimensional input to a lower-dimensional representation, and (ii) the *decoder*, which converts the representation back into the original high-dimensional space. Although they account for a considerable improvement in performance, the training process of these models requires labeled representative data of the target domain—in this case, images of documents and their corresponding binarization. The amount of training data required to obtain acceptable results tends to be vast.

In some cases, such as those dealing with historical documents, there are little labeled data available, which leads to the binarization of documents with models trained on different graphical domains, causing a drop in performance. The varying data distributions might not be accessible initially, but instead, they become available sequentially, one by one. We could, then, (i) re-train the model on all the available data each time a new domain arrives—computationally expensive and inefficient—or (ii) fine-tune the model only on the newly acquired domain. The latter alternative is known to cause forgetting of old domains as new ones are learned, which is referred to as *catastrophic forgetting* [9,15].

Continual Learning (CL) aims to gradually extend acquired knowledge by learning from a (theoretically) infinite stream of data [6]. To our knowledge, there is only one work that has studied CL for binarization [7]. However, the authors considered the use of hypernetworks, which do not scale well to deep neural networks. In contrast, in this work, we explore different alternatives to CL that barely modify the existing state-of-the-art architectures for binarization.

Our experiments on different datasets show that it is possible to incrementally learn a new domain without catastrophically forgetting the previous ones—even when the scenarios considered are very different (e.g., from text to music documents). Interestingly, the results show that the use of transfer learning mechanisms is often as effective as applying state-of-the-art CL approaches, thereby pointing out interesting avenues for future work.

The remainder of the paper is organized as follows: Sect. 2 thoroughly develops the framework for CL in document image binarization; Sect. 3 describes the experimental setup; Sect. 4 presents and analyses the results; and, finally, Sect. 5 concludes the work and discusses possible ideas for future research.

2 Methodology

In this work, we address the document image binarization task from a CL perspective, specifically in the context of domain-incremental learning: in each learning episode, the model is trained with images from a new graphic document domain. We shall now introduce some notations to then formally define this framework.

For training the model on a new domain d, only a specific annotated dataset for the binarization problem in such domain is used (X^d, Y^d), with X^d being the collection of document images from a specific manuscript and Y^d their binarized versions. Note that the images in X^d are in the form of $[0, 255]^{h_j^d \times w_j^d \times c}$, being $h_j^d \times w_j^d$ px. The resolution of the j-th image and c the number of channels, whereas the binarized versions have the shape $\{0, 1\}^{h_j^d \times w_j^d}$.[1] In this work, we used color images as input ($c = 3$), but other considerations may be applied.

For the *baseline* binarization model, we will use a state-of-the-art AE [3]. This is illustrated in Fig. 1. The proposed methodology is to train the *same* baseline AE but apply different CL strategies in the same scenarios to analyze their performances. One can expect to find catastrophic forgetting with the baseline model, yet to avoid partially or totally catastrophic forgetting when a specific CL strategy is applied.

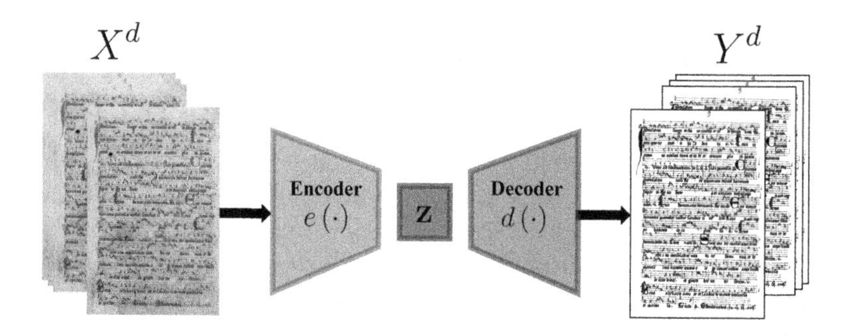

Fig. 1. Graphical summary of the *baseline* methodology. The encoder $e(\cdot)$ compresses the colored input data X^d (continuous-valued image) to a latent representation space $\mathbf{z}$ so that the decoder $d(\cdot)$ maps it to its binarized version Y^d (binary image). In this work, we introduce different constraints either on the encoder $e(\cdot)$ or on the latent space $\mathbf{z}$ to promote continual learning in this *baseline* model.

Hereinafter, the *baseline* model discussed in this section will be referred to as "Base". Below, we discuss the three strategies that will be applied to introduce mechanisms that should favor CL in this baseline model.

2.1 Frozen Encoder

The first strategy to study is based on transfer learning. The encoder $e(\cdot)$ of the baseline model is frozen during all the training process (we will hereafter refer to

[1] Typically, document image sizes are not fixed. Instead, patches of $h_j^d \times w_j^d$ are considered.

this as "Base-F"). Therefore, for each new domain, the model will only update the weights of the decoder $d(\cdot)$ part.

Intuitively, by not allowing the update of the encoder's weights, and keeping the representations learned from previous data, we will provide the model with greater stability, thereby mitigating the effects of catastrophic forgetting.

2.2 Vector Quantized-Variational AutoEncoder

The second proposed strategy is based on imposing a constraint on the latent space $\mathbf{z}$. Specifically, a discretization of this space is proposed by introducing a Vector Quantizer (VQ) layer, converting the AE into a Vector Quantized-Variational AutoEncoder (VQ-VAE) [17]. For this model, the encoder $e(\cdot)$ is kept frozen to follow the strategy used in existing works [16].

The VQ-VAE is an extension of the traditional AE architecture that learns a discrete latent representation of the input data, instead of a continuous one, by training a codebook of discrete latent variables onto which the encoder maps the input data [17]. During training, the encoder $e(\cdot)$ representations are quantized to the nearest code in the codebook, which serves as the (quantized) latent representation $\mathbf{z}^q$. The decoder $d(\cdot)$ then reconstructs the input from the quantized latent representation $\mathbf{z}^q$. Subsequent works proposed using multiple codebooks by splitting the representations between the different codebooks to further improve the performance [2,14].

We believe that the introduction of this component will contribute to additional stability when facing domain changes, thereby leading to less catastrophic forgetting.

2.3 Discrete Key-Value Bottleneck

Our last strategy is the use of a Discrete Key-Value Bottleneck (DKVB). DKVB is a model proposed to address the issue of *catastrophic forgetting* in CL for classification tasks [16]. It builds upon the VQ-VAE architecture but transforms the discretized latent space $\mathbf{z}^q$ of VQ-VAE into a discrete dictionary of keys and values $\mathbf{z}^q_{(k,v)}$. In this regard, the input is encoded and passed through the discrete bottleneck to select the nearest keys, whose corresponding values are fed to the decoder for the current task. Note that VQ-VAE can be seen as a special case of DKVB in which the keys are equal to the values (i.e., $\mathbf{z}^q = \mathbf{z}^q_{(k,v)}$ with $k = v$).

In our case, we introduce the DKVB to discretize the latent space and enable localized and context-dependent model updates. Theoretically, the introduction of this component designed for CL should achieve a considerable reduction of catastrophic forgetting and, possibly, provide other desirable CL capabilities (such as improving general performance).

2.4 Training Considerations

For all strategies, we use an encoder pre-trained from ImageNet using the self-supervised learning method DINO [4].[2] Moreover, Base and Base-F are simply trained sequentially on the different domains, while VQ-VAE and DKVB models follow the training strategy proposed in previous works [16]. Such a process, summarized in Table 1, is described below.

- *Phase 0: Keys initialization.* The keys of both VQ-VAE and DKVB models are initialized using images from the first domain and remain unchanged for the rest of the training process. Thus, the first domain is of special importance for the subsequent ones.
- *Phase 1: Decoder training.* The decoder $d(\cdot)$ is trained to solve the binarization task. For VQ-VAE, this phase represents the remainder of the training process—it is trained for all domains—while for DKVB, the decoder $d(\cdot)$ is kept unchanged for all domains after being trained for the first domain. Therefore, the first domain is even more important for this method.
- *Phase 2: Training values.* This phase is unique to DKVB and involves training the bottleneck values for all domains except the first one. This phase helps activate different keys for different domains, improving the stability of the model's memory.

Table 1. Summary of the training methodology for the different architectures considered.

Name	Freeze encoder $e(\cdot)$?	Latent space	Initialize keys?	Freeze decoder $d(\cdot)$?
Base	No	$\mathbf{z}$	–	No
Base-F	Yes	$\mathbf{z}$	–	No
VQ-VAE	Yes	$\mathbf{z}^q$	Yes	No
DKVB	Yes	$\mathbf{z}^q_{(k,v)}$	Yes	Yes

3 Experimental Setup

This section describes the experimental setup used in our study, including the datasets, CL scenarios, and evaluation metrics considered.

3.1 Datasets

We have considered different types of documents, including text manuscripts and music score images, to assess our methodology. A summary of the features depicted in these corpora is given in Table 2, while some image examples are illustrated in Fig. 2.

[2] We empirically checked that the baseline behaves similarly in our experiments when initialized randomly.

- PHI. A set of scanned images of Persian manuscripts from the Persian Heritage Image Binarization Competition [1]. This dataset represents a text domain.
- DIB. A collection of images of handwritten Latin text documents from the Document Image Binarization Contest (DIBCO) [8] created by combining sets from 2009 to 2016 as a single unified corpus, which represents another text domain.
- SAL and EIN. Two compilations of high-resolution images of scanned documents that contain lyrics and music scores in neumatic notation. Specifically, the images come from Salzinnes Antiphonal (CDM-Hsmu 2149.14).[3] and Einsiedeln, Stiftsbibliothek, Codex 611(89)[4]

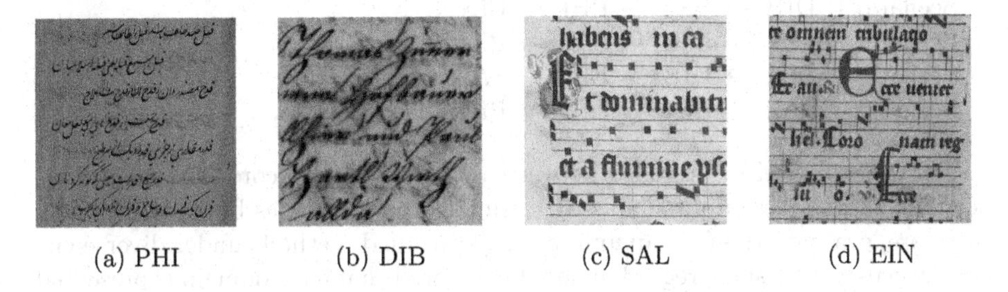

(a) PHI (b) DIB (c) SAL (d) EIN

Fig. 2. Examples of pages from the selected datasets.

Table 2. Details of the corpora. The columns represent the names, the number of page images, the average resolution of the images, and the percentage of ink pixels.

Corpus	Pages	Size	Ink
Text documents			
PHI	15	1 022 × 1 158 px	9.2%
DIB	86	659 × 1 560 px	7.2%
Music documents			
SAL	10	5 100 × 3 200 px	19.2%
EIN	10	5 550 × 3 650 px	20.0%

For the experimentation, the images from the corpora have been configured for a 5-fold cross-validation. Each dataset is independently split into three partitions for training, validating, and testing with 60%, 20%, and 20% of the whole collections, respectively.

[3] https://cantus.simssa.ca/manuscript/25/.
[4] https://cantus.simssa.ca/manuscript/74/.

Note that all the approaches process images by patches of size 256×256. For the evaluation of the model, we process full images with a sliding window. The performance of the model is measured by averaging the result for each image. For each domain of each scenario, we save the model with the best performance in the validation set of the newly introduced domain.

3.2 CL Scenarios

In this work, we study several scenarios that vary in the order in which the different domains are presented sequentially. Due to the high number of possibilities, we restrict ourselves to a series of scenarios that can show meaningful results, as in previous works [7]:

- Scenario 1: DIB $\rightarrow$ SAL $\rightarrow$ PHI $\rightarrow$ EIN
- Scenario 2: EIN $\rightarrow$ DIB $\rightarrow$ SAL $\rightarrow$ PHI
- Scenario 3: PHI $\rightarrow$ DIB $\rightarrow$ EIN $\rightarrow$ SAL
- Scenario 4: SAL $\rightarrow$ EIN $\rightarrow$ DIB $\rightarrow$ PHI

Scenarios 1 and 2 alternate between text and music score domains, while Scenarios 3 and 4 group domains with similar characteristics by pairs. We can, therefore, evaluate the CL capabilities of the studied methods under diverse circumstances. In the same regard, it should be noted that each domain is presented at least once as a first domain, given the importance of this issue in some of the strategies.

3.3 Metrics

Binarization is a binary classification problem: pixels are either classified as foreground or background. However, as seen in Table 2, the evaluation calls for the use of metrics that are not biased toward the dominant class—in this case, background. For this, we consider the *F-measure* (F_1), which is defined as

$$\mathrm{F}_1 = \frac{2 \cdot \mathrm{TP}}{2 \cdot \mathrm{TP} + \mathrm{FP} + \mathrm{FN}}, \tag{1}$$

where TP (*True Positives*) represents ink pixels classified as such, FP (*False Positives*) denotes background pixels classified as ink, and FN (*False Negatives*) stands for ink pixels classified as background.

3.4 Implementation Details

All the models presented in Sect. 2 share the same neural architecture for the encoder and the decoder:

- Encoder. As in [16], we use a standard ResNet-50 configuration without the last two residual blocks in order to keep high dimensional representations in the output. The input size is fixed to $256 \times 256 \times 3$.

- Decoder. Four stacked convolutional blocks with transposed convolution of size 2×2 with stride of 2 for upsampling the image, followed by two convolutions of 3×3 with batch normalization and *ReLU* as activation. The number of filters of the decoder blocks are 256, 128, 64, and 32, respectively. The last block is followed by a last convolutional layer with kernel size 1×1 and 1 output channel, and *sigmoid* activation, whose output represents the probability for each pixel to be "ink".

Furthermore, as mentioned in Sect. 2.2, previous works mention that using multiple codebooks helps both the performance of the system and alleviates catastrophic forgetting in the case of DKVB [16]. We, therefore, explored the use of different numbers of codebooks in preliminary experiments, resulting in the use of 128 codebooks for both VQ-VAE and DKVB.

During training, all the models are optimized through gradient descent using mean squared error as the loss criteria and Adam optimizer [13], with a fixed learning rate of $3 \cdot 10^{-4}$, and a batch size of 32 patches. The training procedure comprises a maximum of 300 epochs with 50 steps per epoch allowing early stopping with a patience of 30 epochs.

4 Results

In this section, we analyze the performance of the studied methods for the different scenarios. We study how the model performs after including a new domain (step) of the scenario to determine its robustness to domain changes. Similarly, we analyze the ability of the model to retain knowledge, as well as its ability to eventually learn all involved domains.

The first point to analyze is the model's ability to gradually acquire knowledge (forward transfer) and maintain successful performance in the domains seen up to each step. In Fig. 3, the performance curves after each step of the process are shown. The F_1 reported is the average considering all domains seen up to each step (including the current one), with shading including the minimum and maximum F_1.

Before analyzing each scenario, it should be noted that the three alternatives to CL show clearly superior results to the baseline, for which a very unstable performance caused by catastrophic forgetting is observed. This is interesting because some of these alternatives, such as Base-F, have not been reported in the literature as sufficient for CL. On the other hand, it can be observed that the instability depends on the scenario and the specific domain of each step. This is because some domains are likely to share features with each other; that is, there is no performance drop due to the change of domain.

In Scenario 1, the effect of forgetting is notorious in the Base model because domains of different types are interspersed (text, music, text, music). Therefore, the Base model tends to be biased towards the last domain provided at all times. Instead, the models proposed for CL maintain a successful performance in general. This is rather equal in Scenario 2, which has similar conditions. In the case of Scenario 3, it can be seen that while the first two domains are of the same typology (text), there is no drop in performance because they share many features. However, when the last two (music) are introduced, the effect of catastrophic forgetting is observed in the base case. On the other hand, Scenario 4 is an example of an order in which a performance drop is not observed in any case. This is probably because text domains are useful for the binarization of music scores as well, but not the other way around (Scenario 3).

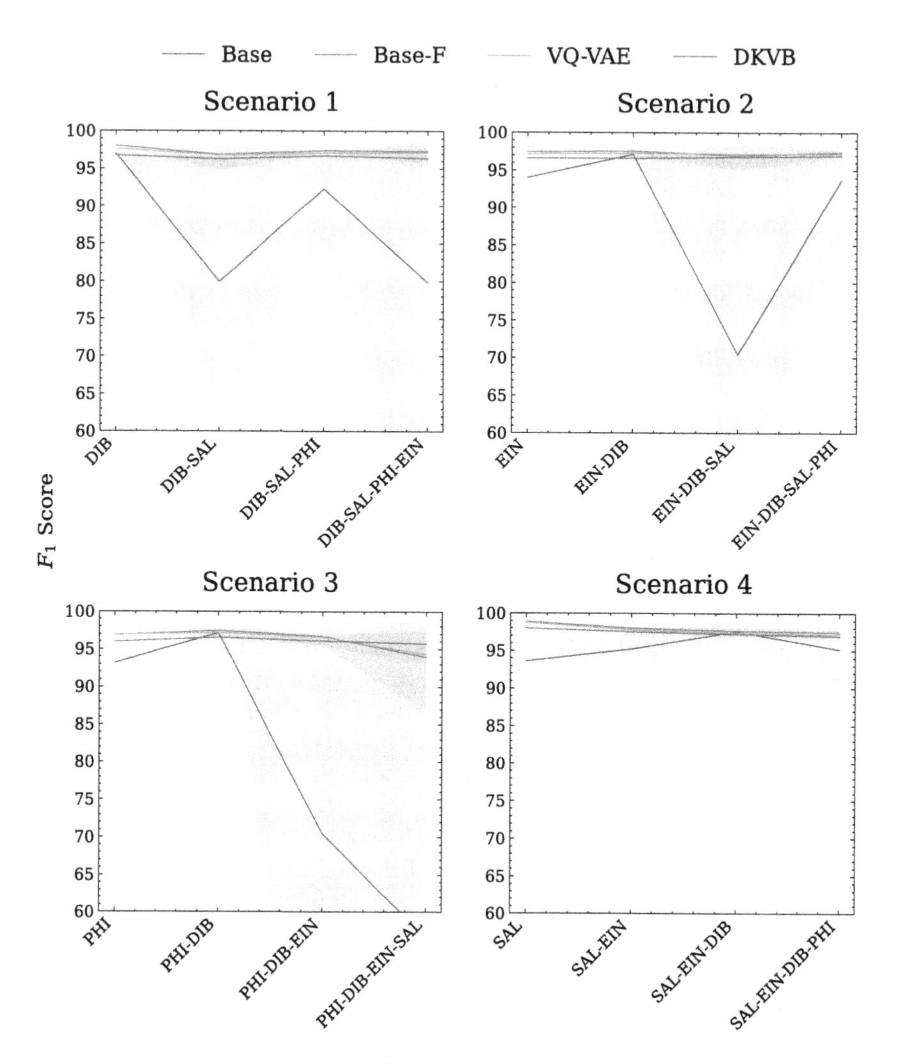

Fig. 3. Performance in terms of F_1 (%) for all the domains provided up to each step. The line represents the average performance, while the shadow area covers the interval between the best and worst performance in the different domains.

Furthermore, Table 3 reflects the average performance on each dataset after completing all domains in the scenario. This is useful to measure the ability of each strategy to maintain awareness at the end of the process, and if being the case, improve performance on previously learned domains (backward transfer). The results reinforce the superiority of CL methods over the baseline, and interestingly, the simplest method (Base-F) achieves the highest F_1 rates in 70% of the cases on average.

Table 3. Mean F_1(%) values obtained on each domain for each scenario after the final learning episode. Bold figures denote the best results for each dataset for a given scenario.

Scenario 1 DIB → SAL → PHI → EIN					Scenario 2 EIN → DIB → SAL → PHI				
	PHI	DIB	SAL	EIN		PHI	DIB	SAL	EIN
Base	39.6	85.1	98.8	95.9	Base	85.0	95.1	97.9	96.7
Base-F	**95.7**	**97.1**	**98.6**	**97.8**	Base-F	**97.5**	**97.6**	97.6	**96.9**
VQ-VAE	**95.7**	96.8	98.4	97.6	VQ	97.1	97.3	97.5	96.7
DKVB	95.3	95.9	97.7	95.9	DKVB	96.2	96.7	**98.0**	**96.9**
Scenario 3 PHI → DIB → EIN → SAL					Scenario 4 SAL → EIN → DIB → PHI				
	PHI	DIB	SAL	EIN		PHI	DIB	SAL	EIN
Base	5.9	62.4	97.3	60.8	Base	90.4	96.0	97.7	96.7
Base-F	86.1	94.4	**98.8**	96.5	Base-F	**97.4**	**97.8**	97.7	**96.9**
VQ-VAE	86.7	94.8	98.7	**96.9**	VQ-VAE	97.0	97.5	97.6	96.6
DKVB	**93.1**	**95.3**	98.0	96.4	DKVB	96.3	96.9	**97.9**	96.7

Regarding backward transfer, we find very few examples of it in the results for these methods, not being able to draw significant conclusions in this regard. We theorize that these methods that constraint the freedom of the architecture to learn by freezing the encoder and introducing a bottleneck may help the stability of the neural network, but in contrast, it may affect its learning plasticity, possibly limiting the backward transfer.

Finally, to visually demonstrate the performance of the various approaches, Fig. 4 presents an example of the binarization results on the second-to-last domain in Scenario 3 (EIN) after completing the scenario. The considered CL strategies clearly enhance the baseline binarization by reducing smudging due to over-detection. Notably, even the simplest CL method (Base-F) exhibits comparable performance to VQ-VAE and outperforms DKVB, which was specifically designed for CL. In this particular example, DKVB fails to fill in the letter S correctly.

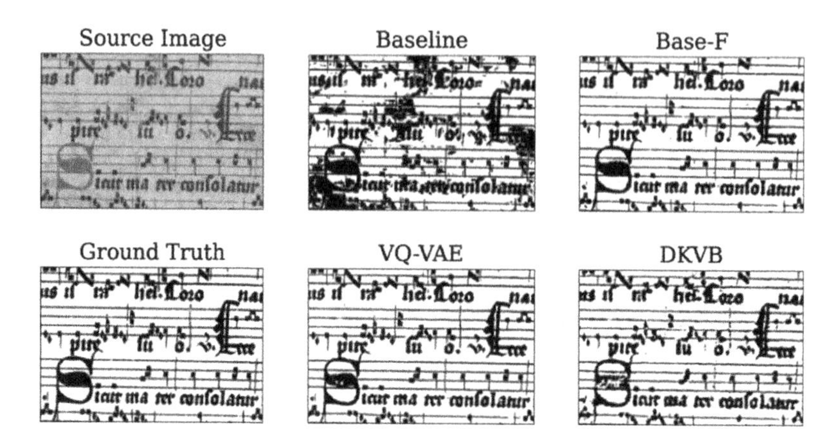

Fig. 4. Examples of binarization from all studied methods for the dataset EIN after completing all domains of Scenario 3.

5 Conclusions

The context of DIA is characterized by the scarcity of labeled data from heterogeneous domains, which quite often become available sequentially. CL helps improve training efficiency, avoiding training models from scratch every time, and retaining knowledge from domains that become unavailable.

In this work, we study three strategies for CL of document binarization: freezing the encoder of the baseline model, VQ-VAE, and DKVB. For the experiments, several scenarios were proposed depending on the order in which the different domains are provided. Our experimentation verified that the three strategies considered avoid catastrophic forgetting and attain positive forward transfer. Interestingly, even a seemingly simple method (freezing the encoder) is capable of achieving similar effects to the approach specially designed for this purpose (DKVB). This is analogous to phenomena occurring in other fields of meta-learning such as Domain Generalization, where rather baseline methods are similar in performance to specific strategies [10].

An interesting future line of this work would be to check the scope of our conclusions in other CL tasks and scenarios. We also plan to broaden the scope of our research to deal with other DIA tasks beyond classification, such as sequential modeling for Handwritten Text Recognition.

Acknowledgments. This paper is part of the project I+D+i PID2020-118447RA-I00 (MultiScore), funded by MCIN/AEI/10.13039/501100011033. The second author is supported by grant FPU19/04957 from the Spanish Ministerio de Universidades.

References

1. Ayatollahi, S.M., Nafchi, H.Z.: Persian heritage image binarization competition (PHIBC 2012). In: Iranian Conference on Pattern Recognition and Image Analysis, pp. 1–4. IEEE (2013)

2. Baevski, A., Zhou, Y., Mohamed, A., Auli, M.: wav2vec 2.0: a framework for self-supervised learning of speech representations. Adv. Neural. Inf. Process. Syst. **33**, 12449–12460 (2020)
3. Calvo-Zaragoza, J., Gallego, A.J.: A selectional auto-encoder approach for document image binarization. Pattern Recognit. **86**, 37–47 (2019)
4. Caron, M., et al.: Emerging properties in self-supervised vision transformers. In: IEEE Conference on Computer Vision and Pattern Recognition, pp. 9650–9660 (2021)
5. Christlein, V., Nicolaou, A., Seuret, M., Stutzmann, D., Maier, A.: ICDAR 2019 competition on image retrieval for historical handwritten documents. In: International Conference on Document Analysis and Recognition (ICDAR), pp. 1505–1509. IEEE (2019)
6. De Lange, M., et al.: A continual learning survey: defying forgetting in classification tasks. IEEE Trans. Pattern Anal. Mach. Intell. **44**(7), 3366–3385 (2021)
7. Garrido-Munoz, C., Sánchez-Hernández, A., Castellanos, F.J., Calvo-Zaragoza, J.: Continual learning for document image binarization. In: International Conference on Pattern Recognition, pp. 1443–1449. IEEE (2022)
8. Gatos, B., Ntirogiannis, K., Pratikakis, I.: ICDAR 2009 document image binarization contest (DIBCO 2009). In: International Conference on Document Analysis and Recognition, pp. 1375–1382. IEEE (2009)
9. Goodfellow, I.J., Mirza, M., Xiao, D., Courville, A., Bengio, Y.: An empirical investigation of catastrophic forgetting in gradient-based neural networks. arXiv preprint arXiv:1312.6211 (2013)
10. Gulrajani, I., Lopez-Paz, D.: In search of lost domain generalization. In: International Conference on Learning Representations (2021)
11. Hinton, G.E., Zemel, R.: Autoencoders, minimum description length and Helmholtz free energy. Adv. Neural Inf. Process. Syst. **6** (1993)
12. Kang, S., Iwana, B.K., Uchida, S.: Complex image processing with less data—document image binarization by integrating multiple pre-trained U-Net modules. Pattern Recogn. **109**, 107577 (2021)
13. Kingma, D.P., Ba, J.: Adam: a method for stochastic optimization. arXiv preprint arXiv:1412.6980 (2014)
14. Ling, S., Liu, Y.: Decoar 2.0: deep contextualized acoustic representations with vector quantization. arXiv preprint arXiv:2012.06659 (2020)
15. McCloskey, M., Cohen, N.J.: Catastrophic interference in connectionist networks: the sequential learning problem. In: Psychology of Learning and Motivation, vol. 24, pp. 109–165. Elsevier (1989)
16. Träuble, F., et al.: Discrete key-value bottleneck. arXiv preprint arXiv:2207.11240 (2022)
17. Van Den Oord, A., Vinyals, O., et al.: Neural discrete representation learning. Adv. Neural Inf. Process. Syst. **30** (2017)

Test-Time Augmentation for Document Image Binarization

Adrian Rosello, Francisco J. Castellanos, Juan P. Martinez-Esteso, Antonio Javier Gallego(✉), and Jorge Calvo-Zaragoza

Department of Software and Computing Systems, University of Alicante, Alicante, Spain
{adrian.rosello,juan.martinez11}@ua.es,
{fcastellanos,jgallego,jcalvo}@dlsi.ua.es

Abstract. Document binarization is a well-known process addressed in the document image analysis literature, which aims to isolate the ink information from the background. Current solutions use deep learning, which requires a great amount of annotated data for training robust models. Data augmentation is known to reduce such annotation requirements, and it can be used in two ways: during training and during prediction. The latter is the so-called Test Time Augmentation (TTA), which has been successfully applied for general classification tasks. In this work, we study the application of TTA for binarization, a more complex and specific task. We focus on cases with a severe scarcity of annotated data over 5 existing binarization benchmarks. Although the results report certain improvements, these are rather limited. This implies that existing TTA strategies are not sufficient for binarization, which points to interesting lines of future work to further boost the performance.

Keywords: Binarization · Test-Time Augmentation · Data augmentation · Convolutional Neural Networks

1 Introduction

Digital transcription is the process that enables retrieving the content of scanned documents and encoding them into a digital format which can be processed by a computer [9]. This process opens up the opportunity to preserve and disseminate this cultural heritage, but there is a major hurdle to overcome: the countless number of documents waiting to be transcribed together with their huge heterogeneity. Traditionally, this digitization process has been manually carried out, but it is a very costly and time-consuming process to be applied on a large scale. This situation led to the need for the development of automatic digitization systems since they represent a scalable solution. The literature covers a multitude of works around this topic [3,7].

This work was supported by the I+D+i project TED2021-132103A-I00 (DOREMI), funded by MCIN/AEI/10.13039/501100011033.

A. Pertusa et al. (Eds.): IbPRIA 2023, LNCS 14062, pp. 158–169, 2023.
https://doi.org/10.1007/978-3-031-36616-1_13

Because of the heterogeneity and variety of documents, digitization can be considered a very challenging and complex process, and it is usually divided into different steps, each one specialized in a specific task. One of the most well-known steps is binarization, which is an image-processing task with the purpose of detecting the relevant information in the image—i.e. the ink of the document. This process simplifies the input image to a binary representation in which the background is represented with a color—typically white—and the ink with another color—typically black. A lot of document image workflows make use of binarization somehow or other, so that is an important process worth to be studied [11,13,15].

Conventional binarization strategies are based on heuristic algorithms to find a global or local threshold, and the literature has a lot of references to perform it [17,19]. Although heuristic methods demonstrated to provide high-performance results, it is also true that these algorithms exploit specific features in the images. However, the heterogeneity of documents makes these algorithms not suitable for real-world scenarios.

The use of deep learning has recently been considered for the development of more generalizable solutions, currently representing the state of the art in binarization. The literature contains a lot of related works using deep neural networks, such as that by Pastor-Pellicer et al. [18], which makes use of a Convolutional Neural Network (CNN) to classify each pixel of the image, that by Afzal et al. [1], in which a Long Short-Term Memory [12] was used to perform this task, or the work by Calvo-Zaragoza and Gallego [5], which proposes the use of a U-net architecture for a more efficient and robust binarization.

Although neural networks have been demonstrated to be a potential solution, their use is associated with an important bottleneck: the need for a great amount of annotated data. In the context of binarization, these methods require manually labeling a set of images to be used as a reference in the training process. Because of the concerns involved, a previous work minimizes this problem by addressing binarization through a domain adaptation technique [8]. Even so, it needs an annotated collection, so an important manual effort is necessary in any case.

For supervised scenarios, since this manual annotation is expensive—both economically and temporally—a possible solution is to apply data augmentation, which has been successfully used in the literature in a lot of contexts including the binarization field [14]. Data augmentation is a technique used to increase the variability of data by applying different transformations on the input, such as rotations, skew, scale, and contrast adjustments, among many others. This process is commonly used during model training, however, and although it is less known in the literature, it can also be applied during the inference stage, technique so-called Test-Time Augmentation (TTA) [16]. TTA was successfully implemented for general classification tasks [6]. Since binarization can be assumed as a classification at the pixel level, we find it interesting to study its application for this task.

In this work, we study the influence of TTA for document binarization. Our experiments consider 5 binarization benchmarks, with data of different nature.

The results report an improvement in the use of TTA, especially in cases of data scarcity. However, we will see that there is still a lot of room for improvement and that the typical TTA strategies in classification do not translate with the same success to more complex tasks such as binarization.

2 Methodology

In this work, a strategy is proposed to apply TTA in binarization tasks. The image to be binarized will be disturbed following data augmentation strategies. Each augmented version will be independently binarized and then the results will be aggregated to give a more robust result. Due to the nature of the process, typical classification strategies must be adapted, given that binarization is a process that affects the pixel level, and therefore the possible geometrical disturbances must be reverted before aggregating the different predictions. Figure 1 shows an overview of the proposed framework.

Let $\mathcal{C}$ be a collection of images in which the i-th image $\mathcal{I}_i^{h_i \times w_i}$ has a w_i width and h_i height. Given a supervised neural network-based model trained for the binarization task with data extracted from $\mathcal{C}$, the objective of this work is to study the influence of the TTA technique in the inference stage.

For the binarization task, we considered a state-of-the-art method [5], which uses a U-net architecture, so-called Selectional Auto-Encoder (SAE). It processes each $\mathcal{I}_i^{h_i \times w_i}$ by sequentially extracting patches of a size of $w \times h$ pixels. The SAE model provides a probabilistic map as a result for each individual patch. That is, for each pixel, the model obtains a numeric value that represents the degree of confidence of belonging to the ink or background class. Then, a threshold ρ can be applied to take a final decision. Note that ρ might be fixed to 0.5 in a typical case, but other criteria also could be considered. In our case, we consider the best ρ value by optimizing a considered metric—F-measure—on a validation partition data. Note that the input data for the model in training and testing will be a sample or patch x.

From an input data x—sample—in the inference stage, TTA is a technique that applies a number of transformations T to obtain different variations of x, whose set is defined as $\mathcal{X} = (x_0, x_1, x_2, x_3, \ldots, x_T)$ with the aim of being individually processed. Note that x_0 represents the original input without any transformation. A binarization model processes each input variant and provides a probabilistic map as a result, obtaining thus a set of results $\hat{\mathcal{X}} = (\hat{x}_0, \hat{x}_1, \hat{x}_2, \hat{x}_3, \ldots, \hat{x}_T)$. Then, a global decision is made by using the results $\hat{\mathcal{X}}$. In the binarization context, where there is a pixel-wise classification in which each pixel is categorized according to two possible classes $\omega \in \Omega$—ink and background—the results have to be combined to give a final result for each pixel. However, to make it possible, it is necessary to undo the previously applied transformations. For this, each individual result has to be transformed in order to, given a specific pixel, all the results for the coordinates of that pixel represent exactly the result for the same pixel of the original input x. Once done this, the results $\hat{\mathcal{X}}' = (\hat{x}'_0, \hat{x}'_1, \hat{x}'_2, \hat{x}'_3, \ldots, \hat{x}'_T)$ can be combined by using a specific combination policy and obtaining thus a final result $\hat{y}$.

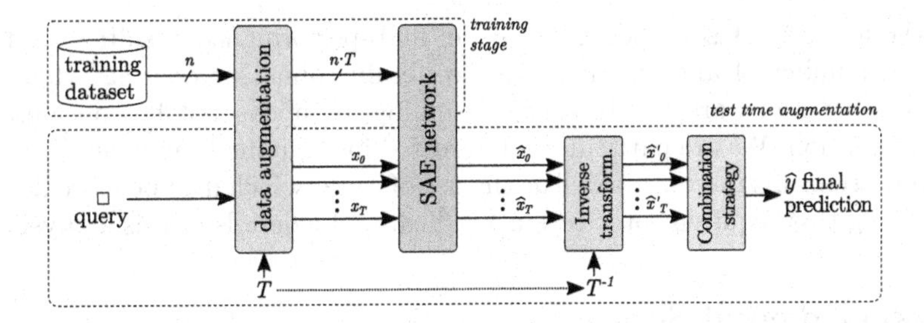

Fig. 1. General scheme of training and testing procedures on SAE network.

The policy employed for the combination of the results can take advantage of the confidence values in one way or another. Since the combination decision may drastically impact the final results, we considered studying several options. For the description of the different policies, we define (x, y) as the coordinate of a pixel, and $\hat{x}_i'^{(x,y)}$ the result of the prediction of the i-th transformed sample from $\hat{\mathcal{X}}'$ in that coordinate. Let also $\hat{y}$ the final result after the policy combination. With this, the combination policies considered are the following ones:

- **Vote:** it counts the number of cases in which a pixel was predicted as ink according to $\hat{\mathcal{X}}'$. It requires a previous thresholding for the individual results in order to determine if a pixel is ink or background. We considered three versions of this policy to select the ink class: when the number of ink votes overcomes 50%; when it overcomes 75% of votes; and when all the results in $\hat{\mathcal{X}}'$ indicate that the pixel is ink, that is 100% of votes. Henceforth, these cases will be denoted as ***vote50***, ***vote75*** and ***vote100***, respectively. For the formulation, $[\![\cdot]\!] \rightarrow \{0, 1\}$ represents the Iverson bracket, which outputs the unit value when the condition in the argument is met and zero otherwise.

$$\hat{y}_i^{(x,y)} = \begin{cases} \omega_1 & \textit{if } \sum\limits_{i=0}^{T} [\![\arg\max\limits_{\omega \in \Omega} P(\omega \mid \hat{x}_i'^{(x,y)}) = \omega_1]\!] \geq \lambda T \\ \omega_0 & \textit{otherwhise} \end{cases}$$

  where $\omega_0 = 0$ represents the background and $\omega_1 = 1$ stands for the ink.
- **Average confidence:** it computes the average confidence between all the results in $\hat{\mathcal{X}}'$. Then, a threshold ρ is applied to determine if a pixel is ink or background. This policy will be referenced as ***avg-policy***.

$$\hat{y}_i^{(x,y)} = \arg\max_{\omega \in \Omega} \frac{\sum_{i=0}^{T} P(\omega \mid \hat{x}_i'^{(x,y)})}{T+1}$$

- **Maximum confidence:** it takes the maximum confidence value for each pixel in $\hat{\mathcal{X}}'$ and then, the resulting probabilistic map is processed by a threshold ρ. This combination criterion is denoted as ***max-policy***.

$$\hat{y}_i^{(x,y)} = \arg\max_{\omega \in \Omega} \left(\max_{0 \leq i \leq T} P(\omega \mid \hat{x}_i'^{(x,y)}) \right)$$

Although supervised models provide high-performance results [5], they require a number of annotated images to obtain robust predictions. However, manual annotations are highly time-consuming, so it can not be considered a scalable solution. We are particularly interested in studying the influence of partial annotations in training, as well as the use of TTA, which may be more flexible by reducing this requirement. Section 4 indicates the details of this analysis.

3 Experimental Setup

In this section, we explain the setup considered for the experiments. We describe the corpora with which we evaluate our method, the metrics used to empirically validate the results, and the implementation details of our neural architecture.

3.1 Corpora

In order to evaluate our method, we considered several manuscripts of different natures commonly used for document binarization. Some examples can be found in Fig. 2, while Table 1 shows a short description of each one.

- *Phi*: the Persian Heritage Image Binarization Competition [2] is a set of 15 pages of Persian documents. 3 pages were used as test and 12 for training.
- *Dibco*: the Document Image Binarization Contest is a contest held for several years from 2009 [10] with a collection of printed and handwritten images. As in previous works [5], we used the images from the 2014 edition as the test set and the remaining editions until 2016 for the training set.
- *Einsiedeln*: a collection of 10 high-resolution images of music scores written in mensural notation, extracted from Einsiedeln, Stiftsbibliothek, Codex 611(89).[1] As in previous works, we use 8 pages for training and 2 for test.
- *Salzinnes*: a collection of 10 high resolution scanned pages of mensural music notation, extracted from Salzinnes Antiphonal (CDM-Hsmu 2149.14).[2] of high resolution. We considered 8 pages for training and 2 for test.
- *Palm*: a set of 100 scanned Balinese palm leaf manuscripts that were part of a binarization contest [4]. We choose one of the annotated versions—specifically the version labeled as the first ground truth. 50 images were used for training and the 50 remaining for test.

Although the number of pages may seem limited, the total number of annotated pixels is large (see the resolution of the images in Table 1). These corpora are typically used in other binarization works or challenges, making them valuable benchmarks to be considered.

[1] http://www.e-codices.unifr.ch/en/sbe/0611/.
[2] https://cantus.simssa.ca/manuscript/133/.

Table 1. Summary description of each corpus considered in this work.

Dataset	Pages	Image size (px)
PHI	15	$1\,022 \times 1\,158$
DIBCO	86	$659 \times 1\,560$
EINSIEDELN	10	$5\,550 \times 3\,650$
SALZINNES	10	$5\,100 \times 3\,200$
PALM	100	$492 \times 5\,116$

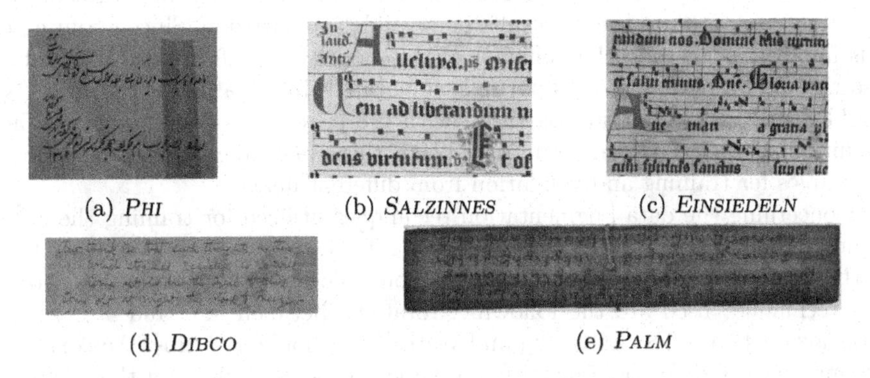

(a) *PHI* (b) *SALZINNES* (c) *EINSIEDELN*

(d) *DIBCO* (e) *PALM*

Fig. 2. Examples of excerpts from the corpora considered.

3.2 Metrics

Binarization is a two-class problem in which each pixel is classified as ink or background. It presents an imbalanced scenario, in which the number of pixels of each class does not match. Since this is the case, we consider a metric that does not bias to the majority class, presumably the background: the F-measure metric (F_1). This metric, widely used for binarization evaluation, can be computed as

$$F_1 = \frac{2\,\mathrm{TP}}{2\,\mathrm{TP} + \mathrm{FP} + \mathrm{FN}},$$

where TP means *True Positives* or ink pixels classified as ink, FP stands for *False Positives* or background pixels predicted as ink, and FN indicates the *False Negatives* or the ink pixels predicted as background.

3.3 Implementation Details

For the experiments, we considered the use of a SAE architecture, based on previous work [5]. This model receives as input a patch of 256×256 pixels in color format and returns an image of the same size with the confidence of each pixel belonging to the ink class according to the neural network. The input image is normalized in the form $\mathcal{I}' = \frac{255-\mathcal{I}}{255}$, where $\mathcal{I}$ is the input image and $\mathcal{I}'$ the normalized version with values between 0 and 1.

The architecture is divided into two stages: an encoder and a decoder stage. The encoder is composed of 4 convolutional blocks in which each block contains a convolutional layer of 32 filters of a size of 3×3, down-sampling of 2×2 pixels, a batch normalization, a Rectified Linear Unit activation function and a dropout of 0.4. Then, the decoder inverts these operations by a series of 4 blocks with the same configuration except changing down-sampling by up-sampling of 2×2 pixels. The architecture includes a final convolutional layer with a set of neurons that predict values between 0 and 1 through the sigmoid activation function, with the confidence of the model in which a pixel is part of the ink information.

The models were trained up to 200 epochs with an early stopping of 10 epochs of patience to optimize the F_1 results on a validation partition. Since in our case it is important to study the behavior of the model with a few annotated data, also the validation partition is partially annotated. In a real-world scenario, we would not have a validation partition with full-page annotations, whereas the training images are partially annotated. For this, we extract the same number of samples for training and validation from different images.

Concerning the data augmentation techniques utilized for training the augmented models as well as the ones applied for TTA, they were randomly applied in the form that several transformations could be used to the selected patch. The techniques used are the following: rotations between -45° and 45°; zoom variations between 0.8x and 1.2x; and vertical and horizontal flips. We consider making up to 10 different transformed versions of the input data for these experiments. This decision was made in previous experimentation that showed that the results did not vary from this amount. Note that TTA applies the same data-augmentation techniques, but in the prediction process.

A lot of combinations are possible by modifying the number of pages and samples extracted. We report the average results between the scenarios with the same training size considering 1, 2, 4, 8, and all the images available for training. In the case where the datasets do not have sufficient images, we selected the maximum possible.

4 Results

Concerning the experiments, in addition to studying different combination policies for TTA (see Sect. 2), we consider also evaluating the influence of TTA in two scenarios:

- **Non-augmented scenario:** the model is trained with patches of 256 × 256 px. sequentially extracted from the images without any data augmentation.
- **Augmented scenario:** the model is trained with augmented patches. We consider random transformations as described in Sect. 3.3.

First, we focus on studying the improvement rate of the augmented model with respect to non-augmented in terms of F_1. As we can see in Fig. 3, all the considered criteria follow the same trend. With a few samples, the behavior seems unstable, while a significant improvement occurs with a relatively small number of samples between 16 and 64. In that range, we obtain up to about 30% of improvement of F_1 when the model is trained with data augmentation, being the best case *max-policy* with 32.7%. As the number of samples increased, the improvement became smaller, obtaining up to 5% of worsening for 256 samples. In this case, the worst criterion was *max-policy* together with the models without TTA, while the criteria with the smallest deterioration were *vote100* and *vote75*. This means that data augmentation on training is particularly beneficial when there is a scarcity of annotated data.

Concerning TTA, Fig. 4 shows the improvement rate of the models when this technique is applied. As reported in Fig. 4a and a previous work [6], the former case—no augmentation in training—indicates that TTA improves the results with a few samples for training, whereas the performance gets worse with a lot of samples. In the range of 32 to 128 samples, *vote100* and *vote75* obtain an improvement, being higher for 64 samples (2.75% and 1.27% respectively). For 256 samples, *vote100* was the only one to get a very slight improvement of 0.22%, while the *max-policy* showed the worst performance. These results also report an instability when there are 8 samples or lower for training the model.

When the model is trained with data augmentation (Fig. 4b), we can see a different behavior. For 16 samples, *vote50*, *vote75* and *max-policy* obtain an improvement between 1% and 2%. These policies get worse from 64 samples, and we can observe that *vote50* keep the same result with TTA and no TTA from that point. However, a different pattern can be found in *vote100*. With a few samples, this policy does not provide any improvement, but when the model is trained with greater than 128 samples, we observe a significant increase in the F_1 results. Therefore, if the model is trained with data augmentation and scarcity of annotated samples, *vote50*, *vote75* and *max-policy* seem the best options, but when there are a lot of annotated samples for training the model, the most restrictive voting criteria *vote100* is the policy to be considered.

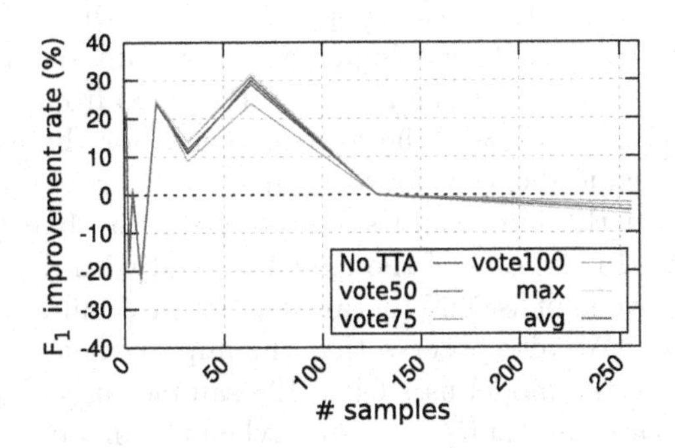

Fig. 3. F_1 improvement rate of using data augmentation in training.

For further analysis, Table 2 reports a comparison between the base case—without TTA—and the use of TTA. Note that these values are those obtained by the best model for each number of samples employed for training. On average,

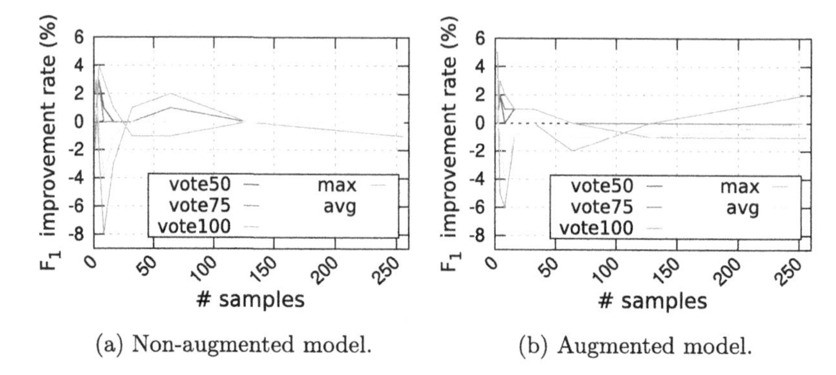

(a) Non-augmented model.

(b) Augmented model.

Fig. 4. F_1 improvement rate of using TTA with respect to no using it for the non-augmented model (a) and the augmented model (b).

Table 2. Average results in terms of F_1 and improvement rate of TTA.

Scenario	# samples									
	1	2	4	8	16	32	64	128	≥256	AVG.
Base case (F_1 %)										
Non-augmented model	20.7	18.2	22.7	26.8	25.8	28.5	38.7	56.9	61.2	33.3
Augmented model	24.7	14.7	23.3	20.7	32.1	31.8	50.6	57.0	58.3	34.8
Improvement rate of TTA (%)										
Non-augmented model	2.7	3.3	4.5	3.4	1.1	1.5	2.8	0.3	0.2	1.7
Augmented model	5.3	5.6	3.1	2.3	1.9	1.1	-0.2	0.0	2.7	1.8

we observe that TTA increases 1.7% the F_1 metric when the model is trained without augmentation and 1.8% on the opposite case. Using TTA with a model trained without data augmentation improves the results for all the cases considered. It particularly improves the results in the cases from 1 to 8 samples in training, obtaining up to 4.5% in the best of them. For the rest of the cases, the improvement is reduced except for 64 samples, which yields an improvement of F_1 of 2.8%. When the number of samples is greater than 64, TTA barely increases the results. In the case of the model trained with data augmentation, the maximum improvement is 5.6% on average when 2 annotated samples are available for training. We also observe that the improvement is extended up to 32 samples, but when the model uses 64 or 128 samples in the training process, TTA does not improve the results. However, when the model uses 256 samples, the improvement increases again, obtaining a 2.7% of improvement rate.

Finally, Fig. 5 shows two selected examples for *Einsiedeln* and *Dibco*. In the former example, shown in Figs. 5a and 5b, we can observe a relevant improvement with respect to no using TTA—green pixels—with *max-policy*. In that case, the model was trained with only 4 samples. It particularly improves the detection of the staff lines, since the texts and music symbols are recognized in both cases—black pixels. If we do not use TTA, we realize that the staff lines are not properly recognized, so the TTA mechanism is definitely necessary in this case. The second example, shown in Figs. 5c and 5d, we can see also an improvement—green pixels—in false positives, but the contribution of TTA is not sufficient to have a high-performance binarized version. However, note that the model used for this example was trained with only 2 samples. That improvement, therefore, is obtained even if the number of annotations is minimum. With the first example, we find an improvement of F_1 from 83% to 91%, while the example of *Dibco* improves from 49% to 60%. Note that in these examples, there are very few false positives caused by TTA.

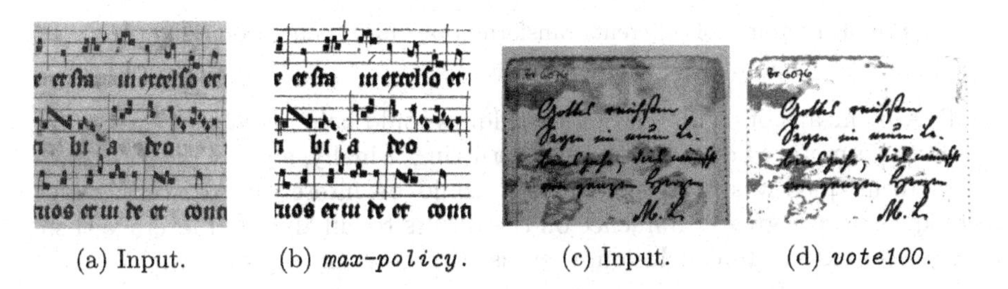

(a) Input. (b) *max-policy*. (c) Input. (d) *vote100*.

Fig. 5. Two selected examples of the results for *Einsiedeln* (a and b) and *Dibco* (c and d). Empirically, for the full page, the former example improves the F_1 from 83% to 91%, while the second example improves from 49% to 60%. The best TTA criteria for each case were applied. Green stands for TTA improvements; blue represents the FN; red stands for FP; and black are TP. (Color figure online)

Even though the results achieved do not represent a great improvement, it should be noted that they do not involve any additional cost or page annotations. Therefore, it is an advantageous procedure to get a slight boost in performance.

Figure 6 shows an example of how an input sample is transformed by data augmentation and then processed by the binarization model. After undoing the transformations, the combination policy—*max-policy* in this case—obtains a combined prediction, which can be processed by a threshold to obtain the final result. In this example, we observe that some pixels are improved with respect to the case of $\hat{x}_o$, which is the prediction of the original data, but when one of the cases obtains high confidence to be ink in a background pixel, *max-policy* will select the maximum value in the combination. For this, it was important to study different criteria combinations.

In general, although TTA improves the results, this occurs in a limited way. This fact implies that the existing strategies of TTA are not sufficient for binarization, but the experiments point out to TTA as an interesting and a potential reference for future research lines to intend to improve the results without additional effort in the manual annotation. Other techniques could be explored in combination with TTA such as meta-learning techniques. Metric learning or Gradient-Based Meta-Learning, or even studying architectures specific for few-shot could be the line to follow in this research. In addition, a specific adaptation

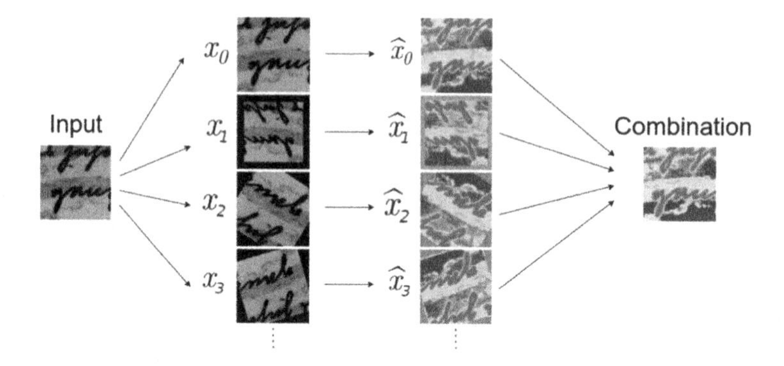

Fig. 6. Example of different transformations for a patch from *Dibco*.

for TTA by means of regularization techniques could help for this task, since the scarcity of annotated data may cause overfitting. Finally, a set of data augmentation techniques was considered, but it could be interesting to study which techniques have a greater influence on the results to enhance and leverage them to a greater extent. In conclusion, there is still room for improvement.

5 Conclusions

This work proposes a study of the use of TTA for the binarization task, focusing on cases with a scarcity of annotated data. TTA is a technique that applies data augmentation in the inference stage. Although this technique was demonstrated in simpler classification tasks, a study was necessary for more complex processes such as image segmentation or binarization.

Our experiments over 5 corpora of document collections reveal a modest improvement by using TTA for binarization. The improvement is greater as the amount of annotated data is reduced, thereby making this technique an interesting solution to be further studied. Among the criteria employed for combining the TTA results, `vote100` is the best option when there are sufficient annotated data, but `max-policy` and `vote50` increases the results with limited data.

One of the most interesting points is the wide range of options for extending this research. The combination of TTA with meta-learning techniques such as Metric learning or Gradient-based Meta-Learning could be interesting to be explored. Also, few-shot-based architectures could boost the quality of binarization in cases with scarce annotated data. In addition, we have employed generic data augmentation transformations, but, because it is a key role in this research, it could be extended to explore other augmentation options or even to enhance those techniques that further influence the results.

References

1. Afzal, M.Z., Pastor-Pellicer, J., Shafait, F., Breuel, T.M., Dengel, A., Liwicki, M.: Document image binarization using LSTM: a sequence learning approach. In: Proc.

of the 3rd International Workshop on Historical Document Imaging and Processing, pp. 79–84. New York, NY, USA (2015)
2. Ayatollahi, S.M., Nafchi, H.Z.: Persian heritage image binarization competition (PHIBC 2012). In: 2013 First Iranian Conference on Pattern Recognition and Image Analysis (PRIA), pp. 1–4. IEEE (2013)
3. Bainbridge, D., Bell, T.: The challenge of optical music recognition. Comput. Humanit. **35**(2), 95–121 (2001)
4. Burie, J.C., et al.: ICFHR 2016 competition on the analysis of handwritten text in images of balinese palm leaf manuscripts. In: 15th International Conference on Frontiers in Handwriting Recognition (ICFHR), pp. 596–601 (2016)
5. Calvo-Zaragoza, J., Gallego, A.J.: A selectional auto-encoder approach for document image binarization. Pattern Recogn. **86**, 37–47 (2019)
6. Calvo-Zaragoza, J., Rico-Juan, J.R., Gallego, A.J.: Ensemble classification from deep predictions with test data augmentation. Soft. Comput. **24**, 1423–1433 (2020)
7. Campos, V.B., Toselli, A.H., Vidal, E.: Natural language inspired approach for handwritten text line detection in legacy documents. In: Proc. of the 6th Workshop on Language Technology for Cultural Heritage, Social Sciences, and Humanities (LaTeCH 2012), pp. 107–111 (2012)
8. Castellanos, F.J., Gallego, A.J., Calvo-Zaragoza, J.: Unsupervised neural domain adaptation for document image binarization. Pattern Recogn. **119**, 108099 (2021)
9. Doermann, D., Tombre, K.: Handbook of Document Image Processing and Recognition. Springer, London (2014). https://doi.org/10.1007/978-0-85729-859-1
10. Gatos, B., Ntirogiannis, K., Pratikakis, I.: ICDAR 2009 document image binarization contest (DIBCO 2009). In: 2009 10th International Conference on Document Analysis and Recognition, pp. 1375–1382. IEEE (2009)
11. Giotis, A.P., Sfikas, G., Gatos, B., Nikou, C.: A survey of document image word spotting techniques. Pattern Recogn. **68**, 310–332 (2017)
12. Greff, K., Srivastava, R.K., Koutník, J., Steunebrink, B.R., Schmidhuber, J.: LSTM: a search space odyssey. IEEE Trans. Neural Netw. Learn. Syst. **28**(10), 2222–2232 (2017)
13. He, S., Wiering, M., Schomaker, L.: Junction detection in handwritten documents and its application to writer identification. Pattern Recogn. **48**(12), 4036–4048 (2015)
14. Huang, X., Li, L., Liu, R., Xu, C., Ye, M.: Binarization of degraded document images with global-local U-Nets. Optik **203**, 164025 (2020)
15. Louloudis, G., Gatos, B., Pratikakis, I., Halatsis, C.: Text line detection in handwritten documents. Pattern Recogn. **41**(12), 3758–3772 (2008)
16. Nalepa, J., Myller, M., Kawulok, M.: Training- and test-time data augmentation for hyperspectral image segmentation. IEEE Geosci. Remote Sens. Lett. **17**(2), 292–296 (2020)
17. Otsu, N.: A threshold selection method from gray-level histograms. IEEE Trans. Syst. Man Cybern. **9**(1), 62–66 (1979)
18. Pastor-Pellicer, J., España-Boquera, S., Zamora-Martínez, F., Afzal, M.Z., Castro-Bleda, M.J.: Insights on the use of convolutional neural networks for document image binarization. In: Rojas, I., Joya, G., Catala, A. (eds.) IWANN 2015. LNCS, vol. 9095, pp. 115–126. Springer, Cham (2015). https://doi.org/10.1007/978-3-319-19222-2_10
19. Sauvola, J., Pietikäinen, M.: Adaptive document image binarization. Pattern Recogn. **33**(2), 225–236 (2000)

A Weakly-Supervised Approach for Layout Analysis in Music Score Images

Eric Ayllon, Francisco J. Castellanos(✉), and Jorge Calvo-Zaragoza

Department of Software and Computing Systems, University of Alicante, Alicante, Spain
eap56@alu.ua.es, {fcastellanos,jcalvo}@dlsi.ua.es

Abstract. In this paper, we propose a data-efficient holistic method for layout analysis in the context of Optical Music Recognition (OMR). Our approach can be trained by just providing the number of staves present in the document collection at issue (weak label), thereby making it practical for real use cases where other fine-grained annotations are expensive. We consider a Convolutional Recurrent Neural Network trained with the Connectionist Temporal Classification loss function, which must retrieve a pretext sequence that encodes the number of staves per page. As a by-product, the model learns to relate every image row according to the presence or not of a staff. We demonstrate that our approach achieves performances close to the full supervised scenario on two OMR benchmarks, according to the eventual performance of the full transcription pipeline. We believe that our work will be useful for researchers working on music score recognition, and will open up new avenues for research in this field.

Keywords: Optical Music Recognition · Layout Analysis · Weakly-Supervised Learning

1 Introduction

Optical Music Recognition (OMR) is the field that studies how to automatically read music notation from music score images [3]. The digitization of music documents is a challenging task that typically involves the supervision of humans. Growing progress in technology opens up the opportunity to develop automatic high-performance techniques. However, designing the most accurate models is not the unique concern. The intervention of humans in building annotated corpora for training current methods can be assumed as an important bottleneck, and any attempt to improve this situation can lighten the workload to become more efficient on a large-scale basis.

This paper is part of the I+D+i project PID2020-118447RA-I00 (MultiScore), funded by MCIN/AEI/10.13039/501100011033.

A. Pertusa et al. (Eds.): IbPRIA 2023, LNCS 14062, pp. 170–181, 2023.
https://doi.org/10.1007/978-3-031-36616-1_14

Recent contributions to the OMR field point out to the use of deep learning techniques. They have demonstrated high performance and efficiency in multiple contexts such as speech recognition, object detection or image processing, among many others. Nevertheless, the need for annotated data, obtained typically by hand, is often a problem that requires a great deal of effort and time by experts in the topic. The vast heterogeneity in music documents does not help to this issue: the specific music notation, the era, the author style, the engraving mode or the manner in which the images of the documents have been captured are examples of this variety.

The state of the art in OMR trends to the use of end-to-end techniques to obtain the music sequence of individual staves, previously detected from the music score images by a document layout analysis process. In the music context and for the needs of the end-to-end approaches, this process detects the position of the different staves within the image. Recent research locates deep learning techniques as a high-performance and efficient strategy for solving document analysis tasks, but they need previous annotations of the position of reference images, used for training the models [8,12,13].

Music manuscripts are typically written on templates,[1] and most of the pages within a piece contain the same distribution and number of staves. Although this characteristic may be useful to train staff-retrieval models, it is a fact that has not been exploited so far for this task, and state-of-the-art strategies tend to ignore this information.

In this paper, therefore, we contribute with an innovative deep-learning method based on Convolutional Recurrent Neural Network (CRNN) and Connectionist Temporal Classification (CTC) to detect the staves in a music score image without annotating their positions. Instead, the method takes advantage of the layout template for each collection and it only uses the number of staves depicted on each music score image to train a staff-retrieval model. This same idea has been carried out previously using Hidden Markov Models [2], whose modeling allows a similar approach. However, how to do this using state-of-the-art technology with deep learning has not yet been explored.

In addition to the performance for locating staves, we study how this method affects the final transcription. Results eventually demonstrate that our method is a potential solution for document analysis tasks in the music context, removing the ground-truth requirements that would be needed by the state-of-the-art approaches.

2 Methodology

While historical music documents are highly heterogeneous in many aspects, most of them were written using templates. These templates follow a uniform structure within the same collection. This means that while the music content is varying, the number of staves n remains the same. The specific value n can

[1] Even in the case of handwritten sources, the staff lines are printed or drawn with a ruling pen.

be easily determined by observing one image, thus being practically effortless to provide if compared with annotating the bounding boxes of the staves for a set of reference images. We propose to exploit this fact with the aim of training a model to extract music staves with just the specific value of n (weak label).

Typically templates of music scores depict the staves within each page positioned one below another with a clear space between them (either empty or containing lyrics). From the top-down order, we can see this vertical structure as a sequence of two elements: non-staff and staff, henceforth $\mathtt{N}$ and $\mathtt{S}$, respectively. As an example, given a template with three staves—$n = 3$—we may represent this structure as a sequence in the form $\mathcal{Q} = \{\mathtt{N}, \mathtt{S}, \mathtt{N}, \mathtt{S}, \mathtt{N}, \mathtt{S}, \mathtt{N}\}$. The first $\mathtt{N}$ stands for the upper margin and the last $\mathtt{N}$ for the bottom one. Note that the number of different types of elements may be extended but this binary categorization will be considered in this work.

Since $\mathcal{Q}$ is a sequence, we can formulate the layout analysis problem with sequential modeling. For this, we propose to use the well-known CRNN architecture [9] trained by using CTC [10], since, in the literature, it is considered as state-of-the-art in sequence transcription without position annotations of the elements to be recognized. The scheme of the proposal can be found in Fig. 1.

Note that the unique information required from humans to represent a music score image as a sequence $\mathcal{Q}$ is the number of staves n within the images. In addition, the method does not need any alignment information between the image and the sequence. Training the model with CTC enables the possibility to skip manual annotations of the bounding boxes, and the model automatically learns to find staves. This process is carried out at full page, and the model processes rows of the image—frames—to determine the presence of a staff, i.e. no bounding boxes are detected, but frames with music information.

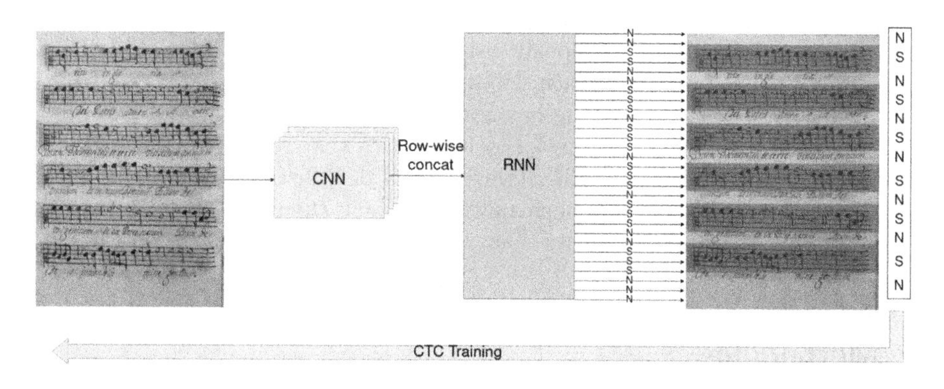

Fig. 1. Scheme of our proposal. We train a CRNN with CTC to eventually predict the sequence that represents the template depicted in the music piece. Note that no positional annotations are required to train the model.

3 Experimental Setup

In this section, we describe the experiments that have been carried out to evaluate our proposal. Section 3.1 presents the corpora; Metrics to evaluate the results are explained in Sect. 3.2; Sect. 3.3 enumerates the different scenarios to be explored in the experiments; and Sect. 3.4 indicates the implementation considerations.

3.1 Corpora

In our experiments, we considered the following corpora:

- CAPITAN containing 129 pages of a *Missa* dated in the second half of the 17th century with annotations of the position of the staves and the music sequence within them [5].
- SEILS [11] including 140 pages of music scores from the 16th century anthology of Italian madrigals *Il Lauro Secco*.

Both corpora have ground-truth information specific for OMR, including bounding boxes of staves and their content, and were used for very related experiments [6]. Table 1 summarizes the description of the CAPITAN and SEILS manuscripts.

Table 1. Summary description of the corpora considered for the experimentation.

Description	CAPITAN	SEILS
Pages	129	140
Staves per page	6	9
Number of symbols	322	184
Number of running symbols	17 016	30 689
Avg. resolution ($w \times h$)	$1\,517 \times 2\,110$ px.	537×813 px.
Avg. height of staves	212 px.	61 px.
Avg. height of non-staff regions	121 px.	27 px.

3.2 Metrics

Since we need to evaluate the quality of the retrieved staves with respect to the expected ones, we considered using the Jaccard index as a reiterative metric in the literature of related fields such as Optical Character Recognition [1]. The Jaccard index intends to measure the goodness of the prediction by the overlapping between the prediction and the expected result. It can be calculated as

$$\mathcal{J}(\mathrm{S}, \mathrm{S}') = \frac{\mathrm{S} \cap \mathrm{S}'}{\mathrm{S} \cup \mathrm{S}'}, \tag{1}$$

where S, in our case, represents the expected position of the staves and S' stands for the respective predictions. Note that each staff involves two 1D coordinates within the image, the start and the end of the staff in the vertical layout. To obtain this measure, it is first necessary to match each expected staff with the predicted one by calculating $\mathcal{J}$ for every possible combination of ground truth and predicted staff and selecting the prediction with the highest value.

For further analysis of the predicted staves, we also considered the F-measure ($\mathcal{F}_1$), which provides more information about the number of staves whose quality overcomes a minimum quality. We used the Jaccard index with a threshold of 0.5 and 0.7 for this purpose. These thresholds were determined in a previous work to analyze the layout analysis results [8]. $\mathcal{F}_1$ is computed as

$$\mathcal{F}_1 = \frac{2 \cdot \mathrm{TP}}{2 \cdot \mathrm{TP} + \mathrm{FP} + \mathrm{FN}}, \tag{2}$$

where TP or *True Positives* indicates the number of predicted staves that overcome the threshold in terms of Jaccard index, FP or *False Positives* represents the number of predicted staves which cannot be matched with any expected one or they did not overcome the Jaccard threshold, and FN or *False Negatives* being those expected staves with no prediction.

Finally, intending to increase the analysis of the results, we include a transcription metric—Symbol Error Rate (SER)—widely used in OMR [4]. This metric is useful to determine if the predicted staves are proper for further transcription, which is the final goal in the OMR field. Let H be the set of hypothesized or predicted sequences of music symbols and G the set of ground-truth ones, the SER metric can be computed as

$$\mathcal{SER}(H, G) = \frac{\sum_{i=0}^{n} d(\mathbf{g}_i, \mathbf{h}_i)}{\sum_{i=0}^{n} |\mathbf{g}_i|}, \tag{3}$$

where $d(\cdot, \cdot)$ stands for the edit distance[2] between a hyphotesis sequence $\mathbf{h}_i$ and its respective ground-truth sequence $\mathbf{g}_i$, and $|\mathbf{g}_i|$ is the amount of tokens of a ground-truth sequence. Note that this metric only can be used for predictions that match with any ground-truth staff, i.e. TP.

3.3 Experiments

The experiments were designed to address two evaluation scenarios: a first evaluation of the quality of the staff detection, and a second evaluation of the influence of these predictions for end-to-end transcriptions. For both scenarios, a 10-fold cross-validation was employed, with 80% of pages for training, 10% for validation, and 10% for testing.

Evaluation Scenario I: Staff Detection. Concerning the first evaluation scenario, a CRNN trained with CTC is used for the staff detection following the

[2] The edit-distance function is implemented as the Levenshtein distance.

methodology described in Sect. 2. The only information required is the number of staves n of the template used for the specific corpus. Our approach encodes n as a sequence $\mathcal{Q}$ and it is used for training the model. We use the Jaccard index and $\mathcal{F}_1$ metrics to evaluate these results and compare them with the state-of-the-art approach in layout analysis without annotations for the target manuscript [7]. The implementation details are described in Sect. 3.4.

Evaluation Scenario II: End-to-End Recognition. Concerning the second evaluation scenario, the objective is to evaluate the influence of staff detection in the final transcription. For this, we make use of the state-of-the-art end-to-end approach, which uses a CRNN trained with CTC to predict the music sequence.

For training the end-to-end transcription model, we considered using data augmentation to increase variability. In this case, we randomly extend the vertical size of the staves employing up to 30 pixels of upper and bottom trims—60 in total as maximum—fixed with preliminary experiments. Figure 2 shows some examples of this augmentation.

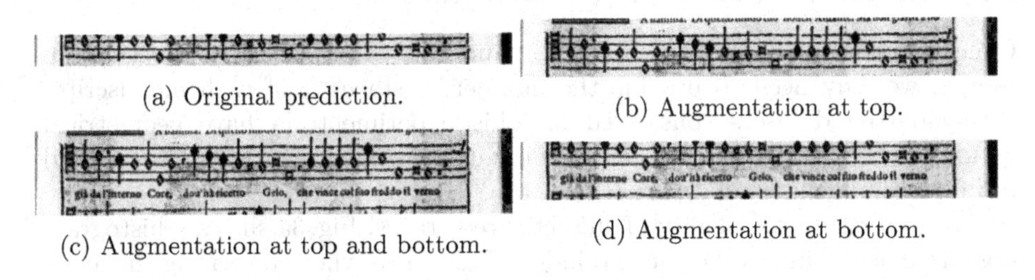

(a) Original prediction.

(b) Augmentation at top.

(c) Augmentation at top and bottom.

(d) Augmentation at bottom.

Fig. 2. Example of different data augmentations cases for a predicted staff.

3.4 Implementation Details

For both evaluation scenarios, we used a CRNN with similar architecture. They have two convolutional layers of 64 filters of 5×5 pixels of size, followed by two more of 128 filters of 3×3 pixels. Each one is followed by batch normalization, a Leaky Rectified Linear Unit with 0.2 of slope, and a 2D max pooling, henceforth referenced as a convolutional block. The detection model is configured with 2×2 pooling in the first convolutional block and 1×2 pooling for the remaining, whereas the transcription model uses 2×2 pooling in the first two convolutional blocks, while it uses 2×1 for the rest.

After the four convolution blocks, both models have a recurrent part with two Bidirectional Long Short-Term Memory (BLSTM) layers with 256 units each and a dropout of 0.5 after them. The models finish the architecture with a fully connected layer with softmax activation.

The staff-detection model is trained for a maximum of 30 epochs with an early stopping of 5 epochs of no improvement in terms of SER on the validation partition, while the transcription model is trained with 40 epochs. It is important to highlight that $\mathcal{F}_1$-based metrics are not viable for validating this model on the training process since they would require ground-truth information about the position of the staves—which we intend to avoid with our method—but SER does not require it.

Note that the images are normalized in the form $x' = \frac{255-x}{255}$, being x each pixel of the image and x' the normalized pixel. In addition, the staff-detection model requires a previous transpose of the input image—the full page—in order to process the image from the left to the right in the training and prediction processes.

4 Results

Once the experiments are defined, we present the results from the point of view of a staff retrieval method first, and then from the point of view of the transcription as a goal-directed experiment.

4.1 Evaluation Scenario I: Staff Detection

Concerning our staff-detection approach, and following the scheme described in Sect. 2, we only need to provide the number of staves n of each manuscript. Although the datasets considered for this experiment do have geometrical ground-truth information, this will be only used in the evaluation protocol (and not for training the model).

For evaluating the quality of the retrieved staves, Fig. 3a shows a histogram organized according to the Jaccard index $\mathcal{J}$ obtained when comparing the automatically detected staves with the ground truth. The figure indicates that most of the staves were predicted with a $\mathcal{J} > 50\%$. This means that at least half of the region of the staves is properly recognized. We can see two relevant local maximum values in the curve: with $\mathcal{J} = 55\%$ with a representation of near to 10% of the staves, and with $\mathcal{J} = 75\%$ with about 11% of the staves. Note that, according to previous works [8], 70% of the Jaccard index can be considered as a value that almost guarantees a good transcription.

To complement the previous histogram, Fig. 3b indicates the number of staves that overcomes a Jaccard index threshold, identified as the x-axis. For example, when the threshold is 50%, the curve represents the number of predicted staves with 50% or more of the Jaccard index. We can observe that there are over 40% of the staves that reach 70% of $\mathcal{J}$, while around 80% of the staves have $\mathcal{J} \geq 50\%$. This fact demonstrates that the method can detect staves with a reasonable quality to be then processed by an end-to-end transcription model.

For further analysis, Table 2 reports the $\mathcal{F}_1$ metric for the staff-retrieval process using a Jaccard threshold of 0.5 ($\mathcal{F}_1^{\text{th}=0.5}$) and 0.7($\mathcal{F}_1^{\text{th}=0.7}$). As a reference, our approach is compared with the results reported on the unsupervised domain adaptation approach [7] by selecting the cases with Capitan and Seils. This work proposed a domain adaptation technique to learn features from an annotated manuscript—source domain—to process another non-annotated

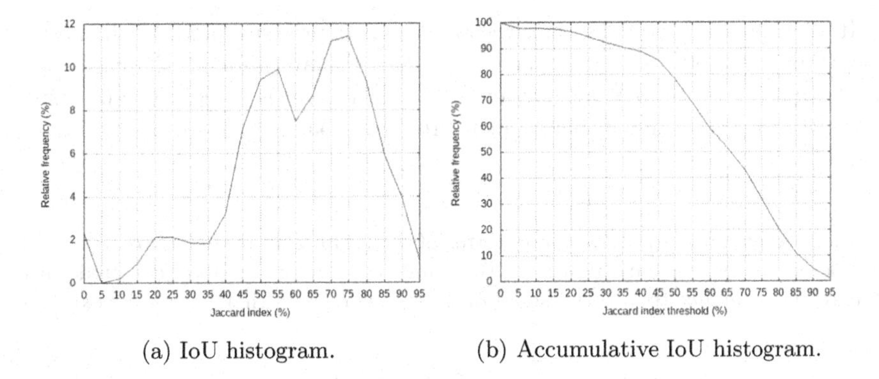

(a) IoU histogram. (b) Accumulative IoU histogram.

Fig. 3. Figure (a) shows the histogram of the number of predicted staves in terms of percentage according to the Jaccard index obtained for both datasets. Figure (b) indicates the number of predicted staves overcoming a Jaccard index threshold.

one—target domain. To evaluate CAPITAN—target—SEILS would be the source domain, and to evaluate SEILS, the opposite. Despite not following the same assumptions (it does use annotated corpora, yet not from the target domain), we believe that this is a good reference of the quality that can be achieved with limited information from the corpus to be transcribed.

Considering, therefore, only the datasets used in our experiments, when the threshold is 0.7, the unsupervised domain adaptation method obtains 78.6% and 20.9% in experiments on CAPITAN and SEILS, respectively. Our approach reports $\mathcal{F}_1^{\text{th}=0.7} = 58.9\%$ for CAPITAN and $\mathcal{F}_1^{\text{th}=0.7} = 19.1\%$ for SEILS. The number of staves predicted with $\mathcal{J} \geq 70\%$ is of course lower with our approach, but note that our model never receives positional information of the staves. In addition, we can observe that SEILS experiences a slight decrease in $\mathcal{F}_1^{\text{th}=0.7}$, from 20.9% to 19.1%. In that case, the results only reduce less than 2 points of this figure. In CAPITAN, the loss is higher, from 78.6% to 58.9%. Even so, a great number of staves have been detected with high quality, and these results would be much higher with a more flexible Jaccard threshold.

In addition, Table 2 also reports the $\mathcal{F}_1$ metric with a Jaccard threshold of 0.5. Figure 3 demonstrates that our approach is able to detect near to 80% of the staves when $\mathcal{J} = 50\%$. Focusing on these results, the unsupervised domain adaptation reports $\mathcal{F}_1^{\text{th}=0.5} = 95.2\%$ for CAPITAN and $\mathcal{F}_1^{\text{th}=0.5} = 33.9\%$ for SEILS. In this case, our approach improves both results, from 95.2% to 95.7% and from 33.9% to 55.3% for CAPITAN and SEILS, respectively. Although in CAPITAN is observed a slight improvement, note the high results for both approaches. Basing on the $\mathcal{F}_1$ results, from the point of view of the error we are obtaining a relative improvement of over 10%. In the case of SEILS, the improvement is much more noticeable, representing over 32% of relative error improvement. Therefore, these results demonstrate that our method is also proper to detect staves with $\mathcal{J} \geq 50\%$.

It is interesting to note that we also studied precision and recall metrics. They report stable results very similar to the $\mathcal{F}_1$ values for all the cases. This means that our method is not particularly biased towards detecting non-existent staves nor missing some of them, but it provides a balanced result regarding this aspect.

Table 2. Average ± std. deviation in terms of F_1 obtained from the 10-fold CV on the staff detection experiment to evaluate the number of staves that were detected with a Jaccard index higher than a threshold of 0.5 and 0.7. Note that values are expressed in %.

Method Corpus	$\mathcal{F}_1^{th=0.5}$ (%)	$\mathcal{F}_1^{th=0.7}$ (%)
Unsupervised Domain Adaptation [7]		
CAPITAN	95.2	78.6
SEILS	33.9	20.9
Our approach		
CAPITAN	95.7 ± 0.1	58.9 ± 0.6
SEILS	55.3 ± 1.9	19.1 ± 0.6

4.2 Evaluation Scenario II: End-to-End Recognition

Concerning the goal-directed experiment, we evaluate our predictions through a staff-level end-to-end recognition model. Table 3 shows the results in terms of SER—the lower, the better. We compare the transcription results of the staves predicted by our approach with the results obtained by the staves of the state-of-the-art in full-page OMR [6], with the same datasets considered in our experiments.

As expected, the augmentation is particularly beneficial for SEILS. This is because the staff-retrieval predictions on SEILS are much less accurate—see Table 2—in both, the reference and our approach. Compared with the reference, if the end-to-end model is trained with ground-truth staves, CAPITAN yields a SER of 16.8%, while SEILS obtains 5.2%. In the case of CAPITAN, the augmented model used on our predicted staves improves the results from 16.8% to 15.5%. However, if the reference model uses (supervised) predicted staves, SER obtains 14.8%, improving slightly our case. In the case of SEILS, our best case obtains 8.6% of SER, while the ground-truth reference model reports 5.2%. Using predicted staves by the reference reduces this metric to 4.4%. This supposes a slightly worst transcription than the reference. Note that the reference results are extracted from previous work [6].

Table 3. Transcription results in terms of SER (%) with standard deviation. The reference method includes results reported in a previous work [6] on predicted staves in two cases: by training the model with ground-truth staves and by training it with predicted staves. Concerning our approach, the table reports the results with and without data augmentation.

Method Case	CAPITAN	SEILS
End-to-end on reference [6]		
Ground-truth staves	16.8 ± 3.7	5.2 ± 1.4
Predicted staves	14.8 ± 3.6	4.4 ± 0.5
End-to-end on our approach		
No augmentation	17.0 ± 1.8	21.0 ± 3.9
With augmentation	15.5 ± 1.4	8.6 ± 2.2

In general, the staves detected automatically with our method allow attaining a competitive performance in the final transcription with respect to the case of both having ground-truth staves or automatically-detected staves from a model trained with positional annotations. In addition to its competitiveness, it is important to highlight the huge cost reduction of our proposal, which reduces almost to the minimum the amount of ground-truth information required.

4.3 Overview of the Pipeline Quality

So far, only empirical results have been analyzed. Because of the importance of the layout analysis for the final transcription, in this section, we show some qualitative and representative examples of the results obtained by our proposal. Figure 4 includes a series of examples with different degrees of quality according to $\mathcal{J}$.

Figures 4a and 4b show two examples of staves extracted by our approach for CAPITAN. In the first example, the staff can be considered as correctly retrieved. Graphically, we can observe that the image contains the complete staff without missing music content for the further transcription. In this case, the Jaccard index is 98.7% with respect to the ground truth, while the transcription obtains a SER of 7.1%. Given the high value of $\mathcal{J}$, the transcription achieves a result better than the global result reported in Table 3—SER = 15.5% in the best of our cases. The second example, which is Fig. 4b, depicts a recurrent problem when the $\mathcal{J}$ metric yields low figures: missing part of the content of the staff. In this case, there are symbols that are not included within the image, increasing the difficulty of correctly transcribing the sequence. In this case, $\mathcal{J}$ reports a result of 47.3%, while SER is 20.8%. However, it should be noted that the same $\mathcal{J}$ figure may be obtained by shifting the prediction to the top and missing part of the information at the bottom. In such a case, since the symbols are in the top part of the staff, the transcription may be successfully made. This demonstrates the importance of layout analysis in the OMR field.

Concerning SEILS, a similar pattern can be found. There are cases in which the staff can be considered as correctly detected, as the example shown in Fig. 4c with $\mathcal{J} = 96.9\%$ and SER $= 7.4\%$, slightly improving the global SER result—8.6%. On the other hand, there are cases with poor $\mathcal{J}$ figures, such as the case shown in Fig. 4d. That example contains a staff in which part of the information at the top and the bottom is removed. There are some symbols affected by this miss-detection, but most of the note-heads are visible. While $\mathcal{J}$ is 44%, which could be considered as a poor retrieval, the end-to-end model obtains 10% of SER. This reinforces the idea of a low $\mathcal{J}$ value may not be totally correlated with the transcription, demonstrating the need of evaluating layout analysis in goal-directed experiments.

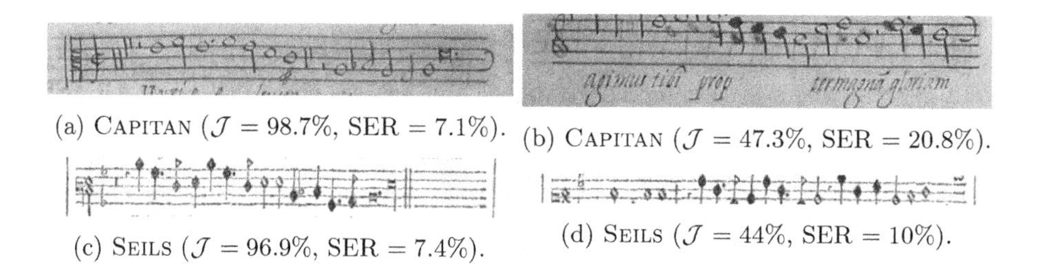

(a) CAPITAN ($\mathcal{J} = 98.7\%$, SER $= 7.1\%$). (b) CAPITAN ($\mathcal{J} = 47.3\%$, SER $= 20.8\%$).

(c) SEILS ($\mathcal{J} = 96.9\%$, SER $= 7.4\%$). (d) SEILS ($\mathcal{J} = 44\%$, SER $= 10\%$).

Fig. 4. Selected examples of retrieved staves and a comparison between the Jaccard index and SER.

5 Conclusions

In this work, we propose an innovative deep-learning approach based on CRNN and CTC to detect the staves from music score images. This method is able to work without stating positional information of the staves, but it only needs to be provided by the number of staves included in each image. This fact can be applied in OMR because a lot of manuscripts are written using templates.

Experiments were divided into two evaluation scenarios: the former validates the quality of our approach for staff detection, and the latter studies the influence of our layout analysis method in the eventual transcription, which is the final goal of OMR. The results support the suitability of our method with results comparable to an existing unsupervised domain adaptation method for this task. The Jaccard index and $\mathcal{F}_1$ metrics point out that the results for high-quality staves (Jaccard ≥ 0.7) are somewhat lower than those of the existing method, but if we increase the flexibility (Jaccard ≥ 0.5), our approach yields better results. On the other hand, the SER metric indicates that our predicted staves obtain transcriptions comparable with a supervised state-of-the-art method. This is particularly interesting since the staves used for training the transcription model are also predicted ones, and the ground truth of bounding boxes has not been used at any time, except for evaluation purposes.

For future work, we plan to extend the experiments to other types of template-based documents such as reports or tablatures. We find it particularly interesting to study the possibility of removing the need of providing the number of staves of the template through clustering techniques in order to detect staves in documents with a non-uniform template.

References

1. Alberti, M., Bouillon, M., Ingold, R., Liwicki, M.: Open evaluation tool for layout analysis of document images. In: 2017 14th IAPR International Conference on Document Analysis and Recognition (ICDAR), vol. 04, pp. 43–47 (2017)
2. Bosch, V., Calvo-Zaragoza, J., Toselli, A.H., Vidal-Ruiz, E.: Sheet music statistical layout analysis. In: 15th International Conference on Frontiers in Handwriting Recognition, ICFHR 2016, Shenzhen, China, 23–26 October 2016, pp. 313–8 (2016)
3. Calvo-Zaragoza Jr., J., J.H., Pacha, A.: Understanding optical music recognition. ACM Comput. Surv. **53**(4), 77:1–77:35 (2021)
4. Calvo-Zaragoza, J., Toselli, A.H., Vidal, E.: Handwritten music recognition for mensural notation with convolutional recurrent neural networks. Pattern Recogn. Lett. **128**, 115–121 (2019)
5. Calvo-Zaragoza, J., Toselli, A.H., Vidal, E.: Handwritten music recognition for mensural notation: Formulation, data and baseline results. In: 14th IAPR International Conference on Document Analysis and Recognition, ICDAR 2017, Kyoto, Japan, 9–15 November 2017, pp. 1081–1086 (2017)
6. Castellanos, F.J., Calvo-Zaragoza, J., Inesta, J.M.: A neural approach for full-page optical music recognition of mensural documents. In: ISMIR, pp. 558–565 (2020)
7. Castellanos, F.J., Gallego, A.J., Calvo-Zaragoza, J., Fujinaga, I.: Domain adaptation for staff-region retrieval of music score images. Int. J. Doc. Anal. Recognit. (IJDAR) 1–12 (2022)
8. Castellanos, F.J., Garrido-Munoz, C., Ríos-Vila, A., Calvo-Zaragoza, J.: Region-based layout analysis of music score images. Expert Syst. Appl. **209**, 118211 (2022)
9. Graves, A.: Supervised sequence labelling with recurrent neural networks. Ph.D. thesis, Technical University Munich (2008)
10. Graves, A., Fernández, S., Gomez, F., Schmidhuber, J.: Connectionist temporal classification: labelling unsegmented sequence data with recurrent neural networks. In: Proceedings of the 23rd International Conference on Machine Learning, ICML 2006, pp. 369–376. ACM, New York (2006)
11. Parada-Cabaleiro, E., Batliner, A., Schuller, B.W.: A diplomatic edition of il lauro secco: Ground truth for OMR of white mensural notation. In: Proceedings of the 20th International Society for Music Information Retrieval Conference, ISMIR 2019, Delft, The Netherlands, 4–8 November 2019, pp. 557–564 (2019)
12. Quirós, L., Toselli, A.H., Vidal, E.: Multi-task layout analysis of handwritten musical scores. In: Morales, A., Fierrez, J., Sánchez, J.S., Ribeiro, B. (eds.) IbPRIA 2019. LNCS, vol. 11868, pp. 123–134. Springer, Cham (2019). https://doi.org/10.1007/978-3-030-31321-0_11
13. Waloschek, S., Hadjakos, A., Pacha, A.: Identification and cross-document alignment of measures in music score images. In: 20th International Society for Music Information Retrieval Conference, pp. 137–143 (2019)

ResPho(SC)Net: A Zero-Shot Learning Framework for Norwegian Handwritten Word Image Recognition

Aniket Gurav[1(✉)], Joakim Jensen[1], Narayanan C. Krishnan[2], and Sukalpa Chanda[1]

[1] Department of Computer Science and Communication, Østfold University College, Halden 1757, Norway
aniketag@hiof.no, sukalpa@ieee.org

[2] Department of Data Science, Indian Institute of Technology Palakkad, Kerala, India
ckn@iitpkd.ac.in

Abstract. Recent advances in deep Convolutional Neural Networks (CNNs) have established them as a premier technique for a wide range of classification tasks, including object recognition, object detection, image segmentation, face recognition, and medical image analysis. However, a significant drawback of utilizing CNNs is the requirement for a large amount of annotated data, which may not be feasible in the context of historical document analysis. In light of this, we present a novel CNN-based architecture ResPho(SC)Net, to recognize handwritten word images in a zero-shot learning framework. Our method proposes a modified version of the Phosc(Net) architecture with a much lesser number of trainable parameters. Experiments were conducted on word images from two languages (Norwegian and English) and encouraging results were obtained.

Keywords: Zero-Shot Learning · Word Image Recognition · Zero-Shot Word Image Recognition · Deep Learning · OCR

1 Introduction

Digital archives of historical documents are important for preserving historical documents and making them accessible to researchers, historians, and the general public. Digitization helps to preserve a society's cultural heritage by providing access to its historical records. It plays an important role in revealing unknown facts about the past society. Earlier end-users of digital archives used to deploy Optical Character Recognition (OCR) to get transcriptions of the manuscripts to extract information of their interest. In this approach, the performance of the OCR used to be greatly affected due to erroneous segmentation of characters in cursive handwritten text. As a historical document is often characterized by several adversities like - ink bleed-through effect, poor background foreground contrast, and faded text ink, such challenges can result in a

A. Pertusa et al. (Eds.): IbPRIA 2023, LNCS 14062, pp. 182–196, 2023.
https://doi.org/10.1007/978-3-031-36616-1_15

loss of valuable information and lead towards erroneous character segmentation and consequently erroneous transcription of the text. Moreover, it was observed that the end-users of such systems (historians and paleographers in particular) are not interested in the full transcription of the text, rather they are often interested in accessing some particular portion of the text where a specific word has occurred. This led to word recognition and word spotting techniques that enabled the end-user to index and search a specific document in a huge digital archive. In the recent past, CNNs have shown impressive performance on various classification tasks. CNNs can automatically learn and extract information and handle the variances from images but it requires the availability of annotated data for all classes during the training process. Annotating historic documents is a time-consuming, costly, and resource-dependent process that needs to deploy knowledge experts (historians and paleographers) in the process. Hence, annotating all possible word classes in a particular script is practically an infeasible task. Word recognition in a ZSL framework can combat this situation, where a word image can be correctly classified even though that particular class was never present in the training set. ZSL has the ability to perform under the limited availability of labeled data. Historical texts are often scarce and can have a large vocabulary, and annotation in this case is a tedious and expensive task. ZSL helps to overcome the problem of the lack of labeled data by allowing the model to learn from auxiliary data from other domains or languages. ZSL is capable of handling variability in handwriting due to different writers with varying handwriting styles. Moreover, it has been observed in [4] that if the data is from multiple writers, then CTC-based approach is not good for zero-shot word recognition, it is very effective to recognize unseen classes in a zero-shot word recognition problem if the dataset is from a single author/handwriting. ZSL is also adaptable to a wide range of domains in historical texts such as literature, science, and law. ZSL can be adapted to new domains by fine-tuning the model on a small set of labeled data from the target domain. These reasons make ZSL a prime candidate over other word recognition and spotting techniques.

The method proposed in the work, referred to as ResPho(SC)Net, incorporates the use of Pyramidal Histograms of Characters (PHOC) and Pyramidal Histograms of Shapes (PHOS) to facilitate the recognition of words by using a CNN of ResNet-18 architecture. The network is trained to learn the PHOC and PHOS representation of word images from various classes present in the training set. The proposed ResPho(SC)Net is capable of recognising Norwegian as well as English handwritten words. Figure 1 depicts sample images of Norwegian handwritten data indicating problems involved with this type of data. The images show variations in writing styles, text size, and text orientation. The differences in handwritten images shown in the top two rows illustrate handwriting deviation across individuals. The bottom three rows highlight the difficulties of intra-class variability in images across writers. The Norwegian words displayed in the third, fourth, and fifth rows, are namely "meget", "norske", and "være", respectively. Although words are the same due to the existence of intra-class variations there is a difference in their visual appearance which can be observed in corresponding columns. This can result in difficulties in accurately recognizing and classifying the written words. In this work, our contributions are as follows:

- A novel ResNet18-based word recognition architecture ResPho(SC)Net is proposed that clubs the benefits of a residual model and the recently proposed Pho(SC)Net [14] representation for word recognition.
- A significant improvement in terms of fast convergence during training is reported, as the model converges much faster compared to the original Pho(SC)Net [14] and gives better results. The proposed architecture exhibits generalization ability across Norwegian and English handwritten word images.
- The Norwegian word image dataset will be freely available for future research and benchmarking purpose.

Fig. 1. Sample of Norwegian Handwritten Text

2 Related Work

Indexing a document page with particular content in a digital archive can be achieved by either word spotting or by word recognition. In word spotting, a query in the form of an image or as a string of text characters is given as input to the word spotting algorithm which then searches for image regions containing similar words [6]. On the other hand, word recognition could retrieve a particular document image from a corpus after identifying the word boundaries where the recognizer has recognized the desired word class within the bounding box Word spotting is often used in the field of document analysis to locate specific words in large digital archives of document images. This approach, initially proposed by Manmatha et al. in [10,11], has led to many significant research in the field like [16,19]. It is difficult to identify the optimal hand-crafted feature that could render discriminating attribute for a word, character, or patch, as outlined in [10,13,18]. With the advent of CNN, the availability of large-scale annotated data, and the increased computing power provided by Graphical Processing Units (GPUs) the recent trend is to shift towards feature learning. In [19], the author utilized CNNs to extract features from word images and trained a Pyramidal Histogram of Characters (PHOC) representation and PHOCNet architecture. They were first to use CNNs for end-to-end word spotting. Similarly, a number of architectures utilizing CNN networks have been proposed, which incorporate PHOC-defined textual embedding spaces, as seen in the works [8,21]. In the work, [13] the authors employ a strategy that utilizes multiple parallel fully connected layers, built upon a modified version of VGGNet, for the purpose of recognizing PHOC attributes. Two different AttributeCNN architectures were introduced in [21], one incorporating conventional layers and the other incorporating a new layer called Temporal Pyramid Pooling (TPP), both of which are end-to-end trainable. The study in [8] demonstrates the use of HWNet to enhance the discriminative capability of deep CNN features through learning a PHOC-based attribute representation. The architecture of the HWNet includes five convolutional layers, along with two fully connected layers and a multinomial logistic regression loss function. The network is trained on synthetic handwritten word images initially, followed by fine-tuning on real-world data. The second fully connected layer's activation features are utilized as a comprehensive representation for word images. In [9], a modified version of [8] is proposed for efficiently learning a detailed representation of handwritten words through a combination of advanced deep convolutional neural networks and standard classification techniques. The method also includes the use of synthetic data for pre-training and utilizes a modified version of the ResNet-34 architecture with the region of interest pooling for improved effectiveness. Shape attributes of characters were first proposed for zero-shot word recognition problem in [5]. Later, utilizing the shape attribute information the authors in [4,14] proposed an advanced approach that utilizes a combination of PHOC and Pyramidal Histogram of Shapes (PHOS) which is capable of recognizing previously unseen words in datasets in a zero-shot learning framework. In [1] the authors obtain highly discriminatory embeddings of handwritten words, using the encoding component of a Sequence-to-Sequence (Seq2Seq) recognition network, and performs word spotting. In [22] the Triplet-CNN model is proposed for word spotting in a segmentation-based scenario. Triplet-CNN uses three identical CNNs sharing weights. The model is learned on triplets of images, with the similarity between two positive images and one negative image being selectively backpropagated. The performance achieved by the triplet-CNN is comparable to that achieved by the PHOCNet. In the work presented in [3] the authors explore different approaches, including computer-generated text, unsupervised adaptation, and using recognition hypotheses on test sets, to minimize manual annotation effort. The research in [20] builds upon the ideas and concepts presented in [19], it also proposes a new CNN architecture named TPP-PHOCNet for the purpose of word spotting. The experimental evaluation in this work demonstrates significant performance improvements compared to the original PHOCNet.study conducted in [15] emphasizes the importance of using strategies like TPP, augmentation, class balancing, and ensemble learning in keyword spotting. The results showed that these strategies can significantly enhance the performance of the network. Additionally, the study also concluded that using fixed-size images for training is more beneficial compared to using images of variable sizes. In [17], a retrieval list is formed by sorting all the word images on the probability of their PHOC representation. The query representation is obtained directly word string. The authors build upon the TPP-PHOCNet architecture by incorporating both an SPP and a TPP layer, and the output from these two layers is combined as input to the Multi-Layer Perceptron (MLP) part of the CNN. The resulting Attribute CNN shows significant improvement over the results obtained in [20]. The end-to-end learning for segmentation-free word spotting is done in [23]. In their work, they combined a Region Proposal Network (RPN) with a heuristic approach to predict potential regions in a document image. These regions are then used to predict a PHOC or Discrete Cosine Transform of Words (DCToW) representation.

3 Methodology

3.1 Pipeline and Network Design

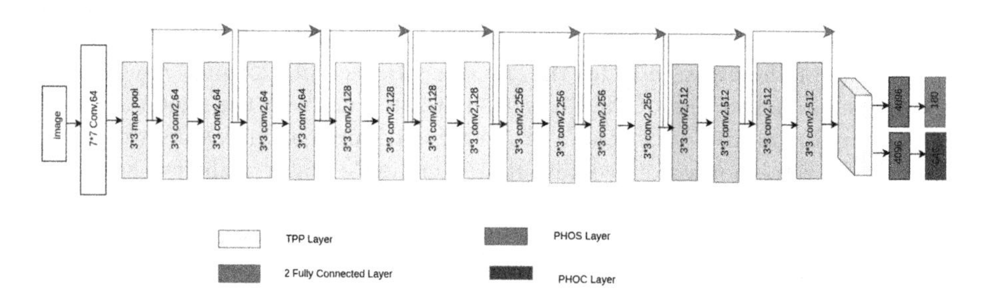

Fig. 2. ResPhoscNet Architecture

The architecture of our model, as depicted in Fig. 2, is comprised of three crucial components, namely, the base feature extractor, the Temporal Pyramid Pooling (TPP) layer, and the PHOS and PHOC layers. The base feature extractor, ResNet-18, adept in processing a 3-channel input image, produces a feature map comprising 512 convolution filters. Subsequently, this feature map is fed through the TPP layer, which effectively decreases the temporal dimensions of the feature map, generating a feature vector of 4096 dimensions. This feature vector undergoes additional processing through dense layers to provide a PHO(SC) representation. Further details regarding our model's architecture and training can be found in Sect. 3.4.

3.2 Problem Description

A collection of examples known as the dataset S is utilized for training the ResPHO(SC) model. This S is represented as triples $(x_i, y_i, c(y_i))$. The S denotes the set of training classes/word labels to which the model will be subjected during training. In $(x_i, y_i, c(y_i))$, x_i is an image from a set of images X, y_i is the corresponding word label from a set of seen word labels Y^S. and $c(y_i)$ is the PHO(SC) attribute signature that belongs to a set C. During the evaluation, the model is expected to handle the unseen classes. These are disjoint sets of unseen labels referred to as $Y^U = \{u_1, u_2, \ldots, u_l\}$. The PHO(SC) attribute signature set for these unseen labels is $U = \{u_l, c(u_l)\}_{l=1}^{L}$, where $c(u_l) \in C$. This attribute signature set is available during testing, but the corresponding class images are not provided during training. In the ZSL it is a common assumption that not all classes of interest will be available during the training phase, and the model should be able to generalize to unseen classes Y^U based on the information provided. The model needs to generalize the learning from the seen data S to identify and classify unseen data. To perform zero-shot word recognition, the aim is to train a classifier $f_{zsl} : X \rightarrow Y^U$, using S and U datasets. In generalized zero-shot word recognition, the goal is to learn the classifier $f_{gzsl} : X \rightarrow Y^U \cup Y^S$. The above problem described is similar to [4].

3.3 PHOS and PHOC Representation

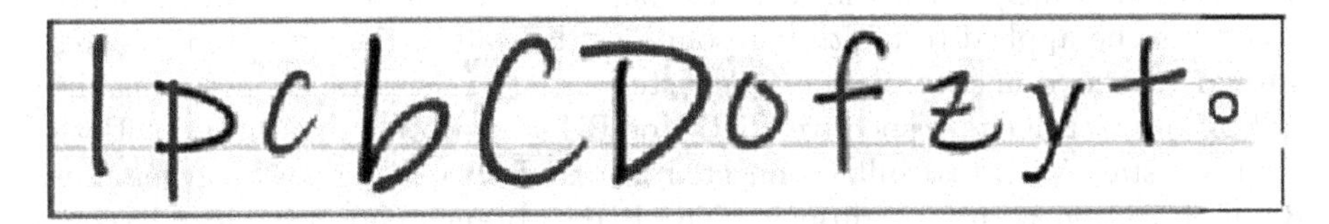

Fig. 3. 12 primary shape attributes: ascender, descender, left small semi-circle, right small semicircle, left large semi-circle, right large semi-circle, circle, vertical line, diagonal line, diagonal line at a slope of 135°, horizontal line, and small circle, This diagram is adopted from [14], and customized to suit our needs.

In this study, we adopt the Pho(SC) attribute representation proposed in [14] for the English and the Norwegian language. The original Pho(SC) representation comprises PHOC and PHOS vectors of 604 and 165 length respectively. The PHOC level representation is based on the [19] where the binary vector is represented using 2, 3, 4, and 5 splits of the word. Furthermore, at the last level, the top 50 most commonly occurring bigrams are encoded in PHOC vector. The PHOC vector length for the English language is 604. For handling the Norwegian data we have modified the vocabulary of PHOC by adding characters not present in the original vocabulary. The characters specific to the Norwegian language - and å,ø,æ are included in each level of the 5-level PHOC representation, due to which the size of the PHOC vector becomes 646.

The PHOS vector is utilized by [14] to establish a resilient connection between seen and unseen words. The PHOS considers the basic shapes consisting of the ascender, descender, left small semi-circle, right small semi-circle, left large semi-circle, right large semi-circle, circle, vertical line, diagonal line, the diagonal line at a slope of 135°, and horizontal line, To cater to the Norwegian language, we have added one extra shape, which is a small circle, to the original PHOS shapes. This original and additional shape is reflected in the last position of the Fig. 3 with red colour. The current version of Fig. 3 in this work is a modification of Fig. 2 from the [14]. Due to these changes in PHOC and PHOS, the total shape of the modified Pho(SC) vector becomes 826 for the Norwegian language. Throughout all experiments and explanations in all sections, it is important to note that the modified PHOSC vector is utilized for the Norwegian language only, while the original PHOSC setting is employed for the English language. This distinction is crucial in ensuring accurate and valid results for each respective language.

3.4 ResPhoscNet Architecture

The Pho(SC)Net is a deep CNN designed for recognizing handwritten words in images [14]. Pho(SC)Net uses a series of convolutional layers followed by ReLu activation and max pooling layers. Having processed the image through the convolutional part of the network, the extracted feature maps are processed by a 3-level Spatial Pyramid Pooling (SPP) layer. The SPP-layer [7,19] is a parameter-free method, which pools the features maps to a constant size, enabling the network to be applied to multiple resolutions. Hereafter, the extracted features are fed to two branches of fully connected layers. Each branch performs the task of outputting its respective PHOS or PHOC vector representation. Both branches start with two fully connected layers of 4096 nodes. Each layer uses ReLu activation with 50% dropout. The PHOC-branch ends with a fully connected layer of 604 nodes with sigmoid activation for English language, and 646 nodes for the Norwegian language. PHOS-branch ends with a fully connected layer of 165 for the English language, and 180 for the Norwegian language, with ReLu activation. For fair comparison on Norwegian data, we modified the last fully connected layer of the Pho(SC)Net architecture to have 646 and 180 nodes, respectively. The original Pho(SC)Net was designed only for the English language. The difference in output layer activation stems from the fundamental difference between PHOS and PHOC. PHOS encapsulates the count of shapes, while PHOC encapsulates the presence of a character (the character is present or not). The Pho(SC)Net has 134.44 million trainable parameters. A large number of trainable parameters in the Pho(SC)Net model makes it susceptible to be biased towards seen class examples and fails to recognize word images from an unseen class in the generalized ZSL setup. In light of a large number of trainable parameters in the Pho(SC)Net model, we introduce a residual version of the Pho(SC)Net, the ResPho(SC)Net. This network shares few components with its predecessor but with some key differences. Instead of using the convolutional architecture of the Pho(SC)Net, the ResNet-18 backbone architecture has been introduced for the convolutional operations [2]. Like ResNet-18, we use a series of residual blocks, doubling the number of feature maps after every second residual block, but at the same time also downsampling the feature maps. Each residual block consists of 2 convolutional layers, each followed by batch normalization and ReLu. For downsampling, the first convolutional layer of the residual block uses stride = 2. For the shortcut of the residual block, the identity function is used whenever downsampling isn't performed. However, to make sure the dimensions match whenever downsampling has been performed the shortcut consists of a single convolutional projection layer followed by batch normalization. The secondary change we made in comparison to Pho(SC)Net [14] was to replace the SPP with a TPP layer. The network branches into a PHOS and PHOC branch which individually, outputs their respective output vectors, however, in the ResPho(SC)Net we have removed a fully connected layer from each of the branches. The ResPho(SC)Net shows improved performance over its predecessor in both seen and unseen GZSL. The residual blocks allow for easier training and better feature extraction. The TPP layer pools the feature maps in a more logical way in context to handwritten words in images. Additionally, we removed a fully connected layer from each branch without affecting the performance. We hypothesize that this is possible because of better feature extraction from the residual blocks.

Temporal Pyramid Pooling. As in [20], the motivation behind TPP is that it might be more useful to subdivide the image along the horizontal axis, completely ignoring the vertical axis, if the images are of sequential nature. Word images tend to be of sequential nature along the horizontal axis, making the TPP layer a good fit for our data. In our approach, we use TPP as proposed in [20]. TPP just like SPP is used to sample the feature maps into a constant size before feeding into the MLP [7]. By having this layer, the feature maps will always be of constant size before going into the MLP, enabling the network to take images of arbitrary sizes. The method contains no extra parameters and is therefore not affected by the back-propagating of loss. In comparison to SPP, which subdivides its feature maps into equal amounts of vertical and horizontal bins, the TPP only subdivides the image along the horizontal axis [20]. Like SPP, a pre-defined number of levels α and a pre-defined divider number β for each level is chosen. In SPP, β divides the feature map along the x-axis and y-axis to create equal size spatial bins per level. However, TPP only subdivides along the horizontal axis creating β temporal bins of equal size. Having the feature maps split into bins, some pooling method is performed on each of the 2D feature maps. Min, max, and average pooling are all available operations for the bins. We perform TPP on 3 levels with β [1, 2, 5], and pool our bins with max pooling.

In comparison to Pho(SC)Net, our proposed ResPho(SC)Net model has 48.12 million parameters, which is almost 3 times less. This reduction in the number of parameters allows for a more computationally efficient model, making it more feasible to run on limited hardware resources. Additionally, a smaller number of parameters also means that the model is less prone to overfitting, allowing for better generalization to new unseen data.

3.5 Loss Function

The final loss function of the proposed ResPho(SC)Net is a linear combination of two different loss types and is defined as follows:

$$L = \sum_{i=1}^{B} \lambda_c L_c(\varphi_c(x_i), c_c(y_i)) + \lambda_s L_s(\varphi_s(x_i), c_s(y_i)) \tag{1}$$

where the term $L_c(\varphi_c(x_i), c_c(y_i))$ represents the cross-entropy loss between the predicted and actual PHOC representations. In which $\varphi_c(x_i)$ represents the predicted PHOC representation of the input image x_i. $c_c(y_i)$ is the true class label in the PHOC representation space for the given word label y_i. The term $L_s(\varphi_s(x_i), c_s(y_i))$ represents the mean squared error loss between the predicted and actual PHOS attributes in which $\varphi_s(x_i)$ represents the predicted PHOS representation of the input image x_i. $c_s(y_i)$ is the true class label in the PHOS representation space for the given word label y_i.

4 Experiments and Results

4.1 Training Set-Up

For training the ResPho(SC)Net model, Norwegian handwritten and IAM datasets are used. For every experiment mentioned in the results, the model is trained for 100 epochs using an Adam optimizer with a batch size of 256. The input images are first resized to a shape of (250,50) before being passed into the network. The learning rate is set to 0.00001. The learning rate scheduler is used to adjust the learning rate based on performance. If the performance does not improve for 5 epochs, the learning rate is reduced by a factor of 0.25. This reduction continues until either the learning rate reaches a minimum value of 1e-12 or the performance improves beyond a specified threshold of 0.0001. This configuration allows the model to effectively learn and recognize variations in handwriting styles and scripts while maintaining high accuracy. In the loss function, parameters λ_c and λ_s are used to determine the weightage given to PHOC and PHOS loss values. The values for λ_c and λ_s are set to 4.5 and 1.5, respectively. All the models are trained on Nvidia V100/32 GB GPU.

4.2 Datasets

In this work, we have used 2 different datasets, to prove the effectiveness of the proposed ResPho(SC)Net model in recognizing seen/unseen words in historical documents. The first one is the Norwegian handwritten dataset procured by the National Library of Norway. The Norwegian handwritten dataset is a collection of handwritten text from historical documents written in the Norwegian language. There are total 7052 samples of handwritten image crops from total 2166 different words or classes. The dataset contains text written by multiple people, with different writing styles. The quality of the documents in the dataset can also be poor with smudges, tears, faded regions, errors, and other distortions, further adding to the difficulty of recognition. During training and evaluation the Norwegian data is divided into 5 folds, each fold contains 70% train data, 10% validation, and 20% test data. Following the aforementioned data split, the training dataset comprises 1766 distinct classes or words. The test dataset is comprised of 400 seen classes and 400 unseen classes, where the former denotes the words that are also present in the training dataset, while the latter indicates the words that are absent in the training dataset. The validation dataset, on the other hand, encompasses a total of 279 classes. The second dataset is the IAM dataset [12]. The dataset contains 115,320 images of handwritten text from 657 writers, including words, paragraphs, and lines of text. The images in the dataset are taken from various sources, including historical documents, books, and forms. The text transcriptions for each image are provided in the dataset, making it a valuable resource for training and evaluating handwriting recognition models. The dataset is also widely used in the research community for developing and evaluating different techniques for handwriting recognition, such as handwriting segmentation, character recognition, and word recognition. For comparison with Pho(SC)Net we have used the standard and ZSL IAM splits mentioned in [14]. It is worth mentioning that the IAM dataset consists of handwritten text from multiple authors. As our Norwegian dataset consists of handwritten text from multiple authors, we preferred the IAM dataset over other benchmark datasets like George Washington as used in [4] for a fair comparison with Pho(SC)Net [4].

Data Augmentation. In [14], limited augmentation techniques were used in the form of adding random noise and shearing. To improve upon our result we added more augmentation methods like random perspective, erosion, dilation, blur, and sharpness to the images in our datasets, in addition to shearing and random noise. We apply a maximum of two methods to an augmented image, the degree of augmentation by one method is decided at random. By adding the improved augmentation we also see improved performance of the model.

4.3 Results on Norwegian Data

Table 1 shows the results of ResPho(SC)Net, on the Norwegian dataset. As the dataset is small to ensure the generalization ability of the proposed method it has been divided into 5 different folds for evaluation, and for each fold, the model's unseen accuracy and seen accuracy are reported. In the context of zero-shot learning, K-Fold cross-validation can be used to evaluate both Zero-shot Learning (ZSL) and Generalized Zero-shot Learning (GZSL) accuracy. In the GZSL scenario, for an input image, the cosine similarity between the predicted vector for that input image and the PHOSC representation of all classes (in the most strict setup) in the training, testing, and validation sets are computed, and the predicted vector is assigned to the class which has got maximum cosine similarity with the predicted vector. In the ZSL scenario, the cosine similarity between the predicted vector and only the unseen classes in the test set is computed and the seen classes are completely ignored. Here again, the predicted vector is assigned to the class which has got max. cosine similarity with the predicted vector. It can be observed from Table 1 that the model performs better on seen words as compared to unseen words, with the highest seen accuracy being 76.85%and the highest unseen accuracy being 39% for GZSL. In the case of ZSL, the maximum accuracy for seen words is 82.59%, while the highest unseen accuracy is 70.5%. The ZSL accuracy outperforms GZSL, due to the fact that GZSL during the testing phase involves a larger number of word classes than ZSL.

Table 1. ResPho(SC)Net GZSL and ZSL results for Norwegian Data in %

fold	GZSL		ZSL	
	Seen Accuracy	Unseen Accuracy	Seen Accuracy	Unseen Accuracy
0	76.45	36	82.59	66.25
1	74.28	39	80.71	67.5
2	72.5	37.75	79.42	70.5
3	75.96	31.75	81.8	68.5
4	76.85	35.0	81.6	68.75

4.4 Comparison of Results on Norwegian Data and IAM Data

Result Comparison on Norwegian Data. Table 2 illustrates the performance of Pho(SC)Net from [14] on the Norwegian 5-fold dataset. The Pho(SC)Net architecture has undergone necessary adaptations to cater to the Norwegian language. A comparison between Table 2 and Table 1 indicates that ResPhosNet has demonstrated superior performance with regard to ZSL and GZSL in both seen and unseen accuracy compared to Pho(SC)Net.

Result Comparison on IAM Dataset. For the sake of fair comparison of our method with another state-of-the-art method [14] we considered the IAM dataset and the corresponding splits of the dataset as it is provided in [4,14].

Table 2. Pho(SC)Net GZSL results for Norwegian Data in %

fold	GZSL		ZSL	
	Seen Accuracy	Unseen Accuracy	Seen Accuracy	Unseen Accuracy
0	42.23	1.5	52.027	10.0
1	52.52	6.5	63.00	31.5
2	52.52	6.5	63.00	31.5
3	54.4	6.0	66.76	33.5
4	47.08	3.75	59.54	15.75

Our Norwegian dataset is a multi-author dataset hence we considered IAM as the dataset in the English language to prove the generalization ability of the proposed method. In [14] the authors have reported ZSL and GZSL accuracy in two different custom-made variants/splits of the IAM dataset namely ZSL split and standard split, we have reported our method's accuracy accordingly on those accordingly in Table 3 and Table 4 with row heading ResPho(SC)Net. For the sake of simplicity in comparison, we have extracted the ZSL and GZSL accuracy values for the aforementioned splits from Table 3 and Table 4 in the paper [14] and mentioned it in Table 3 and Table 4 with row heading Pho(SC)Net. The corresponding higher results are indicated in bold font.

Table 3. Accuracy of ResPho(SC)Net and Pho(SC)Net on IAM ZSL split (in %) with test set details.

Method	Accuracy	Unseen	Seen	Test Set Details
ResPho(SC)Net	ZSL	**91.82**	NA	509 (unseen)
	GZSL	**87.91**	**96.11**	509 (unseen), 748(seen)
Pho(SC)Net	ZSL	86.0	NA	509 (unseen)
	GZSL	77.0	93.0	509 (unseen), 748(seen)

Table 3 shows the accuracy of two models, ResPho(SC)Net and Pho(SC)Net, on the IAM dataset. It can be observed that ResPho(SC)Net performs better than Pho(SC)Net in terms of ZSL and GZSL accuracy. Specifically, ResPho(SC)Net has a 5.82% better ZSL accuracy and a 10.91% better GZSL accuracy for unseen classes. Moreover, when we compare the GZSL accuracy for seen classes, ResPho(SC)Net is 3.11% better than Pho(SC)Net. Interestingly, ResPho(SC)Net achieves better performance even though it has almost three times fewer parameters than Pho(SC)Net.

The results for the standard split are in Table 4. It can be observed that in the case of GZSL ResPho(SC)Net seen and unseen accuracies are higher. The unseen accuracy is 6.36% high and seen accuracy is 3.13% more. In the case of the ZSL split the ResPho(SC)Net accuracy is 84.6% and the corresponding

Table 4. Accuracy of ResPho(SC)Net and Pho(SC)Net on IAM standard split (in %) with test set class details.

Method	Accuracy	Unseen	Seen	Test Set Details
ResPho(SC)Net	ZSL	84.6	NA	1071 (unseen)
	GZSL	**76.36**	**93.13**	1071 (unseen), 1355 (seen)
Pho(SC)Net	ZSL	**93**	NA	1071 (unseen)
	GZSL	70.0	90.0	1071 (unseen), 1355 (seen)

Pho(SC)Net accuracy is 93.0%. Thus, it is evident that ResPho(SC)Net outperforms Pho(SC)Net except for the unseen word ZSL accuracy on IAM standard split. The working code for this research will be made available at https://github.com/aniketntnu/ResPho-SC-Net-ZSL.

5 Conclusion

Based on experiments on Norwegian and English data, it can be observed that the proposed ResPho(SC) performs better in terms of unseen accuracy and seen accuracy compared to the Pho(SC)Net model for Norwegian data. For English data, the model performs at a similar or higher level, with only a few exceptions. These results suggest that the proposed ResPho(SC) model is better suited for zero-shot text recognition tasks and can effectively handle variations in writing style, script, and language with less data. Additionally, the ResPho(SC) model has less parameters compared to Pho(SC)Net, which makes it more computationally efficient and easier to deploy in resource-constrained environments. In summary, the proposed ResPho(SC) model, specifically the ResPho(SC)-18 variant, is a highly efficient, advanced, and lightweight deep learning model that has been developed for zero-shot text recognition tasks. This model has been shown to exhibit performance and faster convergence compared to other models in the field, making it an ideal choice for real-world applications in this domain.

Acknowledgement. We would like to thank Andre Kåsen from the National Library of Norway for providing the Norwegian Handwritten word images used in this study, We are grateful to Simula HPC cluster, The research presented in this paper has benefited from the Experimental Infrastructure for Exploration of Exascale Computing (eX3), which the Research Council of Norway financially supports under contract 270053, We are grateful to Professor Marius Pedersen from Department of Computer Science at Norwegian University of Science and Technology, Gjøvik, for his informative comments, invaluable support, and encouragement throughout this work. This work is supported under the aegis of the Hugin-Munin project funded by the Norwegian Research Council (Project number 328598).

References

1. The 32nd British Machine Vision (Virtual) Conference 2021 : From Seq2Seq Recognition to Handwritten Word Embeddings. https://www.bmvc2021-virtualconference.com/conference/papers/paper_1481.html
2. Deep Residual Learning for Image Recognition. https://www.computer.org/csdl/proceedings-article/cvpr/2016/8851a770/12OmNxvwoXv
3. Ahmad, I., Fink, G.A.: Training an Arabic handwriting recognizer without a handwritten training data set. In: 2015 13th International Conference on Document Analysis and Recognition (ICDAR), pp. 476–480, August 2015. https://doi.org/10.1109/ICDAR.2015.7333807
4. Bhatt, R., Rai, A., Chanda, S., Krishnan, N.C.: Pho(SC)-CTC—a hybrid approach towards zero-shot word image recognition. Int. J. Doc. Anal. Recognit. (IJDAR) (2022)
5. Chanda, S., Baas, J., Haitink, D., Hamel, S., Stutzmann, D., Schomaker, L.: Zero-shot learning based approach for medieval word recognition using deep-learned features. In: 16th International Conference on Frontiers in Handwriting Recognition, ICFHR 2018, Niagara Falls, NY, USA, 5–8 August 2018, pp. 345–350 (2018)
6. Chanda, S., Okafor, E., Hamel, S., Stutzmann, D., Schomaker, L.: Deep learning for classification and as tapped-feature generator in medieval word-image recognition. In: 13th IAPR International Workshop on Document Analysis Systems, DAS 2018, Vienna, Austria, 24–27 April 2018, pp. 217–222. IEEE Computer Society (2018)
7. He, K., Zhang, X., Ren, S., Sun, J.: Spatial pyramid pooling in deep convolutional networks for visual recognition. In: Fleet, D., Pajdla, T., Schiele, B., Tuytelaars, T. (eds.) ECCV 2014. LNCS, vol. 8691, pp. 346–361. Springer, Cham (2014). https://doi.org/10.1007/978-3-319-10578-9_23
8. Krishnan, P., Dutta, K., Jawahar, C.: Deep feature embedding for accurate recognition and retrieval of handwritten text. In: 2016 15th International Conference on Frontiers in Handwriting Recognition (ICFHR), pp. 289–294, October 2016
9. Krishnan, P., Jawahar, C.V.: HWNet v2: an efficient word image representation for handwritten documents. Int. J. Doc. Anal. Recognit. (IJDAR) **22**(4), 387–405 (2019)
10. Manmatha, R., Han, C., Riseman, E.M., Croft, W.B.: Indexing handwriting using word matching. In: Proceedings of the First ACM International Conference on Digital Libraries, pp. 151–159. DL '96, Association for Computing Machinery, New York, NY, USA, April 1996
11. Manmatha, R., Han, C., Riseman, E.: Word spotting: a new approach to indexing handwriting. In: Proceedings CVPR IEEE Computer Society Conference on Computer Vision and Pattern Recognition, pp. 631–637, June 1996
12. Marti, U.V., Bunke, H.: The IAM-database: an English sentence database for offline handwriting recognition. Int. J. Doc. Anal. Recognit. **5**(1), 39–46 (2002). https://doi.org/10.1007/s100320200071, https://doi.org/10.1007/s100320200071
13. Poznanski, A., Wolf, L.: CNN-N-Gram for hand writing word recognition. In: 2016 IEEE Conference on Computer Vision and Pattern Recognition (CVPR), pp. 2305–2314, June 2016. https://doi.org/10.1109/CVPR.2016.253
14. Rai, A., Krishnan, N.C., Chanda, S.: Pho(SC)Net: an approach towards zero-shot word image recognition in historical documents. In: Lladós, J., Lopresti, D., Uchida, S. (eds.) ICDAR 2021. LNCS, vol. 12821, pp. 19–33. Springer, Cham (2021). https://doi.org/10.1007/978-3-030-86549-8_2

15. Retsinas, G., Sfikas, G., Stamatopoulos, N., Louloudis, G., Gatos, B.: Exploring critical aspects of CNN-based keyword spotting. A PHOCNet study. In: 2018 13th IAPR International Workshop on Document Analysis Systems (DAS), pp. 13–18, April 2018. https://doi.org/10.1109/DAS.2018.49
16. Rothfeder, J.L., Feng, S., Rath, T.M.: Using corner feature correspondences to rank word images by similarity. In: 2003 Conference on Computer Vision and Pattern Recognition Workshop, vol. 3, p. 30, June 2003. https://doi.org/10.1109/CVPRW.2003.10021
17. Rusakov, E., Rothacker, L., Mo, H., Fink, G.A.: A probabilistic retrieval model for word spotting based on direct attribute prediction. In: 2018 16th International Conference on Frontiers in Handwriting Recognition (ICFHR), pp. 38–43, August 2018. https://doi.org/10.1109/ICFHR-2018.2018.00016
18. Rusiñol, M., Aldavert, D., Toledo, R., Lladós, J.: Efficient segmentation-free keyword spotting in historical document collections. Pattern Recogn. **48**(2), 545–555 (2015). https://doi.org/10.1016/j.patcog.2014.08.021
19. Sudholt, S., Fink, G.A.: PHOCNet: a deep convolutional neural network for word spotting in handwritten documents. In: 2016 15th International Conference on Frontiers in Handwriting Recognition (ICFHR), pp. 277–282, October 2016. https://doi.org/10.1109/ICFHR.2016.0060
20. Sudholt, S., Fink, G.A.: Evaluating word string embeddings and loss functions for CNN-based word spotting. In: 2017 14th IAPR International Conference on Document Analysis and Recognition (ICDAR), vol. 01, pp. 493–498, November 2017. https://doi.org/10.1109/ICDAR.2017.87
21. Sudholt, S., Fink, G.A.: Attribute CNNs for word spotting in handwritten documents. Int. J. Doc. Anal. Recognit. (IJDAR) **21**(3), 199–218 (2018). https://doi.org/10.1007/s10032-018-0295-0
22. Wilkinson, T., Brun, A.: Semantic and verbatim word spotting using deep neural networks. In: 2016 15th International Conference on Frontiers in Handwriting Recognition (ICFHR), pp. 307–312, October 2016. https://doi.org/10.1109/ICFHR.2016.0065
23. Wilkinson, T., Lindström, J., Brun, A.: Neural Ctrl-F: Segmentation-free query-by-string word spotting in handwritten manuscript collections. In: 2017 IEEE International Conference on Computer Vision (ICCV), pp. 4443–4452, October 2017. https://doi.org/10.1109/ICCV.2017.475

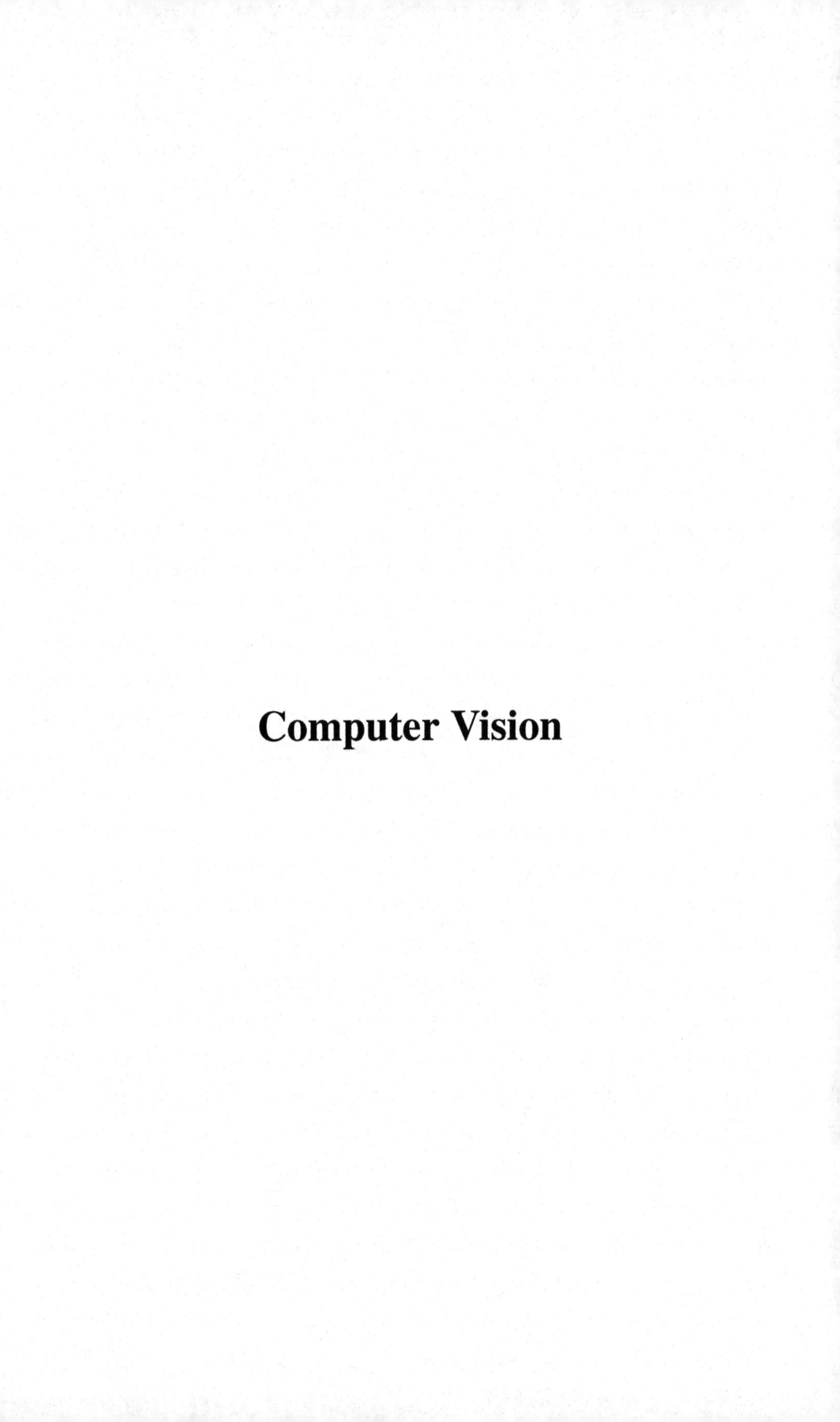

Computer Vision

DeepArUco: Marker Detection and Classification in Challenging Lighting Conditions

Rafael Berral-Soler[1](✉), Rafael Muñoz-Salinas[1,2], Rafael Medina-Carnicer[1,2], and Manuel J. Marín-Jiménez[1,2]

[1] Department of Computing and Numerical Analysis, University of Córdoba, Córdoba, Spain
{rberral,rmsalinas,rmedina,mjmarin}@uco.es

[2] Maimonides Institute for Biomedical Research of Córdoba (IMIBIC), Córdoba, Spain

Abstract. Detection of fiducial markers in challenging lighting conditions can be useful in fields such as industry, medicine, or any other setting in which lighting cannot be controlled (e.g., outdoor environments or indoors with poor lighting). However, explicitly dealing with such conditions has not been done before. Hence, we propose DeepArUco, a deep learning-based framework that aims to detect ArUco markers in lighting conditions where the classical ArUco implementation fails. The system is built around Convolutional Neural Networks, performing the job of detecting and decoding ArUco markers. A method to generate synthetic data to train the networks is also proposed. Furthermore, a real-life dataset of ArUco markers in challenging lighting conditions is introduced and used to evaluate our system, which will be made publicly available alongside the implementation.

Code available in GitHub: https://github.com/AVAuco/deeparuco/.

Keywords: Fiducial Markers · Deep Neural Networks · Marker Detection · CNNs

1 Introduction

Square planar markers, such as ArUco [2], AprilTag [8] or ARTag [1], are popular computer vision tools for object pose estimation and detection, and have been applied for different applications such as healthcare [12,13] and robotics [11,14].

These methods assume adequate lighting and are inflexible since they depend on tailored classical image processing techniques. For example, ArUco relies on

Supported by the MCIN Project TED2021-129151B-I00/AEI/10.13039/ 501100011 033/ European Union NextGenerationEU/PRTR, project PID2019-103871GB-I00 of the Spanish Ministry of Economy, Industry and Competitiveness, and project PAIDI P20_00430 of the Junta de Andalucía, FEDER. We thank H. Sarmadi his contribution to the data preparation.

A. Pertusa et al. (Eds.): IbPRIA 2023, LNCS 14062, pp. 199–210, 2023.
https://doi.org/10.1007/978-3-031-36616-1_16

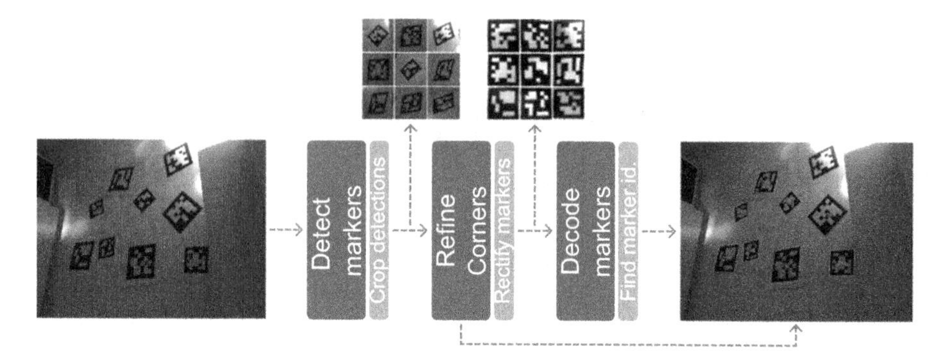

Fig. 1. DeepArUco framework. The marker detector receives as input a color image and returns a series of bounding boxes, which are used to obtain crops of the input image. Then, the corner regressor model is applied over each crop to refine the position of the corners. The detected markers are then rectified with the refined corners, and used as input for the marker decoder. Finally, the marker is assigned the ID with the least Hamming distance w.r.t. the decoded bits. (Best viewed in digital format)

adaptive thresholding, contour and polygon extraction, binary thresholding, and bit lookup to detect and classify. In this case, a blurry image can easily break contour and polygon extraction, and inadequate lighting can also freely break the marker's thresholding and decoding.

Since the success of AlexNet [4], there has been a surge in the usage of deep convolutional neural networks (CNNs) in computer vision. Deep CNNs have retained their status as the dominant image-based detection and recognition approach to this day, showing high flexibility by performing in different applications. Inspired by this, we propose a system (Fig. 1) based on deep CNNs to detect and decode square fiducial planar markers. In our implementation and experiments, we target the ArUco markers (hence the name DeepArUco). However, we claim that our method could be validly adapted for any square planar markers because of their high structural similarities. We needed many samples with challenging illumination to train our deep neural networks, which were not publicly available. Hence, we created a synthetic dataset that can be used to train our networks. Furthermore, we created a real-life dataset of ArUco markers under challenging illumination to evaluate our approach.

Our contributions are the following. First, we introduce a framework containing a marker detector, corner regressor and marker classifier based on deep CNNs, that can detect and decode the ArUco markers in challenging lighting conditions, outperforming both ArUco and DeepTag, while keeping a competitive throughput; using separate models allows for a simpler approach to the problem, and favors modularity. Second, we introduce a method to produce synthetic datasets for training a detector and identifier for square fiducial markers, used to generate a synthetic dataset. Third, we created a real-life dataset of ArUco markers in challenging lighting conditions. Finally, the implementation of our methods and the created datasets will be published online for public use (see the URL provided in the Abstract).

The rest of the paper is organized as follows. First, Sect. 2 summarises some related work. The datasets created for training and testing our proposal are introduced in Sect. 3. Then, we describe our proposed approach in Sect. 4. Section 5 presents our experiments and discusses their results. Finally, we conclude the paper in Sect. 6.

2 Related Works

ArUco [10], AprilTag [15], and ARTag [1] are some well-known planar markers employed in the industry and academia. Specifically, ArUco [2] is present in the well-known computer vision library OpenCV. They all have similar structures: the shape of a square with black borders and a code pattern inside (a grid of black and white square cells representing bit values). All these classical image processing-based methods assume favourable image conditions. Other classical-based works focus on real-world problems [7,9]. However, they either do not address challenging illumination at all [9] or do not perform detection and localization while also not addressing low illumination [7]. On the other hand, a CNN-based method for detecting boards of markers under low-light and high-blur conditions is proposed in [3]. This approach, however, is engineered based on a fixed configuration of certain markers on a board (called ChArUco) and hence cannot be used to identify or detect different individual markers, i.e. the goal of our work. Authors in [5] introduced a YOLO-based method to detect ArUco markers under occlusions. This work only focuses on detecting the bounding box of the markers and does not address the estimation of corner positions or marker identification. Furthermore, they do not consider difficult lighting conditions, as is in our case. Even more recently, a new work has been published [16], using a neural network based on a MobileNet backbone to detect fiducial markers and iteratively refine their corners; however, their work focuses only on sensor noise and motion blur and thus it does not tackle the problem presented in this work.

3 Datasets

To train our models, we generated a synthetic dataset dubbed "FlyingArUco". We tested our method on a real-life dataset of markers in challenging lighting conditions, dubbed "Shadow-ArUco".

3.1 Flying-ArUco Dataset

Our synthetic dataset is divided into two sets. The first consists of markers from the ArUco DICT_6X6_250 (OpenCV) in different configurations overlaid on random backgrounds sampled from the COCO [6] dataset's training set. First, a random camera field of view is chosen for the image. Then, between 1 and 10 randomly sampled markers are generated. White borders with the width of the marker's cell size are added. The contrast of each marker, along with the

Fig. 2. Flying ArUco dataset. Using images from COCO dataset as background, new challenging samples are created to train our models. Top: detection dataset. Bottom: refinement/decoding dataset. (Best viewed in digital format)

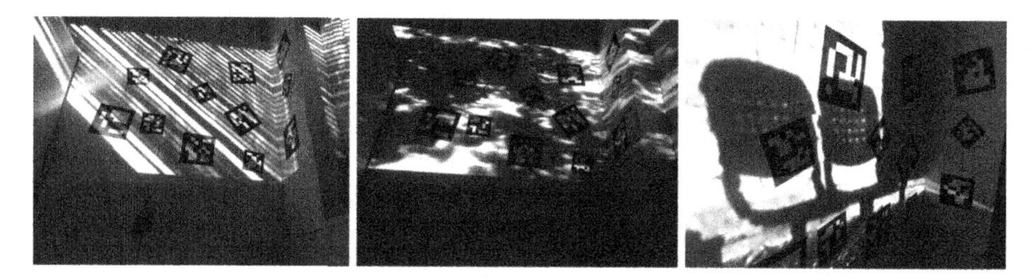

Fig. 3. Shadow-ArUco dataset. Given a static camera, a video is projected on the walls containing various markers; the resulting patterns make their detection difficult for the classical detection approaches. (Best viewed in digital format)

illumination's intensity and color, plus color noise, are randomly sampled and applied to the marker. We also generate a shadow map by creating a random 3×3 image and up-scaling it to the marker's original size using the bi-cubic interpolation, which is helpful in simulating shadow cast on the markers. After that, for each marker, position, distance from the camera, out-of-plane rotation, and in-plane rotation are randomly sampled; then the marker is warped into its position using a corresponding perspective transformation. Gaussian smoothing is applied beforehand to remove aliasing. This dataset contains 2500 images (2000 for training and 500 for validation), containing 13788 markers. Some examples are shown in Fig. 2.

The second set, built by sampling markers from the ARUCO_MIP_36h12 (official) dictionary but with different parameters, such as lighting, was used to train our corner refinement and decoder models. The ground-truth corners of the markers on this dataset were used as training data for our corner refinement network, and the encoded IDs were used to train our marker decoder. 50k markers were used for training and 4.9k for validation.

3.2 Shadow-ArUco Dataset

We have also prepared a dataset for real-world testing of ArUco detectors in challenging lighting conditions. In this dataset, different ArUco markers are attached to the walls in the corner of a room. The room is darkened by closing the windows and turning off the lights. A projector has projected a video[1] of moving shadows with different shapes on the wall with the markers attached to it. We have captured videos of this scene from different angles. For each viewing angle, the camera is fixed in its place. We employed a semi-manual method to estimate the ground-truth labels of the markers and their corner positions. First, we process the sequence frames using the traditional ArUco method, setting the error correction bits (ECB) parameter equal to zero to get only the detections with the highest confidence. This leads us to have a lower number of detections. However, we take advantage of the fact that the camera is fixed while recording the video sequence. For each frame where a marker is not detected, we get the ground-truth detection information (corner positions and marker ID) from the closest frame where we have a detection of the marker. In cases where a marker is not detected in any frame of the sequence, we add the detection information manually and copy it for all of the frames in the sequence. Six video sequences were recorded from six different points of view. A total of 8652 frames with a resolution of 800×600 are captured for the dataset. Some examples of these frames from two different video sequences (v1 and v4) are depicted in Fig. 3.

4 Proposed method

The DeepArUco framework is summarised in Fig. 1. Our marker detector is composed of three parts. The first part detects the marker and its bounding box (see Sect. 4.1). This is done without determining its ID, i.e., it detects the marker without decoding it. The second part determines the corners of the marker (see Sect. 4.2), allowing us to determine the precise position of the marker and rectify it for easier decoding. Finally, the values for each bit in the marker are extracted, and its ID is obtained by comparing it against the ArUco codes dictionary (see Sect. 4.3).

4.1 Bounding-Box-Level Detection

As the first step in our pipeline, we need to find the coarse bounding boxes of the markers appearing in a picture. To do so, we have trained a model based on the YOLOv8[2] object detector, in its "medium" size variant (YOLOv8m). We have trained the model starting from a pre-trained model. As a form of offline data augmentation, we have increased the number of training and validation samples by introducing 10 additional variants for each original image in the detection dataset. These variants are obtained by multiplying the original image by a synthetic gradient with an arbitrary angle, with minimum and maximum values also randomized, ranging from 0.0 to 2.0; this is done aiming to simulate a broader range of lighting conditions. Online data augmentation is done using

[1] https://www.youtube.com/watch?v=nQJopt_M5mE.

[2] https://github.com/ultralytics/ultralytics.

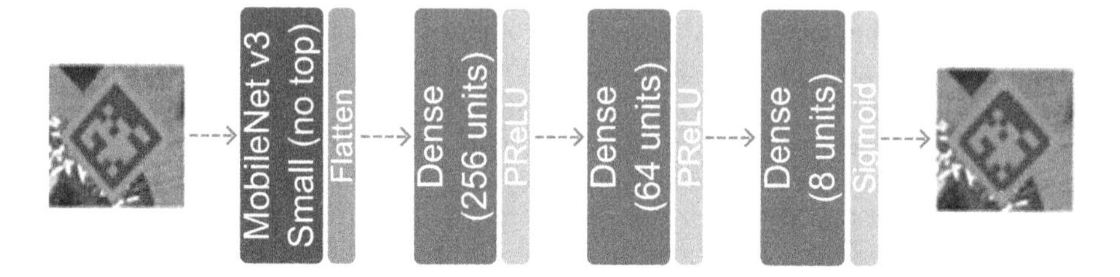

Fig. 4. Corner regressor architecture. The model's input is a 64×64 color image containing a marker; the output is a vector containing 8 values, corresponding to the X and Y coordinates of the four corners of the marker in clockwise order.

the Albumentations library[3], which can be easily integrated with the YOLOv8 training. Other parameters (e.g., loss function) have been kept as default.

4.2 Corner Regression

From the boxes obtained in the detection step, we can refine the corners to fit better the real shape and dimensions of the detected marker. To do so, we have built a model around a pre-trained MobileNetV3 backbone (in its "small" variant); we have removed the fully-connected top of the original model and kept the rest as a feature extractor; then, we have added our custom top, composed of two fully-connected layers (sizes 256 and 64) with a PReLU activation function, and another additional fully-connected layer (size 8, 2 coordinates for each corner) at the output, using a sigmoid activation function. As an input to the network, we expect a 64×64 pixels color image; the model will output four pairs of coordinates, with values in the range [0, 1] (i.e., the position of the corners is represented using values relative to the width/height of the input image). A summary of the architecture appears in Fig. 4.

We have trained the model on the FlyingArUco regression dataset described in Sect. 3.1. As a form of online data augmentation, we introduce changes in lighting using the same method as in Sect. 4.1. To further increase the number of samples (in the absence of a simple integration such as Albumentations in Sect. 4.1), we also rotate images by a multiple of 90° and mirror them horizontally or vertically; multiple transformations can be applied to the same sample. The associated "labels" (i.e., the corners of the image) must also be transformed, keeping consistency with the transformations applied to the corresponding input crops. We used the mean average error (MAE) as the loss function between the predicted coordinates and the ground-truth corners' positions.

On inference time, the bounding boxes obtained in the detection step are expanded, adding a margin of 20% over the original width at both left and right and a margin of 20% over the original height at both top and bottom; the original picture is cropped using the expanded bounding box, and the input is resized to a size of 64×64. Then, the corner regressor is applied over this crop.

[3] https://albumentations.ai/.

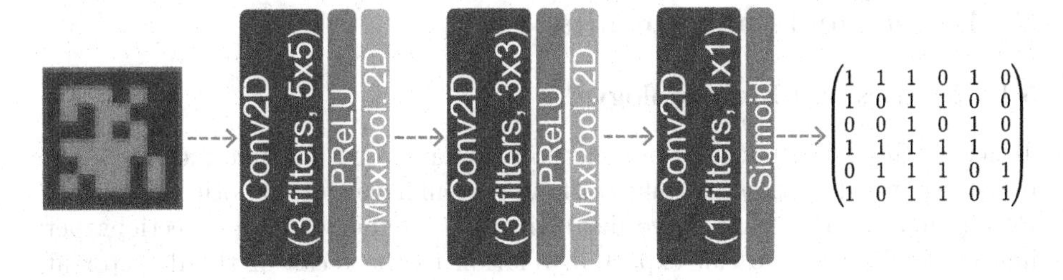

Fig. 5. Marker decoder architecture. The model's input is a 32×32 grayscale image containing a rectified marker; the output is a matrix of size 6×6, containing the prediction for each bit in the input picture, in the range [0–1].

4.3 Marker Decoding

As a final step, we must decode the detected marker (i.e., extract the encoded bits and find a match in the ArUco dictionary). We have resorted to a simple architecture using three convolutional layers to do this. As input, the network expects a 32×32 pixels black and white picture, normalized w.r.t. its maximum and minimum values in the [0–1] range. At its output, it returns a 6×6 array of values in the range [0, 1], which after rounding can be interpreted as the decoded bits. A summary of the architecture appears in Fig. 5.

For training, we have used the pictures in the FlyingArUco regression dataset. From the ground truth coordinates, we find a homography that allows us to rectify the region defined by the four corners into a perfect square, and we crop the picture to obtain a marker. During training, as a form of online data augmentation, ground-truth corner coordinates (both X and Y values), normalized in the range [0–1] against the width/height of the crop, are shifted by up to 0.05 over a maximum value of 1; this aims to simulate "imperfect" corner detections, making our model more robust. Then, the marker is extracted, and we apply lighting changes, rotations and mirroring as described in Sect. 4.2. As loss function, we used the MAE between the predicted values and the ground-truth encoded bits.

On inference time, the corners predicted by the corner regressor are used to extract the markers from the crops obtained at the bounding-box detection step, using a homography to rectify the markers, which are normalized and used as input for the decoder network. The sequence of values at the network output is then rounded to the nearest integer (i.e., 0 or 1) and compared with every possible code in the ArUco dictionary. As there are four possible orientations for the marker, the comparison is performed 4 times, rotating 90° each time (i.e., the output 6×6 matrix is rotated, then is flattened into a vector, and finally, it is compared against every code). The marker is then assigned the ID corresponding to the ArUco code with the smallest Hamming distance w.r.t. the encoded bits in any orientation.

5 Experiments and Results

5.1 Metrics and Methodology

To assess the performance of our method we have used multiple metrics. As a measure of the detection capabilities of our system have plotted both a Precision-Recall curve and another curve displaying the number of false detections per image (FDPI) that we can expect in relation to the recall of the detector at varying minimum confidence thresholds, dubbed *FDPI curve*; as a criterion to determine whether a detection is correct or not, we use the *Intersection-over-Union* (IoU) metric, expecting at least an IoU of 0.5 with any of the ground-truth markers in the image. In a similar vein, to measure the performance of the whole system, including both detection and marker decoding, we have obtained a third curve, showing the number of false detections plus wrongly classified markers that we can expect in relation to the ratio of correctly detected and classified markers (w.r.t. the number of ground-truth markers in the image), dubbed *BCPI curve* in this work (as in bad classifications per image, BCPI).

Furthermore, the performance of each component of the system has been numerically summarised with a single metric: the area under the Precision-Recall curve (AUC-PR) for the YOLO detector, the *Mean Average Error* (MAE) over the individual components of each corner for the corner regressor (considering every *correct* detection in the dataset), and the *accuracy* (ratio of correctly classified markers w.r.t. the number of *correct* detections in the dataset) for the marker decoder. Finally, as a measure of the expected throughput of the system, we have measured the processing time per frame, both as a whole and for each component.

5.2 Implementation Details

To train our marker detector, we have used YOLOv8; both our corner regressor and marker decoder have been implemented using TensorFlow. Our experiments were run on a laptop with an AMD Ryzen(TM) 7 5800HS CPU and an NVIDIA GeForce RTX 3060(R) (6 GB, 80W max. TDP). For brevity, we refer the reader to the public source code of DeepArUco (see Abstract) for further implementation details (optimizer, learning rate, etc.).

5.3 Experimental Results and Analysis

We have applied the methodology described in 5.1 to our method DeepArUco, and to better identify the weaknesses and strengths of our system, we have also evaluated the classic ArUco implementation[4] (version 3.1.15) and the DeepTag method[5]. The metrics have been obtained over the Shadow-ArUco dataset. The plots in Fig. 6 compare DeepArUco against ArUco and DeepTag. We can observe

[4] https://sourceforge.net/projects/aruco/.

[5] https://github.com/herohuyongtao/deeptag-pytorch.

Table 1. Quantitative comparison on Shadow-ArUco dataset. DeepArUco, DeepTag and ArUco methods have been considered. Note that while DeepArUco yields the highest MAE value, it also yields the highest AUC-PR, encompassing difficult detections whose corners are harder to pinpoint.

	DeepArUco	DeepTag	ArUco*
MAE-corners ↓	3.56	2.23	**0.57**
Accuracy-ID ↑	**0.93**	0.91	0.84
AUC-PR ↑	**0.82**	0.42	0.53
FPS ↑	**17.78**	4.25	14.82

*: Method running on CPU.

in (a) that our method is the one that manages to recover the most markers from each frame at any required precision value, performing better than both ArUco and DeepTag. In (b), we can see that from a detection standpoint, our method can recover the most markers while introducing the least amount of false positives, when compared to both ArUco and DeepTag. The previous conclusion can be extended to the entire system, as shown in (c), with our method producing the least false detections + misclassifications for each required classification rate (i.e., the ratio between good detections and ground-truth markers in the image). For each curve, IoU values (used to match ground truth and detected markers) have been computed using the corners provided by the corner refinement model.

In Table 1, we compare the performance of the three methods in each step. The MAE metric is the aggregate over the individual component of each corner for every correct detection in the dataset (in pixels); from this, we can see how ArUco performs the best while our method is the least performant. This could be related to the far higher number of detections obtained with our method, encompassing "difficult" detections in which it is difficult to identify with high precision *where* the corners are. The Accuracy metric has been computed similarly, considering only good detections: from here, we attest the superior decoding capabilities of our method, an ever greater feat considering the far wider range of detected markers. The AUC-PR metric confirms our conclusions from Fig. 6: our method recovers the most markers at any required precision value. Finally, the FPS metric shows the good throughput of our method, achieving 17.78 FPS on GPU (with corner refinement and marker decoding running on-batch).

In Table 2, we have a breakdown of the time required per frame by each of the components of our system. In the last row, "Others" refer to all the auxiliary tasks in our execution loop needed to use our models, such as picture cropping, homography finding, marker ID matching, etc. From the time allocated for such tasks, we can realize that many improvements could be made to reduce the time per frame.

5.4 Working Examples and Limitations

Figure 7 contains two frames of the ShadowArUco dataset, processed with DeepAruco (a and d), DeepTag (b and e), and classic ArUco (c and f). Notice

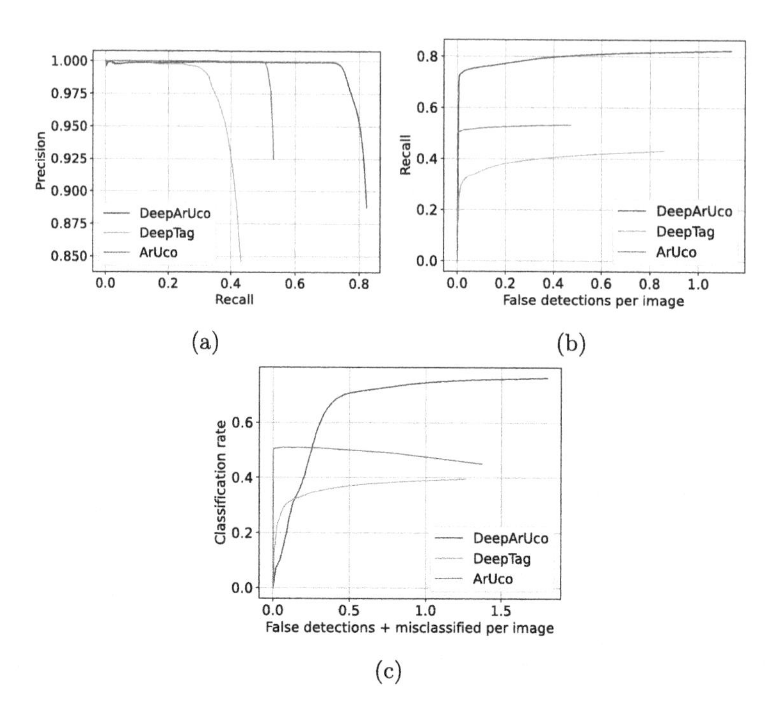

Fig. 6. Quantitative evaluation on Shadow-ArUco dataset. (a) Precision-Recall curve, **(b)** FDPI curve and **(c)** BCPI curve for DeepArUco, DeepTag and ArUco methods. (Best viewed in digital format)

Table 2. Time breakdown of our DeepArUco. In milliseconds per frame.

Task	Time per frame ↓
Marker detection	24.22
Corner refinement	6.42
Marker decoding	2.01
Others	23.59

how our method outperforms the rest in poor lighting conditions (top row), detecting nearly every marker, although producing a few false positives. Also, our method is resilient to complex lighting patterns (bottom row); while DeepTag can recover more markers, it introduces a lot of false positives and wrong identifications. Overall, while our method seems to perform "well enough" in the comparison, there is ample room for improvement, aiming to improve the detection capabilities with complex lighting and/or markers not directly in front of the camera (as in the right wall in d).

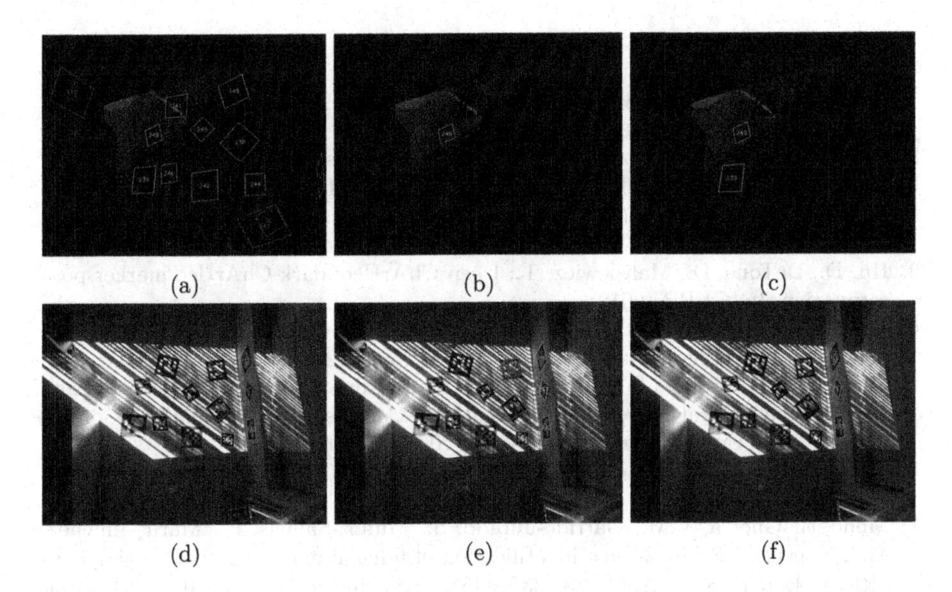

Fig. 7. Qualitative evaluation on Shadow-ArUco dataset. We have considered DeepArUco (a, d), DeepTag (b, e) and classic ArUco (c, f). Red outlines indicate bad detections. Red IDs indicate wrong identifications. (Best viewed in digital format)

6 Conclusions and Future Work

We have presented a methodology to tackle marker detection and identification on challenging lighting conditions, using neural networks, in a modular fashion. To train and test our method, we have created two datasets: a fully synthetic one (FlyingArUco dataset) and another obtained by recording physical markers under varying lighting settings (Shadow-ArUco dataset).

Overall, the results of the work suggest that neural networks are well suited for this task, with our approach matching or outperforming tested state-of-the-art methods in marker detection and identification. Furthermore, the decomposition of the task in multiple steps allows for improvements at the level of individual modules, which could be replaced if a better alternative is available.

As future work, our methodology could be extended to a broader range of situations and lighting conditions, training on a larger dataset of fiducial markers in the wild instead of relying on synthetic data. Also, it could be interesting to assess the impact of marker size, distance, angle, and blur in the results. Another possible variation would be to replace the direct corner regression with a heatmap-based corner detector to later compute the exact position of the corners algorithmically. Finally, much optimization could be applied to achieve a better throughput making it suitable for a broader range of tasks and/or deploying on low-power hardware.

References

1. Fiala, M.: Designing highly reliable fiducial markers. IEEE PAMI **32**(7), 1317–1324 (2010). https://doi.org/10.1109/TPAMI.2009.146
2. Garrido-Jurado, S., Muñoz-Salinas, R., Madrid-Cuevas, F.J., Marín-Jiménez, M.J.: Automatic generation and detection of highly reliable fiducial markers under occlusion. Pattern Recogn. **47**(6), 2280–2292 (2014). https://doi.org/10.1016/j.patcog.2014.01.005
3. Hu, D., DeTone, D., Malisiewicz, T.: Deep ChArUco: dark ChArUco marker pose estimation. In: CVPR (2019)
4. Krizhevsky, A., Sutskever, I., Hinton, G.E.: ImageNet classification with deep convolutional neural networks. In: NeurIPS, vol. 25 (2012)
5. Li, B., Wu, J., Tan, X., Wang, B.: ArUco marker detection under occlusion using convolutional neural network. In: International Conference on Automation, Control and Robotics Engineering (CACRE), pp. 706–711 (2020)
6. Lin, T.Y., et al.: Microsoft COCO: common objects in context. arXiv:1405.0312 [cs] (2015)
7. Mondéjar-Guerra, V.M., Garrido-Jurado, S., Muñoz-Salinas, R., Marín-Jiménez, M.J., Carnicer, R.M.: Robust identification of fiducial markers in challenging conditions. Expert Syst. Appl. **93**, 336–345 (2018). https://doi.org/10.1016/j.eswa.2017.10.032
8. Olson, E.: AprilTag: a robust and flexible visual fiducial system. In: 2011 IEEE International Conference on Robotics and Automation, pp. 3400–3407 (2011). https://doi.org/10.1109/ICRA.2011.5979561. iSSN: 1050-4729
9. Ramírez, F.J.R., Muñoz-Salinas, R., Carnicer, R.M.: Tracking fiducial markers with discriminative correlation filters. Image Vis. Comput. **107**, 104094 (2021). https://doi.org/10.1016/j.imavis.2020.104094
10. Romero-Ramirez, F.J., Muñoz-Salinas, R., Medina-Carnicer, R.: Speeded up detection of squared fiducial markers. Image Vis. Comput. **76**, 38–47 (2018). https://doi.org/10.1016/j.imavis.2018.05.004
11. Sani, M.F., Karimian, G.: Automatic navigation and landing of an indoor AR. Drone quadrotor using ArUco marker and inertial sensors. In: 2017 International Conference on Computer and Drone Applications (IConDA), pp. 102–107 (2017). https://doi.org/10.1109/ICONDA.2017.8270408
12. Sarmadi, H., Muñoz-Salinas, R., Álvaro Berbís, M., Luna, A., Medina-Carnicer, R.: 3D Reconstruction and alignment by consumer RGB-D sensors and fiducial planar markers for patient positioning in radiation therapy. Comput. Methods Programs Biomed. **180**, 105004 (2019). https://doi.org/10.1016/j.cmpb.2019.105004
13. Sarmadi, H., Muñoz-Salinas, R., Álvaro Berbís, M., Luna, A., Medina-Carnicer, R.: Joint scene and object tracking for cost-Effective augmented reality guided patient positioning in radiation therapy. Comput. Methods Programs Biomed. **209**, 106296 (2021). https://doi.org/10.1016/j.cmpb.2021.106296
14. Strisciuglio, N., Vallina, M.L., Petkov, N., Muñoz Salinas, R.: Camera localization in outdoor garden environments using artificial landmarks. In: IWOBI, pp. 1–6 (2018)
15. Wang, J., Olson, E.: AprilTag 2: efficient and robust fiducial detection. In: 2016 IEEE/RSJ International Conference on Intelligent Robots and Systems (IROS), pp. 4193–4198 (2016). https://doi.org/10.1109/IROS.2016.7759617
16. Zhang, Z., Hu, Y., Yu, G., Dai, J.: DeepTag: a general framework for fiducial marker design and detection. IEEE Trans. Pattern Anal. Mach. Intell. **45**(3), 2931–2944 (2023). https://doi.org/10.48550/arXiv.2105.13731

Automated Detection and Identification of Olive Fruit Fly Using YOLOv7 Algorithm

Margarida Victoriano[1,2](✉), Lino Oliveira[1], and Hélder P. Oliveira[1,2]

[1] INESC TEC, Porto, Portugal
{ana.victoriano,lino.oliveira,helder.f.oliveira}@inesctec.pt
[2] FCUP, Porto, Portugal
https://www.inesctec.pt

Abstract. The impact of climate change on global temperature and precipitation patterns can lead to an increase in extreme environmental events. These events can create favourable conditions for the spread of plant pests and diseases, leading to significant production losses in agriculture. To mitigate these losses, early detection of pests is crucial in order to implement effective and safe control management strategies, to protect the crops, public health and the environment. Our work focuses on the development of a computer vision framework to detect and classify the olive fruit fly, also known as Bactrocera oleae, from images, which is a serious concern to the EU's olive tree industry. The images of the olive fruit fly were obtained from traps placed throughout olive orchards located in Greece. The approach entails augmenting the dataset and fine-tuning the YOLOv7 model to improve the model performance, in identifying and classifying olive fruit flies. A Portuguese dataset was also used to further perform detection. To assess the model, a set of metrics were calculated, and the experimental results indicated that the model can precisely identify the positive class, which is the olive fruit fly.

Keywords: Image processing · Object detection and classification · Pest management · Olive orchards

1 Introduction

Climate change can have a significant impact on the spread of pests and diseases in agriculture by affecting temperature and precipitation patterns, pest behaviour and life cycle, crop maturity, and natural pest enemies. This can lead to an imbalance between the phenology of crops and insect populations, increasing crop damage [6]. So, early pest detection and precision in pesticide administration are crucial for reducing crop losses and ensuring human and environmental safety.

This work is financed by National Funds through the Portuguese funding agency, FCT - Fundação para a Ciência e a Tecnologia, within project LA/P/0063/2020.

A. Pertusa et al. (Eds.): IbPRIA 2023, LNCS 14062, pp. 211–222, 2023.
https://doi.org/10.1007/978-3-031-36616-1_17

Our research concentrates on the olive fruit fly, which is one of the major threats to olive trees. The olive fruit fly is widespread in the European Union, particularly in Spain and Italy which are home to the largest olive tree cultivation areas [2].

Monitoring the number of pests captured in a trap is a crucial step in confirming an outbreak and assessing olive orchard fly infestations. Manual trap monitoring by professionals is the usual method, but it can be expensive and time-consuming in large and dispersed orchards [9]. Nonetheless, effective pest management is essential for maintaining the health and productivity of olive trees and the quality of their by-products, such as olive oil and table olives.

Our goal is to develop a framework that can detect and classify olive fruit flies, also known as Bactrocera oleae in Portuguese olive orchards using trap images. To achieve this, we used trap images collected in Corfu, Greece, and fine-tuned the YOLO (You Only Look Once)v7 model. We employed three-fold cross-validation with hyperparameter tuning to ensure optimal performance. Finally, we retrained the model on the full training set with the best hyperparameters.

To evaluate the model's performance, we measured several metrics. These models may be used to develop a reliable and efficient system that can control olive fruit fly populations and protect olive crops.

2 Related Work

Object detection is an important research field in computer vision that focuses on detecting objects in images or videos. With the emergence of deep learning techniques, there has been significant progress in this area. Insect detection and classification is a specific application of object detection that has important uses in pest control, agriculture, and disease prevention. In this section, we will review the latest approaches to insect detection and classification, including image processing, machine learning, and deep learning.

2.1 Image Processing Techniques

Image processing techniques are used to detect objects in images since they allow for the extraction of meaningful features and patterns from images that can be used to identify objects [10,13].

Wang et al. [13] developed a vegetable detection algorithm for complex backgrounds. The algorithm used median filtering and gray scale conversion to distinguish between the leaves and the background. The OTSU algorithm was used to extract the foreground leaves, while morphological operations were applied to remove the effects of insects and veins. The overlapped objects were extracted using an opening operation. Disease spots area was calculated by estimating the proportion of yellow to green in each pixel of the image. The method was tested on a set of images, and the authors concluded that it usually works well, although the recognition accuracy may decrease in some complex cases.

Thenmozhi and Reddy [10] use image processing techniques to detect the shape of the sugarcane crop. Firstly, the image is converted from RBG into gray scale, to preserve the brightness of the image and to be segmented later. Then, the obtained gray scale image is divided into smaller regions to detect its edges. The edges were detected using the Sobel edge detection method, since it usually performs well, and it is more resistant to noise. The next step consisted in applying morphological operations. Foreground objects are expanded or thickened by the dilation operation, and the structuring element is employed to regulate the thickening of pixel values. Then, by joining and filling gaps that are smaller than the structural element, the closure operation is used to smooth the contours of the objects. Finally, a filling operation fills the holes in the insect image. Following the segmentation step, shape features are extracted from the border, surface, and texture of the insect. These characteristics are shown quantitatively so that different objects can be compared. Nine geometric shape features are extracted from insects' shape feature analysis, including area, perimeter, major axis length, minor axis length, eccentricity, circularity, solidity, form factor and compactness. These features allowed the authors to identify the various shapes of insects. The type of insect will then be classified and determined using these finite feature vectors based on attributes related to shape.

2.2 Learning Techniques

Deep learning is used to detect objects because it has shown great success in solving complex problems, such as object detection, that were previously difficult to solve with traditional machine learning methods. Deep learning models, such as convolutional neural networks, can automatically learn and extract high-level features from raw image data, making them well-suited for image analysis tasks [1,4,5,14].

Kasinathan et al. [4] conducted insect classification and detection using two datasets of five and nine classes, applying image processing techniques, to reduce noise and increase sharpness, and machine learning algorithms like ANN, SVM, KNN, and NB. The SVM classifier achieved a higher accuracy of 79.9% and 71.8% for the five and nine insect classes, respectively, compared to the other classifiers. Data augmentation techniques and shape features were used to improve accuracy and prevent overfitting, and the machine learning algorithms were evaluated using 9-fold cross-validation.

Boukhris et al. [1] proposed a deep learning architecture using transfer learning to detect damage in leaves and fruits, classify their severity levels, and located them. The model used data augmentation techniques and was trained using a modified version of Mask-RCNN. Transfer learning was applied using a pre-trained model approach, and several models were tested with VGG16 performing the best. Resnet 50 was ultimately chosen as the best model for the object detection task based on its metric results.

Xia et al. [14] proposed an improved convolutional neural network based on VGG19 to locate and classify insects. The model combined the VGG19 and RPN models and was trained using stochastic gradient descent. The proposed method achieved the highest mean average precision among other methods, and a dataset of 739 images of stored-product insects with or without foreign materials was created and used to improve the accuracy of the Faster R-CNN method. The inception network was used to extract feature maps, and the RPN network located potential insect-containing areas. These areas were then classified by the improved inception network and corrected simultaneously. Finally, Non Maximum Suppression was used to merge overlapping target boxes.

The study by Le et al. [5] evaluated three single-stage object detection models (SSD-MobileNetV1, SSD-MobileNetV2, and YOLOv4-tiny) using a dataset of fruit flies on sticky traps. Synthetic datasets were also used to model real-world disturbances. The YOLOv4-tiny model performed the best overall in all areas with and without synthetic disturbances, maintaining strong F1-score and localization, followed by SSD-MobileNetV2. SSD-MobileNetV1 performed well at lower IoU thresholds but was hindered at the highest IoU threshold. The study concluded that YOLOv4-tiny is the best model for this case.

3 Proposed Framework

We present a framework for detecting the olive fruit fly, a pest with a relative negative impact for the olive oil industry, using YOLOv7, a state-of-the-art single-stage object detector. The framework involves several stages, including preparing the dataset for training, performing data augmentation techniques to simulate real-world scenarios, and training using YOLOv7 architecture. To evaluate the model's performance, we performed three-fold cross-validation with hyperparameter tuning and tested it on two different test sets. One of the test sets belongs to the DIRT dataset [3], which was the dataset that the model was trained on, while the other is a dataset collected in Portugal [7] and it is a new and unseen dataset to evaluate the model's generalization ability. Our proposed framework aims to achieve high accuracy in insect detection and classification and is designed to be adaptable to other object detection tasks.

3.1 Dataset Description

The DIRT dataset consists of 848 images, mostly showing olive fruit flies captured in McPhail traps from 2015 to 2017 in various locations in Corfu, Greece. The images were mainly collected through the e-Olive smartphone app, which enables users to report captures and upload images; however, the images were taken with various devices (smartphones, tablets, and cameras), making the dataset non-standardized. The images include expert-annotated spatial identification of Dacuses in each image and were annotated using the labelImg tool [11], where bounding boxes were manually drawn around objects and labelled.

The DIRT dataset, as stated above, consists of images mostly showing olive fruit flies; however, it also contains several other insects that do not represent olive fruit flies, so experts from our team labelled a set of the negative occurrences with the labelImg tool. These insects are of varied species, but all of them were labelled as the negative class, non-dacus, so the different species will not be differentiated. Figure 1 shows examples of images from the DIRT dataset, containing the original labels and other examples containing labels of both classes.

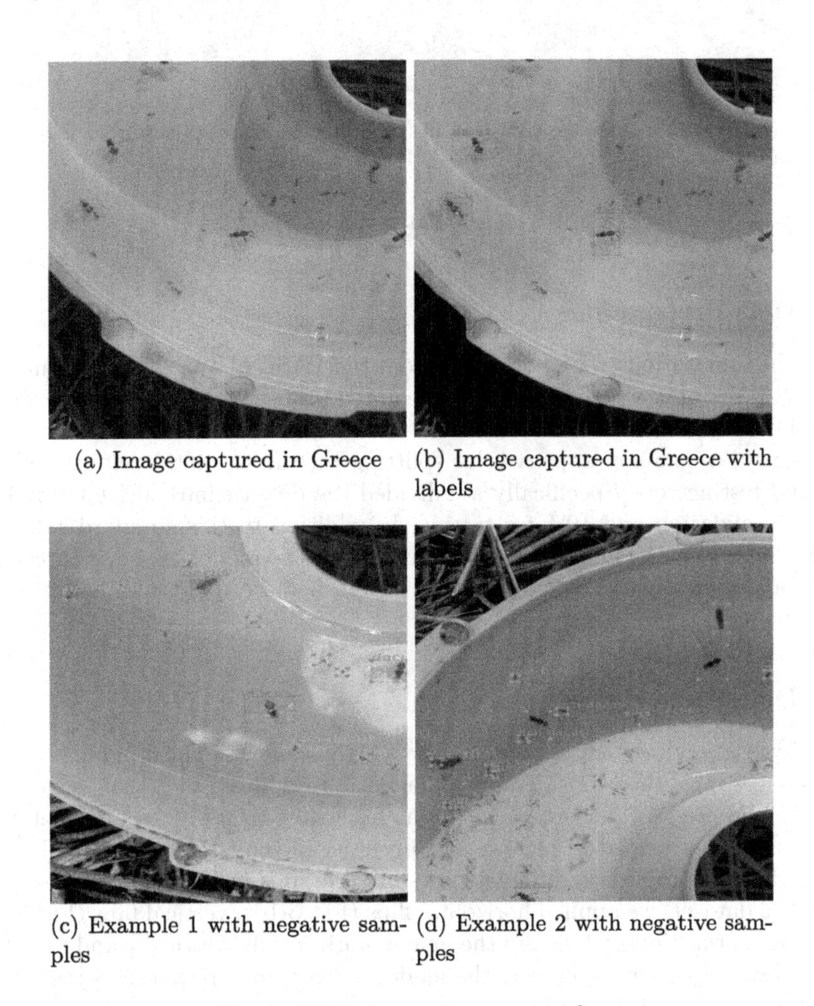

(a) Image captured in Greece (b) Image captured in Greece with labels

(c) Example 1 with negative samples (d) Example 2 with negative samples

Fig. 1. DIRT dataset images samples

To perform further detection, after training the model, we will be using a different dataset that contains 321 images, collected in Portuguese olive groves, to assess the model's ability to adapt to a set of new and unseen images. This dataset contains images of yellow sticky traps, which is a different trap from the one depicted in the DIRT dataset. The yellow sticky traps also contain the olive fruit fly, which is the pest under study. Figure 2 shows two images from the Portuguese dataset.

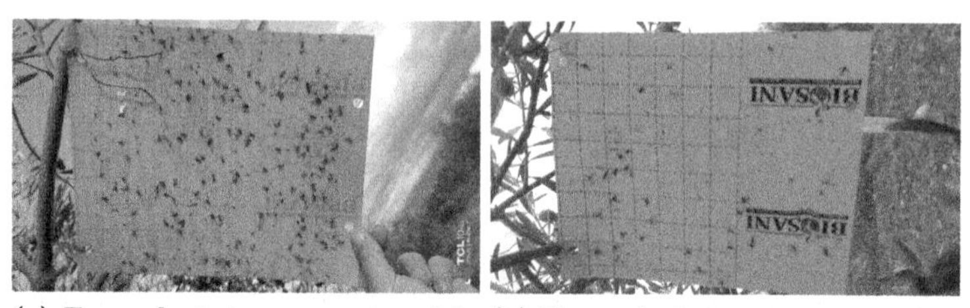

(a) Example 1: image captured in Portugal

(b) Example 2: image captured in Portugal

Fig. 2. Portuguese dataset images samples

3.2 Data Preparation

First, the annotations were converted from the PASCAL VOC XML format to the YOLO format, which is the format required for use with the YOLOv7 object detection model.

We used a common approach for splitting the dataset into training, validation and testing sets. Specifically, we divided the dataset into 80% for training, 10% for validation and 10% for testing. In addition to this standard split, we further improved the accuracy and robustness of the model by using three-fold cross validation on the training set, to obtain a more reliable estimate of its performance.

3.3 Data Augmentation

In real-world environments, there can be a significant amount of noise, and the trap images may vary considerably. Therefore, to address these challenges and improve the model's performance, it was necessary to use data augmentation techniques. This helped to increase the diversity of the training data and make the model more robust to different types of variations in the field.

In the dataset, we applied horizontal flips (Fig. 3(b)), vertical flips (Fig. 3(c)) and random rotations, which flip the images horizontally, vertically and to different angles, respectively, allowing the model to learn to detect insects regardless of the orientation of the image.

We also applied random brightness contrast (Fig. 3(g)), which adjusts the brightness and contrast of the images to help the model learn to recognize insects under different lighting conditions. Random brightness contrast, in this specific context, is vital since capturing images on the field usually involves capturing them under varying and unpredictable lighting conditions, such as sunlight or cloudy skies. These different lighting conditions can greatly affect the appearance of the insects in the images and make it more challenging for the model to recognize them. By applying this technique on the training data, we can artificially alter the brightness and contrast of the images, generating new training samples that more accurately reflect the range of lighting conditions the model may encounter when applied in real-world scenarios, and to generalize better to new, unseen data.

We also applied multiplicative noise (Fig. 3(f)), which adds noise to the images to help the model learn to recognize insects under noisy conditions. Additionally, we applied a blurring transformation (shown in Fig. 3(d)) to intentionally blur the images. This technique aimed to help the model learn to recognize insects under blurry conditions, which commonly occur during field image acquisition due to factors such as camera shake, motion blur, or poor focusing.

We applied JPEG compression (Fig. 3(h)), which compresses the images, helping the model learn to recognize insects under low-resolution conditions, since capturing pictures in the field frequently entails taking pictures that would need to be compressed to make them smaller for transmission or storage, resulting in poorer quality. Finally, we also applied cutout (Fig. 3(e)), which cuts out a portion of the image, helping the model learn to focus on the most important features of the insect, thus learning better the main object features.

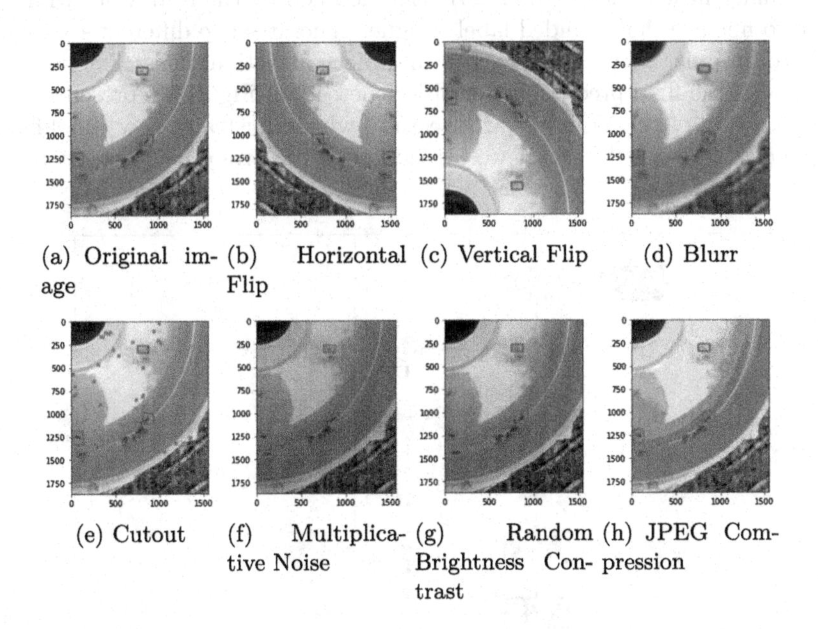

(a) Original image (b) Horizontal Flip (c) Vertical Flip (d) Blurr

(e) Cutout (f) Multiplicative Noise (g) Random Brightness Contrast (h) JPEG Compression

Fig. 3. Data augmentation techniques applied.

3.4 YOLOv7 Architecture

YOLO (You Only Look Once) [8] is an object detection system that employs a single deep neural network to detect objects in an image. YOLOv7 is the seventh version of the YOLO architecture, and it is a quick and accurate object detector that use a single neural network to predict bounding boxes and class probabilities for items in an image.

An input image is divided into a grid of cells by the YOLO architecture, and each cell is in charge of predicting a series of bounding boxes. In order to handle several objects in a single cell, the network uses anchor boxes, each of which has a corresponding class probability. The center coordinates, width, and height of each bounding box are predicted by the network, together with the class probabilities [8].

YOLOv7 [12] is an object detection architecture divided in three main components: the backbone, the neck and the head. The backbone network, composed of convolutional and pooling layers, extracts features for object detection. YOLOv7 introduces E-ELAN, which enhances the learning capability of the backbone network without destroying the original gradient path. The authors also introduce model scaling to produce models at various scales for different inference speeds. Additionally, the network uses deep supervision by adding an auxiliary head in the middle layers of the network with an assistant loss as a guide. The lead head guided label assigner method generates soft labels to train both heads, allowing the auxiliary head to learn the information learned by the lead head, while the coarse-to-fine lead head guided label assigner generates two different sets of soft labels to optimize recall and precision in object detection tasks. The detection head makes the final predictions for object detection. Figure 4 represents a proposed diagram of the YOLOv7 architecture, which illustrates how the different components of the architecture work together to perform object detection.

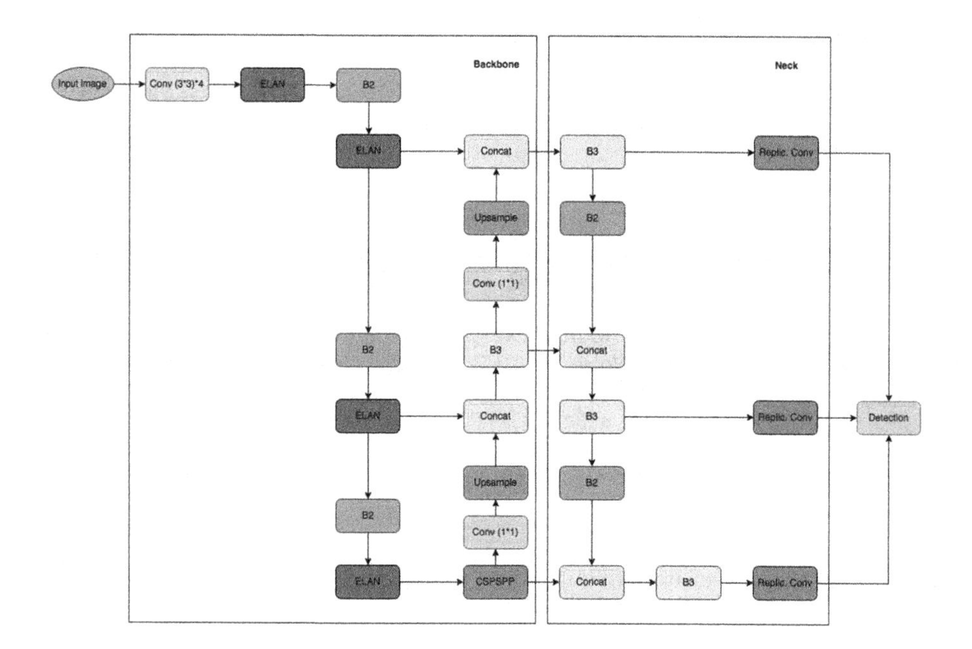

Fig. 4. YOLOv7 architecture

3.5 YOLOv7 Training and Testing

We used the training data that was previously split and performed three-fold cross-validation with parametrization. This allowed us to evaluate our model's performance on multiple subsets of the training data and to tune our hyperparameters. The parametrization was chosen randomly for several hyperparameters, including the number of epochs, batch size, weight decay, and learning rate range. We varied the parameters between 50 and 100 epochs, the batch size between 16 and 20, the learning rate ranged between 0.0001 and 0.1, and the weight decay assumed values between 0.0005 and 0.001.

After performing the cross-validation, we evaluated the performance of our model using precision, recall, and mean average precision. We then retrained the model using the parameters that obtained the best results in terms of these metrics. The parametrization that obtained the best results was defined with 100 epochs, a batch size of 16, weight decay of 0.0005 and learning rate that ranged between 0.01 and 0.1.

Once the model was retrained, we tested it on the 10% set, that we split previously and on the Portuguese dataset, that was not used for training. This allowed us to evaluate the performance of the model on data that it had not seen before, and to ensure that it was able to generalize well.

4 Results Evaluation

The proposed object detection and identification model was evaluated to detect and identify the olive fruit fly, represented by the class dacus. The results of the evaluation are presented in Table 1, and they demonstrate the effectiveness of the model in detecting the class dacus and distinguishing it from other objects, represented as non-dacus.

The model was evaluated using three-fold cross-validation, and then it was retrained with the best parameters obtained. The mean precision of the model was 0.916 for all classes, 0.868 for non-dacus, and 0.943 for dacus in three-fold cross-validation, which shows that the model has a high accuracy in identifying dacus. After testing with the best parameters, the precision improved to 0.958 for all classes, 0.951 for non-dacus, and 0.966 for dacus, indicating that the model's performance improved, and it may be further improved by fine-tuning its parameters.

The recall of the model was 0.801 for all classes, 0.707 for non-dacus, and 0.896 for dacus in three-fold cross-validation. This means that the model can identify most of the dacus instances in the images, but has a lower recall rate for non-dacus objects. After testing with the best parameters, the recall increased to 0.900 for all classes, 0.826 for non-dacus, and 0.975 for dacus, which indicates that the model can better identify dacus instances, which are the target class, while reducing false negatives.

The mAP@.5 of the model was 0.855 for all classes, 0.726 for non-dacus, and 0.984 for dacus in three-fold cross-validation, which indicates that the model can accurately detect dacus with a high confidence score. After testing with the best parameters, the mAP@.5 improved to 0.938 for all classes, 0.885 for non-dacus, and 0.991 for dacus, indicating that the model can better distinguish dacus from non-dacus objects.

The mAP@.5:.95 of the model was 0.525 for all classes, 0.298 for non-dacus, and 0.777 for dacus in three-fold cross-validation, which shows that the model has a lower performance in detecting dacus with higher confidence scores. After testing with the best parameters, the mAP@.5:.95 improved to 0.630 for all classes, 0.486 for non-dacus, and 0.774 for dacus, indicating that the model can better detect dacus with high confidence scores after fine-tuning.

Table 1. Experimental results obtained

Metric	3-fold Cross Validation			Testing with Best Parameters		
	All Classes	Non Dacus	Dacus	All Classes	Non Dacus	Dacus
Precision	0.916	0.868	0.943	0.958	0.951	0.966
Recall	0.801	0.707	0.896	0.900	0.826	0.975
mAP@.5	0.855	0.726	0.984	0.938	0.885	0.991
mAP@.5:.95	0.525	0.298	0.777	0.630	0.486	0.774

The detection results on the test set, obtained from the DIRT dataset, showed that the proposed object detection and identification model performed well, as shown in Fig. 5.

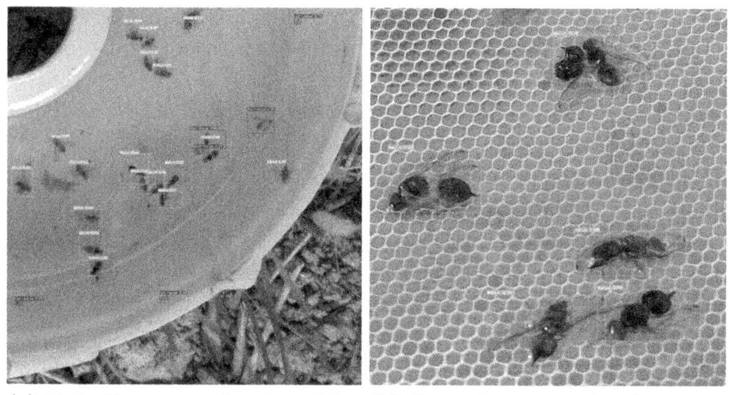

(a) Detection example of positive (dacus) and negative (non dacus) class

(b) Detection example of positive class (dacus)

Fig. 5. DIRT dataset detection images samples

Figure 6 shows the detection results on the Portuguese dataset, that was not used for training, so it comprised a set of new, unseen data. The confidence

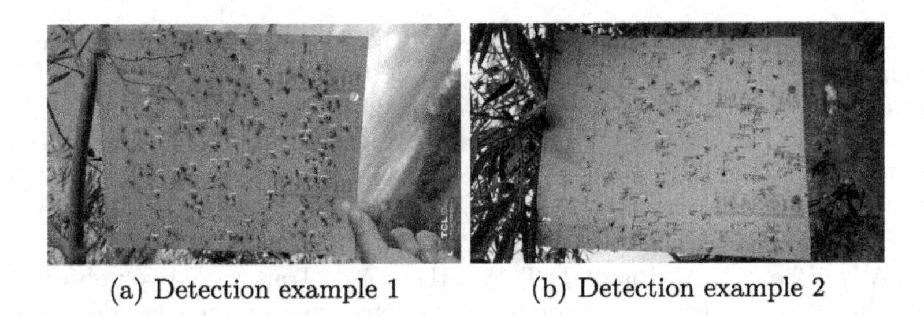

(a) Detection example 1 (b) Detection example 2

Fig. 6. Portuguese dataset detection images samples

associated with each prediction is lower, comparing to the results obtained previously.

In conclusion, the proposed object detection and identification framework has demonstrated high accuracy in detecting and identifying dacus, and the performance of the model can be further improved by fine-tuning its parameters. The evaluation results show that the model can be a valuable tool for identifying and controlling the olive fruit fly in agricultural settings.

5 Conclusion and Future Work

In conclusion, this research aimed to develop a framework for the detection and classification of olive fruit flies, which are a major threat to olive production in the European Union. The framework involved augmenting the dataset and fine-tuning the YOLOv7 model with the augmented data. The results showed that the model was able to accurately detect the positive class (olive fruit fly). To obtain more precise results, acquiring a larger dataset is necessary, but this can be a costly and time-consuming activity. This framework is going to be integrated in a web-based information system, that receives trap images captured from electronic traps placed on olive groves, and proceeds to analyse the image and return the detection results.

Therefore, a future work proposal is to develop a method using segmentation techniques to artificially generate new datasets with existing data to overcome these challenges and make the models more accurate.

References

1. Boukhris, L., Ben Abderrazak, J., Besbes, H.: Tailored deep learning based architecture for smart agriculture. In: 2020 International Wireless Communications and Mobile Computing (IWCMC), pp. 964–969 (2020). https://doi.org/10.1109/IWCMC48107.2020.9148182
2. Eurostat, E.: Agricultural production - orchards: olive trees. Statistics Explained (2019). https://ec.europa.eu/eurostat/statistics-explained/index.php?title=Agricultural_production_-_orchards#Olive_trees. Accessed 2 Feb 2023

3. Kalamatianos, R., Karydis, I., Doukakis, D., Avlonitis, M.: DIRT: the dacus image recognition toolkit. J. Imaging **4**, 129 (2018). https://doi.org/10.3390/JIMAGING4110129, https://www.mdpi.com/2313-433X/4/11/129/htm
4. Kasinathan, T., Singaraju, D., Uyyala, S.R.: Insect classification and detection in field crops using modern machine learning techniques. Inf. Process. Agric. **8**(3), 446–457 (2021). https://doi.org/10.1016/j.inpa.2020.09.006, https://www.sciencedirect.com/science/article/pii/S2214317320302067
5. Le, A.D., Pham, D.A., Pham, D.T., Vo, H.B.: AlertTrap: a study on object detection in remote insects trap monitoring system using on-the-edge deep learning platform (2021). https://doi.org/10.48550/ARXIV.2112.13341, https://arxiv.org/abs/2112.13341
6. Nazir, N., et al.: Effect of climate change on plant diseases. Int. J. Curr. Microbiol. Appl. Sci. **7**, 250–256 (2018). https://doi.org/10.20546/IJCMAS.2018.706.030
7. Pereira, J.A.: Yellow sticky traps dataset _ olive fly (Bactrocera Oleae) (2023). https://doi.org/10.34620/dadosipb/QFG85C
8. Redmon, J., Divvala, S., Girshick, R., Farhadi, A.: You only look once: unified, real-time object detection (2015). https://doi.org/10.48550/ARXIV.1506.02640, https://arxiv.org/abs/1506.02640
9. Shaked, B., et al.: Electronic traps for detection and population monitoring of adult fruit flies (diptera: Tephritidae). J. Appl. Entomol. **142**(1–2), 43–51 (2017). https://doi.org/10.1111/jen.12422
10. Thenmozhi, K., Reddy, U.S.: Image processing techniques for insect shape detection in field crops. In: 2017 International Conference on Inventive Computing and Informatics (ICICI), pp. 699–704 (2017). https://doi.org/10.1109/ICICI.2017.8365226, https://ieeexplore.ieee.org/document/8365226/
11. Tzutalin: Labelimg (2015). https://github.com/tzutalin/labelImg
12. Wang, C.Y., Bochkovskiy, A., Liao, H.Y.M.: YOLOv7: trainable bag-of-freebies sets new state-of-the-art for real-time object detectors. arXiv preprint arXiv:2207.02696 (2022)
13. Wang, K., Zhang, S., Wang, Z., Liu, Z., Yang, F.: Mobile smart device-based vegetable disease and insect pest recognition method. Intell. Autom. Soft Comput. **19**, 263–273 (2013). https://doi.org/10.1080/10798587.2013.823783, https://www.tandfonline.com/doi/abs/10.1080/10798587.2013.823783
14. Xia, D., Chen, P., Wang, B., Zhang, J., Xie, C.: Insect detection and classification based on an improved convolutional neural network. Sensors **18**, 4169 (2018). https://doi.org/10.3390/S18124169, https://www.mdpi.com/1424-8220/18/12/4169/htm

Learning to Search for and Detect Objects in Foveal Images Using Deep Learning

Beatriz Paula[1] and Plinio Moreno[1,2](✉)

[1] Instituto Superior Técnico, Univ. Lisboa, 1049-001 Lisbon, Portugal
[2] Institute for Systems and Robotics (ISR/IST), LARSyS, 1049-001 Lisbon, Portugal
plinio@isr.tecnico.ulisboa.pt

Abstract. The human visual system processes images with varied degrees of resolution, with the fovea, a small portion of the retina, capturing the highest acuity region, which gradually declines toward the field of view's periphery. However, the majority of existing object localization methods rely on images acquired by image sensors with space-invariant resolution, ignoring biological attention mechanisms. As a region of interest pooling, this study employs a fixation prediction model that emulates human objective-guided attention of searching for a given class in an image. The foveated pictures at each fixation point are then classified to determine whether the target is present or absent in the scene. Throughout this two-stage pipeline method, we investigate the varying results obtained by utilizing high-level or panoptic features and provide a ground-truth label function for fixation sequences that is smoother, considering in a better way the spatial structure of the problem. Additionally, we present a novel dual task model capable of performing fixation prediction and detection simultaneously, allowing knowledge transfer between the two tasks. We conclude that, due to the complementary nature of both tasks, the training process benefited from the sharing of knowledge, resulting in an improvement in performance when compared to the previous approach's baseline scores.

Keywords: Visual Search · Foveal Vision · Deep Learning

1 Introduction

A fundamental difference between the human visual system and current approaches to object search is the acuity of the image being processed [1]. The human eye captures an image with very high resolution in the fovea, a small region of the retina, and a decrease in sampling resolution towards the periphery of the field of view. This biological mechanism is crucial for the real-time image processing of the rich data that reaches the eyes (0.1–1 Gbits), since visual attention prioritizes interesting and visually distinctive areas of the scene, known as salient regions, and directs the gaze of the eyes. In contrast, image sensors, by default, are designed to capture the world with equiresolution in a homogeneous

A. Pertusa et al. (Eds.): IbPRIA 2023, LNCS 14062, pp. 223–237, 2023.
https://doi.org/10.1007/978-3-031-36616-1_18

space invariant lattice [2], and current solutions to vision system performance rely on the increase of the number of pixels. This limits real-time applications due to the processing bottleneck and the excessive amount of energy needed by state-of-the-art technologies.

The Convolutional Neural Network (CNN) [3,5], a very sucessful Deep Learning (DL) technique, is inspired by the human visual processing system. In the ImageNet Challenge, the winner, Alex Krizhevsky, introduced a CNN [4] that showed its massive power as a feature learning and classification architecture. Although AlexNet is very similar to LeNet [5] (published in 1998), the by scaling up of both the data and the computational power brought large performance improvements. Nevertheless, it remains challenging to replicate and model the human visual system. Recent advances combine DL with image foveation and saliency detection models. In [6], a foveated object detector has performance similar to homogeneous spatial resolution, while reducing computational costs.

In the context of foveated image search, our work aims to utilize goal-guided scanpath data for object detection in foveated images. Our object search approach receives as input an image and an object category, then indicates the presence or absence of instances of that category in the scene while adjusting the acuity resolution to mimic the human visual system.

Our contributions include: (i) Benchmark of recent approaches based on DL, which are able to predict fixations, on a recent large-scale dataset; (ii) a ground-truth label function for fixation sequences that is smoother, considering in a better way the spatial structure of the problem; (iii) evaluation of two alternative visual representations (conventional high-level features from VGG and a more elaborate multi-class presence description); and (iv) the introduction of a novel dual task approach that simultaneously performs fixation and target detection.

2 Related Work

Human attention is driven by two major factors, bottom-up and top-down factors [7]. While bottom-up is driven by low-level features in the field of vision, which means saliency detection is executed during the pre-attentive stage, top-down factors are influenced by higher level features, such as prior knowledge, expectations and goals [8]. Depending on the goal/task description, the distribution of the points of fixation on an object varies correspondingly [9].

In Computer Vision, Gaze Prediction models aim to estimate fixation patterns made by people in image viewing. These models can have a spatial representation, in fixation density maps, and an added temporal representation when predicting scanpaths. In this area of study, most work focuses on free-viewing, which, as mentioned, is led by bottom-up attention.

In [10], CNNs are used for feature extraction and feature maps compilation, which are then used in a Long Short Term Network (LSTM) responsible for modeling gaze sequences during free-viewing. LSTM [11] was proposed as a solution to the vanishing gradient problem of RNNs. LSTM networks have a more complex structure that tweak the hidden states with an additive interaction, instead of a linear transformation, which allows the gradient to fully backpropagate all the way to the first iteration.

However, human scanpaths during search tasks vary depending on the target items they are trying to gain information from, therefore guided search cannot be predicted based on free-viewing knowledge. Goal-directed attention is additionally relevant due to the human search efficiency in complex scenes that accounts for scene context and target spatial relations [12].

In [13], a guided search approach inspired in the architecture of [10] shows promising results. Their approach relies on a Convolutional Long Short Term Memory (ConvLSTM) architecture, and introduced a foveated context to the input images on top of an additional input encoding the search task, which found human fixation sequences to be a good foundation for object localization. The ConvLSTM had been previously introduced in [14] as a variant of LSTMs better suited for 3-dimensional inputs, such as images. This adaptation still contains the same two states: a hidden state, h, and a hidden cell state, c; and the same four intermediate gates: the input gate i, forget gate f, output gate o and candidate input $\tilde{c}$; as the LSTM architecture. However, a convolution is performed during the computation of the gates instead of the previous product operations, as seen in the following equations:

$$
\begin{aligned}
i_t &= \sigma(W_i * x_t + U_i * h_{t-1}) & f_t &= \sigma(W_f * x_t + U_f * h_{t-1}) \\
o_t &= \sigma(W_o * x_t + U_o * h_{t-1}) & \tilde{c}_t &= \tanh(W_c * x_t + U_c * h_{t-1}) \\
c_t &= f_t \odot c_{t-1} + i_t \odot \tilde{c}_t & h_t &= o_t \odot tanh(c_t)
\end{aligned}
$$

where $\odot$ denotes an element wise product, $*$ denotes a convolution, and W and U are the weight matrices of each gate that operate over the hidden states.

The limited amount of available data containing human scanpaths in visual search was, however, identified as a significant obstacle in [13]. Since then a new large-scale dataset has been introduced in [15], which has shown promising results in [16], where an inverse reinforcement learning algorithm was able to detect target objects by predicting both the action (fixation point selection) and state representations at each time step, therefore replicating the human attention transition state during scanpaths. This approach additionally utilized features extracted from a Panoptic Feature Pyramid Network (Panoptic FPN) model [18], that performs panoptic segmentation which is the unification of "the typically distinct tasks of semantic segmentation (assign a class label to each pixel) and instance segmentation (detect and segment each object instance)" [17].

3 System Overview

In this section, we present the architecture of the two strategies used in this study: a two-stage pipeline system consisting of a gaze fixations predictor and an image classifier, and a dual-task model that conducts scanpath prediction and target detection simultaneously.

3.1 Fixation Prediction Module

We consider the same network architecture for the two types of features of the fixation model: (i) High-level feature maps and (ii) panoptic image features. At each time-step $T = t$, the Input Transformation Section aggregates the features of the foveated pictures at each fixation location from the beginning of the gaze sequence, $T \in 0, ..., t$, as well as the task encoding of the target object. This combined input is then sent to the Recurrent Section, which uses ConvLSTM layers to emulate human-attention through its hidden states. The Recurrent Section then outputs its final hidden state h_{t+1} to the model's Output Section, which predicts the next scanpath fixation as a discrete location in an image grid with dimensions $H \times W$. We now present the architecture of each model in detail.

Fixation Prediction from High-Level Features. In this model, we utilized the high-level features retrieved from the ImageNet-trained VGG16 model [19, 20] with dimensions $H \times W \times Ch$, and it is composed of the following sections:

- **Input Transformation:** To condition the image feature maps on the task, we perform an element-wise multiplication of these inputs. In addition, depending on its format, the task encoding may be transmitted through a Fully Connected (FC) Layer with Ch units and a tanh activation, followed by a Dropout Layer with a rate of $r_{Dropout}$ in order to prevent overfitting.
- **Recurrent Section:** This portion mainly consists of a ConvLSTM layer with F filters (dimensionality of its output), a kernel size of K x K, a stride of S, and a left and right padding of P. The ConvLSTM has a tanh activation, and the recurrent step utilizes a hard sigmoid activation[1]. Subsequently, to prevent overfitting we perform batch normalization, where the features are normalized with the batch mean and variance. During inference, the features are normalized with a moving mean and variance.
- **Output Section:** We perform a flattening operation to each temporal slice of the input with the help of a Time Distributed wrapper. The flattened array is then fed to a FC layer and has $H \times W$ units and a softmax activation function.

Fixation Prediction from Panoptic Features. To compute these new features we resorted to the Panoptic FPN model in [18]. The belief map computes a combination of high and low resolution belief maps:

$$B(t) = M_t \odot H + (1 - M_t) \odot L, \tag{1}$$

where H and L are the belief maps for the high and low resolution images, respectively, and M_t is the binary fixation mask at time step t, of size $H \times W$, where every element is set to 0 except the grid cells at an euclidean distance shorter than r from the current fixation point.

[1] a piece-wise linear approximation of the sigmoid function, for faster computation.

To duplicate the belief maps in [16], the task encoding is a one-hot encoding with dimensions $H \times W \times Cl$, where each row of the axis Cl corresponds to an object class and the two-dimensional map $H \times W$ is all set to one for the target class and zero for the others. This input is subsequently transmitted to the Input Transformation stage, where it is concatenated with the image feature maps.

The recurrent section of the model is composed of d ConvLSTM layers, each followed by a Batch Normalization layer. Every ConvLSTM is constructed with the same hyper-parameters: each one has F_{LSTM} filters with a kernel size of $K_{LSTM} \times K_{LSTM}$, a stride of S_{LSTM}, a padding of P_{LSTM}, a tanh activation function and a hard sigmoid activation during the recurrent step.

Finally, in the output section, we conducted experiments over two different setups. The first one is composed of a 3d-Convolutional layer with a sigmoid activation followed by a time distributed flattening operation. The second setup is comprised of the same 3d-Convolutional layer, but with a ReLu activation, and a flattening layer followed by a FC layer with softmax activation.

3.2 Target Detection Module

In the last stage of the pipeline, the model evaluates at each time-step, if the fixation point coincides with the location of the target object. To detect the target, we rely on VGG16 architecture, and develop 18 binary classifiers, one for each task. In addition, as a baseline, we utilize a complete VGG16 trained on the ImageNet dataset to perform classification on our data.

To fine-tune the already pre-trained VGG16 model, we substituted its classification layers, with three fully connected layers, FC_i with $i \in \{1, 2, 3\}$, each with U_i units, where FC_1 and FC_2 were followed by a ReLu activation function and FC_3 was followed by a sigmoid action function. During training, only the parameters of these last FC layers were updated.

3.3 Dual Task Model

We aim to both predict the fixation point and localize the target object, by sharing the internal states on two LSTMs branches. We consider three different architectures: A, B and C. All models receive as input the high-level feature maps, with dimensions $H \times W \times Ch$ and a one-hot task encoding array, of size Cl, which are then aggregated. Similar to Sect. 3.1, the grouping of both of these inputs is accomplished by passing the task encoding through a FC layer with Ch units and tanh activation, and conducting an element-wise multiplication with the foveated image's feature maps. After this shared module, the models branch off to complete each specific task using the following architectures:

- **Architecture A (fixation-first)** After performing the input transformation, where we aggregate the feature maps and task encoding, the array x_t is fed to two ConvLSTM layers. Then, following each iteration of the fixation prediction recurrent module, its internal states h_t^{fix} and c_t^{fix} are passed to the detection branch as the internal states, h_{t-1}^{det} and c_{t-1}^{det}, of the preceding time step, as illustrated in Fig. 1a. In the first branch, a temporal flattening operation is performed to h^{fix}, followed by an output layer consisting of a FC layer with softmax activation. In the second we classify each temporal slice of h^{det} by employing the same structure of three FC layers $FC_i \in \{1, 2, 3\}$, with U_i units, respectively, where the first two layers have a ReLu activation while the output layer has a sigmoid activation.

- **Architecture B (detection-first)** Similar to the previous architecture, each task branch employs ConvLSTM layer. The sole difference is that we now conduct the iterations of the detection module first, and send the internal states h_t^{det} and c_t^{det} to the fixation prediction module for the preceding time step $t-1$.
- **Architecture C** The fixation prediction is a copy from architecture A. In difference, the target detection branch no longer has a ConvLSTM layer. Instead, at each time step t, the combined input x_t computed by the shared module is concatenated with the output of the ConvLSTM layer, h_{t+1}^{fix} of the fixation prediction task, as illustrated in Fig. 1b. Finally, this concatenation is followed by the same three FC layers utilized by the previous architectures.

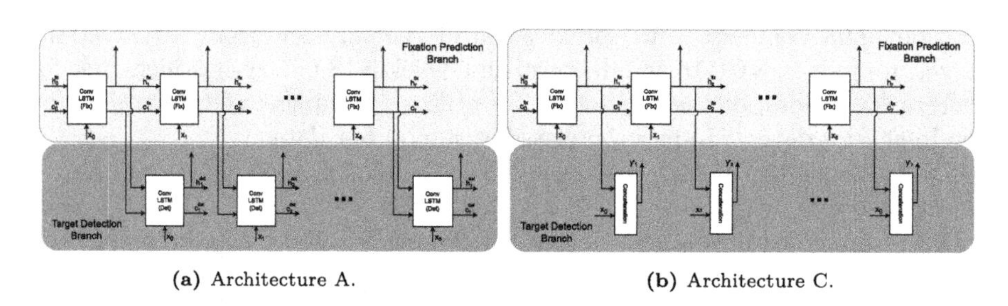

(a) Architecture A. **(b)** Architecture C.

Fig. 1. Information flow across the fixation prediction and target detection branch in architectures A, on the left, and C, on the right.

4 Implementation

4.1 Dataset

We use the COCO-Search18 dataset [15]. This dataset consists of 6,202 images from the Microsoft COCO dataset [22], evenly split between target-present and target-absent images, of 18 target categories[2], with eye movement recordings from 10 individuals. As humans were able to fixate the target object within their first six saccades 99% of the time, fixation sequences with length greater than that were discarded. Additionally, the sequences were padded with a repeated value of the last fixation point to achieve a fixed length of 7, including the initial

[2] bottle, bowl, car, chair, analogue clock, cup, fork, keyboard, knife, laptop, microwave, mouse, oven, potted plant, sink, stop sign, toilet and tv.

center fixation. This was done to replicate the procedure of a similar work [21], where participants were instructed to fixate their gaze on the target object, once they found them, during search tasks. To train the fixation prediction module and the dual task model, we used a random dataset split of 70% train, 10% validate and 20% test over each class category and all images were resized to 320×512 which resulted in feature maps with 10×16 spacial dimensions.

4.2 Training

During the training phase, all our models were optimized with the Adam algorithm [23] and a learning rate of $lr = 0.001$, for a maximum of 100 epochs with an early stopping mechanism activated when the validation loss stops improving after a duration of 5 epochs. Additionally, every dropout is performed with $r_{Dropout} = 0.5$ and we use a batch size of 256 in every module apart from the fixation prediction performed with high-level features.

Fixation Prediction from High-Level Features. We estimate the weights and bias parameters that minimize the loss between the predicted output $\hat{y}$ and the ground truth label y, with the cross entropy function computed for every fixation time step t for every sequence s of each mini-batch:

$$L_{CE} = -\sum_{s=1}^{S}\sum_{t=0}^{T}\sum_{i=1}^{H\times W} y_i * log(\hat{y}_i), \tag{2}$$

where S corresponds to the batch size, T to the sequence length which is set to 6 (in addition to the initial fixation point at $t = 0$) and $H \times W$ to the output size which is set to 160. We set with $F = 5$ filters, a kernel size of $K = 4$ and a stride of $S = 2$, and varied the batch size between 32, 64, 128 and 256. We conduct an ablation study over these additional hyper-parameters and settings: (i) Fovea size: We use the same real-time foveation system as in [13], considering three fovea sizes: 50, 75 and 100 pixels. (ii) Task encoding: We consider two representations. The first is a one-hot encoding array of size 18. The second is a normalized heat map of fixations made during the observations of that same task, compiled exclusively with training data. (iii) Ground truth function: We consider both a one-hot encoding representation of the ground-truth label and a two dimensional Gaussian function with the mean set to the cell coordinates of the actual fixation location and the variance set to 1.

Fixation Prediction from Panoptic Features. We want to minimize the loss function in Eq. 2. Additionally, the feature maps used have dimension $10 \times 16 \times 134$ in order to replicate the scale of our grid shaped output. The configuration of the ConvLSTM: $F_{LSTM} = 10$ filters with square kernels of size $K_{LSTM} = 3$, a stride of $S_{LSTM} = 1$ and a padding of $P_{LSTM} = 1$ to maintain the features spatial resolution. In the output section, $F_{Conv} = 1$ for the 3d-Convolutional layer to have a kernel size of $K_{Conv} = 2$, a stride of $S_{Conv} = 1$ and padding of $P_{Conv} = 1$. The second setup of this section is configured to have a Fully Connected layer with 160 units.

For this model, we additionally varied the depth of the recurrent section with $d \in \{1, 3, 5\}$, and altered the structure of the output section to utilize both a sigmoid and softmax as the final activation function. Concerning the data representation, we once again evaluated the impact of having a one-hot or a Gaussian ground truth encoding, and explored several belief maps settings: we varied the radius r of the mask, M_t, with values $r \in \{1, 2, 3\}$ (each to emulate a corresponding fovea size of 50, 75 and 100); and experimented with a cumulative mask configuration, M_t', where the binary mask utilized in Eq. 1, in addition to the information of the current time step, accumulates the high acuity knowledge of all previous time steps. All panoptic feature maps were computed with a low resolution map L extracted from a blurred input image with a Gaussian filter with radius $\sigma = 2$.

Target Detection. The binary classifiers were implemented with each fully connected layer having $U_1 = 512$, $U_2 = 256$ and $U_3 = 1$ units, a dropout rate of $r_{Dropout} = 0.5$, and we varied the fovea size between 50, 75 and 100 pixels. They were trained with a loss function defined as:

$$L_{BCE} = -\frac{1}{N}\sum_{i=1}^{N} y_i \cdot log(\hat{y}_i) + (1 - y_i) \cdot log(1 - \hat{y}_i) \tag{3}$$

Dual Task. In this case, the ConvLSTM layers are configured with $F = 5$ filters of size $K = 4$ to execute the convolutional operations with stride $S = 2$, while the fully connected layers of the detection branch are configured with $U_1 = 64$, $U_2 = 32$ and $U_3 = 1$ units. The loss function of the dual task L_{Dual} is:

$$L_{Dual} = w_{fix} \cdot L_{fix} + (1 - w_{fix}) \cdot L_{det}, \tag{4}$$

where L_{fix} and L_{det} correspond to the loss of the fixation and detection prediction, respectively. L_{fix} corresponds to the categorical cross entropy, like in (2), while L_{det} is a weighted binary cross entropy as follows in (5):

$$w = y \cdot w_1 + (1 - y) \cdot w_0; \quad L_{det} = w \cdot [y * log(\hat{y}) + (1 - y) \cdot log(1 - \hat{y})]. \tag{5}$$

Since the target is absent in half of the images and appears in a limited section of the scanpath sequence, (5) includes sample-based weights. Due to the high imbalance of the detection data we add the weights $w_1 = 1.6$ and $w_0 = 0.7$. In the case of w_1 we compute the multiplicative inverse of the ratio of positive detections on the total number of detections, and dividing it by 2. In the case of w_0, the inverse ratio of negative detections divided by 2.

To determine the optimal configuration for each model's architecture, an ablation study was done over the fovea size (50, 75 and 100 pixels) and the degree of importance w_{fix} (0.10, 0.25, 0.50, 0.75, 0.90) in Eq. (4).

4.3 Prediction

During the testing phase, the single and dual task models aim to predict a scanpath sequence of fixed length $l = 7$, based on the training data, and the fixation point at $t = 0$ as the center cell of the discretized grid. We apply the beam search algorithm, which selects the best m fixation points at each time step. Then, the selected points are appended to the sequences they were generated from, while saving the target detection prediction in the case of the dual task model. In the next time step, the model runs for each of these m predicted sequences, to select the next best m predictions. In our experiments, $m = 20$.

In regards to the target presence detector, all models were deployed on the scanpaths produced by the highest performing scanpath predictor. The baseline classifier was tested similarly to the binary models, but cropping the images to 224×224. Due to the mismtaching classes between the datasets, we adapt the ImageNet classes to our targets by grouping some sibling sub-classes[3], and remove the non-exisiting classes in ImageNet. The target is present when the ground-truth class has the highest classification score and as absent otherwise.

5 Results

5.1 Two Stage-Pipeline

Fixation Prediction. The evaluation metrics include: (i) **Search Accuracy** which is computed as the ratio of sequences in which a fixation point selects a grid cell that intersects the target's bounding box, (ii) **Target Fixation Cumulative Probability (TFP)**, which is plotted in Fig. 3, and presents the search accuracy attained by each time step. On the TFP, we compute the **TFP - Area Under Curve (TFP-AUC)** and the **Probability Mismatch**, which is the sum of absolute differences between the model's TFP and the human's observable data. Finally, (iii) the **Scanpath Ratio** as the ratio between the sum of euclidean distances between each fixation point and the distance from the initial fixation to the center of the target's bounding box.

Through the ablation study we conducted for the two stage pipeline, we found that the high-level features scanpath predictor achieves top search accuracy scores (0.69) when using a one-hot task encoding and a Gaussian ground-truth, as seen in Fig. 2, where the search scores are depicted in box-plots grouped by each training setting.

Figure 3 shows the TFP curve of several highest model configurations, as well as human search behavior, and a random scanpath baseline model[4]. We noticed: (i) A decrease in performance across time for all models through the slopes of each function, (ii) all the models but the random one were able to detect the most of the targets by the second fixation step and (iii) our models barely detected any new targets on the last four fixation points. The right side of Fig. 3 shows that fixations converge across time steps, but seems to be a behavior copycat from the training set.

[3] e.g. the task bowl corresponds to the joint sub-classes mixing bowl and soup bowl.

[4] We select a human sequence randomly from the train split for the same search target class on the testing set.

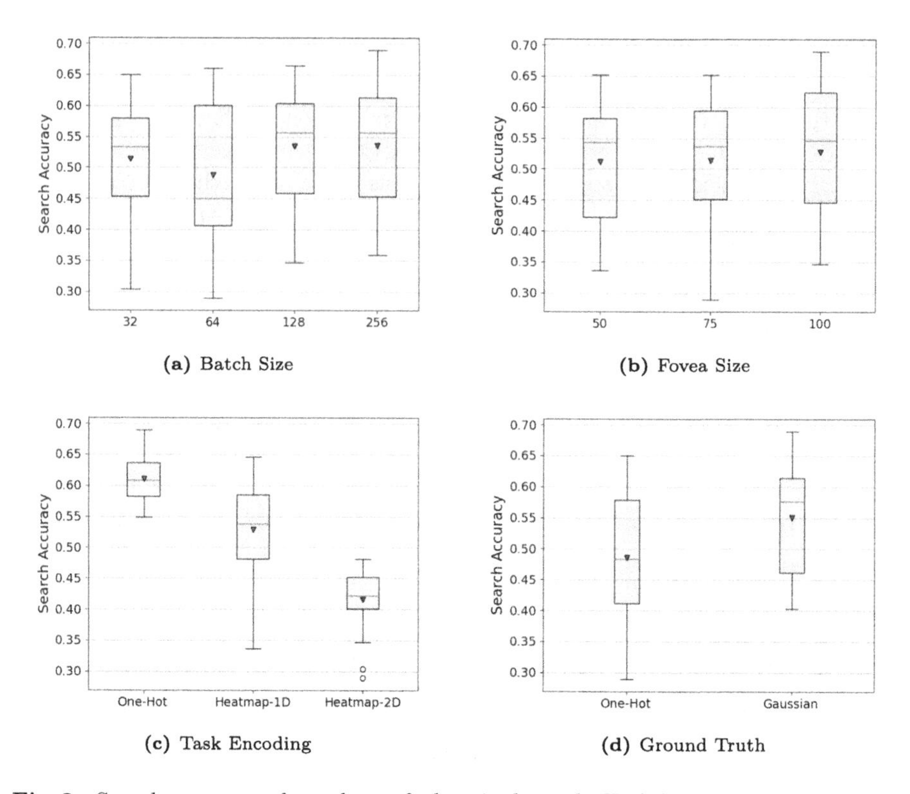

Fig. 2. Search accuracy box-plots of the single task High-level features' models, grouped by training settings, with the mean values (triangles) and outliers (circles).

Regarding the ablation study of fixation prediction with the panoptic features shown in Fig. 4, the highest search accuracy score of 0.686 was attained with a cumulative mask of radius $r = 1$. We note that using a sigmoid activation improves the model's results because the ground truth is Gaussian. In contrast, when a final softmax activation is utilized, the model turns the scores of the last hidden layers into class probabilities. Due to the ground truth encoding, there is a saturation of the loss when every grid cell is considered, as opposed to only examining the probability of the true class in a one-hot encoding configuration, resulting in the model's poor performance.

By comparing the high-level vs. panoptic features on Fig. 3, we see that the models with high-level feature maps fixate targets much sooner. However, the panoptic features lead to a similar search accuracy. The panoptic-based model is much less efficient as the scanpaths travel a much greater distance, as seen in Fig. 3, leading to a scanpath ratio score of only 0.463.

Target Detection. For the target detection task we used **accuracy**, **precision** and **recall** as metrics. The fine-tuned classifiers with foveation radius of 50 pixels had the maximum performance for all measures, with a mean accuracy, precision

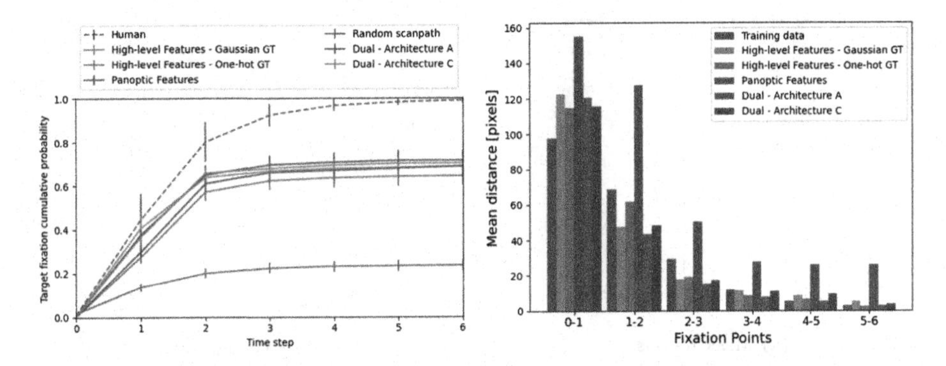

Fig. 3. Left side: Search accuracy per model along scanpaths, with means and standard errors computed over target classes. The Human TFP refers to the human behavior observed in the entire COCO-Search18 dataset. The remaining TFP curves were computed for the test data split. Right side: Euclidean distances between fixation points.

Table 1. Performance evaluation of best performing models (rows) based on Fixation Prediction metrics (columns). The ↑ indicates higher is better, and ↓ lower is better.

	Search Accuracy ↑	TFP-AUC ↑	Probability Mismatch ↓	Scanpath Ratio ↑
Human	0.990	5.200	–	0.862
High-Level Features - One-hot GT	0.650	3.068	1.727	0.753
High-Level Features - Gaussian GT	0.690	3.413	1.360	0.727
Panoptic Features	0.686	3.259	1.514	0.463
Dual - Architecture A	0.719	3.496	1.263	0.808
Dual - Architecture C	0.701	3.446	1.320	0.791
IRL	N/A	4.509	0.987	0.826
BC-LSTM	N/A	1.702	3.497	0.406
Random Scanpath	0.235	1.150	3.858	–

and recall of 82.1%, 86.9% and 75.4%, respectively, whereas the configuration of 75 pixels achieved the lowest accuracy and recall. In turn, the baseline pre-trained model has a very large variance across metrics for the vast majority of the classes, as seen in Fig. 5b. Performance increases slightly with larger foveation radius, reaching the highest accuracy, precision and recall scores (64.0%, 85.7% and 34.3%) for the 100 pixels fovea size.

5.2 Dual Task

Figure 6a shows that architecture A outperforms more than half of the models of architecture C on search accuracy (top value of 73.4% with a fovea size of 100

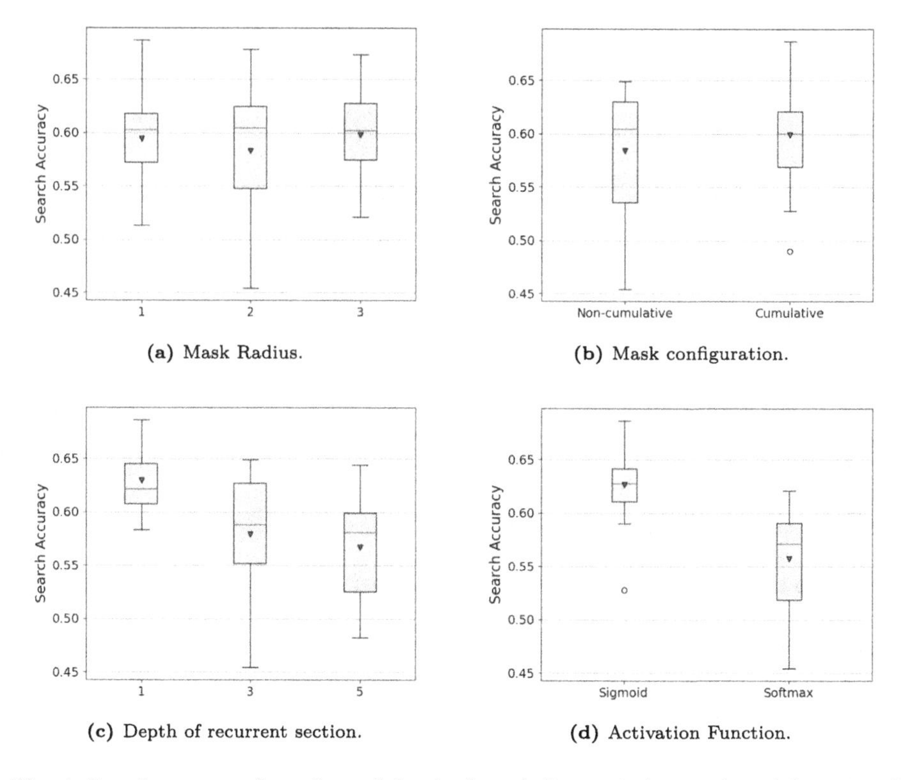

(a) Mask Radius. **(b)** Mask configuration. **(c)** Depth of recurrent section. **(d)** Activation Function.

Fig. 4. Search accuracy box-plots of the single task Panoptic features' models grouped by their training settings, with the mean values (trinagles) and outliers (circles).

pixels and $w_{fix} = 0.75$). Regarding detection performance, the dual task model with architecture C, a fovea size of 50, and $w_{fix} = 0.9$ achieved the highest target presence detection rate of 68.7%. In Fig. 6c, considering the quartiles and upper limit of its performance, the architecture of design C is deemed to be the most effective. Regarding the weight of the fixation loss, we can also see that models trained with bigger values obtained a larger interquartile range than models trained with smaller values.

In addition, note that the best scanpath prediction model achieved a detection accuracy of 49.7% while the best target presence predictor achieved a search accuracy of 63.9%. To have a single value for evaluation, we also considered the average of both metrics. The majority of the time, design A earned a higher score than design C, while design B ranked the lowest. Regarding the remaining parameters, we observe that a bigger fovea radius led to higher average scores, and a higher fixation loss' weight resulted in a better top score, with the exception of setting $w_{fix} = 0.25$. The model configured with architecture C, a fovea size of 75 pixels, and $w_{fix} = 0.9$ achieved the top score of 67.7% with search and detection accuracies of 70.1% and 65.3%, respectively.

Finally, the dual approach led to higher search accuracy, as seen in Table 1, resulting in a higher TFP-AUC score of 3.496 and 3.446 and a lower probability

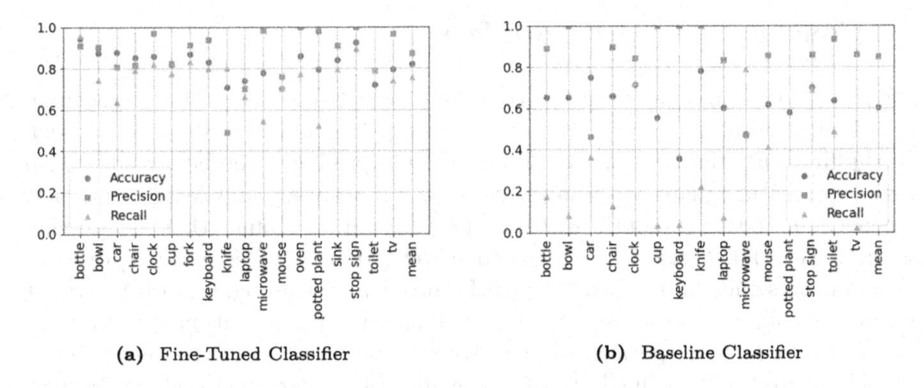

(a) Fine-Tuned Classifier **(b)** Baseline Classifier

Fig. 5. Performances of the fine-tuned and baseline target detectors in terms of accuracy, precision and recall, for the fovea size setting of 50 pixels on the left side and 100 pixels on the right side.

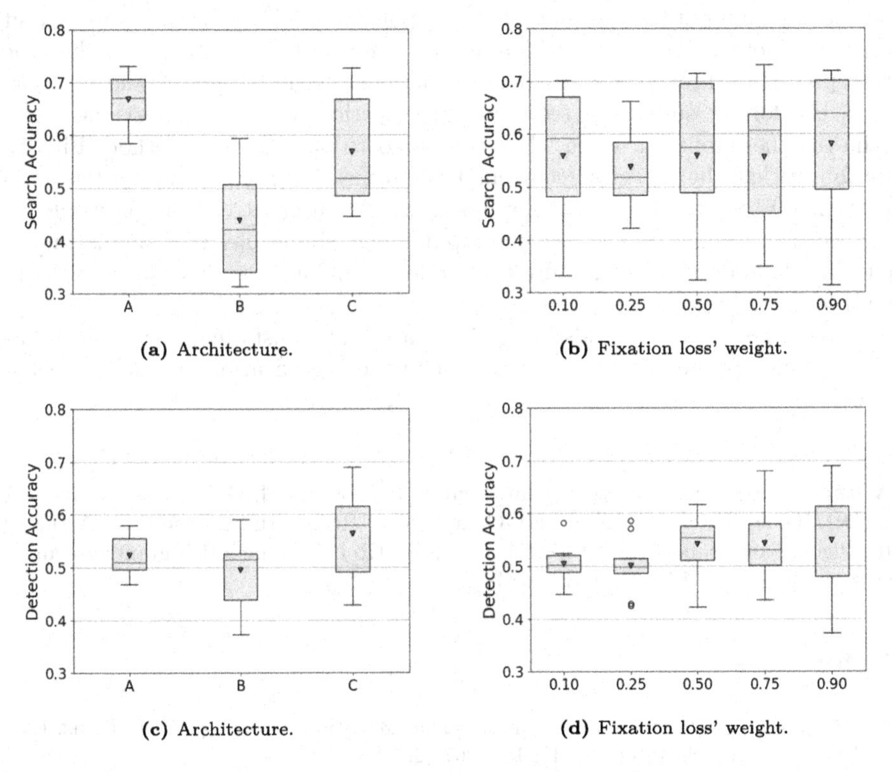

(a) Architecture. **(b)** Fixation loss' weight.

(c) Architecture. **(d)** Fixation loss' weight.

Fig. 6. Box-plots of the Search accuracy, on the top, and Detection accuracy, on the bottom, grouped by configuration, with mean values (triangles) and outliers (circles).

mismatch of 1.263 and 1.320 for the overall best models with architectures A and C, respectively. In addition, the two models exhibit a higher search efficiency with scanpath ratios of 0.808 and 0.791, respectively.

6 Conclusions and Future Work

We present two methods for predicting the presence or absence of a target in an image with foveated context: (i) a two-stage pipeline and (ii) a dual task model. In the first one, the fixation prediction module produced the best results with high-level feature maps, both in terms of search accuracy and search efficiency, when compared to panoptic features. In addition, we found that a Gaussian ground truth label encoding, enhanced search accuracy. This novel representation captures the spatial structure of the problem, encouraging both the exact discretized human fixation positions as well as attempts to cells near these locations. Two classifiers performed the target presence of the two-stage pipeline model, where the fine-tuned classifiers for multiple binary tasks performed better than a pre-trained VGG-16.

The final contribution of this work is a dual-task model that executes both tasks concurrently while enabling information sharing between them by executing a common input transformation and establishing linking channels throughout each task branch. This multi-task approach improved search precision when the task prediction branch initiated the predictions, i.e. in designs A and C. However, the former suffered a reduction in detection accuracy, whilst the latter achieved the maximum score when compared to our baseline method. Finally, we found that the use of a recurrent layer biases the model towards temporal patterns of target detection of the dual-task. An alternative solution would be for the task branch to generate the input image simultaneously, simulating an encoder-decoder, so as to require the model to maintain its knowledge of high-level features in its hidden states.

Future research should also investigate a visual transformer-based design, as it has shown promising results in similar image classification and goal-directed search tasks.

Acknowledgements. Work partially supported by the LARSyS - FCT Project [UIDB/50009/2020], the H2020 FET-Open project Reconstructing the Past: Artificial Intelligence and Robotics Meet Cultural Heritage (RePAIR) under EU grant agreement 964854, the Lisbon Ellis Unit (LUMLIS).

References

1. Borji, A., Itti, L.: State-of-the-art in visual attention modeling. IEEE Trans. Pattern Anal. Mach. Intell. **35**(1), 185–207 (2012)
2. Bandera, C., Scott, P.D.: Foveal machine vision systems. In Conference Proceedings. In: IEEE International Conference on Systems, Man and Cybernetics, pp. 596–599. IEEE (1989)
3. LeCun, Y., Bottou, L., Bengio, Y., Haffner, P.: Gradient-based learning applied to document recognition. Proc. IEEE **86**(11), 2278–2324 (1998)
4. Krizhevsky, A., Sutskever, I., Hinton, G.E.: ImageNet classification with deep convolutional neural networks. Commun. ACM **60**(6), 84–90 (2017)

5. Fukushima, K.: Neocognitron: a hierarchical neural network capable of visual pattern recognition. Neural Netw. **1**(2), 119–130 (1988)
6. Akbas, E., Eckstein, M.P.: Object detection through search with a foveated visual system. PLoS Comput. Biol. **13**(10), e1005743 (2017)
7. James, W.: The Principles of Psychology, vol. 1. Henry Holt and Co. (1890)
8. Corbetta, M., Shulman, G.L.: Control of goal-directed and stimulus-driven attention in the brain. Nat. Rev. Neurosci. **3**(3), 201–215 (2002)
9. Yarbus, A.L.: Eye Movements and Vision. Springer, Heidelberg (2013)
10. Ngo, T., Manjunath, B.S.: Saccade gaze prediction using a recurrent neural network. In: 2017 IEEE International Conference on Image Processing (ICIP), pp. 3435–3439. IEEE (2017)
11. Graves, A.: Long short-term memory. In: Graves, A. (ed.) Supervised Sequence Labelling with Recurrent Neural Networks, vol. 385, pp. 37–45. Sprnger, Heidelberg (2012). https://doi.org/10.1007/978-3-642-24797-2_4
12. Kreiman, G., Zhang, M.: Finding any Waldo: zero-shot invariant and efficient visual search (2018)
13. Nunes, A., Figueiredo, R., Moreno, P.: Learning to search for objects in images from human gaze sequences. In: Campilho, A., Karray, F., Wang, Z. (eds.) ICIAR 2020. LNCS, vol. 12131, pp. 280–292. Springer, Cham (2020). https://doi.org/10.1007/978-3-030-50347-5_25
14. Shi, X., Chen, Z., Wang, H., Yeung, D.Y., Wong, W.K., Woo, W.C.: Convolutional LSTM network: A machine learning approach for precipitation nowcasting. Adv. Neural Inf. Process. Syst. **28** (2015)
15. Chen, Y., Yang, Z., Ahn, S., Samaras, D., Hoai, M., Zelinsky, G.: COCO-search18 fixation dataset for predicting goal-directed attention control. Sci. Rep. **11**(1), 1–11 (2021)
16. Yang, Z., et al.: Predicting goal-directed human attention using inverse reinforcement learning. In: Proceedings of the IEEE/CVF Conference on Computer Vision and Pattern Recognition, pp. 193–202 (2020)
17. Kirillov, A., He, K., Girshick, R., Rother, C., Dollár, P.: Panoptic segmentation. In: Proceedings of the IEEE/CVF Conference on Computer Vision and Pattern Recognition, pp. 9404–9413 (2019)
18. Kirillov, A., Girshick, R., He, K., Dollár, P.: Panoptic feature pyramid networks. In: Proceedings of the IEEE/CVF Conference on Computer Vision and Pattern Recognition, pp. 6399–6408 (2019)
19. Simonyan, K., Zisserman, A.: Very deep convolutional networks for large-scale image recognition. arXiv preprint arXiv:1409.1556 (2014)
20. Deng, J., Dong, W., Socher, R., Li, L.J., Li, K., Fei-Fei, L.: ImageNet: a large-scale hierarchical image database. In: 2009 IEEE Conference on Computer Vision and Pattern Recognition, pp. 248–255. IEEE (2009)
21. Cabarrão, B.: Learning to search for objects in foveal images using deep learning, Master's thesis, Universidade de Lisboa - Instituto Superior Técnico (2022)
22. Lin, T.-Y., et al.: Microsoft COCO: common objects in context. In: Fleet, D., Pajdla, T., Schiele, B., Tuytelaars, T. (eds.) ECCV 2014. LNCS, vol. 8693, pp. 740–755. Springer, Cham (2014). https://doi.org/10.1007/978-3-319-10602-1_48
23. Kingma, D.P., Ba, J.: Adam: a method for stochastic optimization. arXiv preprint arXiv:1412.6980 (2014)

Relation Networks for Few-Shot Video Object Detection

Daniel Cores[1(✉)], Lorenzo Seidenari[2], Alberto Del Bimbo[2], Víctor M. Brea[1], and Manuel Mucientes[1]

[1] Centro de Investigación en Tecnoloxías Intelixentes (CiTIUS), Universidade de Santiago de Compostela, Santiago de Compostela,, Spain
{daniel.cores,victor.brea,manuel.mucientes}@usc.es

[2] Media Integration and Communication Center (MICC), University of Florence, Firenze, Italy
{lorenzo.seidenari,alberto.delbimbo}@unifi.it

Abstract. This paper describes a new few-shot video object detection framework that leverages spatio-temporal information through a relation module with attention mechanisms to mine relationships among proposals in different frames. The output of the relation module feeds a spatio-temporal double head with a category-agnostic confidence predictor to decrease overfitting in order to address the issue of reduced training sets inherent to few-shot solutions. The predicted score is the input to a long-term object linking approach that provides object tubes across the whole video, which ensures spatio-temporal consistency. Our proposal establishes a new state-of-the-art in the FSVOD500 dataset.

Keywords: few-shot object detection · video object detection

1 Introduction

The paradigm of few-shot object detection aims to address the issue of large training sets in modern object detectors, which shows up mainly in high annotation costs, or, which is worse, eventually in useless deep learning models, simply because there might not be enough data to train them. This makes few-shot object detection a timely topic.

Video object detectors leverage spatio-temporal information to tackle challenges such as motion blur, out of focus, occlusions or high changes in an object appearance to increase detection precision [6,7,9], which is not straightforward for image object detectors working on isolated frames. Similarly, the issue of large training sets as a need for a high precision in video object detectors calls for few-shot video object detectors, which today are less abundant in the literature than their image few-shot object detector counterpart.

Few-shot becomes even more challenging when referred to attention mechanisms, which today have become widely accepted for modeling object proposal relationships in video object detection [24], increasing precision, but at the cost

A. Pertusa et al. (Eds.): IbPRIA 2023, LNCS 14062, pp. 238–248, 2023.
https://doi.org/10.1007/978-3-031-36616-1_19

of multiple video frames in each training iteration. On the contrary, a few-shot video object detector should learn from a limited number of labeled instances per object category while exploiting spatio-temporal information at inference time. In a proper few-shot framework, whole videos are not available for training. This leads to a data gap between train and test sets that must be addressed.

All the above leads us to design a new few-shot object detector to take advantage of spatio-temporal information in videos. Our solution comprises a relation model with attention mechanisms and a classification score optimization component for spatio-temporal feature aggregation and consistency.

The main contributions of this work are:

- A method to bridge the data gap between training and test sets through synthetic frame augmentation, which permits the relation module to match new objects with proposals from the current image in the training phase, and avoids the degradation of our solution in the single image setting.
- A new spatio-temporal double head for spatial and spatio-temporal information with a spatial branch for object location and classification in the current frame, and a spatio-temporal branch to improve the classification and predict the overlap among detections and ground truth as a detection precision metric.
- Object tube generation from the predicted overlap among detections and ground truth to link detections in successive frames. This benefits from the generally high spatio-temporal redundancy of videos to modify the classification confidence of object detections.

Our approach outperforms both single image and previous video object detectors in FSVOD500, a specific dataset for few-shot video object detection.

2 Related Work

General object detectors based on deep neural networks fall mainly into two main categories: one- and two-stage detectors. Two-stage detectors [2,21] generate a set of object proposals with high probability of containing an object of interest, and then perform a bounding box refinement and object classification. In contrast, one-stage detectors [16,20] directly calculate the final detection set by processing a dense grid of unfiltered candidate regions.

More recently, object detection frameworks specifically designed to work with videos [6,7,9] were introduced to take advantage of spatio-temporal information, improving detection precision. The main idea behind these methods is to aggregate per-frame features throughout time, achieving more robust feature maps. This aggregation can be done at pixel- [12] or at object-level [6,7,9], linking object instances through neighboring frames.

Recently, the few-shot object detection problem has drawn significant attention in order to replicate the success achieved in the image classification field. Learning new tasks from just a few training examples is very challenging for most machine learning algorithms, especially for deep neural networks.

Current few-shot object detectors follow two main approaches: meta-learning and fine-tuning.

Meta-learning based methods are one of the research lines to address the few-shot object detection problem. They learn a similarity metric, so a query object can be compared with support sets from different categories. FSRW [14] proposes a framework based on YOLOv2 [20] including a feature reweighting module and a meta-feature extractor. Alternatively, a two-stage approach based on Faster/Mask R-CNN is proposed on Meta R-CNN [26]. This work proposes a new network head that applies channel-wise soft attention between RoI features given by a Region Proposal Network (RPN), and category prototypes to calculate the final detection set. The authors in [11] also propose to modify the RPN of a two-stage architecture, including an attention RPN that takes advantage of the support information to increase the RPN recall. A key component in all meta-learning based object detectors is to compute category prototypes from a set of annotated objects that are general enough to represent each object category. DAnA [5] implements a background attenuation block to minimize the effect of background in the final category prototype and also proposes a new method to summarize the support information in query position aware (QPA) support vectors.

TFA [25] proves that a simple transfer-learning approach, in which only the last layers of existing detectors are fine-tuned with new scarce data, can achieve results comparable to those of the state-of-the-art of meta-learning based detectors. The fine-tuning approach is further developed in DeFRCN [19], including a gradient decoupled layer (GDL) that modulates the influence of the RPN and the network classification and localization head in the training process and a prototypical calibration layer (PCB) to decouple the classification and localization tasks. Few-shot detectors that follow a fine-tuning approach suffer from a non-exhaustive training process, in which objects from novel categories in the base images are not annotated. Therefore, the model learns to treat novel categories as background in the pre-training stage. This issue is addressed in [3] by mining annotations from novel categories in base images. An automatic annotation framework was also proposed in [15] to expand the annotations available for the novel categories through label mining in unlabeled sets.

To the best of our knowledge, there is only one attempt to consider spatio-temporal information for few-shot object detection [10]. They are also the first to propose a dataset specifically designed for few-shot video object detection. Their method is based on a meta-learning approach, including a tube proposal network to associate objects across frames and a Tube-based Matching Network to compare tube features with support prototypes. Alternatively, we propose a fine-tuning based video object detector that implements feature aggregation at object level to enhance per-frame proposal features.

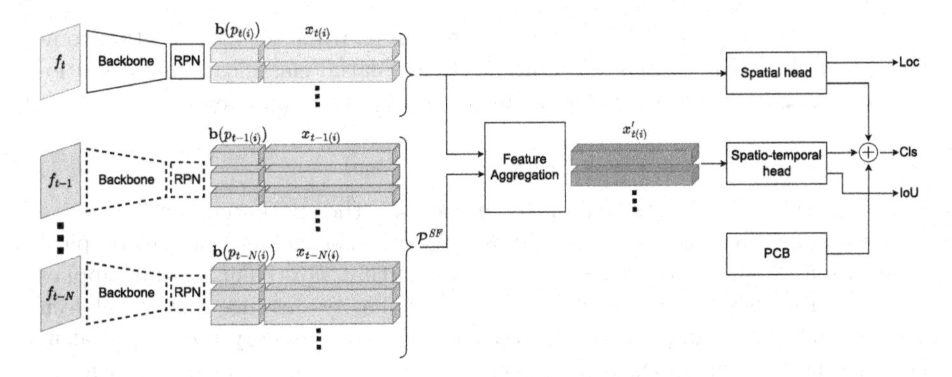

Fig. 1. FSVDet architecture overview, without the Confidence Score Optimization (CSO) module.

3 Proposed Method

3.1 Problem Definition

We follow the standard few-shot object detection setting established in previous works [11,14,19,25]. The whole dataset is divided into $\mathcal{D}_{base}$, with several annotated objects of the base classes $\mathcal{C}_{base}$, and $\mathcal{D}_{novel}$, with only a few annotations of the novel classes $\mathcal{C}_{novel}$ for training, with $\mathcal{C}_{base} \cap \mathcal{C}_{novel} = \emptyset$. The detection problem is defined as K-shot, where K is the number of objects annotated for each novel category $\mathcal{C}_{novel}$. Typically, the number of annotations K used to evaluate few-shot detectors in the literature ranges from 1 to 30 examples.

Video object detection aims to localize and classify objects in every frame of an input video. Object detection at each time step t is performed by considering information from the reference frame f_t and N previous frames $f_{t-N}, ..., f_{t-1}$. In the proposed framework, we also include a long-term optimization that considers information from the complete video to optimize the confidence of the detections. Although we propose a video object detector, annotations are object instances in single video frames. Thus, $\mathcal{D}_{novel}$ follows the same definition as for single-image approaches.

3.2 FSVDet: Few-Shot Video Object Detection

We propose FSVDet, a two-stage video object detection framework that can be trained with few labeled examples. The selected spatial baseline is DeFRCN [19], a modification of the original Faster R-CNN [21] able to perform a quick adaptation from a set of base categories $\mathcal{C}_{base}$ to a new domain $\mathcal{C}_{novel}$.

Figure 1 shows an overview of the proposed network architecture. First, per-frame image features and object proposals are independently calculated. Then, proposal boxes $\mathbf{b}(p_{t(i)})$ and the corresponding features $\phi(p_{t(i)})$, extracted through RoI Align for proposal $p_{t(i)}$ in the current frame f_t, are fed to the spatial branch of the double head. Object localization is exclusively performed with information extracted from the current frame, while the classification combines spatial and spatio-temporal information. Spatial classification and spatio-temporal classification scores for each category are combined as:

$$s = s_{tmp} + s_{spt}(1 - s_{tmp}) \tag{1}$$

being s_{spt} the spatial classification score and s_{tmp} the spatio-temporal score.

As our spatial baseline is DeFRCN [19], we also include the Prototypical Calibration Block (PCB) originally proposed in that work. This module applies a classification score refinement based on a similarity distance metric between network detections and category prototypes. For the category prototype calculation, annotated images from $\mathcal{D}_{novel}$ are fed to a CNN pre-trained on ImageNet to extract deep image features. Then, a RoI Align layer extracts object features with ground truth boxes, calculating K feature maps per each category in $\mathcal{C}_{novel}$. The final category prototype is calculated by an average pooling operator over the K feature maps of each category. The final classification score is calculated as follows:

$$s' = \beta \cdot s + (1 - \beta) \cdot s^{\cos} \tag{2}$$

being $s^{\cos}$ the cosine distance used as similarity metric between category prototypes and detection features extracted by the same CNN as the category prototypes. The hyperparameter β sets the tradeoff between the two confidence scores.

Finally, we define the Confidence Score Optimization method (CSO) that links object detections throughout the video and updates their classification scores. The linking method is based on the category agnostic scores and the overlap between detections in consecutive frames. The goal of this method is to modify the classification score of detections in each tube, ensuring spatio-temporal consistency. Further details of this method are given in Sect. 3.6.

3.3 Single Image Spatio-Temporal Training

Traditional video object detectors randomly select support frames in the input video sequence for each reference frame for training [6,7,9]. However, few-shot object detectors are trained with a limited number of annotated instances per category. Hence, the training set is composed of single images rather than fully annotated videos.

Our model overcomes this issue with the generation of a set $\mathcal{F}$ of synthetic support frames f_q for each training image I_t by inserting transformations of objects from I_t in different positions of f_q. In so doing, each annotated object $\alpha_{t(j)}$ from I_t is subject to horizontal random flipping and inserted L times in f_q in random positions within the image boundaries. The relative size of the object with respect to the whole image sets an upper bound for L. Undesirable artifacts from a naive insertion of a cropped object in a different position from the original object in an image is addressed with a seamless cloning operator [18]. Figure 2 shows a set of synthetically generated support frames.

Fig. 2. Examples of real reference frames (left) and their corresponding synthetic support frames (right). The synthetic support frames might appear flipped.

Reusing the original image background to place new objects decreases the probability of inconsistencies between background and foreground features. Reducing these background mismatches is crucial as the detector learns not only the object features but also the context features [1,4]. Previous methods for position selection to insert new objects in video frames rely on spatio-temporal consistency and many annotated objects to select valid positions [1,4]. However, the data availability restrictions of few-shot training makes the application of these methods infeasible.

3.4 Proposal Feature Aggregation

Relations among objects in the same image were successfully explored in [13], defining an object relation module based on the multi-head attention introduced in [24]. Similar approaches were also successfully applied in video object detection, mining relations between objects in different frames [7,9,13].

The goal of the feature aggregation module is to compute M relation features $\mathbf{r}_R^m$ for each object proposal $p_{t(i)} \in \mathcal{P}_t$ in the reference frame f_t:

$$\mathbf{r}_R^m(p_{t(i)}, \mathcal{P}^{SF}) = \sum_{r=1}^{R} \sum_{j=1}^{|\mathcal{P}_r|} w_{t(i),r(j)}^m \, (W_V \, \phi(p_{r(j)})), \quad m = 1, ..., M \tag{3}$$

where $\mathcal{P}^{SF} = \{\mathcal{P}_1, \mathcal{P}_2, ..., \mathcal{P}_R\}$ contains object proposals in the R support frames. In training, $\mathcal{P}_t$ contains the proposals extracted from the input image I_t while $\mathcal{P}^{SF}$ contains proposals from the synthetic frames set $\mathcal{F}$ (Sect. 3.3). $w_{t(i),r(j)}^m$ is a pairwise relation weight between $p_{t(i)}$ and each proposal $p_{r(j)}$ in the support frame f_r based on appearance and geometry similarities [13].

Following previous work [7,9], we implement a multi-stage relation module with a basic stage and an advanced stage. The basic stage enhances proposal features in the current frame $\mathcal{P}_t$ by mining relationships with the proposals in the support frames $\mathcal{P}^{SF}$, generating $\mathcal{P}_t'$. In the advanced stage, the proposals in $\mathcal{P}^{SF}$ are first improved by aggregating them with the top-ϵ% scoring proposals in $\mathcal{P}^{SF}$. This enhanced support proposal features are finally used in this advanced stage to aggregate with proposals in $\mathcal{P}_t'$, generating the final proposal set.

3.5 Loss Function

We propose a double objective optimization loss for the training of the spatio-temporal branch of the network double head: a classification loss and an overlap prediction. On the one hand, the classification loss function is implemented as a cross-entropy loss, following the standard approach to train a multi-class classifier. On the other hand, the overlap prediction is based on estimating the overlap of each detection with the actual objects in the image. This loss function is implemented as a binary cross-entropy loss. Hence, the final loss for the spatio-temporal branch is defined as:

$$\mathcal{L} = \mathcal{L}_{CLS} + \mathcal{L}_{IoU} \tag{4}$$

3.6 Confidence Score Optimization (CSO)

Confidence prediction plays a fundamental role in object detection. First, redundant detections are removed by means of Non-Maximum Suppression (NMS). Then, a confidence threshold is usually applied to discard low confidence detections. Commonly, the classification score is used as detection confidence.

Long tube generation has been successfully applied to optimize detection confidence, leveraging long-term spatio-temporal consistency [8,9]. This idea is the basis of our Confidence Score Optimization (CSO) method, which first builds long tubes from detections of one object and, then, increases the classification scores based on the detections of the tube. We propose to calculate the link score ls between a detection $d_{t(i)}$ in frame f_t and a detection $d_{t'(j)}$ in $f_{t'}$ as follows:

$$ls(d_{t(i)}, d_{t'(j)}) = \hat{\psi}(d_{t(i)}) + \hat{\psi}(d_{t'(j)}) + 2 \cdot IoU(d_{t(i)}, d_{t'(j)}) \quad (5)$$

being $\hat{\psi}(d_{t(i)})$ the predicted IoU score described in Sect. 3.5. This increases the probability of linking detections with greater predicted overlap with the ground truth.

To build the object tubes we apply the Viterbi algorithm, using as association scores the values generated by Eq. 5 [6,7,9]. Then, detections belonging to each tube are updated, setting their classification score to the mean classification score of the top-20% detections of the tube.

4 Experiments

We have evaluated our proposal on the FSVOD-500 dataset [10]. It contains 2,553 annotated videos with 320 different object categories for $\mathcal{D}_{base}$, and 949 videos with 100 object categories for $\mathcal{D}_{novel}$[1]. Object categories in $\mathcal{C}_{base}$ and $\mathcal{C}_{novel}$ are completely different. Following [10], we experiment with different partitions of $\mathcal{D}_{novel}^{train}$ and $\mathcal{D}_{novel}^{test}$ for fine-tuning. Thus, we randomly divide $\mathcal{D}_{novel}$ into two subsets ($\mathcal{D}_{novel}^{train}$ and $\mathcal{D}_{novel}^{test}$), keeping the same distribution of videos per object category. The subsets are interchanged, so each video is in $\mathcal{D}_{novel}^{test}$ once. We repeat this whole process 5 times —with different random splits—, and the reported results include the mean and standard deviation of these 5 executions.

We use ResNet-101 pretrained on ImageNet [22] as backbone. For training, we first learn the spatial part of FSVDet on $\mathcal{D}_{base}$. Then, we fine-tune both the spatial and spatio-temporal parts on $\mathcal{D}_{novel}^{train}$ —the spatio-temporal weights are randomly initialized. Training of the spatial part is done with a batch size of 16, with a learning rate of 2×10^{-2} for the first 20K iterations, reducing it to 2×10^{-3} for the next 5K iterations, and to 2×10^{-4} for the last 5K iterations. For fine-tuning the learning rate is 1×10^{-2} for the first 9K iterations, reducing it to 1×10^{-3} for the last 1K iterations. The spatio-temporal training is performed for 40K iterations with 1 image per batch and an initial learning rate of 2.5×10^{-4},

[1] FSVOD500 also contains a validation set with 770 annotated videos with 80 object categories. We do not use this set in the experimentation.

Table 1. Results on the few-shot video object detection FSVOD-500 dataset.

Type	Method	AP_{50}
Obj. Det.	Faster R-CNN [21]	$26.4_{\pm 0.4}$
Few-shot Obj. Det.	TFA [25]	$31.0_{\pm 0.8}$
	FSOD [11]	$31.3_{\pm 0.5}$
	DeFRCN [19]	$37.6_{\pm 0.5}$
Vid. Obj. Det.	MEGA [6]	$26.4_{\pm 0.5}$
	RDN [9]	$27.9_{\pm 0.4}$
Mult. Obj. Track.	CTracker [17]	$30.6_{\pm 0.7}$
	FairMOT [27]	$31.0_{\pm 1.0}$
	CenterTrack [28]	$30.5_{\pm 0.9}$
Few-shot Vid. Obj. Det.	FSVOD [10]	$38.7_{\pm 0.7}$
	FSVDet	$\mathbf{41.9}_{\pm 2.0}$

reducing it to 2.5×10^{-5} after the first 30K iterations. The loss function is the cross entropy with label smoothing regularization [23]. For training, 2 synthetic support frames are generated for each input image with a maximum number of $\gamma = 5$ new objects into each support frame. For test, 15 support frames are used for each video frame to mine object relations. The hyperparameter β that modulates the influence of the Prototypical Calibration Block on the classification score is set to 0.5. For the relation module, the ratio of proposals selected for the advanced stage ϵ is 20%.

Table 1 shows the results of several state-of-the-art approaches on the FSVOD-500 dataset for a shot size of $K = 5$ —the one tested in [10]. As the few-shot video object detection problem remains almost unexplored with only one previous work, for comparison purposes we also include single image few-shot object detectors, traditional video object detectors and methods based on multiple object tracking (MOT). The results for traditional video object detectors and MOT-based methods were originally reported in [10]. FSVDet outperforms previous approaches, improving our single image baseline (DeFRCN) by 4.3 points, and previous few-shot video object detectors by 3.2 points. Traditional video object detectors that include attention mechanisms for mining proposal relationships [6,9] fail to perform few-shot object detection, falling behind single image few-shot detectors. This proves the need for algorithms specifically designed for few-shot video object detection.

5 Conclusions

We have proposed FSVDet, a new few-shot video object detection framework that, first, applies attention mechanisms to mine proposals relationships between different frames. Then, a spatio-temporal double head classifies object proposals leveraging spatio-temporal information, and it also predicts the overlap of each proposal with the ground truth. Finally, overlapped predictions are used in an object linking method to create long tubes and optimize classification scores. Moreover, we have defined a new training strategy to learn from single images while considering a group of input frames at inference time. FSVDet outperforms previous solutions by a large margin, and establishes a new state-of-the-art result for the FSVOD-500 dataset for few-shot video object detection.

Acknowledgment. This research was partially funded by the Spanish Ministerio de Ciencia e Innovación (grant number PID2020-112623GB-I00), and the Galician Consellería de Cultura, Educación e Universidade (grant numbers ED431C 2018/29, ED431C 2021/048, ED431G 2019/04). These grants are co-funded by the European Regional Development Fund (ERDF).

References

1. Bosquet, B., Cores, D., Seidenari, L., Brea, V.M., Mucientes, M., Bimbo, A.D.: A full data augmentation pipeline for small object detection based on generative adversarial networks. Pattern Recognit. **133**, 108998 (2022)
2. Cai, Z., Vasconcelos, N.: Cascade R-CNN: delving into high quality object detection. In: IEEE Conference on Computer Vision and Pattern Recognition (CVPR) (2018)
3. Cao, Y., Wang, J., Lin, Y., Lin, D.: Mini: Mining implicit novel instances for few-shot object detection. arXiv preprint arXiv:2205.03381 (2022)
4. Chen, C., et al.: RRNet: a hybrid detector for object detection in drone-captured images. In: IEEE International Conference on Computer Vision Workshops (ICCV) (2019)
5. Chen, T.I., et al.: Dual-awareness attention for few-shot object detection. IEEE Trans. Multimed. **25**, 291–301 (2021)
6. Chen, Y., Cao, Y., Hu, H., Wang, L.: Memory enhanced global-local aggregation for video object detection. In: IEEE Conference on Computer Vision and Pattern Recognition (CVPR), pp. 10337–10346 (2020)
7. Cores, D., Brea, V.M., Mucientes, M.: Short-term anchor linking and long-term self-guided attention for video object detection. Image Vis. Comput. **110**, 104179 (2021)
8. Cores, D., Brea, V.M., Mucientes, M.: Spatiotemporal tubelet feature aggregation and object linking for small object detection in videos. Appl. Intell. **53**, 1205–1217 (2023)
9. Deng, J., Pan, Y., Yao, T., Zhou, W., Li, H., Mei, T.: Relation distillation networks for video object detection. In: IEEE International Conference on Computer Vision (ICCV), pp. 7023–7032 (2019)
10. Fan, Q., Tang, C.K., Tai, Y.W.: Few-shot video object detection. In: European Conference on Computer Vision (ECCV) (2022)
11. Fan, Q., Zhuo, W., Tang, C.K., Tai, Y.W.: Few-shot object detection with attention-RPN and multi-relation detector. In: IEEE Conference on Computer Vision and Pattern Recognition (CVPR), pp. 4013–4022 (2020)
12. Guo, C., et al.: Progressive sparse local attention for video object detection. In: IEEE International Conference on Computer Vision (ICCV), pp. 3909–3918 (2019)

13. Hu, H., Gu, J., Zhang, Z., Dai, J., Wei, Y.: Relation networks for object detection. In: IEEE Conference on Computer Vision and Pattern Recognition (CVPR), pp. 3588–3597 (2018)
14. Kang, B., Liu, Z., Wang, X., Yu, F., Feng, J., Darrell, T.: Few-shot object detection via feature reweighting. In: IEEE International Conference on Computer Vision (ICCV), pp. 8420–8429 (2019)
15. Kaul, P., Xie, W., Zisserman, A.: Label, verify, correct: a simple few shot object detection method. In: IEEE Conference on Computer Vision and Pattern Recognition (CVPR), pp. 14237–14247 (2022)
16. Lin, T.Y., Goyal, P., Girshick, R., He, K., Dollár, P.: Focal loss for dense object detection. In: IEEE Conference on Computer Vision and Pattern Recognition (CVPR) (2017)
17. Peng, J., et al.: Chained-tracker: chaining paired attentive regression results for end-to-end joint multiple-object detection and tracking. In: European Conference on Computer Vision (ECCV), pp. 145–161 (2020)
18. Pérez, P., Gangnet, M., Blake, A.: Poisson image editing. In: ACM SIGGRAPH 2003 Papers, pp. 313–318 (2003)
19. Qiao, L., Zhao, Y., Li, Z., Qiu, X., Wu, J., Zhang, C.: DeFRCN: decoupled Faster R-CNN for few-shot object detection. In: IEEE International Conference on Computer Vision (ICCV), pp. 8681–8690 (2021)
20. Redmon, J., Farhadi, A.: YOLO9000: better, faster, stronger. In: IEEE Conference on Computer Vision and Pattern Recognition (CVPR), pp. 7263–7271 (2017)
21. Ren, S., He, K., Girshick, R., Sun, J.: Faster R-CNN: towards real-time object detection with region proposal networks. In: Advances in Neural Information Processing Systems (NIPS) (2015)
22. Russakovsky, O., et al.: Imagenet large scale visual recognition challenge. Int. J. Comput. Vis. **115**(3), 211–252 (2015). https://doi.org/10.1007/s11263-015-0816-y
23. Szegedy, C., Vanhoucke, V., Ioffe, S., Shlens, J., Wojna, Z.: Rethinking the inception architecture for computer vision. In: IEEE Conference on Computer Vision and Pattern Recognition (CVPR), pp. 2818–2826 (2016)
24. Vaswani, A., et al.: Attention is all you need. In: Advances in Neural Information Processing Systems (NIPS), pp. 5998–6008 (2017)
25. Wang, X., Huang, T., Gonzalez, J., Darrell, T., Yu, F.: Frustratingly simple few-shot object detection. In: International Conference on Machine Learning (ICML), pp. 9919–9928 (2020)
26. Yan, X., Chen, Z., Xu, A., Wang, X., Liang, X., Lin, L.: Meta R-CNN: towards general solver for instance-level low-shot learning. In: IEEE International Conference on Computer Vision (ICCV), pp. 9577–9586 (2019)
27. Zhang, Y., Wang, C., Wang, X., Zeng, W., Liu, W.: A simple baseline for multi-object tracking, p. 6 (2020). arXiv preprint arXiv:2004.01888
28. Zhou, X., Koltun, V., Krähenbühl, P.: Tracking objects as points. In: Vedaldi, A., Bischof, H., Brox, T., Frahm, J.-M. (eds.) ECCV 2020. LNCS, vol. 12349, pp. 474–490. Springer, Cham (2020). https://doi.org/10.1007/978-3-030-58548-8_28

Optimal Wavelength Selection for Deep Learning from Hyperspectral Images

S. Dehaeck[1(✉)], R. Van Belleghem[2], N. Wouters[2], B. De Ketelaere[2], and W. Liao[1]

[1] Flanders Make, Oude Diestersebaan 133, 3920 Lommel, Belgium
sam.dehaeck@flandersmake.be
[2] KU Leuven, Department of Biosystems, MeBioS, Kasteelpark Arenberg 30, 3001 Leuven, Belgium

Abstract. While hyperspectral images typically contain two orders of magnitude more information about the detailed colour spectrum of the object under study compared to RGB images, this inevitably leads to larger storage costs and larger bandwidth requirements. This, in consequence, typically makes it challenging to use such images for real-time industrial defect detection. A second challenge of working with images having tens to hundreds of colour channels is that most available deep-learning (DL) networks are designed for images with three colours. In this paper, we will demonstrate that training typical DL segmentation networks on the full hyperspectral colour channel stack performs differently than when training on only three channels. We will also test two different approaches (Greedy Selection and Bayesian Optimization) to select the N-best wavelengths. For a given use-case, we demonstrate how optimal 3-wavelength images can outperform baseline wavelength selection methods as RGB and PCA (principal component analysis) as well as outperform networks trained with the full colour channel stack.

Keywords: Hyperspectral · Segmentation · Wavelength selection

1 Introduction

Vision-based inspection is an important tool to improve quality control in industry. This field is being revolutionized by deep learning and the improved defect detection performance this brings. At the same time, the advent of high performance hyperspectral imagers also allows to visualize defects that are impossible to see with regular RGB cameras.

In the present article, we will focus on the best way to combine deep learning models with hyperspectral images. We illustrate our approach using a semantic segmentation task applied to an industrial object with a small dataset (90 samples). As we will show, feeding the network with all colour channels (211 in the present case) leads to a network where training convergence is not as easily reached as when a selection of a few wavelengths is used. This is understandable

A. Pertusa et al. (Eds.): IbPRIA 2023, LNCS 14062, pp. 249–260, 2023.
https://doi.org/10.1007/978-3-031-36616-1_20

as most deep learning models are directly optimized for three channel images. In addition, each hyperspectral image requires 70x more space to store and 70x more bandwidth for streaming than a regular three channel image, which will clearly have an impact on the maximum framerate that can be achieved with a real-time inspection system.

What will be demonstrated in the present paper, is how the training stability and final performance of the deep-learning (DL) network can be improved by performing a reduction of the total amount of wavelengths to only three, for the present application. Both a greedy selection procedure as a Bayesian Optimization (BO) approach can improve upon typical baseline techniques as RGB, PCA (principal components analysis) and full stack training.

2 Related Work

Hyperspectral images consist of hundreds of contiguous spectral bands, yet most of these bands are highly correlated to each other. Many applications demonstrated that relatively few bands extracted from the hyperspectral image can achieve similar performances as using all spectral bands (e.g. [4]). This process of finding the most interesting bands is referred to as 'wavelength selection' and has drawn great research attention because fewer wavelengths can mean faster algorithms or more affordable sensors [3,8].

Within the field of chemometrics, wavelength selection often revolves around variable selection of a Partial Least Square (PLS) model [10]. Although already in use for a long time, these methods still perform excellent in the case of spectroscopy data, as demonstrated in an experiment by Sonobe et al. [13] for chlorophyll prediction in leafs. In a paper by Schumacher et al. [11], the authors demonstrated that Bayesian Optimization can be used to select the optimal filter central wavelengths and widths for a variable amount of filters for such PLS models. The final reduced model was clearly capable of improving the performance compared to equidistant filters. Now, while these PLS methods can be applied to hyperspectral images on a pixel level, convolutional neural networks (CNNs) have been proposed recently to extract both spectral and spatial features at the same time [12]. Therefore, recent advances are more focused on how to perform wavelength selection in hyperspectral image analysis with CNNs.

Liu et al. [9] explored using the weights of the first convolution layer to determine the most influential wavelengths. This way they could reduce the input from 181 to 15 wavelengths, while obtaining similar performance in classifying bruised from healthy strawberries. On the other hand, Varga et al. [15] examined the integrated gradients after backpropagation of a trained model to see which parts of the spectrum were most influencing. Their case consisted of classifying avocados and kiwis in different ripeness classes. Although the integrated gradients highlighted those parts of the spectrum that were biologically logical, the authors did not test the performance for training a model only on the most influential wavelengths.

The common approach in these papers for wavelength selection for CNN, is the use of techniques from the field of Explainable AI (XAI) applied to a fully trained network using all channels as input. One possible issue, in our opinion, is that such an approach could never adequately reduce the amount of wavelengths to a truly small number. Imagine that a wavelength range from 500 to 550 nm provides good contrast between two components. This entire region might get flagged by XAI-techniques, yet how to select the best 3 wavelengths from this range? What is the best spacing to use? It is impossible to get this information from querying the DL-network trained on the full channel stack, as the network never had to 'optimize' for this situation. In the current paper, we will immediately address the question "if you only have 3 wavelengths to pick, what would be the best selection?" and solving for it with different global optimization techniques. While this does require many training runs, we will show that efficient optimization techniques such as Bayesian Optimization are able to get results outperforming all common baselines, within 15 h of calculation (for the presented use-case).

3 Experimental Study

3.1 Experimental Setup

Hyperspectral images were collected with a FX10e linescan camera (SPECIM, Oulu, Finland) on a conveyor belt. This linescan camera has a spatial resolution of 1024 pixels and allows capturing 224 wavelengths in the range from 400 nm to 1000 nm, with a bandwidth (FWHM) of 5.5 nm per channel. Due to a low signal to noise ratio in the blue region, only the 211 channels between 430 nm and 1000 nm are used for further processing. Raw images were translated to relative reflectance values between 0 and 1 with a white reference (empty conveyor belt) and a dark reference (closed camera shutter) in the following manner: $Reflectance = (Raw - Dark)/(White - Dark)$

Due to the fact that hyperspectral imaging requires a light source that emits photons across all measured wavelengths, halogen spots are often used. However the direct use of halogen spots resulted in highly reflective gloss spots on the samples, and therefore a custom diffuse illumination was designed to allow for an even distribution of the light (see Fig. 1[I]). The design consists of 8 halogen spots facing upward in a reflective dome, such that the light is spread and reaches the sample from all directions evenly.

3.2 Deep Learning Vision Task

In order to investigate a generic deep learning vision task, we have chosen to perform semantic segmentation. The physical objects under study, are washing pods that can be freely obtained in commerce (see Fig. 1[II]). These pods contain 3 different soap compartments. While it is possible to distinguish these compartments by inspection of the RGB-image shown, it is clear that hyperspectral imaging might improve the contrast between the two 'blue' soaps and the

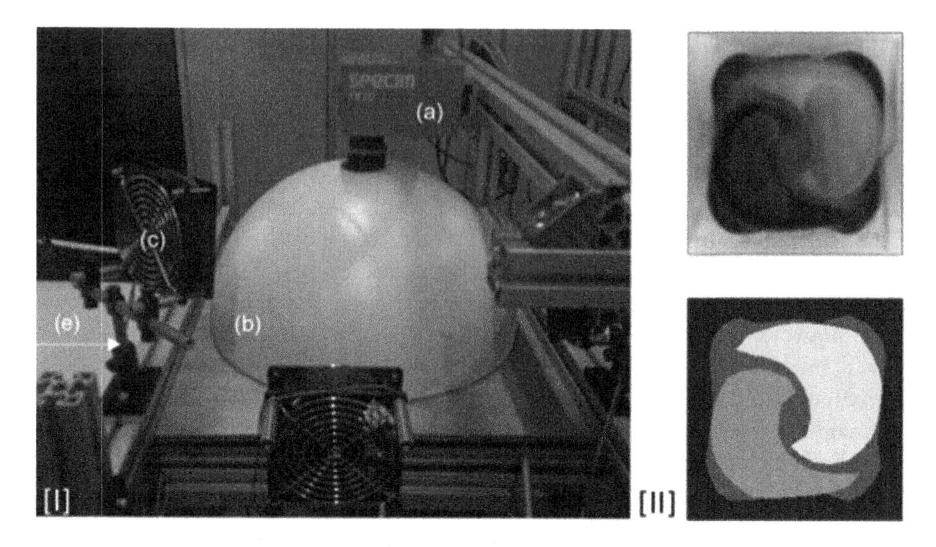

Fig. 1. [I] Hyperspectral measurement setup with hyperspectral linescan camera (a), custom diffuse dome illumination (b), cooling system for diffuse dome (c) and white conveyor belt (e) with arrow indicating moving direction. [II] Example washing pod with the 3 different soap masks.

white soap versus the background. A dataset with 90 pods was hand-annotated, of which 60 were used for training and 30 for testing (no validation set). While the task seems random, we imagine it to be a good proxy for detecting soap leaks from the pod, which could be an interesting industrial use case.

To perform the multi-class semantic segmentation, the 'Feature Pyramid Network' architecture from [7] was used on top of an efficientnet-b2 backbone [14]. The implementation present in [5] was used. A dedicated data-loader was written, which takes the relevant wavelengths from the hypercubes and feeds the newly formed 'image' into the model. As the models from [5] can accommodate an arbitrary amount of channels in the input image, there was no issue in having less or more than 3 channels. All training hyperparameters (except for training epochs) were kept constant for the different training runs performed in this paper (i.e. batch size 8, Adam Optimizer with learning rate 0.0001 and Lovasz-loss).

4 Experimental Analysis

4.1 Baseline Results

Three baseline results are considered for comparison. The first baseline is the performance on reconstructed RGB-images. This is obtained by simply extracting the channels with wavelengths around 450 nm, 550 nm and 650 nm. Feeding these images into the network, leads to a final intersection-over-union score (IOU) of 93.82%, when averaged over 20 repetitions.

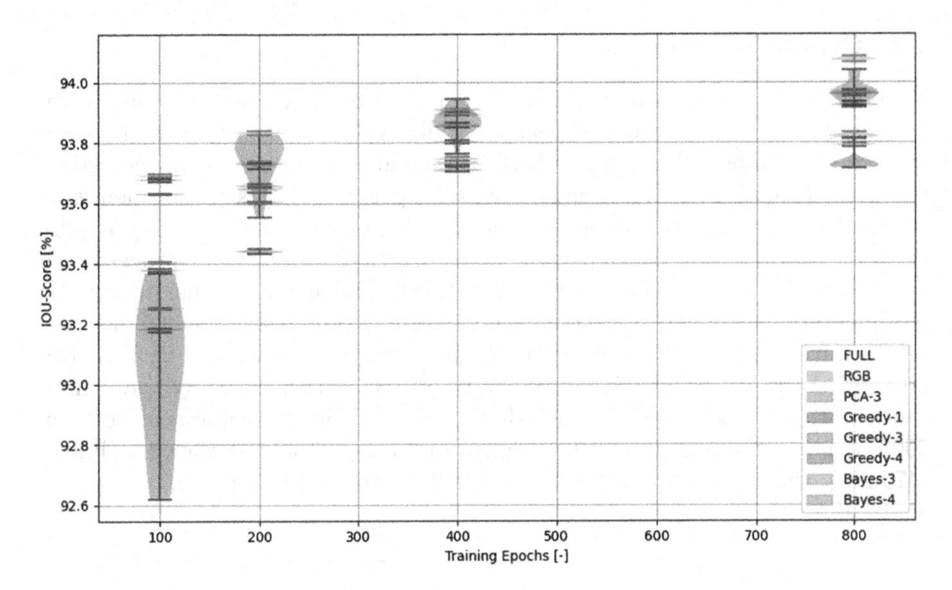

Fig. 2. Violin plot of 20 repetitions of IOU-score versus training epoch.

The second baseline is using all 211 wavelengths directly as input to the neural network. A violin plot of the obtained test IOU-scores in 20 repetitions versus the number of training epochs is shown in Fig. 2. When comparing these results to the RGB-baseline, it is immediately clear that using all channels leads to a much larger variance on the score, even after 800 epochs. Another interesting fact is that the full stack initially leads to a worse performance, but continued training eventually leads to an increased performance as compared to the RGB-baseline. The final average score after 800 epochs is 93.97%.

The third baseline, is using Principal Component Analysis (PCA) [6]. This is one of the most popular unsupervised methods to reduce dimensionality by capturing the maximum variance in the data. In our experiments, we use Python Sklearn library for PCA implementation (with 3 extracted components, and other parameter settings as default)[1]. We compute the PCA model from samples randomly selected from the training dataset, and then we apply the model to the test samples for dimension reduction. Here, we have selected the three first principal components and the results are shown in Fig. 2 as 'PCA-3'. This reveals that PCA cannot improve the performance with respect to normal RGB-imaging (or full-stack training) *for the present application.* The final performance obtained is 93.79%.

[1] https://scikit-learn.org/stable/modules/generated/sklearn.decomposition.PCA.html.

4.2 Greedy Selection

Next, we will examine a first application-specific technique for selecting wavelengths. As a full grid search is clearly prohibitive with respect to the required amount of trainings ($211^{N_{wavs}}$ with N_{wavs} the amount of chosen wavelengths), one possible approach for selecting wavelengths is a greedy selection heuristic. In such an approach, the selection is made progressively and will only require $211 * N_{wavs}$ runs (without repetitions). In practice, we start the process with the first selection and trying out all 211 options (repeated 3 times to average out fluctuations but only for 100 epochs because of time restrictions). The score versus wavelength based on such one-channel images is shown in Fig. 3. As can be seen, the peak to peak performance is only 3.5% and thus any wavelength can lead to reasonable results. Yet, there are clear optima visible near 500 nm, 700 nm and in the near infrared. We (greedily) choose the best wavelength here (479 nm), which leads to an IOU-score of 92.7% (after 100 epochs).

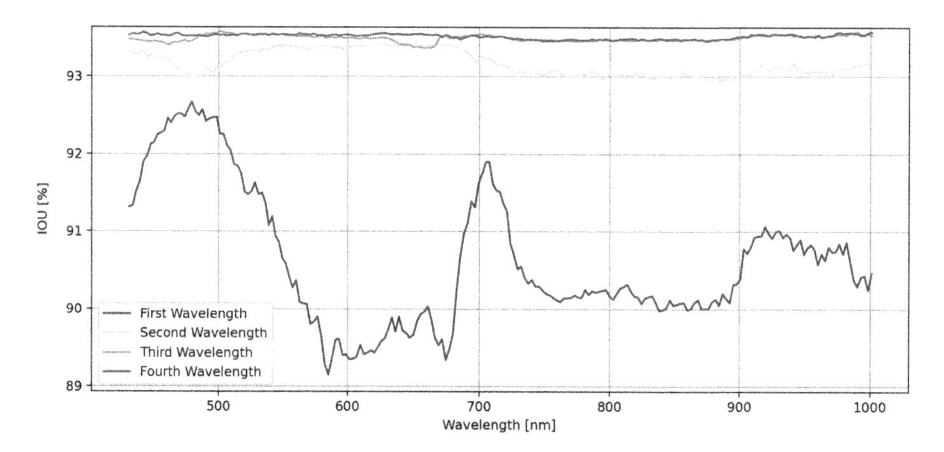

Fig. 3. IOU-score versus wavelength for the different stages in the greedy selection process.

In order to select the second wavelength, we fix the wavelength found above as the first one and run a loop over all wavelengths to choose the second wavelength (also shown in Fig. 3). Note that adding a second wavelength to the image will *always* improve the performance of the system, no matter which wavelength. Second important point is that the peak-to-peak difference between a 'good' and a 'bad' choice has decreased significantly to 0.5%. Nevertheless, a clear optimum for the second wavelength is found at 669 nm, which leads to an improvement in IOU-score of 0.7% (to 93.4% for 100 epochs). A curious observation is that there is a performance gap when using two times the same wavelength as compared to only a single time. This is perhaps an artefact due to not training for enough epochs and needs to be investigated further.

Repeating the cycle for the third wavelength, we notice for the first time that a bad choice might actually decrease performance as compared to using 2-channel images. There are 2 nearly equivalent choices to be made for the third wavelength. Choosing a wavelength of 501 nm leads to a score of 93.59% and using 990 nm as the third wavelength leads to a score of 93.58% (for 100 epochs). Thus, including a third wavelength leads to a relative improvement of only 0.18%. Also the peak-to-peak score has decreased to 0.2%.

Finally, running the loop for the fourth time, we notice that we cannot find a wavelength, which will improve the IOU-score. Adding a fourth channel leads to a net decrease of 0.1%, even in the best case (1001 nm). For completeness, we also mention that the peak-to-peak difference is 0.12%. We did not continue searching for extra wavelengths, as it is becoming highly questionable whether the peak-to-peak difference is not getting below the noise-level.

Having performed our selection, we now run the training with a larger amount of epochs and plot the results next to our previous baselines in Fig. 2. We notice that the same order is respected; i.e. 1 wav $<$ 4 wavs $<$ 3 wavs. Thus, there seems to be a regression in the IOU-score if we include too many wavelengths. Compared to the baselines, we notice that the greedy 3 wavelength selection performs similar to the (average) full stack training and better than the RGB and PCA baselines.

4.3 Bayesian Optimization

Of course, the plots shown by the greedy selection strategy do not cover the full search space. For each new search phase, the parameters found so far are locked in place. Theoretically, it is still possible that a better 4-wavelength selection could be made, which can outperform the 3-channel image performance.

To explore this, we have also employed Bayesian Optimization to optimize the wavelength selection [1]. This technique is often used in deep learning literature to perform hyperparameter optimization and can thus deal with noisy measurements. It is also geared to obtaining results with a minimum of experiments, in contrast to other global optimization techniques such as Particle Swarm Optimization or Genetic algorithms [2].

For the experiments, we have used the Ax-platform[2] with 200 optimization iterations and each training having 200 training epochs (as it reduces the variation in the test results). As global optimization techniques have random elements in them, it is typically advised to repeat such runs to see the possible variance in results. Therefore, the optimization for a given amount of wavelengths was repeated 7 times and the combination yielding the best model performance was kept for the final training performance.

To start, we will examine the IOU-scores of the best wavelength combinations trained for 800 epochs. The best results for a 3 and 4-wavelength selection run are added to Fig. 2. As can be seen, using the three suggested wavelengths by BO leads to 94.07%, which is a clear increase in performance as compared to the

[2] https://ax.dev/.

RGB-baseline. Actually, the result is even outperforming the training with the full stack, though we suspect that this is due to not having enough repetitions in the training with the full stack to cover all possible cases (and hence underestimating the variance). Also notice that the training is quite fast and only has a small variance, signalling that the network quickly finds the information it needs to perform the task at hand. Quite unintuitively, using 4 wavelengths leads to a worse performance than using 3 wavelengths when trained with 800 epochs. This is in-line to what was found from the Greedy selection procedure and confirms those results. Comparing this result to the Greedy selection approach, we note that Bayesian Optimization was able to make a better selection of wavelengths, both for the 3-wavelength optimization problem as for 4 wavelengths.

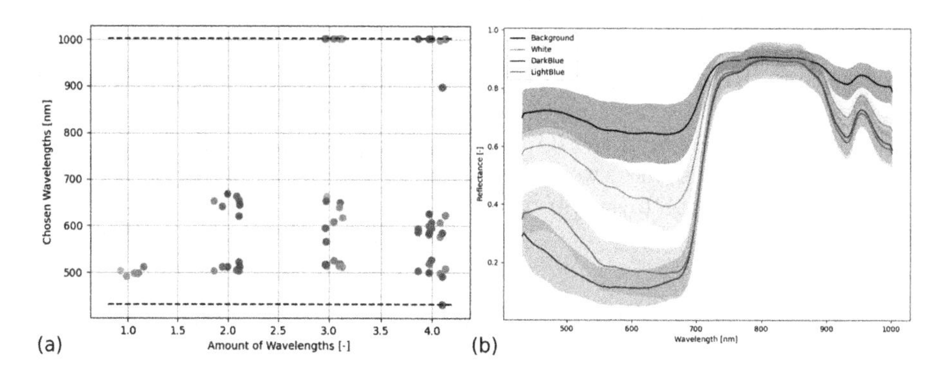

Fig. 4. (a) Visualization of chosen wavelengths versus the amount of wavelengths. Some jitter in the x-axis is added to the points in order to allow discerning the combinations made in the different optimization runs. (b) Average reflectance spectra and standard deviation for the different components being Background and the 3 different soaps (White, DarkBlue and LightBlue). Spectra where extracted from a single pod with labels based on the segmentation mask as provided during annotation (see also Fig. 1)

Next, we plot out the individual combinations found by each of the 7 Bayesian Optimization runs performed for a given amount of wavelengths (1 to 4). In Fig. 4(a), the chosen wavelengths are displayed versus the amount of wavelengths. Note that we have placed some jitter on the x-axis of the results to allow discerning which combinations were given by the seven runs (to be clear; we never selected 3.9 wavelengths). Interestingly, we notice the clear clustering for the first two wavelengths, indicating a wavelength near 500 nm and one near 650 nm is the best starting selection. For the third wavelength choice, we notice the same ambiguity as in the Greedy selection approach. In most of the cases, the third wavelength is near the end of the spectrum around 1000 nm nm. However, in one case (out of 7), all three wavelengths were in the range 500 nm–700 nm. When there are 4 wavelengths, the clustering is less precise, but the dominant tendency seems to be to have one wavelength at 500 nm, two wavelengths in the

580–620 nm range and the last one in the far infrared. We have not performed optimization for larger amounts of wavelengths as the reported accuracy drops below the performance for 3 wavelengths.

One interesting pattern that we noticed in the obtained parameter set for 4 wavelengths, is that the algorithm suggested using two wavelengths very close to each other, e.g. 504, *588, 593* and 1001 nm. This could indicate that the network is able to use the extra information coming from a local gradient operation between the two nearby channels. Another interesting fact that can be seen in Fig. 4(a), is that the chosen wavelengths around 620 nm are distinctly shifted downwards when comparing 2 to 4 wavelengths. This implies that there is some interaction between the chosen wavelengths, which a greedy selection procedure is not able to grasp.

5 Discussion

5.1 Expert Interpretation of Optimal Wavelengths

When looking at the spectra for the different soaps and the background in Fig. 4(b), it can be seen that the wavelengths around 500 nm are a logical choice if only one wavelength can be selected, because reflectance values are different for all 4 components. This is also the first component that both selection techniques chose (see Fig. 3 and Fig. 4(a)). The second 'cluster' of wavelengths found by both techniques, is located between 550 nm and 680 nm. This can be related to the fact that the difference between the white soap and background is most discriminative for these wavelengths. The third 'cluster' of chosen wavelengths, lies at the far infrared end (1000 nm). Here, all soaps can be distinguished from the background, allowing for an easy foreground-background separation.

Another detail, we might try to propose an explanation for, is the fact that when only two wavelengths were chosen by BO, it was preferred to select a wavelength near 680 nm, whereas the choice was around 580 nm when a wavelength near 1000 nm was already present in the image. We conjecture that this is because the separation between white soap and background has gotten a boost by the 1000 nm wavelength channel, allowing to push the selection to lower values around 580 nm where the contrast between the blue soaps is distinctly better, while still having an acceptable contrast between white soap and background.

From this analysis, it is clear that Bayesian or Greedy Optimization is capable of finding the wavelengths providing maximal contrast between the different components in the system. This is done in a fully data-driven task-specific manner without the need for human intervention. Now, while a human expert could easily highlight the three most interesting regions in the spectrum, it is harder to know for the expert, what the best way is to place e.g. 4 wavelengths in these regions, without trial-and-error.

5.2 Final Comparison of Techniques

The final (averaged) IOU-scores after training the networks with the chosen selection of wavelengths is shown in Table 1. Important to note however, is that we are dealing with a small dataset and only a limited amount of training epochs were used in the optimization loops. Thus, we should be careful not to overgeneralize our results. Nevertheless, as a single Bayesian Optimization run now already takes 15 h with annotation costing another 3 h, we nevertheless feel that the current study already presents a realistic use case of having 1 day to decide on the optimal wavelengths for a particular use case. In addition, it is a conceptual starting point in any kind of global optimization endeavour that it is infeasible to test all possible combinations and, in the same vein, you also cannot train on an infinitely large dataset for an infinite amount of epochs.

Table 1. IOU-score versus amount of wavelengths used for the different techniques.

Technique	1 Wav	2 Wav	3 Wav	4 Wav	211 Wav
RGB	–	–	93.82%	–	–
FULL	–	–	–	–	93.97%
PCA	–	–	93.79%	–	–
Greedy	93.74%	93.92%	93.96%	93.92%	–
BO	93.78%	93.96%	94.07%	93.96%	–

In spite of these practical limitations, this analysis has shown that, for the present use case, we were able to optimize the choice of 3 wavelengths with a result that outperforms three common baselines, i.e. RGB, PCA and full-stack training. An expert analysis of the different spectra revealed that there is a sound justification for the wavelength selection, further strengthening our believe in the suggested approach.

Important lessons we have learned in the process, is that training with the full spectral range turned out to be quite temperamental and the accuracy after 800 epochs could easily turn out to be smaller than for our optimized 3-wavelength system. The same decrease in performance was found when selecting 4 wavelengths instead of 3. This is most remarkable for the Greedy selection approach, as in that case, the network could have learned to disregard the 4th colour channel, but it did not. In the future, we would like to investigate this further.

Nevertheless, one could already ponder on the question what is determining that 3 channels seems to be the optimal choice for this problem. Is it because there are three components to detect in the image and each class is focusing on one channel (i.e. 3 channels for 3 unknowns)? An alternative explanation could come from the design of the neural networks. Countless hours are spend trying out all possible combinations to see what works best for 3-channel images. Thus, perhaps we should not be surprised that the performance is best for 3 wavelengths. It is an unfortunate side-effect of our experimental design that we have exactly 3 classes to segment into and thus are not able to distinguish between both mentioned drivers. Another topic for future research would be to consider a segmentation problem with more classes.

6 Conclusions

In this paper, we have found that using all information coming from a hyperspectral image and feeding it into a deep learning segmentation network does lead to substantial variation in the final accuracy (at least for moderate training epochs). Reducing the complexity of the input image through a Bayesian (or Greedy) optimization technique to select a few wavelengths (optimally 3 here) can be beneficial for the final test score. This has the added benefit that we can, either reduce the bandwidth needed to process images coming from a hyperspectral imager, or design a dedicated multispectral imager capable of snapshot images. Both techniques should allow to use these kinds of systems in industry with similar or even improved performance w.r.t. a full-stack processing.

Acknowledgements. This research is supported by Flanders Make, the strategic research Centre for the Manufacturing Industry and the Flemish Innovation and Entrepreneurship Agency through the research project 'DAP2CHEM' (project number: HBC.2020.2455). The authors would like to thank all project's partners for their inputs and support to make this publication.

References

1. Balandat, M., et al.: BoTorch: a framework for efficient Monte-Carlo Bayesian optimization (2019). https://doi.org/10.48550/ARXIV.1910.06403
2. Gendreau, M., Potvin., J.: Handbook of Metaheuristics, Springer, New York (2019). https://doi.org/10.1007/978-1-4419-1665-5, ISBN 978-3-319-91085-7
3. Genser, N., Seiler, J., Kaup, A.: Camera Array for Multi-Spectral Imaging. In: IEEE Transactions on Image Processing, vol. 29 (2020)
4. Ghamisi, P., et al.: Advances in hyperspectral image and signal processing: a comprehensive overview of the state of the art. IEEE Geosci. Remote Sens. Mag. **5**(4), 37–78 (2017)
5. Iakubovskii, P.: Segmentation Models Pytorch. https://github.com/qubvel/segmentation_models.pytorch. Accessed 13 Feb 2023
6. Jolliffe, I.T: Principal Component Analysis. Springer Series in Statistics. Springer-Verlag, New York (2002). https://doi.org/10.1007/b98835
7. Lin, T.Y., Dollár, P., Girshick, R., He, K., Hariharan, B., Belongie, S.: Feature pyramid networks for object detection (2016). https://doi.org/10.48550/ARXIV.1612.03144
8. Liu, D., Sun, D.-W., Zeng, X.-A.: Recent advances in wavelength selection techniques for hyperspectral image processing in the food industry. Food Bioprocess Technol. **7**(2), 307–323 (2014). https://doi.org/10.1007/s11947-013-1193-6
9. Liu, Y., Zhou, S., Han, W., Liu, W., Qiu, Z., Li, C.: Convolutional neural network for hyperspectral data analysis and effective wavelengths selection. Anal. Chim. Acta **1086**, 46–54 (2019) https://doi.org/10.1016/j.aca.2019.08.026
10. Mehmood, T., Liland, K.H., Snipen, L., Sæbø, S.: A review of variable selection methods in partial least squares regression. Chemom. Intell. Lab. Syst. **118**, 62–69 (2012). https://doi.org/10.1016/j.chemolab.2012.07.010

11. Schumacher, P., Gruna, R., Längle, T., Beyerer, J.: Problem-specific optimized multispectral sensing for improved quantification of plant biochemical constituents. In: 2022 12th Workshop on Hyperspectral Imaging and Signal Processing: Evolution in Remote Sensing (WHISPERS), pp. 1–6, Rome, Italy (2022). https://doi.org/10.1109/WHISPERS56178.2022.9955113
12. Signoroni, A., Savardi, M., Baronio, A., Benini, S.: Deep learning meets hyperspectral image analysis: a multidisciplinary review. J. Imaging **5**(5), 52 (2019). https://doi.org/10.3390/jimaging5050052
13. Sonobe, R., et al. Hyperspectral wavelength selection for estimating chlorophyll content of muskmelon leaves. Eur. J. Remote Sens. **54**(1), 513–524 (2021). https://doi.org/10.1080/22797254.2021.1964383
14. Tan, M., Le, Q.V.: EfficientNet: rethinking model scaling for convolutional neural networks (2019). https://doi.org/10.48550/ARXIV.1905.11946
15. Varga, L.A., Makowski, J., Zell, A.: Measuring the ripeness of fruit with hyperspectral imaging and deep learning. In: Proceedings of the International Joint Conference on Neural Networks (2021). https://doi.org/10.1109/IJCNN52387.2021.9533728
16. Zhu, S., Zhou, L., Gao, P., Bao, Y., He, Y., Feng, L.: Near-infrared hyperspectral imaging combined with deep learning to identify cotton seed varieties. Molecules (Basel, Switzerland) **24**(18), 3268 (2019). https://doi.org/10.3390/molecules24183268

Can Representation Learning for Multimodal Image Registration be Improved by Supervision of Intermediate Layers?

Elisabeth Wetzer(✉), Joakim Lindblad, and Nataša Sladoje

Centre for Image Analysis, Department of Information Technology, Uppsala University, Uppsala, Sweden
elisabeth.wetzer@it.uu.se

Abstract. Multimodal imaging and correlative analysis typically require image alignment. Contrastive learning can generate representations of multimodal images, reducing the challenging task of multimodal image registration to a monomodal one. Previously, additional supervision on intermediate layers in contrastive learning has improved biomedical image classification. We evaluate if a similar approach improves representations learned for registration to boost registration performance. We explore three approaches to add contrastive supervision to the latent features of the bottleneck layer in the U-Nets encoding the multimodal images and evaluate three different critic functions. Our results show that representations learned without additional supervision on latent features perform best in the downstream task of registration on two public biomedical datasets. We investigate the performance drop by exploiting recent insights in contrastive learning in classification and self-supervised learning. We visualize the spatial relations of the learned representations by means of multidimensional scaling, and show that additional supervision on the bottleneck layer can lead to partial dimensional collapse of the intermediate embedding space.

Keywords: Contrastive Learning · Multimodal Image Registration · Digital Pathology

1 Introduction

Multimodal imaging enables capturing complementary information about a sample, essential for a large number of diagnoses in digital pathology. However, directly co-aligned data can only be provided if an imaging device hosts multiple imaging modalities, otherwise individually acquired data have to be registered by image processing. Different sensors may produce images of very different

Supported by VINNOVA (projects 2017-02447, 2020-03611, 2021-01420) and the Centre for Interdisciplinary Mathematics (CIM), Uppsala University.

A. Pertusa et al. (Eds.): IbPRIA 2023, LNCS 14062, pp. 261–275, 2023.
https://doi.org/10.1007/978-3-031-36616-1_21

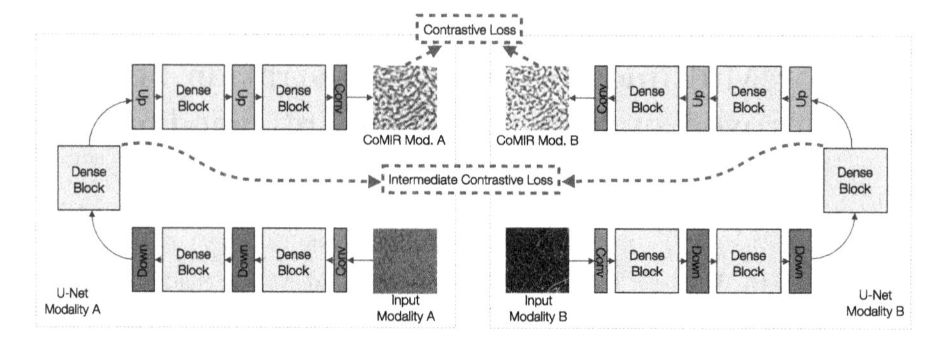

Fig. 1. Overview of the representation learning approach: Two U-Nets sharing no weights are trained in parallel to learn dense representations (CoMIRs) through a contrastive loss on the final output layers of the networks. Additional contrastive losses acting on the features of the bottleneck (BN) layers are evaluated with an aim to improve CoMIRs for registration.

appearance, making automated multimodal registration a very challenging task. Consequently, the registration is often performed manually; a difficult, labor- and time consuming, and hence expensive approach. Reliable automated multimodal registration can reduce the workload, allowing for larger datasets to be studied in research and clinical settings. Numerous methods are available for automated monomodal registration, however, multimodal image registration is both more difficult and with fewer tools available. Recently, a modality-independent representation learning technique was introduced in [33]; the method generates Contrastive Multimodal Image Representations (CoMIRs) for Registration, which can be registered using monomodal (intensity-based and feature-based) registration methods. CoMIR-based multimodal registration [33] has been successfully applied to various datasets: brightfield (BF) & second harmonic generation (SHG) images in [33], remote sensing images (RGB & near-infrared), and correlative time-lapse quantitative phase images (QPI) & fluorescence microscopy (FM) images, and magnetic resonance images (MRI T1 & T2) in [25,29].

Based on a recent study [19] which has shown that representations learned contrastively performed better in the downstream task of biomedical image classification when additional contrastive losses were used to supervise intermediate layers in the network, we investigate if such additional supervision can be applied to further improve CoMIRs for the downstream task of registration. A schematic overview of the considered approach is given in Fig. 1.

Our contributions are as follows: (i) We evaluate three approaches of including contrastive losses on U-Net bottleneck (BN) features when generating CoMIRs of BF and SHG images; for each we test three different similarity/distance measures. We observe that *leaving intermediate features unconstrained* results in CoMIRs leading to more successful rigid multimodal image registration. This differs from what was previously observed for features used in biomedical image classification. (ii) We confirm that unconstrained intermediate features result in better registration on a second dataset of QPI and FM images, using the best performing similarity measure observed on the BF and SHG dataset in each approach. (iii) We investigate the reasons for the drop in registration performance and observe that contrastive training on the BN features can lead to a partial dimensional collapse of the feature space. (iv) Furthermore, we evaluate the relation between the quality of the generated representations and the downstream task of image registration, using several image distance/similarity measures.

2 Background

2.1 Representation Learning

Several recent approaches find common representations for images of different modalities to enable monomodal registration methods. The common representation may be one of the involved modalities, e.g., learned by GAN-based Image-To-Image (I2I) translation, where a generator and discriminator are trained in a zero-sum game, generating representations in line with the appearance and statistical distribution of one modality given an image of another modality [3,34]. Another strategy is to map both of the considered image modalities into a common space: In [33], the contrastive loss InfoNCE [30] was used to produce dense, image-like representations – called CoMIRs – which proved useful for multimodal image registration on a variety of different datasets [25,29,33]. CoMIRs are generated by two U-Nets trained in parallel, sharing no weights, connected by the contrastive loss InfoNCE [12,14], which maximizes a lower bound of mutual information (MI). Contrastive Learning is also used in [32] to perform I2I, which inspired the development of ContraReg [7] to contrastively learn to perform unsupervised, deformable multimodal image registration.

A number of I2I approaches [5,16,21,46] were evaluated in [25,33] for multimodal registration, but the representations lacked similarity in structures or intensity needed for registration. CoMIRs [33], however, performed well in combination with both intensity-based [28] and with feature-based registration methods.

Additional Supervision on Intermediate Layers
Recently it was shown in [19] that a momentum contrastive (MoCo) learning-based framework benefits from additional supervision imposed on intermediate layers. The features are more similar earlier in the network, resulting in better performance in biomedical image classification. In [10], a three stream architecture (TS-Net) is proposed combining Siammese and pseudo-Siammese networks for patch matching of multimodal images. An additional contrastive loss on intermediate features improves the matching performance on three multimodal datasets. Additional supervision on intermediate layers in U-Nets has mainly been applied to the BN layer. In [23], additional supervision on the BN features is provided for liver and tumor segmentation. The authors argue that supervision on the BN layer can reduce information loss, due to its highly condensed low-dimensional information of label maps. In [20], a Tunable U-Net (TU-Net) uses a tuning parameter acting on the BN features, to change the output without the requirement of multiple trained models. In [1], the effect of fine-tuning

different sets of layers of a pretrained U-Net for ultrasound and X-ray image segmentation is studied. The authors show that the choice of layers is critical for the downstream tasks and differs significantly between segmentation and classification. This is based on the assumption that low-level features, which are associated with shallow layers, are shared across different data sets. The authors observe that freezing the BN has equivalent segmentation performance as fine-tuning the entire network, highlighting the importance of that particular latent representation.

Similarly, we focus on evaluating additional losses constraining the BN layer in the U-Nets which generate CoMIRs for alignment of multimodal images.

2.2 Contrastive Learning

Contrastive Learning (CL) has been successfully employed in many tasks - from self-supervised pretraining, to image retrieval and classification [7,10,19,25,33, 36,37]. The goal of CL is to learn a representation s.t. similar samples are mapped close to each other, while dissimilar samples are not. *Similar* can refer to the same class, or different views or augmentations of one particular sample. Usually, the learned representation is a 1D vector. CoMIRs differ in that they are 2D image-like representations intended to preserve spatial structures, not only optimized for their distinction to other samples. Learning CoMIRs is closely related to self-supervised learning (SSL). While CoMIR training currently relies on a labelled training set in form of aligned image pairs, the two images in different modalities of one sample act as different views of one sample in SSL.

The features of multiple views in SSL depend strongly on data augmentation (DA), as it decouples the correlations of features between the representations of positive samples [42]. The importance of view selection is also addressed in [37,45]. The authors in [37] find that the MI between views should be reduced while information relevant for the downstream task has to be preserved. While [37,42] argue for extensive DA, in [6] it was shown that extensive DA can yield the projector head of an SSL network invariant to the DA to a higher degree than the encoder itself, resulting in a projection to a low-dimensional embedding space. This so-called *dimensional collapse* (DC) has been observed in different CL settings [4,6,15,18,40] and is currently subject of research. The phenomenon has been first studied in [18] and [15]. DC can occur if the variance caused by DA is larger than the variance in the data distribution, implicit regularization to favor low-rank solutions, or strong overparametrization [18]. For non-contrastive SSL methods such as SimSiam or BYOL, underparametrization of the model can cause at least partial collapse in [22].

Following these recent observations, we inspect if DC of the feature space is the cause of representations which are unsuitable for feature-based registration.

3 Method

CoMIRs, as originally proposed, are learned by two U-Nets [17], sharing no weights, connected by a contrastive loss given as follows:

For $\mathcal{D} = \{(\boldsymbol{x}_i^1, \boldsymbol{x}_i^2)\}_{i=1}^n$ an i.i.d. dataset containing n data points, $\boldsymbol{x}^j$ is an image in modality j, and $f_{\boldsymbol{\theta}_j}$ the network processing modality j with respective parameters $\boldsymbol{\theta}_j$ for $j \in \{1, 2\}$. Given an arbitrary image pair $\boldsymbol{x} = (\boldsymbol{x}^1, \boldsymbol{x}^2) \in \mathcal{D}$ and a set of negative samples $\mathcal{D}_{neg}$, the loss is given by

$$\mathcal{L}(\mathcal{D}) = -\frac{1}{n}\sum_{i=1}^{n}\left(\log \frac{e^{h(\boldsymbol{y}_i^1, \boldsymbol{y}_i^2)/\tau}}{e^{h(\boldsymbol{y}_i^1, \boldsymbol{y}_i^2)/\tau} + \sum_{\boldsymbol{y}_j^1, \boldsymbol{y}_j^2 \in \mathcal{D}_{neg}} e^{h(\boldsymbol{y}_j^1, \boldsymbol{y}_j^2)/\tau}}\right). \tag{1}$$

$\mathcal{L}(\mathcal{D})$ is named InfoNCE as described in [31]. The exponential of a similarity function (called critic) $h(\boldsymbol{y}^1, \boldsymbol{y}^2)$ computes a chosen similarity between the representations $\boldsymbol{y}^1 = f_{\boldsymbol{\theta}_1}(\boldsymbol{x}^1)$ and $\boldsymbol{y}^2 = f_{\boldsymbol{\theta}_2}(\boldsymbol{x}^2)$ for the scaling parameter $\tau > 0$.

3.1 Additional Supervision of the BN Latent Representation

Different Critic Functions. We consider three types of similarity functions in Eq. 1 for the supervision of the BN latent representations:

- A Gaussian model with a constant variance $h(\boldsymbol{y}^1, \boldsymbol{y}^2) = -||\boldsymbol{y}^1 - \boldsymbol{y}^2||_2^2$ which uses the L^2 norm, i.e. mean squared error (MSE) as a similarity function;
- A trigonometric model $h(\boldsymbol{y}^1, \boldsymbol{y}^2) = \frac{\langle \boldsymbol{y}^1, \boldsymbol{y}^2 \rangle}{||\boldsymbol{y}^1|| \, ||\boldsymbol{y}^2||}$ relating to cosine similarity;
- A model using the L^1 norm as a similarity $h(\boldsymbol{y}^1, \boldsymbol{y}^2) = -||\boldsymbol{y}^1 - \boldsymbol{y}^2||_1$.

We investigate the following approaches for supervision of the BN:

Approach I: Alternating loss $\mathcal{L}(\mathcal{D})$ given in Eq. (1) is computed in an alternating way on the final output representations in the network as $\mathcal{L}_C(\mathcal{D})$ and on the BN latent features as $\mathcal{L}_{BN}(\mathcal{D})$, taking turns every iteration. A hyperparameter ensures losses of the same magnitude for stable training.

Approach II: Summed loss As proposed in [19], the loss in Eq. (1) is calculated in each iteration on the final representations in the network and on the BN latent features. The two losses are combined in a weighted sum, requiring an additional hyperparameter α to ensure that the two losses are of the same magnitude: $\mathcal{L}_{Sum}(\mathcal{D}) = \mathcal{L}_C(\mathcal{D}) + \alpha\mathcal{L}_{BN}(\mathcal{D})$

Approach III: Pretraining using a contrastive loss on the BN layers For this approach, the networks are trained with $\mathcal{L}_{BN}(\mathcal{D})$ for the first 50 epochs on the BN latent features. After this pretraining of the networks, $\mathcal{L}_C(\mathcal{D})$ is computed acting only on the final layer for 50 more epochs (no layers frozen).

3.2 Implementation Details

All models are trained for 100 epochs, except the baseline model denoted "Baseline 50" in Fig. 3, which is trained for 50 epochs. In all experiments, $\mathcal{L}_C(\mathcal{D})$ uses MSE as $h(\boldsymbol{y}^1, \boldsymbol{y}^2)$ (as suggested in [33]), while the similarity functions acting on the BN layers are varied between none, MSE, cosine similarity, and L^1-norm as described in Sect. 3.1. In all cases 1-channel CoMIRs are generated with identical, random data augmentation consisting of random flips, rotations, Gaussian blur, added Gaussian noise and contrast adjustments. More detailed implementation information, including all chosen hyperparameters, can be found in appendix Sec. [9.2] in [43].

4 Evaluation

4.1 Datasets

SHG and BF Dataset
The dataset is a publicly available, manually aligned registration benchmark [8] consisting of SHG and BF crops of tissue micro-array (TMA) image pairs. The training set consists of 40 image pairs of size 834×834 px cropped from original TMA images [9] to avoid any border affects due to transformations. The test set comprises 134 pairs of synthetically, randomly rotated and translated images of up to $\pm 30°$, and ± 100 px.

QPI and FM Dataset
The dataset consists of simultaneously acquired correlative time-lapse QPI [38] and FM images [39] of three prostatic cell lines, captured as time-lapse stacks at seven different fields of view while exposed to cell death inducing compounds. It is openly available for multimodal registration [26] and is used as in [25]. The images are of size 300×300 px, with 420 test samples in each of three evaluation folds. All images originating from one cell line are used as one fold in 3-folded cross-validation. The test set was created by synthetic random rotations of up to $\pm 20°$ and translations of up to ± 28 px.

4.2 Evaluation Metrics

Registration Performance: We register CoMIRs by extracting Scale-Invariant Feature Transform (SIFT, [44]) features, and match them by Random Sample Consensus (RANSAC [11]). The registration error is calculated as $err = \frac{1}{4}\sum_{i=1}^{4} ||C_i^{Ref} - C_i^{Reg}||_2$, where C_i are the corner points of the reference image C_i^{Ref}, and the registered image C_i^{Reg} respectively. The registration success rate (RSR) is measured by the percentage of test images which are successfully registered, whereas success is defined by an error below a dataset-dependent threshold. The implementation details, including parameter choices, are reported in appendix Sec. [9.2] in [43].

Measuring Representation Quality: Intuitively, the higher similarity in appearance the images have, the more successful their registration. Therefore, the "goodness" of CoMIRs can be evaluated in two ways: by evaluating (i) the similarity/distance between the CoMIRs of the corresponding images; and (ii) the success of their registration. We correlate both types of evaluation in this study. For comparing CoMIRs we utilize several approaches common for quantifying image similarity/distance. More precisely, we evaluate the following:

- the pixelwise measures **MSE** and **correlation**,
- the perceptual similarity measure **structural similarity index measure** (SSIM, [41]),
- a distance measure which combines intensity and spatial information, namely **α-Average Minimal Distance** (α-AMD, [24,28]),
- a distance measure comparing two distributions and popular choice to evaluate GAN generated representations [27], namely **Fréchet Inception Distance** (FID, [13]).

The respective definitions are given in appendix Sec. [9.1] in [43].

5 Results

Registration Performance

Fig. 2a shows the RSR, computed as the percentage of the test set which was registered with an error less than 100 px on the BF & SHG dataset. The methods are grouped color-wise w.r.t. the similarity function used for the BN supervision. In green the baseline results are shown for CoMIRs as introduced in [33], using no additional loss on intermediate layers. Detailed results of each run per experiment and examples of CoMIRs for the tested approaches can be found in appendix Sec. [9.3] in [43].

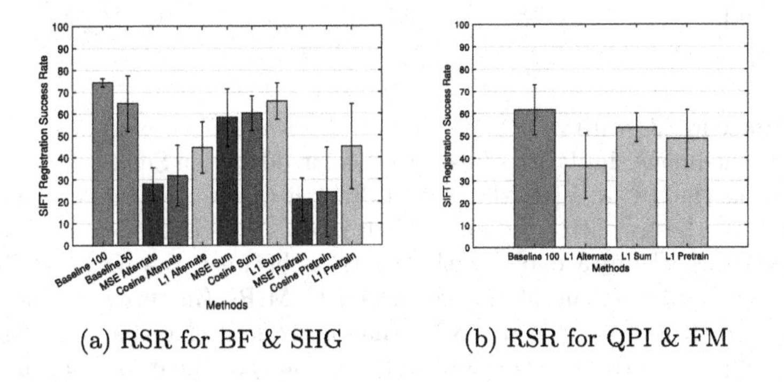

(a) RSR for BF & SHG (b) RSR for QPI & FM

Fig. 2. Evaluation of the CoMIRs w.r.t. to registration performance.

It should be considered that the Approaches I and III only spend half of all epochs on evaluating $\mathcal{L}_C(\mathcal{D})$, ensuring pixel-wise intensity similarity in the final CoMIRs. To confirm that this does not cause the subsequent lower registration performance, a second set of baseline experiments was run for only 50 instead of 100 epochs. Hence, the plots include the baseline trained for 100 epochs ("Baseline 100") and also for 50 epochs ("Baseline 50").

The best performing critic function, w.r.t. the registration performance on the BF & SHG Dataset (i.e. L^1), is used to assess the three approaches on the second multimodal dataset, containing QPI and FM images of cells. This dataset confirms our observations that no supervision on intermediate layers results in CoMIRs more suitable for registration. The RSR (averaged over three folds) corresponding to the percentage of the test set with a relative registration error of less than 2% of the image size, as used in [25], is shown in Fig. 2b (detailed results are listed in the appendix in [43]). We observe that on both datasets the Baseline approach outperforms the considered alternatives, producing CoMIRs which lead to the highest RSR.

Table 1. The medians of all the considered performance measures, computed on the test set and averaged over three runs, for the different approaches to include supervision of the BN latent representations. Arrows indicate if a high (↑) or low (↓) value correspond to a good result.

Approach	Intermediate Loss	MSE ↓	SSIM ↑	Correlation ↑	α-AMD ↓	FID ↓	Reg. Success Rate ↑
baseline 100	none	4,771	0.53	**0.66**	1.86	**93.26**	**74.38**
alternate	MSE	4,534	0.46	0.36	2.21	137.89	27.86
alternate	Cosine	3,949	0.50	0.42	**1.32**	123.31	31.84
alternate	L1	7,603	0.37	0.47	5.68	190.26	44.53
sum	MSE	4,674	**0.57**	0.60	2.33	129.31	58.21
sum	Cosine	5,776	0.48	0.52	7.35	219.76	60.20
sum	L1	**3,894**	**0.57**	0.64	1.70	106.09	65.67
pretrain	MSE	4,815	0.37	0.32	3.46	179.98	20.90
pretrain	Cosine	8,139	0.35	0.28	7.31	181.93	24.13
pretrain	L1	5,668	0.44	0.48	2.86	160.34	45.02

Representation Quality

We report the image similarities/distances as introduced in Sect. 4.2 for CoMIRs produced on the BF & SHG dataset. Table 1 lists the performance measures computed for the test set (median over all images where applicable), averaged over three runs. Figure 3 explores the relations between the mean of the considered quality assessment of the generated CoMIRs (in terms of their similarity/distance), and their RSR. Note that horizontal error-bars in Fig. 3 can only be compared intraplot-wise, but not inter-plot, as the evaluation measures are of different ranges. The Pearson correlation coefficient (PCC) is reported to assess the linear relationship between the measures and RSR. The PCC ranges in $[-1, 1]$ with -1 corresponding to a direct, negative correlation, 0 representing no correlation, and 1 representing a direct, positive correlation. Values close to ± 1 indicate a high agreement between a similarity/distance measure and RSR.

6 Exploration of the Embedding Space

To investigate the reasons of the reduced performance when using additional supervision on the BN features, we inspect the training features on the BF & SHG dataset.

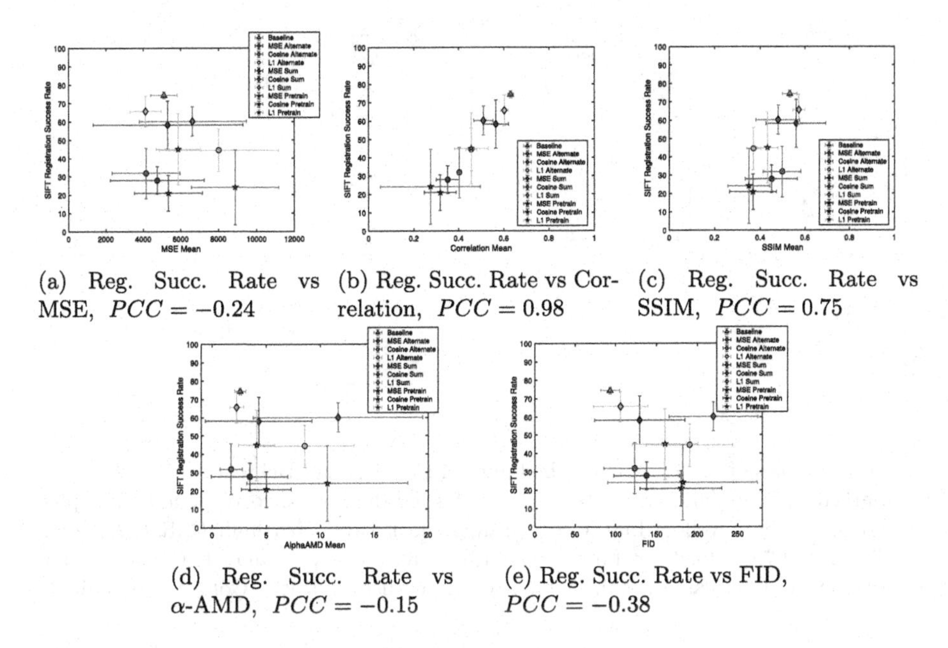

(a) Reg. Succ. Rate vs MSE, $PCC = -0.24$

(b) Reg. Succ. Rate vs Correlation, $PCC = 0.98$

(c) Reg. Succ. Rate vs SSIM, $PCC = 0.75$

(d) Reg. Succ. Rate vs α-AMD, $PCC = -0.15$

(e) Reg. Succ. Rate vs FID, $PCC = -0.38$

Fig. 3. Relation between RSR and image (dis-)similarity measures. Marker style indicates the approach (alternating, summed or pretraining loss), color of the marker and bars correspond to the critic functions used in $\mathcal{L}_{BN}(\mathcal{D})$ (none, MSE, cosine sim. and L^1), error-bars correspond to the standard deviations computed over 3 runs. The PCC between each measure and the registration performance is reported. Best appreciated zoomed in.

We compute the relative (dis-)similarities between all CoMIR pairs, e.g. w.r.t. MSE. Based on dissimilarities of data points, Multidimensional Scaling (MDS, [2]), used in dimensionality reduction and visualization, maps high dimensional data into a low dimensional metric space by numerical optimization, preserving the relative dissimilarities between the samples.

We perform metric MDS on the dissimilarity matrix $\Delta = (d_{ij}) \in \mathbb{R}_+^{n \times n}$ to find 2D points whose distances $\bar{\Delta} = (\bar{d}_{ij})$ approximate the dissimilarities in Δ. The resulting optimization problem minimizes an objective function called stress. We empirically observe that using Sammon's non-linear stress criterion [35] results in the best fit and lowest stress to embed CoMIR and BN features, given by

$$\text{Stress} = \frac{1}{\sum_{i<j} d_{ij}} \sum_{i<j} \frac{(d_{ij} - \bar{d}_{ij})^2}{d_{ij}} \tag{2}$$

where d_{ij} denotes the distance between sample i and j in the high-dimensional space and $\bar{d}_{ij}$ the distance in the 2D projection space. Δ contains the pairwise MSE between all features, either BN or CoMIR features.

Figure 4 shows the MDS embeddings of BN and CoMIR features of the training set for a subset of runs for each approach. Visualizations for all runs are in

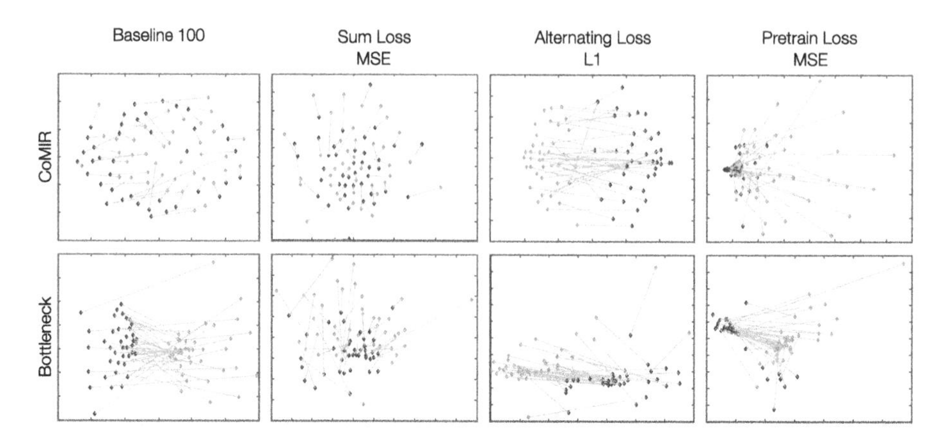

Fig. 4. Examples of Metric MDS embeddings of CoMIRs and BN features. BF features are marked by blue diamonds, SHG features by red diamonds, corresponding samples are connected by a yellow line. Green frames mark runs with high RSR (72.4% for Baseline 100, 67.9% for Sum Loss using MSE), the magenta frames low RSR (32.1% for Alternating Loss using L^1, 9.7% for Pretraining using MSE). (Color figure online)

appendix Sec. [9.3] in [43]. Features resulting from the BF images are marked by blue diamonds, from SHG images by red diamonds. All corresponding samples, i.e. the features resulting from a multimodal image pair, are connected by a yellow line. The green frames mark runs with high RSR, magenta frames runs with low RSR. We observe in Fig. 4 that CoMIRs without intermediate loss on BN layers ("Baseline 100") are spatially spread. The distances between corresponding samples are reasonably small, though using a $\mathcal{L}_{BN}(\mathcal{D})$ with a MSE critic in a sum with $\mathcal{L}_C(\mathcal{D})$ can further reduce the distance between corresponding pairs. This embedding configuration is the best among all experiments w.r.t. pairwise distances, however the RSR is below the baseline.

For some experiments with low RSR, we observe that the embeddings have a tendency to cluster w.r.t. the modalities, as for example using an alternating loss with L^1 in Fig. 4. Furthermore, we observe that the pretraining approach, can lead to a cluster of SHG samples, contracted to a single point. Additionally, we visualize the embeddings of the BN features for each of these approaches (visualizations for all runs are in the appendix in [43]). We observe that training CoMIRs without intermediate supervision results in BN features which are clustered by the modality they originate from.

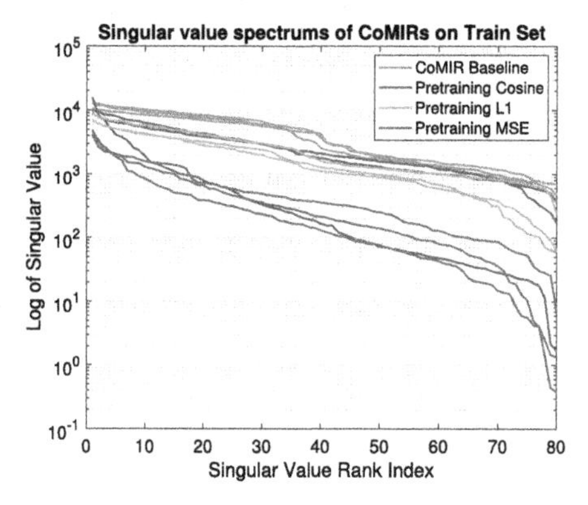

Fig. 5. Log-Plot of the SVs of CoMIRs learned after pretraining with a loss on the BN, and the baseline. Three runs were performed for each experiment.

We observe that CoMIRs of SHG images learned after pretraining the BN layer can collapse to a point. To quantify the dimensionality of the embedding space - or its collapse - the authors in [18] compute the covariance matrix of the embedding vectors and perform its singular value decomposition (SVD). Inspecting the spectrum of the resulting singular values (SVs) by plotting their log value in sorted order, shows how many collapse to zero, corresponding to collapsed dimensions. This approach to quantify DC in CL has also been adopted in [6,22].

We plot the SV spectrum of the covariance matrix of CoMIRs in Fig. 5. While we do not observe a complete DC, we see that the SVs decay much faster for approach III using pretraining, i.e. they are closer to a low-rank solution. The square of a SV is proportional to the variance explained by its corresponding singular vector, i.e. the smaller the SVs the less variance in the data is explained and more likely corresponds to noise.

7 Discussion

Our results shown in Fig. 2a and 2b indicate that CoMIRs generated using a loss acting only on the final features yield the best representations for the downstream task of registration on both observed datasets, outperforming all evaluated attempts to impose similarity of the features earlier in the networks. We further observe on both datasets that Approach II leads to a higher RSR than Approach I and III. On the BF & SHG Dataset, using L^1 as a similarity in the contrastive loss produces, on average, better representations w.r.t. RSR, compared to cosine similarity, which in turn performs better than using MSE. However, the error bars indicate a wide spread in the results. On the QPI & Fluorescence Dataset, Approach III results in CoMIRs with higher RSR compared to Approach I (alternating the losses), unlike in the case of BF & SHG images.

Figure 3 presents the relation between similarity/distance of generated CoMIR pairs and their RSR. This connects our two stated evaluation objectives: the quality of appearance of the representations, and their applicability for registration. For the distance measures (MSE, α-AMD, FID) a well performing measure would be located in the upper left corner (high RSR, and low distance between the image pairs), while for the similarity measures (Correlation, SSIM) it would be in the upper right corner. We observe that the mean correlation between CoMIR pairs (Fig. 3b) correlates strongly with the RSR of these representations (PCC 0.98), while for the other measures the relations are less clear. As intuitively expected, the PCC indicates (weak) negative correlation between RSR and distance measures and positive correlation for the similarity measures. We observe that for the BF & SHG Dataset the baseline approach performs best in terms of RSR, also exhibiting stability (low standard deviation), high similarity and low distance between the corresponding image pairs. Among the experiments using intermediate losses on the BN, L^1-based loss generated the most correlated CoMIRs within any loss fusion approach and the highest RSR.

Furthermore, we investigate the reasons of the lower RSR by visualizing the embeddings of the train set in Sect. 6. We observe that additional supervision on the BN can lead to clustering by modality in the embedding space, and in turn to less discriminative features among the corresponding CoMIR pair. The embedding of the BN features shows that the network learns features at the BN which are modality-specific rather than similar across modalities when left unregulated. This is in line with other observations that suggest pseudo-Siammese networks are prefered over Siammese networks for multimodal tasks in which the modalities differ strongly from each other [10].

We observe that pretraining on the BN features can lead to partial DC from which the training cannot recover in the subsequent training of the final layer. This observation may be related to observations in [22], which connect partial DC to models which are too small relative to the dataset size. However, this was observed for SimSiam, a non-contrastive SSL approach not relying on negatives.

8 Conclusions

Contrastive learning can generate common representations for multimodal images, called CoMIRs, which are similar in intensity, structures and features, allowing the utilization of monomodal registration methods. This reduces the very challenging task of multimodal registration to a, typically easier, monomodal one. In this study, we explore three approaches to add supervision to the latent representations in the bottleneck of the U-Net used to generate CoMIRs, imposing similarity of the features earlier in the network. For each approach we test three critic functions for the loss evaluated on these latent features. Our results show that CoMIRs learned with no additional supervision perform best in the downstream task of registration. More so, we show that without additional supervision, the BN features tend to extract modality-specific information and the shared features are likely extracted in the decoder of the U-Net. Additional supervision on the BN features often propagates the tendency to modality-wise clusters in the feature space into the final CoMIRs, making them less useful for registration. We observe partial DC of the learned features when the contrastive loss is only applied to the BN during pretraining and see that this collapse is maintained during the subsequent training on the final layers.

We address the quality quantification of learned image representations. We relate commonly used image distance and similarity measures to the representations' usefulness for registration. We show that correlation corresponds highly to registration performance. Representations useful for registrations also score reasonably well with respect to SSIM, MSE, α-AMD, and FID, however the reverse does not necessarily hold. The study explores the behavior of CoMIRs with BN supervision on a publicly available multimodal dataset of BF and SHG images. We confirm the main finding that CoMIRs without additional intermediate supervision are more useful for registrationon on a second multimodal dataset of correlative time-lapse QPI and fluorescence images.

Our study indicates the importance to develop representation learning approaches with a particular application on mind, as we show that concepts applicable for biomedical image classification do not necessarily generalize to registration tasks in the same domain and learning context, in this case contrastive learning. Multimodal image registration remains in general a very difficult task and we believe it is important to explore and further improve upon suitable learning strategies, which we will continue in our future work.

References

1. Amiri, M., Brooks, R., Rivaz, H.: Fine-tuning U-Net for ultrasound image segmentation: different layers, different outcomes. IEEE Trans. Ultrason. Ferroelectr. Freq. Control **67**(12), 2510–2518 (2020). https://doi.org/10.1109/TUFFC.2020.3015081
2. Borg, I., Groenen, P.J.F.: Modern Multidimensional Scaling - Theory and Applications. Springer, New York (2005). https://doi.org/10.1007/0-387-28981-X
3. Chen, Z., Wei, J., Li, R.: Unsupervised multi-modal medical image registration via discriminator-free image-to-image translation (2022). https://doi.org/10.48550/ARXIV.2204.13656
4. Chi, Z., et al.: On the representation collapse of sparse mixture of experts (2022). https://doi.org/10.48550/ARXIV.2204.09179
5. Choi, Y., Uh, Y., Yoo, J., Ha, J.W.: StarGAN v2: diverse image synthesis for multiple domains. In: CVPR, pp. 8185–8194 (2020). https://doi.org/10.1109/CVPR42600.2020.00821
6. Cosentino, R., et al.: Toward a geometrical understanding of self-supervised contrastive learning (2022). https://doi.org/10.48550/ARXIV.2205.06926
7. Dey, N., Schlemper, J., Salehi, S.S.M., Zhou, B., Gerig, G., Sofka, M.: ContraReg: contrastive learning of multi-modality unsupervised deformable image registration (2022). https://doi.org/10.48550/ARXIV.2206.13434
8. Eliceiri, K., Li, B., Keikhosravi, A.: Multimodal Biomedical Dataset for Evaluating Registration Methods (patches from TMA Cores), June 2020. https://doi.org/10.5281/zenodo.3874362
9. Eliceiri, K., Li, B., Keikhosravi, A.: Multimodal biomedical dataset for evaluating registration methods (full-size TMA cores), February 2021. https://doi.org/10.5281/zenodo.4550300
10. En, S., Lechervy, A., Jurie, F.: TS-NET: combining modality specific and common features for multimodal patch matching. In: ICIP, pp. 3024–3028 (2018). https://doi.org/10.1109/ICIP.2018.8451804
11. Fischler, M.A., Bolles, R.C.: Random sample consensus: a paradigm for model fitting with applications to image analysis and automated cartography. Commun. ACM **24**(6), 381–395 (1981). https://doi.org/10.1145/358669.358692
12. Gutmann, M., Hyvärinen, A.: Noise-contrastive estimation: a new estimation principle for unnormalized statistical models. In: Proceedings of the International Conference on Artificial Intelligence and Statistics, vol. 9, pp. 297–304. PMLR (2010)
13. Heusel, M., Ramsauer, H., Unterthiner, T., Nessler, B., Hochreiter, S.: GANs trained by a two time-scale update rule converge to a local Nash equilibrium. In: NeurIPS, vol. 30 (2017). Proceedings.neurips.cc/paper/2017/file/8a1d694707eb0fefe65871369074926d-Paper.pdf
14. Hjelm, R.D., et al.: Learning deep representations by mutual information estimation and maximization. In: ICLR (2019). https://openreview.net/forum?id=Bklr3j0cKX

15. Hua, T., Wang, W., Xue, Z., Ren, S., Wang, Y., Zhao, H.: On feature decorrelation in self-supervised learning. In: ICCV, pp. 9598–9608, October 2021
16. Isola, P., Zhu, J.Y., Zhou, T., Efros, A.A.: Image-to-image translation with conditional adversarial networks. In: CVPR (2017)
17. Jégou, S., Drozdzal, M., Vazquez, D., Romero, A., Bengio, Y.: The one hundred layers tiramisu: fully convolutional densenets for semantic segmentation. In: CVPR Workshops, pp. 11–19 (2017)
18. Jing, L., Vincent, P., LeCun, Y., Tian, Y.: Understanding dimensional collapse in contrastive self-supervised learning. arXiv preprint arXiv:2110.09348 (2021)
19. Kaku, A., Upadhya, S., Razavian, N.: Intermediate layers matter in momentum contrastive self supervised learning. In: NeurIPS, vol. 34, pp. 24063–24074 (2021). https://Proceedings.neurips.cc/paper/2021/file/c9f06258da6455f5bf50c5b9260efeff-Paper.pdf
20. Kang, S., Uchida, S., Iwana, B.K.: Tunable U-Net: controlling image-to-image outputs using a tunable scalar value. IEEE Access **9**, 103279–103290 (2021). https://doi.org/10.1109/ACCESS.2021.3096530
21. Lee, H.Y., et al.: DRIT++: diverse image-to-image translation via disentangled representations. Int. J. Comput. Vis. **128**, 2402–2417 (2020). https://doi.org/10.1007/s11263-019-01284-z
22. Li, A.C., Efros, A.A., Pathak, D.: Understanding collapse in non-contrastive Siamese representation learning. In: Avidan, S., Brostow, G., Cisse, M., Farinella, G.M., Hassner, T. (eds.) Computer Vision – ECCV 2022. ECCV 2022. Lecture Notes in Computer Science, vol. 13691, pp. 490–505. Springer, Cham (2022). https://doi.org/10.1007/978-3-031-19821-2_28
23. Li, S., Tso, G.K., He, K.: Bottleneck feature supervised U-Net for pixel-wise liver and tumor segmentation. Expert Syst. Appl. **145**, 113131 (2020). https://doi.org/10.1016/j.eswa.2019.113131
24. Lindblad, J., Sladoje, N.: Linear time distances between fuzzy sets with applications to pattern matching and classification. TIP **23**(1), 126–136 (2014). https://doi.org/10.1109/TIP.2013.2286904
25. Lu, J., Öfverstedt, J., Lindblad, J., Sladoje, N.: Is image-to-image translation the panacea for multimodal image registration? A comparative study. PLOS ONE **17**(11), 1–33 (2022). https://doi.org/10.1371/journal.pone.0276196
26. Lu, J., Öfverstedt, J., Lindblad, J., Sladoje, N.: Datasets for Evaluation of Multimodal Image Registration, April 2021. https://doi.org/10.5281/zenodo.5557568
27. Morozov, S., Voynov, A., Babenko, A.: On self-supervised image representations for GAN evaluation. In: ICLR (2021). https://openreview.net/forum?id=NeRdBeTionN
28. Öfverstedt, J., Lindblad, J., Sladoje, N.: Fast and robust symmetric image registration based on distances combining intensity and spatial information. TIP **28**(7), 3584–3597 (2019). https://doi.org/10.1109/TIP.2019.2899947
29. Öfverstedt, J., Lindblad, J., Sladoje, N.: Fast computation of mutual information in the frequency domain with applications to global multimodal image alignment. Pattern Recogn. Lett. **159**, 196–203 (2022). https://doi.org/10.1016/j.patrec.2022.05.022
30. van den Oord, A., Li, Y., Vinyals, O.: Representation learning with contrastive predictive coding. CoRR (2018). http://arxiv.org/abs/1807.03748
31. Oord, A.v.d., Li, Y., Vinyals, O.: Representation learning with contrastive predictive coding. arXiv preprint arXiv:1807.03748 (2018)

32. Park, T., Efros, A.A., Zhang, R., Zhu, J.-Y.: Contrastive learning for unpaired image-to-image translation. In: Vedaldi, A., Bischof, H., Brox, T., Frahm, J.-M. (eds.) ECCV 2020. LNCS, vol. 12354, pp. 319–345. Springer, Cham (2020). https://doi.org/10.1007/978-3-030-58545-7_19
33. Pielawski, N., et al.: CoMIR: contrastive multimodal image representation for registration. In: NeurIPS, vol. 33, pp. 18433–18444 (2020). https://Proceedings.neurips.cc/paper/2020/file/d6428eecbe0f7dff83fc607c5044b2b9-Paper.pdf
34. Qin, C., Shi, B., Liao, R., Mansi, T., Rueckert, D., Kamen, A.: Unsupervised deformable registration for multi-modal images via disentangled representations. In: Chung, A.C.S., Gee, J.C., Yushkevich, P.A., Bao, S. (eds.) IPMI 2019. LNCS, vol. 11492, pp. 249–261. Springer, Cham (2019). https://doi.org/10.1007/978-3-030-20351-1_19
35. Sammon, J.W.: A nonlinear mapping for data structure analysis. Trans. Comput. **C-18**(5), 401–409 (1969)
36. Tian, Y., Krishnan, D., Isola, P.: Contrastive multiview coding. CoRR (2019). http://arxiv.org/abs/1906.05849
37. Tian, Y., Sun, C., Poole, B., Krishnan, D., Schmid, C., Isola, P.: What makes for good views for contrastive learning? In: NeurIPS, vol. 33, pp. 6827–6839 (2020). https://Proceedings.neurips.cc/paper/2020/file/4c2e5eaae9152079b9e95845750bb9ab-Paper.pdf
38. Vicar, T., Raudenska, M., Gumulec, J., Masarik, M., Balvan, J.: Quantitative phase microscopy timelapse dataset of PNT1A, DU-145 and LNCaP cells with annotated caspase 3,7-dependent and independent cell death, March 2019. https://doi.org/10.5281/zenodo.2601562
39. Vicar, T., Raudenska, M., Gumulec, J., Masarik, M., Balvan, J.: Fluorescence microscopy timelapse dataset of PNT1A, DU-145 and LNCaP cells with annotated caspase 3,7-dependent and independent cell death, February 2021. https://doi.org/10.5281/zenodo.4531900
40. Wang, F., Liu, H.: Understanding the behaviour of contrastive loss. In: CVPR, pp. 2495–2504, June 2021
41. Wang, Z., Bovik, A., Sheikh, H., Simoncelli, E.: Image quality assessment: from error visibility to structural similarity. TIP **13**(4), 600–612 (2004). https://doi.org/10.1109/TIP.2003.819861
42. Wen, Z., Li, Y.: Toward understanding the feature learning process of self-supervised contrastive learning, vol. 139, pp. 11112–11122. PMLR, 18–24 July 2021
43. Wetzer, E., Lindblad, J., Sladoje, N.: Can representation learning for multimodal image registration be improved by supervision of intermediate layers? (2023). https://doi.org/10.48550/ARXIV.2303.00403
44. Wu, W., Yang, J.: Object fingerprints for content analysis with applications to street landmark localization. In: Proceedings of the ACM International Conference on Multimedia, pp. 169–178 (2008)
45. Xiao, T., Wang, X., Efros, A.A., Darrell, T.: What should not be contrastive in contrastive learning. In: ICLR (2021). https://openreview.net/forum?id=CZ8Y3NzuVzO
46. Zhu, J.Y., Park, T., Isola, P., Efros, A.A.: Unpaired image-to-image translation using cycle-consistent adversarial networks. In: ICCV (2017)

Interpretability-Guided Human Feedback During Neural Network Training

Pedro Serrano e Silva[1,2,3], Ricardo Cruz[2,3], A. S. M. Shihavuddin[4], and Tiago Gonçalves[2,3](✉)

[1] NILG.AI, Porto, Portugal
pedro.serrano@nilg.ai
[2] Faculty of Engineering, University of Porto, Porto, Portugal
{rpcruz,tiagofs}@fe.up.pt
[3] INESC TEC, Porto, Portugal
[4] Green University, Dhaka, Bangladesh
shihav@eee.green.edu.bd

Abstract. When models make wrong predictions, a typical solution is to acquire more data related to the error: an expensive process known as active learning. Our supervised classification approach combines active learning with interpretability so the user can correct such mistakes during the model's training. At the end of each epoch, our training pipeline shows examples of mistaken cases to the user, using interpretability to allow the user to visualise which regions of the images are receiving the model's attention. The user can then guide the training through a regularisation term in the loss function. This approach differs from previous works where the user's role was to annotate unlabelled data since, in this proposal, the user directly influences the training procedure through the loss function. Overall, in low-data regimens, the proposed method returned lower loss values in the predictions made for all three datasets used: 0.61, 0.47, 0.36, when compared with fully automated training methods using the same amount of data: 0.63, 0.52, 0.41, respectively. We also observed higher accuracy values in two datasets: 81.14% and 92.58% over the 78.41% and 92.52% seen in fully automated methods.

Keywords: artificial neural networks · active learning · explainable artificial intelligence · human feedback · interpretability

1 Introduction

The performance of deep neural networks has challenged human performance in many cases [14]; yet, they can fail spectacularly at surprisingly simple cases [25]. When a neural network makes a wrong prediction, a popular solution is to retrain it with more related cases. However, this approach may present high optimisation

Supported by the Portuguese Foundation for Science and Technology — FCT within PhD grant 2020.06434.BD.

A. Pertusa et al. (Eds.): IbPRIA 2023, LNCS 14062, pp. 276–287, 2023.
https://doi.org/10.1007/978-3-031-36616-1_22

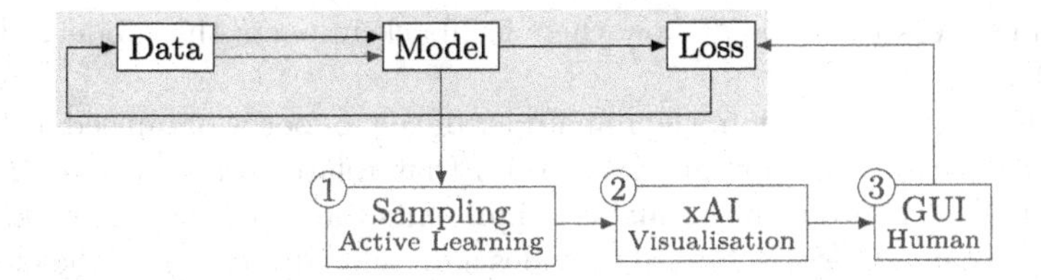

Fig. 1. Overview of the proposed pipeline: **black** lines represent the traditional model and **blue** lines represent the proposal. (Color figure online)

costs, while obtaining new data may not be trivial nor guarantee better results. Therefore, to overcome such problems, an alternative solution may be to let the model efficiently choose the data samples that may positively benefit its training process. In machine learning (ML) literature, this strategy is known as *active learning* [18]. The intuition behind this approach assumes that the model will react better to a given problem if we provide more examples of that instance during training. Recently, the advantages of active learning have been studied and discussed in the literature of deep learning (DL), where the main goal is to retain the powerful learning capabilities of DL while reducing the cost of sample annotation [16]. Moreover, instead of relying solely on a curated set of data samples provided by an active learning approach, one may also integrate human experts in the training loop and benefit from their domain knowledge, in line with other Human-in-the-Loop (HITL) work [6,28]. These frameworks focus on having the human improve the model [4] (e.g., new human annotations [10], reinforcement learning policies that mimic human behaviour [15]).

In this work, we propose more direct human participation by having the human guide the training process itself. The field of explainable artificial intelligence (xAI) (or ML interpretability) generally splits into three main research lines: *pre-*, *in-*, and *post-model* strategies [7]. Pre-model strategies aim to understand the data before making any ML model (e.g., exploratory data analysis). In-model strategies focus on inherently interpretable algorithms through rules or constraints [20]. Post-model strategies aim to produce explanations after the model's training (e.g., saliency maps on the image showing the locations that contributed most to the model's output [19,24]). Interestingly, integrating human users into the optimisation processes of interpretability is already part of the literature on xAI [9], thus motivating our work and showing that this is a timely opportunity to contribute towards this research line.

Our proposal, illustrated in Fig. 1, uses post-model explanations to show the user which parts of the image seem to contribute more to the prediction. The user interface shows rectangles around the most salient regions, and the user clicks on the image regions that should not be relevant to the model's prediction. We integrate this user input into the loss function as a regularisation term. Our proposal uses sampling strategies inspired by active learning methods to decide which images to show the user. This framework aims to promote transparency by allowing users to visualise how their feedback impacts the model's decision iteratively.

Besides this Introduction, the remainder of this paper is organised as follows: Sect. 2 discusses other approaches of HITL that relate to our proposal; Sect. 3 extensively describes our proposal; Sect. 4 presents the experimental work; Sect. 5 shows the obtained results; Sect. 6 provides a detailed discussion of the outcomes of our proposal; and Sect. 7 concludes the paper and suggests future work directions.

2 Related Work

Liu et al. [11] proposed a reinforcement learning method based on HITL, which avoids pre-labelling steps and keeps the model upgrading with progressively collected data. Their strategy uses a *deep reinforcement active learning* method to guide an agent in selecting training samples by an oracle. In this case, the uncertainty value of each human who picked a training sample works as a reinforcement reward. Uehara et al. [27] developed a novel process for object detection in satellite images that uses active learning combined with Gradient-weighted Class Activation Mapping (Grad-CAM) [17] to select the best images from an unlabelled pool of samples to train a feature extractor and classifier efficiently. Using Grad-CAM, this application queries users about the features in a given image, who must select the regions they consider to be related to a given object. Le et al. [10] proposed a simple and efficient *interactive self-annotation* framework to cut down time and human labour costs for video object bounding box annotation. The interactive recurrent annotation relies on a human annotator that gives feedback to the bounding-box detector. Later, Adhikari and Huttunen [1] improved on this framework by modifying the role of humans, which transitioned from performing full annotations to correcting errors.

On autonomous driving, Rajendran et al. [15] use HITL learning for vehicle self-exploration with HITL learning, in two phases. In the first phase (non-exploratory), the AI system learns from available historical data and the imitation of an oracle. In the second phase (exploratory), the AI system interacts with the environment, updating its policy through a long short-term memory network (LSTM), restricted by the anomaly detector trained in the first phase with the aid of a human annotator. Smailagic et al. [23] innovated the task of medical image classification by introducing MedAL, a novel way of selecting relevant samples from an unlabelled image pool. This method maximises the average distance to all training set examples in a learned feature space. The main goal is to reduce the amount of data required for state-of-the-art results by generating an optimal initial training set (i.e., with the most information about the overall data distribution) without needing a trained model. Regarding interpretability-guided frameworks, we highlight the works developed by Silva et al. [21,22], where they proposed the use of interpretability methods to localise relevant regions of images, promoting more focused feature representations and, consequently, improve medical image retrieval; Mahapatra et al. [13], where they proposed an interpretability-guided method that enforces that learned features yield more distinctive and spatially consistent saliency maps for different class labels of trained models; and, lastly, Mahapatra et al. [12], where they presented a sample selection approach based on graph analysis to identify informative samples in a multi-label setting. While this literature served as inspiration, in this work, we propose a more direct approach at HITL by having the user guide the model directly through its loss function.

3 Proposal

In our supervised classification proposal, we approach the neural network and the human as two entities that speak different languages. This way, xAI methods establish the bridge between these two entities (i.e., a translator). From active learning and HITL, we can learn the most effective way of mediating this communication and make it more efficient by finding the least amount of information required to convey any ideas in the conversation. The pipeline is summarised in Algorithm 1 and includes the following steps that are detailed in the following subsections:

1. Active learning sampling to find the most egregious model mistakes (line 3).
2. Use xAI to detect which regions of the images are being considered by the model (line 4).
3. The user feedback is integrated into the loss function (lines 5–6).

Algorithm 1. Pseudocode of our training method.

1: **function** TRAIN(model, images, labels)
2: preds $\leftarrow$ model(images)
3: images $\leftarrow$ EntropySample(images)
4: saliencies $\leftarrow$ xAI(images, preds)
5: $W_{i,j} \leftarrow$ UserInterface(saliencies)
6: loss $\leftarrow$ CE(preds, labels) $+ \sum_{i,j} \lambda W_{i,j} \frac{\partial \text{preds}}{\partial \text{images}_{i,j}}$
7: **end function**

3.1 Sampling

Data sampling and querying play a crucial role in the proposed workflow. In active learning, one may use several traditional *uncertainty sampling* methods to identify the most relevant data points in a dataset. The logic behind these sampling methods is that the data points which are hard to classify must contain relevant information in an active learning workflow. Although several uncertainty sampling exists (e.g., *least confident strategy*, *margin sampling*, and *entropy sampling*), Settles [18] concluded that entropy sampling is more suited if the objective is to decrease logarithmic loss. Since one of our goals is to minimise model overfitting and considering that the solution should generalise for tasks with any number of classes, we selected entropy sampling as the best option (x^*_H), as described in Eq. 1:

$$x^*_H = \arg\max_x - \sum_i P_\theta(y_i \mid x) \log P_\theta(y_i \mid x), \tag{1}$$

where y_i is the label for observation i, x is the image, and P_θ is the probability of y_i given x, predicted by the model. Cases with high entropy are relevant since those are the cases where the model has higher uncertainty of its prediction. We then select the top-N observations for user feedback.

3.2 Interpretability

In this work, we use DeepLIFT [19] to generate the saliency maps the human user will assess. This xAI method decomposes the output prediction of a neural network on a specific input by performing backpropagation of the contributions [3] of all neurons in the network to every feature of the input signal (i.e., the image). It defines *importance* as a function of differences from a reference state, where the reference is chosen based on the problem at hand. In this work, we used the variation DeepLIFT-Rescale, included in the Captum library for Python programming language [8]. DeepLIFT has two optional, although important, parameters to be defined: the target and the reference. The target is related to which class (i.e., label) DeepLIFT will compute the pixel attributions. As the HITL process should happen in a clean, single-image display, we must execute DeepLIFT for a single target class. Our proposal aims to fix the model's prediction. Hence, it is logical to target the label predicted by the model since we are basing human interaction on this prediction. The reference refers to the reference image previously explained. The default reference is a white image of the same size as the input image. However, Shrikumar et al. [19] recommend a different strategy for more complex datasets such as the ones used in our experiments: using a blurred version of the input image as the reference to obtain more well-defined explanations.

3.3 User Feedback

After selecting the most relevant samples, it is necessary to show information to the annotator (i.e., the user) in a human-understandable manner. Simply showing the DeepLIFT saliency map to the annotator may be too cryptic, and overlaying them on top of the original can damage the perception of the image. To overcome these visualisation issues, rectangles over the regions with higher attribution scores are identified by dividing the saliency map in a grid (see Fig. 2). Besides, to allow for a more informed decision, our interface also shows the ground-truth label of the image, the model's prediction, and the image index.

The human feedback requires the following steps:

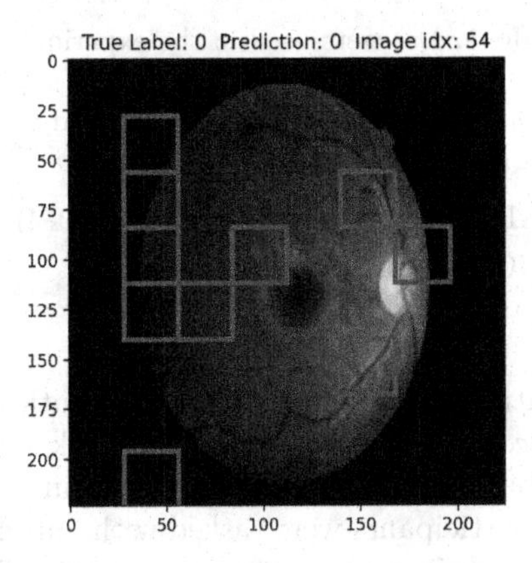

Fig. 2. Example of a query when training the model on the APTOS2019 dataset. The interface displays the sampled image, its ground-truth label, the model's prediction, and the image index. The users can click on the squared regions they perceive as less relevant for the classification task. After finishing their selection, the users can submit it and close the image.

1. The annotator must select which of the presented regions the model should disregard by clicking the square.
2. When clicked, the rectangle is highlighted, and a second time can deselect it.
3. After selecting all the desired squares, the user can close the image. We store this information in the form of a weight tensor $W_{i,j} \in \{0,1\}$ related to the locations of the selected pixels i,j. This tensor is 1 for the pixels chosen by the user and 0 otherwise.

To backpropagate the errors identified by the user, we penalise the selected pixels (that is, where $W_{i,j}$ is 1). A regularisation term is added to the loss function that penalises the model if the pixels are being used to produce the model output. That is, the derivative of the output relative to the input ($\frac{\partial \hat{y}}{\partial x_{i,j}}$) should be zero for those cases. The final loss is described in Eq. 2:

$$L(\hat{y},y) = \mathrm{CE}(\hat{y},y) + \lambda \sum_{i,j} W_{i,j} \left(\frac{\partial \hat{y}}{\partial x_{i,j}} \right)^2, \tag{2}$$

where λ is a hyper-parameter weighting the new loss term.

4 Experiments

4.1 Data

The three datasets used by the experiments are described below with a summary in Table 1. **ISIC2017** is a skin cancer dataset from the ISIC2017 Challenge [5][1],

[1] https://challenge.isic-archive.com/landing/2017/.

Table 1. Datasets used in the experiments.

Dataset	N	Classes used
ISIC2017	2,000	1 vs 2
APTOS2019	3,600	multiclass (0–4)
NCI	74,820	0 vs 1

that had participants attempt to build automated solutions for the diagnosis of melanoma from dermoscopic images. We have used Task 3: Disease Classification. **APTOS2019** is a diabetic retinopathy dataset[2] originates from a competition held on Kaggle, where participants were tasked with automating the identifying of different stages of diabetic retinopathy, to detect indications of blindness. **NCI** is a cervical cancer dataset from the American National Cancer Institute (NCI). Besides the medical image data (i.e., cervigrams), the dataset includes the patient's age, HPV test, and histology results. The labels for each cervigram are related to the neoplasia progression level and divided into the following categories: normal, CIN1, CIN2, CIN3, and cancer. This dataset is not publicly available.

4.2 Model Architecture

After preliminary experiments, we decided to use EfficientNet [26] as the model's architecture. We initialised this model with the weights from the training on the ImageNet dataset. In this proposal, the weight tensor $W_{i,j}$ only stores information about the non-augmented version of the images. Hence, to include data augmentation methods during training, we must compute the backward pass twice: the first pass computes the ordinary loss for the augmented images, and the second pass computes our regularisation term using the original image.

4.3 Training Phase

Data Augmentation: We used the following data transformations: vertical and horizontal flipping, 10% cropping, free 360° rotation, and 10% colour brightness. We resized all images to 224×224.

Baseline: We start by training a baseline model for 10 epochs without human feedback (i.e., we optimised this hyper-parameter through preliminary studies). From the progression of the entropy values, we noticed that most data points shared the same values, meaning high entropy sampling would work as random sampling. By running more epochs, we are letting the model *figure out* which data points are harder to classify, and at this point, we can correctly apply high entropy sampling. The learning rate for the baseline was 1.0×10^{-4}, the optimiser was adaptive moment estimation (Adam), and the batch size was 4.

[2] https://www.kaggle.com/c/aptos2019-blindness-detection.

HITL Models: The HITL models were initialised with the baseline's weights and trained for 20 epochs (i.e., 10 with human input and 10 without). The number of queries was 20, which amounts to roughly 7% of the training data, and the λ value was 10^6. To compare the HITL training with automated training, we trained another model without human feedback for 20 epochs. Both models used a learning rate of 10^{-4}, Adam as the optimiser, and a batch size of 4.

The code related to the experimental work is publicly available in a GitHub repository[3].

5 Results

Table 2. Results of the proposed framework.

Dataset	Method	Accuracy	Min Loss
ISIC2017	Baseline 10%	**85.91**	0.52
	HITL 10%	83.87	**0.47**
	Baseline 100%	88.59	0.34
APTOS2019	Baseline 10%	78.41	0.63
	HITL 10%	**81.14**	**0.61**
	Baseline 100%	85.11	0.47
NCI	Baseline 0.5%	92.52	0.41
	HITL 0.5%	**92.58**	**0.36**
	Baseline 50%	**93.18**	0.23
	HITL 50%	93.16	**0.22**

The performance of the proposed pipeline relates to the prediction accuracy and the minimum loss value achieved by the model. The test scenarios use only a percentage of the dataset so that the impact of the method is made more visible. Models trained with the proposed human feedback approach (i.e., HITL) are contrasted with models trained without human feedback (i.e., baseline). Table 2 shows the raw performance of the methods for the most relevant experiments performed in this work. Regarding the ISIC2017 and APTOS2019 datasets, we achieved the best performance when training on 10% of the data. We also present the baseline results when using 100% of the data. Concerning the NCI dataset, we achieved the best performance when training on 0.5% of the data. To confirm the low impact of the HITL method in scenarios with high data availability, we performed other experiments as well. These tests were performed on 50% of the data, using models pre-trained on 0.5%, with the HITL method, and compared with no feedback training on 50% of the dataset. We achieved the lowest min-loss using the proposed HITL training method in all tests. Regarding accuracy

[3] https://github.com/PedroSerran0/ig-human-feedback-nn.

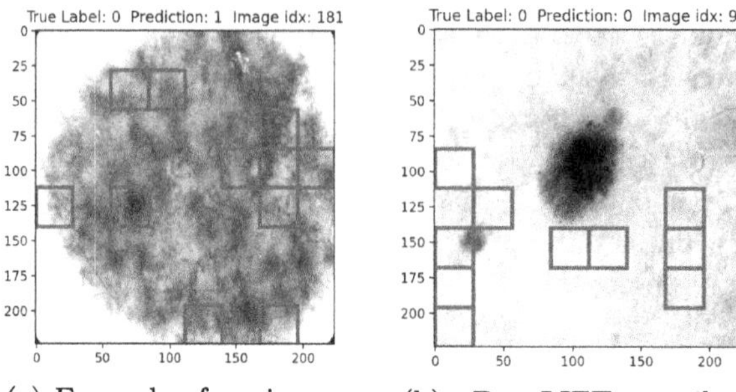

(a) Example of an image hard to interpret.

(b) DeepLIFT attributions poorly summarized.

(c) Presence of hair in the image.

Fig. 3. Example of the problems encountered when annotating query images while training the HITL approach on the ISIC2017 dataset.

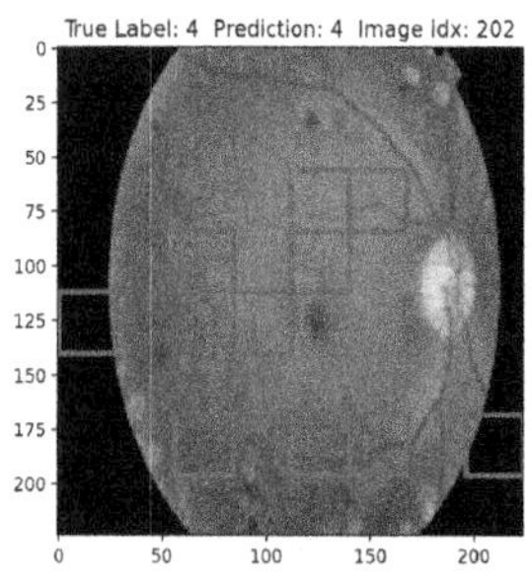

(a) Correct prediction with unclear attributions.

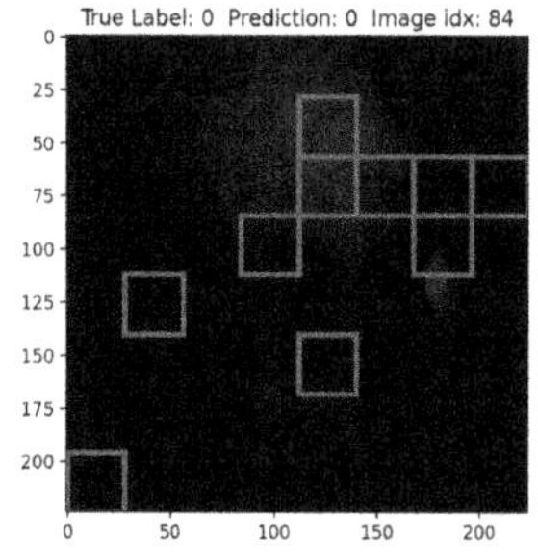

(b) Darkness hiding important features.

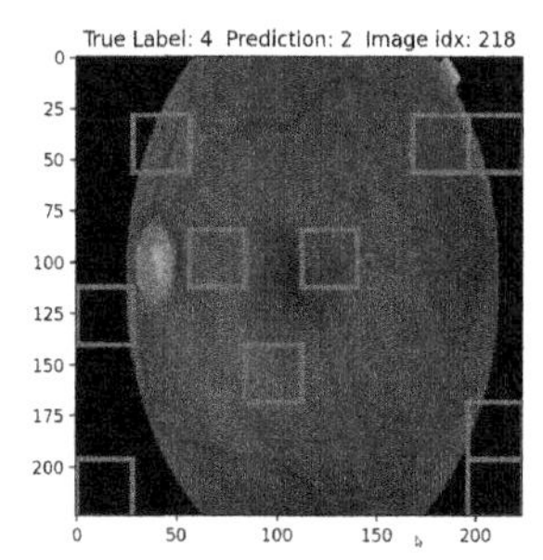

(c) Possibly wrong label.

Fig. 4. Examples of the problems encountered when annotating query images while training the HITL approach on the APTOS2019 dataset.

values, the HITL-trained model outperformed the baseline, in the same task, on two different occasions. Moreover, models trained with HITL and low data availability could not achieve similar accuracy values to models trained with 100% or 50% of the data. Testing on 100% and 50% of the datasets showed that using feedback or not achieves similar results, albeit when using feedback, there is more stochasticity in the training progression. These results suggest a delay in the model overfitting, resulting, by hypothesis, in lower min-loss values. Interestingly, we detected a similar pattern in most low-data training scenarios.

6 Discussion

Figure 3 presents examples of some of the difficulties found when working with the ISIC2017 dataset. From these, we realise that there are cases where the model focuses on irrelevant features (e.g., higher attribution values outside the area of the lesion). Therefore, we argue that these are the situations where the training of the models may benefit from the HITL framework. One of the drawbacks of working with this dataset is the difficulty in understanding the type of features the model should ignore or pay attention to (e.g., the presence of hair or the existence of multiple lesions). Naturally, this process would benefit from the aid of a domain expert. Oppositely to other use cases, the APTOS2019 dataset seems intuitive to manipulate, regarding the features that the model should ignore (i.e., the features that are always present, independently of the stage of the disease, such as background, fovea, the optic disc, and the blood vessels) or pay attention (i.e., features that should relate to the stages of the disease) when classifying an image. Hence, the HITL framework is essential in guiding the model's training to focus on the relevant objects.

Nevertheless, as shown in Fig. 4, we also identified some difficulties when working with the APTOS2019 dataset. We encountered several queries where the image was too dark (making it very hard to analyse while returning a high entropy prediction), ambiguity on the features (i.e., the features where the model should focus were not easy) or on the labels (i.e., model's prediction made more sense than the ground-truth annotation), and where the attributions did not motivate the model's prediction.

Regarding the NCI dataset, we followed the annotation rationale described by Albuquerque et al. [2], which uses a Gaussian kernel for regularisation, forcing the model to focus on a more central part of the images, as most tools are near the peripheries. Hence, assuming this prior knowledge, the HITL framework serves as a flexible refined method of this state-of-the-art approach. Besides, as with APTOS2019, we noted that this dataset was intuitive. However, we also registered some abnormal behaviour during training (e.g., the model focusing on irrelevant features such as the instruments or the textual description of the cervigram). Interestingly, after a few training epochs, when the model started to focus on relevant features (e.g., the endocervix), we could still register some cases of abnormal behaviour. We do not provide a figure with the results obtained, as this dataset is not publicly available.

7 Conclusion

In this work, we proposed a HITL framework that helps comprehend the images that models find harder to classify and where they seem to fail the most. It also succeeded in further exploring the requirements for effective interaction between neural networks and people, and its limitations, with several cases highlighted during testing. All of this information can play a big part in finding better ways to improve the training process and fine-tune models while increasing the transparency of neural networks. Experiments performed in three datasets showed some loss reduction – 0.61, 0.47, and 0.36 for the proposed pipeline versus 0.63, 0.52, and 0.41 for the baseline, respectively, for each dataset.

Further work should be devoted to fine-tuning hyper-parameters such as the number of training epochs in which we ask humans for feedback, the space between those same epochs, the number of queries per epoch, the type of sampling, the threshold for entropy sampling, the number of squares to display, the shape of those squares, the optimal learning rate or the optimal λ for the regularisation term in the proposed loss function. Concerning user interface improvements, we aim to explore other geometric shapes besides the current squared grid. Moreover, we intend to perform additional experiments on clustering methods, connecting high-value pixels in the prediction explanations or showing the personalised feature shape to the user. Furthermore, it would be interesting to extrapolate this methodology into other tasks (e.g., segmentation or object detection since there are multiple outputs).

References

1. Adhikari, B., Huttunen, H.: Iterative bounding box annotation for object detection. In: 2020 25th International Conference on Pattern Recognition (ICPR), pp. 4040–4046. IEEE (2021)
2. Albuquerque, T., Cardoso, J.S.: Embedded regularization for classification of colposcopic images. In: 2021 IEEE 18th International Symposium on Biomedical Imaging (ISBI), pp. 1920–1923. IEEE (2021)
3. Ancona, M., Ceolini, E., Öztireli, C., Gross, M.: Towards better understanding of gradient-based attribution methods for deep neural networks. arXiv preprint arXiv:1711.06104 (2017)
4. Budd, S., Robinson, E.C., Kainz, B.: A survey on active learning and human-in-the-loop deep learning for medical image analysis. Med. Image Anal. **71**, 102062 (2021)
5. Codella, N.C., et al.: Skin lesion analysis toward melanoma detection: a challenge at the 2017 international symposium on biomedical imaging (ISBI), hosted by the international skin imaging collaboration (ISIC). In: 2018 IEEE 15th international symposium on biomedical imaging (ISBI 2018), pp. 168–172. IEEE (2018)
6. Fischer, M., Kobs, K., Hotho, A.: NICER: aesthetic image enhancement with humans in the loop. arXiv preprint arXiv:2012.01778 (2020)
7. Kim, B., Doshi-Velez, F.: Interpretable machine learning: the fuss, the concrete and the questions. ICML Tutor. Interpret. Mach. Learn. (2017)
8. Kokhlikyan, N., et al.: Captum: A unified and generic model interpretability library for pytorch. arXiv preprint arXiv:2009.07896 (2020)
9. Lage, I., Ross, A., Gershman, S.J., Kim, B., Doshi-Velez, F.: Human-in-the-loop interpretability prior. Adv. Neural Inf. Process. Syst. **31**, 10180–10189 (2018)
10. Le, T.N., Sugimoto, A., Ono, S., Kawasaki, H.: Toward interactive self-annotation for video object bounding box: recurrent self-learning and hierarchical annotation based framework. In: Proceedings of the IEEE/CVF Winter Conference on Applications of Computer Vision, pp. 3231–3240 (2020)
11. Liu, Z., Wang, J., Gong, S., Lu, H., Tao, D.: Deep reinforcement active learning for human-in-the-loop person re-identification. In: Proceedings of the IEEE/CVF International Conference on Computer Vision, pp. 6122–6131 (2019)
12. Mahapatra, D., Poellinger, A., Reyes, M.: Graph node based interpretability guided sample selection for active learning. IEEE Trans. Med. Imaging **42**(3), 661–673 (2022)
13. Mahapatra, D., Poellinger, A., Reyes, M.: Interpretability-guided inductive bias for deep learning based medical image. Med. Image Anal. **81**, 102551 (2022)
14. McKinney, S.M., et al.: International evaluation of an AI system for breast cancer screening. Nature **577**(7788), 89–94 (2020)

15. Rajendran, P.T., Espinoza, H., Delaborde, A., Mraidha, C.: Human-in-the-loop learning for safe exploration through anomaly prediction and intervention. In: Proceedings of SafeAI, AAAI (2022)
16. Ren, P., et al.: A survey of deep active learning. ACM Comput. Surv. **54**(9), 1–40 (2021)
17. Selvaraju, R.R., Cogswell, M., Das, A., Vedantam, R., Parikh, D., Batra, D.: Grad-CAM: visual explanations from deep networks via gradient-based localization. In: Proceedings of the IEEE International Conference on Computer Vision (2017)
18. Settles, B.: Active learning literature survey. Computer Sciences Technical report 1648, University of Wisconsin-Madison (2009)
19. Shrikumar, A., Greenside, P., Kundaje, A.: Learning important features through propagating activation differences. In: International Conference on Machine Learning, pp. 3145–3153. PMLR (2017)
20. Silva, W., Fernandes, K., Cardoso, M.J., Cardoso, J.S.: Towards complementary explanations using deep neural networks. In: Stoyanov, D., et al. (eds.) MLCN/DLF/IMIMIC -2018. LNCS, vol. 11038, pp. 133–140. Springer, Cham (2018). https://doi.org/10.1007/978-3-030-02628-8_15
21. Silva, W., et al.: Computer-aided diagnosis through medical image retrieval in radiology. Sci. Rep. **12**(1), 20732 (2022)
22. Silva, W., Poellinger, A., Cardoso, J.S., Reyes, M.: Interpretability-guided content-based medical image retrieval. In: Martel, A.L., et al. (eds.) MICCAI 2020. LNCS, vol. 12261, pp. 305–314. Springer, Cham (2020). https://doi.org/10.1007/978-3-030-59710-8_30
23. Smailagic, A., Costa, P., Noh, H.Y., Walawalkar, D., Khandelwal, K., et al.: MedAL: accurate and robust deep active learning for medical image analysis. In: 2018 17th IEEE International Conference on Machine Learning and Applications (ICMLA). IEEE (2018)
24. Sundararajan, M., Taly, A., Yan, Q.: Axiomatic attribution for deep networks. In: International Conference on Machine Learning, pp. 3319–3328. PMLR (2017)
25. Szegedy, C., et al.: Intriguing properties of neural networks. arXiv preprint (2013)
26. Tan, M., Le, Q.: EfficientNet: rethinking model scaling for convolutional neural networks. In: International Conference on Machine Learning. PMLR (2019)
27. Uehara, K., Nosato, H., Murakawa, M., Sakanashi, H.: Object detection in satellite images based on active learning utilizing visual explanation. In: 2019 11th International Symposium on Image and Signal Processing and Analysis (ISPA), pp. 27–31. IEEE (2019)
28. Zhang, L., Wang, X., Fan, Q., Ji, Y., Liu, C.: Generating manga from illustrations via mimicking manga creation workflow. In: Proceedings of the IEEE/CVF Conference on Computer Vision and Pattern Recognition, pp. 5642–5651 (2021)

Calibration of Non-Central Conical Catadioptric Systems from Parallel Lines

James Bermudez-Vargas(✉), Jesus Bermudez-Cameo, and Jose J. Guerrero

Instituto de Investigación en Ingeniería de Aragón, I3A, Universidad de Zaragoza, Zaragoza, Spain
792069@unizar.es

Abstract. Most camera calibration approaches require the use of calibration patterns like checkerboards. This also applies to Omnidirectional Catadioptric Systems. In this paper, we propose a new calibration method for Non-Central Conical Catadioptric Systems which only requires long and parallel lines which are plentiful in man-made environments. It is based on the extraction of lines from a single omnidirectional image, where we impose parallelism restrictions to obtain the analytical solution of the calibration parameters. However, the calibration of this kind of non-central catadioptric systems is noise sensitive. The proposal has been tested with simulations and then with synthetic and real images to evaluate the accuracy of the proposed method.

Keywords: Catadioptric cameras · Non-Central projection · Omnidirectional vision

1 Introduction

The methods used in cameras calibration are sensitive to the type of omnidirectional camera it will be used with: Central or Non-Central. These systems are mainly differentiated by the way of projecting the elements of the scene and therefore the resulting images. In central systems any projection ray passes through a common viewpoint that we know as the optical center. Therefore, the projection surface of a 3D line is a plane that passes through the optical center of the camera. This causes some loss of information since any 3D line in this plane will be projected as the same line image.

On the contrary, in non-central systems the projection rays do not pass through the same viewpoint [9], since it is generally tangent to the caustic [2,15], which is an enveloping surface of the projection rays. When the non-central system is axial and has revolution symmetry, the projection rays pass through the axis of symmetry and the projection surface of a 3D line is formed by skew lines that form a ruled surface.

Our goal is to propose a new calibration method for Non-Central Catadioptric Systems, specifically those that use a conical mirror. And our main contribution

A. Pertusa et al. (Eds.): IbPRIA 2023, LNCS 14062, pp. 288–299, 2023.
https://doi.org/10.1007/978-3-031-36616-1_23

(a) Real conical catadioptric image containing the projection of parallel lines.

(b) Conical Catadioptric System

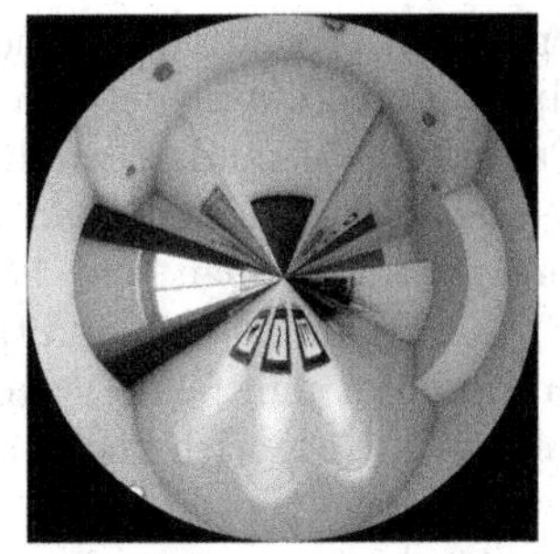
(c) Synthetic Image with parallel lines

Fig. 1. Real and Synthetic images from a Conical Catadioptric System.

is to be able to make this calibration from images acquired in man made environments (like corridors, sidewalks or train tracks, where we can find long and parallel lines), avoiding the use of traditional calibration pattern (chessboard type) that is currently used, as in [13], or the Charuco in [4], for example.

A Conical Catadioptric System is one with revolution symmetry in which we have a conventional camera aligned to the vertex of a conical mirror through a rotation axis, and that is sometimes called a cone sensor [18]. In Fig. 1 we can see two examples of omnidirectional images, one Real (Fig. 1(a)) and one Synthetic (Fig. 1(c)), obtained by a Conical Catadioptric System (Fig. 1(b)), and both of them contain parallel lines.

1.1 Related Work

In [27], they apply the definition of non-central as the misalignment between the optical center of the camera and the axis of rotation of the mirror. Image projection in Non-Central Catadioptric Systems [2,3] differs a lot from Central Catadioptric Systems [6]. Focusing in line projections, in Central Catadioptric Systems lines are projected onto conics, while in non-central are projected in higher degree curves. The projection of the final image in non-central systems is typically defined by the geometry of the reflection on the mirror and it depends on the particular mirror used.

In our case, we work with a conical mirror, and its projective geometry has been previously studied in [17], highlighting the variable position of the viewpoint (Unitary Torus Model) of the system, position that in turn is conditioned by the aperture angle of the mirror, which is explained in [5]. In [25], we can see a deep analysis of the properties of the caustic and its close relationship with the viewpoint of the conical catadioptric systems. Also, we can observe in [10] a new method for the automatic extraction of line images from those circular panoramas. This method later is generalized in [9] for non-central catadioptric systems with revolution symmetry and the analytical solution of the forward and back projection model of a line on the omnidirectional image is presented.

Calibration of central catadioptric systems has been previously addressed in numerous works and summarized in [23]. However, calibration of non-central cameras is still an open field. For example, in [1] where they make use of two or more spherical mirrors, in addition to two or more refractive lenses to obtain the calibration of a multi-axis system from a single 2D image, which consists of a calibration pattern with known parameters. In [20,21] they use a calibration pattern with known coordinates and an internally calibrated spherical catadioptric camera to obtain the extrinsic system calibration parameters.

In [16] a new method to solve the problem of pose estimation of a calibrated multi-camera system is proposed and that is represented by a set of non-central rays (in Plucker coordinates) that pass, each one of them, through the optical center of its corresponding camera. As for the extraction of lines, [14,24,26] describe how to calculate one or several lines that are formed from the intersection of four lines in 3D space, while [12] focuses on non-central cameras with symmetry of revolution, addressing the problem of 3D line estimation from single projection.

Previous studies have improved the reconstruction of 3D lines reducing the degrees of freedom of the problem, such are the cases of [11,22] who focused only on horizontal lines, [19] taking advantage of the cross-ratio properties, or [7] imposing additional constraints.

2 Background

Henceforth, we will use the following nomenclature: P, represents the 3D points; ρ, represents the line-images; r, represents the distance from each point to the principal point; L, represents the Plüker coordinates of a line in 3D; l, represents the direction vector of the 3D line; $\bar{l}$, represents the moment vector of the 3D line; $_i$, represents the index of an array of points; $_j$, represents the index of an array of lines; and $_k$, represents the index of an element within the coordinates. For example, we use $\rho_{j,k}$ with $j = 2$ and $k = 3$ for the 3nd element of the projection of the 2nd line-image. In other words, for each line ($_j$) we will have a $\rho_j = (\rho_{j,1}, \rho_{j,2}, \rho_{j,3}, \rho_{j,4}, \rho_{j,5}, \rho_{j,6})^T$.

2.1 Plücker Coordinate Lines

Traditionally, we use the equation $y = mx + b$, to define a line in $\mathbb{R}^2$, and its vector notation $(x, y, z)^T = (x_0, y_0, z_0)^T + \lambda(u_x, u_y, u_z)^T$ to define a line in $\mathbb{R}^3$, however to better working with 3D lines and their projections, we use Plücker coordinates to define our lines.

The Plücker coordinates of a 3D line is a homogeneous representation of a line $L \in \mathbb{P}^5$. This representation can be organized in such a way that they form two vectors, $L = (l, \bar{l})^T = (l_1, l_2, l_3, \bar{l}_1, \bar{l}_2, \bar{l}_3)^T$. One, the director vector

(l), which represents the direction of the line; the other, the moment vector ($\bar{l}$), which represents the normal to the plane formed by the line and the origin of the reference system.

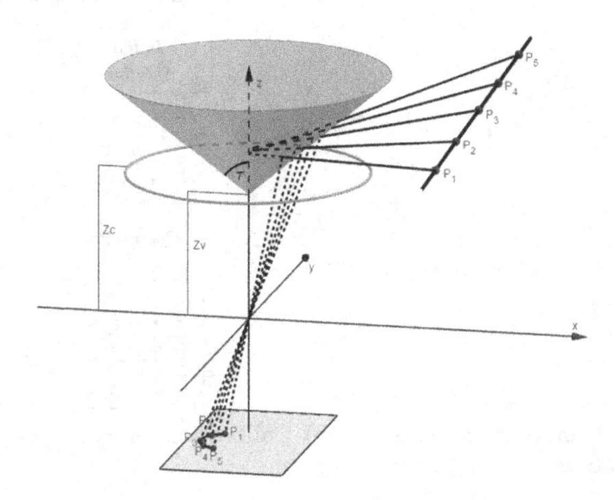

Fig. 2. Projection of a line in Conical Catadioptric System. Being a non-central system, the projection centers are different for each point, however they all pass through the axis of symmetry of the mirror.

2.2 Lines in Conical Mirror Systems

In the non-central projection of the conical mirror, each ray corresponding to a point in the scene passes through the conical mirror, crosses its axis of symmetry at a specific point (see Fig. 2), until it hits the circle (representing the point of view of the system), on its opposite side from where the lightning came from. The part of the ray that is reflected by the mirror surface passes through the optical center of the camera until it reaches the image plane.

As presented in [8], when particularizing the line image equation, expressed in polar coordinates ($x_n = r\cos\theta, y_n = r\sin\theta$), to the conical catadioptric system, it can be written in a compact form with 6 parameters that encapsulate the Plucker coordinates of the line and the parameters from the system mirror.

This allows us to linearly calculate the line image with five or more points without knowing neither the aperture angle of the mirror 2τ or the distance to the vertex of the mirror Z_v, solving

$$(r_i x_{ni}, r_i y_{ni}, {r_i}^2, x_{ni} z_{ni}, y_{ni} z_{ni}, r_i z_{ni})\omega = 0, \; for \; i = 1..5 \tag{1}$$

After knowing the line image $\omega \in \mathbb{P}^5$, it is easy to calculate τ from $\tan 2\tau = \frac{\omega_3}{\omega_6}$. It is important to clarify that in conical catadioptric systems, if the distance from the camera to the mirror Z_v is unknown, it is not possible to reconstruct the scale of a scene from line images in a single image. Notice that ω is defined in a normalized plane.

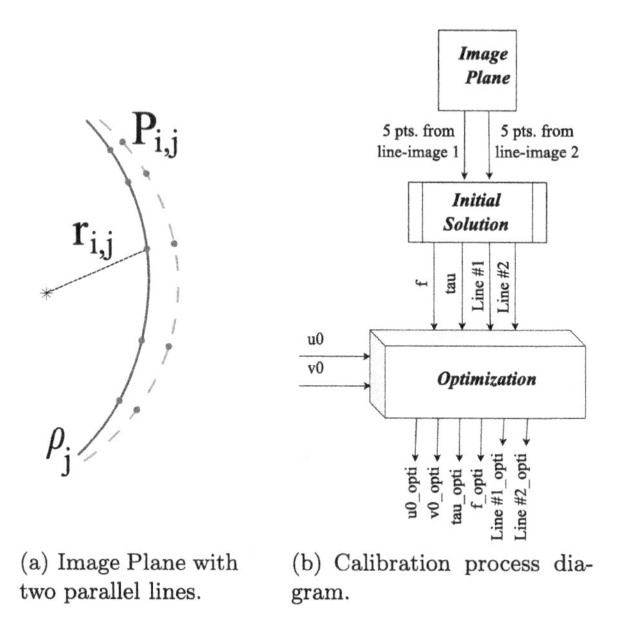

(a) Image Plane with two parallel lines.

(b) Calibration process diagram.

Fig. 3. Lines on the Image Plane and the Calibration Process. (Color figure online)

On the other hand, we define ρ as the representation of ω in a new reference on the image plane whose origin coincides with the projection of the optical center, called "aligned reference".

3 Calibration Method

The input of our proposal is a set of points supporting line projections on a 2D image, which is obtained by projecting this lines via an uncalibrated conical catadioptric system. In Fig. 3(a) we can see an example of a 2D image that contains two line-images (solid blue and dotted green) which correspond to two parallel lines in 3D.

The first step is to identify two parallel lines within the omnidirectional 2D image, then we take at least five points from each one of them (see Fig. 3(b)) to start with their respective adjustment.

As a next step, we take the transformation matrix of the projection model (without skew parameter, s) and decompose it, in T_{n1} and T_{n2}, to separate the focal length (f) from the main point (x_{p0}, y_{p0}) and be able to work it separately in calibration:

$$\begin{pmatrix} x_p \\ y_p \\ 1 \end{pmatrix} = \underbrace{\begin{pmatrix} 1 & 0 & x_{p0} \\ 0 & 1 & y_{p0} \\ 0 & 0 & 1 \end{pmatrix}}_{Tn_2} \underbrace{\begin{pmatrix} f & 0 & 0 \\ 0 & f & 0 \\ 0 & 0 & 1 \end{pmatrix}}_{Tn_1} \begin{pmatrix} x_n \\ y_n \\ z_n \end{pmatrix} \tag{2}$$

With Tn_1 we can project a 3D point on the "aligned reference", which is a reference lying on the image plane whose origin coincides with the projection of the optical center.

$$\begin{pmatrix} \hat{x}_p \\ \hat{y}_p \\ \hat{z}_p \end{pmatrix} = \begin{pmatrix} f & 0 & 0 \\ 0 & f & 0 \\ 0 & 0 & 1 \end{pmatrix} \begin{pmatrix} x_n \\ y_n \\ z_n \end{pmatrix} \tag{3}$$

Now, with the 2D coordinates of each point on the image plane (x_p, y_p) and of principal point (x_{p0}, y_{p0}) previously calculated, we get the coordinates of each point in the aligned reference $(\hat{x}_p, \hat{y}_p, \hat{z}_p)$. With these coordinates we calculate the corresponding radius $(\hat{r}_{i,j})$ of each point in the aligned reference.

We know, from [8], that with five points we can linearly calculate the line-image by applying the Moore-Penrose Pseudoinverse (*pinv*) and solving for:

$$\rho_j = -\hat{r}_{i,j} pinv(\hat{r}_{i,j}\hat{x}_{n(i,j)}, \hat{r}_{i,j}\hat{y}_{n(i,j)}, \hat{r}^2_{i,j}, \hat{x}_{n(i,j)}, \hat{y}_{n(i,j)}), \quad for\ i = 1..5,\ and\ j = 1..2. \tag{4}$$

From (2) and (3) we can also clearly see how f is introduced into the system model through: $x_n = \frac{\hat{x}_p}{f}$; $y_n = \frac{\hat{y}_p}{f}$; $z_n = \hat{z}_p = 1$; $\hat{r}^2 = \frac{\hat{x}_p^2 + \hat{y}_p^2}{f^2}$. And, if we substitute in (4), we get our new line-image (ρ_j) represented by $(\frac{\omega_1}{f^2}, \frac{\omega_2}{f^2}, \frac{\omega_3}{f^2}, \frac{\omega_4}{f}, \frac{\omega_5}{f}, \frac{\omega_6}{f})^T$.

Therefore, with the point coordinates $(\hat{x}_p, \hat{y}_p)$ and radius $(\hat{r})$ on the aligned reference, and assuming homogeneous coordinates $(\hat{z}_n = 1, \rho_{j,6} = 1)$, we can compose a system of linear equations, which is expressed as,

$$\begin{pmatrix} 0 & s & -\frac{1}{f^2} & 0 & 0 \\ -s & 0 & 0 & -\frac{1}{f^2} & 0 \\ 0 & 0 & 0 & 0 & \frac{t}{f} \\ 0 & t & t & 0 & 0 \\ -t & 0 & 0 & t & 0 \\ 0 & 0 & 0 & 0 & \frac{1}{f} \end{pmatrix} \begin{pmatrix} l_{j,1} \\ l_{j,2} \\ \bar{l}_{j,1} \\ \bar{l}_{j,2} \\ \bar{l}_{j,3} \end{pmatrix} = \begin{pmatrix} \rho_{j,1} \\ \rho_{j,2} \\ \rho_{j,3} \\ \rho_{j,4} \\ \rho_{j,5} \\ \rho_{j,6} \end{pmatrix} \tag{5}$$

where $s = \frac{1-\cos 2\tau}{f^2 \cos 2\tau} Z_v$; $t = \frac{\tan 2\tau}{f} Z_v$.

Then solving our system of equations in (5), we get L_j (depending on ρ_j) in Plücker coordinates, such that

$$\boldsymbol{L_j} = \begin{pmatrix} l_{j,1} \\ l_{j,2} \\ l_{j,3} \\ \bar{l}_{j,1} \\ \bar{l}_{j,2} \\ \bar{l}_{j,3} \end{pmatrix} = \begin{pmatrix} -\frac{f^2 \sin 2\tau \rho_{j,2} + f \cos 2\tau \rho_{j,5}}{Z_v tan 2\tau} \\ \frac{f^2 \sin 2\tau \rho_{j,1} + f \cos 2\tau \rho_{j,4}}{Z_v tan 2\tau} \\ \frac{f^2 \cos 2\tau (\rho_{j,2}\rho_{j,4} - \rho_{j,1}\rho_{j,5})}{Z_v \tan 2\tau \rho_{j,6}} \\ \frac{f(1-\cos 2\tau)\rho_{j,4} - f^2 \sin 2\tau \rho_{j,1}}{\tan 2\tau} \\ \frac{f(1-\cos 2\tau)\rho_{j,5} - f^2 \sin 2\tau \rho_{j,2}}{\tan 2\tau} \\ f\rho_{j,6} \end{pmatrix} \tag{6}$$

We know that two lines are parallel if and only if their direction vectors are linearly dependent $\left(\frac{l_{1,1}}{l_{2,1}} = \frac{l_{1,2}}{l_{2,2}} = \frac{l_{1,3}}{l_{2,3}}\right)$.

Based on first equality $\left(\frac{l_{1,1}}{l_{2,1}} = \frac{l_{1,2}}{l_{2,2}}\right)$, replacing by the values of (6) and solving we get a quadratic equation with two solutions:

$$f \tan 2\tau = \frac{-b \pm \sqrt{b^2 - 4ac}}{2a} \tag{7}$$

where $a = \rho_{1,1}\rho_{2,2} - \rho_{1,2}\rho_{2,1}$; $b = \rho_{1,1}\rho_{2,5} - \rho_{1,5}\rho_{2,1} + \rho_{1,4}\rho_{2,2} - \rho_{1,2}\rho_{2,4}$; and $c = \rho_{1,4}\rho_{2,5} - \rho_{1,5}\rho_{2,4}$.

From the second equality $\left(\frac{l_{1,1}}{l_{2,1}} = \frac{l_{1,3}}{l_{2,3}}\right)$ and from the third equality $\left(\frac{l_{1,2}}{l_{2,2}} = \frac{l_{1,3}}{l_{2,3}}\right)$ we get respectively:

$$f \tan 2\tau = -\frac{\rho_{1,5}\rho_{1,6}(u) - \rho_{2,5}\rho_{2,6}(v)}{\rho_{1,2}\rho_{1,6}(u) - \rho_{2,2}\rho_{2,6}(v)}; \tag{8}$$

$$f \tan 2\tau = -\frac{\rho_{1,4}\rho_{1,6}(u) - \rho_{2,4}\rho_{2,6}(v)}{\rho_{1,1}\rho_{1,6}(u) - \rho_{2,1}\rho_{2,6}(v)} \tag{9}$$

where $u = \rho_{2,1}\rho_{2,5} - \rho_{2,2}\rho_{2,4}$; and $v = \rho_{1,1}\rho_{1,5} - \rho_{1,2}\rho_{1,4}$.

Also, from ρ_1 and ρ_2 we can deduce that $\frac{\rho_{j,3}}{\rho_{j,6}} = \frac{tan2\tau}{f}$.

Then, with these equations we can compute the missing calibration parameters such as f and τ.

4 Experimental Results

In this section, we present the experimental results of applying our proposed calibration method. First, we show simulations to study the behavior of the calibration. Then we apply the method on synthetic and real images to obtain the calibration parameters and compare the results with the known ground truth. The use of simulations and synthetic images allows us to evaluate the full algorithms without doubts about the accuracy of the ground truth. The Gaussian noise is added to the image points. In the simulations, the distance from the camera to the vertex of conical mirror $Z_v = 0.1$ m, the aperture angle of the mirror $2\tau = 120°$, the size of the image is 1024×1024 px. and the focal distance is $f = 512$ px. The tests were performed with 1000 combinations of 5 points randomly selected.

Knowing the ground truth of the parameters and the position of lines in 3D space, we can determine the corresponding errors in the solution calculated for our method. The error in the calculation of the aperture angle of the mirror (τ) measured in radians and compared with the ground truth is shown in the boxplot of Fig. 4(a), where we can see how it grows slightly when we gradually increase, from 0 to 1, the standard deviation of the added Gaussian noise. Finally we show in the boxplot of Fig. 4(b) the error in pixels of calculating the focal distance (f), in which we can observe a similar behavior to the previous aperture angle, growing slightly when we increase the standard deviation of the added Gaussian noise from 0 to 1.

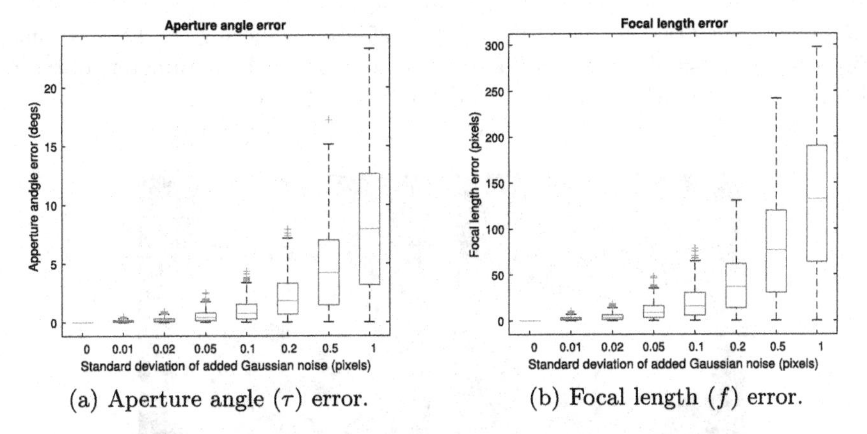

(a) Aperture angle (τ) error. (b) Focal length (f) error.

Fig. 4. Error in the calibration parameters obtained by applying Gaussian noise to 1000 cases in simulation.

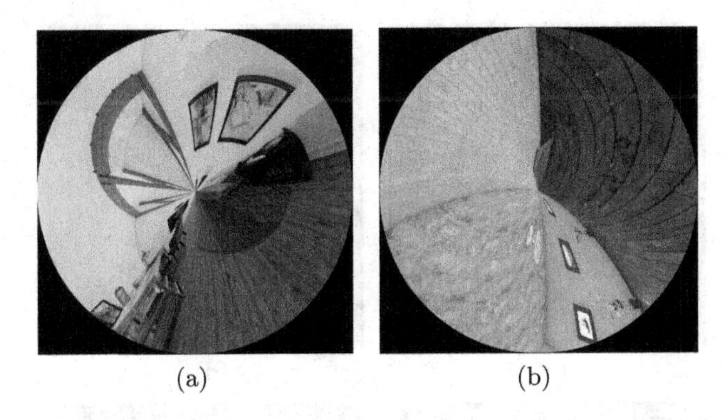

(a) (b)

Fig. 5. Synthetic images with parallel lines, used for calibration.

4.1 Real and Synthetic Images

Table 1. Calibration parameters obtained from the Synthetic Images of Fig. 5.

Image	τ	f
Figure 5(a)	64.2429	270.3770
Figure 5(b)	55.5587	256.1342

In Table 1 we show the calibration parameters obtained, both the aperture angle of the mirror (τ) and the focal length (f) of the camera, after applying our proposed method to the synthetic images of the Fig. 5. The ground truth of the first image (Fig. 5(a)) is $\tau = 60°$ and $f = 256px.$, while for the second image (Fig. 5(b)) it is $\tau = 55°$ and $f = 256px$. And comparing the parameters obtained with the ground truth, we can see that the values are very close.

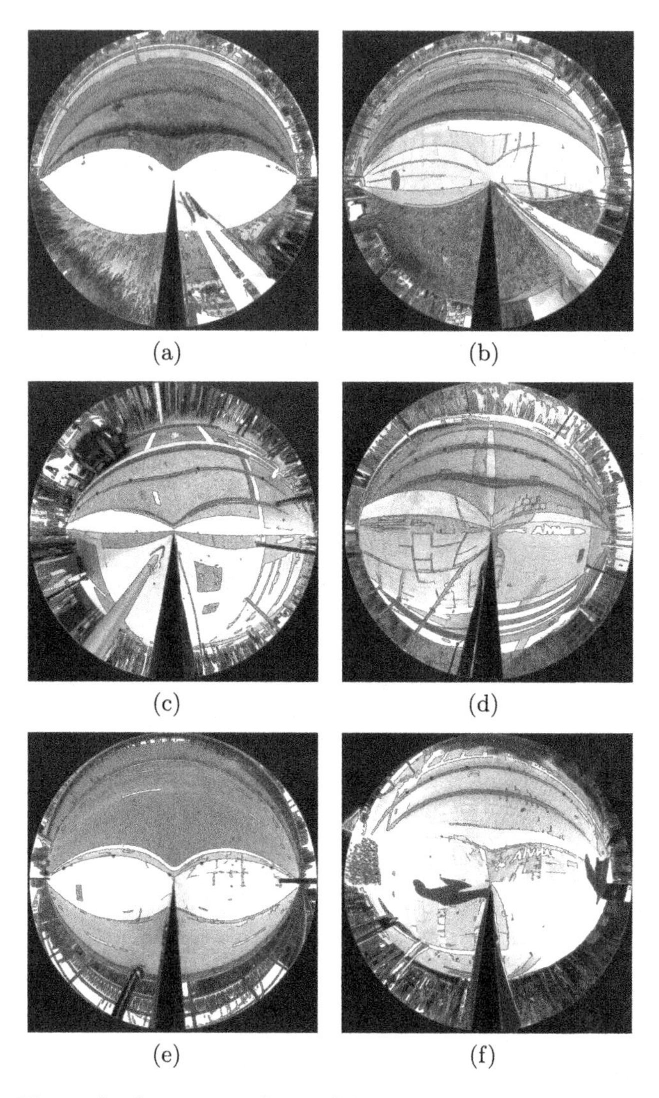

Fig. 6. Real images with parallel lines, used for calibration.

Table 2. Calibration parameters obtained from the Real Images of Fig. 6.

Image	τ	f
Figure 6(a)	67.3479	413.0064
Figure 6(b)	65.7288	718.8974
Figure 6(c)	62.2185	524.7578
Figure 6(d)	68.5955	397.7260
Figure 6(e)	60.1300	676.7471
Figure 6(f)	64.0100	458.0764

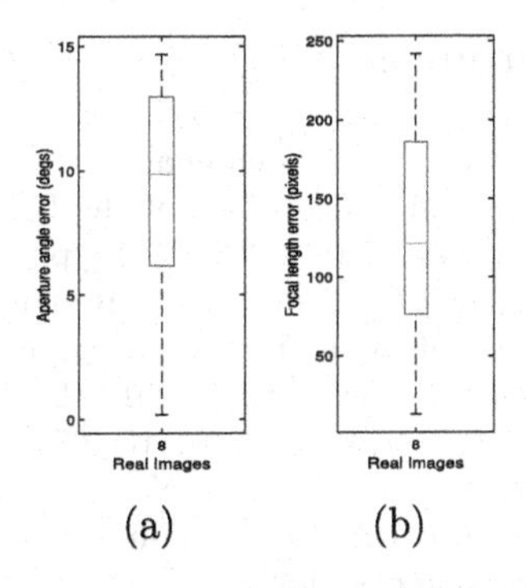

Fig. 7. Error in the aperture angle and focal length with Real Images

On the other hand, in Table 2 we show the calibration parameters obtained, both τ and f, after applying our proposed method, now to the real images that we show in Fig. 6. The ground truth of those images, from 6(a) to 6(f), is $\tau = 55°$ and $f = 512px$.. And if we compare the values of Table 2 with the ground truth, we can see that they are also close.

In addition, in Fig. 7(a) we can see a boxplot with the mean error and standard deviation of the mirror aperture angle obtained after having applied our proposed method on eight real images produced by the conical catadioptric system shown in Fig. 1(b) and that contain parallel lines. In the same way, we have obtained the mean error and the standard deviation of the focal length of the camera that we show in the boxplot of Fig. 7(b), from the eight images produced by the system in Fig. 1(b).

In this system, a fisheye camera (uEye UI-148xSE-C) is implemented, aligned to the rotation axis of a conical mirror with a total aperture angle of 110° ($\tau = 55°$).

5 Conclusion

In this paper we propose a new method for calibrating conical catadioptric cameras from the projection of two parallel lines. We have shown that working in a selected projection it is possible to decouple the focal length of the camera f and the aperture angle of the mirror τ in an analytical procedure. However, the obtained results suggest that the proposal is very sensitive to noise. As future work, we will consider new parameterizations that perform better in calibration estimation.

Acknowledgments. This work was supported by projects: TED2021-129410B-I00 (MCIN/AEI/10.13039/501100011033 and NextGenerationEU/PRTR) and JIUZ-2021-TEC-01 and the government of Panama under the IFARHU-SENACYT scholarship program for PhD studies.

References

1. Agrawal, A., Ramalingam, S.: Single image calibration of multi-axial imaging systems. In: Proceedings of the IEEE Conference on Computer Vision and Pattern Recognition (CVPR 2013), pp. 1399–1406 (2013)
2. Agrawal, A., Taguchi, Y., Ramalingam, S.: Analytical forward projection for axial non-central dioptric and catadioptric cameras. In: Daniilidis, K., Maragos, P., Paragios, N. (eds.) ECCV 2010. LNCS, vol. 6313, pp. 129–143. Springer, Heidelberg (2010). https://doi.org/10.1007/978-3-642-15558-1_10
3. Agrawal, A., Taguchi, Y., Ramalingam, S.: Beyond Alhazen's problem: analytical projection model for non-central catadioptric cameras with quadric mirrors. In: Computer Vision and Pattern Recognition (CVPR 2011), pp. 2993–3000. IEEE (2011)
4. An, G.H., Lee, S., Seo, M.W., Yun, K., Cheong, W.S., Kang, S.J.: Charuco board-based omnidirectional camera calibration method. Electronics **7**(12), 421 (2018)
5. Baker, S., Nayar, S.K.: A theory of single-viewpoint catadioptric image formation. Int. J. Comput. Vis. **35**(2), 175–196 (1999)
6. Barreto, J.P., Araujo, H.: Geometric properties of central catadioptric line images and their application in calibration. IEEE Trans. Pattern Anal. Mach. Intell. **27**(8), 1327–1333 (2005)
7. Bermudez-Cameo, J., Barreto, J.P., Lopez-Nicolas, G., Guerrero, J.J.: Minimal solution for computing pairs of lines in non-central cameras. In: Cremers, D., Reid, I., Saito, H., Yang, M.-H. (eds.) ACCV 2014. LNCS, vol. 9003, pp. 585–597. Springer, Cham (2015). https://doi.org/10.1007/978-3-319-16865-4_38
8. Bermudez-Cameo, J., Lopez-Nicolas, G., Guerrero, J.J.: Line-images in cone mirror catadioptric systems. In: 2014 22nd International Conference on Pattern Recognition, pp. 2083–2088. IEEE (2014)
9. Bermudez-Cameo, J., Lopez-Nicolas, G., Guerrero, J.J.: Fitting line projections in non-central catadioptric cameras with revolution symmetry. Comput. Vis. Image Underst. **167**, 134–152 (2018)
10. Bermudez-Cameo, J., Saurer, O., Lopez-Nicolas, G., Guerrero, J.J., Pollefeys, M.: Exploiting line metric reconstruction from non-central circular panoramas. Pattern Recogn. Lett. **94**, 30–37 (2017)
11. Chen, W., Cheng, I., Xiong, Z., Basu, A., Zhang, M.: A 2-point algorithm for 3d reconstruction of horizontal lines from a single omni-directional image. Pattern Recogn. Lett. **32**(3), 524–531 (2011)
12. Gasparini, S., Caglioti, V.: Line localization from single catadioptric images. Int. J. Comput. Vis. **94**(3), 361–374 (2011)
13. Gong, X., Lv, Y., Xu, X., Jiang, Z., Sun, Z.: High-precision calibration of omnidirectional camera using an iterative method. IEEE Access **7**, 152179–152186 (2019)
14. Griffiths, P., Harris, J.: Principles of Algebraic Geometry. John Wiley & Sons, Hoboken (2014)
15. Ieng, S.h., Benosman, R.: Geometric construction of the caustic curves for catadioptric sensors. In: 2004 International Conference on Image Processing, 2004. ICIP 2004, vol. 5, pp. 3387–3390. IEEE (2004)
16. Lee, G.H., Li, B., Pollefeys, M., Fraundorfer, F.: Minimal solutions for the multi-camera pose estimation problem. Int. J. Robot. Res. **34**(7), 837–848 (2015)
17. Lopez-Nicolas, G., Sagues, C.: Unitary torus model for conical mirror based catadioptric system. Comput. Vis. Image Underst. **126**, 67–79 (2014)

18. Marhic, B., Mouaddib, E.M., Pegard, C.: A localisation method with an omnidirectional vision sensor using projective invariant. In: Proceedings. 1998 IEEE/RSJ International Conference on Intelligent Robots and Systems, pp. 1078–1083 (1998)
19. Perdigoto, L., Araujo, H.: Reconstruction of 3d lines from a single axial catadioptric image using cross-ratio. In: Proceedings of the 21st International Conference on Pattern Recognition (ICPR2012), pp. 857–860. IEEE (2012)
20. Perdigoto, L., Araujo, H.: Calibration of mirror position and extrinsic parameters in axial non-central catadioptric systems. Comput. Vis. Image Underst. **117**(8), 909–921 (2013)
21. Perdigoto, L., Araujo, H.: Estimation of mirror shape and extrinsic parameters in axial non-central catadioptric systems. Image Vis. Comput. **54**, 45–59 (2016)
22. Pinciroli, C., Bonarini, A., Matteucci, M.: Robust detection of 3d scene horizontal and vertical lines in conical catadioptric sensors. In: Proceedings of the 6th Workshop on Omnidirectional Vision (2005)
23. Puig, L., Bermudez, J., Sturm, P., Guerrero, J.J.: Calibration of omnidirectional cameras in practice: a comparison of methods. Comput. Vis. Image Underst. **116**(1), 120–137 (2012)
24. Semple, J.G., Kneebone, G.T.: Algebraic Projective Geometry. Oxford University Press, Oxford (1998)
25. Swaminathan, R., Grossberg, M.D., Nayar, S.K.: Non-single viewpoint catadioptric cameras: geometry and analysis. Int. J. Comput. Vis. **66**(3), 211–229 (2006)
26. Teller, S., Hohmeyer, M.: Determining the lines through four lines. J. Graph. Tools **4**(3), 11–22 (1999)
27. Xiang, Z., Dai, X., Gong, X.: Noncentral catadioptric camera calibration using a generalized unified model. Opt. Lett. **38**(9), 1367–1369 (2013)

S^2-LOR: Supervised Stream Learning for Object Recognition

César D. Parga[1](✉), Gabriel Vilariño[2], Xosé M. Pardo[1], and Carlos V. Regueiro[3]

[1] Centro Singular de Investigación en Tecnoloxías Intelixentes (CiTIUS), Universidade de Santiago de Compostela, Santiago de Compostela, Spain
{cesardiaz.parga,xose.pardo}@usc.es

[2] Universidade de Santiago de Compostela, Santiago de Compostela, Spain
gabriel.vilarino@rai.usc.es

[3] Universidade da Coruña, CITIC, Computer Architecture Group, A Coruña, Spain
carlos.vazquez.regueiro@udc.es

Abstract. In a stream learning scenario, where new data arrives at a slow pace, it is crucial to leverage new knowledge at the same rate without losing prior knowledge, and without assuming data stationarity. This scenario presents a significant challenge for incremental learning, particularly for tasks such as object recognition in video streams. In this paper, a novel approach is proposed that uses a set of weak classifiers that evolves into ensembles to enhance the generalization power of the system, as new video subsequences of the same instances are presented. We evaluate the efficiency of our approach and compare with state-of-the-art methods using a benchmark dataset. The code is available at https://github.com/vilaB/object_recognition.

Keywords: Stream Learning · Ensemble learning · Incremental Learning

1 Introduction

Humans have the ability to continually learn and adapt to new experiences throughout their lives without the need to accumulate large amounts of examples and counterexamples. They learn from non-stationary streams of samples, presented one at a time, and can immediately use the new knowledge to enhance their understanding of visual scenes [6]. This learning method is called stream learning, where model training and testing are entwined operations [8].

Deep neural networks are usually trained from scratch using large batches of data, and when new batches are available, they can be retrained either from scratch, or by passing the new batches through additional epochs of learning. However, if data is non-stationary, gradient descent updates cause the networks

C. D. Parga and G. Vilariño—These authors contributed equally to this work.

A. Pertusa et al. (Eds.): IbPRIA 2023, LNCS 14062, pp. 300–311, 2023.
https://doi.org/10.1007/978-3-031-36616-1_24

to forget previously knowledge [13]. This means that, for instance, the adaptation to perform the same task under new environmental conditions could be at the expense of degrading task performance under previously learned conditions.

Most incremental object recognition systems are not suitable for stream learning [18,22]. Instead of updating in real-time with the current sample, these systems typically update using large batches of samples, which slows down the updating process and is not practical for use in devices with limited resources [1].

In this work, we tackle the problem of instance object recognition in a stream learning scenario. Our approach is able to process data in small batches, as short as a few frames of video, as they arrive. We use ensembles of weak classifiers in order to easy adaptation based on few examples. Besides, our approach can deal with memory size constrains, as it does not rely on keeping a large reservoir of video frames to preserve the generalization power and avoid the catastrophic forgetting. It tries to organize global knowledge into individual pieces, or facets, that can be added or removed without any impact on the rest of the facets. In applications with scarce resources, it is useful to identify redundancy to make room for new insights and make feasible to identify and eliminate errors in the incremental learning process.

This approach scales linearly with the number of entities to be recognized.

The rest of the paper is organized as follows: in Sect. 2, we survey the related work; then, in Sect. 3 our method is described, and Sects. 4 and 5 present the results and main conclusions.

2 Related Work

Incremental learning is a paradigm that deals with the adaptation of models to new information over time. This is a challenging task, particularly in the case of stream data, where information is received at low pace and in small amounts. Most of the approaches to incremental learning are based on batches of samples [23,25,26], which have to be gathered before new epochs of training can be performed. However, in stream learning settings, where knowledge has to be updated as soon as possible, only mini-batches of new samples, when not single samples, must be used in each updating step. iCarl [21] is an example of the use of mini-batches, where new information is provided progressively at the same time as the model is incrementally updated.

Typically, adaptive models have difficulties in maintaining previous knowledge, when dealing with non-stationary data distributions. Some techniques can be used to preserve the knowledge of the previous tasks while learning new ones. EwC [13] makes it possible to regularize the learning of the model by penalizing changes in the most important weights of a neural network. GEM [17] is based on a replay mechanism, since it stores a subset of learning samples, called episodic memory, which it uses to revisit that information during training and find the optimal updates that maintain the balance between new and old tasks. LwF [14] uses a slightly different approach based on distillation mechanisms to transfer knowledge from the old model to the new one, allowing the system to consolidate the acquired knowledge into new streams over time.

Streaming Learning [11,12,20] adds a new constraint to the training process, namely that data samples can only be observed once [9]. It comprises two phases: a first phase where the system is initialized and the operation phase where the data stream is received. In a supervised context, the operation mixes query samples for prediction with labeled samples for training. Under this paradigm, it also becomes key to preserve prior knowledge while training with incoming samples. For this purpose, solutions based on replaying previous relevant information stand out above all. GDM [19] uses two types of memory, a short-term memory, where the information about the new stream received is stored, and a long-term memory where the knowledge of the previous data streams is stored, combining both in the training of the system. RODEO [1] uses a single memory buffer, but makes use of compressed representations in an attempt to mimic the sample repetition mechanism performed by the human brain. REMIND [10] use the same methodology as the previous one, incorporating a samples indexing mechanism inspired on biological replay and reducing the instance number and the buffer size used during the replay.

Nevertheless, these techniques have several limitations when they are applied in a streaming context. Regularization schemes are weak in the face of significant changes in context because most of these techniques assume static data distributions [19]. Replay mechanisms can be effective in certain scenarios, but may not work well in certain cases of continuous learning. Specifically, the main limitation lies in the additional computation and storage required for keeping the selection of representative samples for memory reply. Solutions to this involve limiting the size of the buffers, but this decreases the ability of the set of exemplars to represent the original distribution of the data. In addition, the use of raw samples may involve privacy issues [4].

Ensemble methods have good properties to tackle the problem of stream learning, as they can be combined with strategies to deal with non-stationary data distributions, and selective addition and removal of weak classifiers [2,3,16, 24]. AAEDS [7] makes use of a great diversity of base classifiers, which include Naive Bayes, k-Nearest Neighbors, Decision Trees, and SVMs, adjusting the size and diversity of the ensemble according to the changes in the data streams.

In this paper, we will focus on solving this streaming learning problem by using a dynamic ensemble approach. This approach could help us to 1) provide more robust and specialized evaluations of samples, and 2) facilitate the incorporation of new knowledge as streams are received. Also, it allows us to deal with the problem of representativeness and the number of samples used, showed by the replay techniques. This works aims to reduce the amount of stored data by providing good performance at a low memory cost.

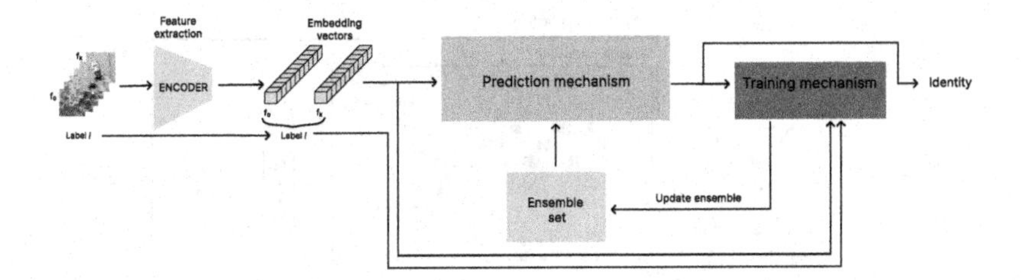

Fig. 1. Sketch of the system, depicting high-level components.

3 Supervised Stream Learning for Object Recognition(S^2-LOR)

In this section, we present our approach as an ensemble-based system that allows learning over time based from non-stationary data streams. This system combines learning operations with query/prediction operations. If the samples received are not labelled, the system only predicts their category. In the case of labeled samples, the system use them to learn.

A scheme of the system is depicted in Fig. 1, where four different parts can be distinguished. When a video sequence arrives, an **encoder** generates their embeddings. The encoder is made up of the hidden layers of the ResNetv2-50, pretrained on Imagenet database [5]. Then, the **prediction mechanism** is feed with these embedding vectors to determine the category the video subsequence belongs to. The classification is provided by a **set of ensembles**, which members are exemplar SVM's. This processing chain correspond to the operation/testing mode.

When video subsequences are provided with category/identity labels, the system enters learning mode. In this case, the output of the prediction mechanism is used to decide whether a training stage is initiated or not. The **training mechanism** is in charge of incorporating new knowledge by dynamically updating the set of ensembles.

3.1 Initialization Setup

The frame embeddings in the first video subsequence of each object are used to train the initial member of each ensemble. As new subsequences are received in this stream setting, ensembles (one per object) are updated with new weak classifiers. When the maximum number of classifiers in an ensemble is reached, the addition of a new classifier implies the replacement of an old one. However, as we will show later, our findings demonstrate that an effective selection strategy for adding classifiers to an ensemble can significantly outperform the alternative approach of adding multiple classifiers and then requiring pruning steps, while requiring fewer computational resources.

From those initial subsequences, we will also grab a subset of frames from each object. This will become a pool of negative samples from we will later take frames to be able to feed the new members with positive and negative samples.

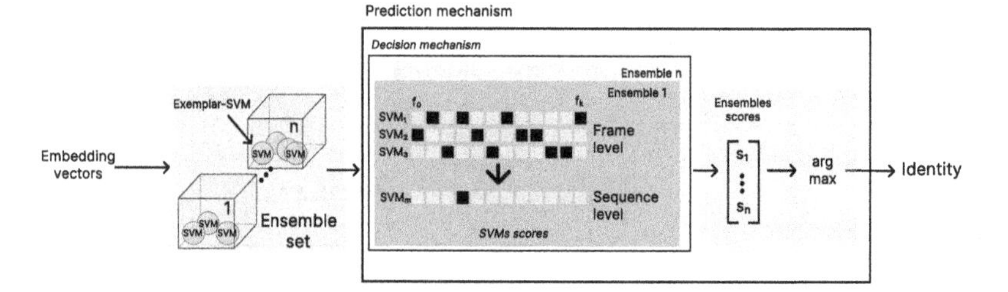

Fig. 2. Embeddings generated by a feature encoder are presented to the Ensembles set. Then, the SVMs scores are given to the decision mechanism, which evaluate the score for every sequence frame (f_i), given at last one representative score per each ensemble. Finally, the system selects the best score to assign an identity to the full sequence.

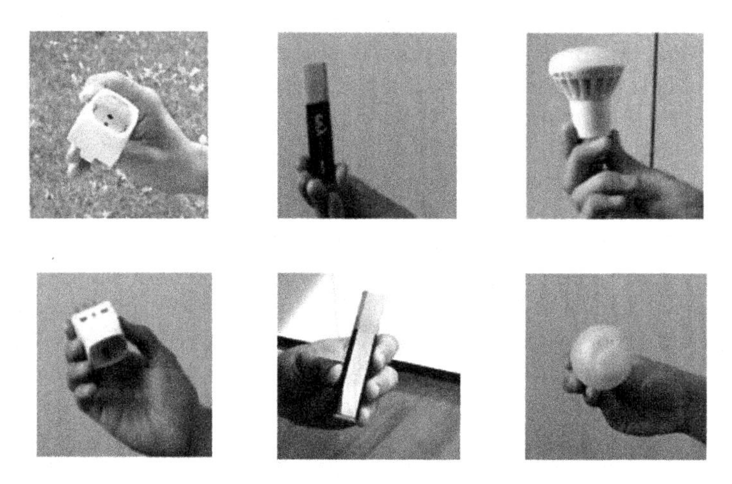

Fig. 3. Camera viewpoints on different backgrounds and objects.

3.2 Operation Mode

After the initial setup, the system will start receiving new sequences to be processed. For every sequence, the system will try to classify the sequence (Fig. 2). The classification occurs in two levels: a) frame level and b) sequence level.

It is important to note that objects' appearance in frames can change dramatically due to a number of factors (Fig. 3). Some of them are the conditions of video recording, the relative pose of object respect to the camera, the brightness and contrast of the scene or the influence of the background, among others. Ideally, weak classifiers, members of the same ensemble, would give stronger responses to specific poses of an object. This would imply that only the activation of a few members of an ensemble for a given frame could be enough to detect the object, this could be represented by a small percentile of all classifiers scores.

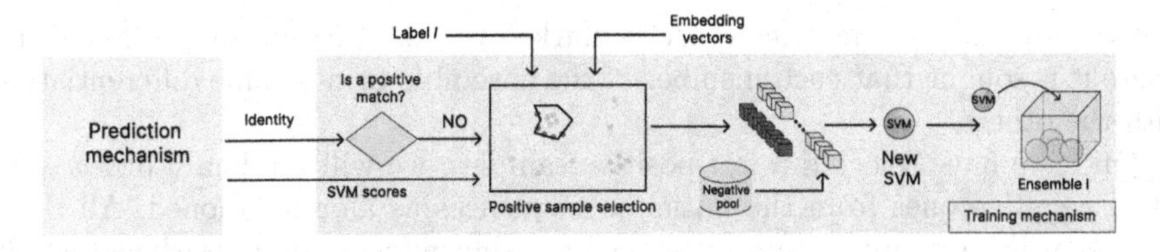

Fig. 4. Once the sequence has been assessed, the learning process begins. First, it is evaluated whether the selected identity corresponds to the true label (l). If it does not, the positive and negative learning samples are selected and the new SVM classifier is generated. Finally, it is added to the true entity ensemble l.

Decision at sequence level is made from the set of scores at frame level. A simple and effective method to exploit the temporal cohesion is to use the median. This technique is useful because it allows balancing situations where high ensemble scores are not obtained for all the frames of the sequence. The median allows us to filter out outliers and obtain a representative score for the sequence. The decision criterion is based on the existence of a few members of the ensemble who recognize the object during, at least, half of the sequence.

Once a score is obtained for a given sequence, the system will evaluate whether it considers the given sequence as a positive match or not. If the system already scores it as a positive, it will not incorporate any new classifier, since the system is able to recognize the given sequence. As in humans, once you already have some knowledge, keep repeating it usually does not improve the capacity to recognize/perform something, since you already have the knowledge.

In the case the current response was considered as a misclassification, the system would add a new member to the ensemble to enhance the recognition ability, regarding its associate object identity (Fig. 4). To join this new classifier to the ensemble, the system will start by selecting a few frames of the subsequence to perform the training. We have seen that the selection of these frames heavily affects the performance of the system. Initially, we used the worst scoring frames from the given sequence as in [16]. However, this criterion has proven to be ineffective to recognize objects in a real-world scenario. This is because selecting the worst frames from a sequence often results in frames that lack any discernible features for object recognition. Instead, using those frames that are hard positives at a given point in time works a lot better.

This approach enables a more realistic form of learning, acknowledging the difficulty of assimilating completely novel information. Specifically, the aim is to extract insights from features that are situated near the boundary of positive recognition (hard positives), and thereby enhance learning in a more targeted fashion. The aim of this approach is to generate specific ensembles that do not combine very different views. This composition makes it possible to maintain a set of weak SVM classifiers. The main objective is that the evaluation of all predictions serves to determine whether the object belongs to a certain class, i.e., to compose a knowledge boundary. This way of operating allows a more flexible update policy to determine when a classifier is no longer relevant and should be replaced by another one. Relevance is marked by the diversity of the classifiers, where it is sought that each member of the ensemble brings some differentiation with the others.

Once we have selected a few positive samples, we will randomly pick a subset of negative ones from the initial subsequences as aforementioned. All these samples will be used to train a new member that will be added to the ensemble the positive frames belong to. It is important to note that only SVM classifiers are trained. The embedding extraction model remains frozen.

4 Experiments

In this section, we will first describe the benchmark used to assess the abilities of our approach to learn and adapt to non-stationary streams of data, and then we will compare its results with other state-of-the-art approaches.

4.1 The Benchmark: CORe50

The CORe50 (Continual Object Recognition [15]) is a benchmark developed for testing Continual Learning systems against a set of test well-defined to compare different approaches to the same problem. The CORe50 contains both: 1) a dataset of frames where objects can be seen, and 2) a set of rules of how to process that dataset of frames to measure the performance of the solution.

The dataset can be used in different ways. The harder one is to identify the objects at an instance level. The agent will have to know to which object instance belongs every query sequence. The other approach could be to recognize the category the object belongs to.

The dataset is composed by **50 objects from 5 different categories**. For each one of the objects, there are **11 sequences**: 8 indoors and 3 outdoors. Each of the sequences is about 300 frames long, and includes both an RGB image and a depth image. The dataset provides not only the complete frames, but also cropped objects of interest.

The benchmark is split into three scenes for testing and eight for training. All the sequences were captured under different environment conditions.

4.2 Benchmark Configuration for S^2-LOR

How we Use the Benchmark? We will apply the measures used in the benchmark to compare our approach with others. In accordance with the benchmark specifications, we will perform ten runs using the proposed protocols.

We will Split Every Sequence in Equally Distributed Sub-sequences. We will use the same split ratio (for example, divide each sequence by 3) for the testing and learning scenarios. Different ratios will result in different lengths of the sequences presented. The system needs sequences of an intermediate size: 1) if the sequence is too short, for example one frame, the system won't be able to

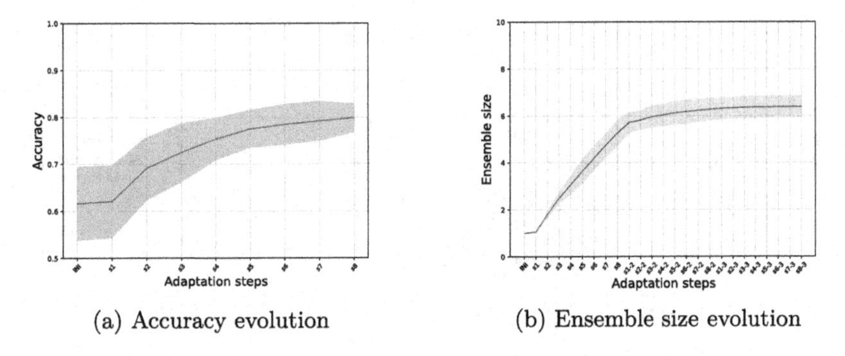

(a) Accuracy evolution

(b) Ensemble size evolution

Fig. 5. Performance results (averaged on 10 runs) for S^2-LOR system. Colored areas represent the standard deviation of each curve.

exploit the temporal cohesion of the sequences, but 2) if the sequence is too long, it will need to wait too much time, and leading to a system that operates in batch mode and awaits accumulation of data. We want a near real-time performance, so we want to get a middle point between these both extremes. **We split the original sequences (300 frames) into 15 subsequences**, which means **a size of 20 frames per subsequence**. As stated on the benchmarks' website[1], the recordings have been made at 20 fps, which means that **we are using just a second of video to query the system.**

Experiments are organized in accordance with the rules described in CORe50 benchmark. We process the 8 training scenario in the exact same order as indicated there. For every training scenario, we process the 50 objects one by one. We divided the original sequences into 15 subsequences of equal size (around 20 frames per sequence, or 1 s of footage). After ending the feeding of the 15 subsequences of every object on one scenario, we run a test step, where the system is queried using the 3 testing scenarios. We provide to the system with those scenes (also divided in 15 subsequences) and measure the performance. During this step, the system is not allowed to gain any knowledge.

To measure the model's performance, we use Accuracy, which is the metric proposed by the benchmark. Since the sample presentation order can affect the final result, the benchmark proposes to compute the average accuracy over 10 runs with sequences randomly shuffled.

4.3 Experiment Results over CORE50 Dataset

In Fig. 5a, it can be observed how the accuracy grows continuously during the whole operation phase. The system keeps on improving its performance over time, which means the agent is not suffering from catastrophic forgetting. Besides, Fig. 5b, illustrates the fact that when the system is queried with samples very similar to those already seen, it does not learn any more and the size of the

[1] https://vlomonaco.github.io/core50/.

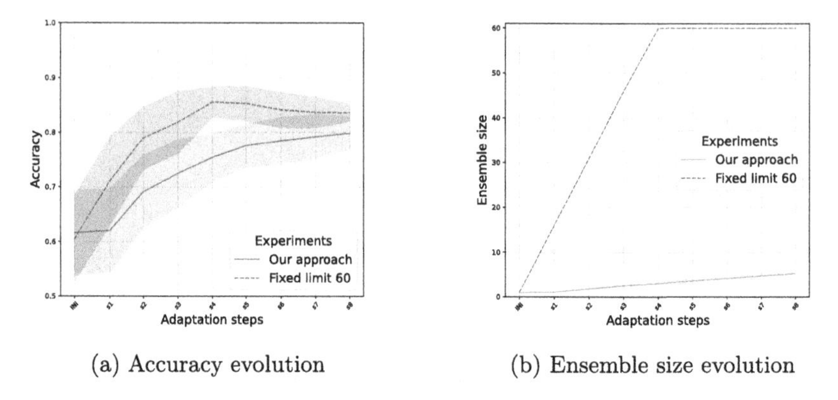

(a) Accuracy evolution

(b) Ensemble size evolution

Fig. 6. Performance about learning mechanism (averaged on 10 runs) for S^2-LOR system. Colored areas represent the standard deviation of each curve.

ensembles stabilizes. Indeed, the final mean size of the ensembles is low, **about six members per object category**. As Fig. 6 illustrates, it is possible to reach high performance even with a small set of classifiers per ensemble. We impose the criterion of adding a new member to the corresponding ensemble only when the new classifier enhances ensemble performance, i.e., it recognizes the object only after the addition of the new classifier. This **prevents the model from the inclusion of ensemble members that are redundant**. Previously, we had attempted to periodically identify redundant members. However, we found this to be impractical due to the following reasons: (1) revisiting previous decisions is more cumbersome than directly avoiding the addition of such members, as it requires re-examining all members on each iteration rather than just the new ones, and (2) conducting an overlap assessment prior to the addition of a new member is more efficient than doing so later on, as it reduces memory requirements. Furthermore, as Fig. 6a shows, with sizes of ensembles up 10 times the maximum value reached applying the aforementioned criterion (adding classifiers only when necessary) (Fig. 6b), the results tend to be on a par.

As we can see on Table 1, S^2-LOR achieves an overall accuracy of 80.41%, outperforming all the others solutions on the conditions set out in [15], which makes our approach the best current solution for the benchmark, without the need to use replay techniques.

On a second iteration (repeated data), our system performs pretty much the same as on the end of the first one. On the Fig. 5b we can see that the system keeps almost the same mean ensemble size from the end of the first iteration. This shows that the system was capable of add most of the useful knowledge on the first iteration and how new iterations simply add some small changes, but mostly the system stays the same, since it can not get useful knowledge from something it already saw. Furthermore, it also shows that our system can quickly adapt to new conditions, not needing several iterations over the same information to get all the useful knowledge. However, we were not able to get

Table 1. Benchmark results for comparison purposes

Strategy	Accuracy	
	Mean (%)	Std. Dev (%)
REMIND (with replay) [10]	88.08	1.33
GDM (with replay) [19]	87.94	1.72
Ours	**80.41**	**1.89**
GDM [19]	74.87	2.54
Cumulative [15]	65.15	0.66
LwF [14]	59.42	2.71
EWC [13]	57.40	3.80
Naive [15]	54.69	6.18

the same levels of accuracy as GDM or REMIND with replay, which means that we are not able to extract as much useful information as their solutions.

In contrast to GDM and REMIND, the storage of prior information is done in a simpler way. In this case, the only information stored is the set of initial samples that are used to build the set of negatives used in the creation of the different classifiers. In the case of GDM, the architecture employed makes use of a complex mechanism of reactivation of the previously stored neural patterns, adding an overhead to the learning process. In REMIND, it is shown how the results are obtained by buffering 4465 samples or images. Our approach, for the CORE50 case, simply needs to store 500 samples or embedding vectors (compressed representations), approximately 9 times less information than in the previous solution, which are used directly in the learning process without the need to include an additional phase to the process.

5 Conclusions

We have developed a system for object recognition in video sequences that is able to adapt to dynamic conditions. It extracts the relevant knowledge from the sequences and uses a filtering mechanism to determine the usefulness of the new data, minimizing the necessary updates and, consequently, the resources used during sample processing. This system has been able to outperform all current solutions without replay on the specific conditions set out in Core50 benchmark. While on a two iterations process, our approach does not beat GDM. This shows that our system is able to get all the possible knowledge it can extract quicker than GDM, but is not capable of achieving the same level of accuracy by re-iterate over the dataset. It also demonstrates that it is able to achieve good performance, storing 9 times less information than REMIND.

We have proven that ensemble methodologies are able to beat other solutions by making easier to understand what knowledge is kept by each part of the system, which allow to easily remove/add new knowledge without affecting to the rest of the already settled knowledge. Lastly, we also developed a way to effectively add knowledge to the system, which avoids waste of computational resources while helps the system to only learn useful knowledge, eliminating or decreasing a lot the need for a purge method where we re-evaluate the current ensembles and remove the duplicate knowledge.

Acknowledgments. This work has received financial support from the Spanish government (project PID2020-119367RB-I00); from the Xunta de Galicia, Consellaría de Cultura, Educación e Ordenación Universitaria (accreditations 2019-2022 ED431G-2019/04 and ED431G 2019/01, and reference competitive groups 2021-2024 ED431C 2021/48 and ED431C 2021/30), and from the European Regional Development Fund (ERDF/FEDER).

References

1. Acharya, M., Hayes, T.L., Kanan, C.: Rodeo: replay for online object detection. In: The British Machine Vision Conference (2020)
2. Bian, Y., Chen, H.: When does diversity help generalization in classification ensembles? IEEE Trans. Cybern. **52**(9), 9059–9075 (2022)
3. Coop, R., Mishtal, A., Arel, I.: Ensemble learning in fixed expansion layer networks for mitigating catastrophic forgetting. IEEE Trans. Neural Netw. Learn. Syst. **24**(10), 1623–1634 (2013)
4. De Lange, M., et al.: A continual learning survey: defying forgetting in classification tasks. IEEE Trans. Pattern Anal. Mach. Intell. **44**(7), 3366–3385 (2022)
5. Deng, J., Dong, W., Socher, R., Li, L.J., Li, K., Fei-Fei, L.: Imagenet: a large-scale hierarchical image database. In: 2009 IEEE Conference on Computer Vision and Pattern Recognition, pp. 248–255. IEEE (2009)
6. Ditzler, G., Roveri, M., Alippi, C., Polikar, R.: Learning in nonstationary environments: a survey. IEEE Comput. Intell. Mag. **10**(4), 12–25 (2015)
7. Gomes, H.M., Read, J., Bifet, A.: Streaming random patches for evolving data stream classification. In: 2019 IEEE International Conference on Data Mining (ICDM), pp. 240–249 (2019)
8. Gomes, H.M., Read, J., Bifet, A., Barddal, J.P., Gama, J.A.: Machine learning for streaming data: state of the art, challenges, and opportunities. SIGKDD Explor. Newsl. **21**(2), 6–22 (2019)
9. Hayes, T.L., Cahill, N.D., Kanan, C.: Memory efficient experience replay for streaming learning. In: 2019 International Conference on Robotics and Automation (ICRA). pp. 9769–9776 (2019)
10. Hayes, T.L., Kafle, K., Shrestha, R., Acharya, M., Kanan, C.: REMIND your neural network to prevent catastrophic forgetting. In: Vedaldi, A., Bischof, H., Brox, T., Frahm, J.-M. (eds.) ECCV 2020. LNCS, vol. 12353, pp. 466–483. Springer, Cham (2020). https://doi.org/10.1007/978-3-030-58598-3_28
11. Hayes, T.L., Kanan, C.: Lifelong machine learning with deep streaming linear discriminant analysis. In: 2020 IEEE/CVF Conference on Computer Vision and Pattern Recognition Workshops (CVPRW), pp. 887–896 (2020)
12. Hulten, G., Spencer, L., Domingos, P.: Mining time-changing data streams. In: ACM SIGKDD International Conference on Knowledge Discovery and Data Mining, pp. 97–106. Association for Computing Machinery, New York, NY, USA (2001)

13. Kirkpatrick, J., et al.: Overcoming catastrophic forgetting in neural networks. Proc. Natl. Acad. Sci. **114**(13), 3521–3526 (2017)
14. Li, Z., Hoiem, D.: Learning without forgetting. IEEE Trans. Pattern Anal. Mach. Intell. **40**(12), 2935–2947 (2018)
15. Lomonaco, V., Maltoni, D.: Core50: a new dataset and benchmark for continuous object recognition. In: Levine, S., Vanhoucke, V., Goldberg, K. (eds.) Proceedings of the 1st Annual Conference on Robot Learning. Proceedings of Machine Learning Research, vol. 78, pp. 17–26. PMLR, 13–15 November 2017
16. Lopez-Lopez, E., Pardo, X.M., Regueiro, C.V.: Incremental learning from low-labelled stream data in open-set video face recognition. Pattern Recogn. **131**, 108885 (2022)
17. Lopez-Paz, D., Ranzato, M.: Gradient episodic memory for continual learning. Adv. Neural Inf. Process. Syst. **30**, 6470–6479 (2017)
18. López Lobo, J., Laña, I., Del Ser, J., Bilbao, N., Kasabov, N.: Evolving spiking neural networks for online learning over drifting data streams. Neural Netw. **108**, 1–19 (2018)
19. Parisi, G.I., Tani, J., Weber, C., Wermter, S.: Lifelong learning of spatiotemporal representations with dual-memory recurrent self-organization. Front. Neurorobotics **12** (2018)
20. Read, J., Bifet, A., Holmes, G., Pfahringer, B.: Scalable and efficient multi-label classification for evolving data streams. Mach. Learn. **88**, 243–272 (2012)
21. Rebuffi, S.A., Kolesnikov, A., Sperl, G., Lampert, C.H.: iCaRL: incremental classifier and representation learning. In: IEEE Conference on Computer Vision and Pattern Recognition (CVPR) (2017)
22. Sahoo, D., Pham, Q., Lu, J., Hoi, S.C.H.: Online deep learning: learning deep neural networks on the fly. In: Proceedings of the 27th International Joint Conference on Artificial Intelligence, pp. 2660–2666. IJCAI'18, AAAI Press (2018)
23. Wang, M., Deng, W.: Mitigating bias in face recognition using skewness-aware reinforcement learning. In: 2020 IEEE/CVF Conference on Computer Vision and Pattern Recognition (CVPR), pp. 9319–9328 (2020)
24. Wankhade, K.K., Jondhale, K.C., Dongre, S.S.: A clustering and ensemble based classifier for data stream classification. Appl. Soft Comput. **102**, 107076 (2021)
25. Yan, J., Jin, D., Lee, C.W., Liu, P.: A comparative study of off-line deep learning based network intrusion detection. In: 2018 Tenth International Conference on Ubiquitous and Future Networks (ICUFN), pp. 299–304 (2018)
26. Zhang, K., Zhang, Z., Li, Z., Qiao, Y.: Joint face detection and alignment using multitask cascaded convolutional networks. IEEE Signal Process. Lett. **23**(10), 1499–1503 (2016)

Evaluation of Regularization Techniques for Transformers-Based Models

Hugo S. Oliveira[1,2(✉)], Pedro P. Ribeiro[1,2], and Helder P. Oliveira[1,2]

[1] Faculty of Sciences, Computer Science Department (DCC), University of Porto, Porto, Portugal
hugo.soares@fe.up.pt
[2] Instituto de Engenharia de Sistemas e Computadores, INESC TEC, Porto, Portugal

Abstract. In recent years the great success of transformers-based models initially employed in Natural Language (NLP) tasks has led to the development of several transformers variations to be employed in a wide range of domains, such as vision. With the correct amount of training data and proper training, transformers can perform excellently compared to the Convolution Neural Networks (CNN) counterpart in the vision tasks. However, the main drawback of transformers concerns the know memory requirements that often exceed the available training platform, growing in a quadratic form regarding the input image size, and a great tendency to overfit.

Several works address the memory problem by relaxing the model architecture versions, but mainly with reduced prediction capabilities. In this work, we evaluate Random Patch erasing among the image patch level of the transformer model as a regularization technique to reduce overfitting while at the same time alleviating training time. The evaluated regularization technique achieves competitive results on several image classification medical datasets. The evaluated Visual Transformers (ViT) models allow to be trained in a single GPU, reaching similar results to CNN counterparts, obtaining an accuracy 91.2%, 79.2% in two competitive image datasets, and reducing the training time on average by 22% on the transformers models.

Keywords: Transformers · Regularization · Vision Transformers

1 Introduction

Transformers [1] are state-of-the-art models in NLP tasks [2,3] and have revolutionized the way sequence modeling is performed.

The model architecture consists of an encoder that maps an input sequence of symbol representations to a sequence of continuous representations independent of their position and a decoder that generates an output sequence of symbols one at a time.

Recently, transformers have been adapted to be used in vision problems by employing ViT [4,5], where the input image is decomposed into smaller patches

A. Pertusa et al. (Eds.): IbPRIA 2023, LNCS 14062, pp. 312–319, 2023.
https://doi.org/10.1007/978-3-031-36616-1_25

and combined with a learnable positional encoding, represented in Fig. 1 for each batch used as input by the transformer. However, for transformers to be effective, they require large amounts of training data and GPU resources complemented with data augmentation techniques.

Often works that employ transformers explore knowledge distillation in a hybrid combination with CNN (e.g., Retina classification [6]) models, leveraged by transfer learning allows them to achieve interesting results. However, directly training transformers on image inputs often poses several difficulties.

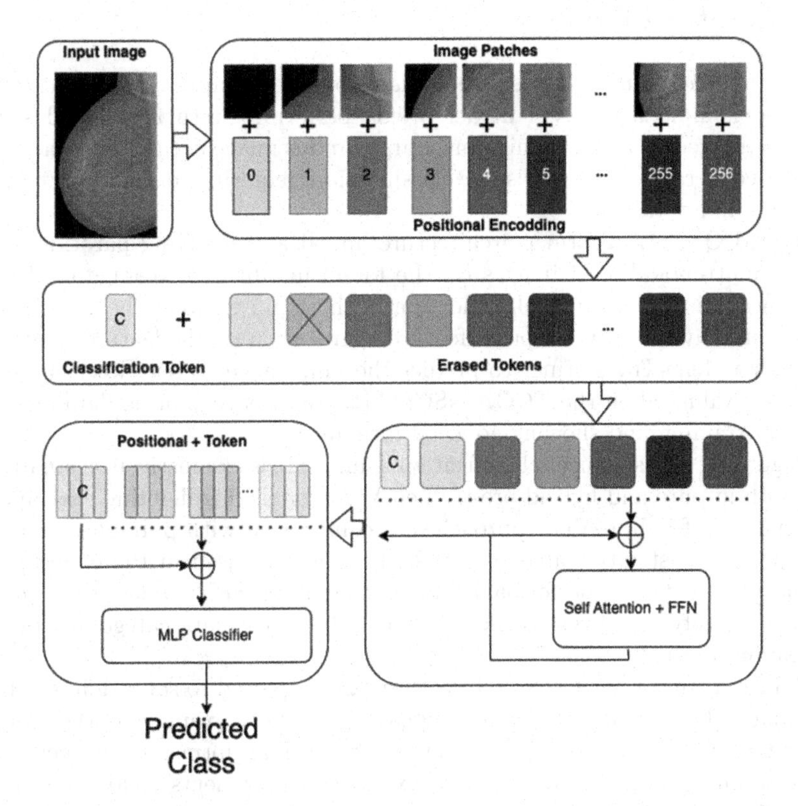

Fig. 1. Random Patch Erasing for Transformers + Classification Token in INbreast Dataset

One of the main difficulties associated with training Transformers with larges medical images is related to the space patch representation, which encompasses a large image space often with a very sparse representation. Decomposing larger images or alternative features into patches leads to long sparse sequence representations. Training sparse representations requires lower learning rates while maintaining redundant information in memory, harming the training process. This sparsity raises problems, namely by a large set of activation that biases the learning process, difficulties convergence and generalization, requiring proper regularization at the input level.

In this work, we evaluate the regularization techniques' effectiveness by randomly erasing images patch components while disentangling the transformer's positional encoding [7,8], improving the generalization of the trained models while effectively reducing memory requirements. We also investigate the reduction of training complexity while attaining the model's overall performance in the specified classification tasks with detailed taxonomy with CNN models counterparts.

2 Literature Review

The use of Transformers in medical domains often encompassed different strategies. For example, a MIL-ViT model [5,9] to performing retina disease classification is leveraged by a pre-training on a large fundus image dataset and fine-tuned in the specific task of retinal disease classification, requiring considerable training resources and datasets.

Hybrid CNN transformer architecture aimed at classifying parotid tumors among multi-modal MRI images can be found in [10]. A two-stage model [11] to perform glioma sub-type classification in the brain explores the use of transformers in a hybrid way to sparse feature aggregation on the feature space. [12] proposes a Gene-Transformer to predict the lung cancer categories. The model has been evaluated on the TCGA-NSCLC [13] dataset, exploring data augmentation in training and showing good performance.

Regarding breast cancer classification using ViT, is exhaustively evaluated by [13], both in pure and hybrid pre-trained ViT models, highlighting the superior performance of hybrid ViT approaches in comparison with pure ones. The two datasets of breast ultrasound datasets, [14] and [15] used in the experiments, evidence the superior performance when compared with models that employ CNN to classify the given medical images into three main categories, benign, malign and normal.

Hybrid Transformer-CNN architectures are employed to solve different medical image classification tasks in various organs. One example is the work of [12], which proposes a Gene classifier based on Transformer to correctly predict the lung cancer categories, with extensive experiments conducted on the TCGA-NSCLC [16] datasets, surpassing the performance of the CNN available baselines. [17] focuses on diagnosing gastric cancer in the stomach by employing a multi-scale GasHis-Transformer. Hybrid models are proposed by [18] to correctly diagnose acute lymphocytic leukemia using a Siamese symmetric cross-entropy loss function.

In the following section, we present the methodology to be employed, clearly explaining the research model design and methods employed in the study.

3 Methodology

The use of ViT in a pure form often leads to extensive image patch sequences that increase the complexity of the transformer models, also aggravated by the fact that many medical images have feature regions aggregated in a particular region. In contrast, the remaining parts of the image often correspond to the background, leading to a sparse representation. We evaluate the performance of Random Patch erasing in transformer models applied to medical image domains using an Bidirectional Encoder representation of Transformers (BERT) like architecture [19,20].

Greatly inspired by the Dropout technique [21], the Random Patch erasing the main objective is to enforce the transformer model to be trained in incomplete input sequences to perform image classification. The patch image generation starts by sampling with overlap from the input image (Fig. 1). During training, the random parts of the sequence patch image are also randomly dropped, effectively reducing the training sequence information and leading to a regularizing effect of the Transformer model. The original input sequence is preserved in the transformer model's original form during inference. We ensure in training that subsequential patches don't get erased, avoiding complete erasing or contiguous information.

Multi-head attention layers [1] receive the computed distance between each pair of positions in the input sequence as attention matrix A, the complexity grows in a quadratic way $O(n^2)$, regarding the n the input sequence size.

Longer complete input sequences using a high degree of patch overlapping in ViT, combined with high-resolution images, greatly increase the memory requirements, challenging the training and deployment of Transformers for image classification. Given an input b samples with the dimension $c \times e$, with b being the batch size and n defining the sequence length, e the embedding size, with each attention layer projecting each input sample into the h query Q, key K and value V matrices, with h being the number of attention heads [1].

4 Setup and Experiments

We evaluated several CNN model variations (Inception, Resnet, DenseNet) and Transformers for comparison purposes. We trained them individually on the following datasets ISIC 2019 [22] for skin cancer and APTOS 2019 [23] with images from diabetic retinopathy, with characteristics summarized in Table 1. For CNN models architecture, we set the dropout rate to 20% among the fully connected layer.

Table 1. Datasets Summary

Dataset	Organ	# Examples	# Classes	% Inbalance
APTOS 2019 [23]	Eye	35,126	5	30%
ISIC 2019[22]	Skin	25,331	10	20%

We trained models from scratch and also fine-tuned from ImageNet toward the final objective. We balance the datasets using importance sampling by assigning a sampling weight to each class proportional to the inverse frequency. In addition to normalization, we perform random horizontal flips and random cropping as additional data augmentation during training. We train and evaluate using an 80%/20% stratified split, resulting in a 70%/10% for training and validation, respectively, and the remainder 20% for testing, using cross-validation with 5 k-CV folders to obtain the optimal set of hyper-parameters for each model evaluated.

As an optimizer, we evaluate Adam [24] and AdamW [25] optimizers with weight decay of 10^{-4} for all models, with a maximum learning rate of 10^{-5}. Linear learning rate decay is used from epoch 15 to 70, dropping the learning rate to 10^{-7}, and all models for each task are fine-tuned for a further 30 epochs, completing a total of 100 epochs.

The performance of all model variations is listed in Table 2, attaining only the best model variation regarding each architecture after exhaustive evaluation. We mostly exploit pre-training from Imagenet in all models and fine-tune toward the task objective. we employ EarlyStopping with patience of 5, with all images resized to 224×224, with patch sizes set to 16×16, resulting in a final sequence of 256 patches.

Table 2. Performance Summary of the proposed models.

Model	DA	DA	PT-I	PT-E	Metrics
Dataset: Isic 2019[22] - 10 classes					
Inception + AdamW	✓	✓	✓	–	acc: 87.8%, AUC: 91.8%, mf1: 60.8%, Mf1: 61.3%
ResNet + AdamW	✓	✓	✓	–	acc: 89.3%, AUC: 93.7%, mf1: 61.9%, Mf1: 62.2%
DenseNet + Adam	✓	✓	✓	–	acc: 89.0%, AUC: 93.5%, mf1: 61.1%, Mf1: 61.9%
Transformer + AdamW (0%)	✓	✓	✓	✓	acc: 87.8%, AUC: 90.9%, mf1: 59.4%, Mf1: 60.5%
Transformer + AdamW (25%)	✓	✓	✓	–	acc: 90.7%, AUC: 94.0%, mf1: 64.2%, Mf1: 65.7%
Transformer + Adam (25%)	–	✓	✓	✓	acc: 91.2%, AUC: 94.3%, mf1: 69.7%, Mf1: 69.9%
Dataset: APTOS 2019 [23] - 5 classes					
Inception Adam	✓	✓	✓	–	acc: 76.1%, AUC: 87.5%, mf1: 50.3%, Mf1: 50.8%
ResNet + Adam	✓	✓	✓	–	acc: 77.3%, AUC: 88.3%, mf1: 51.5%, Mf1: 52.3%
DenseNet + Adam	✓	✓	✓	–	acc: 78.5%, AUC: 88.4%, mf1: 52.9%, Mf1: 53.5%
Transformer + AdamW (0%)	✓	✓	✓	–	acc: 76.4%, AUC: 84.4%, mf1: 50.1%, Mf1: 50.9%
Transformer + AdamW (25%)	✓	✓	✓	–	acc: 77.6%, AUC: 86.2%, mf1: 51.5%, Mf1: 52.0%
Transformer + AdamW (25%)	–	✓	✓	✓	acc: 79.2%, AUC: 89.2%, mf1: 53.0%, Mf1: 54.1%

DO: dropout; DA: data augmentation; PT-I: Pre-trained ImageNet; PT-E: Patch Erase; acc: accuracy; AUC: area under the receiver operating characteristic curve; mf1: micro f1score; Mf1: macro f1score;.

5 Results and Discussion

Table 2 presents the Accuracy, Area Under the Curve (AUC) micro and macro F1 scores results on the ISIC 2019 [22], APTOS 2019 [23] datasets. A detailed analysis shows that Transformers in all variations achieved similar results when directly compared to the CNN counterparts, with one model slightly superior. Fine-tuning from Imagenet proved to be fundamental in all models, allowing faster convergence and superior performance results in all transformer and CNN models.

The Random Patch erasing employed in transformers models increased training performance regarding transformers models trained without Random Patch erasing by around 2.0% in accuracy compared with transformers models that employ dropout. The degree of patch overlap also affected the Transformer's overall performance. Without any degree of overlap, Transformer presented slightly weaker results compared to CNN models in datasets containing a small number of samples. In contrast, the larger ones presented slightly superior results. After exhaustive model experimentation, the degree of overlap was set to 25% in all evaluated datasets.

Other transformers models variations experiments without any genre fine-tuned from other domains had lower performance, and results were not reported, empathizing the need to employ pre-trained models to overcome the limitation in the number of training examples in medical datasets.

Regarding CNN models, pre-trained weights and data augmentation (rotation, mirror, crop) improved model performance as expected, with models trained from scratch also not listed due to lower performance. APTOS 2019 showed lower performance on all evaluated models, evidencing the impact in the size of the images, with challenging classes having difficulties correctly classifying. Fine-tuning the transformer and CNN models is vital in all considered datasets.

We replace the Multi Layer Perceptron (MLP) classifiers in the pre-trained models to perform fine-tuning toward the dataset class objective. The Random Patch erasing optimal random probability was set to 30% of the input sequence during training. We repeated each experiment 5 times to reach an average result. The top-performer models align with those reported in the literature, meaning that further fine-tuning and detailed analyses could easily improve the model's performance.

Random Patch erasing as an additional benefit allowed faster training times in the order of 22% when compared without using Random Erasing. This is mainly due to the $O(n^2)$ quadratic complexity exhibited by transformers. An interesting fact is related to the size of the datasets, as the number of samples grows, the margin between ViT and CNN grows as well in favour of ViT, with the benefit of built-in high-resolution saliency maps that can be useful to understand the model's decisions better.

6 Conclusions

We evaluate a regularization method to effectively train transformers for image classification, obtaining similar performance when compared with CNN counterparts in the same evaluated datasets. The evaluated Patch Erasing method avoids overfitting and marginally reduces the computational complexity required to train transformers from in-domain and transfer the domain, outperforming CNN models in larger datasets of samples.

In the future, we aim to employ a variation of Random Erase at the embedding space to access the regularization capabilities when using pre-trained backbones as feature embeddings, enabling it to be fine-tuned on several downstream tasks simultaneously.

Acknowledgements. This work is Co-financed by Component 5 - Capitalization and Business Innovation, integrated in the Resilience Dimension of the Recovery and Resilience Plan within the scope of the Recovery and Resilience Mechanism (MRR) of the European Union (EU), framed in the Next Generation EU, for the period 2021–2026, and by National Funds through the Portuguese funding agency, FCT-Foundation for Science and Technology Portugal, a PhD Grant Number 2021.06275.

References

1. Vaswani, A., et al.: Attention is all you need. Adv. Neural Inf. Process. Syst. **30** (2017)
2. Tenney, I., Das, D., Pavlick, E.: Bert rediscovers the classical nlp pipeline, arXiv preprint arXiv:1905.05950 (2019)
3. Liu, Y., Lapata, M.: Text summarization with pretrained encoders, arXiv preprint arXiv:1908.08345 (2019)
4. Nguyen, C., Asad, Z., Huo, Y.: Evaluating transformer-based semantic segmentation networks for pathological image segmentation, arXiv preprint arXiv:2108.11993 (2021)
5. Dosovitskiy, A., et al.: An image is worth 16 × 16 words: transformers for image recognition at scale, arXiv preprint arXiv:2010.11929 (2020)
6. Yang, H., Chen, J., Xu, M.: Fundus disease image classification based on improved transformer. In: 2021 International Conference on Neuromorphic Computing (ICNC), pp. 207–214. IEEE (2021)
7. Chu, X., et al.: Conditional positional encodings for vision transformers, arXiv preprint arXiv:2102.10882 (2021)
8. Amir, S., Gandelsman, Y., Bagon, S., Dekel, T.: Deep vit features as dense visual descriptors, vol. 2, no. 3, p. 4 (2021). arXiv preprint arXiv:2112.05814
9. Yu, S., et al.: MIL-VT: multiple instance learning enhanced vision transformer for fundus image classification. In: de Bruijne, M., et al. (eds.) MICCAI 2021. LNCS, vol. 12908, pp. 45–54. Springer, Cham (2021). https://doi.org/10.1007/978-3-030-87237-3_5
10. Dai, Y., Gao, Y., Liu, F.: Transmed: transformers advance multi-modal medical image classification. Diagnostics **11**(8), 1384 (2021)
11. Lu, M., et al.: Smile: sparse-attention based multiple instance contrastive learning for glioma sub-type classification using pathological images. In: MICCAI Workshop on Computational Pathology, pp. 159–169. PMLR (2021)
12. Khan, A., Lee, B.: Gene transformer: transformers for the gene expression-based classification of lung cancer subtypes, arXiv preprint arXiv:2108.11833 (2021)
13. Gheflati, B., Rivaz, H.: Vision transformer for classification of breast ultrasound images (2021). arXiv preprint arXiv:2110.14731
14. Al-Dhabyani, W., Gomaa, M., Khaled, H., Fahmy, A.: Dataset of breast ultrasound images. Data Brief **28**, 104863 (2020)
15. Shah, S.M., Khan, R.A., Arif, S., Sajid, U.: Artificial intelligence for breast cancer detection: trends & directions, arXiv preprint arXiv:2110.00942 (2021)

16. Sandy, N., Plevritis Sylvia, K.: Nsclc radiogenomics: initial stanford study of 26 cases. the cancer imaging archive (2014)
17. Chen, H., et al.: Gashis-transformer: a multi-scale visual transformer approach for gastric histopathology image classification, arXiv preprint arXiv:2104.14528 (2021)
18. Jiang, Z., Dong, Z., Wang, L., Jiang, W.: Method for diagnosis of acute lymphoblastic leukemia based on ViT-CNN ensemble model. Computational Intelligence and Neuroscience, vol. 2021 (2021)
19. Devlin, J., Chang, M.-W., Lee, K., Toutanova, K.: Bert: pre-training of deep bidirectional transformers for language understanding (2018)
20. Koutini, K., Schlüter, J., Eghbal-zadeh, H., Widmer, G.: Efficient training of audio transformers with patchout, arXiv preprint arXiv:2110.05069 (2021)
21. Srivastava, N., Hinton, G., Krizhevsky, A., Sutskever, I., Salakhutdinov, R.: Dropout: a simple way to prevent neural networks from overfitting. J. Mach. Learn. Res. **15**(1), 1929–1958 (2014)
22. Nahata, H., Singh, S.P.: Deep learning solutions for skin cancer detection and diagnosis. In: Jain, V., Chatterjee, J.M. (eds.) Machine Learning with Health Care Perspective. LAIS, vol. 13, pp. 159–182. Springer, Cham (2020). https://doi.org/10.1007/978-3-030-40850-3_8
23. Bejnordi, B.E., et al.: Diagnostic assessment of deep learning algorithms for detection of lymph node metastases in women with breast cancer. JAMA **318**(22), 2199–2210 (2017)
24. Kingma, D.P., Ba, J.: Adam: a method for stochastic optimization (2014)
25. Loshchilov, I., Hutter, F.: Decoupled weight decay regularization (2017)

3D Computer Vision

Guided Depth Completion Using Active Infrared Images in Time of Flight Systems

Amina Achaibou[1,2](✉), Nofre Sanmartín-Vich[1,2], Filiberto Pla[2], and Javier Calpe[1]

[1] Analog Devices Inc., 46980 Paterna, Spain
{amina.achaibou,nofre.sanmartinvich,javier.calpe}@analog.com
[2] Institute of New Imaging Technologies, University Jaume I, 12071 Castellón de la Plana, Castellón, Spain
pla@uji.es

Abstract. Depth information has been successfully used in many computer vision applications, but depth imaging sensors frequently provide missing values, mainly around objects boundaries. These invalid values and image gaps cause serious problems in some applications. In order to estimate missing depth values and fill gaps in depth images (D), we propose a new algorithm for depth completion based on belief propagation. The rationale of the proposed technique is based on the idea that missing values must be estimated by taking into account object boundaries, mainly those related with depth discontinuities. Time of Flight (ToF) cameras provide depth information and some additional data, such as active infrared (IR) brightness images. Therefore, object boundaries information for depth missing areas can be reconstructed by using auxiliary IR information or by RGB images in RGB-D systems. These auxiliary images are used as a guidance for the depth completion, also known as depth inpainting. Experimental results show that our algorithm is very simple to implement, fast and produces better results than other more complex, and usually slower, existing methods.

1 Introduction

Depth sensing has become extensively used in many computer vision applications such as autonomous driving, augmented reality, and scene reconstruction. Despite recent advances in depth sensing technology, RGB-D cameras like Microsoft Kinect, Intel RealSense, Google Tango, and other ToF based systems still produce depth images with missing data when surfaces are too glossy (bright), thin, close, or far from the camera (low reflectance). These problems appear when distances are large, surfaces are shiny, or there is strong ambient lighting. Therefore, depth sensing devices are unable to correctly estimate depth data in some cases due to the limit of working distance, occlusion, or reflective surface angles, leading to missing values regions and unstable boundaries in depth maps. In our study we use the ADTF3175 module from Analog Devices Inc with 1024 × 1024 pixels resolution based on a VCSEL laser light at 940nm

A. Pertusa et al. (Eds.): IbPRIA 2023, LNCS 14062, pp. 323–335, 2023.
https://doi.org/10.1007/978-3-031-36616-1_26

[1]. This module provides less noise comparing to the other technologies, and it consists of an IR light projector and a depth image sensor. Image completion or image inpainting, is the technique for reconstructing damaged or missing areas of an image. A long-standing and analogous challenge to the depth filling problem has been to complete a color image after a selected object or region is removed or alternatively to create a plausible synthesis of the image over a larger spatial area [10,12,16,20]. Although extensive literature on that subject already exists, in this work we focus on methods that can potentially be used for depth filling, more specifically in the guided depth completion algorithms [6,7,17,21], and we avoid deep learning methods which require unavailable large data-sets with ground-truth. Moreover, due to our final aim to embed the algorithm in dedicated hardware, we propose a guided completion based on optimization, which better fits sensors and hardware resources requirements.

In this paper we propose a novel depth inpainting approach which takes advantage from IR images captured simultaneously with depth maps from a ToF system, using the IR information as guidance. The presented method outperforms some methods of depth completion found in the literature and shows an improvement for edges recovery.

2 Related Work

There is an extensive literature about methods that have been proposed for filling gaps of missing values in depth maps of RGB-D images. Those methods can be roughly categorized into two groups: guided-inpainting and self-inpainting. Techniques of the first group consider the structural similarity between depth images and auxiliary color images to help the depth image gap filling and are the most researched depth completion methods. Guided inpainting include a large variety of algorithms such as adaptive joint bilateral filter [17], methods based on color image segmentation [7], guided fast marching methods [6], color image guided anisotropic diffusion model [8], and image guided auto-regressive model [19]. This group of methods also includes background surface extrapolation [13], color-depth edge alignment [21], low-rank matrix completion [18], Mumford-Shah functional optimization [9], joint optimization with other properties of intrinsic images [4], and patch-based image synthesis [5].

Recently, deep learning has also led to the development of depth completion methods based on neural networks. Some of them have been proposed for inpainting color images with auto-encoders [14] and GAN architectures [15]. However, most of these deep-learning works have not been used for depth image inpainting. This problem is more difficult due to the absence of strong features in depth images and the lack of large real training datasets.

There are few studies on self-inpainting methods based on depth images such as exemplar depth inpainting [3] and coherence transport method [12]. Zang et al. applied deep depth completion networks for a single depth image [22], using conventional networks to predict surface normal direction and occlusion boundaries from color images and filled missing gaps by a global optimization using original damaged depth images. Junbo Mao et al. [11] used a U-Net network for self-inpainting depth, which is trained comparing the depth image inpainting result with the input and ground truth images.

For the best of our knowledge, apart from the methods using color images as a guidance and a few studies based on self-inpainting depth, there are no methods based on IR images in ToF cameras for guided depth inpainting, as it is addressed in this paper.

3 Proposed Approach

The regions of missing depth values, hereafter referred to as image depth holes, cause serious problems in applications that use image depth sensors. These depth holes often straddle part of the background and foreground objects. Therefore, the recovery of depth values in these regions must take into account the possible depth discontinuities lost in the depth estimation process provided by the camera. However, objects boundaries inside holes can be estimated in ToF cameras by the active IR image captured simultaneously with the depth map. Thus, this paper proposes a guided depth completion method based on the objects border information extracted from the IR image. The rationale of the proposed method consists of filling in the depth holes starting from the hole boundary and propagating the valid depth values, converging from both sides to the object borders. Figure 1 shows the intuitive idea of how depth information is propagated in the proposed method.

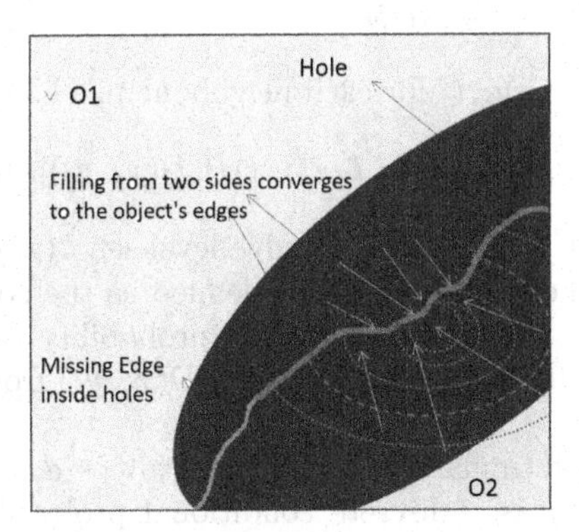

Fig. 1. Propagation of estimated depth values to recover object edges inside a depth hole (blue region). O1 represents object 1 and O2 represents object 2. (Color figure online)

The guidance information is based on the active IR image edges. Therefore, let us define the objects edge map $WE(x, y)$ as

$$WE(x, y) = \exp\left(-k\, GI(x, y)\right) \tag{1}$$

where k is a constant and GI(x,y) is the module of the gradient of the IR image smoothed with a Gaussian kernel $G_\sigma(x, y)$ with standard deviation σ in the image plane.

$$GI(x, y) = ||G_\sigma(x, y) * \nabla IR(x, y)|| \quad (2)$$

Note that $WE(x, y) \in [0, 1]$, and the nearer to 0 the stronger the border. In a further step, the edge map is represented as a level set

$$L(x, y, t) = \begin{cases} WE(x, y);\ WE(x, y) > s(t) \\ 0;\ otherwise \end{cases} \quad (3)$$

let $s(t)$ be the threshold of level set to define the level $t \in 1, ..., N$. This level set $L(x, y, t)$ is used as the guidance to determine at each level t what are the missing value pixels $V(x, y) = 0$ in depth holes to indicate where to propagate the depth information from their neighbours with a valid depth value $V(x, y) = 1$. That is, the propagation of the depth values into the depth hole from the neighbouring valid pixels is performed by means of an iterative process going through the level set of the edge map $L(x, y, t)$, where at each level the estimated depth for a missing depth pixel is calculated as follows,

$$D_m(x, y) = \frac{1}{NW(x, y)} \sum_{i \in N(x,y)} [V(x_i, y_i)WP(x, y, x_i, y_i, t)D(x_i, y_i)] \quad (4)$$

Where NW is the normalization term,

$$NW(x, y) = \sum_{j \in N(x,y)} [V(x_j, y_j)WP(x, y, x_j, y_j, t)] \quad (5)$$

and weights $WP(x, y, x_j, y_j, t)$ for each neighbour pixel are defined as

$$WP(x, y, x_j, y_j, t) = L(x, y, t)P(D(x_j, y_j)|IR(x, y)) \quad (6)$$

that is, as the product of the guidance edge level set $L(x, y, t)$ and the belief to be propagated from the neighbour pixels defined as the conditional probability $P(D(x_j, y_j)|IR(x, y))$, which stands for the probability of a given pixel (x, y) with IR value $IR(x, y)$ to be assigned a depth $D(x_j, y_j)$ from its neighbour pixel (x_j, y_j).

Note that, in this formulation, depth map values d must be discretized in order to define $P(d|ir)$ as a discrete conditional probability. In practice, this conditional probability is computed from the joint probability using the chain rule as $P(d|ir) = P(d, ir)/P(ir)$, and prior $P(ir) = \sum_d P(d, ir)$ as the marginal of the joint probability $P(d, ir)$. The joint probability $P(d, ir)$ is estimated by computing the normalized joint histogram between the initial depth map D and the corresponding active IR image.

The whole belief propagation algorithm for depth completion can be summarized in the following iterative process using the IR edge level set $L(x, y, t)$ as a guidance, Note that the estimation of the missing depth values is done in an iterative way according to the edge level set guidance, propagating at each iteration the beliefs of the neighbours depths according to the conditional probabilities $P(D(x_j, y_j)|IR(x, y))$ of the neighboring depths $D(x_j, y_j)$ given the target pixel IR value $IR(x, y)$. The iterative nature of this algorithm produces the recovering of missing depth pixel values as described and represented in Fig. 1, starting from high values of the edge map (W_E) and propagating the depth values towards the object borders with lower values of the edge map.

Algorithm 1. Guided depth completion through belief propagation

Compute IR edge weights WE
Compute conditional probabilities $P(d|ir)$
Initialize V, $V = 0$ for missing depth pixels and $V = 1$ for initial valid depth pixels
Initialize the belief propagation rate $bs = constant$
Initialize iteration number $t = 0$
Initialize IR edge level set threshold $s(t) = 1$
while $s(t) > 0$ **do**
  Compute $L(x, y, t)$
  for all pixels (x, y) with missing depth values $V = 0$ **do**
    Compute $D_m(x, y)$ according to equation in Eq. (4)
  end for
  Update valid depth pixels V
  Next threshold of the IR level set $s(t + 1) = s(t) - bs$
  Next level $t = t + 1$
end while

4 Experiments and Datasets

4.1 Experiments Based on the ADTF3175 Dataset

To validate the proposed approach, both quantitative and qualitative experiments have been carried out. For qualitative experiments, a set of real depth maps are acquired using the ADTF3175 module for different scenes with distance ranging from 0.2 to 4 m. Simultaneously, IR images of each scene are also captured by the same sensor. The IR images and depth maps D are already co-registered as they are obtained from the same sensor. Image resolution is 1024×1024 pixels. Figure 4 and Fig. 5 illustrate some representative examples where second column, first row is the captured depth map and first column, first row shows the corresponding IR image. The scenarios are chosen to include both narrow and large missing depth holes caused by occlusion, dark or reflective surfaces, or other factors. More than 10 scenarios captured with the ADTF3175 camera were used for the experiments to test how the methods recover the missing edges and how they can reconstruct large depth missing regions. These scenarios contain different sizes of holes, where the complexity of the processing depends on the number of invalid or missing depth pixels for each scene.

4.2 Experiments Based on RGB-D Middlebury Dataset

Since the ground truth depth for ToF IR and depth real-world scenes cannot be easily obtained, we used the Middlebury dataset [2], consisting of RGB-D images with ground truth depth data for quantitative analysis. Moreover, we adapted the proposed guided depth inpainting approach based on IR to color guidance by pre-processing the RGB images. This pre-processing consists of using the gray level intensity corresponding to the RGB image as a proxy of the IR image. In the case of this dataset, missing depth regions were simulated by defining some masks on specific areas to analyze the behaviour of the depth completion algorithms regarding missing values around object borders or low reflectance areas. Several depth holes sizes were simulated to assess the capability of the depth inpainting algorithms to recover small and large regions. Five different scenarios from the Middlebury dataset were used to evaluate the study quantitatively (Plastic, Arts, Baby, Aloevera and Flowerpots images). The simulated depth holes for each scenario have different shapes with different number of invalid pixels. Holes included objects boundaries, transition edges (two different objects). Average metrics on the five images are computed to evaluate the missing depth completion process, and due to the large variability among the scenes, we also computed the standard deviation to better understand the stability of the metrics for each algorithm.

4.3 Selected Baseline Methods

The proposed method is compared with guided depth inpainting state-of-the-art approaches based on different principles, such as the guided Anisotropic Diffusion approach (AD) [8], Adaptive Joint Bilateral Filtering (AJBLF) [17], Guided Fast Marching Method based on color images (GFMM) [6] and guided inpainting based on color segmentation [7]. This set of methods were originally proposed for depth filling based on color image guidance. In the present work these algorithms are adapted for guided depth inpainting based on IR images, to be compared with the proposed guided depth completion approach.

5 Results and Discussion

The different methods are quantitatively assessed using several metrics on data supported with ground truth. In particular the Root Mean Squared Error (RMSE), Structural Similarity (SSIM), Peak Signal-to-Noise Ratio (PSNR), and computational time are used to evaluate and compare the performance of the depth completion methods tested. These indicators are the commonly accepted evaluation metrics for depth completion and are categorized into errors and accuracy estimators. The RMSE is a global metric based on the distance between the estimated D and ground truth $\hat{D}$. Smaller RMSE means the estimated values are closer to the ground truth. The SSIM index assesses the loss and distortion between the estimated depth and the image ground truth. In this case, values closer to 1 indicate better image quality. PSNR indicates the ratio of the square peak signal value (maximum depth range) and the MSE between the estimated and the ground truth depth, the larger PSNR of the estimated depth the closer to the ground truth, indicating higher accuracy in the estimated values. Regarding computational time, lower computational times are preferred for the sake of computational efficiency. Experiments were carried out with a PC with Intel Core i7, CPU @ 2.90 GHz, and 16 GB memory.

5.1 Comparison to Baseline Methods

In order to compare the proposed approach to the baseline reference methods for guided depth inpainting described in Sect. 4.3, the RGB-D images from the Middlebury data set were used. Table 1 shows the results of the comparison among the considered methods regarding the metrics using the ground truth depths in this dataset. The results for each metric are expressed as the mean and standard deviation over the five images selected from this dataset. Figure 2 and Fig. 3 show some qualitative results. For instance, Fig. 2 shows, for the depth holes around the objects boundaries (shown in red box), the results obtained by the state-of-the-art guided inpainting methods, where they produce some blurring at edges, while the result of the proposed method (shown in Fig. 2 last column) provides less blurring effects. From a quantitative point of view, results shown from the RGB-D dataset reveal that the proposed approach deals better when filling in depth holes of all sizes and results are closer to the ground truth values, as shown in Table 1, where the proposed method significantly outperforms the rest of state-of-the-art methods used in the comparison.

Figure 4 and Fig. 5 show some examples of actual depth images captured with the ADTF3175 ToF sensor. In this case, the objective is to analyze qualitatively how our proposed technique compares to the selected state-of-art methods perform in the context of IR-based guidance in ToF real data sets. Therefore, to analyze the ability of the different methods to fill in depth holes of different sizes. As an example, let us observe the image in Fig. 4 with large and small holes from the testing data set. The large depth holes have thousands of invalid pixels, and the small depth holes only have at most hundreds of invalid pixels. The results obtained from the tested methods show that some of these techniques can fill in small holes but cannot estimate completely some large holes, for instance in the case of the AJBLF method. By contrast, the result of the proposed method could successfully deal with large depth holes.

Concerning the GFMM method, it fails to fill in large missing areas when using a small radius (<10 pixels) as the parameter of this algorithm. This method can only fill in large holes when big radius values (>20 pixels) are chosen, but increasing the value of the radius produces large blurring effects around boundaries, such as those observed from the details in Fig. 5. Note also in this figure the result obtained by GAD and the inpainting based on IR segmentation, which shows that these two methods can fill in some large areas and around edges but they are not stable in case of complex scenarios, producing in these cases some blurring effects and artifacts. However, the proposed method is more stable regarding to variability of the different scenarios used in the experiments and provide less artifacts and blurring effects, according also to the standard deviation results of the metrics shown in Table 1.

Table 1. Comparative results of guided depth completion methods on RGB-D Middlebury data set.

Methods	Metrics			
	RMSE	SSIM	PSNR	Run time
Adaptive bilateral filtering [17]	16.286±8.725	0.475±0.0427	26.077±8.454	19.677±12.731
Fast marching [6]	2.551±1.586	0.915±0.053	41.983±7.150	1,904±1,381
Guided color segmentation [7]	2.174±0.756	0.946±0.024	41.744±3.356	45.669±23.669
Anisotropic diffusion [8]	1.524±0.971	0.947±0.023	42.278±6.676	47.041±27.660
Proposed belief propagation	**1.308±0.855**	**0.954±0.019**	**43.950±5.485**	**12.773±0.628**

5.2 Edge Recovery Analysis

One of the main challenges in inpainting algorithms is their ability to recover image discontinuities. As already described in Sect. 3, the proposed belief propagation algorithm based on IR edge map guidance is aimed at filling in the missing depth regions in such a way that they propagate the depth information from the depth hole boundary to object edges. Therefore, analyzing how the depth inpainting algorithms recover depth near discontinuities is a key feature to assess the performance of such algorithms.

Analyzing the results obtained when using the RGB-D dataset, Fig. 2 and Fig. 3 show in the bottom row some zoomed patches to see what happens when recovering missing object edges. In Fig. 2, the last column of the third row and of the last row show the edges recovery obtained by the proposed approach. The rest of the columns are the results based on the reference methods corresponding to AJBLF, GFMM, color segmentation and GAD, whilst the first column is the ground truth depth patch. As we can see, the results obtained from the depth inpainting reference methods add some blurring effect around the edges and add more artifacts, as shown in the depth filling of the pencils parts in the zoomed patch. Note how the estimation of pencils edges with the proposed method is more accurate than the other approaches, with less blurring and less artifacts.

We can observe the same quality of recovering from the second patch shown in the last row, where it was more challenging to recover the missing part of the red pencil, and the missing part of the orange cone. It is worth noting that the proposed method readily reconstructs these two missing parts by leveraging the IR edges information, which is exploited as guidance during the belief propagation process.

The test in case of Fig. 3 consisted of adding large holes to the stripes mask around two challenging areas. The first area masked an intersection of three objects (see region in red box), and the second area masked a corner of an object located inside another object, as shown in the blue box. Observing the results shown in the first and second rows, and the zoomed patches in the third row of Fig. 3, our approach better recovered the object borders comparing to the other methods used in the comparison. Note how the AJBLF algorithm still fails to fill in some large regions and produces blurring around edges, and the GFMM method provides more artifacts when it works with large radius to try to recover larger depth holes. The method based on color segmentation often provides more artifacts and also produces some blurring. Finally, the result of the GAD model is better than the other reference methods, but less stable than the proposed approach, as shown in Table 1, where the proposed algorithm provides results with less variance, taking into account the variability of the different scenarios tested based on the RGB-D Middlebury dataset.

From the experiments based on the ADTF3175 data set, where the guidance employed is the IR image, it can also be noticed the stability of the proposed approach concerning contrast edges recovering compared to the other reference algorithms tested. For instance, as it is shown in Fig. 5, and the zoomed parts shown in the last row of this figure. The zoomed patch in the last row of Fig. 4 shows how the result of the proposed approach produces sharper edges, while the rest of the baseline methods produce some smoothing in the depth discontinuity between the object border and the background.

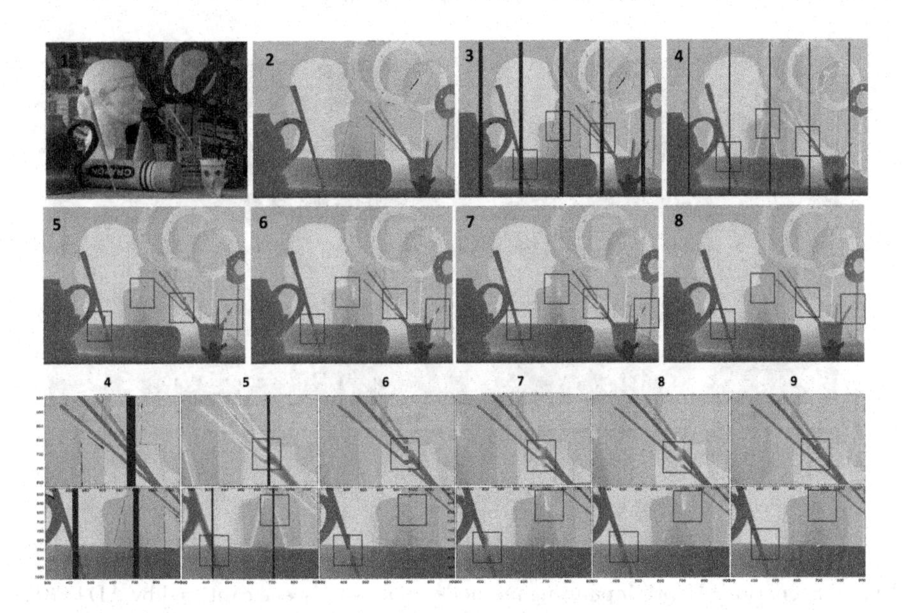

Fig. 2. Comparison of depth inpainting methods for Middlebury data (Arts image). First and second rows: from left to right: 1) IR image, 2) Depth map, 3) Map of missing depth values, 4) AJBLF, 5) GFMM, 6) color Segmentation, 7) GAD, 8) Our approach. Two last rows: two zoomed patches of the pencils and cone parts.

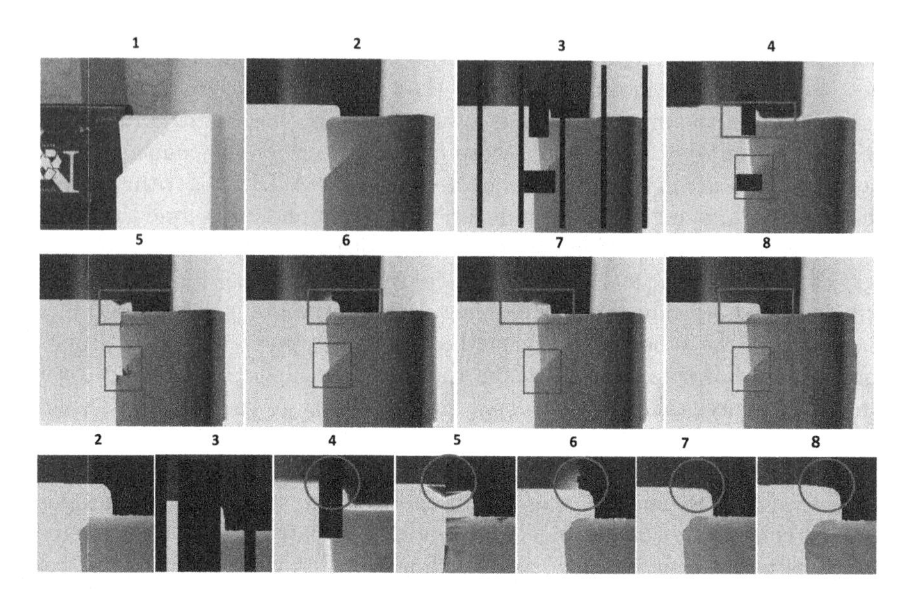

Fig. 3. Comparison of depth inpainting methods for Middlebury data (Plastic image). From left to right: 1) IR image, 2) Depth map, 3) Map of missing depth values, 4) AJBLF, 5) GFMM, 6) color Segmentation, 7) GAD, 8) Our approach. Last row is the zoomed patch of the red box part. (Color figure online)

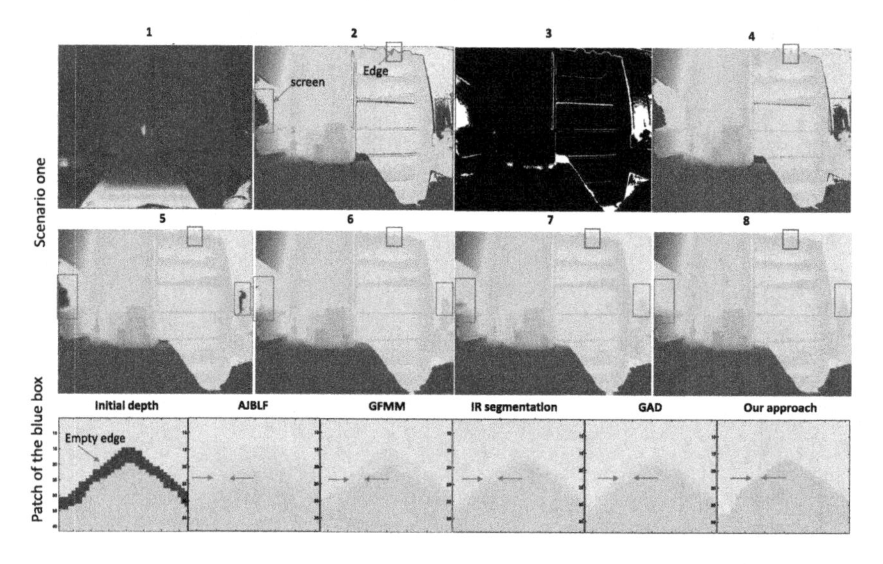

Fig. 4. Scenario 1: Depth inpainting methods results of a scene captured by ADTF3175 module. From left to right : 1) IR image, 2) Depth map, 3) Map of missing depth values, 4) AJBLF, 5) GFMM, 6) IR Segmentation, 7) GAD, 8) Our approach. Blue box patch shows the transition of recovered edges between two objects. (Color figure online)

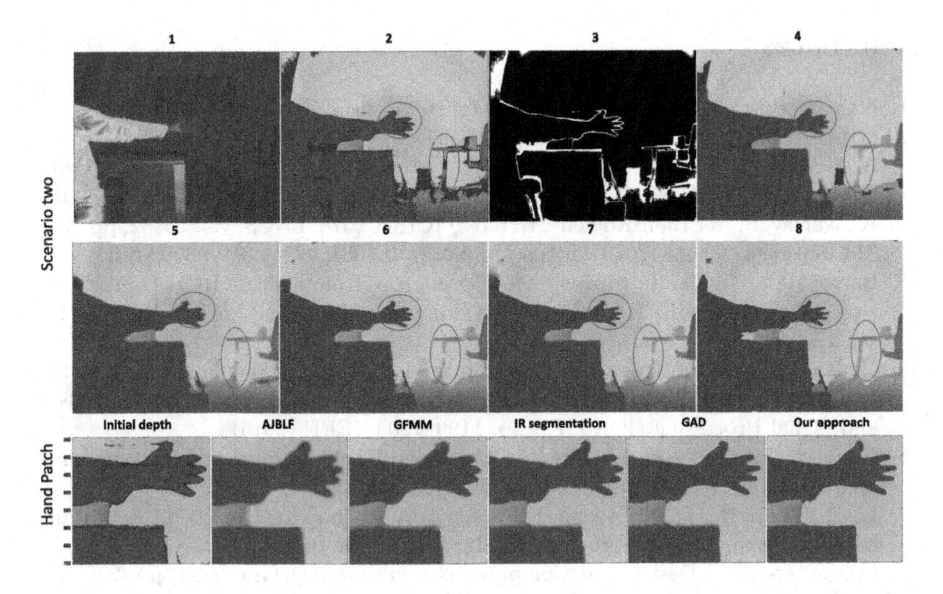

Fig. 5. Scenario 2: Depth inpainting methods results of a scene captured by ADTF3175 module. From left to right: 1) IR image, 2) Depth map, 3) Map of missing depth values, 4) AJBLF, 5) GFMM, 6) IR Segmentation, 7) GAD, 8) Our approach. Hand patch shows a comparison of the blurring effects around the recovered edges.

6 Conclusion

In this work, a guided depth inpainting method for ToF image data has been proposed, which is based on the active IR gradient, the level-set map information and a belief propagation strategy. According to the experiments carried out on real data and the comparative study with other state-of-the-art guided depth inpainting methods, the proposed algorithm yields more enhanced recovered depth maps around edges and can recover satisfactorily large missing depth areas. These results support that the proposed depth recovery method reduces the effect of leakage in large missing depth regions close to depth discontinuities. After analyzing the obtained results, we conclude that the proposed algorithm recovers edges with less artifacts and less blurring effects, it is computationally more efficient than the reference methods used in the comparison. The performance is also better in terms of RMSE and PSNR with respect to ground truth values in the case of the RGB-D dataset used for the quantitative analysis. Further work is directed to integrate the belief propagation strategy in a holistic IR-D fusion framework to recover missing depth values and provide more consistent depth maps mainly at object borders.

Acknowledgments. This work was partially supported by Analog Devices, Inc. and by the Agencia Valenciana de la Innovacion of the Generalitat Valenciana under program "Plan GEnT. Doctorados Industriales. Innodocto"

References

1. https://www.analog.com/en/products/adtf3175.html#product-overview
2. https://vision.middlebury.edu/stereo/data/
3. Atapour-Abarghouei, A., Breckon, T.P.: Extended patch prioritization for depth filling within constrained exemplar-based RGB-D image completion. In: Campilho, A., Karray, F., ter Haar Romeny, B. (eds.) ICIAR 2018. LNCS, vol. 10882, pp. 306–314. Springer, Cham (2018). https://doi.org/10.1007/978-3-319-93000-8_35
4. Barron, J.T., Malik, J.: Intrinsic scene properties from a single RGB-D image. In: Proceedings of the IEEE Conference on Computer Vision and Pattern Recognition, pp. 17–24 (2013)
5. Ciotta, M., Androutsos, D.: Depth guided image completion for structure and texture synthesis. In: 2016 IEEE International Conference on Acoustics, Speech and Signal Processing (ICASSP), pp. 1199–1203. IEEE (2016)
6. Gong, X., Liu, J., Zhou, W., Liu, J.: Guided depth enhancement via a fast marching method. Image Vis. Comput. **31**(10), 695–703 (2013)
7. Li, H., Chen, H., Tu, C., Wang, H.: A recovery method for Kinect-like depth map based on color image segmentation. In: 2015 14th International Conference on Computer-Aided Design and Computer Graphics (CAD/Graphics), pp. 230–231. IEEE (2015)
8. Liu, J., Gong, X.: Guided depth enhancement via anisotropic diffusion. In: Huet, B., Ngo, C.-W., Tang, J., Zhou, Z.-H., Hauptmann, A.G., Yan, S. (eds.) PCM 2013. LNCS, vol. 8294, pp. 408–417. Springer, Cham (2013). https://doi.org/10.1007/978-3-319-03731-8_38
9. Liu, M., He, X., Salzmann, M.: Building scene models by completing and hallucinating depth and semantics. In: Leibe, B., Matas, J., Sebe, N., Welling, M. (eds.) ECCV 2016. LNCS, vol. 9910, pp. 258–274. Springer, Cham (2016). https://doi.org/10.1007/978-3-319-46466-4_16
10. Lu, W., et al.: Diverse facial inpainting guided by exemplars. arXiv preprint arXiv:2202.06358 (2022)
11. Mao, J., Li, J., Li, F., Wan, C.: Depth image inpainting via single depth features learning. In: 2020 13th International Congress on Image and Signal Processing, BioMedical Engineering and Informatics (CISP-BMEI), pp. 116–120. IEEE (2020)
12. März, T.: Image inpainting based on coherence transport with adapted distance functions. SIAM J. Imaging Sci. **4**(4), 981–1000 (2011)
13. Matsuo, K., Aoki, Y.: Depth image enhancement using local tangent plane approximations. In: Proceedings of the IEEE Conference on Computer Vision and Pattern Recognition, pp. 3574–3583 (2015)
14. Van den Oord, A., Kalchbrenner, N., Espeholt, L., Vinyals, O., Graves, A., et al.: Conditional image generation with pixelcnn decoders. Adv. Neural Inf. Process. Syst. **29** (2016)
15. Pathak, D., Krahenbuhl, P., Donahue, J., Darrell, T., Efros, A.A.: Context encoders: feature learning by inpainting. In: Proceedings of the IEEE Conference on Computer Vision and Pattern Recognition, pp. 2536–2544 (2016)
16. Wan, Z., Zhang, J., Chen, D., Liao, J.: High-fidelity pluralistic image completion with transformers. In: Proceedings of the IEEE/CVF International Conference on Computer Vision, pp. 4692–4701 (2021)
17. Wang, D., Chen, X., Yi, H., Zhao, F.: Hole filling and optimization algorithm for depth images based on adaptive joint bilateral filtering. Chin. J. Lasers **46**(10), 294–301 (2019)

18. Xue, H., Zhang, S., Cai, D.: Depth image inpainting: improving low rank matrix completion with low gradient regularization. IEEE Trans. Image Process. **26**(9), 4311–4320 (2017)
19. Yang, J., Ye, X., Li, K., Hou, C., Wang, Y.: Color-guided depth recovery from RGB-D data using an adaptive autoregressive model. IEEE Trans. Image Process. **23**(8), 3443–3458 (2014)
20. Yu, Y., Zhang, L., Fan, H., Luo, T.: High-fidelity image inpainting with GAN inversion. In: Avidan, S., Brostow, G., Cissé, M., Farinella, G.M., Hassner, T. (eds.) Computer Vision – ECCV 2022. ECCV 2022. LNCS, vol. 13676, pp. 242–258. Springer, Cham (2022). https://doi.org/10.1007/978-3-031-19787-1_14
21. Zhang, H.T., Yu, J., Wang, Z.F.: Probability contour guided depth map inpainting and superresolution using non-local total generalized variation. Multimed. Tools Appl. **77**(7), 9003–9020 (2018)
22. Zhang, Y., Funkhouser, T.: Deep depth completion of a single RGB-D image. In: Proceedings of the IEEE Conference on Computer Vision and Pattern Recognition, pp. 175–185 (2018)

StOCaMo: Online Calibration Monitoring for Stereo Cameras

Jaroslav Moravec(✉) and Radim Šára

Department of Cybernetics, Faculty of Electrical Engineering, Czech Technical University in Prague, Prague, Czech Republic
{moravj34,sara}@fel.cvut.cz

Abstract. Cameras are the prevalent sensors used for perception in autonomous robotic systems, but initial calibration may degrade over time due to dynamic factors. This may lead to the failure of the downstream tasks, such as simultaneous localization and mapping (SLAM) or object recognition. Hence, a computationally light process that detects the decalibration is of interest.

We propose StOCaMo, an online calibration monitoring procedure for a stereoscopic system. StOCaMo is based on epipolar constraints; it validates calibration parameters on a single frame with no temporal tracking. The main contribution is the use of robust kernel correlation, which is shown to be more effective than the standard epipolar error.

StOCaMo was tested on two real-world datasets: EuRoC MAV and KITTI. With fixed parameters learned on a realistic synthetic dataset from CARLA, it achieved 96.2% accuracy in decalibration detection on EuRoC and KITTI. In the downstream task of detecting SLAM failure, StOCaMo achieved 87.3% accuracy, and its output has a rank correlation of 0.77 with the SLAM error. These results outperform a recent method by Zhong et al., 2021.

Keywords: Autonomous Robots · Stereo Cameras · Calibration Monitoring

1 Introduction

Visual sensor calibration is critical in autonomous robotic systems like self-driving cars or aerial vehicles. A decalibration (an unexpected or gradual change of the calibration parameters, such as the relative translation and rotation between a pair of stereo cameras) may degrade the performance of a downstream visual task [6], which may lead to an accident of the vehicle. Therefore, the sensor system has to be equipped with a self-assessment mechanism. Beyond hardware failure,

This work was supported by the OP VVV MEYS grant CZ.02.1.01/0.0/0.0/16 019/0000765 and by the Czech Technical University in Prague grant SGS22/111/OHK3/2T/13. We thank the anonymous reviewers for their remarks that helped improve the final version.

A. Pertusa et al. (Eds.): IbPRIA 2023, LNCS 14062, pp. 336–350, 2023.
https://doi.org/10.1007/978-3-031-36616-1_27

the calibration also needs to be monitored. This paper considers the problem of calibration monitoring for a pair of cameras in a stereoscopic system.

Let us assume that the reference calibration is obtained, for example, in a calibration room with known targets. *Calibration monitoring* aims to provide a validity stamp for the calibration parameters on the current data. By the definition of the task, the monitoring process needs to have real-time performance and a small computational overhead. The most important performance metrics are low false positive (false decalibration alarms) and false negative (not reporting actual decalibration) rates. Unlike in automatic calibration, when all parameters need to be found, it is not important to monitor all the parameters, as the decalibration typically manifests in all degrees of freedom.

2 Related Work

The standard approach to camera calibration uses predefined targets of known dimensions, e.g. [23]. The targets are captured from various viewing angles, and parametric constraints are derived from the geometric relationships of the corresponding points. These methods offer high precision at the cost of longer and/or inconvenient execution. In this work, we focus on monitoring the extrinsic calibration parameters (the relative camera orientation).

Automatic Targetless Calibration. Targetless (or self-) calibration has been studied for over three decades. We divide the methods into several groups.

Correspondence-based. These methods use detected (or hand-picked) matched features to optimise the calibration parameters either alone (e.g., by utilising epipolar geometry) [24] or together with the 3D structure of the scene (calibration from infrastructure) [9]. The former is very fast but lacks precision. On the other hand, the latter is very precise, but the optimisation over thousands of parameters (using bundle adjustment) can be computationally very expensive. Thus, some combination of the two is often used [7].

Odometry-based These methods employ visual odometry estimation in each sensor separately, and then they find the transformation between them by solving the hand-eye (HE) calibration problem [2]. Although it does not require any common field of view of the sensors or any initial guess, the precision of the parameters is usually low. HE methods require sufficiently complex motions in 6D [10]. This is often infeasible in the automotive domain, where the vehicle is bound to the ground (plane).

End-to-end Learning-based. The fast development of deep learning (DL) also considerably impacts the stereo self-calibration task. These methods differ in the way they employ the DL. For example, they can estimate the fundamental matrix [21] or optimise the consistency between monocular depth and inferred stereo depth [13]. They achieve good results at the cost of long training and inference and/or the need for high-performance hardware.

Online Calibration Tracking. Precise and stable approaches for targetless calibration are usually computationally expensive and can hardly be run during the operation of the sensor system. Hence, online calibration tracking methods that follow unpredictable changes in calibration parameters were proposed.

Following up on their previous work [7], the authors of [8] studied three geometric constraints for calibration parameters tracking with robust iterated extended Kalman Filter (IEKF). Using a reduced bundle adjustment, they achieved very accurate results in real-time, although dynamic objects hurt the stability of the approach in some environments. They found that combining epipolar constraints for instantaneous coarse calibration with their reduced bundle adjustment method yields stable and accurate results.

Lowering the effect of the dynamic objects in the scene was then studied in [18], where they used CNN to segment the pixels. Only static points were used in the optimisation. Besides epipolar geometry, the homographies induced by the ground plane also provide calibration information [19]. This requires image segmentation, too. Such segmentation could be obtained as a by-product of the downstream task preprocessing.

Epipolar constraints for online calibration were further studied in [14]. They aggregated detected keypoints over several frames to optimise the epipolar error using Kalman Filter. Hence, in contrast to [8], they did not require temporal matching of the features. They show that the epipolar error is a sufficient constraint for accurate online recalibration (tested downstream for visual odometry and reconstruction).

Online Calibration Monitoring. Online calibration tracking methods provide good results, but changing the calibration during the operation might be unsafe. For example, in the automotive domain, if the data fusion does not work, the vehicle should undergo some authorised service rather than rely on parameters tracked during the drive. Online tracking is also quite expensive and cannot be run constantly as the computing resources are needed elsewhere. A more lightweight system could instead detect a miscalibration and trigger the calibration procedure only when needed.

Calibration monitoring as a research problem (specifically in the automotive domain) was introduced in [16]. They detected the LiDAR-Camera system extrinsic miscalibration by examining the alignment between image edges and projected LiDAR corners (points with large radial distance differences with neighbours). As the method had no memory, it could validate the calibration parameters on a single frame without any temporal tracking of the parameters.

The detection of decalibration for intrinsic camera parameters was studied in [5]. They employed end-to-end learning of the average pixel position difference (APPD) using convolutional neural networks (CNN). A similar method for extrinsic calibration monitoring of stereo cameras was introduced in [25]. They trained CNN to output the extrinsic parameters' actual miscalibration. They presented the effectiveness of their monitoring system on ORB-SLAM2 [20] failure detection. Even though the method showed promising results, it had a low

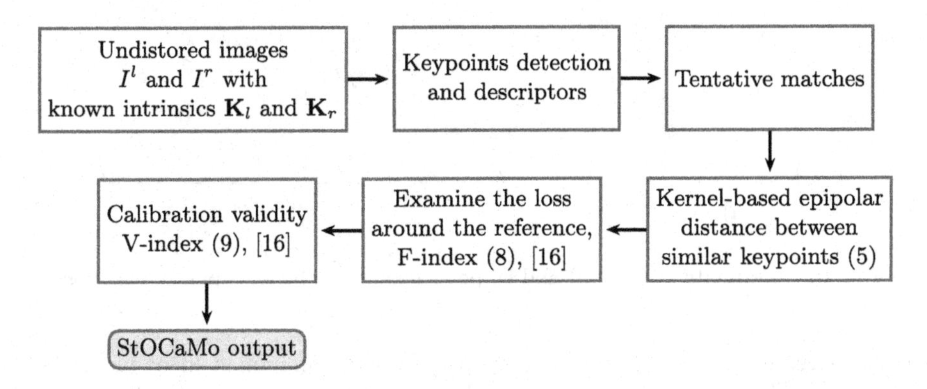

Fig. 1. A diagram of the proposed StOCaMo method, which outputs a decision whether the visual system is decalibrated.

recall (many undetected SLAM failures). We use this method as a baseline in our downstream experiment (see Sect. 4.4).

Contributions. In this work, we focus on extrinsic calibration monitoring and propose an online method for stereo cameras, which we call StOCaMo. It is based on the idea of minimising epipolar error. However, instead of one-to-one matching, we employ the kernel correlation principle [22] to get more accurate results. Our main contributions are: (1) introduction of a lightweight monitoring process using epipolar constraints without one-to-one matching, nor temporal tracking, (2) adoption of kernel correlation principle for implicitly robust loss function used in the optimisation, and (3) a thorough experimental evaluation of the method on synthetic and real data in synthetic and real-world scenarios.

3 Methods

StOCaMo is based on the examination of epipolar error between similar keypoints. Instead of using some robust optimisation technique (RANSAC, LMedS), we employ the kernel correlation principle, which is implicitly robust [22]. Because of the low time-to-detection requirement, we use the method from [16] to validate the calibration on a single frame. Figure 1 summarises the method: First, we extract keypoints [1] and their descriptors [4] from each stereo image. Second, we find tentative left-to-right and right-to-left matches guided by the descriptor similarity, which we then use in the kernel-based epipolar error [22]. Third, we evaluate the error function over a grid of small parameter perturbations around the reference. These are the primary measurements for the monitoring task. Finally, a probability distribution maps the primary measurements to the probabilities of correct and incorrect calibration. These are combined to what we term the calibration validity index, as in [16].

We assume that sensors are global-shutter and that both cameras' intrinsic parameters (calibration matrices and distortion coefficients) are known. Our

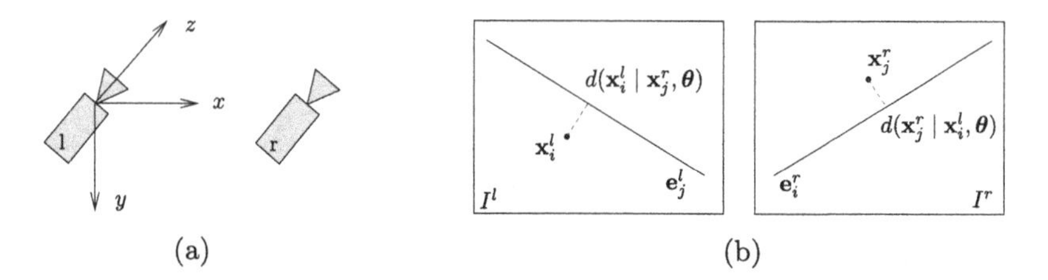

Fig. 2. Coordinate system (a) and epipolar geometry (b) for a pair of cameras.

method requires the cameras to have a sufficient common field of view. The sensors should also be synchronised well, so that there is a minimal effect of the relative latency in the data.

3.1 The StOCaMo Method

Let us assume we are given two undistorted images, I^l and I^r, captured by two cameras with known camera matrices, $\mathbf{K}_l$ and $\mathbf{K}_r$. Our goal is to decide whether the given extrinsic calibration parameters $\boldsymbol{\theta}^{\text{ref}}$ are correct.

Keypoint Detection and Feature Extraction. As already stated, our method minimises the epipolar distance of tentative matches. Therefore, we need to detect suitable keypoints and extract descriptors for matching.

Let $\mathcal{I}^l$ be the set of keypoint indices in the left image and $\mathcal{I}^r$ be the set of keypoints in the right image. Each left-image keypoint $i \in \mathcal{I}^l$ has a location $\mathbf{p}_i^l \in \mathbb{P}^2$ (hence, it has three 'homogeneous' coordinates in the projective space $\mathbb{P}^2$) and descriptor $\mathbf{f}_i^l \in \mathbb{R}^c$. Analogically, we have $\mathbf{p}_j^r$, $\mathbf{f}_j^r$ for $j \in \mathcal{I}^r$ in the right image.

In our experiments, we use the STAR keypoint detector [1] and the BRIEF descriptor [4], which provides descriptor vectors of dimension $c = 32$. These choices are not critical for the statistical performance of StOCaMo.

Epipolar Geometry. As both images are captured by calibrated cameras, we first transform the keypoints by the 3×3 calibration matrices $\mathbf{K}_l$, $\mathbf{K}_r$:

$$\mathbf{x}_i^l = \mathbf{K}_l^{-1}\mathbf{p}_i^l \quad \text{and} \quad \mathbf{x}_j^r = \mathbf{K}_r^{-1}\mathbf{p}_j^r, \quad \mathbf{x}_i^l, \mathbf{x}_j^r \in \mathbb{P}^2. \tag{1}$$

Let now $\mathbf{R} \in \mathbb{R}^{3\times3}$ be the matrix of the relative rotation between the cameras, and $\mathbf{t} \in \mathbb{R}^3$ be the relative translation vector. The rotation matrix is represented via Rodrigues' formula by the axis-angle vector $\boldsymbol{\omega} \in \mathbb{R}^3$, which describes a rotation by angle $||\boldsymbol{\omega}||$ around the rotation axis $\boldsymbol{\omega}$. The extrinsic parameters of dimension six are then composed of two three-element vectors $\boldsymbol{\theta} = (\mathbf{t}, \boldsymbol{\omega})$.

Due to the epipolar constraint [15], given a left-image keypoint $i \in \mathcal{I}^l$, the corresponding right-image keypoint $j \in \mathcal{I}^r$ needs to lie on the epipolar line $\mathbf{e}_i^r$ in the right image I^r, $\mathbf{x}_j^r \in \mathbf{e}_i^r$ (see Fig. 2b), such that

$$\mathbf{e}_i^r = \mathbf{E}(\boldsymbol{\theta})\mathbf{x}_i^l, \qquad \mathbf{e}_i^r = (e_{i,1}^r, e_{i,2}^r, e_{i,3}^r) \in \mathbb{P}^2,$$

and, analogically, given the right-image keypoint $j \in \mathcal{I}^r$, the left-image keypoint $i \in \mathcal{I}^l$ must lie on the line

$$\mathbf{e}_j^l = \mathbf{E}^\top(\boldsymbol{\theta})\mathbf{x}_j^r\,, \qquad \mathbf{e}_j^l = (e_{j,1}^l, e_{j,2}^l, e_{j,3}^l) \in \mathbb{P}^2\,,$$

where $(\cdot)^\top$ is matrix transposition. As the intrinsic parameters are known, the map $\mathbf{E}(\boldsymbol{\theta})$ is given by a 3×3 rank-deficient essential matrix

$$\mathbf{E}(\boldsymbol{\theta}) = [\mathbf{t}]_\times \mathbf{R}, \tag{2}$$

in which $[\cdot]_\times$ is the 3×3 skew-symmetric matrix composed of the elements $\mathbf{t}$.

If the locations $\mathbf{x}_i^l$, $\mathbf{x}_j^r$ are imprecise, one expresses a pair of *epipolar errors*, each defined as the distance of a point from the corresponding epipolar line it belongs to:

$$d(\mathbf{x}_j^r \mid \mathbf{x}_i^l, \boldsymbol{\theta}) = \frac{\left|(\mathbf{x}_j^r)^\top \mathbf{E}(\boldsymbol{\theta})\,\mathbf{x}_i^l\right|}{\sqrt{(e_{i,1}^r)^2 + (e_{i,2}^r)^2}}, \quad d(\mathbf{x}_i^l \mid \mathbf{x}_j^r, \boldsymbol{\theta}) = \frac{\left|(\mathbf{x}_i^l)^\top \mathbf{E}^\top(\boldsymbol{\theta})\,\mathbf{x}_j^r\right|}{\sqrt{(e_{j,1}^l)^2 + (e_{j,2}^l)^2}}. \tag{3}$$

where $\mathbf{a}^\top \mathbf{b}$ stands for the dot product of two vectors. Note that the numerators are the same, but the denominators differ.[1] Note also, that due to the calibration (1), this error is essentially expressed in angular units.

Average Epipolar Error. The standard approach to estimate the quality of a calibration (used also as a baseline in [25]) would minimise the average epipolar error over the nearest neighbours in the descriptor space:

$$\mathrm{AEE}(\boldsymbol{\theta}) = \frac{1}{n} \sum_{i \in \mathcal{I}^l} \sum_{j \in \mathrm{NN}_i^r} d(\mathbf{x}_j^r \mid \mathbf{x}_i^l, \boldsymbol{\theta}) + \frac{1}{n} \sum_{j \in \mathcal{I}^r} \sum_{i \in \mathrm{NN}_j^l} d(\mathbf{x}_i^l \mid \mathbf{x}_j^r, \boldsymbol{\theta}), \tag{4}$$

where $n = |\mathcal{I}^l| + |\mathcal{I}^r|$ and NN_i^r is the nearest neighbour of $\mathbf{f}_i^l$ in the set $\{\mathbf{f}_j^r\}_{j \in \mathcal{I}^r}$ and NN_j^l is the nearest neighbour of $\mathbf{f}_j^r$ in $\{\mathbf{f}_i^l\}_{j \in \mathcal{I}^l}$. Lower $\mathrm{AEE}(\boldsymbol{\theta})$ is better.

In order to reduce false matches, one can use the so-called Lowe's ratio [17]: The descriptor distance to two nearest neighbours is computed, and the match (i, j) is removed from (4) if their ratio is lower than some threshold.

Robust Loss Function. Even though Lowe's ratio lowers the number of outliers to the epipolar geometry, there may still be undetected false matches (e.g., due to repetitive objects in the scene). Hence, we use a different model, which considers the uncertainty of a certain match. Specifically, we employ the kernel correlation principle, studied in the context of registration problems, e. g. in [22].

Let us define a kernel function for a point and a line (between subspaces of dimension zero and one). As the distance between these is symmetric, the

[1] Alternatively, we could use a single (symmetric) distance of a correspondence $(\mathbf{x}^l, \mathbf{x}^r)$ from the epipolar manifold $\mathbf{x}^r \mathbf{E} \mathbf{x}^l = 0$. Since there is no closed-form solution for this distance, it can be approximated with the Sampson error [15]. We prefer the epipolar error that needs no approximation.

kernel will be a symmetric function of the distance as well. The uncertainty of the keypoint location is expressed through a predefined variance of the kernel. Hence, keypoints that have a large distance from their corresponding epipolar line (see (3)) will have a small effect on the resulting loss. Specifically, we use the Gaussian kernel and evaluate the error on the $k\,n$ tentative matches in the feature space:

$$KC(\boldsymbol{\theta}) = -\frac{1}{n}\sum_{i\in\mathcal{I}^l}\sum_{j\in\mathrm{kNN}_i^r}\exp\left[-\frac{d^2(\mathbf{x}_j^r \mid \mathbf{x}_i^l,\boldsymbol{\theta})}{2\sigma^2}\right] - \frac{1}{n}\sum_{j\in\mathcal{I}^r}\sum_{i\in\mathrm{kNN}_j^l}\exp\left[-\frac{d^2(\mathbf{x}_i^l \mid \mathbf{x}_j^r,\boldsymbol{\theta})}{2\sigma^2}\right], \tag{5}$$

where kNN_i^r are k nearest neighbours of $\mathbf{f}_i^l$ in the set $\{\mathbf{f}_j^r\}_{j\in\mathcal{I}^r}$ and kNN_j^l are the k nearest neighbours of $\mathbf{f}_j^r$ in $\{\mathbf{f}_i^l\}_{j\in\mathcal{I}^l}$ and n is as in (4). Lower $KC(\boldsymbol{\theta})$ is better.

There are two model parameters: k and σ. Using $k > 1$ nearest neighbours helps the performance, as non-matching descriptors usually have larger distances and do not contribute to the loss too much. We set $k = 5$ in our experiments. The σ parameter depends on the *calibration tolerance* denoted as δ. This value should be provided by the user of the monitoring system and it should represent the tolerable deviation of the calibration parameters from the true ones (i.e., the system's sensitivity to the decalibration). In this work, we assume that the decalibration of $\delta = 0.005$ rad in the rotation around the x axis (Fig. 2a) is within the tolerance. Then we set $\sigma = \delta$.

Calibration Monitoring. As the percentage of inlier correspondences changes from frame to frame (not to say from dataset to dataset), we cannot simply decide the calibration validity based on the loss function value (5) itself. The time-to-detection requirement also prefers to make the decision on a single frame, without any previous memory needed, as in [25]. We, therefore, use the approach introduced in [16], based on the examination of the loss function around the reference, in a small perturbation grid defined in the parameter space.

Each of the six extrinsic parameters will have its own 1D grid step constant. For example, in the case of the translation in x it is defined as follows:

$$\boldsymbol{\Theta}_{\mathrm{tx}}^{\mathrm{pert}} = \left\{\theta_{\mathrm{tx}}^{\mathrm{ref}} - e_{\mathrm{tx}}, \theta_{\mathrm{tx}}^{\mathrm{ref}}, \theta_{\mathrm{tx}}^{\mathrm{ref}} + e_{\mathrm{tx}}\right\}, \tag{6}$$

where e_{tx} is the grid constant, and for the remaining parameters, it is defined analogically. The loss function (5) will then be evaluated on the Cartesian product of such 6D decalibrations:

$$\boldsymbol{\Theta}^{\mathrm{pert}} = \boldsymbol{\Theta}_{\mathrm{tx}}^{\mathrm{pert}} \times \boldsymbol{\Theta}_{\mathrm{ty}}^{\mathrm{pert}} \times \boldsymbol{\Theta}_{\mathrm{tz}}^{\mathrm{pert}} \times \boldsymbol{\Theta}_{\mathrm{rx}}^{\mathrm{pert}} \times \boldsymbol{\Theta}_{\mathrm{ry}}^{\mathrm{pert}} \times \boldsymbol{\Theta}_{\mathrm{rz}}^{\mathrm{pert}}, \tag{7}$$

which we call the *perturbation grid.* This can yield up to $3^6 = 729$ evaluations when all the parameters are perturbed. If the calibration is correct, the decalibrations should yield a higher loss value than in $\boldsymbol{\theta}^{\mathrm{ref}}$. Hence, inspired by [16], we define a quality measure called an *F-index*:

$$F\left(\boldsymbol{\theta}^{\mathrm{ref}}\right) = \frac{1}{|\boldsymbol{\Theta}^{\mathrm{pert}}|}\sum_{\boldsymbol{\theta}\in\boldsymbol{\Theta}^{\mathrm{pert}}} \mathbb{1}\left[KC(\boldsymbol{\theta}^{\mathrm{ref}}) \le KC(\boldsymbol{\theta})\right], \tag{8}$$

where $\mathbb{1}[\cdot]$ is the indicator function. If the calibration parameters are correct, the F-index should be close to the unit value and smaller (about 0.5) otherwise.

In [16], they proposed to learn two different probability distributions over the $F(\boldsymbol{\theta}^{\text{ref}})$ random values for: (1) small noise within the calibration tolerance $\boldsymbol{\theta}^{\text{ref}} + \delta$, denoted as p_c, and (2) large decalibration well behind the tolerance $\boldsymbol{\theta}^{\text{ref}} + \Delta$, denoted as p_d, with $\Delta > \delta$. Then, they defined a validity index (called *V-index* here) as a posterior probability with equal priors:

$$V(\boldsymbol{\theta}^{\text{ref}}) = \frac{p_c(\boldsymbol{\theta}^{\text{ref}})}{p_c(\boldsymbol{\theta}^{\text{ref}}) + p_d(\boldsymbol{\theta}^{\text{ref}})}. \tag{9}$$

In the original paper [16], they suggested using the normal distribution for p_c and p_d. This selection is not optimal, as the random values F are from the (discretised) interval $[0, 1]$. Instead, we use the empirical distributions (histograms) of F for p_c and p_d.

4 Experiments

In this work, experiments are performed on one synthetic (CARLA [11]) and two real-world datasets (KITTI [12], EuRoC [3]). All the parameters are learned on the synthetic dataset and then used on the real-world ones without any modification. This illustrates a good generalisation of StOCaMo.

CARLA. [11] is a simulator based on the Unreal Engine, hence it provides highly realistic scenes in several pre-created worlds. Although we use it in this work to simulate the stereo pair of cameras, it provides a plethora of other sensors (LiDARs, depth cameras, Radars, etc.). We simulate a stereo pair of two cameras with 70° horizontal and 24° vertical fields of view with a resolution of 1241 × 376 pixels. Both cameras are front-facing and have a baseline equal to 0.4 m. We use the default map TOWN10HD_OPT with 100 autopilot vehicles to simulate the traffic. We recorded 155 sequences (from different spawn points) with 200 frames each.

KITTI. is one of the most popular public datasets in the automotive domain [12]. It contains data for several different tasks, such as odometry estimation, optical flow estimation, object recognition, or object tracking. We will be using the rectified and synchronised data, specifically Sequence 10 from the odometry evaluation dataset with 1201 frames. There are two grayscale, global-shutter cameras with 70° horizontal and 29.5° vertical field of view and a resolution of 1226 × 370 pixels.

EuRoC. MAV [3] dataset was captured on-board on a micro aerial vehicle (MAV). Hence it provides unique fast movements, which are not present in the automotive data. In the experiments, we use Sequence 01 from the Machine Hall set, with 3682 frames from the stereo, global-shutter cameras. The horizontal field of view of unrectified images is 79° and the vertical field of view is 55°, the resolution 752 × 480 and the baseline 0.11 m.

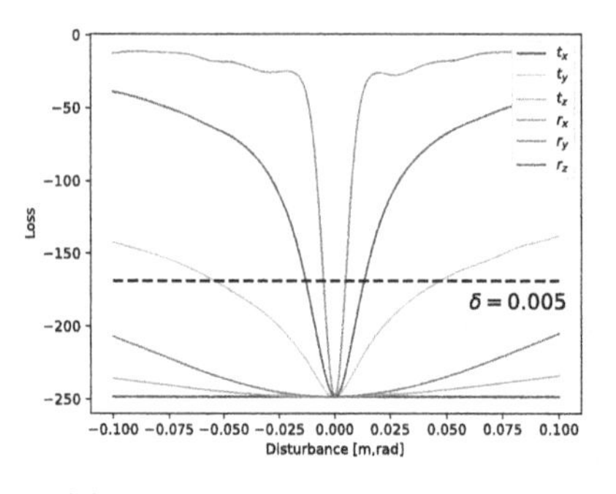

(a) Proposed kernel correlation loss

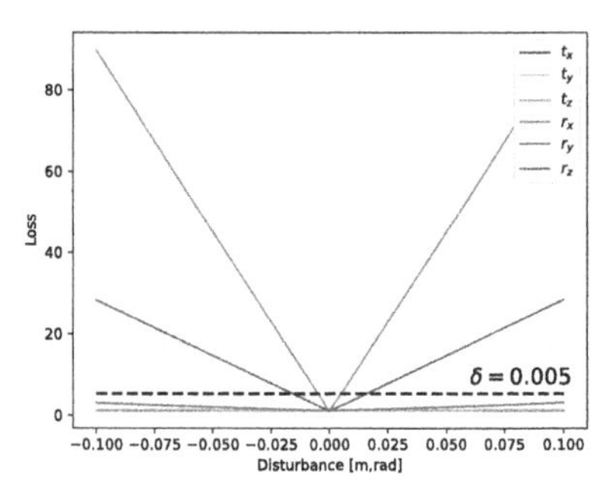

(b) Average epipolar error

Fig. 3. Evaluation of the proposed KC loss (5) (a) and the standard AEE loss (4) (b). Translations in the x direction (blue) are unobservable in epipolar error. The translation plot in the y direction (orange) (almost) coincides with the rotation around the y axis (purple) in (b). This comparison shows increased relative sensitivity in the translation in y by our proposed loss (a).

4.1 Loss Function Shape

We first examined the shape of the proposed KC loss (5) in a similar way as in [25]. We took one sequence from the CARLA dataset (200 frames) and evaluated the average loss. The perturbations were selected as in [25], from $[-0.1, 0.1]$ rad in rotation and $[-0.1, 0.1]$ m in translation. The loss function should be minimal for the reference parameters and quickly increase for large perturbations. As both cameras are front-facing, the change in the baseline length (translation in x) has no impact, because the translation does not change the $\mathbf{E}(\boldsymbol{\theta})$ in (3).

The result of the KC loss (5) is shown in Fig. 3a. All the parameters except translations in the x direction have clear minima in the reference, but some have a higher impact on the KC loss than others. Specifically, the x-axis rotation has the most prominent influence because it corrupts all the epipolars identically (either moves them up or down). The rotation around the z (optical) axis and the y-axis translation have a similar effect on the epipolars, and the decalibration is observable. One can see (black dashed line) that the decalibration of 0.005 rad in the x-axis rotation (set as the calibration tolerance δ) has a similar effect on the KC loss as 0.012 rad decalibration in the z-axis rotation and 0.05 m decalibration in the y-axis translation. The other two observable degrees of freedom (y-axis rotation and z-axis translation) are less apparent than the others. To increase the sensitivity to those, one would probably need to control the distribution of keypoints in the image (preferring the periphery, for instance).

We also performed the same experiment with the average epipolar error (4) with Lowe's ratio on matched STAR features using the BRIEF descriptor. The result is shown in Fig. 3b, and it replicates the corresponding results from [25], where they used SIFT instead of our combination STAR+BRIEF. KC loss (5) has a greater sensitivity to translation in y than AEE loss (4). Moreover, KC loss also exhibits robustness to larger errors in the rotation in the x axis (red).

In order to estimate the F-index (8), we need to set the grid step e in $\boldsymbol{\Theta}^{\text{pert}}$ (6). In this work, we will detect the decalibration on the three most observable degrees of freedom (DoFs) from Fig. 3a, i.e., rotations around the x and z axes and the translation in y. Decalibration in these parameters will have the largest impact on the epipolar geometry (and all subsequent computer vision tasks). The choice of the grid step for rotations around x is based on the calibration tolerance δ so that it is outside of the basin of attraction: $e_{\text{rx}} = 3 \cdot \delta = 0.015$ rad. The perturbations of the other two DoFs are set so that their relative change of the loss with respect to the rotation around the x axis is the same. Based on Fig. 3a, they are: $e_{\text{ry}} = 3 \cdot 0.012 = 0.036$ rad and $e_{\text{ty}} = 3 \cdot 0.015 = 0.045$ m.

4.2 Decalibration Detection

By the derivations in Sect. 3.1, a decalibration should manifest in the F-index (8). Nevertheless, the magnitude of the change is also important. A small perturbation should not deviate the $F(\theta^{\text{ref}})$ values much from the unit value, while larger decalibrations should make it smaller, with low variance. This sensitivity depends on the selected calibration tolerance δ and thus on the σ and $\boldsymbol{\Theta}^{\text{pert}}$ (7).

In this experiment, we evaluate the F-index on all 155 CARLA sequences (200 frames each) with random (uniform) decalibration of several magnitudes:

$$\{0, 0.0025, 0.005, 0.01, 0.02, 0.05, 0.075\}, \tag{10}$$

meters or radians in all six degrees of freedom. Due to the selected calibration tolerance $\delta = 0.005$ rad, the F-index should be close to the unit value up to this decalibration magnitude. After that, it should quickly decrease to about 0.5. This behaviour can be seen in Fig. 4, where the bar corresponds to the average $F(\theta^{\text{ref}})$ over all frames and the errorbars show the 15% and 85% quantiles for that decalibration magnitude. Small decalibrations up to 0.005 (m or rad) have high $F(\theta^{\text{ref}})$ values around 0.98. With increasing the decalibration magnitude, this value drops to 0.55 for the 0.05 and 0.075 magnitudes.

To estimate the posterior probability of the reference calibration parameters (see V-index in (9)), we need to learn the distribution p_c for a small noise within the calibration tolerance δ around the reference and p_d for large decalibrations Δ. We use the actual histograms (as opposed to the normal distribution in [16]) of the F-index values on two magnitudes. The p_c is learned from the magnitude equal to the calibration tolerance, i.e. 0.005 (m or rad). For the p_d, we pick the magnitude $\Delta = 0.05$, as the 0.01 and 0.02 are too close to the reference, and 0.075 yields similar values. One can see both histograms in Fig. 5.

StOCaMo yields a value between zero and one (see (9)). Hence, we also need to select a decision threshold for reporting decalibration. As it is the posterior probability, we set the threshold to 0.5. StOCaMo then reports that reference calibration parameters are incorrect if the value is below this threshold.

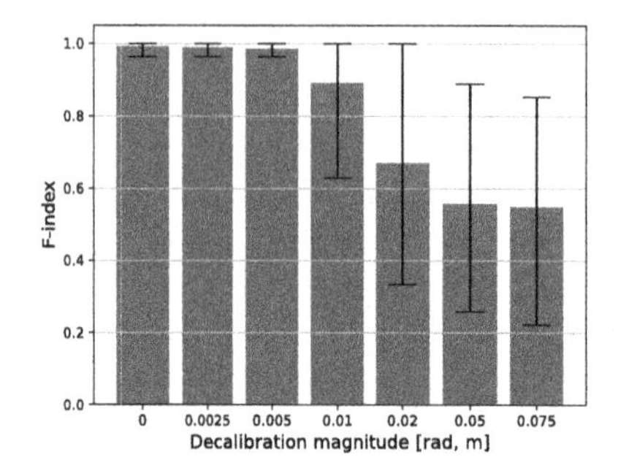

Fig. 4. F-index (8) as a function of decalibration on the CARLA dataset.

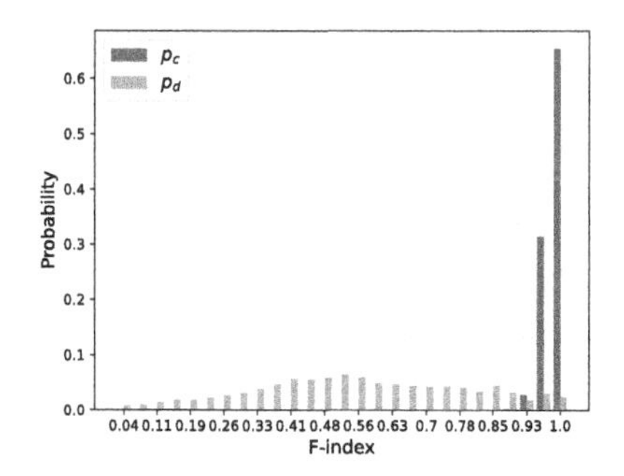

Fig. 5. Histograms of p_c and p_d from (9).

Table 1. Results of our calibration monitoring procedure on synthetic decalibration on KITTI and EuRoC datasets.

	TP	FN	TN	FP	Precision	Recall	Accuracy
KITTI	11854	156	11666	344	97.2	98.7	97.9
EuRoC	33240	3580	36321	499	98.5	90.3	94.5

4.3 Calibration Monitoring on Synthetic Decalibration

With the model learned and the precision parameters set, we can evaluate the performance of StOCaMo on the two real-world datasets. Based on the results from Sect. 4.2, we should be able to identify small decalibrations as valid, but larger ones should be detected. We hence investigate two decalibration magnitudes: (1) a small one: $[-\delta, \delta]$ m or rad, and (2) a large one: $[-4\delta, -2\delta] \cup [2\delta, 4\delta]$ m or rad. The first decalibration is rather small (within the calibration tolerance), and it should be labelled as calibrated by the monitoring method. Therefore, we investigate the true negative (no decalibration) and false positive (false alarm) metrics. The second decalibration magnitude is already large enough to be detected. We use it to estimate the true positive (detected decalibration) and false negative (undetected decalibration) rates of StOCaMo.

On each frame of the KITTI (1,201) and EuRoC (3,682) sequences, we perform ten decalibrations of each of the two kinds. This gives us 12,010 negative examples (without decalibration) and 12,010 positive examples (with decalibration) for the KITTI dataset (and analogically, 36,820 examples of each kind for the EuRoC). We then evaluated StOCaMo on these stereo images with perturbed parameters, as shown in Table 1.

On both datasets, we achieved a very good accuracy of 97.9 and 94.5%, respectively (96.2% on average). This shows that StOCaMo is not only able to correctly distinguish decalibration magnitude on CARLA (see F-index evaluation in Sect. 4.2), but also generalises well on different, real-world data. Interestingly, the difference in recall shows that the decalibration magnitude has a different impact in each of the sensor sets. As the EuRoC has a much wider vertical field of view than both KITTI and CARLA, the decalibration is smaller relative to the sensor resolution. This could be compensated by having different decalibration magnitudes or σ^2 variances for different datasets (sensor setups). However, in the next practical (downstream) experiment, this will not cause any damage to the accuracy.

4.4 Long-Time SLAM Stability with Calibration Monitoring

Incorrect calibration parameters may have a negative effect on the downstream computer vision tasks. In [25], they investigated the reliability of their calibration monitoring system in detecting the ORB-SLAM2 [20] failure. Given a sequence from a dataset, the SLAM was considered failing if the root mean squared error (RMSE) from the ground truth was higher than some threshold (or if it did not finish at all). The threshold was set to 75 m RMSE for KITTI and 2 m RMSE for EuRoC (we kept these thresholds for our experiment). Their monitor was then used as a predictor of the SLAM failure, and they achieved 62% accuracy on the KITTI and 86% on the EuRoC dataset.

We performed a similar experiment with synthetic decalibration of six magnitudes from (10) to investigate the behaviour of our monitoring method fully. The experiment is supposed to investigate the ability of StOCaMo to predict the SLAM failure (or even regress the SLAM error). From each magnitude, we uniformly drew 100 perturbations and these were given to the ORB-SLAM2. Based on the failure of the SLAM method, we obtained positive and negative examples. We executed StOCaMo on frames with the corresponding decalibration and let it decide based on a single frame, whether there is a decalibration or not. For each perturbation, we have drawn ten random frames from the sequence, for more precise statistics. This yields 1,000 predictions for each decalibration magnitude. Full results are shown in Table 2, for each decalibration magnitude individually.

Results on both datasets are very similar (see Table 2a and Table 2b). There are very few false alarms for decalibration magnitudes within the tolerance (0.0025 and 0.005). Then the accuracy drops with borderline decalibrations (0.01 and 0.02) on the decision boundary between calibrated and decalibrated parameters. The accuracy for large decalibration magnitudes (0.05 and 0.075) is very high again for both datasets. Interestingly, for the 0.01 borderline decalibration magnitude, there are around 80% of the sequences in the SLAM tolerance for EuRoC, but only 50% for KITTI (compare TN+FP vs. TP+FN). This again shows that the decalibration magnitude has a different impact on each dataset, which differ in the field of view and resolution. However, the overall accuracy is similar, with 87.1% on the KITTI and 87.6% on the EuRoC (87.3% on average). These results highly depend on the selected thresholds on RMSE, which we took from [25]. Nevertheless, we still achieved much better results on the KITTI dataset and slightly better ones on the EuRoC (Table 2).

Besides the accuracy, we have also evaluated Spearman's rank correlation (see ρ in Table 2) between the V-index (9) and the RMSE of SLAM. One can see that StOCaMo achieved a higher correlation than [25] on both datasets, demonstrating a better ability to predict the SLAM error based on the V-index.

Table 2. Evaluation metrics on the downstream experiment on KITTI (a) and EuRoC (b) for six decalibration magnitudes. The horizontal lines separate decalibrations within tolerance, borderline decalibrations and large decalibrations. The ρ is Spearman's correlation. Results from [25] are shown for comparison.

Dec.	TP	FN	TN	FP	Acc.	ρ
0.0025	0	0	990	10	99.0	
0.005	0	0	959	41	95.9	
0.01	262	218	376	144	63.8	
0.02	699	191	82	28	78.1	
0.05	924	66	3	7	92.7	
0.075	928	72	0	0	92.8	
Avg.					**87.1**	**-0.76**
[25]					62	0.44

(a) KITTI

Dec.	TP	FN	TN	FP	Acc.	ρ
0.0025	0	0	1000	0	100.0	
0.005	0	0	970	30	97.0	
0.01	141	69	600	190	74.1	
0.02	604	186	163	47	76.7	
0.05	862	108	11	19	87.3	
0.075	905	95	0	0	90.5	
Avg.					**87.6**	**-0.78**
[25]					86	0.59

(b) EuRoC

For a more detailed analysis, Fig. 6 shows the breakdown of the results from Table 2 to individual pairs (V-index, RMSE SLAM). Each point is the mean V-index (over the ten random frames) and RMSE from ORB-SLAM2 for each decalibration. Colours correspond to decalibration magnitudes. We also visualise thresholds and the names of the corresponding metrics in each quadrant, where this dot falls.

Overall, the results correspond to those in Table 2. One can see that (when averaged over ten frames), StOCaMo has no problem with small decalibrations (blue and orange dots), which our method indicates as calibrated (top left quadrant). Similarly, substantial decalibrations (purple and brown dots) are correctly denoted as decalibrated (bottom right quadrant) most of the time. The most problematic decalibration magnitudes are 0.01 and 0.02 (green and red dots), which yield many false negatives (top right quadrant) and false positives (bottom left quadrant). Nevertheless, these results are still quite good, as it is not clear what kind of calibration errors influence ORB-SLAM2 most prominently. This can vary from downstream method to method and from task to task.

It is also important to note that, in StOCaMo, the decision boundary can be easily tightened/loosened up, based on the user's selection of the calibration tolerance δ. Therefore, the ratio between FPs and FNs can be changed, based on the metric (precision or recall) preferred by the user.

4.5 Algorithmic Efficiency

A direct comparison with neuronal nets [25] is difficult to make because of the different hardware architecture. StOCaMo runs on a laptop CPU[2], and our Python implementation needs only 70 ms per frame. The most time-consuming parts are the kNN search in the KDTree (30 ms) and the keypoint detection and descriptor construction (17 ms), which can be recycled from the image preprocessing for the SfM downstream task.

[2] AMD Ryzen 7 5800H.

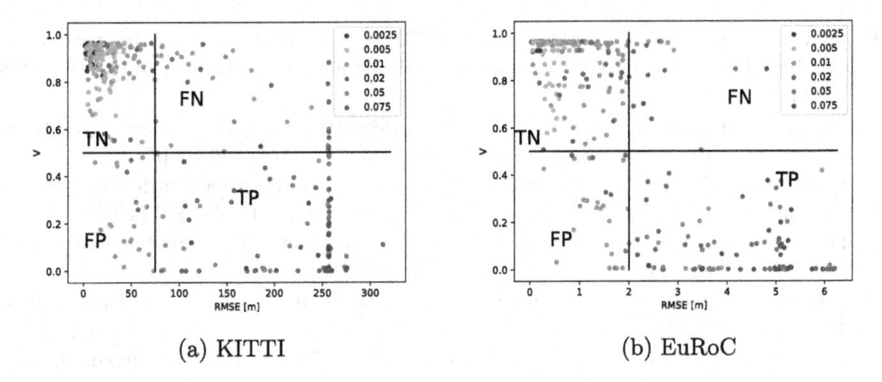

(a) KITTI

(b) EuRoC

Fig. 6. Visualisation of statistical metrics on 100 decalibrations for the downstream performance experiment (averaged over ten frames) of six magnitudes (only converged ORB-SLAM2 runs).

5 Conclusion

We have described an efficient, robust and single-frame method for monitoring the calibration quality of a stereoscopic vision system. The proposed calibration validity index has a greater sensitivity to decalibration than the standard epipolar error. As a result, monitoring synthetic decalibrations on two real-world datasets achieve statistical accuracy of 96.2% (Table 1).

Monitoring results on downstream performance experiments with ORB-SLAM achieve statistical accuracy of 87.3% (Table 2), which exceeds the recent results [25] in the same experiment by more than 13% with a method that is computationally more lightweight. The SLAM error prediction, expressed as a rank correlation value, achieved 0.77, which is by a margin of 0.25 better result than in [25].

References

1. Agrawal, M., Konolige, K., Blas, M.R.: CenSurE: center surround extremas for realtime feature detection and matching. In: Forsyth, D., Torr, P., Zisserman, A. (eds.) ECCV 2008. LNCS, vol. 5305, pp. 102–115. Springer, Heidelberg (2008). https://doi.org/10.1007/978-3-540-88693-8_8
2. Brookshire, J., Teller, S.: Extrinsic calibration from per-sensor egomotion. In: Proceedings Robotics: Science and Systems Conference, pp. 504–512 (2013)
3. Burri, M., Nikolic, J., Gohl, P., et al.: The EuRoC micro aerial vehicle datasets. Int. J. Robot. Res. **35**, 1157–1163 (2016)
4. Calonder, M., Lepetit, V., Strecha, C., Fua, P.: BRIEF: binary robust independent elementary features. In: Daniilidis, K., Maragos, P., Paragios, N. (eds.) ECCV 2010. LNCS, vol. 6314, pp. 778–792. Springer, Heidelberg (2010). https://doi.org/10.1007/978-3-642-15561-1_56
5. Cramariuc, A., Petrov, A., Suri, R., et al.: Learning camera miscalibration detection. In: IEEE International Conference on Robotics and Automation, pp. 4997–5003 (2020)

6. Cvišić, I., Marković, I., Petrović, I.: SOFT2: stereo visual odometry for road vehicles based on a point-to-epipolar-line metric. IEEE Trans. Robot. **39**(1), 273–288 (2023)
7. Dang, T., Hoffmann, C., Stiller, C.: Self-calibration for active automotive stereo vision. In: IEEE Intelligent Vehicles Symposium, pp. 364–369 (2006)
8. Dang, T., Hoffmann, C., Stiller, C.: Continuous stereo self-calibration by camera parameter tracking. IEEE Trans. Image Process. **18**(7), 1536–1550 (2009)
9. Dang, T., Hoffmann, C.: Stereo calibration in vehicles. In: IEEE Intelligent Vehicles Symposium, pp. 268–273 (2004)
10. Daniilidis, K.: Hand-eye calibration using dual quaternions. Int. J. Robot. Res. **18**(3), 286–298 (1999)
11. Dosovitskiy, A., Ros, G., Codevilla, F., et al.: CARLA: an open urban driving simulator. In: Proceedings of the Annual Conference on Robot Learning, pp. 1–16 (2017)
12. Geiger, A., Lenz, P., Stiller, C., et al.: Vision meets robotics: the KITTI dataset. Int. J. Robot. Res. **32**, 1231–1237 (2013)
13. Gil, Y., Elmalem, S., Haim, H., et al.: Online training of stereo self-calibration using monocular depth estimation. IEEE Trans. Comput. Imaging **7**, 812–823 (2021)
14. Hansen, P., Alismail, H., Rander, P., et al.: Online continuous stereo extrinsic parameter estimation. In: IEEE Conference on Computer Vision and Pattern Recognition, pp. 1059–1066 (2012)
15. Hartley, R., Zisserman, A.: Multiple View Geometry in Computer Vision. Cambridge University Press, Cambridge (2003)
16. Levinson, J., Thrun, S.: Automatic online calibration of cameras and lasers. In: Proceedings Robotics: Science and Systems Conference, vol. 2, p. 7 (2013)
17. Lowe, D.G.: Distinctive image features from scale-invariant keypoints. Int. J. Comput. Vis. **60**, 91–110 (2004)
18. Mueller, G.R., Wuensche, H.J.: Continuous stereo camera calibration in urban scenarios. In: International Conference on Intelligent Transportation Systems, pp. 1–6 (2017)
19. Mueller, R.G., Burger, P., Wuensche, H.J.: Continuous stereo self-calibration on planar roads. In: IEEE Intelligent Vehicles Symposium, pp. 1755–1760 (2018)
20. Mur-Artal, R., Tardós, J.D.: ORB-SLAM2: an open-source SLAM system for monocular, stereo, and RGB-D cameras. IEEE Trans. Robot. **33**(5), 1255–1262 (2017)
21. Poursaeed, O., Yang, G., Prakash, A., et al.: Deep fundamental matrix estimation without correspondences. In: Proceedings of the European Conference on Computer Vision Workshops, pp. 485–497 (2018)
22. Tsin, Y., Kanade, T.: A correlation-based approach to robust point set registration. In: European Conference on Computer Vision, pp. 558–569 (2004)
23. Zhang, Z.: A flexible new technique for camera calibration. IEEE Trans. Pattern Anal. Mach. Intell. **22**(11), 1330–1334 (2000)
24. Zhang, Z., Luong, Q.T., Faugeras, O.: Motion of an uncalibrated stereo rig: self-calibration and metric reconstruction. IEEE Trans. Robot. Autom. **12**(1), 103–113 (1996)
25. Zhong, J., Ye, Z., Cramariuc, A., et al.: CalQNet-detection of calibration quality for life-long stereo camera setups. In: IEEE Intelligent Vehicles Symposium, pp. 1312–1318 (2021)

Smart-Tree: Neural Medial Axis Approximation of Point Clouds for 3D Tree Skeletonization

Harry Dobbs(✉), Oliver Batchelor, Richard Green, and James Atlas

UC Vision Research Lab, University of Canterbury, Christchurch 8041, New Zealand
harry.dobbs@pg.canterbury.ac.nz, {oliver.batchelor,richard.green,
james.atlas}@canterbury.ac.nz
https://ucvision.org.nz/

Abstract. This paper introduces Smart-Tree, a supervised method for approximating the medial axes of branch skeletons from a tree point cloud. Smart-Tree uses a sparse voxel convolutional neural network to extract the radius and direction towards the medial axis of each input point. A greedy algorithm performs robust skeletonization using the estimated medial axis. Our proposed method provides robustness to complex tree structures and improves fidelity when dealing with self-occlusions, complex geometry, touching branches, and varying point densities. We evaluate Smart-Tree using a multi-species synthetic tree dataset and perform qualitative analysis on a real-world tree point cloud. Our experimentation with synthetic and real-world datasets demonstrates the robustness of our approach over the current state-of-the-art method. The dataset (https://github.com/uc-vision/synthetic-trees) and source code (https://github.com/uc-vision/smart-tree) are publicly available.

Keywords: Tree Skeletonization · Point Cloud · Metric Extraction · Neural Network

1 Introduction

Digital tree models have many applications, such as biomass estimation [13,14,22], growth modelling [7,31,33], forestry management [3,25,35], urban microclimate simulation [36], and agri-tech applications, such as robotic pruning [2,38], and fruit picking [1].

A comprehensive digital tree model relies on the ability to extract a skeleton from a point cloud. In general, a skeleton is a thin structure that encodes an object's topology and basic geometry [6]. Skeleton extraction from 3D point clouds has been studied extensively in computer vision and graphics literature [30] for understanding shapes and topology. Applied to tree point clouds, this is a challenging problem due to self-occlusions, complex geometry, touching branches, and varying point densities.

Supported by University of Canterbury, New Zealand.

A. Pertusa et al. (Eds.): IbPRIA 2023, LNCS 14062, pp. 351–362, 2023.
https://doi.org/10.1007/978-3-031-36616-1_28

There are many existing approaches for recovering tree skeletons from point clouds (recent survey [5]). These approaches can be categorized as follows; neighbourhood graph, medial axis approximation, voxel and mathematical morphology, and segmentation.

Neighbourhood graph methods use K-nearest neighbours (usually within a search radius) or Delaunay triangulation to create an initial graph from the point cloud. Multiple implementations [11,19,34,37], then use this graph to quantize the points into bins based on the distance from the root node. The bin centroids are connected based on constraints to create a skeleton. Livny et al. [24] use the neighbourhood graph to perform global optimizations based on a smoothed orientation field. Du et al. [12] use the graph to construct a Minimum Spanning Tree (MST) and perform iterative graph simplification to extract a skeleton. However, these methods have some drawbacks. Gaps in the point cloud, such as those caused by occlusions, can lead to a disconnected neighbourhood graph. Additionally, false connections may occur, especially when branches are close to one another.

Medial axis approximation works by estimating the medial surface and then thinning it to a medial axis. An approach in the *L1-Medial Skeleton* method [20] was proposed for point clouds by iterative sampling and redistributing points to the sample centre. Similarly, Cao et al. [4] proposed using Laplacian-based contraction to estimate the medial axis. However, these methods are sensitive to irregularities in the point cloud and require a densely sampled object as input.

Voxel and mathematical morphology were implemented in [16] and later refined in [15]. The point cloud is converted into a 3D voxel grid and then undergoes opening and closing, resulting in a thinned voxel model. The voxel spatial resolution is a key parameter, as a too-fine resolution will lead to many holes. In contrast, a larger resolution can lose significant detail as multiple points become a single voxel (different branches may all become connected). A further limitation of this method is the required memory and time, which increases with the third power of the resolution. This method only aims to find prominent structures of trees rather than finer branches.

Segmentation approaches work by segmenting points into groups from the same branch. Raumonen [28] et al. create surface patches along the tree and then grow these patches into branches. However, this method assumes that local areas of the tree have a uniform density.

Recently a deep learning segmentation approach was proposed in TreePartNet [23]. This method uses two networks, one to detect branching points and another to detect cylindrical representations. It requires a sufficiently sampled point cloud as it relies on the ability to detect junctions accurately and embed local point groups. However, it struggles to work on larger point clouds due to the memory constraints of PointNet++ [27]. Numerous network architectures can process point clouds [18]; however, point clouds in general, but in particular, trees are spatially large and sparse, containing fine details - for this reason, we utilise submanifold sparse CNNs [8,9,17,32].

We propose a deep-learning-based method to estimate the medial axis of a tree point cloud. A neighbourhood graph approach is then used to extract the skeleton. This method mitigates the effects of errors commonly caused when neighbouring branches get close or overlap, as well as improving robustness to common challenges such as varying point density, noise and missing data. Our contributions are as follows:

- A synthetic point cloud generation tool was developed for creating a wide range of labelled tree point clouds.
- A demonstration of how a sparse convolutional neural network can effectively predict the position of the medial axis.
- A skeletonization algorithm is implemented that can use the information from the neural network to perform a robust skeletonization.
- The method is evaluated against the state-of-the-art automatic skeletonization method.
- The method's ability to generalize is demonstrated by applying it to real data.

2 Method

2.1 Dataset

We use synthetic data for multiple reasons. First, it has a known ground truth skeleton for quantitative evaluation. Second, we can efficiently label the point clouds - allowing us to generate data for a broad range of species.

We create synthetic trees using a tree modelling software called SpeedTree [21]. For evaluation, we generate the tree meshes without foliage, which otherwise increase the level of occlusion. In the general case, we remove foliage using a segmentation step. Generating point clouds from the tree meshes is done by emulating a spiral drone flight path around each tree and capturing RGBD images at a resolution of 2.1 megapixels (1920 $\times$ 1080 pixels).

We randomly select a sky-box for each tree for varying lighting conditions. The depth maps undergo augmentations such as jitter, dilation and erosion to replicate photogrammetry noise. We extract a point cloud from the fused RGBD images. We remove duplicate points by performing a voxel downsample at a 1cm resolution. The final point clouds have artefacts such as varying point density, missing areas (some due to self-occlusion) and noise.

We select six species with SpeedTree models. We produce 100 point clouds (600 total) for each of the six tree species, which vary in intricacy and size. Of these, 480 are used for training, while 60 are reserved for validation and 60 for testing. We show images of the synthetic point cloud dataset in the results section (Sect. 3). Future revisions will include point clouds with foliage and a wider variety of species.

2.2 Skeletonization Overview

Our skeletonization method comprises several stages shown in Fig. 2. We use labelled point clouds to train a sparse convolutional neural network to predict each input point's radius and direction toward the medial axis (ground truth skeleton). Using these predictions, we then translate surface points to estimated medial axis positions and construct a constrained neighbourhood graph. We extract the skeleton using a greedy algorithm to find paths from the root to terminal points. The neural network predictions help to avoid ambiguities with unknown branch radii and separate points which would be close in proximity but from different branches.

2.3 Neural Network

Our network takes an input set of N arbitrary points $\{Pi|i = 1, ..., N\}$, where each point Pi is a vector of its (x, y, z) coordinates plus additional features such as colour (r, g, b). Each point is voxelized at a resolution of 1cm. Our proposed network will then, for each voxelized point, learn an associated radius $\{Ri|i = 1, ..., N\}$ where Ri is a vector of corresponding radii and a direction vector $\{Di|i = 1, ..., N\}$ where Di is a normalized direction vector pointing towards the medial axis.

The network is implemented as a submanifold sparse CNN using SpConv [9], and PyTorch [26]. We use regular sparse convolutions on the encoder blocks for a wider exchange of features and submanifold convolutions elsewhere for more efficient computation due to avoiding feature dilation. The encoder block uses a stride of 2. Each block uses a kernel size of $3 \times 3 \times 3$, except for the first sub-manifold convolution and the fully connected blocks, which use a kernel size of $1 \times 1 \times 1$.

The architecture comprises a UNet backbone [29] with residual connections, followed by two smaller fully connected networks to extract the radii and directions. A ReLU activation function and batch normalization follow each convolutional layer. We add a fully-connected network head when branch-foliage segmentation is required.

A high-level overview of the network architecture is shown in Fig. 1.

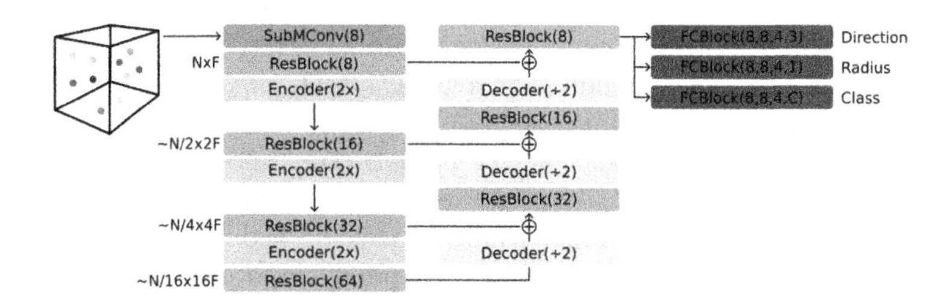

Fig. 1. Network architecture diagram.

A block sampling scheme ensures the network can process larger trees. During training, for each point cloud, we randomly sample (at each epoch) a $4\,\text{m}^3$ block and mask the outer regions of the block to avoid inaccurate predictions from the edges. We tile the blocks during inference, overlapping the masked regions.

We estimate a logarithmic radius to handle the variation in branch radii, which spans several orders of magnitude [10], which provides a relative error. The loss function (Eq. 3) comprises two components. Firstly we use the L1-loss for the radius (Eq. 2) and the cosine similarity (Eq. 1) for the direction loss. We use the Adam optimizer, a batch size of 16, and a learning rate of 0.1. The learning rate decays by a factor of 10 if the validation data-set loss does not improve for 10 consecutive epochs.

$$L_D = \sum_{i=0}^{n} \frac{Di \cdot \hat{Di}}{||Di||_2 \cdot ||\hat{Di}||_2} \quad (1) \qquad L_R = \sum_{i=0}^{n} |\ln(Ri) - \hat{Ri}| \quad (2)$$

$$Loss = L_R + L_D \quad (3)$$

2.4 Skeletonization Algorithm

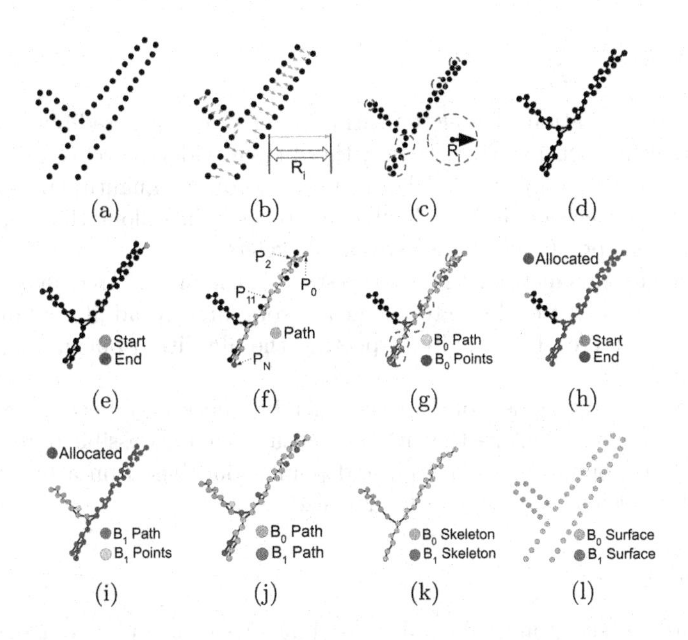

Fig. 2. Skeletonization Algorithm: (a) Input points, (b) Medial axis approximation, (c) Neighbourhood radius search, (d) Neighbourhood graph, (e) B_0 Farthest point, (f) B_0 Trace path,(g) B_0 Allocated points, (h) B_1 Farthest (unallocated) point, (i) B_1 Trace path and allocated points, (k) Branch skeletons, (i) Corresponding surface points.

Once $\{Ri\}$ and $\{Di\}$, have been predicted by the network. We use this information to project the input surface points $\{Pi\}$ (2a) onto the medial axis (2b).

We form a neighbourhood graph (2d) where points are connected to neighbours with weight equal to the distance between points and restricted to edges with a distance less than the predicted radius (2c).

As the point cloud has gaps due to self-occlusion and noise, we end up with multiple connected components. Each connected component we call a sub-graph. We process each sub-graph sequentially. For each sub-graph:

1. A distance tree is created based on the distance from the root node (the lowest point in each sub-graph - shown in red in (2e)) to each point in the sub-graph.
2. We assign each point a distance based on a Single Source Shortest Path (SSSP) algorithm. A greedy algorithm extracts paths individually until all points are marked as allocated (steps e to j).
3. We select a path to the furthest unallocated point and trace its path back to either the root (2e) or an allocated point (2i).
4. We add this path to a skeleton tree (2f).
5. We mark points as allocated which lie within the predicted radius of the path (2g).
6. We repeat this process until all points are allocated (2i, 2j)

3 Results

Performance evaluation of skeletonization algorithms is incredibly challenging and remains an open problem. Hence, there is no widely accepted metric used for evaluation. We compare our algorithms skeleton for quantitative evaluation using a form of precision and recall which matches points along the ground truth skeleton against points along the estimated skeleton.

We evaluate our method against the state-of-the-art AdTree algorithm [12]. As our metrics do not evaluate topological errors directly, additional qualitative analysis is conducted by visually inspecting the algorithm outputs against the ground truth.

Due to augmentations, some of the finer branches may become excessively noisy or disappear. To ensure the metrics measure what is possible to reconstruct, we prune the ground truth skeleton and point cloud based on a branch radius and length threshold - respective to tree size.

3.1 Metrics

We evaluate our skeletons using point cloud metrics by sampling our skeletons at a 1mm resolution. We use the following metrics to assess our approach: f-score, precision, recall and AUC. For the following metrics, we consider $p \in \mathcal{S}$ points along the ground truth skeleton and $p^* \in \mathcal{S}^*$ points obtained by sampling the output skeleton. p_r is the radius at each point. We use a threshold variable t, which sets the distance points must be within based on a factor of the ground truth radius. We test this over the range of $0.0 - 1.0$. The f-score is the harmonic mean of the precision and recall.

Skeletonization Precision. To calculate the precision, we first get the nearest points from the output skeleton $p_i \in \mathcal{S}$ to the ground truth skeleton $p_j^* \in \mathcal{S}^*$, using a distance metric of the euclidean distance relative to the ground truth radius r_j^*. The operator $[\![.]\!]$ is the Iverson bracket, which evaluates to 1 when the condition is true; otherwise, 0.

$$d_{ij} = ||p_i - p_j^*|| \tag{4}$$

$$P(t) = \frac{100}{|S|} \sum_{i \in \mathcal{S}} [\![d_{ij} < t\, r_j^* \wedge \underset{k \in \mathcal{S}}{\forall} d_{ij} \leq d_{kj}]\!] \tag{5}$$

Skeletonization Recall. To calculate the recall, we first get the nearest points from the ground truth skeleton $p_j^* \in \mathcal{S}^*$ to the output skeleton $p_i \in \mathcal{S}$. We then calculate which points fall inside the thresholded ground truth radius. This gives us a measurement of the completeness of the output skeleton.

$$R(t) = \frac{100}{|S^*|} \sum_{j \in \mathcal{S}^*} [\![d_{ij} < t\, r_j^* \wedge \underset{k \in \mathcal{S}^*}{\forall} d_{ij} \leq d_{ik}]\!] \tag{6}$$

3.2 Quantitative Results

We evaluate our method on our test set of sixty synthetic ground truth skeletons (made up of six species). Our results are summarized in Table 1 and illustrated in Fig. 3. We compute the AUC for F1, precision, and recall using the radius threshold t ranging from 0 to 1.

Table 1. Skeletonization Results.

Metric	Smart-Tree	AdTree
Precision AUC	0.53	0.21
Recall AUC	0.40	0.38
F1 AUC	0.45	0.26

Smart-Tree achieves a high precision score, with most points being close to the ground truth skeleton (Fig. 3a). Compared to AdTree, Smart-Tree has lower recall at the most permissive thresholds, and this is due to Smart-Tree's inability to approximate missing regions of the point cloud, making it prone to gaps. AdTree, on the other hand, benefits from approximating missing regions. However, this also leads to AdTree having more topological errors and lower precision. Smart-Tree consistently achieves a higher F1 score and AUC for precision, recall, and F1, respectively.

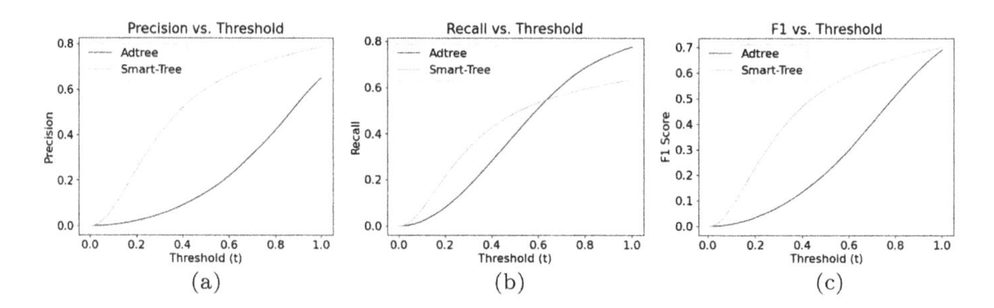

Fig. 3. Left to right: Precision Results (a), Recall Results (b), F1 Results (c).

3.3 Qualitative Results

In Fig. 5, we show qualitative results for each species in our synthetic dataset. We can see that Smart-Tree produces an accurate skeleton representing the tree topology well. Adtree produces additional branches that would require post-processing to remove.

AdTree often fails to capture the correct topology of the tree. This is due to Adtree using Delaunay triangulation to initialise the neighbourhood graph. This can lead to branches that have no association being connected.

Smart-Tree, however, does not generate a fully connected skeleton - but rather one with multiple sub-graphs. The biggest sub-graph can still capture the majority of the major branching structure, although to provide a full topology of the tree with the finer branches, additional work is required to connect the sub-graphs by inferring the branching structure in occluded regions.

To demonstrate our method's ability to work on real-world data. We test our method on a tree from the Christchurch Botanic Gardens, New Zealand. As this tree has foliage, we train our network to segment away the foliage points and then run the skeletonization algorithm on the remaining points. In Fig. 4c, we can see that Smart-Tree can accurately reconstruct the skeleton.

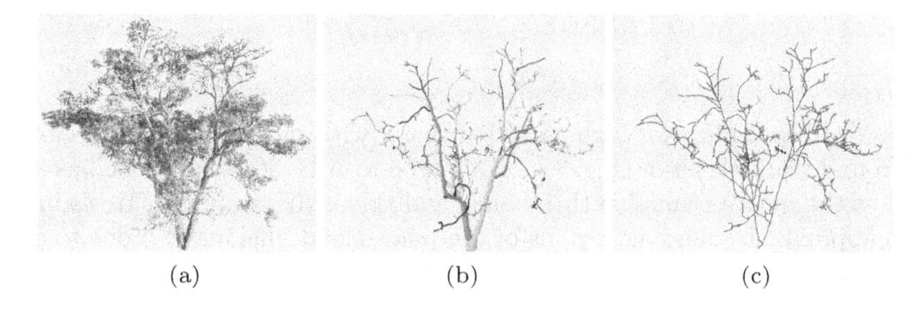

Fig. 4. Left to Right: Input point cloud (a), Branch meshes (b), Skeleton Sub-graphs (c).

Fig. 5. Several results of skeletonization of synthetic point clouds. From left to right: synthetic point cloud, ground truth skeleton, Smart-Tree skeleton, and Adtree skeleton. From top to bottom (species): cherry, eucalyptus, walnut, apple, pine, ginkgo

4 Conclusion and Future Work

We proposed Smart-Tree, a supervised method for generating skeletons from tree point clouds. A major area for improvement in the literature on tree point-cloud skeletonization is quantitative evaluation, which we contribute towards with a synthetic tree point cloud dataset with ground-truth and error metrics.

We demonstrated that using a sparse convolutional neural network can help improve the robustness of tree point cloud skeletonization. One novelty of our work is that a neighbourhood graph can be created based on the radius at each region, improving the accuracy of our skeleton.

We used a precision and recall-based metric to compare Smart-Tree with the state-of-the-art AdTree. Smart-Tree is generally much more precise than AdTree, but it currently does not handle point clouds containing gaps (due to occlusions and reconstruction errors). AdTree has problems with over-completeness on this dataset, with many duplicate branches.

In the future, we would like to work towards robustness to gaps in the point cloud by filling gaps in the medial-axis estimation phase. We plan to train our method on a wider range of synthetic and real trees; to do this, we will expand our dataset to include more variety, trees with foliage, and human annotation on real trees. We are also working towards error metrics which better capture topology errors.

Acknowledgment. This work was funded by the New Zealand Ministry of Business, Innovation and Employment under contract C09X1923 (Catalyst: Strategic Fund).

This research/project is supported by the National Research Foundation, Singapore under its Industry Alignment Fund - Pre-positioning (IAF-PP) Funding Initiative. Any opinions, findings and conclusions or recommendations expressed in this material are those of the author(s) and do not reflect the views of the National Research Foundation, Singapore.

References

1. Arikapudi, R., Vougioukas, S.G.: Robotic tree-fruit harvesting with telescoping arms: a study of linear fruit reachability under geometric constraints. IEEE Access **9**, 17114–17126 (2021)
2. Botterill, T., et al.: A robot system for pruning grape vines. J. Field Robot. **34**(6), 1100–1122 (2017)
3. Calders, K., et al.: Terrestrial laser scanning in forest ecology: Expanding the horizon. Remote Sens. Environ. **251**, 112102 (2020)
4. Cao, J., Tagliasacchi, A., Olson, M., Zhang, H., Su, Z.: Point cloud skeletons via laplacian based contraction. In: 2010 Shape Modeling International Conference, pp. 187–197. IEEE (2010)
5. Cárdenas-Donoso, J.L., Ogayar, C.J., Feito, F.R., Jurado, J.M.: Modeling of the 3D tree skeleton using real-world data: a survey. IEEE Trans. Vis. Comput. Graph. (2022)
6. Chaudhury, A., Godin, C.: Skeletonization of plant point cloud data using stochastic optimization framework. Front. Plant Sci. **11**, 773 (2020)

7. Chaudhury, A., et al.: Machine vision system for 3D plant phenotyping. IEEE/ACM Trans. Comput. Biol. Bioinf. **16**(6), 2009–2022 (2018)
8. Choy, C., Gwak, J., Savarese, S.: 4D spatio-temporal convnets: Minkowski convolutional neural networks. In: Proceedings of the IEEE/CVF Conference on Computer Vision and Pattern Recognition, pp. 3075–3084 (2019)
9. Contributors, S.: Spconv: Spatially sparse convolution library (2022). https://github.com/traveller59/spconv
10. Dassot, M., Fournier, M., Deleuze, C.: Assessing the scaling of the tree branch diameters frequency distribution with terrestrial laser scanning: methodological framework and issues. Ann. For. Sci. **76**, 1–10 (2019)
11. Delagrange, S., Jauvin, C., Rochon, P.: PypeTree: a tool for reconstructing tree perennial tissues from point clouds. Sensors **14**(3), 4271–4289 (2014)
12. Du, S., Lindenbergh, R., Ledoux, H., Stoter, J., Nan, L.: AdTree: accurate, detailed, and automatic modelling of laser-scanned trees. Remote Sens. **11**(18), 2074 (2019)
13. Fan, G., et al.: A new quantitative approach to tree attributes estimation based on lidar point clouds. Remote Sens. **12**(11), 1779 (2020)
14. Fan, G., Nan, L., Dong, Y., Su, X., Chen, F.: AdQSM: a new method for estimating above-ground biomass from TLS point clouds. Remote Sens. **12**(18), 3089 (2020)
15. Gorte, B.: Skeletonization of laser-scanned trees in the 3D raster domain. In: Abdul-Rahman, A., Zlatanova, S., Coors, V. (eds.) Innovations in 3D Geo Information Systems. Lecture Notes in Geoinformation and Cartography, pp. 371–380. Springer, Heidelberg (2006). https://doi.org/10.1007/978-3-540-36998-1_29
16. Gorte, B., Pfeifer, N.: Structuring laser-scanned trees using 3D mathematical morphology. Int. Arch. Photogram. Remote Sens. **35**(B5), 929–933 (2004)
17. Graham, B., Van der Maaten, L.: Submanifold sparse convolutional networks. arXiv preprint arXiv:1706.01307 (2017)
18. Guo, Y., Wang, H., Hu, Q., Liu, H., Liu, L., Bennamoun, M.: Deep learning for 3D point clouds: a survey. IEEE Trans. Pattern Anal. Mach. Intell. **43**(12), 4338–4364 (2020)
19. Hackenberg, J., Spiecker, H., Calders, K., Disney, M., Raumonen, P.: Simpletree-an efficient open source tool to build tree models from TLS clouds. Forests **6**(11), 4245–4294 (2015)
20. Huang, H., et al.: L1-medial skeleton of point cloud. ACM Trans. Graph. **32**(4), 65–1 (2013)
21. Interactive Data Visualization, I.: The standard for vegetation modeling and middleware. https://store.speedtree.com/
22. Kankare, V., Holopainen, M., Vastaranta, M., Puttonen, E., Yu, X., Hyyppä, J., Vaaja, M., Hyyppä, H., Alho, P.: Individual tree biomass estimation using terrestrial laser scanning. ISPRS J. Photogramm. Remote. Sens. **75**, 64–75 (2013)
23. Liu, Y., Guo, J., Benes, B., Deussen, O., Zhang, X., Huang, H.: TreePartNet: neural decomposition of point clouds for 3D tree reconstruction. ACM Trans. Graph. **40**(6) (2021)
24. Livny, Y., Yan, F., Olson, M., Chen, B., Zhang, H., El-Sana, J.: Automatic reconstruction of tree skeletal structures from point clouds. In: ACM SIGGRAPH Asia 2010 papers, pp. 1–8. ACM (2010)
25. Molina-Valero, J.A., et al.: Operationalizing the use of TLS in forest inventories: the R package FORTLS. Environ. Modell. Softw. **150**, 105337 (2022)
26. Paszke, A., et al.: PyTorch: an imperative style, high-performance deep learning library. Adv. Neural Inf. Process. Syst. **32** (2019)
27. Qi, C.R., Yi, L., Su, H., Guibas, L.J.: PointNet++: deep hierarchical feature learning on point sets in a metric space. Adv. Neural Inf. Process. Syst. **30** (2017)

28. Raumonen, P., et al.: Fast automatic precision tree models from terrestrial laser scanner data. Remote Sens. **5**(2), 491–520 (2013)
29. Ronneberger, O., Fischer, P., Brox, T.: U-net: convolutional networks for biomedical image segmentation. In: Navab, N., Hornegger, J., Wells, W.M., Frangi, A.F. (eds.) MICCAI 2015. LNCS, vol. 9351, pp. 234–241. Springer, Cham (2015). https://doi.org/10.1007/978-3-319-24574-4_28
30. Saha, P.K., Borgefors, G., di Baja, G.S.: A survey on skeletonization algorithms and their applications. Pattern Recogn. Lett. **76**, 3–12 (2016)
31. Spalding, E.P., Miller, N.D.: Image analysis is driving a renaissance in growth measurement. Curr. Opin. Plant Biol. **16**(1), 100–104 (2013)
32. Tang, H., Liu, Z., Li, X., Lin, Y., Han, S.: TorchSparse: efficient point cloud inference engine. Proc. Mach. Learn. Syst. **4**, 302–315 (2022)
33. Tompalski, P., et al.: Estimating changes in forest attributes and enhancing growth projections: a review of existing approaches and future directions using airborne 3D point cloud data. Curr. For. Rep. **7**, 1–24 (2021)
34. Verroust, A., Lazarus, F.: Extracting skeletal curves from 3D scattered data. In: Proceedings Shape Modeling International 1999. International Conference on Shape Modeling and Applications, pp. 194–201. IEEE (1999)
35. White, J.C., Wulder, M.A., Vastaranta, M., Coops, N.C., Pitt, D., Woods, M.: The utility of image-based point clouds for forest inventory: a comparison with airborne laser scanning. Forests **4**(3), 518–536 (2013)
36. Xu, H., Wang, C.C., Shen, X., Zlatanova, S.: 3D tree reconstruction in support of urban microclimate simulation: a comprehensive literature review. Buildings **11**(9), 417 (2021)
37. Xu, H., Gossett, N., Chen, B.: Knowledge and heuristic-based modeling of laser-scanned trees. ACM Trans. Graph. (TOG) **26**(4), 19-es (2007)
38. Zahid, A., Mahmud, M.S., He, L., Heinemann, P., Choi, D., Schupp, J.: Technological advancements towards developing a robotic pruner for apple trees: a review. Comput. Electron. Agric. **189**, 106383 (2021)

A Measure of Tortuosity for 3D Curves: Identifying 3D Beating Patterns of Sperm Flagella

Andrés Bribiesca-Sánchez[1](✉), Adolfo Guzmán[2], Alberto Darszon[1], Gabriel Corkidi[1], and Ernesto Bribiesca[3]

[1] Instituto de Biotecnología, Universidad Nacional Autónoma de México, Cuernavaca, Morelos, Mexico
javier.bribiesca@ibt.unam.mx

[2] Centro de Investigación en Computación, Instituto Politécnico Nacional, Mexico City, Mexico

[3] Instituto de Investigaciones en Matemáticas Aplicadas y en Sistemas, Universidad Nacional Autónoma de México, Mexico City, Mexico

Abstract. Tortuosity is an intrinsic property of 3D curves, related to their meanders, turns, and twists. In this work, we present a measure of tortuosity for 3D polygonal curves. These curves are composed of straight-line segments of different lengths. The proposed measure of tortuosity is computed from the slope change (scaled to lie between 0 and 1) and torsion (scaled to lie between 0 and 0.5) at each vertex of the polygonal curve. This descriptor is invariant under translation, rotation, symmetry, starting point, and scaling (change of size). Additionally, a normalized measure of tortuosity that ranges from 0 to 1 is introduced, allowing the comparison of curves with varying numbers of vertices. These properties of the tortuosity make it useful for shape comparison, description, and classification, potentially contributing to the analysis of 3D tubular structures in nature such as veins, optic nerves, proteins, and DNA chains. As examples of applications, we show how it can be used to quantitatively analyze and rank closed curves. Finally, we show that 3D tortuosity is a key feature in identifying 3D flagellar beat patterns of human sperm that correlate with the cell's fertility, making it an important feature to include in clinical and scientific software.

Keywords: 3D tortuosity · 3D polygonal curve · 3D sperm flagellar beat

1 Introduction

The tortuosity of curves is an important subject matter in pattern recognition, computer vision, and visualization. It is related to the twists and turns of curves;

This project is supported by CONACYT (PhD scholarship) and DGAPA PAPIIT IN105222. We thank the anonymous reviewers for their helpful feedback. We gratefully acknowledge the support of Paul Hernandez-Herrera in image segmentation.

A. Pertusa et al. (Eds.): IbPRIA 2023, LNCS 14062, pp. 363–374, 2023.
https://doi.org/10.1007/978-3-031-36616-1_29

to how crooked the curve is. The classical measure of tortuosity is the ratio between the curve's arc length and the length of the underlying chord, however, the abundance and complexity of lines and tubular structures in nature and engineering have promoted the development of more robust tortuosity measures. These have particularly contributed to biomedical sciences, where links between the tortuosity of blood vessels and the severity of eye diseases, aneurysms, and tumors have been found [1,3,6,8]. Bribiesca [1] defined a tortuosity measure for curves in the 2D domain, based on a manner to codify 2D changes in slope, called slope chain code (SCC). The application of his 2D tortuosity to retinal images assisted in the diagnosis of retinopathy of prematurity (ROP). A review of measures of tortuosity can be found in [7].

In this paper, we propose a tortuosity measure for 3D polygonal curves, based on the definition of 2D tortuosity introduced by Bribiesca [1]. 3D polygonal curves are composed of 3D straight-line segments of different lengths. This measure of tortuosity is based on the computation of the slope changes and torsions between contiguous straight-line segments, scaled to a continuous range from 0 to 1. The proposed measure of tortuosity is valid for open and closed curves; it is invariant under translation, rotation, scaling, reflection, and starting point, and different resolutions. We show a real-world application of this descriptor by using it to identify 3D flagellar beat patterns of human sperm that are necessary for fertilization.

The main differences between the measure proposed by Bribiesca [1] and the present one are the following: • The tortuosity measure proposed here is for 3D curves, while the former was for the 2D domain. This opens the way to analyze real-world 3D curves, permitting more applications (identifying 3D flagellar beat patterns of human sperm, atherosclerotic changes in arteries, and inspecting aircraft routes, to name a few), not just those pertaining to flat curves. • When a curve is confined to a plane (2D), the rotation between a pair of connected segments can be expressed as a single angle (slope change). 3D lines have an extra degree of freedom, which allows the next vertex to tilt above or below the plane containing the previous pair of segments, a phenomenon called torsion. Thus, the 2D measure of tortuosity is based solely on the sum of the slope changes, while the 3D measure considers both, the slope change and the torsion at each vertex. • The SCC code can reproduce the shape of the 2D curve, while we now only compute the 3D tortuosity. • Finally, the slope change between a pair of segments in a 2D curve is positive if the rotation is clockwise and negative if it is counterclockwise, whereas for 3D curves it's always positive, regardless of the relative direction of the segments, which facilitates the computation 3D tortuosity. As a consequence, the concept of accumulated slope describes different properties for 2D and 3D curves. Because it includes positive and negative angles, the 2D accumulated slope is helpful in identifying whether a curve is closed or open, as well as analyzing its concavity. On the other hand, the 3D accumulated slope is a measurement of how much a polygonal curve deviates at each vertex from a forward linear path.

2 Definitions

It is assumed that, once a 3D curve has been isolated from an image, it is converted to a 3D polygonal curve, formed by straight-line segments.

Definition 1 (Polygonal curve). *A 3D polygonal curve is a 3D curve formed by a sequence of straight-line segments, connecting consecutive vertices.*

Figure 1(a) illustrates an example of a 3D polygonal curve. We refer to vertices connecting a pair of segments as inner vertices and vertices connected to only one segment as outer vertices. A 3D polygonal curve is formed by m straight-line segments and n inner vertices. Thus, $n = m-1$ for open curves. Closed curves inherently do not have outer vertices so in that case $n = m$. Notice that the segments may be of different lengths. Later, for scale independence (Sect. 3.4), we will require them to be of the same size.

Definition 2 (Displacement vector). *The displacement vector j is the difference between the position of vertex j and $j+1$. It is denoted as $\boldsymbol{v}_j$.*

Definition 3 (Slope change). *The slope change at vertex j is the angle formed by the line segments connected at this vertex, normalized to lie in [0,1). We denote the slope change vertex j as α_j.*

In simple terms, the slope change describes how much two consecutive segments deviate from a straight-line path, ranging from 0 (both segments have the same slope) to 1 (the segments have opposite slopes), as shown in Fig. 1(b).

Definition 4 (Torsion). *The torsion at vertex j (which connects a pair of segments s_j and s_{j+1}) is the angle that a previous segment s_i connected to vertex j through a straight line path must roll along s_j to be in the same plane as s_{j+1} instead. The torsion is normalized to lie in between 0 (no rolling is needed) and 0.5 (s_{j+1} produces a 90° tilt). We denote the torsion at vertex j as β_j.*

In other words, torsion measures the elevation angle of a sequence of segments. Vertex i is the most recent corner before j, i.e. the last vertex with $\alpha_i \neq 0$. If $\alpha_{j-1} \neq 0$, i = j − 1; Otherwise, i is selected so that the slope changes of vertices i + 1, i + 2, ..., j − 1, are all 0. The computation of the torsion will be further explained in Sect. 2.2.

Torsion is only present in 3D polygonal curves comprised of 3 or more segments since any curve formed by fewer is flat. Despite being a simple observation, it is helpful to understand that slope change is a property of pairs of contiguous line segments, while torsion is a feature of longer sequences. Therefore, the first inner vertex of an open polygonal curve does not have torsion because there is no prior rotation. For this reason, an open curve has n slope changes $(\alpha_0, \alpha_1, \alpha_2, \ldots, \alpha_n)$ and $n-1$ torsions $(\beta_1, \beta_2, \ldots, \beta_n)$, whereas a closed one has an equal amount of both ($\alpha_1, \alpha_2, \ldots, \alpha_n$ and $\beta_1, \ldots, \beta_n$). We denote the number of torsions of a curve as n_t.

2.1 Computation of the Slope Changes of a 3D Polygonal Curve

To obtain the slope changes of any curve, we first need to select the origin of the 3D polygonal curve. The selected origin of the curve shown in Fig. 1(a) is represented by a sphere. Thus, the first computed slope change between the contiguous segments is illustrated in Fig. 1(c). In general, the slope change between a pair of segments is obtained by means of the following equation:

$$\alpha_j = \arccos(\frac{\boldsymbol{v}_j \cdot \boldsymbol{v}_{j+1}}{||\boldsymbol{v}_j||\,||\boldsymbol{v}_{j+1}||}) \tag{1}$$

where $\boldsymbol{v}_j \cdot \boldsymbol{v}_{j+1}$ is the dot-product of these vectors ($\boldsymbol{v}_j \cdot \boldsymbol{v}_{j+1} = u_j u_{j+1} + v_j v_{j+1} + w_j w_{j+1}$ and $||\boldsymbol{v}_j||$ is the magnitude of the displacement vector. The slope change at that vertex of the curve is considered as $\frac{\alpha}{\pi}$. Slope changes equal to 1 are not considered since that would imply that two consecutive segments coincide. The next slope change is computed considering the next 3D straight-line segments, and so on. Figure 1(d) shows all the slope changes of the curve.

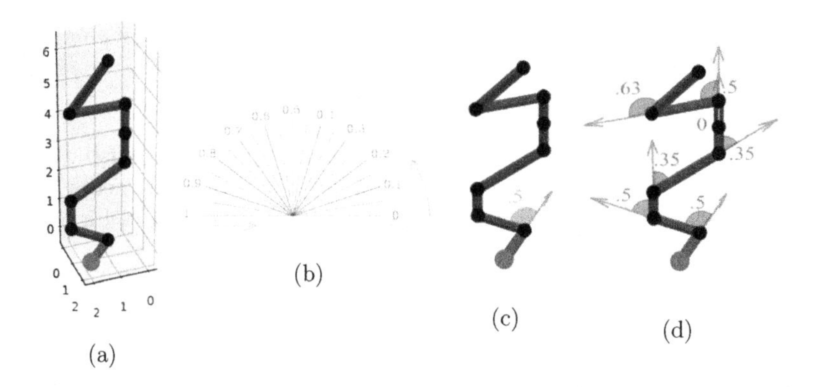

Fig. 1. Slope change computation: (a) an example of a 3D polygonal curve, (b) range of slope changes [0, 1), (c) first step in the computation of the slope changes, and (d) computation of all the slope changes of the curve. Notice that the segments of the curve are not of the same length.

2.2 Computation of the Torsions of a 3D Polygonal Curve

The torsions of a 3D polygonal curve can be obtained by computing the angle between planes containing segments connected at vertex i and vertex j. This is equivalent to computing the angle between vectors that are perpendicular to these planes. This can be expressed as:

$$\beta_j = \arccos(\frac{|\boldsymbol{q}_i \cdot \boldsymbol{q}_j|}{||\boldsymbol{q}_i||\,||\boldsymbol{q}_j||}), \tag{2}$$

where $\boldsymbol{q}_i$ and $\boldsymbol{q}_j$ are perpendicular vectors to the segments intersecting vertex i and j respectively, obtained through the cross product ($\boldsymbol{q}_i = \boldsymbol{v}_i \times \boldsymbol{v}_{i+1}$, $\boldsymbol{q}_j = \boldsymbol{v}_j \times \boldsymbol{v}_{j+1}$) and $|\boldsymbol{q}_i \cdot \boldsymbol{q}_j|$ is the absolute value of the dot product. Notice that the argument of Eq. 2 is constrained in $[0, 1]$. This is an important detail because torsion is a measure of non-flatness, so it should be minimal when the sequence is flat and maximal at 90°. Thus, if the angle formed by the planes is between 90 and 180°, the supplementary angle should be taken instead, which the absolute value does implicitly.

Figure 2(a) shows how to compute the first torsion of a 3D polygonal curve. Since the first vertex of an open polygonal curve does not have torsion, we must start by computing the torsion of the second one. Then proceed to the next vertex and so on. On the other hand, for a closed curve, we can easily compute the torsion of the first vertex by looking at the previous ones. Figure 2(b) shows all of the torsions the curve in Fig. 1(a).

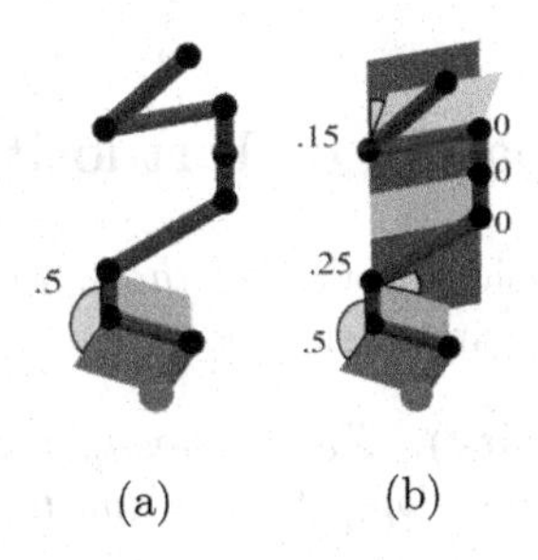

Fig. 2. Torsion computation: (a) First torsion of an open curve; (b) Torsion at every vertex represented as the angle between planes containing each pair of segments.

Slope changes being zero makes the algorithm to compute torsions slightly more complicated because a tangent plane could have any orientation and the frame of reference for torsion would be lost. Thus, the algorithm uses Remark 1 and keeps track of the last corner vertex i to use as a reference for the torsion of vertex j. Also, if the initial vertex of an open curve is zero, an initial plane of reference for torsion cannot be immediately established because the vectors are co-linear. In this case, we set β_1=0. We repeat $\beta_{j+1} = 0$ for every vertex j until finding $\alpha_j \neq 0$, which finally allows us to set the tangent plane. The torsion at vertex $j + 1$ will be the result between the segments vertex j and $j + 1$.

Remark 1. For any vertex j =1,...,n; If $\alpha_j = 0$, then $\beta_j = 0$

2.3 Definition and Computation of Accumulated Angles

Definition 5 (Accumulated slope). *The accumulated slope of a 3D polygonal curve is the sum of its slope changes, denoted as:*

$$S = \Sigma_0^n \alpha_j \quad \text{for open curves} \tag{3}$$
$$S = \Sigma_1^n \alpha_j \quad \text{for closed curves}$$

The accumulated slope of the polygonal curve in Fig. 1(a) is 2.83.

Remark 2. A 3D polygonal curve is a straight line if and only if its accumulated slope is 0.

Definition 6 (Accumulated torsion). *The accumulated torsion of a 3D polygonal curve is the sum of its torsion angles, i.e.:*

$$T = \Sigma_1^n \beta_j \tag{4}$$

The accumulated torsion of the curve in Fig. 2(a) is 0.9.

Remark 3. A polygonal curve is planar if and only if its accumulated torsion is 0.

3 The Proposed Measure of Tortuosity

Our aim is to find an intrinsic property of *the shape* of a 3D polygonal curve; one that is independent of position, size, etc.

Definition 7 (3D Tortuosity). *The tortuosity τ of a 3D polygonal curve is the sum of all the values of the slope changes and torsions, and is defined by*

$$\tau = S + T \tag{5}$$

The accumulated slope and torsion of the 3D polygonal curve shown in Fig. 1(a) are respectively *value* and *value*, thus, the numerical value of its tortuosity is equal to 3.73, i.e. $\tau = 2.83 + 0.9 = 3.73$). The following sections are dedicated to describing and illustrating the properties of this descriptor.

3.1 Independence of Translation, Rotation, and Scale

The proposed measure for 3D polygonal curves is invariant under translation, rotation, and scale (as long as the scale factor is equal for every axis). This is because only relative slope changes and torsion between contiguous straight-line segments are considered. Figure 3 illustrates different rotations, translations, and scales of the 3D polygonal curve presented in Fig. 1(a). The value of the tortuosity for all these transformed curves is 3.73.

3.2 Invariance Under Starting Point

The slope changes and torsions obtained by traveling the curve in one direction are just in the reverse order of the slope changes and torsions obtained by traveling the same curve in the opposite direction, so no matter in what direction we travel the curve, Eq. 5 will give the same value for τ. Thus, the tortuosity is invariant to the starting point (for open curves), or invariant to the direction of travel (for open and closed curves). Figure 4(a) and (b) show the invariance under the starting point for an open curve. Figure 4(c) and (d) show the invariance under the starting point and direction for a closed curve ($\tau = 5.5$).

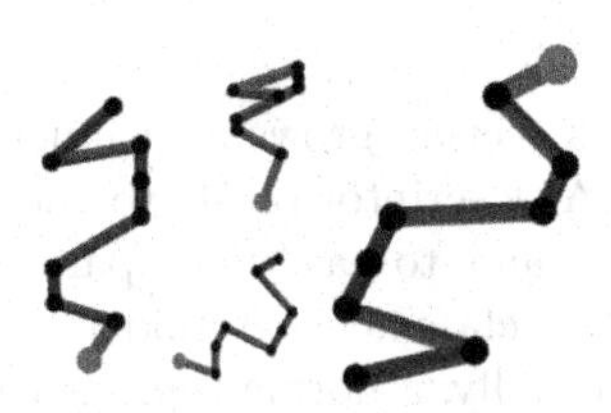

Fig. 3. Rotatated, translated, and scaled polygonal curve. All of them have the same tortuosity.

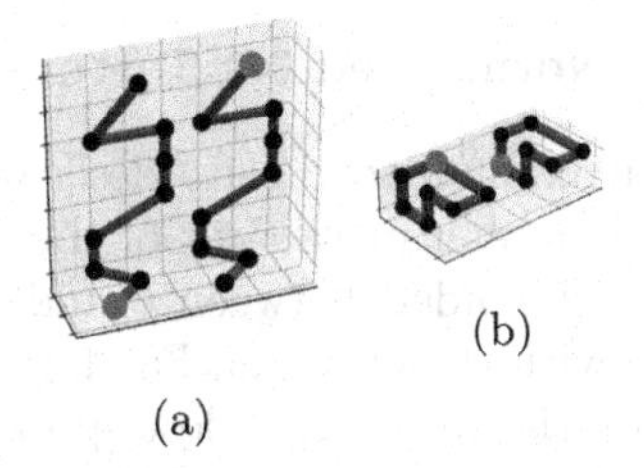

Fig. 4. Invariance under starting point. (a) The 3D polygonal curve shown in Fig. 1(a) is traversed in the direction shown, (b) closed curve with different starting points.

3.3 Invariance Under Mirror Imaging

A mirror does not distort the shape. Thus, a mirror image of a 3D polygonal curve will have all its slope changes unaltered, as seen in Fig 5. Therefore, Eq. 5 will give equal values for the tortuosity τ for the original image and for the mirror image.

3.4 The Extrema of 3D Tortuosity

From this section on, we will consider only 3D polygonal curves having m segments of equal size. A 3D polygonal curve with all its slope changes equal to zero has a tortuosity $\tau = 0$, its minimum value. In contrast, a curve with all slope changes ~1 and all torsion angles of 0.5 will have an accumulated slope of ~n and an accumulated torsion of $\frac{n_t}{2}$. That is the maximum for τ. Thus the tortuosity of any curve lies in $[0, n + \frac{n_t}{2})$.

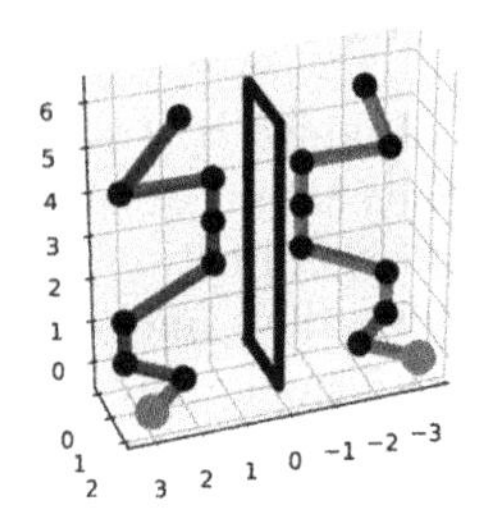

Fig. 5. The proposed measure of tortuosity is invariant under mirror imaging

3.5 Normalized 3D Tortuosity

The maximum tortuosity described in the previous section grows as a function of the number of vertices of the polygonal curve. A descriptor of 3D tortuosity that is bounded between 0 and 1 would be convenient to easily compare any curve with the extrema. Furthermore, regression and classification models often require the values of a descriptor to be bounded. Finally, a normalized measure of tortuosity allows the comparison of curves even if they have a different amount of vertices. Thus, we have the following.

Definition 8 (Normalized tortuosity).
The normalized tortuosity τ_N *of a 3D polygonal curve is equal to the tortuosity of the 3D polygonal curve divided by the upper bound (supremum) of the tortuosity given the number of vertices, i.e.*

$$\tau_N = \frac{\tau}{n + \frac{n_t}{2}}. \tag{6}$$

The range of values of this new measure is $[0, 1)$, varying continuously. Thus, the minimum value of the normalized tortuosity for a shape composed of n slope changes is zero (for example, the left curve in Fig. 6), and its maximum value is approximately one (for instance, the curve on the right side in Fig. 6).

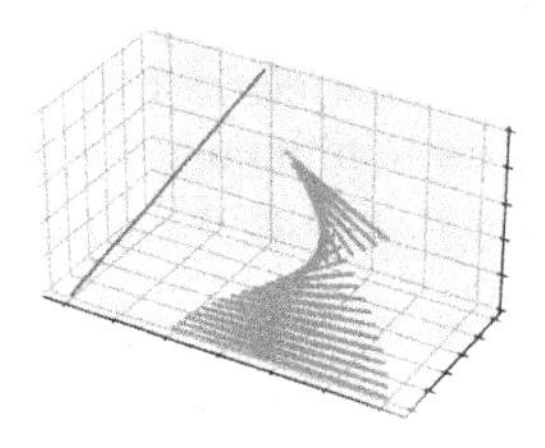

Fig. 6. 3D polygonal curves with the minimal and maximal tortuosity.

Tortuosity and normalized tortuosity are two measures related to the shape of 3D curves, independent of translation, size, etc. Therefore, they can be useful for ranking curves according to their tortuosity, for shape comparison, storage, compression, and classification (see Sect. 4.2), and to gauge the similarity between two 3D curves. In addition, important detection problems are related to the shape of 3D tubular structures and curves such as veins, respiratory airways, arteries, fissures in walls and pillars of buildings, etc.

4 Results

4.1 Analysis of Closed Curves by Means of Their Tortuosity

The study of closed curves has important applications in topology, graph theory, fluid mechanics, chemistry, etc. The presented measures can be helpful in these fields because they help translate qualitative visual observations into quantitative features and contribute to classifying and ranking these patterns. Figure 7 shows some sample closed curves and Table 1 describes their accumulated slope, accumulated torsion, and tortuosity. Figure 7(a) contains a simple knot. It is very smooth so it has the lowest tortuosity. Figure 7. Figures 7(b) has a high accumulated slope and torsion, making it highly tortuous. Figure 7(c) has a tilting, zigzagging pattern that increases the accumulated slope and torsion, making this curve the most tortuous, especially using the normalized measure, which is much higher than in Fig. 7(b) due to the lower number of vertices. Finally, Fig. 7(d) is fairly planar which gives it the lowest accumulated torsion, however, it has a significant tortuosity due to its loops. The ranking from least to most tortuous is: (a), (d), (c), and (b).

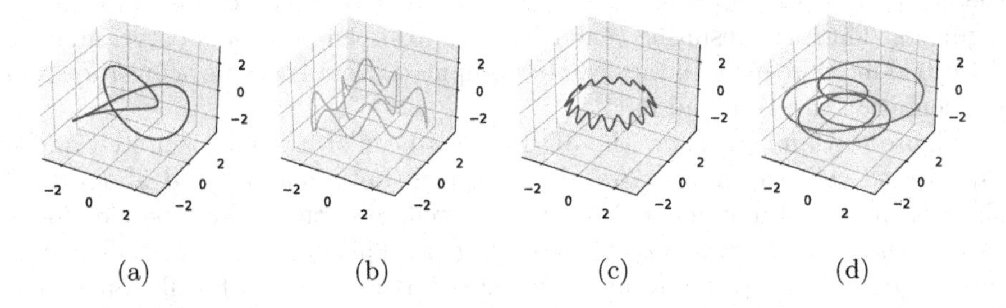

Fig. 7. Sample closed 3D polygonal curves.

4.2 Tortuosity of the 3D Flagellar Beat of Human Sperm

A human spermatozoon propels itself towards the egg through the 3D beat of its tail, known as the flagellum. During its journey, the sperm must undergo *capacitation*, which produces changes in the flagellar beat pattern that are essential for fertilization. Very few cells complete the capacitation process, and identifying them is difficult since it requires the extraction of features of the flagellar

Table 1. Values of the tortuosity of the 3D polygonal curves shown in Fig. 7. The columns describe the number of vertices (n), the accumulated slope (S), the accumulated torsion (T), the 3D tortuosity measure τ, and the normalized tortuosity τ_N.

Curve	n	S	T	τ	τ_N
(a)	200	4.44	1.61	6.05	0.02
(b)	150	15.4	15.6	31.0	0.14
(c)	100	26.13	7.24	33.37	0.22
(d)	100	7.99	0.7	8.69	0.06

beat, which is very fast, complex, and three-dimensional. Furthermore, freely-swimming spermatozoa rotate along their axis of movement as they swim (see Fig. 8(a–b)), which has hindered our ability to analyze freely-swimming sperm, since failing to account for this rotation can produce artifacts in the measurements [2,4]. Therefore, using descriptors that are invariant to the cells' rotation is imperative to ensure an accurate assessment of the flagellar beat pattern.

We monitored the 3D tortuosity of the beating flagellum of not capacitated and capacitated human sperm over short time intervals (~3 s). During each experiment, an individual, freely-swimming cell was recorded using an experimental device comprised of an inverted microscope piezoelectric device that makes the objective vibrate along the z-axis 90 Hz, covering a volume with a 20 μm height, while a high-speed camera records images of the sample at 8000 fps fps. Thus, images are recorded at different heights and arranged to reconstruct the 3D+t recorded volume. The center line of the flagellum was segmented from each 3D frame using a semi-automated path-finding algorithm that connects the head and distal point of the flagellum through a sequence of ~60 3D linear segments. Thus, the resulting center line is a 3D open polygonal curve, and the described measure of tortuosity can be computed. Details on the acquisition and segmentation can be found in [5].

Figure 8(a) and (b) show the 3D flagellar beat of a not capacitated and a capacitated cell elapsing 0.175 s, during which both cells rotated about 180° along their axis of movement. Notice the flagellum of the capacitated cell looks curvier than the other one during the process. This can be confirmed quantitatively by comparing the mean normalized tortuosity of both cells during the elapsed time, which is 0.112 for the capacitated cell and only 0.079 for the other one (see Fig. 8(c)). It is important to note that despite the complex 3D motion of the cell, the form of the flagellum remains similar, evidenced by the fact that the tortuosity isn't increasing or decreasing on average. There seems to be a fluctuation in the tortuosity of the flagellum of the activated cell, correlated with a subtle high-frequency flagellar beat. The capacitated cell has a more irregular pattern.

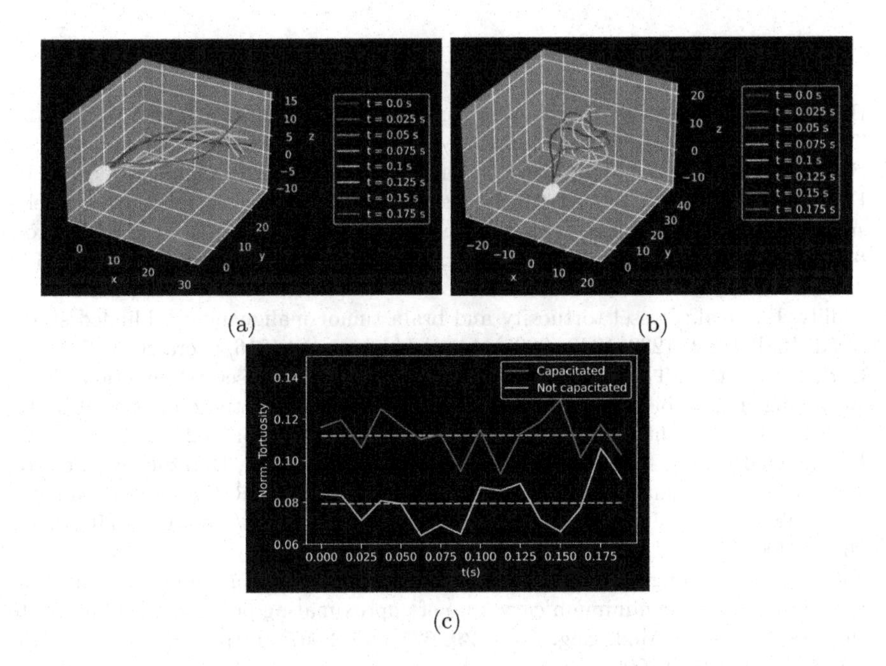

Fig. 8. Normalized 3D tortuosity of the flagellar beat of human sperm. Flagellum of (a) a not capacitated and (b) a capacitated sperm at different times. (c) Tortuosity of the flagellum of the activated sperm (blue) and hyperactivated sperm (red) (Color figure online).

This measure of tortuosity can be used for a variety of applications of open and closed 3D curves, including analyzing knots, airplane routes, flights of insects, etc.

5 Conclusion

A measure of tortuosity for open and closed 3D polygonal curves was defined. The measure has invariance properties that make it suitable for identifying objects despite being rotated, scaled, reflected, etc. Furthermore, the descriptor can be normalized between 0 and 1 which is helpful to compare and classify curves even when comprised of a different amount of segments. This descriptor has potential applications in computer vision and pattern recognition, such as the analysis of long molecules, veins, flights of insects, orbits of celestial bodies, and fissures in buildings. We demonstrated its effectiveness in a real-world application: analyzing the 3D flagellar beat of human sperm. The 3D tortuosity descriptor could be helpful in clinical systems for the analysis of male sub-fertility.

References

1. Bribiesca, E.: A measure of tortuosity based on chain coding. Pattern Recognit. **46**, 716–724 (2013). https://doi.org/10.1016/J.PATCOG.2012.09.017
2. Bribiesca-Sánchez, A., Montoya, F., González-Cota, A.L., Hernández-Herrera, P., Darszon, A., Corkidi, G.: Artifacts generated by the 3D rotation of a freely-swimming human sperm in the measurement of intracellular ca2+. IFMBE Proc. **86**, 355–362 (2023). https://link.springer.com/chapter/10.1007/978-3-031-18256-3_39
3. Bullitt, E., et al.: Vessel tortuosity and brain tumor malignancy: a blinded study. Acad. Radiol. **12**, 1232–1240 (2005). https://doi.org/10.1016/j.acra.2005.05.027
4. Corkidi, G., et al.: The human sperm head spins with a conserved direction during swimming in 3D. bioRxiv, p. 2022.11.28.517870 (2022). https://doi.org/10.1101/2022.11.28.517870, https://www.biorxiv.org/content/10.1101/2022.11.28.517870v1
5. Hernandez-Herrera, P., Montoya, F., Rendón-Mancha, J.M., Darszon, A., Corkidi, G.: 3-D + t human sperm flagellum tracing in low SNR fluorescence images. IEEE Trans. Med. Imaging **37**, 2236–2247 (2018). https://doi.org/10.1109/TMI.2018.2840047
6. Johnson, M.J., Dougherty, G.: Robust measures of three-dimensional vascular tortuosity based on the minimum curvature of approximating polynomial spline fits to the vessel mid-line. Med. Eng. Phys. **29**, 677–690 (2007). https://doi.org/10.1016/j.medengphy.2006.07.008
7. Krestanova, A., Kubicek, J., Penhaker, M.: Recent techniques and trends for retinal blood vessel extraction and tortuosity evaluation: a comprehensive review. IEEE Access **8**, 197787–197816 (2020). https://doi.org/10.1109/ACCESS.2020.3033027
8. Lisowska, A., Annunziata, R., Loh, G.K., Karl, D., Trucco, E.: An experimental assessment of five indices of retinal vessel tortuosity with the ret-tort public dataset. In: 2014 36th Annual International Conference of the IEEE Engineering in Medicine and Biology Society, EMBC 2014, pp. 5414–5417 (2014). https://doi.org/10.1109/EMBC.2014.6944850

The ETS2 Dataset, Synthetic Data from Video Games for Monocular Depth Estimation

David María-Arribas(✉), Alfredo Cuesta-Infante, and Juan J. Pantrigo

Computer Science and Statistics Department, Universidad Rey Juan Carlos, Calle Tulipán s/n, 28933 Móstoles, Madrid, Spain
david.maria@urjc.es

Abstract. In this work, we present a new dataset for monocular depth estimation created by extracting images, dense depth maps, and odometer data from a realistic video game simulation, Euro Truck Simulator 2^{TM}. The dataset is used to train state-of-the-art depth estimation models in both supervised and unsupervised ways, which are evaluated against real-world sequences. Our results demonstrate that models trained exclusively with synthetic data achieve satisfactory performance in the real domain. The quantitative evaluation brings light to possible causes of domain gap in monocular depth estimation. Specifically, we discuss the effects of coarse-grained ground-truth depth maps in contrast to the fine-grained depth estimation. The dataset and code for data extraction and experiments are released open-source.

Keywords: Synthetic dataset · Depth estimation · Video game simulation · Domain gap

1 Introduction

Depth estimation is a key process for a multitude of applications in computer vision where a physical agent must interact in an environment that it perceives through 2D images; e.g. robot-assisted surgery, production processes with closed-loop vision based control, 3D scene reconstruction or autonomous driving. It is also necessary in computer graphics to provide realism or a sense of immersion when the source are 2D real images.

Time-of-flight (TOF) cameras provide accurate estimations but the longer the range, the more expensive they are. Stereo cameras are able to measure the disparity and obtain the depth from it after a calibration process. Monocular cameras are ubiquitous but depth estimation from their images is an ill-posed regression problem.

This research work has been supported by project TED2021-129162B-C22, funded by the Recovery and Resilience Facility program from the NextGenerationEU and the Spanish Research Agency (Agencia Estatal de Investigación); and PID2021-128362OB-I00, funded by the Spanish Plan for Scientific and Technical Research and Innovation of the Spanish Research Agency.

A. Pertusa et al. (Eds.): IbPRIA 2023, LNCS 14062, pp. 375–386, 2023.
https://doi.org/10.1007/978-3-031-36616-1_30

Deep learning has proven its superiority over other computer vision methods when a ground truth is available. Thus, the supervised solution for depth estimation from monocular cameras would be a convolutional neural network (CNN) with a regression top that learns the depth of every pixel. Clearly, there are two severe drawbacks: 1) it is task-dependent because the backbone has learned to extract visual features from the data set, and 2) it requires the use of TOF cameras to provide a ground-truth. One stream of research is focused on reducing, or even eliminating, the dependency on labels leading to semi-supervised, self-supervised and unsupervised approaches. Another one is to generate synthetic ground-truth data sets. This paper follows the latter.

Computer Graphics Imagery (CGI) techniques are able provide hyper-realistic 3D simulations, but it has been shown that training CNNs with them and testing on the real images works much worse than expected. This issue is referred to as *Domain gap*, and has motivated an intense research on synthetic image generation.

Game engines are an inexpensive way to generate very realistic 3D images with lots of additional information in a totally controlled and parameterized world. Specifically, in this paper we use Euro Truck Simulator 2^{TM} (ETS2) to create and openly distribute a data set for training monocular depth estimation CNNs on urban environments. Our experiments show a possible reason for the domain gap specific of this task. We also remark that ETS2 can be used in many other computer vision tasks such as object detection, semantic segmentation, multi-object tracking or scene reasoning, just to mention a few. Thus, this paper makes the following contributions:

- The ETS2 data set, which consists of 18 video-sequences with 2D images, together with depth maps and odometer data.
- Validation of ETS2 on two state-of-the-art depth estimation models, namely Monodepth2 and Densenet, trained with the KITTI ground-truth.
- Discussion about the reasons of the domain gap issue.

ETS2, as well as all the data and tools are publicly available:

- ETS2 data set: https://ets2-dataset.dmariaa.es/
- Extracting data from ETS2: https://github.com/dmariaa/ets2-data-capture
- Data set tools: https://github.com/dmariaa/dataset-tools
- Monodepth2 experiments: https://github.com/dmariaa/monodepth2
- Densedepth experiments: https://github.com/dmariaa/ets2-trainer.

2 Related Works

Over the past few years, a variety of datasets, encompassing both authentic and synthesized data, have been generated to fulfill the data requirements for training neural networks dedicated to depth estimation (refer to Table 1).

Table 1. Related datasets size comparison

	dataset	frames	ground truth
real	Make3D	984	depth
	Kitti	23.488	depth + segmentation
	Cityscapes	25.000	segmentation + depth (SGM)
	NYU Depth	407.024	depth
synthetic	Virtual Kitti	21.260	depth + segmentation + flow
	Kitti Carla	35.000	depth + segmentation
	Synthia	220.000	depth + segmentation
	GTA5	24.966	segmentation
	Sail-vos 3D	111.654	segmentation
	ETS2 (ours)	46.780	depth

2.1 Real World Datasets

One of the earliest studies [20] in this field presents innovative techniques for depth estimation and image reconstruction together with the "Make3D" dataset. Make3D comprises 400 training images and 134 test images, all of which contain low-resolution depth maps acquired using a specialized laser scanner. In a subsequent publication by Saxena et al. [21], a new dataset is introduced, which includes stereo pairs along with depth maps. These depth maps contain raw laser scanner output, resulting in a total of 282 stereo pairs for this dataset.

The KITTI dataset [10] was released in 2012. It is likely the most widely utilized dataset in this field. In addition there is a benchmark that also includes multiple labels for various tasks such as optical flow, depth estimation, depth completion, semantic and instance segmentation, as well as road and lane detection. The entire dataset was generated by driving a vehicle equipped with a custom camera setup, a Velodyne laser scanner, and a GPS for odometry capture. The data was collected by traversing various urban, rural, and highway routes in the city of Karlsruhe, Germany. The benchmark provided by the KITTI dataset is routinely employed by researchers to evaluate and compare the performance of new depth estimation models or algorithms, as well as for other related tasks. The dataset includes 7,480 images in its training set and 7,517 images in its test set. Subsequent works, such as Behley et al. [2], Rashed et al. [17], and Recalibrating KITTI [4], have improved upon this dataset by introducing additional or enhanced labels and/or images.

The Cityscapes dataset [3] is comprised of up to 25,000 images captured using a stereo camera mounted on a moving vehicle that traveled through 50 different cities in Germany and some neighboring countries. The images in this dataset have been rectified and manually annotated with semantic labels, with 5,000 images having dense pixel-level annotations and 20,000 having coarse ones. In addition, this dataset provides instance segmentation labels for vehicles and people. Although depth annotations are not included in this dataset, they can be obtained by applying the SGM algorithm [13] to the rectified stereo pairs.

In the realm of indoor depth estimation, the NYU Depth dataset [22,23], which utilizes a Kinect device to capture images and depth maps of indoor scenes, is among the most commonly used datasets. These datasets consist of roughly 500,000 images with depth maps acquired from the device, with around 4,000 of these images featuring manually annotated instance and semantic labels. The annotations were obtained via Amazon Mechanical Turk services.

2.2 Synthetic Datasets

The significance of the KITTI dataset has inspired numerous projects to generate synthetic data that closely resemble it, with the aim of augmenting the available training data. One of this projects is Virtual KITTI [9], where some of KITTI's real scenes are transformed into virtual worlds under the Unity game development engine, and then a complete new dataset is generated from this virtual sequences with new, dense ground truth data with more than 20,000 frames. Based on the CARLA simulator [7], KITTI Carla [6], which replicates the KITTI setup inside the simulator, provides a synthetic dataset with 35,000 images, about 5 times bigger than the KITTI dataset.

Also using the Unity engine, the Synthia dataset [19] presents a dataset split into two differente sets of images, random frames and video sequences, for about 13,000 images the former and 200,000 the latter, all obtained from a custom made virtual city. All frames include semantic segmentation and depth.

Realistic video-games like GTAv5 also have been used to provide datasets for vision tasks like depth estimation. Works like [16,18] explore this idea, to generate data from this video game using the APIs provided by the game developers to facilitate gamers to extend the game world. More recently, [14] present a dataset which contains more than 230,000 images, both indoor and outdoor, with annotated ground truth labels.

3 Obtaining Data from the Video Game

The goal of this paper is to introduce a new dataset created for training monocular depth estimation deep neural networks. This dataset contains RGB images, dense depth labels and camera odometer information. All this data is extracted from a hyper-realistic driving simulation, ETS2.

ETS2 is a video game designed by SCS Software that simulates the driving experience of various vehicles, with a focus on trucks. The game offers a network of roads, highways, villages, and cities throughout Europe. It also provides tools to modify and improve the game world and its assets. As a result, a community of users interested in building mods has emerged, pushing the game graphics and simulation to an extraordinary level of realism. It is a multi-platform video game available on Windows, macOS, and Linux operating systems, accessible through the Steam gaming platform. The game engine used to develop the video game is Prism3D, which is proprietary to SCS Software. In the Windows platform, DirectX 11 is used to render the game's graphics.

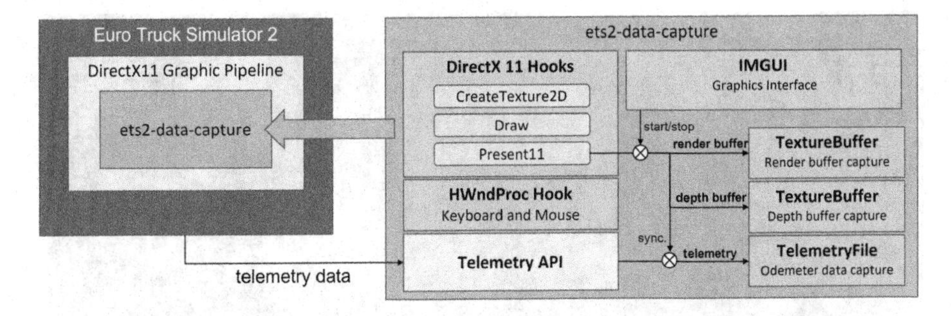

Fig. 1. The ets2-data-data capture software pipeline

Fig. 2. A frame of data in the dataset, game image (top left), depth map (top right) and telemetry data (bottom).

3.1 The Data Capture Tool

To obtain the dataset, we develop a piece of software entitled *ets2-data-capture tool*. A blueprint of this tool is shown in Fig. 1. This software intercepts several DirectX 11 drawing calls in the ETS2 video game to capture its depth and rendering buffers. These buffers are saved to files in real-time, providing images of the game camera and dense depth maps at a user-defined rate. Furthermore, we capture and include the vehicle's linear and angular velocity, world position, and scale factor of the world synchronized for each captured frame. To achieve this, we use the C++ API provided by the video game developers to obtain telemetry of the player's vehicle.

Figure 2 shows the monocular RGB image together with the depth map and the list of measurements. Finally, the video game also supports the simulation of various weather conditions and traffic environments.

During game play, the user must press a configurable key combination to activate our software. Then, the simulation is paused and an interface, developed using the Imgui API, overlays the game window. This interface allows the user to set up several capture settings, such as the frame capture rate or the log level. It also allows the user to start and stop the capture process. A screenshot of this GUI and the capture log are shown in Fig. 3.

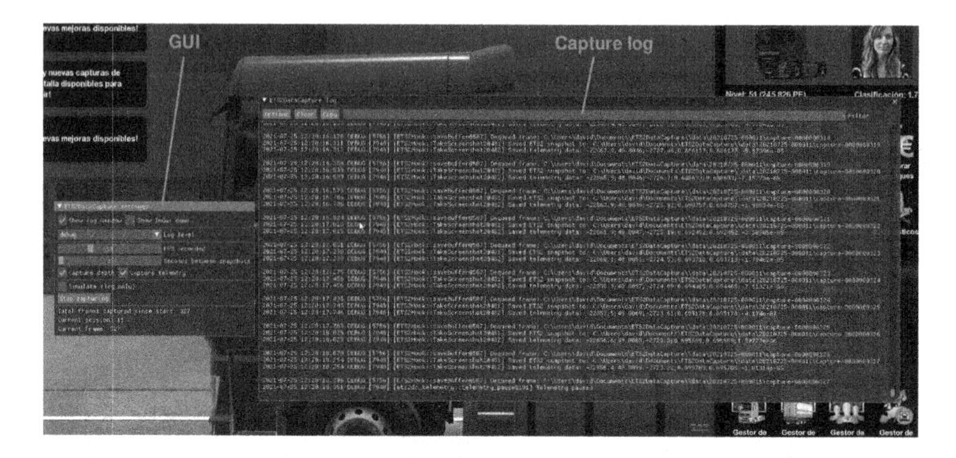

Fig. 3. The ets2-data-capture software GUI overlaying the video game pause screen

After setting up this tool, we play the video game in several sessions in different cities and roads, with diverse traffic, daylight and weather conditions to generate the dataset introduced in this work, the ETS2 Dataset.

4 The ETS2 Dataset

The ETS2 dataset is organized in 18 driving sessions. Each session takes place in a unique environment that is customized by 4 features: road type, time of day, traffic density, and weather conditions. The distribution of environments is shown in Fig. 5 and has been chosen to ensure sufficient variability in the dataset for training neural networks.

The data collected in a session is split in a description file, a telemetry file, and a sequence of paired RGB monocular images and depth maps.

The session description file specifies the date and the identification number of the session, in-game time of day and location, environment features and camera parameters (field of view, near and far planes in game units) stored in JSON format. An illustration of two sessions and their parameters, as they appear in the dataset website, is provided in Fig. 4.

The telemetry file is in CSV format. Each row corresponds to a single capture, including an identification that matches the image and depth file names, vehicle 3D coordinates in the game world, linear and angular velocity, linear and angular acceleration, and world scale factor.

The resolution of both images and depth map is 1440×816. The color information is saved as a 24-bit .bmp file, and the dense depth is saved as a raw file, with each pixel depth represented as a 2-byte short float. The camera simulated in the game has a $fov = 72°$. Assuming that the center of the image corresponds with the center of the camera, the final intrinsic matrix of this virtual camera is:

ETS2-0122-000001.zip (2.53 GB)
Frames: 2568
Date: 22/01/2023 22:30:00
Location: Paris, France
Environment: urban/road
Traffic (1-10): 4
Weather: Sunny
Camera parameters: FOV: 72 Near plane: 0.1 Far plane: 3000

Fig. 4. Example of two sessions with different environmental settings as they appear in the dataset website.

$$\begin{bmatrix} 70 & 0 & 720 \\ 0 & 39.66 & 408 \\ 0 & 0 & 1 \end{bmatrix}$$

The depth *near* and *far* planes are 0.1 and 3000 representing the maximum and minimum depth values.

In summary, the ETS2 dataset contains 46,780 frames of driving images, covering 1 h and 18 min of in-game driving. 1659 frames (3.55%) correspond to intervals during which the truck was stopped such as at a traffic light or a stop sign.

5 Validating the Dataset

The goal of ETS2 is to serve as dataset for training deep neural networks on monocular depth estimation. To assert its validity on real images, we train Monodepth2 [11] and Densedepth [1] on ETS2 and evaluate them on KITTI.

Monodepth2 follows a self-supervised approach, whereas Densedepth employs an encoder network pretrained with the ImageNet dataset [5] and a decoder fully trained in a supervised manner.

5.1 Training the Neural Networks with ETS2

We train both Monodepth2 and DenseDepth with no modifications with respect to their original architecture, using only ETS2, rescaling its images and ground-truth depth maps to the resolution of the input layer.

Monodepth2 consists of a U-Net with a Resnet50 encoder [12] pretrained on Imagenet. Then, both encoder and decoder are updated training 20 epochs with ETS2 and a batch size of 16 images rescaled to 640 × 192. The rest of hyperparameters and the training protocol follow [11].

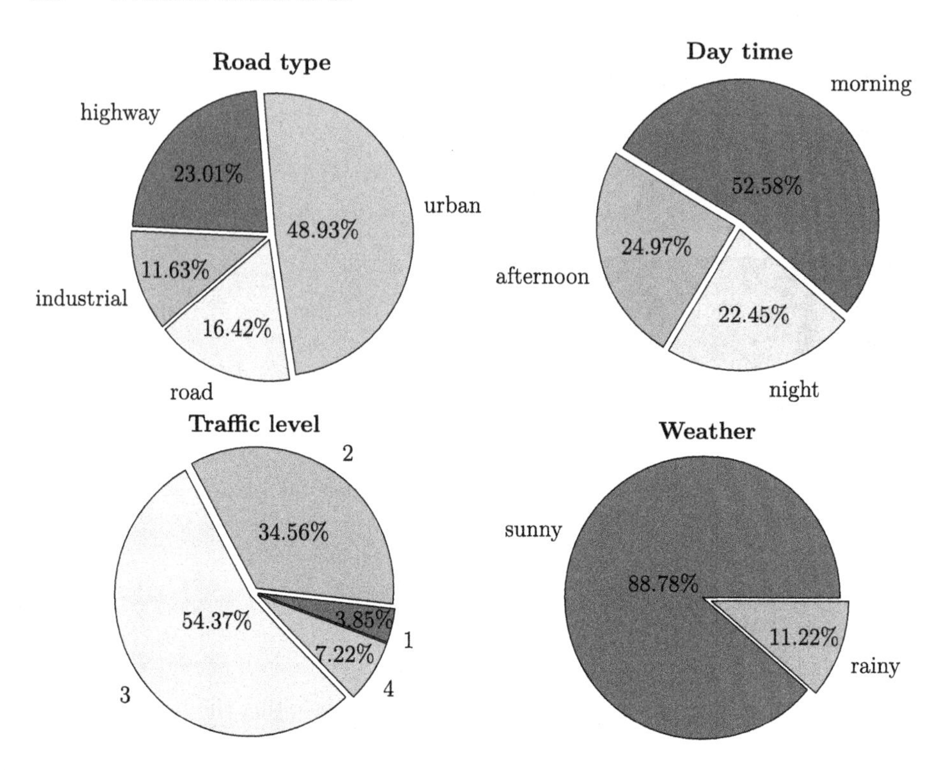

Fig. 5. Distribution of dataset images by road type (top left), day time (top right), traffic level (bottom left) and weather conditions (bottom right)

Regading DenseDepth, the encoder is DenseNet-169 [15], also pretrained on ImageNet. In this case the training takes 10 epochs with a batch-size of 6 images. To match the input and output layers, the images are rescaled to 640×480 and the ground-truth depth maps to 320×240 respectively.

5.2 Results

The models trained above are then evaluated on real data. Specifically, we use the KITTI test-split and the performance metrics proposed by Eigen et al. [8]. For the sake of clarity, we compile here the expressions of such metrics. In all of them, Y represents the ground-truth and $\hat{Y}$ the estimation. Since we are dealing with images, all the operations are pixel-wise.

- **Accuracy under threshold:** $\delta_j = \max\left(\frac{Y'}{Y}, \frac{Y}{Y'}\right) < 1.25^j; \text{for } j = \{1, 2, 3\}$
- **Absolute relative error:** $\text{abs\_rel} = \frac{1}{n}\sum_{i=1}^{n} \frac{|Y_i - \hat{Y}_i|}{\hat{Y}_i}$
- **Square relative error:** $\text{sq\_rel} = \frac{1}{n}\sum_{i=1}^{n} \frac{||Y_i - \hat{Y}_i||^2}{\hat{Y}_i}$

Table 2. Performance comparison for both neural networks, trained on KITTI (real images) and on ETS2 (synthetic images)

Model - Train set	abs_rel ↓	sq_rel ↓	rmse ↓	rmse_log ↓	δ_1 ↑	δ_2 ↑	δ_3 ↑
Monodepth2 - KITTI	**0.115**	**0.903**	**4.863**	**0.193**	**0.877**	**0.959**	**0.981**
Monodepth2 - ETS2 (30K)	0.321	6.186	10.721	0.374	0.609	0.827	0.912
Monodepth2 - ETS2 (50K)	0.253	2.510	7.450	0.317	0.597	0.861	0.949
Densedepth - KITTI	**0.093**	**0.589**	**4.170**	**0.171**	**0.886**	**0.965**	**0.986**
Densedepth - ETS2 (50K)	1.522	7.443	7.628	0.563	0.546	0.742	0.829

- **Root mean squared error:** $\text{rmse} = \sqrt{\frac{1}{n}\sum_{i=1}^{n}||Y - \hat{Y}||^2}$
- **RMSE in log space:** $\text{rmse\_log} = \sqrt{\frac{1}{n}\sum_{i=1}^{n}||\log Y - \log \hat{Y}||^2}$

The quantitative results are shown in Table 2. Models trained with KITTI perform better that those trained using the synthetic dataset. It is also remarkable that the larger ETS2 is needed. The sq_rel and rmse metrics are more sensitive to areas where large deviations between prediction and ground-truth occur. As a consequence, we carried out a detailed analysis of such areas and observed that our predictions are finer-grained compared to the ground-truth. This motivates the discussion in the following section.

5.3 Discussion About Domain Gap

The performance discrepancy observed when running neural networks trained on real data versus synthetic data is a common issue in vision-related tasks. This occurrence, referred to as synthetic to real domain shift, is a widely studied area of research.

Potential reasons for the inferior quantitative performance have been identified through a qualitative evaluation of the results. Figure 6 displays several examples that were utilized for this comparative analysis. At first glance, it is apparent that the predictions generated by the models trained using real-world data and the models trained with synthetic data share many similarities. However, upon closer examination, as depicted in Fig. 7, finer details are observed in the model trained with synthetic data, which are not present in either the model trained with real data or the ground-truth depth.

We believe that the evaluation metrics, which assess the dissimilarity between the predicted and ground-truth depth, are influenced by the finer details exhibited by the model trained on synthetic data. These finer details may result in a larger error value, and hence inferior evaluation results, due to their absence in the ground-truth image.

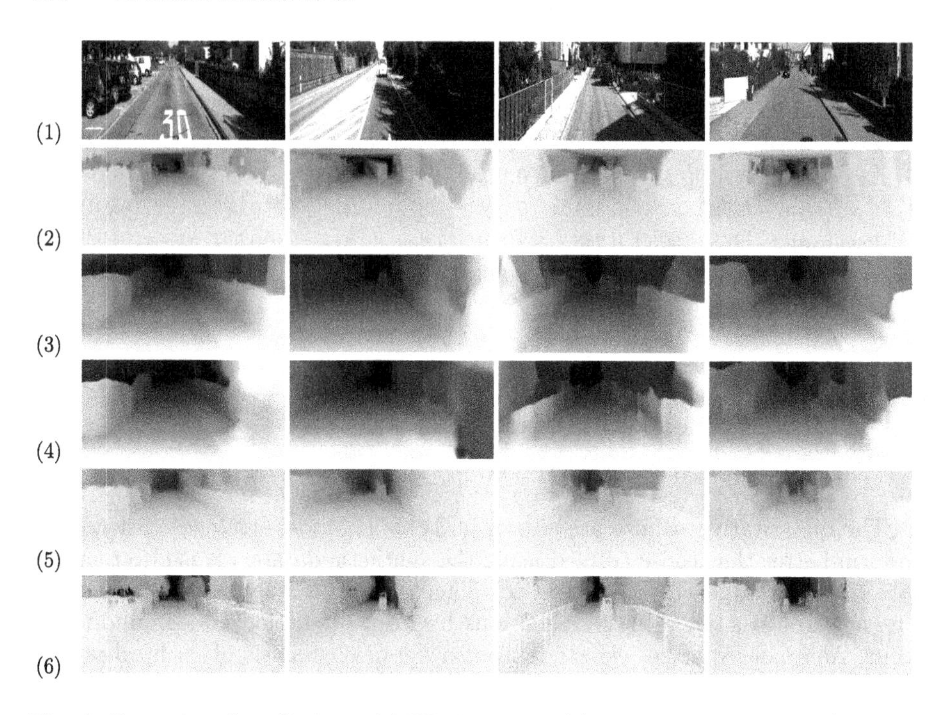

Fig. 6. Examples of predictions. (1) KITTI image, (2) KITTI ground-truth, (3) Monodepth2 trained on KITTI (4) Monodepth2 trained on ETS2 Dataset, (5) Densedepth trained on KITTI (6) Densedepth trained on ETS2 Dataset.

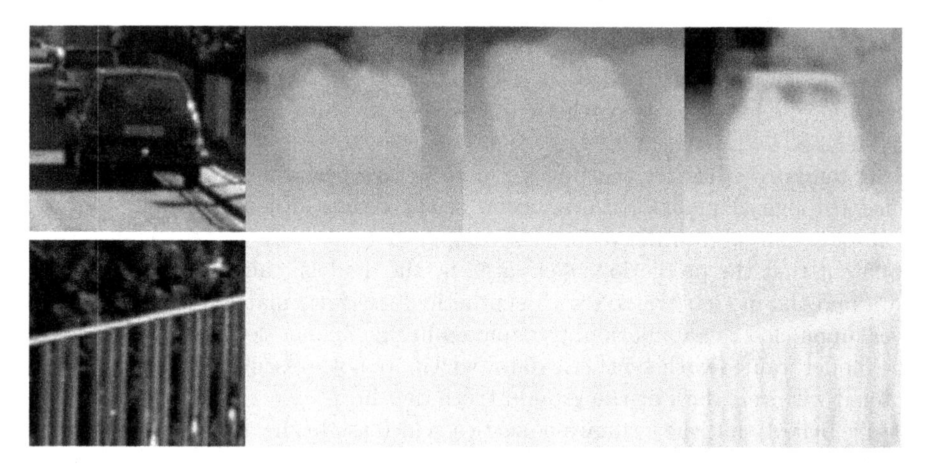

Fig. 7. Details for Densedepth, left to right: image, KITTI ground-truth, predictions of the model when trained on KITTI dataset, predictions when trained on ETS2-Dataset

6 Conclusion

We present a synthetic dataset obtained from a truck driving simulation video game for training neural networks in depth estimation tasks. We also demonstrate that state-of-the-art depth estimation models trained solely on this synthetic data achieved good quantitative and qualitative results. This opens up possibilities for using video game simulations to obtain large amounts of training data at a low cost for related automated driving and vision tasks. Additionally, we perform a qualitative evaluation and observe finer details in predictions made by the models trained on the synthetic dataset. This suggests that the differences in performance between real and synthetic data in this experiments are not due to domain gap but rather a lack of fine-grained ground-truth data.

References

1. Alhashim, I., Wonka, P.: High quality monocular depth estimation via transfer learning abs/1812.11941, arXiv:1812.11941 (2018)
2. Behley, J., et al.: SemanticKITTI: a dataset for semantic scene understanding of lidar sequences. In: IEEE/CVF International Conference on Computer Vision (ICCV), pp. 9296–9306 (2019). https://doi.org/10.1109/ICCV.2019.00939
3. Cordts, M., et al.: The cityscapes dataset for semantic urban scene understanding (2016). https://doi.org/10.1109/CVPR.2016.350, www.cityscapes-dataset.net
4. Cvišić, I., Marković, I., Petrović, I.: Recalibrating the KITTI dataset camera setup for improved odometry accuracy, pp. 1–6 (2021). https://doi.org/10.1109/ECMR50962.2021.9568821
5. Deng, J., Dong, W., Socher, R., Li, L.J., Li, K., Fei-Fei, L.: ImageNet: a large-scale hierarchical image database. In: Proceedings of the IEEE Conference on Computer Vision and Pattern Recognition (CVPR), pp. 248–255 (2009). https://doi.org/10.1109/CVPR.2009.5206848
6. Deschaud, J.E.: KITTI-carla: a kitti-like dataset generated by CARLA simulator (2021). https://doi.org/10.48550/arxiv.2109.00892, https://arxiv.org/abs/2109.00892
7. Dosovitskiy, A., Ros, G., Codevilla, F., Lopez, A., Koltun, V.: CARLA: an open urban driving simulator (2017)
8. Eigen, D., Fergus, R.: Predicting depth, surface normals and semantic labels with a common multi-scale convolutional architecture (2015). https://doi.org/10.1109/ICCV.2015.304
9. Gaidon, A., Wang, Q., Cabon, Y., Vig, E.: Virtual worlds as proxy for multi-object tracking analysis, pp. 4340–4349 (2016)
10. Geiger, A., Lenz, P., Urtasun, R.: Are we ready for autonomous driving? The KITTI vision benchmark suite. In: Proceedings of the IEEE Computer Society Conference on Computer Vision and Pattern Recognition (CVPR), pp. 3354–3361 (2012). https://doi.org/10.1109/CVPR.2012.6248074
11. Godard, C., Aodha, O.M., Firman, M., Brostow, G.: Digging into self-supervised monocular depth estimation (2018). https://doi.org/10.1109/ICCV.2019.00393
12. He, K., Zhang, X., Ren, S., Sun, J.: Deep residual learning for image recognition (2016)
13. Hirschmüller, H.: Accurate and efficient stereo processing by semi-global matching and mutual information. In: Proceedings of the IEEE Computer Society Conference on Computer Vision and Pattern Recognition (CVPR), vol. II, pp. 807–814 (2005). https://doi.org/10.1109/CVPR.2005.56, https://researchcode.com/code/672268296/accurate-and-efficient-stereo-processing-by-semi-global-matching-and-mutual-information/

14. Hu, Y.T., Wang, J., Yeh, R., Schwing, A.: SAIL-VOS 3D: a synthetic dataset and baselines for object detection and 3D mesh reconstruction from video data, pp. 3359–3369 (2021). https://doi.org/10.1109/CVPRW53098.2021.00375
15. Huang, G., Liu, Z., Van Der Maaten, L., Weinberger, K.Q.: Densely connected convolutional networks, pp. 2261–2269 (2017). https://doi.org/10.1109/CVPR.2017.243
16. Huang, Y., Dong, D., Lv, C.: Obtain datasets for self-driving perception from video games automatically. In: 12th International Conference on Reliability, Maintainability, and Safety (ICRMS), pp. 203–207 (2018). https://doi.org/10.1109/ICRMS.2018.00046
17. Rashed, H., Ramzy, M., Vaquero, V., El Sallab, A., Sistu, G., Yogamani, S.: FuseMODNet: real-time camera and LiDAR based moving object detection for robust low-light autonomous driving. In: The IEEE International Conference on Computer Vision (ICCV) Workshops (2019). https://doi.org/10.1109/ICCVW.2019.00293
18. Richter, S.R., Vineet, V., Roth, S., Koltun, V.: Playing for data: ground truth from computer games. In: Leibe, B., Matas, J., Sebe, N., Welling, M. (eds.) ECCV 2016. LNCS, vol. 9906, pp. 102–118. Springer, Cham (2016). https://doi.org/10.1007/978-3-319-46475-6_7
19. Ros, G., Sellart, L., Materzynska, J., Vazquez, D., Lopez, A.M.: The SYNTHIA dataset: a large collection of synthetic images for semantic segmentation of urban scenes. In: Proceedings of the IEEE Computer Society Conference on Computer Vision and Pattern Recognition (CVPR), pp. 3234–3243 (2016). https://doi.org/10.1109/CVPR.2016.352
20. Saxena, A., Chung, S.H., Ng, A.: Learning depth from single monocular images. In: Advances in Neural Information Processing Systems, vol. 18 (2005). https://doi.org/10.5555/2976248.2976394
21. Saxena, A., Schulte, J., Ng, A.: Depth estimation using monocular and stereo cues. In: Proceedings of the 20th International joint conference on Artifical Intelligence (IJCAI) (2007). https://doi.org/10.5555/1625275.1625630
22. Silberman, N., Fergus, R.: Indoor scene segmentation using a structured light sensor. In: Proceedings of the IEEE International Conference on Computer Vision (ICCV) Workshops, pp. 601–608 (2011). https://doi.org/10.1109/ICCVW.2011.6130298
23. Silberman, N., Hoiem, D., Kohli, P., Fergus, R.: Indoor segmentation and support inference from RGBD images. In: Fitzgibbon, A., Lazebnik, S., Perona, P., Sato, Y., Schmid, C. (eds.) ECCV 2012. LNCS, vol. 7576, pp. 746–760. Springer, Heidelberg (2012). https://doi.org/10.1007/978-3-642-33715-4_54, https://www.scinapse.io/papers/125693051

Computer Vision Applications

Multimodal Human Pose Feature Fusion for Gait Recognition

Nicolás Cubero[1(✉)], Francisco M. Castro[1], Julián R. Cózar[1], Nicolás Guil[1], and Manuel J. Marín-Jiménez[2]

[1] University of Malaga, Bulevar Louis Pasteur. 35, 29071 Malaga, Spain
ncubero@uma.es

[2] University of Cordoba, Rabanales campus, 14014 Córdoba, Spain

Abstract. *Gait recognition* allows identifying people at a distance based on the way they walk (i.e. gait) in a non-invasive approach. Most of the approaches published in the last decades are dominated by the use of silhouettes or other appearance-based modalities to describe the Gait cycle. In an attempt to exclude the appearance data, many works have been published that address the use of the human pose as a modality to describe the walking movement. However, as the pose contains less information when used as a single modality, the performance achieved by the models is generally poorer. To overcome such limitations, we propose a multimodal setup that combines multiple pose representation models. To this end, we evaluate multiple fusion strategies to aggregate the features derived from each pose modality at every model stage. Moreover, we introduce a weighted sum with trainable weights that can adaptively learn the optimal balance among pose modalities. Our experimental results show that (a) our fusion strategies can effectively combine different pose modalities by improving their baseline performance; and, (b) by using only human pose, our approach outperforms most of the silhouette-based state-of-the-art approaches. Concretely, we obtain 92.8% mean Top-1 accuracy in CASIA-B.

Keywords: Gait recognition · human pose · surveillance · biometrics · deep learning · multimodal fusion

1 Introduction

Gait-based people identification, or simply *gait recognition* aims at recognizing people by their manner of walking. Unlike other biometrical features, such as iris or fingerprints, gait recognition can be performed at a distance without the subject cooperation. Hence, it owns very potential applications in social security or medical research [19], among others, and many works have been published in this area during the last decades.

Supported by the Junta de Andalucía of Spain (P18-FR-3130, UMA20-FEDERJA-059 and PAIDI P20_00430) and the Ministry of Education of Spain (PID2019-105396RB-I00 and TED2021-129151B-I00). Including European funds, FEDER.

A. Pertusa et al. (Eds.): IbPRIA 2023, LNCS 14062, pp. 389–401, 2023.
https://doi.org/10.1007/978-3-031-36616-1_31

Despite multiple modalities that have been proposed to describe the gait motion, silhouettes are still the most studied modality in literature [3,5,14]. Silhouette holds a binary representation of the human body shape. A sequence of silhouettes reflects the body limbs' movements. However, the silhouette also contains information about the human shape and body contours that is unrelated to the motion of the limbs. In this sense, models may be biased by that appearance-based information and their performance could be penalized.

To remove that appearance-based information, many authors propose to use the human pose as an alternative modality. Human pose describes the positions of the body limbs at every instant and removes any other unnecessary shape information, so it is more robust to that appearance-related bias. Typically, pose-based gait recognition approaches exploit the 2D or 3D coordinates of the joints from the human body [1,11] and extract features from the correlation between the motion of different body joints to predict the identity of the subjects. However, those approaches perform worse than those based on visual descriptors like silhouettes [3,5,14]. This is caused by the less information received from the human pose, *i.e.*a set of 2D/3D coordinates versus a typically 64×64 silhouette image. To overcome such limitations, recent approaches [9,10] propose a multimodal setup that combines the pose information with silhouettes. However, although this multimodal setup reaches a substantial improvement, it brings up again the body shape information, limiting the benefits of the human pose.

In this work, we propose a combination of multiple pose representations in a multimodal setup to overcome the lack of information in every single representation, and hence, to avoid the use of shape descriptors. Moreover, instead of pose coordinates, we use two different representations: *(i)* a set of pose heatmaps images that are extracted from a human pose estimator [23] and; *(ii)* a dense pose representation extracted from DensePose [18] model. These two representations contain richer information than a solely set of coordinates and allow us to build a multimodal model that combines both representations and extracts more valuable features through different fusion strategies.

Therefore, our main contributions are: *(i)* a multimodal setup that exploits information from different pose representation modalities achieving state-of-the-art results on CASIA-B; and, *(ii)* a thorough experimental study comparing different fusion strategies to better combine the information from both pose representations.

The rest of this paper is organized as follows: Sect. 2 presents previous works. Section 3 describes our fusion strategies and Sect. 4 contains the experimental results on CASIA-B. Finally, Sect. 5 concludes the work.

2 Related Work

Recent gait recognition approaches have been mostly dominated by silhouettes or derived descriptors. GaitSet [3] uses a random stack of silhouettes where each frame is handled independently to extract and combine features through a Horizontal Pyramid Pooling (HPP). GaitPart [5] includes a novel part-based model that extracts features from horizontal splits of intermediate convolutional activations. GLN [7] introduces concatenation at intermediate convolutional activations and a compression module that is attached at the end of the model to reduce feature dimensionality. GaitGL [14] applies split convolutions within the convolutional pipeline together with a simplified version of the HPP proposed in [3]. As an alternative to silhouettes, other works use descriptors extracted from alternative sensors like accelerometer [4], floor-sensors [17], wave-sensors [16], or visual modalities [2,15] However, since all those descriptors are based on the human shape, like the silhouette, they are affected by changes in the body shape, illumination, etc.

In an attempt to remove the human shape, many other proposals use human pose descriptors [20] as input modality. Liao *et al.* [11] extract 2D joints from the human body and fed an LSTM and CNN model for gait recognition. In [1], the previous idea was improved by extracting 3D joints instead of 2D joints. In [13], Liao *et al.*compute multiple temporal-spatial features from the joint positions, the joint angles and motion, and limb length from the 3D human pose model. Teepe *et al.* [22] proposes a Graph Convolutional-based model to further exploit the spatial information originated from the 2D joints and their adjacency. Finally, Liao *et al.* [12] extract features from both pose heatmaps and skeleton graph images with a colored joint and limbs skeleton. Although pose-based models have some benefits with respect to visual-based approaches, their performance is lower in comparison with silhouette-based models.

Finally, many works propose multimodal models that exploit pose in combination with other modalities to improve the performance of pose-based models. Li *et al.* [9] propose a multimodal approach that combines pose heatmaps with silhouettes through a set of Transformers blocks that jointly process patches from both modalities. In [10], authors use a 3D pose model inspired by the Human Mesh Recovery Model (HMR), [8] combined with silhouettes that are fed into an ensemble of CNN and LSTM models.

Nevertheless, all these methods, despise proposing genuine strategies for modalities fusion, most of them bring up again the silhouette or another shape-derived, so the models learn the bias inherent to the shape covariates. In contrast, we propose a combination of different pose representation methods, without shape information.

3 Methodology

In this work, we propose multiple fusion strategies to combine and aggregate features extracted from pose (i.e. heatmaps and dense pose images). We start by describing in detail both studied pose representations. Then we describe the key elements of our fusions strategies.

3.1 Pose Representations

Hierarchical Heatmap Representation. *Heatmap* is a feature map generated by a keypoint-based pose extractor network before computing the output

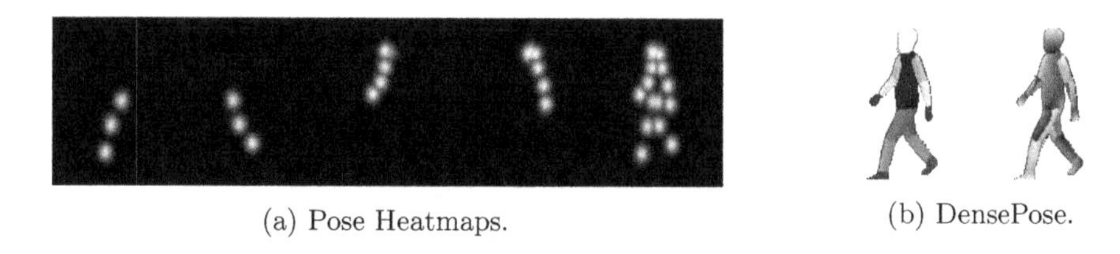

(a) Pose Heatmaps. (b) DensePose.

Fig. 1. Pose representations. Both pose heatmaps and dense pose representations are studied as modalities. For (a), from left to right, each of the joint group channels used as a pose heatmap representation: Right leg and hip, Left leg and hip, Right arm and eye, Left arm and the whole body at last channels. For (b), images I+V are displayed. Best viewed in digital format.

coordinates of each body joint. These maps hold a channel per body joint indicating the probability distribution of the target joint. Thus, a higher value indicates more confident detection while a low value may indicate that the joint is not visible or its estimation is low confident.

Therefore, heatmaps codify richer information than single 2D/3D coordinates and allow the model to be more robust to low-confident joint locations due to noise or occlusions. We regroup the joint heatmaps into the following hierarchical schema, which is composed of five channels: The first four channels contain different parts of the body like the left/right arm and the left/right leg while the last channel contains an image with all the joints of the full body. In this way, this representation allows us to better isolate the movement of each arm or leg from each other while keeping an overall description of the motion of the whole body, in addition, to remove the memory requirements. Heatmap aggrupation is depicted in Fig. 1a

Dense Pose. DensePose [18] is a dense 3D mesh representation of the human body. This mesh indicates information about body segmentation and the 3D location of each body part.

More concretely, the mesh is codified into three channels: *(i)* a body part segmentation map image I, which splits the human body into 24 segments, where each segment is colored in a different shade of gray – for each body part, a texture planar gradient is used to indicate the horizontal and vertical relative coordinates of each point concerning the origin of each body part; *(ii)* a mapping image U with the horizontal gradient coordinates; and, *(iii)* a mapping image V with the vertical gradient.

3.2 Fusion Strategies

We evaluate multiple fusion strategies to aggregate the information from both pose hierarchical heatmaps and the DensePose channels. We first implement strategies for fusing at an *early* stage by aggregating the output layers at a certain depth through an aggregation function. Secondly, we also explore *late* fusion on the final predictions obtained from each modality.

Early Fusion. Model architecture is split into two parallel branches, one per modality, until the fusion stage. At the fusion stage, the gait features obtained from each branch are aggregated into one feature map that is fed to the rest of the model layers to obtain the final prediction. We consider different aggregation strategies among the gait features:

- **Concatenation:** Features produced by the last layer of each parallel branch are concatenated along the filter dimension. After fusion, the input filters of the immediate after layer hold duplicated input filters to accommodate the duplicated size while maintaining the number of output filters.
- **Sum:** The features output by each branch are summed into a single feature map. Hence, we sum the output features produced by the last layer on each branch.
- **Weighted sum:** Features are aggregated through a weighted sum. Weights are learned as the rest of the model parameters during the training process. Thus, the model finds the optimal balance among both modalities.
  Let n the number of modalities and β_i the trainable parameter associated with the i-th modality, the weight w_i associated with the i-th modality is computed through softmax function. β_i is divided by a factor $\sqrt{n}$ so to stabilize values as follows:
  $$w_i = \text{softmax}(\frac{\beta_i}{\sqrt{n}}) \tag{1}$$
  This fusion also adds a layer normalization operator followed by a residual connection to the original feature values. Early fusion with Weighted sum is illustrated in Fig. 2.

Late Fusion. Fusion is performed at inference time on the final prediction output by every single model trained with each modality. At test stage, we compute the final embedding returned by each model for every sequence within gallery and probe sets. Then, softmax probability scores are computed for every sequence, based on the pair distance among the embeddings in the gallery, and the embeddings in the probe. Softmax scores measures, for every sequence, the affinity among the target subject and the subjects in the gallery. Fusion of the softmax probabilities is carried out by means of the following strategies:

- **Product:** Let P_i the set of softmax scores vectors output from the m_i modality. The final softmax score vectors S_{prod} can be computed as follows:
  $$S_{prod}(v = c) = \prod_{i=1}^{n} P_i(m_i = c) \tag{2}$$
  where $S_{prod}(v = c)$ refers to the probability of assigning the score of video v to the subject c and $P_i(m_i = c)$ is the probability of assigning the identity of subject c to that subject in the modality m_i.

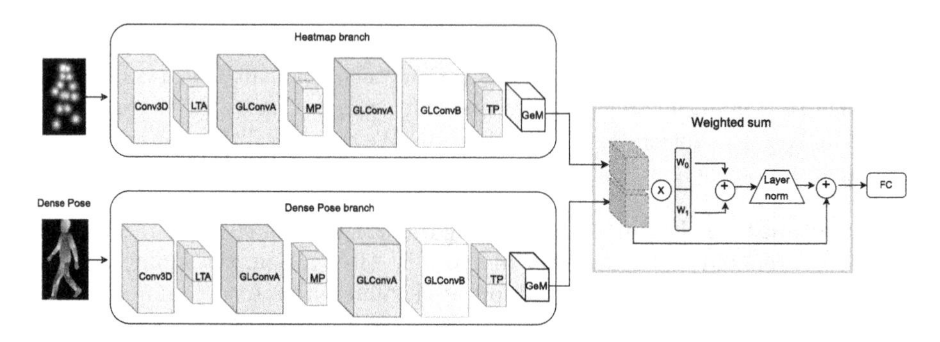

Fig. 2. Scheme of Early fusion with weighted sum. Our model has two parallel branches, one is fed with pose heatmaps and the other one with dense pose. Note that fusion is illustrated over the GaitGL [14] architecture on GeM layer. However, we tested other locations as explained in Sect. 4.4. GLConv refers to Global-Local Convolution, LTA to Local Temporal Aggregation, MP to Max Pooling, TP to Temporal Pooling; GeM to Generalized-Mean pooling and FC to Fully-Connected layer (Best viewed in digital format).

- **Weighted sum:** Final softmax scores S_{ws} are computed from the softmax vector scores P_i output from each modality m_i by a weighted sum as follows:

$$S_{ws}(v = c) = \sum_{i=1}^{n} \lambda_i P_i(m_i = c) \tag{3}$$

where λ_i is the weight assigned to each modality m_i, subject to $\forall i = 1, .., n,\ \lambda_i > 0$ and $\sum_{i=1}^{n} \lambda_i = 1$.
Considering $n = 2$ modalities, we grid values for λ_1 from 0.1 to 0.9 with steps of 0.1 for one modality, and assign $\lambda_2 = 1 - \lambda_1$ to the other modality.

4 Experiments and Results

In this section, we report the experimental results of our fusion approach. Firstly, we describe the datasets and metrics considered to evaluate our models' performance. Then, we report the implementation details of our models. Finally, we report our experimental results and the comparison against the *state of the art*.

4.1 Datasets

We carry out our experimental study on CASIA-B dataset [24]. Note that other popular datasets like OU-MVLP [21] or GREW [25] have not released the original RGB video sequences, so it is not possible to apply the pose estimators on them to extract the pose heatmaps or the dense poses.

CASIA-B collects 124 subjects walking in an indoor environment while they are recorded from 11 different viewpoints (*i.e.*from 0° to 180° in steps of 18°).

Video resolution is 320×240 pixels and fps is 24. For every subject, three walking conditions are considered: normal walking (*NM*), carrying a bag (*BG*) and wearing a coat (*CL*). We follow the *Large-Sample Training* (LT) experimental protocol followed too by [14].

The sequences from the first 74 subjects of all the walking conditions and viewpoints are used for training. For the remaining subjects, the first four *NM* sequences are used as gallery set while the rest of the walking conditions and types are used as probe set.

As evaluation metrics, we use the standard Rank-1 (R1) accuracy to measure the accuracy of our models, *i.e.*the percentage of correctly classified videos: $R1 = \#correct/\#total$.

4.2 Implementation Details

As GaitGL [14] is the current state-of-the-art model in gait recognition using silhouettes we employ it with the two proposed modalities: pose heatmaps and dense pose. Notice that the model is trained from scratch in all our experiments.

Table 1 summarises the training hyperparameters. For training, input samples contain 30 frames to reduce memory requirements, while at test time, we use all video frames to evaluate model accuracy.

Regarding pose image preprocessing, our input data is scaled and cropped so that the subject is always located in the middle of the frame, resulting in an input shape of 64×44. Pose heatmaps are obtained through ViTPose [23], while I-V images are obtained from DensePose [18]. Since image I is represented with 25 gray tones (24 body parts + background), we scale its values to the complete gray scale range ($[0, 255]$). We performed preliminary ablation experiments with every single I, U, and V image and concluded that image U does not provide valuable data. Hence, we only use I+V images. Figure 1b shows the I+V channels.

Table 1. Training hyperparameters. Description of the hyperparameters used to train GaitGL [14].

GaitGL [14] train hyperparameters	
# iterations	$80k$
Batch size (P subjects x K samples)	P: 8, K: 8
Optimizer	lr: 10^{-4} (10^{-5} after iter. $70k$)
Regularization	L2 (Weight decay: $5 \cdot 10^{-4}$)
# of filters per conv. block	32, 64, 128, 128
GeM pooling	p initial value: 6.5
Triplet loss margin	0.2

For early fusion with the weighted sum, weights β_i for every modality are initialized to 1, and we allow the model to find the optimal values during the training process. All the models are developed using OpenGait [6] and PyTorch v1.12.1.

4.3 Baseline Results

Firstly, we train and evaluate the base GaitGL model with each individual pose heatmaps and dense pose modalities. Hence, we obtain the baseline accuracy that can be obtained per each individual modality.

Table 2 reports the accuracy obtained by the base GaitGL model trained on every single modality. It can be observed that Dense Pose representation achieves higher accuracy as it manages richer information than Pose Heatmaps representation.

Table 2. Baseline accuracy. Top-1 accuracy (%) obtained per each single modality at test stage. Mean accuracy per each walking condition (NM, BG and CL) is reported along with the overall accuracy. The best result is highlighted in bold.

Modality	Walking condition			Mean
	NM	*BG*	*CL*	
Hierarchical pose heatmaps	92.7	80.0	70.3	81.0
Dense Pose	96.7	91.6	83.2	**90.5**

4.4 Study of Early Fusion Strategies

In this section, we report a thorough study of the proposed early fusion strategies over each stage of the GaitGL architecture. Thus, we have tested the proposed aggregation strategies at several locations of the model: Conv3D, LTA, first GLConvA layer (called GLConvA0), second GLConvA layer (called GLConvA1) and GLConvB (called GLConvB), TP, GeM and FC.

The mean global top-1 accuracy obtained by each early fusion method over all the fusion stages is summarised in Fig. 3.

It can be observed that fusion strategies based on both concatenation (blue bars) and sum (red bars) obtain worse results than the baseline result achieved by the single Dense Pose modality, indicating that a more complex fusion strategy is necessary.

Thus, we also tested fusion through the weighted sum, where the contribution of each modality must be learned. The results of this fusion strategy (green bars) show an important improvement with respect to previous fusion schemes. It can improve in 2.3% the best result achieved by the Dense Pose baseline when applied at the GeM module's output. This best model holds 4.543 M of parameters, and the average inference time per sample on an NVIDIA Titan Xp GPU is 28 ms.

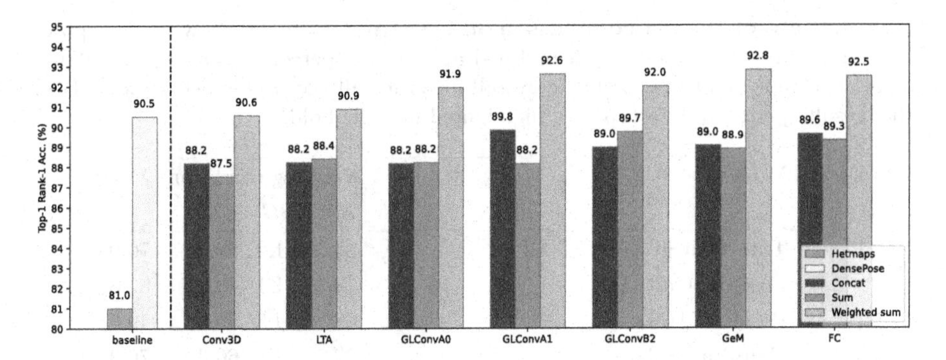

Fig. 3. Results with Early fusion. Mean Top-1 for all the early fusion strategies in all the fusion locations. Note that the accuracy scale is cropped to 80–95%. Best viewed in digital format.

4.5 Study of Late Fusion Strategies

The following tests focus on evaluating our proposed strategies for late fusion: product and weighted sum (w-sum). Thus, Table 3 collects both the mean top-1 accuracy (%) for every walking condition and the global mean for each late fusion strategy.

Table 3. Results with Late fusion, for both Product and Weighted sum strategies. Mean Top-1 accuracy, in percentage, for every walking condition is reported along with the overall mean for all the walking conditions. Note that λ_{hm} refers to the weight associated with heatmaps and λ_{dp} the weight associated with densepose.

Fusion	Weights		Walking condition			Mean
			NM	*BG*	*CL*	
Product			97.5	91.5	83.4	90.8
W-sum	λ_{hm}	λ_{dp}				
	0.1	0.9	97.1	92.2	84.0	91.1
	0.2	0.8	97.4	**92.7**	84.8	91.6
	0.3	0.7	**97.6**	**92.7**	**84.9**	**91.7**
	0.4	0.6	**97.6**	92.3	84.4	91.4
	0.5	0.5	97.5	91.5	83.4	90.8
	0.6	0.4	97.1	90.3	82.0	89.8
	0.7	0.3	96.5	88.6	80.0	88.4
	0.8	0.2	95.7	86.5	77.4	86.5
	0.9	0.1	94.6	83.6	74.1	84.1

It shows that both product and w-sum improve the baseline results. For w-sum, when dense pose modality is weighted under 0.5, the performance gets lower

Table 4. State-of-the-art comparison on CASIA-B. Comparison with other pose-based and shape-based models. Mean Top-1 accuracy, in percentage, for every walking condition is reported along with the overall mean for all the walking conditions. Best other results per each data type are highlighted in italic-bold.

Data	Model	Walking condition			Mean
		NM	*BG*	*CL*	
Pose	TransGait [9] (Pose + STM)	84.5	71.2	54.4	70.0
	PoseMapGait [12]	79.3	61.1	48.1	62.8
	PoseGait [13]	68.7	44.5	39.9	49.7
	GaitGraph [22]	87.7	74.8	66.3	76.3
	End-to-end Pose LSTM [10]	66.1	49.3	37.0	50.8
	End-to-end Pose CNN [10]	91.2	83.9	60.2	***78.4***
Shape	GaitSet [3]	95.0	87.2	70.4	84.2
	GaitPart [5]	96.2	91.5	78.7	88.8
	GaitGL [14]	97.4	94.5	83.6	91.8
	TransGait [9] (Sil + STM)	97.3	92.8	80.6	90.2
	TransGait [9] (Multimodal)	98.1	94.9	85.8	***92.9***
	End-to-end shape model [10]	97.5	90.6	75.1	87.7
	End-to-end ensemble [10]	97.9	93.1	77.6	89.5
Ours	PoseFusionGaitGL (tw-sum after GeM)	98.4	93.4	86.6	**92.8**

than the baseline accuracy. For most of the late fusion strategies, performance gets higher than early fusion by concatenation and sum methods. Finally, the weighted sum with $\lambda = 0.3$ for the pose heatmaps features and $\lambda = 0.7$ for the dense pose features achieves the best performance but does not improve the best results with the best early fusion strategy.

4.6 Comparison to the State of the Art Using Pose

Table 4 compares our best approach with the state-of-the-art pose-based and shape-based models. Our proposal outperforms all the pose-based methods, including 'End-to-end Pose' [10], which has been trained end-to-end with the original RGB frames. Our model obtains the best accuracy in the 'walking with clothes' (CL) condition and improves over the performance of most of the silhouette-based approaches, such as GaitGL [14], or GaitPart [5], and reaches very competitive results with multimodal TransGait [9] which has been trained using both pose and silhouette modalities.

These results show the capability of our fusion strategies to optimally aggregate features from multiple pose representation modalities using little or no shape data.

5 Conclusions

In this work, we presented an experimental study of multiple fusion pipelines for a multimodal framework for gait recognition that exploits various human pose representations. Concretely, we consider two pose representations: (a) pose heatmaps rearranged in a hierarchical decomposition of the human limbs and (b) dense pose.

As fusion strategies, we proposed, on the one hand, early fusion on the output descriptors produced by different layers of the model through concatenation and sum. In addition, we introduced a weighted sum, where the weights are learned during the training process and allow the model to leverage both modalities optimally. And, in the other hand, we proposed multiple late fusions on the final softmax probabilities output by each branch.

We evaluated our pose fusion approaches on the base GaitGL model for silhouette-based gait recognition. We maintained the original GaitGL architecture except for the parallel branches and fusion. We also included a comparison against the baseline performance achieved by every single modality.

Our experimental results show that: (a) Concatenation and sum early fusion methods do not allow one modality to enrich the features of the other modality, so results get poor. By contrast, weighted sum allows the model to learn to combine the features produced by each modality in a more optimal way. (b) Late fusion generally obtains good results too, and improves the baseline performance. And, (c) pose fusion obtains higher results than pose-based models and most shape-based models. Our approach reaches comparable performance to other multimodal fusions proposals based on silhouettes.

In future work, we plan to extend our study to alternative pose representations models that provide new perspectives on gait motion. In addition, we consider studying more elaborated fusion strategies. In our study, all the proposed fusion strategies treat equally the whole modality feature without taking into account useful information that is derived from local regions.

References

1. An, W., Liao, R., Yu, S., Huang, Y., Yuen, P.C.: Improving gait recognition with 3D pose estimation. In: Zhou, J., et al. (eds.) CCBR 2018. LNCS, vol. 10996, pp. 137–147. Springer, Cham (2018). https://doi.org/10.1007/978-3-319-97909-0_15
2. Castro, F.M., Marín-Jiménez, M.J., Guil, N., de la Blanca, N.P.: Multimodal feature fusion for CNN-based gait recognition: an empirical comparison. Neural Comput. Appl. **32**, 1–21 (2020)
3. Chao, H., He, Y., Zhang, J., Feng, J.: GaitSet: regarding gait as a set for cross-view gait recognition. In: Proceedings of the AAAI Conference on Artificial Intelligence (2019)
4. Delgado-Escaño, R., Castro, F.M., Cózar, J.R., Marín-Jiménez, M.J., Guil, N., Casilari, E.: A cross-dataset deep learning-based classifier for people fall detection and identification. Comput. Methods Programs Biomed. **184**, 105265 (2020)

5. Fan, C., et al.: GaitPart: temporal part-based model for gait recognition. In: CVPR, pp. 14225–14233 (2020)
6. Fan, C., Shen, C., Liang, J.: OpenGait (2022). https://github.com/ShiqiYu/OpenGait
7. Hou, S., Cao, C., Liu, X., Huang, Y.: Gait lateral network: learning discriminative and compact representations for gait recognition. In: Vedaldi, A., Bischof, H., Brox, T., Frahm, J.-M. (eds.) ECCV 2020. LNCS, vol. 12354, pp. 382–398. Springer, Cham (2020). https://doi.org/10.1007/978-3-030-58545-7_22
8. Kanazawa, A., Black, M.J., Jacobs, D.W., Malik, J.: End-to-end recovery of human shape and pose. In: Computer Vision and Pattern Recognition (CVPR) (2018)
9. Li, G., Guo, L., Zhang, R., Qian, J., Gao, S.: TransGait: multimodal-based gait recognition with set transformer. Appl. Intell. **53**, 1–13 (2022)
10. Li, X., Makihara, Y., Xu, C., Yagi, Y., Yu, S., Ren, M.: End-to-end model-based gait recognition. In: CVPR (2020)
11. Liao, R., Cao, C., Garcia, E.B., Yu, S., Huang, Y.: Pose-based temporal-spatial network (PTSN) for gait recognition with carrying and clothing variations. In: Zhou, J., et al. (eds.) CCBR 2017. LNCS, vol. 10568, pp. 474–483. Springer, Cham (2017). https://doi.org/10.1007/978-3-319-69923-3_51
12. Liao, R., Li, Z., Bhattacharyya, S.S., York, G.: PoseMapGait: a model-based gait recognition method with pose estimation maps and graph convolutional networks. Neurocomputing **501**, 514–528 (2022)
13. Liao, R., Yu, S., An, W., Huang, Y.: A model-based gait recognition method with body pose and human prior knowledge. Pattern Recognit. **98**, 107069 (2020)
14. Lin, B., Zhang, S., Yu, X.: Gait recognition via effective global-local feature representation and local temporal aggregation. In: ICCV, pp. 14648–14656 (2021)
15. Marín-Jiménez, M.J., Castro, F.M., Delgado-Escaño, R., Kalogeiton, V., Guil, N.: UGaitNet: Multimodal gait recognition with missing input modalities. IEEE Trans. Inf. Forensics Secur. **16**, 5452–5462 (2021)
16. Meng, Z., et al.: Gait recognition for co-existing multiple people using millimeter wave sensing. In: Proceedings of the AAAI Conference on Artificial Intelligence (2019)
17. Nakajima, K., Mizukami, Y., Tanaka, K., Tamura, T.: Footprint-based personal recognition. IEEE Trans. Biomed. Eng. **47**(11), 1534–1537 (2000)
18. Güler, R.A., Neverova, N., Kokkinos, I.: DensePose: dense human pose estimation in the wild. In: CVPR (2018)
19. Sepas-Moghaddam, A., Etemad, A.: Deep gait recognition: a survey. IEEE Trans. Pattern Anal. Mach. Intell. **45**, 264–284 (2022)
20. Shen, C., Yu, S., Wang, J., Huang, G.Q., Wang, L.: A comprehensive survey on deep gait recognition: algorithms, datasets and challenges (2022). https://arxiv.org/abs/2206.13732
21. Takemura, N., Makihara, Y., Muramatsu, D., Echigo, T., Yagi, Y.: Multi-view large population gait dataset and its performance evaluation for cross-view gait recognition. IPSJ Trans. Comput. Vision Appl. **10**(1), 1–14 (2018). https://doi.org/10.1186/s41074-018-0039-6
22. Teepe, T., Khan, A., Gilg, J., Herzog, F., Hormann, S., Rigoll, G.: GaitGraph: graph convolutional network for skeleton-based gait recognition. In: 2021 IEEE International Conference on Image Processing (ICIP). IEEE (2021)
23. Xu, Y., Zhang, J., Zhang, Q., Tao, D.: ViTPose: simple vision transformer baselines for human pose estimation. In: NeurIPS (2022)

24. Yu, S., Tan, D., Tan, T.: A framework for evaluating the effect of view angle, clothing and carrying condition on gait recognition. In: Proceedings of the ICPR, vol. 4, pp. 441–444 (2006)
25. Zhu, Z., et al.: Gait recognition in the wild: a benchmark. In: ICCV, pp. 14789–14799 (2021)

Proxemics-Net: Automatic Proxemics Recognition in Images

Isabel Jiménez-Velasco[1(✉)], Rafael Muñoz-Salinas[1,2], and Manuel J. Marín-Jiménez[1,2]

[1] Department Computing and Numerical Analysis, University of Córdoba, Córdoba, Spain
{isajimenez,rmsalinas,mjmarin}@uco.es

[2] Maimonides Institute for Biomedical Research of Córdoba (IMIBIC), Córdoba, Spain

Abstract. Proxemics is a branch of anthropology that studies how humans use personal space as a means of nonverbal communication; that is, it studies how people interact. Due to the presence of physical contact between people, in the problem of proxemics recognition in images, we have to deal with occlusions and ambiguities, which complicates the process of recognition. Several papers have proposed different methods and models to solve this problem in recent years. Over the last few years, the rapid advancement of powerful Deep Learning techniques has resulted in novel methods and approaches. So, we propose Proxemics-Net, a new model that allows us to study the performance of two state-of-the-art deep learning architectures, ConvNeXt and Visual Transformers (as backbones) on the problem of classifying different types of proxemics on still images. Experiments on the existing Proxemics dataset show that these deep learning models do help favorably in the problem of proxemics recognition since we considerably outperformed the existing state of the art, with the ConvNeXt architecture being the best-performing backbone.

Keywords: Proxemics · Human interactions · Images · Deep Learning

1 Introduction

The Proxemics concept was established in 1963 by the American anthropologist Edward T. Hall [4]. Proxemics is a subcategory of nonverbal communication that studies how people use personal space and the distance they keep between them when communicating. In other words, it studies the proximity or distance between people and objects during an interaction, as well as postures and the presence or absence of physical contact. In this way, this information allows us

Supported by the MCIN Project TED2021-129151B-I00/AEI/10.13039/ 501100011033/European Union NextGenerationEU/PRTR, and project PID2019-103871GB-I00 of the Spanish Ministry of Economy, Industry and Competitiveness, FEDER.

A. Pertusa et al. (Eds.): IbPRIA 2023, LNCS 14062, pp. 402–413, 2023.
https://doi.org/10.1007/978-3-031-36616-1_32

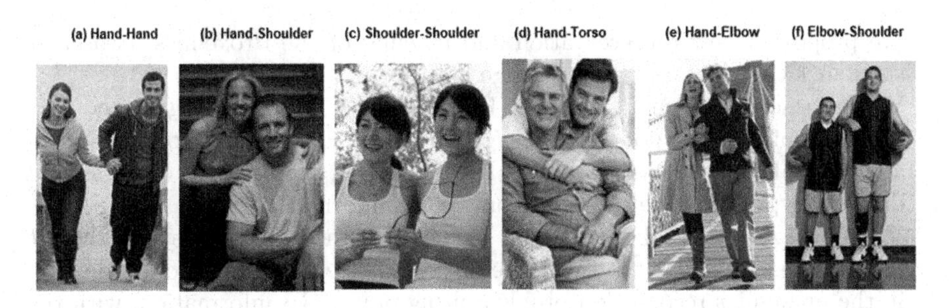

Fig. 1. Touch codes in Proxemics. Images showing the six specific "touch codes" [18] that were studied in this work.

to determine the type of social interaction and the interpersonal relationships between the members present in the interaction, as it will vary greatly depending on whether they are acquaintances, friends or family.

In 2012, inspired by Hall's work, Yang et al. [18] introduced the problem of proxemics recognition in Computer Vision and provided a computational grounding of Hall's work. They characterized proxemics as the problem of recognizing how people physically touch each other, and they called "touch codes" to each type of interaction or proxemics. In particular, they defined "touch codes" as the pairs of body parts (each pair element comes from a different person) in physical contact. After analyzing an extensive collection of images showing two to six people in physical contact, the authors identified that there were six dominant "touch codes", namely, **Hand-Hand** (HH), **Hand-Shoulder** (HS), **Shoulder - Shoulder** (SS), **Hand-Torso** (HT), **Hand-Elbow** (HE) and **Elbow-Shoulder** (ES). Figure 1 shows example images corresponding to these "touch codes".

The authors of [18] stated that the best way to address the problem of recognizing and classifying different types of proxemics was by using **specific detection models**. They argued that other alternatives, such as pose estimation, were significantly affected by ambiguity and occlusion when there was physical interaction between people.

Since then, other methods have been proposed to solve the problem of classifying proxemics in images [1,5]. However, this problem is still far from being solved.

Even so, in the last decade, the increasing popularity of Computer Vision and the rapid advancement of powerful Deep Learning techniques has resulted in novel methods and approaches.

So, the main objective of this work is to analyze if the new Computer Vision methods that have been developed in the last years can help in the problem of proxemic recognition and to obtain better results using only RGB information, instead of using specific models for each type of proxemic as it was proposed in [18].

The main contributions of this work are the following:

- We propose a new representation and labeling of the Proxemics dataset at the pair level instead of at the image level.
- We propose a new model, coined Proxemics-Net, and investigate the performance of two state-of-the-art deep learning architectures, **ConvNeXt** [9] and **Visual Transformers** [2] (as backbones) on the problem of classifying different types of proxemics using only RGB information.
- We show experimental results that outperform the existing state of the art and demonstrate that the two state-of-the-art deep learning models do help in the proxemics recognition problem using only RGB information, with the ConvNeXt architecture being the best-performing backbone.

To do so, after presenting the related work in Sect. 2, we will describe the proposed Proxemics-Net model and the backbones we have used to define it (Sect. 3). Subsequently, we will explain all the experiments' characteristics and the implementation details (Sect. 4). Then, we will show and comment on the results of all the experiments performed. In addition, we will show some failure cases of our model (Sect. 4.4), and finally, we will finish with some conclusions and future work (Sect. 5).

2 Related Work

Human activity recognition (HAR) is one of the most important and challenging problems in computer vision. The main goal of human activity recognition (HAR) is to automatically recognize as many activities from images or videos as possible. It has a large number of applications, such as in video surveillance, human-computer interaction, and other fields [6,11]. Human activity recognition is challenging due to the complex postures of human beings, the number of people in the scene, and the existence of specific challenges, such as illumination variations, clutter, occlusions, and background diversity.

Human interaction recognition (HIR) is one of the categories of human activity recognition. HIR is a challenge due to the participation of multiple people in a scene, the large differences in how actions are performed, and the scene's characteristics, such as lighting, clothing, objects that may cause occlusions, etc.

The difference between HAR and HIR is that HAR focuses on detecting activities in which one or more people are present and which may or may not interact with objects, such as drinking, walking, and cooking. On the other hand, HIR focuses on distinguishing interactions that occur specifically between humans, such as a hug or a kiss.

To address the problem of human-to-human interaction detection, the American anthropologist E.T. Hall [4] introduced the concept of proxemics, a categorization of human individual interactions based on spatial distances. However, Yang et al. [18] were the first to study this problem systematically. In particular, they exploited the concept of "touch code" in proxemics to estimate the body poses of interacting people. In that work [18], the authors proposed a joint model for jointly recognizing proxemic classes and pose estimation for pairs of people.

The model was a pictorial structure [3] consisting of two people plus a spring connecting the body parts of the pair that were in physical contact.

To solve the problem of the existence of ambiguity and occlusion when there is physical interaction between people, Xiao et al. [1] incorporated additional information into the "touch code" concept. Specifically, they defined a complete set of "immediacy" cues like touching, relative distance, body leaning direction, eye contact and standing orientation. The authors built a deep multi-task recurrent neural network (RNN) to model complex correlations between immediacy cues as well as human pose estimation. They showed that by jointly learning all prediction tasks from a full set of immediacy cues with the deep RNN, the performance of each task and pose estimation improved. Thus, this new model improved the state-of-the-art results achieved on the Proxemics dataset by [18].

In 2017, motivated by limitations in the speed, efficiency, and performance of people detectors (unknown scales and orientations, occlusion, ambiguity of body parts, etc.), Jiang et al. [5] proposed a new approach to segment human instances and label their body parts (arms, legs, torso, and head) by assembling regions. Thus, this new approach improved the state of the art on proxemics recognition.

Throughout these years, the problem of detecting proxemics or physical contact between people has been applied to various fields, such as recognition of the role of each person in social events [13], detection of social relations between pedestrians [12], intelligent video surveillance [19], etc. Despite its numerous applications, it has continued to be a major challenge for the computer vision community, especially when working with monocular images since, compared to video [7,8,10], there is not as much information in a single image.

Thus, in this paper, we propose the Proxemics-Net model that consists of novel deep learning architectures as backbones (ConvNeXt [9] and Visual Transformers [2]) and is able to address the proxemics classification problem using only the RGB information of monocular images.

3 Proposed Method

3.1 Overview of the Model: Proxemics-Net

For this work, a new model called Proxemics-Net has been proposed (see Fig. 2). This model makes use of two different deep networks that have been previously pre-trained: ConvNeXt [9] and Vision Transformers (ViT) [2].

Our Proxemics-Net model has three inputs. Two inputs corresponding to the RGB clipping of each of the individuals composing a pair (***p0_branch***) and (***p1_branch***) and a third input corresponding to the RGB clipping showing the pair to be classified (***pair_branch***). The three input branches receive RGB images of 224×224 resolution.

The three branches of our model will have the same type of backbone. The backbone of each branch is responsible for extracting the characteristics of the corresponding input.

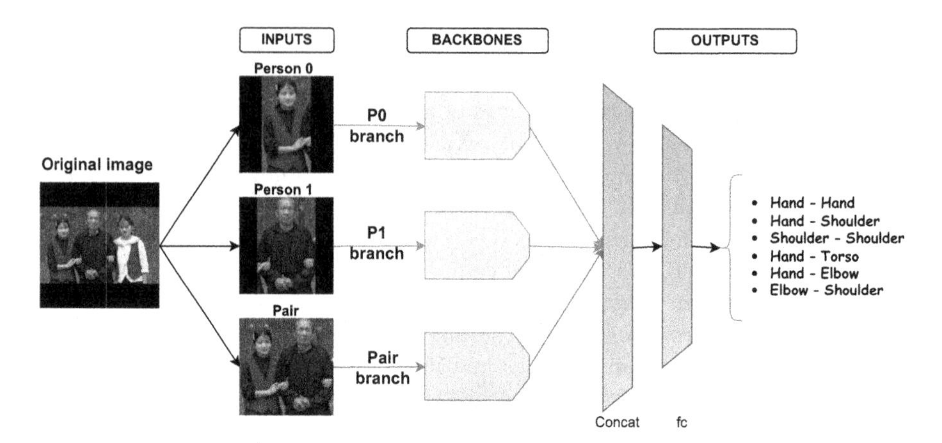

Fig. 2. Our Proxemics-Net model. It consists of the individual branches of each person (*p0_branch* and *p1_branch*) (blue) and the *pair branch* (red) as input. All branches consist of the same type of backbone (ConvNeXt or ViT). The outputs of these 3 branches are merged in a concatenation layer and passed through a fully connected layer that predicts the proxemic classes of the input samples.

The results of the three branches are combined by a concatenation layer and passed through a fully connected layer that predicts the Proxemic classification of the input samples.

Finally, since we are dealing with a multi-label problem in which each image can be classified with more than one type of proxemics, the output layer is a 6-unit Sigmoid layer(one for each class).

3.2 Evaluated Backbones

Visual Transformer. Transformers are deep learning models designed to manipulate sequential data, mainly in the field of natural language processing (NLP) (translation, text prediction, etc.) and Computer Vision. The "Transformer" structure was presented in [15]. This structure is based on a layered encoding and decoding system, and the recurrent layers are replaced by a "self-attention" model. Like recurrent neural networks (RNNs), Transformers are designed to process sequential input data. However, they do not require the data to be ordered. This allows for greater parallelization than RNNs and therefore reduces training times.

The Vision Transformer (ViT) model was proposed in [2]. It is the first paper that successfully trained a Transformer encoder on ImageNet dataset [14], attaining very good results compared to familiar convolutional architectures.

ConvNeXt. The ConvNeXt model was proposed by Liu et al. [9]. ConvNeXt is a pure convolutional model (ConvNet) constructed entirely from standard ConvNet modules and inspired by the design of Vision Transformers.

ConvNeXt models comprise standard layers such as depthwise convolutions, layer normalization, etc., and use standard network topologies. They do not use self-attention or any hybrid approach, unlike the recent architectures such as Vision Transformers. The authors start with a base architecture and gradually refine it to match some of the design choices of Vision Transformers. In the process, they developed a family of models named ConvNeXt, achieving high performance on the ImageNet-1k and ImageNet-21k dataset [14].

4 Experiments

4.1 Dataset: Proxemics

The Proxemics dataset [17] is an annotated database with joint positions and "touch code" labels, introduced in [18]. This dataset is composed of a total of 1178 images. Specifically, 589 images are different, and the other 589 are the same images but flipped.

These images consist of personal photos of family and friends collected from web searches on Flickr, Getty Images, and image search engines on Google and Bing.

All images are in color (RGB) and contain between 2 to 6 people (most images usually have only two) in various poses, scene arrangements, and scales.

For each image in this collection, the authors labeled the locations (coordinates) of the **10 joints** of all the people present in each image (head, neck, right and left shoulder, right and left elbow, right and left wrist, and right and left hand) and all **types of proxemics** between pairs (Hand-Hand (HH), Hand-Shoulder (HS), Shoulder-Shoulder (SS), Hand-Torso (HT), Hand-Elbow (HE) and Elbow-Shoulder (ES) (See Fig. 1)).

Finally, after analyzing all the proxemics labeled in the dataset, the authors found that most of the pairs of people had between 0 (there are none of the six types of proxemics) and 1 "touch codes", but there were also a significant number of images that had two or more "touch codes". This is because, for example, a single arm's elbow and hand may touch another person's body.

4.2 Metrics

We have used the **Average Precision (AP)** metric to evaluate the proposed variants of our Proxemics-Net model.

The average precision is intended to calculate the performance of the model in each of the classes instead of calculating its performance in general. In this way, we will see if it presents a similar behavior in all classes and, therefore, it is a good model or, on the contrary, it has learned to classify only some specific classes of the problem. The closer the value of this metric is to 1, the better our model will be.

Finally, once we have obtained the average precision for each class, we calculate the **mean Average Precision (mAP)**, which is the average of the AP values obtained with the six classes.

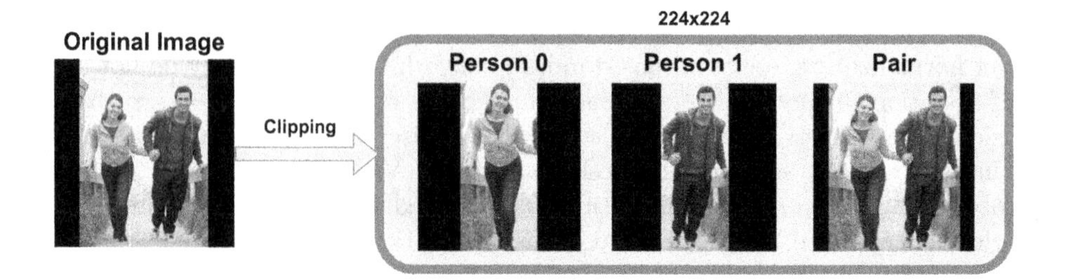

Fig. 3. Input data for Proxemics-Net. Given the target image, the first 2 image crops ('Person 0' and 'Person 1') correspond to the individual clippings of the members of the pair and the third image window corresponds to the clipping of the pair. All clipping images have a resolution of 224 × 224 pixels.

4.3 Implementation Details

Image Preprocessing. To work with the Proxemics dataset, we made clippings of all the images (except the flipped ones) in order to obtain all the possible pairs that appear in each of the images, as well as individual clippings of each member of the pair. Thus, we obtained three clippings for each pair in an image, one for each person in the pair, and a third clipping with both members of the pair (See Fig. 3). Since the deep models (ConvNeXt and ViT) work with 224 × 224 input images, we have preprocessed the clippings by adding padding to make them square and resizing them to 224 × 224 resolution without changing the aspect ratio.

Label Preprocessing. We preprocessed the labels of the Proxemics dataset to adapt them to our experiments since we have approached the problem of proxemics classification at the pair level instead of at the image level as the dataset authors did [18].

In this way, we have obtained the different types of proxemics that appear in each of the pairs that we can find in the original images.

Dataset Partitions. To train and test all the proposed models, we use the same train and test partitions proposed by the Proxemics dataset's authors.

It is important to note that all models have been subjected to a 2-fold cross-validation (set1/set2) as the authors of [18] did. In this way, we can directly compare our results with [18] under the same experimental conditions and see how the model behaves on average.

Finally, in all the experiments of our models and for both sets, we have taken approximately 10% of the samples from the train set to use them as validation. In addition, we have applied data augmentation techniques (flipping images, zooming, changes in brightness, etc.) to our training images in order to train our models with a larger number of different samples.

Table 1. Results of the best models obtained for each of the 3 types of Proxemics-Net variants. The results (average of set1 and set2) show that the model incorporating the ConvNeXt network as a backbone obtains the best results. The %AP results for each type of proxemics and the %mAP are shown.

Backbone	HH	HS	SS	HT	HE	ES	mAP (Set1-Set2)
ViT	45.5	**55.7**	46.4	76.6	56	51.9	55.7
ConvNeXt_Base	54.1	51.7	59.3	82.3	57.2	54.2	59.8
ConvNeXt_Large	**57.7**	54.9	**59.3**	**85.3**	**60.3**	**64.9**	**63.7**

Training Details. First, for training our models with the different backbones, we have used pre-trained models since our Proxemics dataset has a reduced number of samples. In this way, by using transfer learning we avoid having to retrain and relearn specific parameters of our network, thus achieving faster and even more efficient training by being able to extract valuable features that the network has already learned.

In the case of using ViT architecture as backbone, the pre-trained Transformer model with the weights "google/vit-base-patch16-224-in21k" has been used [16] and in the case of using ConvNeXt architecture as backbone, we have used two pre-trained ConvNeXt models of different complexity. Specifically, a *"Base"* model and a *"Large"* model. Thus, our Proxemics-Net model will have three variants (two of type ConvNeXt and one of type ViT). In each variant, the same backbone is used for all 3 branches, no backbones are combined within the same variant.

In addition, the three models have been trained with images of 224×224 resolution so, in order to use these pre-trained networks, the images of the Proxemics dataset had to be preprocessed to rescale them to the required resolution of 224×224.

In order to train and obtain the best results we have tested different batch sizes (6 and 8) and used the Adam and SGD optimizers. Regarding the learning rate, we have varied between $1 \cdot 10^{-2}$, $1 \cdot 10^{-3}$, $1 \cdot 10^{-4}$, $1 \cdot 10^{-5}$, $5 \cdot 10^{-5}$. Finally, we have selected *binary_crossentropy* as the loss function.

4.4 Experimental Results

ConvNeXt vs Vision Transformers. The table 1 shows the best results %mAP obtained for each of the three proposed variants. Specifically, the first row summarizes the results of the best model trained with the architecture that has a Vision Transformer as the backbone, and the two following rows are the results of the best model trained with the architecture that has a ConvNeXt network as the backbone (loading the pre-trained weights of the *Base* and *Large* model).

First of all, if we compare the two variants that incorporate the ConvNeXt network as the backbone, we can observe that the *Large* model obtains better

Table 2. Ablative study: *Pair_branch* vs. *Full model*. Results of %mAP obtained when we disable the Individual branches (*p0_branch* and *p1_branch*) versus using the full model (three branches). These results are shown for the three variants of Proxemics-Net.

Model		mAP (Set1-Set2)		
Pair	**Pair+Individuals**	ViT	ConvNeXt Base	ConvNeXt Large
✓		55.5	58.7	61.2
✓	✓	**55.7**	**59.8**	**63.7**

results in terms of both AP % in all types of proxemics and mAP % on average of set1 and set2 (59.8 mAP% (Base Model) versus 63.7 mAP% (Large Model)). This observation is in line with expectations, as the pre-trained *Large* model incorporated into our architecture demonstrated slightly improved accuracy on ImageNet compared to the *Base* model.

Secondly, if we compare the models obtained with the network incorporating the ConvNeXt backbone with the best model we have obtained with the network incorporating the Vision Transformers backbone, we can observe that the models with the ConvNeXt network obtain better results both in terms of AP % in almost all types of proxemics and in %mAP on average of set1 and set2, with the ConvNeXt_Large model again being the best performer (55.7% of mAP versus 63.7% of mAP).

Thus, we can state that the model incorporating the ConvNeXt network as the backbone obtains the best results and, therefore, is the most optimal method for this type of problem.

Ablative Study. Table 2 shows a comparison between the best %mAP results obtained when we disable the Individual branches of our model and only use as input the RGB information of the pairs (first row), against when we use the full model (RGB information of the pair and of the members that compose it) (second row). Thus, by disabling the branches of the individuals (*p0_branch* and *p1_branch*) we can see how they contribute to the performance of our models.

Focusing on Table 2, we can see that the results of %mAP improve in all three types of backbones as soon as we incorporate the RGB information of the individual members (second row). In addition, the model incorporating the ConvNeXt network (ConvNeXt_Large) as the backbone obtains the best results of the three backbone types.

Thus, we can state that the RGB information of the two persons of a couple (the first two branches of our network architecture) does bring relevant information to our problem, as opposed to using only the RGB information of the couple (third branch). This may be because we provide the network with more information about the members of a pair and help it focus even more on the pair rather than on less important details.

Table 3. Comparison of our three Proxemics-Net variants with the state of the art. Table shows the Average Precision (%) in proxemics recognition: mAP(a) is the average of all classes (Set1 and Set2) and mAP(b) excludes class HT.

Model	HH	HS	SS	HT	HE	ES	mAP (a)	mAP (b)
Yang et al. [18]	37	29	50	61	38	34	42	38
Chu et al. [1]	41.2	35.4	62.2	–	43.9	55	–	46.6
Jiang et al. [5]	59.7	52	53.9	33.2	36.1	36.2	45.2	47.5
Our ViT	48.2	**57.6**	50.4	76.6	57.6	52.8	57.2	53.3
Our ConvNeXt_Base	56.9	53.4	61.4	83.4	68.7	58.8	63.8	59.8
Our ConvNeXt_Large	**62.4**	56.7	**62.4**	**86.4**	**68.8**	**67.9**	**67.4**	**63.8**

Comparison to the State of the Art. Table 3 compares our three best models concerning the existing state of the art. In this Table, two values of %mAP are compared: mAP(a) is the value of mAP explained in the previous sections (the mean of the AP values of the six types of proxemics) and mAP(b) is the mean of the AP values but excluding the Hand-Torso (HT) class as done in [1].

Since our %mAP results are obtained at the pair level rather than at the image level, we had to re-evaluate the best models obtained in Sect. 4.4 to compare our results with the state of the art. For each image, all possible pairs are processed by the target network. The output at image level is computed as the maximum classification score obtained among all image pairs in each proxemics class.

Looking at the Table 3, we can see that our three proposed models (which use three branches as input) perform the best in both comparatives (mAP(a-b)), with the model that uses the ConvNeXt network as a backbone achieving the highest %mAP value (67.4% vs 47.5% mAP(a) and 63.8% vs 47.5% mAP(b)). Thus, we outperformed the existing state of the art by a significant margin, with improvements of up to 19.9% of %AP (mAP(a)) and 16.3% of %mAP (mAP(b)).

Therefore, these results demonstrate that the two state-of-the-art deep learning models (ConvNeXt and Vision Transformers) do help in the proxemics recognition problem since, using only RGB information, they can considerably improve the results obtained by all competing models.

Failure Cases. One detail that caught our attention is the difference in Average Precision that the Hand-Shoulder class has with respect to the other five types of proxemics (see Table 3). So, we took the best model obtained from our experiments (ConvNeXt_Large) and evaluated how it classified each image in the Proxemics dataset.

Figure 4 shows the top 5 images that the model has classified as having the highest Hand-Shoulder score but which do not actually belong to that class, i.e., false positives. It can be seen that they are very similar images in which the elbow-shoulder class predominates and in which the shoulder is not completely

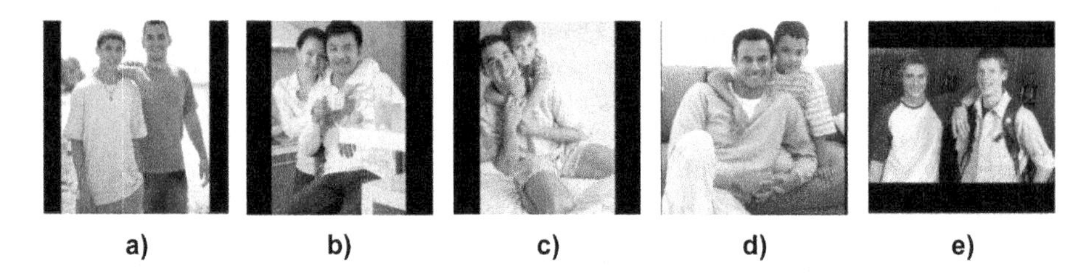

Fig. 4. Failure cases. Samples not annotated as Hand-Shoulder in the dataset, but incorrectly highly scored as such by Proxemics-Net. Note that hands are close to the shoulder in several cases.

visible due to occlusion by other areas of the body (image a), confusion due to the clothing and its color that does not allow to distinguish well each part of the body (image b) or even due to more complex postures (image c).

Thus, one of the limitations of our model is dealing with the occlusion and ambiguity that some parts of the body may have.

5 Conclusions and Future Work

In this work, we have proposed Proxemics-Net, a model capable of addressing the problem of proxemic classification using only RGB information from still images. Proxemics-Net has three input branches (two input branches for the image windows of the individual members of a pair and one for the image window of the pair) and has been evaluated with recent deep learning architectures as backbones. Specifically, three different variations of Proxemics-Net have been proposed (two variations with ConvNeXt models of different complexity and one with a Visual Transformers model).

Our experiments with the Proxemics dataset have shown that state-of-the-art deep learning models such as ConvNeXt and Visual Transformers do help favorably in the problem of proxemics recognition, since using them as backbones in our Proxemics-Net model, we considerably outperformed the results obtained so far by other competent methods.

In addition, after a comparative study, we have been able to check that the variant of our Proxemics-Net model that incorporates the ConvNeXt network as a backbone, obtains the best results in all our different experiments. Therefore, we could affirm that it is the most optimal method for the problem of proxemics recognition in images of all those that have been evaluated in this work.

Future work includes incorporating a more significant number of samples to train our models with a larger number of images and experimenting with other models as backbones. In addition, we could add the pose information of the members to our model and test whether this information reduces the ambiguity and occlusion that some images present and, therefore, further helps in our problem of proxemics recognition in images.

References

1. Chu, X., Ouyang, W., Yang, W., Wang, X.: Multi-task recurrent neural network for immediacy prediction. In: 2015 IEEE ICCV, pp. 3352–3360 (2015)
2. Dosovitskiy, A., et al.: An image is worth 16×16 words: transformers for image recognition at scale. In: ICLR (2021)
3. Felzenszwalb, P., Huttenlocher, D.: Pictorial structures for object recognition. IJCV **61**, 55–79 (2005)
4. Hall, E.T.: A system for the notation of proxemic behavior. Am. Anthropol. **65**(5), 1003–1026 (1963)
5. Jiang, H., Grauman, K.: Detangling people: individuating multiple close people and their body parts via region assembly. In: IEEE CVPR, pp. 3435–3443 (2017)
6. Le, V.T., Tran-Trung, K., Hoang, V.T.: A comprehensive review of recent deep learning techniques for human activity recognition. Comput. Intell. Neurosci. **2022** (2022)
7. Lee, D.G., Lee, S.W.: Human interaction recognition framework based on interacting body part attention. Pattern Recognit. **128** (2022)
8. Li, R., Porfilio, P., Zickler, T.: Finding group interactions in social clutter. In: 2013 IEEE CVPR, pp. 2722–2729 (2013)
9. Liu, Z., Mao, H., Wu, C.Y., Feichtenhofer, C., Darrell, T., Xie, S.: A convnet for the 2020s. In: Proceedings of the IEEE/CVF CVPR (2022)
10. Motiian, S., Siyahjani, F., Almohsen, R., Doretto, G.: Online human interaction detection and recognition with multiple cameras. IEEE Trans. Circuits Syst. Video Technol. **27**(3), 649–663 (2017)
11. Muhamada, A.W., Mohammed, A.A.: Review on recent computer vision methods for human action recognition. ADCAIJ **10**(4), 361–379 (2021)
12. Ouyang, W., Zeng, X., Wang, X.: Modeling mutual visibility relationship in pedestrian detection. In: 2013 IEEE CVPR, pp. 3222–3229 (2013)
13. Ramanathan, V., Yao, B., Fei-Fei, L.: Social role discovery in human events. In: 2013 IEEE CVPR, pp. 2475–2482 (2013)
14. Russakovsky, O., et al.: ImageNet large scale visual recognition challenge. Int. J. Comput. Vision **115**(3), 211–252 (2015). https://doi.org/10.1007/s11263-015-0816-y
15. Vaswani, A., et al.: Attention is all you need. In: Advances in NeurIPS, vol. 30. Curran Associates, Inc. (2017)
16. Wu, B., et al.: Visual transformers: token-based image representation and processing for computer vision (2020)
17. Yang, Y., Baker, S., Kannan, A., Ramanan, D.: PROXEMICS dataset (2012). https://www.dropbox.com/s/5zarkyny7ywc2fv/PROXEMICS.zip?dl=0. Accessed 3 Mar 2023
18. Yang, Y., Baker, S., Kannan, A., Ramanan, D.: Recognizing proxemics in personal photos. In: 2012 IEEE CVPR, pp. 3522–3529 (2012)
19. Ye, Q., Zhong, H., Qu, C., Zhang, Y.: Human interaction recognition method based on parallel multi-feature fusion network. Intell. Data Anal. **25**(4), 809–823 (2021)

Lightweight Vision Transformers for Face Verification in the Wild

Daniel Parres[1(✉)] and Roberto Paredes[1,2]

[1] PRHLT Research Center, Universitat Politècnica de València, Valencia, Spain
{dparres,rparedes}@prhlt.upv.es
[2] Valencian Graduate School and Research Network of Artificial Intelligence, Camí de Vera s/n, 46022 Valencia, Spain

Abstract. Facial verification is an important task in biometrics, particularly as it is increasingly being extended to wearable devices. It is therefore essential to study technologies that can adapt to resource-limited environments. This poses a challenge to current research trends, as the deep learning models that achieve the best results typically have millions of parameters. Specifically, convolutional neural networks (CNN) dominate the state of the art in computer vision, although transformer models have recently shown promise in tackling image problems. Vision transformers (ViT) have been successfully applied to various computer vision tasks, outperforming CNNs in classification and segmentation tasks. However, as ViT models have a high number of parameters, it is crucial to investigate their lightweight variants. The Knowledge Distillation (KD) training paradigm enables the training of student models using teacher models to transfer knowledge. In this study, we demonstrate how to train a lightweight version of ViT using KD to create one of the most competitive lightweight models in the state of the art. Our analysis of ViT models for facial verification sheds light on their suitability for resource-constrained environments such as smartphones, smartwatches, and wearables of all kinds.

Keywords: Face Verification · Transformer · Lightweight Transformers

1 Introduction

Biometric technology involves the analysis of various human characteristics to verify or identify individuals. While fingerprint recognition is the most well-known, other biometric traits such as speech, iris, and palm print also exist. However, these methods typically require active cooperation from the user. In contrast, facial recognition presents an attractive and convenient solution that does not necessitate active user engagement.

Facial recognition is a sub-field of biometrics that aims to identify an individual from an image or video of their face. Facial verification involves analyzing whether a given face image corresponds to a particular person. This requires

A. Pertusa et al. (Eds.): IbPRIA 2023, LNCS 14062, pp. 414–425, 2023.
https://doi.org/10.1007/978-3-031-36616-1_33

to compare the image with a database of facial images of the claimed person obtaining a one-to-one match score.

Facial verification relies on obtaining a high-quality representation or embedding of the face in question. Convolutional neural networks (CNN) are the most widely used approach for generating such representations. By feeding an image of a face into a CNN, the network can extract the most informative features to create a facial embedding.

In this paper, we propose using vision transformer models (ViT) [8] as an alternative to CNNs for producing this embedding. While transformers generally have a high number of parameters, this can present a challenge for face verification when the technology is implemented in memory-limited devices such as wearables. Thus, exploring lightweight transformer architectures that can be deployed on mobile technologies with limited resources is worthwhile.

2 Related Work

CNNs are the most widely employed approach for generating facial embeddings due to their state-of-the-art performance. Usually some previous preprocessing of image is necessary to achieve optimal results. This typically involves detecting and aligning the face, which requires identifying key facial landmarks such as the eyes and nose. There are several methods for performing this preprocessing step, including MTCNN, YOLO, and RetinaFace [5,17,27].

DeepFace [21] was the first successful application of CNNs to the task of face verification. The authors achieved state-of-the-art results at the time by training a CNN in a classification setting, with each individual representing a distinct class. This classification-based approach to facial verification yields facial embeddings that perform well across diverse datasets.

The FaceNet [19] model takes a different approach, eschewing classification-based training in favor of directly learning facial embeddings. To accomplish this, FaceNet trains a CNN using triplet loss, which involves selecting two images for each training image: one image of the same person (positive) and another image of a different person (negative). During training, the CNN is optimized to reduce the distance between an image and its positive example while increasing the distance between the image and its negative example. By taking this approach, FaceNet provides a novel approach to training facial verification models.

In recent years, researchers in facial verification have developed more advanced CNN architectures and enhanced versions of the original DeepFace and FaceNet approaches. Margin-based softmax methods such as SphereFace [13], CosFace [24] and ArcFace [6] are recent techniques that focus on adjusting the softmax loss function. This approach is more efficient than using triplet loss techniques since it performs global comparisons between sample-class pairs, rather than sample-sample pairs. The primary objective of margin-based softmax methods is to maximize class separability, so that each class is represented by an embedding that is adjusted during training.

ArcFace is a margin-based softmax method that stands out for its ability to compute facial embeddings that are more discriminative than those produced by other state-of-the-art approaches. By prioritizing intra-class similarity and inter-class diversity, ArcFace is able to achieve greater separation between classes. The authors of [6] demonstrate how ArcFace outperforms the state-of-the-art on various benchmarks, resulting in wide acceptance by the research community.

3 Proposed Approach

3.1 Vision Transformers

This work investigates the application of vision transformers to facial verification. ViT architectures are designed to maintain the essence of the original transformer proposal by A. Vaswani et al. [23]. In [8], this is achieved by splitting input images into N patches and treating them as sequences.

As described in [8], the number of patches N is a hyperparameter that needs to be set before training the model. Once an input image is fed into the model, it is divided into N patches, which are then flattened and projected to a fixed latent vector size D using a trainable dense layer. This process allows ViT to process image data as if it were a sequence of tokens, enabling the transformer architecture to be applied to image analysis tasks.

In this work, vision transformers are applied using a classification approach with person identities as classes. A classification or class token is used, with the representation of the image contained within it. This representation is called facial embedding and is fed to a multilayer perceptron (MLP) classification head to recognize the identity of the person in the input image. The ViT architecture employed in this work is illustrated in Fig. 1, where an example is shown of an input image being divided into 16 patches. As depicted, the multi-layer perceptron (MLP) plays a crucial role in recognizing the identity of the person, leveraging class embedding.

Table 1 outlines the three variants of Vision Transformer (ViT) proposed in this study for face verification: ViT-B, which has the largest number of parameters and embedding dimension, ViT-Ti, the smallest variant of ViT-B, and ViT-S, a lightweight version of ViT-B. The objective of this investigation is to compare the expressive power of the embeddings of each variant and assess their suitability for different scenarios in face verification.

Table 1. Variants of the vision transformers proposed in this work

Model	Embedding dimension	Num. of params.
ViT-B	768	86M
ViT-S	384	22M
ViT-Ti	192	5M

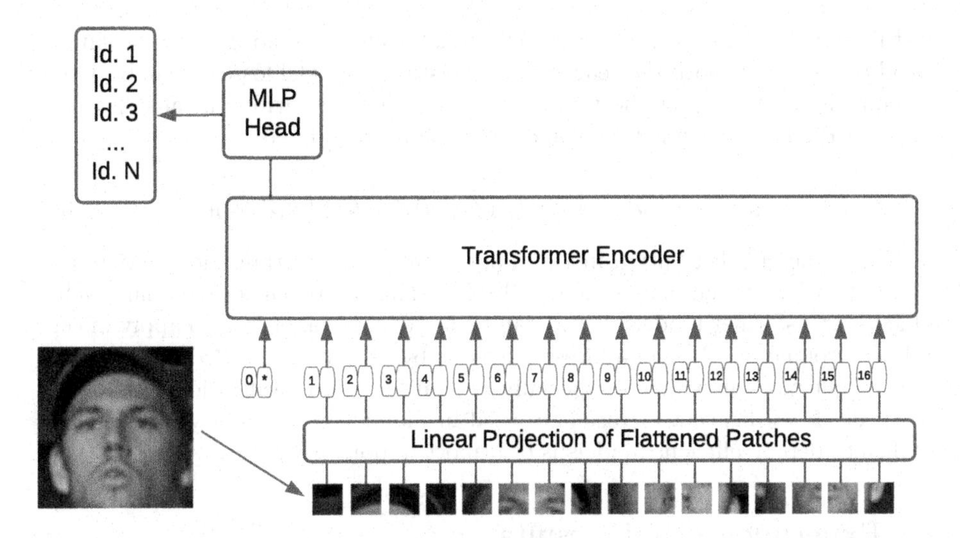

Fig. 1. ViT architecture used in this work.

3.2 Knowledge Distillation

The Knowledge Distillation (KD) training paradigm, introduced in [10], aims to transfer knowledge between models, specifically from a larger teacher model to a smaller student model. In this approach, the teacher model has been trained independently of the student model and typically has a larger capacity. The main objective of KD is to enable the student model to learn from the teacher model's knowledge, which could include generalization capabilities or domain-specific expertise.

In general, the KL divergence (Eq. 1) is employed to train the student model, with the aim of adjusting its parameters so as to minimize the loss function given in Eq. 2. Specifically, the loss function seeks to reduce the difference between the outputs of the teacher and student models, where Z_t and Z_s represent the logits of the teacher and student models, respectively. The function incorporates several parameters, including the temperature parameter τ, the trainable balancing coefficients λ, the cross-entropy $\mathcal{L}_{\mathrm{CE}}$, and the softmax function ψ.

$$D_{\mathrm{KL}}(x, y) = \sum_{n=0}^{N-1} x_n \log\left(\frac{x_n}{y_n}\right) - x_n + y_n \tag{1}$$

$$\mathcal{L}(Z_s, Z_t, y) = (1-\lambda)\mathcal{L}_{\mathrm{CE}}(\psi(Z_s), y) + \lambda\tau^2 D_{\mathrm{KL}}(\psi(Z_{\mathrm{s}}/\tau), \psi(Z_{\mathrm{t}}/\tau)) \tag{2}$$

In this study, we adopt the hard-label (HL) distillation method proposed by [22] for training transformer-based student models. H. Touvron et al. reveal in [22] that this method outperforms the commonly used KL divergence method for transformer-based models. In Eq. 3, we present the formulation of HL distillation, which differs from KL divergence in several key aspects. Notably, HL distillation does not rely on trainable weights and instead employs a simple weighting scheme. Both the teacher and the student model losses are halved and combined without using the temperature parameter τ, which means that the logits of the teacher and student models remain unmodified.

$$\mathcal{L}_{\mathrm{HL}}(Z_s, y_t, y) = \frac{1}{2}\mathcal{L}_{\mathrm{CE}}(\psi(Z_s), y) + \frac{1}{2}\mathcal{L}_{\mathrm{CE}}(\psi(Z_s), y_{\mathrm{t}}) \tag{3}$$

Using the HL distillation method, our study aims to train and compare the lightweight ViT variants presented in Table 1. This approach is particularly valuable for transformer models, which can be large and challenging to apply in certain environments. We can address this issue by exploring knowledge transfer to lightweight models, which is especially critical for facial verification technology. With facial verification increasingly used in mobile devices and wearables, the ability to deploy efficient models is of utmost importance.

4 Experiments and Results

4.1 Datasets

The datasets used in our experiments are detailed in Table 2. For training, we employed a refined version of MS-Celeb-1M [9] known as MS1M-RetinaFace [7]. This dataset was specifically curated for recognition tasks, consisting of 5.1 million face images and 93,000 identities, as shown in Table 2. The main difference between MS-Celeb-1M and MS1M-RetinaFace is the preprocessing step. By utilizing RetinaFace [5], all images are aligned to a size of 112×112, allowing for improved performance. This dataset has been widely used to benchmark state-of-the-art models. Therefore, this training set aims to obtain a good feature extractor for face images. Once the model is trained, the classification layer is removed so that every time an image is forwarded, a facial embedding is obtained, which is used for the verification task.

For evaluating the performance of the models and comparing them to state-of-the-art methods, we used the Labeled Faces in the Wild (LFW) dataset [11]. LFW is widely used in the field of face verification due to its large and diverse collection of face images, making it a suitable benchmark dataset for different types of face recognition problems. To evaluate the performance of a model, we commonly use the Area Under Curve (AUC) metric, which we will refer to as the verification performance in this paper. The verification performance measures the model's ability to distinguish between genuine and impostor face pairs, and it is a widely used metric in face recognition research.

Table 2. Datasets for training and testing.

Dataset	Partition	Num. Images	Num. IDs
MS1M-RetinaFace	Train	5.1M	93k
LFW	Test	13.2k	5.7k

4.2 Training Vision Transformers

In this section, we will outline the methodology used to train transformer models for facial verification. We will focus specifically on the ViT-B, ViT-S, and ViT-Ti models and thoroughly examine our approach.

Weight initialization is a critical step in training vision models, as it can significantly impact the subsequent performance of the model. Over the years, pre-training on the ImageNet [4,18] dataset has emerged as a popular approach for initializing models, followed by fine-tuning on a specific task. This technique enables models to adapt to new data and achieve better results than random weight initialization. In our study, we adopt this approach and initialize our transformer models with ImageNet weights to leverage these benefits.

For this study, the MS1M-RetinaFace database was used as the training dataset, consisting of 5.1 million images. To maximize the use of this dataset, we selected RandAugment [3] as our data augmentation method. RandAugment is an automatic method that can apply up to 14 different transformations to training images, including contrast, equalization, rotation, solarization, color jittering, posterizing, brightness, sharpness, shear in x and y, and translation in x and y. However, negative filters were not applied to the images in this study.

Selecting an appropriate optimizer and learning rate scheduler is critical for training transformer models as they are highly sensitive to these choices. AdamW has emerged as the most popular optimizing algorithm for vision models in recent years. In this study, we utilized cosine annealing [14] as our learning rate scheduler to adjust the learning rate during training. Multiple learning rate values were investigated to determine the best starting weight adjustment, with a value of 1×10^{-4} found to yield the optimal results using batches of size 180. This batch size was chosen as it represents the maximum size our GPUs can handle. For this classification problem, cross-entropy was used as the default loss function to train the transformer models, as each identity corresponds to a class.

As a first experiment in this section, we compared the performance of the ViT-B model on the LFW database using two different loss functions: cross-entropy and a margin-based softmax method. Margin-based softmax methods have gained popularity, and it is interesting to investigate whether they can improve the performance of transformer models over cross-entropy. The margin-based softmax method we used in this study is ArcFace, which is currently the most widely used method. The results are presented in Table 3, where we observed that ViT-B trained with cross-entropy outperformed the ArcFace-based approach in terms of LFW performance.

Table 3. Verification performance (%) of ViT-B on LFW.

Model	Loss	Num. of params.	LFW (%)
ViT-B	ArcFace	86M	99.67
ViT-B	cross-entropy	86M	99.75

Table 4. Verification performance (%) of ViT-B, ViT-S and ViT-Ti on LFW.

Model	Loss	Num. of params.	LFW (%)
ViT-B	cross-entropy	86M	99.75
ViT-S	cross-entropy	22M	99.70
ViT-Ti	cross-entropy	5M	99.50

Given the superior performance of cross-entropy, we trained the ViT-S and ViT-Ti lightweight models under the same conditions as ViT-B. Specifically, we used RandAugment for data augmentation, AdamW [15] for optimization, cosine annealing for learning rate schedule, and a learning rate of 1×10^{-4}. However, due to their different parameter sizes, we used a batch size of 256 for ViT-S and 512 for ViT-Ti. The results are presented in Table 4. The ViT-S model, with 22M parameters, achieved excellent verification results. However, ViT-Ti, with only 5M parameters, achieved worse performance.

This section outlines the training process for transformer models used in face verification. Our results demonstrate that using the cross-entropy loss function leads to superior performance in these models. Additionally, we observed that the lightweight ViT-Ti model performs worse in verification than its base and small versions. To improve its performance, we explore the use of knowledge distillation with ViT-B and its potential to enhance the verification performance of ViT-Ti.

4.3 Improving Lightweight Vision Transformers

The knowledge distillation paradigm enables a student model to learn from a teacher model, making it a popular technique for training lightweight models from more complex ones. With this approach, we aim to enhance the performance of the lightweight ViT-Ti model, as it yielded lower results in our previous experiment (refer to Table 4).

Based on the results obtained in our experiments, the ViT-B model achieved the highest performance with a score of 99.75% on the LFW database. Therefore, we have used ViT-B as the teacher model and ViT-Ti, initialized with ImageNet weights, as the student model. Following the findings in [22], we have used the HL distillation approach, which has been shown to yield better results than the KL method. Hence, we employed the loss function presented in Eq. 3 for this experiment.

To train the lightweight model, we have used RandAugment, AdamW, cosine annealing, a learning rate of 1×10^{-4}, and a batch size of 512. The results are

Table 5. Verification performance (%) of ViT-Ti on LFW.

Model	KD	Num. of params.	LFW (%)
ViT-Ti	no	5M	99.50
ViT-Ti	yes	5M	99.58

presented in Table 5, where it can be seen that ViT-Ti manages to improve its verification performance over LFW using the KD paradigm.

This experiment has demonstrated that the performance of the lightweight model can be enhanced by utilizing ViT-B as a teacher model through knowledge distillation. However, the significance of this improvement needs to be further analyzed and compared to the results obtained by state-of-the-art models to determine if it is competitive.

4.4 Comparison with the State of the Art

In order to assess the competitiveness of the transformers trained in this work, it is imperative to compare them with state-of-the-art models. To this end, we employ LFW, and present a comparative analysis of the results in Table 6.

Table 6. Verification performance (%) of different state of the art methods on LFW sorted by number of parameters

Model	Num. of params	LFW (%)
VGGFace [16]	145M	98.95
Facenet [19]	140M	99.63
DeepFace [21]	120M	97.35
ViT-B (ours)	86M	99.75
ViT-B-ArcFace (ours)	86M	99.67
CurricularFace [12]	65M	99.80
FaceGraph [28]	65M	99.80
ArcFace [6]	65M	99.83
CosFace [24]	48M	99.73
SphereFace [13]	48M	99.42
VGGFace2 [1,12]	28M	99.43
ViT-S (ours)	22M	99.70
MV-Softmax [25]	18M	99.80
ViT-Ti-KD (ours)	5M	99.58
ViT-Ti (ours)	5M	99.50
DeepID2+ [20]	2.5M	99.47
MobileFaceNet [2]	1M	99.55
Center Loss [26]	800k	99.28

Transformer models are highlighted in bold. In general, the models perform exceptionally well, but the highest accuracy is achieved by ArcFace, with a score of 99.83%. Our ViT-B, which achieved a third-best performance of 99.75%, is not far behind. However, it is worth noting that training our ViT-B with ArcFace resulted in slightly lower accuracy compared to the original ViT-B.

Our lightweight models have achieved competitive positions on the table, placing in the middle range. Therefore, we can conclude that we have developed models that perform well compared to state-of-the-art models taking into account the number of parameters. For a better visualization of the comparison, we present Fig. 2.

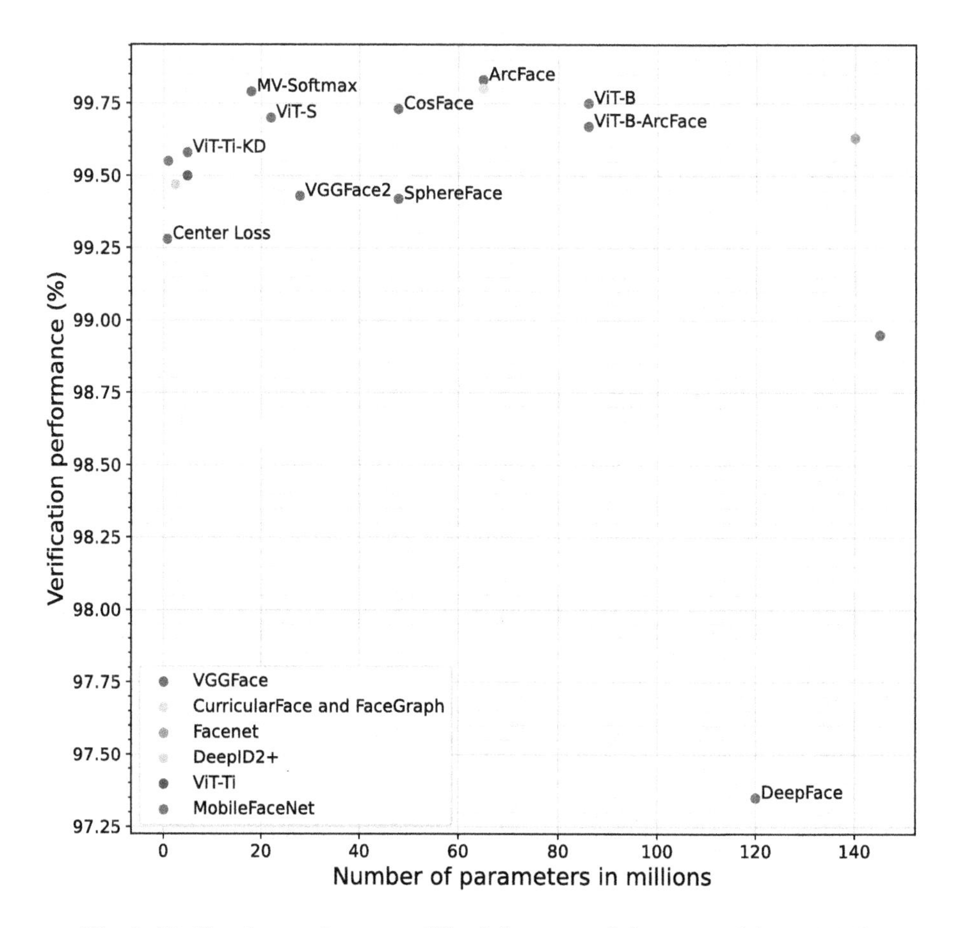

Fig. 2. Verification performance (%) of the state-of-the-art models on LFW.

In Fig. 2, ViT-B stands among the most competitive models. There is an interesting zone in the table, which could be called the "lightweight zone". The Center Loss, DeepID2+, MobileFaceNet, ViT-Ti, and ViT-Ti-KD models are located in this zone. These models have in common that they are the lightest models in state of the art. The model achieving the best verification performance in this area is ViT-Ti-KD. ViT-Ti trained conventionally is below ViT-Ti-KD and MobileFaceNet. Therefore, it can be concluded that the improvement provided by training with the KD paradigm is significant for lightweight models. Overall, our lightweight models have achieved competitive positions on the table, staying in the middle.

As demonstrated, vision transformers have achieved impressive results in facial verification tasks. Specifically, the base model attained the third-best performance among the state-of-the-art models, and training ViT-Ti with the KD paradigm resulted in a significant improvement for the lightweight models.

5 Conclusions

This paper has demonstrated how to train competitive transformer models in face verification. Our experiments show that using cross-entropy as the loss function yields competitive results for ViT base and small models, which have 86M and 22M parameters, respectively. However, more than conventional training is required for the lightweight ViT-Ti model to be competitive with other state-of-the-art lightweight models. On the other hand, when the KD paradigm is applied, ViT-Ti achieves the highest competitiveness among state-of-the-art lightweight models for facial verification on LFW. This is particularly important for adopting vision transformers in mobile and wearable devices.

Acknowledgements. Work partially supported by the Universitat Politècnica de València under the PAID-01-22 programme, by the support of valgrAI - Valencian Graduate School and Research Network of Artificial Intelligence and the Generalitat Valenciana, and co-funded by the European Union.

References

1. Cao, Q., Shen, L., Xie, W., Parkhi, O.M., Zisserman, A.: VGGFace2: a dataset for recognising faces across pose and age. In: Proceedings of the 13th IEEE International Conference on Automatic Face & Gesture Recognition, pp. 67–74 (2018)
2. Chen, S., Liu, Y., Gao, X., Han, Z.: MobileFaceNets: efficient CNNs for accurate real-time face verification on mobile devices. In: Zhou, J., et al. (eds.) CCBR 2018. LNCS, vol. 10996, pp. 428–438. Springer, Cham (2018). https://doi.org/10.1007/978-3-319-97909-0_46
3. Cubuk, E.D., Zoph, B., Shlens, J., Le, Q.V.: RandAugment: practical automated data augmentation with a reduced search space. In: Proceedings of the IEEE/CVF Conference on Computer Vision and Pattern Recognition Workshops, pp. 702–703 (2020)
4. Deng, J., Dong, W., Socher, R., Li, L.J., Li, K., Fei-Fei, L.: ImageNet: a large-scale hierarchical image database. In: Proceedings of the IEEE Conference on Computer Vision and Pattern Recognition, pp. 248–255 (2009)

5. Deng, J., Guo, J., Ververas, E., Kotsia, I., Zafeiriou, S.: RetinaFace: single-shot multi-level face localisation in the wild. In: Proceedings of the IEEE/CVF Conference on Computer Vision and Pattern Recognition, pp. 5203–5212 (2020)
6. Deng, J., Guo, J., Xue, N., Zafeiriou, S.: ArcFace: additive angular margin loss for deep face recognition. In: Proceedings of the IEEE/CVF Conference on Computer Vision and Pattern Recognition, pp. 4690–4699 (2019)
7. Deng, J., Guo, J., Zhang, D., Deng, Y., Lu, X., Shi, S.: Lightweight face recognition challenge. In: Proceedings of the IEEE/CVF International Conference on Computer Vision Workshops (2019)
8. Dosovitskiy, A., et al.: An image is worth 16×16 words: transformers for image recognition at scale. arXiv preprint arXiv:2010.11929 (2020)
9. Guo, Y., Zhang, L., Hu, Y., He, X., Gao, J.: MS-Celeb-1M: a dataset and benchmark for large-scale face recognition. In: Leibe, B., Matas, J., Sebe, N., Welling, M. (eds.) ECCV 2016. LNCS, vol. 9907, pp. 87–102. Springer, Cham (2016). https://doi.org/10.1007/978-3-319-46487-9_6
10. Hinton, G., Vinyals, O., Dean, J.: Distilling the knowledge in a neural network. arXiv preprint arXiv:1503.02531 (2015)
11. Huang, G.B., Mattar, M., Berg, T., Learned-Miller, E.: Labeled faces in the wild: a database for studying face recognition in unconstrained environments. In: Proceedings of the Workshop on faces in 'Real-Life' Images: Detection, Alignment, and Recognition (2008)
12. Huang, Y., et al.: CurricularFace: adaptive curriculum learning loss for deep face recognition. In: Proceedings of the IEEE/CVF Conference on Computer Vision and Pattern Recognition, pp. 5901–5910 (2020)
13. Liu, W., Wen, Y., Yu, Z., Li, M., Raj, B., Song, L.: SphereFace: deep hypersphere embedding for face recognition. In: Proceedings of the IEEE Conference on Computer Vision and Pattern Recognition, pp. 212–220 (2017)
14. Loshchilov, I., Hutter, F.: SGDR: stochastic gradient descent with warm restarts. arXiv preprint arXiv:1608.03983 (2016)
15. Loshchilov, I., Hutter, F.: Decoupled weight decay regularization. arXiv preprint arXiv:1711.05101 (2017)
16. Parkhi, O., Vedaldi, A., Zisserman, A.: Deep face recognition. In: Proceedings of the British Machine Vision Conference, pp. 1–12 (2015)
17. Pebrianto, W., Mudjirahardjo, P., Pramono, S.H.: YOLO method analysis and comparison for real-time human face detection. In: Proceedings of the 11th Electrical Power, Electronics, Communications, Controls and Informatics Seminar, pp. 333–338 (2022)
18. Russakovsky, O., et al.: ImageNet large scale visual recognition challenge. Int. J. Comput. Vision **115**, 211–252 (2015)
19. Schroff, F., Kalenichenko, D., Philbin, J.: FaceNet: a unified embedding for face recognition and clustering. In: Proceedings of the IEEE Conference on Computer Vision and Pattern Recognition, pp. 815–823 (2015)
20. Sun, Y., Wang, X., Tang, X.: Deeply learned face representations are sparse, selective, and robust. In: Proceedings of the IEEE Conference on Computer Vision and Pattern Recognition, pp. 2892–2900 (2015)
21. Taigman, Y., Yang, M., Ranzato, M., Wolf, L.: DeepFace: closing the gap to human-level performance in face verification. In: Proceedings of the IEEE Conference on Computer Vision and Pattern Recognition, pp. 1701–1708 (2014)
22. Touvron, H., Cord, M., Douze, M., Massa, F., Sablayrolles, A., Jegou, H.: Training data-efficient image transformers & distillation through attention. In: Proceedings of the 38th International Conference on Machine Learning, pp. 10347–10357 (2021)

23. Vaswani, A., et al.: Attention is all you need. In: Proceedings of the Advances in Neural Information Processing Systems, pp. 5998–6008 (2017)
24. Wang, H., et al.: CosFace: large margin cosine loss for deep face recognition. In: Proceedings of the IEEE Conference on Computer Vision and Pattern Recognition, pp. 5265–5274 (2018)
25. Wang, X., Zhang, S., Wang, S., Fu, T., Shi, H., Mei, T.: Mis-classified vector guided softmax loss for face recognition. In: Proceedings of the AAAI Conference on Artificial Intelligence, pp. 12241–12248 (2020)
26. Wen, Y., Zhang, K., Li, Z., Qiao, Yu.: A discriminative feature learning approach for deep face recognition. In: Leibe, B., Matas, J., Sebe, N., Welling, M. (eds.) ECCV 2016. LNCS, vol. 9911, pp. 499–515. Springer, Cham (2016). https://doi.org/10.1007/978-3-319-46478-7_31
27. Zhang, K., Zhang, Z., Li, Z., Qiao, Y.: Joint face detection and alignment using multitask cascaded convolutional networks. IEEE Signal Process. Lett. **23**, 1499–1503 (2016)
28. Zhang, Y., et al.: Global-local GCN: large-scale label noise cleansing for face recognition. In: Proceedings of the IEEE/CVF Conference on Computer Vision and Pattern Recognition, pp. 7731–7740 (2020)

Py4MER: A CTC-Based Mathematical Expression Recognition System

Dan Anitei[1(✉)], Joan Andreu Sánchez[1], and José Miguel Benedí[1,2]

[1] Pattern Recognition and Human Language Technologies Research Center, Universitat Politècnica València, Valencia, Spain
{danitei,jandreu,jmbenedi}@prhlt.upv.es

[2] Valencian Graduate School and Research Network of Artificial Intelligence, Camí de Vera s/n, 46022 Valencia, Spain

Abstract. Mathematical expression recognition is a research field that aims to develop algorithms and systems capable of interpreting mathematical content. The recognition of MEs requires handling two-dimensional symbol relationships such as sub/superscripts, matrices and nested fractions, among others. The prevalent technology for addressing these challenges are based on encoder-decoder architectures with attention models. In this paper we propose the Py4MER system, based on Convolutional Recurrent Neural Network (CRNN) models, for the recognition and transcription of MEs into LaTeX mark-up sequences. This model is proposed as an alternative to encoder-decoder approaches, as CRNN models trained through Connectionist Temporal Classification (CTC) implicitly model the dependencies between symbols and do not suffer from under/over parsing of the input image, generating more consistent LaTeX mark-up.

The proposed model is evaluated on the Im2Latex-100k data set based on both textual and image-level metrics, showing a remarkable improvement from other CTC-based approaches. Recognition results are analyzed for different ME lengths and ME structures. Furthermore, a study based on the edit distance is performed, showing a considerable improvement in precision when up to 5 edit operations are considered. Finally, we show that CTC-based CRNN models can adapt to non left-to-right ordering of ME elements, warranting more research for this approach.

Keywords: Mathematical Expression Recognition · Deep Learning · Convolutional Recurrent Neural Network · Connectionist Temporal Classification

1 Introduction

Mathematical Expression Recognition (MER) is a research field that studies the development of algorithms and systems capable of interpreting handwritten or typed Mathematical Expressions (MEs) [36]. These expressions can include numbers, symbols, and operations, which, when combined, create different results,

A. Pertusa et al. (Eds.): IbPRIA 2023, LNCS 14062, pp. 426–441, 2023.
https://doi.org/10.1007/978-3-031-36616-1_34

from simple formulae to complex structures. MEs can be found in a variety of sources, such as textbooks, research papers and educational materials.

MER systems can be divided into: online and offline recognition [5]. In online recognition, the MEs to be interpreted are handwritten through electronic devices and are represented as sets of strokes that define the pen trajectory. On the other hand, in offline recognition systems, the MEs are represented as images of the handwritten or typeset expression. In this paper, we focus on offline recognition of typeset MEs.

Current MER research is based on image processing, pattern recognition, and machine learning techniques. Unlike traditional Optical Character Recognition (OCR) tasks, recognition of MEs must consider the mathematical language's two-dimensional nature. Different approaches have been developed to address the challenges of recognizing MEs, such as complex structures, ambiguity at both symbol and mark-up language levels, font and style choices, and variability in notation depending on the field of application. In addition, a crucial aspect to consider is that ME structural information is associated with spatial relationships that do not have a visual representation. This information has to be inferred from the position of symbols in relation to other parts of the ME.

Research in this field is active and growing with the recent advancements in deep learning techniques. Sequence-to-sequence architectures are typically used to solve complex language problems like machine translation [6], speech [7] and handwriting [13] recognition, image captioning [25], and scene text recognition [12]. These techniques have been successfully used in MER tasks and shown to improve the performance of recognition systems [4,18,24].

In this context, this paper presents a Convolutional Recurrent Neural Network (CRNN) model for the recognition of typeset LaTeX MEs. The system, `Py4MER`, is trained end-to-end by minimizing the Connectionist Temporal Classification (CTC) loss function [10]. Our approach is segmentation-free and does not suffer from coverage or under/over parsing problems which are common for attention-based encoder-decoder architectures [26,30]. In addition, we demonstrate that CTC with bidirectional RNN is able to align the input image with the output sequence even when non left-to-right reading order is used. We propose this architecture as an alternative to the widely used attention-based encoder-decoder models, showing that CTC-based architectures are adequate for tackling the MER task.

The proposed model is analyzed at different levels, addressing the problem of generating structurally correct LaTeX mark-up, and studying the recognition performance for different ME lengths and ME structure types. In addition, the results are analyzed based on the number of edit operations needed to correctly interpret the ground-truth. The proposed model was evaluated on the public `Im2Latex-100k` data set [8]. The results are compared with state-of-the-art MER systems based on textual and image-level metrics, showing a remarkable improvement from other CTC-based approaches [8].

This paper is organized as follows: in Sect. 2, we explore the current state-of-the-art for MER; Sect. 3 addresses the problem of the image to LaTeX mark-up generation; Sect. 4 presents the characteristics of the employed data set; Sect. 5 introduces the neural model architecture used to tackle the MER problem; then Sect. 6 presents the evaluation protocol and a detailed analysis of the experimental results. Finally, Sect. 7 summarizes the work presented.

2 Related Work

Mathematical Expression Recognition is an active field that has been the focus of researchers since 1967, when [2] proposed a syntax-directed recognition system for handwritten MEs. Much research on handwritten and typeset MER has been published, stemming from mathematical notations presenting critical information in many fields. The problem of offline MER poses three significant challenges: symbol segmentation, symbol recognition, and structural analysis. Two main approaches address these challenges. In the first approach, all three challenges are dealt with sequentially, with symbol segmentation performed before symbol recognition and structural analysis [28,36]. This approach propagates segmentation and recognition errors and affects the system's performance. Therefore, this paper focuses on the second approach, in which all steps are executed and optimized simultaneously [3,35].

Structural and Grammatical Models. Structural and grammatical models are excellent techniques for representing the syntactic relationships between mathematical symbols. However, these models have a significant manual component requiring a substantial effort to design them in which expert knowledge is essential. In [22], the `Infty` OCR system is proposed, in which the structure of MEs is represented as trees, and structure analysis is performed using a minimum-cost spanning-tree algorithm. In [28], a method is proposed in which baseline structure trees are employed to describe the two-dimensional arrangement of symbols. [16,35] proposed an ME recognition system (`Seshat`) based on two-dimensional probabilistic context-free grammars (2D-PCFG). A notable feature of this system is that it is possible to obtain not only the 1-best interpretation but also a hypergraph that generalizes a set of N-best interpretations [17].

Encoder-Decoder Models. Sequence-to-sequence (encoder-decoder) neural models have greatly improved the performance of MER systems, given their ability to handle variable-length input and output sequences. However, the problem of obtaining syntactic relations between the different parts of an ME has yet to be researched for neural models [27], although it is directly addressed in syntactic models. This fact is crucial when considering MER as a precursor to ME indexing for ME retrieval systems.

In [8], the authors propose an encoder-decoder architecture with an attention model. This system (`Im2Tex`) includes an RNN row encoder over the extracted features to preserve the relative positions of symbols in the source image. In [32], an MER encoder-decoder system (`DoubleAtt`) is presented, which uses a double attention mechanism to improve the interpretation of the structure of MEs. The authors also propose doubling the input images' size to improve the recognition of fuzzy or small symbolic feature information. [26] introduces an entirely convolutional encoder-decoder architecture (ConvMath), allowing it to perform parallel computation and obtain much better efficiency than RNN-based systems. In [20], the MER problem is tackled by adding a Graph Neural Network (GNN) to model the spatial information of mathematical symbols. In this approach, connected components are computed from the formula image from which a graph is generated, recording symbols' relative coordinates and positional information. [4] presents an encoder-decoder architecture that uses a fine-grained feature extractor (FGFE). The authors use unevenly shaped max pooling to slow the reduction in the size of the input images and, thus, capture the fine-grained structures of the MEs. The system also uses an LSTM to encode the rows of the feature maps. [18] propose a transformer-based encoder (GCBN) inserted between the feature extractor and decoder to encode symbol relationships better. The authors also employ a mask attention-based decoder to deal with the input image's over-parsing and under-parsing. In [24], the authors propose the MI2LS system in which they augment the extracted feature maps by adding sinusoidal positional encoding, which results in richer modeling of symbol interrelationships. In addition, a sequence-level objective function based on Bleu [19] is used to further aid in modeling the interrelationship among different tokens in the LaTeX sequences.

CTC-Based Approaches. CTC has been employed for training recognition systems for handwritten MEs [9,31], or as an auxiliary learning method to form a multi-task learning approach, which is shown to improve the performance of encoder-decoder models [15]. However, the capabilities of CTC to reproduce naturally occurring rendered LaTeX mark-up, have not been thoroughly tested on very large collections of real-world MEs, in which variability in notation is significant (more than 500 different LaTeX tokens in Im2Latex). CTC does not suffer from over-parsing and under-parsing of the input image. CTC implicitly models the dependencies between symbols and allows the training of a neural network directly with unsegmented data. This motivates us to demonstrate that CTC can be employed successfully for the recognition of typeset MEs, providing an alternative to encoder-decoder architectures.

3 Problem Formulation

The Image-to-Mark-up problem consists of converting two-dimensional images into one-dimensional structured sequences. In the case of Image-to-LaTeX for typeset MEs, the input $\mathbf{x}$ consists of a typeset image representation of a ME, and the target $\mathbf{y}$ is a sequence of LaTeX tokens that describe both the ME content and its layout. Following the definition given by [8], we consider a token to be a minimal meaningful LaTeX character, symbol, bracket, special character, or a command such as a function, accent, environment, etc. The problem of Image-to-LaTeX can be addressed by trying to solve the following optimization problem:

$$\hat{\mathbf{y}} = \arg\max_{\mathbf{y}} p(\mathbf{y} \mid \mathbf{x}) \tag{1}$$

In this formulation, the input $\mathbf{x}$ is represented as a sequence of T frames (x_1^T) and the target as a sequence of LaTeX tokens of length U (y_1^U), with the input and output sequences generally of different lengths ($T \neq U$). For CTC-based approaches, as the one presented, it is important to mention that $U \leq T$. In this approach, a CTC classifier outputs a LaTeX token or a CTC blank symbol [10] for every frame T of the input image. Thus, Eq. (1) can be rewritten as:

$$\hat{\mathbf{y}} = \arg\max_{\mathbf{y} \in L^{\leq T}} p(\mathbf{y} \mid \mathbf{x}) \tag{2}$$

$L^{\leq T}$ denotes the set of LaTeX sequences of length less than or equal to T, in which the CTC outputs have been collapsed, and the CTC blank symbol has been removed. Given that the actual probability $p(\mathbf{y} \mid \mathbf{x})$ is unknown, the solution is approximated by a parametric distribution $p_\Theta(\mathbf{y} \mid \mathbf{x})$. The training of the CTC system is done by tuning the model's parameters on a training set D by minimizing the negative log-likelihood:

$$\sum_{(\mathbf{x},\mathbf{y}) \in D} - \log\, p_\Theta(\mathbf{y} \mid \mathbf{x}) \tag{3}$$

$p_\Theta(\mathbf{y} \mid \mathbf{x})$ is computed by summing all the CTC alignments that collapse into the given ground-truth sequence $\mathbf{y}$.

The model performance is assessed by considering the pair $(\hat{\mathbf{y}}, \mathbf{y})$, in which a strict LaTeX token recognition is tested (see *Bleu* and *ExStr* metrics in Sect. 6.1). In addition, evaluation is also performed on the pair $(\hat{\mathbf{x}}, \mathbf{x})$, in which $\hat{\mathbf{x}}$ is the rendered representation obtained by compiling the LaTeX sequence $\hat{\mathbf{y}}$. This second evaluation (see *ExIm* and *ExIm-ws* metrics in Sect. 6.1) tests symbol and structure recognition while suppressing mark-up level ambiguity, given that the rendering of LaTeX MEs is a many-to-one function as different ME definitions can have the same visual representation.

4 Data Set Characteristics

The ME recognition experiments presented in this paper are performed on the `Im2Latex-100k` public data set [8]. This collection consists of 103 556 MEs in LaTeX format, for which rendered images are provided. The corpus was built by parsing the LaTeX sources of documents from tasks I and II of the 2003 KDD cup. The resulting data are provided in three separate partitions: training set (83 883 MEs), validation set (9 319 MEs), and test set (10 354 MEs). Figure 1 shows the distribution of these MEs as a function of the number of LaTeX tokens. These MEs have a mean of 63 and a median of 53 tokens.

An analysis of the MEs based on their structure and components, shows that 95.3% of these MEs feature either sup or sub-script expressions, making these two-dimensional relationships between symbols the most common. These sup/sub script structures are defined by special LaTeX characters such as "^ " and "_ ". In addition, the special characters "{ " and "} " are used to enclose

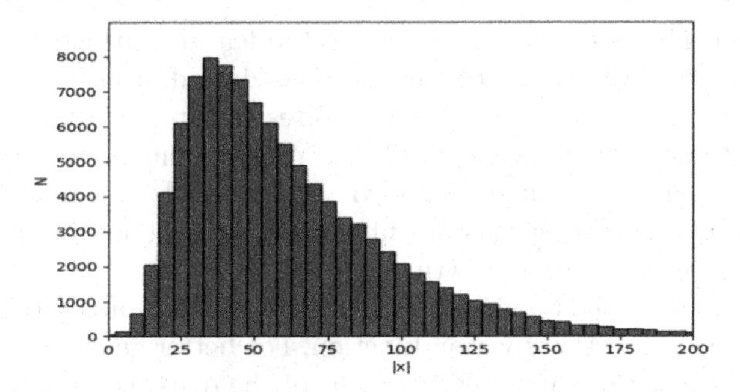

Fig. 1. MEs distribution as a function of the number of LaTeX tokens.

groups of tokens. All four of these characters are part of the LaTeX syntax and, thus, are not rendered. However, they represent 43.6% of the total number of tokens of the MEs. Correct detection and recognition of these special characters is crucial for the performance of MER systems.

In addition, we found that 48.3% of MEs contain fractions and 6.5% of the MEs contain multi-line expressions such as matrices (6.3%) or vertically aligned sub-expressions (0.2%). These complex vertical relationships go beyond conventional OCR approaches, which makes the MER task particularly difficult.

5 Model Architecture

For the experiments presented in this paper, we designed a CTC-based CRNN model. The CRNN architecture has been successfully employed in many Natural Language Processing tasks, such as Handwritten Text Recognition (HTR) [1,21] and Speech Recognition [11,34]. This segmentation-free approach allows us to train the model with only the underlying LaTeX mark-up ME definition with no additional information about the segmentation of individual symbols.

The proposed model consists of 3 components: the convolutional blocks for feature extraction, the recurrent blocks for capturing long-term dependencies, and finally, the CTC transcription block for generating the LaTeX mark-up. The design of this model takes into account the following aspects that are critical for the performance of CTC-based architectures in the case of Image-to-LaTeX tasks: 1) as stated in Sect. 4, more than 43% of the LaTeX tokens are not rendered and thus do not occupy frames in the corresponding ME images; 2) structures like fractions and matrices feature mathematical symbols that are vertically aligned and thus share frame positions; 3) as indicated in Eq. (2), CTC produces a target sequence with at most as many tokens as frames in the input [10]. Taking into account the ratio of frames to tokens, is crucial as excessively reducing the dimensionality of the data can incur losing training data as sequences that cannot be generated cannot contribute to minimizing the CTC loss. An overview of the architecture is shown in Fig. 2. This system is implemented through the publicly available PyLaia[1] software.

[1] https://github.com/jpuigcerver/PyLaia.

Convolutional Blocks. The densely connected feature extractor component is inspired by the work of [23,29], taking the CTC data dimensionality restriction in mind. The initial convolutional layer features 48 filters with a (3×3) kernel and a stride of (1×1), followed by a (2×2) max pooling layer with a stride of (2×2). Next, we defined a single dense convolutional block with a depth $D = 8$ convolutional layers starting with 48 filters for the first layer. Each following convolutional layer features a growth rate of $K = 24$, resulting in new features being added on top of the previously extracted ones. We employ Batch Normalization and a LeakyReLU activation function. For better generalization, we also use Dropout with a value of 0.2. The output of the dense convolutional block is transformed through an Adaptive Average Pooling layer to sequence the image for further RNN processing.

Recurrent Blocks. Dependencies between tokens of LaTeX MEs are captured using 3 Bidirectional Long Short-Term Memory (BiLSTM) layers, each with $U = 512$ units. These recurrent layers process the feature sequence column-wise in left-to-right and right-to-left order. The output of the two directions is concatenated depth-wise [21], resulting in a $2 \cdot U$ channel depth. To prevent overfitting of the model, we apply a Dropout of 0.5 before each LSTM layer.

Transcription Block. A Linear Softmax output layer forms the final transcription block with the number of units (L) equal to the number of different LaTeX tokens that form the alphabet plus the CTC blank symbol. Thus, the output of the recurrent block is transformed from a dimension of $2 \cdot U$ to obtain per-frame predictions over L. This output is then decoded with CTC to generate the final LaTeX transcription. To increase the generalization capabilities of the model, we apply a Dropout of 0.5 before the Linear layer.

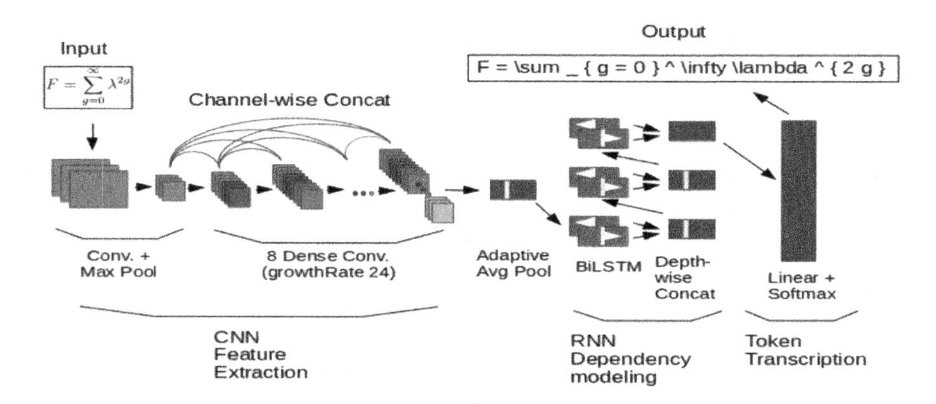

Fig. 2. Overview of the CRNN model showing the three components of the system.

While dense convolutional layers are not publicly available in PyLaia, for reproducibility purposes, in this paper, we also report results with standard CNN architecture, available by default. In this architecture, dense connections are replaced with sequential connections. The number of convolutional layers and filters was kept the same.

6 Evaluation

In this section the training protocol is first presented, followed by an introduction of the employed evaluation metrics, which are widely used for testing ME recognition systems. As mentioned in Sect. 3, the evaluation is performed at both mark-up and image level, highlighting the models' capabilities for symbol recognition as well as structural interpretation.

The proposed model is evaluated for its capabilities of producing syntactically correct mark-up sequences, studying the performance under different ME lengths as well as different ME types. A study is performed based on the number of edit operations required to transform hypotheses into corresponding references. Lastly, we present several recognition examples, illustrating the capabilities of the model.

6.1 Training and Evaluation Protocol

The model's training was done by minimizing the CTC loss function with the AdamW [14] optimizer. The hyperparameters of the model were tuned on the validation set. We used a batch size of 8 samples and a learning rate of 0.0001. For computational purposes, ME samples with a height of more than 110 pixels were discarded from the training set. Training was stopped once the token error rate on the validation set showed no further improvement for 20 epochs. Results obtained by the proposed model are compared with state-of-the-art ME recognition systems based on the following string and image based metrics:

- **Bleu** (*Bleu*): mark-up language level metric that measures the similarity between the model's best prediction and the associated ME reference in accordance with n-gram accuracy [19]. Scores up to 4-grams were considered for this metric.
- **Exact Match String** (*ExStr*): mark-up language level metric that measures the percentage of hypotheses that coincide precisely with the ground-truth LaTeX ME. Note that this metric is sensitive to the length of the MEs and, thus, tends to be overly pessimistic in the case of larger MEs. In addition, both this metric and *Bleu* are affected by mark-up level ambiguity in which LaTeX synonyms with the same visual representation are not considered.
- **Exact Match Image** (*ExIm*): image level metric that measures the percentage of rendered hypotheses that match the corresponding rendered references. Image pairs are considered to match exactly if they display a misalignment of less than 5 pixels [8].
- **Exact Match Image -ws** (*ExIm-ws*): similar to *ExIm*. This image level metric measures the percentage of rendered hypotheses that match the corresponding rendered references. However, unlike *ExIm*, both predicted, and reference images are horizontally compressed by removing whitespace columns [8].

6.2 Results

Table 1, shows the ME recognition results of the Py4MER system in comparison to state-of-the-art (SOTA) recognition systems. Results are reported for two versions of the model, in which dense connections show an improvement to the strictly sequential convolutional connections. While outperforming other CTC-based recognition models and structural and grammatical approaches by a large margin, there is still room for improvement compared to advanced encoder-decoder architectures. To provide more insight for future improvements of the CRNN CTC-based model, experiments are conducted for different ME structures, lengths, and required edit operations for correcting recognition errors.

Table 1. ME recognition results for the Im2Latex-100k, comparing the proposed Py4MER model with state-of-the-art recognition systems. The system is evaluated under two architectures: *Conv* and *Dense*. The first employs strictly sequential connections between convolutional layers, while the second employs dense connections.

Model	Bleu	ExStr	ExIm	ExIm-ws
		SOTA		
CTC [8]	30.4	–	7.6	9.2
INFTY [22]	66.7	–	15.6	26.7
Seshat [16]	81.3	–	–	34.6
Im2Tex [8]	87.7	41.2	77.5	79.9
DoubleAtt [32]	88.4	–	79.8	–
ConvMath [26]	88.3	–	–	83.4
GNN [20]	90.2	–	81.8	–
FGFE [4]	90.3	46.8	–	84.3
GCBN [18]	89.7	–	82.1	–
MI2LS [24]	90.3	–	82.3	84.8
		OURS		
Py4MER-*Conv*	89.3	41.4	76.8	78.9
Py4MER-*Dense*	89.4	43.6	79.1	81.5

Mark-up Consistency. An essential requirement of recognition systems addressing the problem of Image-to-LaTeX is generating syntactically correct (consistent) mark-up. To address this problem, Table 2 shows a consistency experiment in which up to 50-best hypotheses were employed to find correctly generated mark-up sequences. In this experiment, different hypotheses for the same ME were tested in a sequential manner until compilation reported no error, or no consistent hypothesis could be found. With this approach, both Py4MER architectures see a reduction in the number of inconsistent MEs compared to considering only the 1-best hypothesis. It is important to note that *Bleu* and *ExStr* scores do not depend on MEs being consistent. However, the majority of compilation errors were due to few errors such as unmatched opening/closing of groups, which have very little effect on the *Bleu* score, but considerably affect all exact-match based scores. From this point forward, all results are reported based on the `Py4MER`-*Dense* model with consistency correction, in which MEs that do not compile are considered errors from an exact match point of view.

Table 2. LaTeX consistency results for the **Py4MER** recognition system, based on up to 50-best hypotheses for obtaining compilable mark-up.

Model	Consist.	CompErr (%)	Bleu	ExStr	ExIm	ExIm-ws
Py4MER-Conv	No	3.6	89.3	41.0	76.2	78.2
Py4MER-Conv	Yes	2.0	89.3	41.4	76.8	78.9
Py4MER-Dense	No	2.8	89.4	42.4	77.3	80.1
Py4MER-Dense	Yes	1.6	89.4	43.6	79.1	81.5

Mathematical Structure Recognition. Table 3 shows recognition results based on three type of structures contained in the MEs: vertically aligned structures like multi-line (M) expressions and fractions (F), and MEs that do not contain such structures (R). The proposed architecture struggles to correctly interpret the majority of multi-line expressions, partially due to the low representation of this type of MEs. However, the model achieves a much better precision for MEs containing fractions, showing that CTC-based models are capable of correctly interpreting top-to-bottom reading order of symbols. In the case of less complex structures, the model achieves a high score for all metrics. In Table 4 we can see that the model is capable of capturing long-term symbol relationships and correctly interpret nested fractions and fractions in which the numerators and denominators have different number of symbols. Furthermore, the model is able to recognize complex structures like matrices, for which the model generates tokens that have no visual representation but are required for defining the environment, the number of columns and the alignment type. In addition, the model is able to place each element in the correct position. However, due to the low representation of MEs containing matrices, we cannot draw firm conclusions on the modelling capability of the system for this type of MEs.

Table 3. Recognition results based on the type of components contained in the MEs. The results are reported at ME level, not at component level, as the latter would require a LaTeX ME parser. The M and F sets are not disjoint, while R contains no element from the other sets.

Type	Freq. (%)	Bleu	ExStr	ExIm	ExIm-ws
Multi-line (M)	6.5	52.2	14.3	26.2	29.1
Fractions (F)	48.3	88.1	31.6	72.9	75.0
Rest (R)	47.3	93.7	59.2	88.0	91.2

Table 4. Examples of correctly recognized MEs, with various levels of complexity: long-term symbol relationships (green); different token length numerators and denominators (light-red and purple, respectively); nested fractions (orange for outer frac., blue for inner frac.); and vertically aligned math content (red).

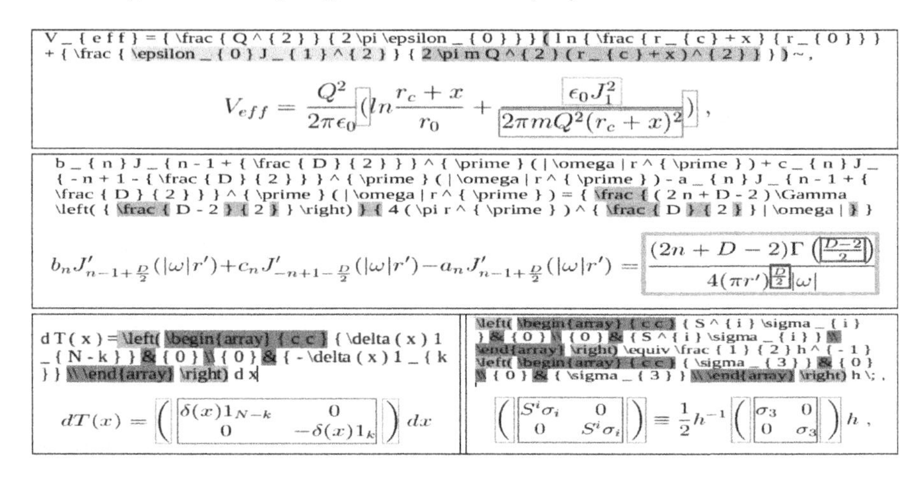

ME Length Impact. From a different point of view, Fig. 3 shows the performance of the Py4MER model when evaluated on MEs with different number of tokens. The precision of the model steadily drops for MEs of increasing length, showing that the *ExStr* score is highly sensitive to the length of the LaTeX sequence. This is an expected behavior, that affects all recognition systems [4,24], considering the difficulty of correctly transcribing very large MEs, where a large percentage of tokens have no visual representation.

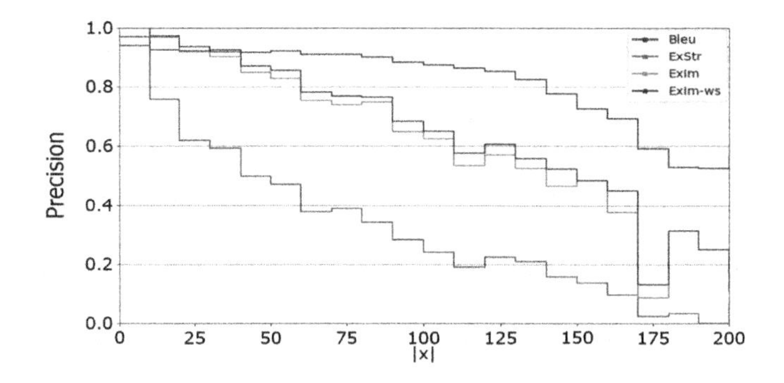

Fig. 3. *ExStr*, *Bleu*, *ExIm*, and *ExIm-ws* precision based on ME length measured in LaTeX tokens. The MEs are grouped in consecutive intervals of size 10. Due to low representation, MEs of more than 200 tokens have been excluded from this chart.

ME Error Assessment. In Fig. 4 the performance of the model is shown when up to 5 errors are considered. The *ExStr* score shows a remarkable improvement as we increase the number of edit operations required to transform the hypotheses into the corresponding references. This behavior highlights the mark-up level ambiguity present in the data set, as the same jump in precision is not seen in the *ExIm* and *ExIm-ws* scores. This implies that LaTeX tokens are substituted with mark-up synonyms that have the same visual representation, leading to a minor improvement in the other scores. However, with two edit operations, the model surpasses state of the art recognition results in all metrics (*Bleu*= 90.8, *ExStr*= 63.2, *ExIm*= 83.7, *ExIm-ws*= 85.3). This is a reasonable correction, since on average, MEs consist of 63 tokens. Figure 5 shows the performance of the model for different ME lengths, simulating queries in ME retrieval systems. It is reasonable to assume that queries are generally short expressions [16]. For MEs indexed by their LaTeX definition (simulated in Fig. 5 left) the model offers good precision for all contemplated ME lengths. However, to mitigate the mark-up ambiguity, the edit distance can be employed to increase the retrieval capabilities of the model. On the other hand, for MEs indexed by their visual representation (simulated in Fig. 5 right), the Py4MER system offers high precision, with editing operations to correct the underlying mark-up language offering minor performance improvement.

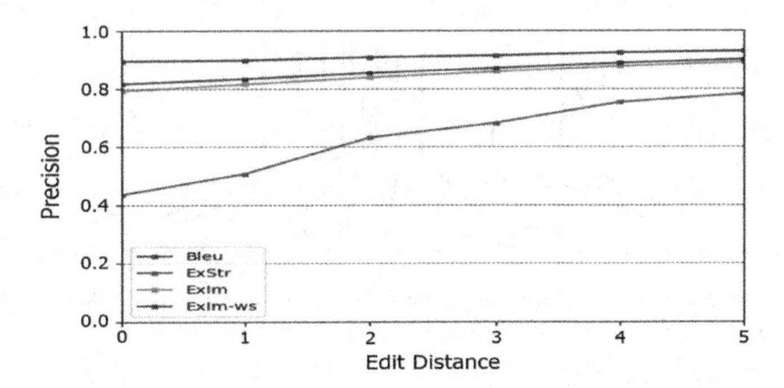

Fig. 4. *Bleu*, *ExStr*, *ExIm*, and *ExIm-ws* precision based on the number of edit operations needed to transform hypotheses into the corresponding references.

CTC Behavior Analysis. Finally, we study the behavior of CTC where the reading order is not strictly left-to-right. As studied in [10,33], CTC predicts labels as a series of localized spikes, separated by 'blanks', or null predictions. Figure 6 shows an example of a ME containing a fraction, which CTC has correctly interpreted. The image was processed frame-wise, highlighting the frames where CTC predicted a LaTeX token. Upon careful examination we can see that for each frame, CTC is able to make an informed decision based on the passed and future context provided by the BiLSTM layers. In this example, CTC has completely reinterpreted the frames of the input image, to generate a LaTeX sequence that would otherwise be in a non left-to-right reading order. These results show that CTC-based CRNN models, are capable of interpreting complex structures, obtaining competitive results when certain aspects are considered, such as: 1)

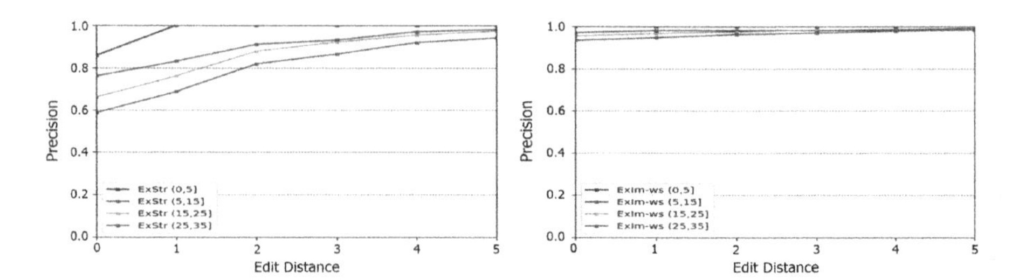

Fig. 5. *ExStr* (left) and *ExIm-ws* (right) precision based on the number of edit operations needed to transform ME hypotheses into the corresponding references. MEs are grouped into intervals based on the number of LaTeX tokens.

using a less aggressive dimensionality reduction approach, ensuring that sufficient frames are left for the CTC to adapt its behavior to the analysed structure, given that a large percentage of tokens have no visual representation; 2) modelling context bidirectionally; 3) and to a lesser degree, using dense connections to model symbol relationships at different abstraction levels.

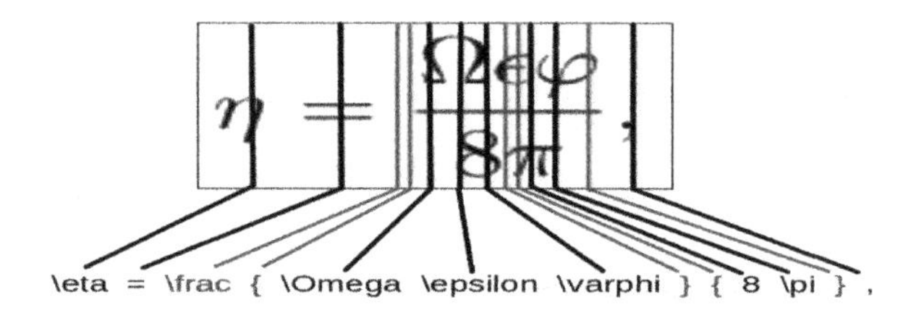

Fig. 6. CTC alignments showing the assigned frame of each ME token. Tokens that have no visual representation are colored in red. (Color figure online)

7 Conclusion

This paper presents the `Py4MER` system for Mathematical Expression Recognition. This system is based on a CRNN architecture trained with the CTC objective function. The model has been implemented in the open source PyLaia toolkit, which facilitates the reproducibility of the experiments presented.

The proposed model has been trained and tested on the public `Im2Latex-100k` data set. Results are reported based on the type of substructures present in MEs and based on the length of MEs measured in LaTeXtokens. A study has been conducted in which the performance of the system is evaluated based on the editing distance required to transform the hypotheses into the corresponding references. A remarkable increase in performance is achieved when up to 5 editing operations are considered. The results presented show that CTC based approaches are a good alternative to encoder-decoder architectures and are adequate for the recognition of MEs. However certain CTC and LaTeX ME properties have to be taken into account in order to obtain strong results.

Acknowledgment. This work has been partially supported by MCIN/AEI/10.13039/501100011033 under the grant PID2020-116813RB-I00 (SimancasSearch); the Generalitat Valenciana under the FPI grant CIACIF/2021/313; and by the support of valgrAI - Valencian Graduate School and Research Network of Artificial Intelligence and the Generalitat Valenciana, and co-funded by the European Union.

References

1. Ameryan, M., Schomaker, L.: A limited-size ensemble of homogeneous CNN/LSTMs for high-performance word classification. Neural Comput. Appl. **33**(14), 8615–8634 (2021). https://doi.org/10.1007/s00521-020-05612-0
2. Anderson, R.H.: Syntax-directed recognition of hand-printed two-dimensional mathematics, p. 436–459 (1967)
3. Awal, A., Mouchère, H., Viard-Gaudin, C.: A global learning approach for an online handwritten mathematical expression recognition system. PRL **35**, 68–77 (2014)
4. Bender, S., Haurilet, M., Roitberg, A., Stiefelhagen, R.: Learning fine-grained image representations for mathematical expression recognition. In: ICDARW, vol. 1, pp. 56–61 (2019)
5. Blostein, D., Grbavec, A.: Recognition of mathematical notation. In: Handbook of Character Recognition and Document Image Analysis, p. 557–582 (1997)
6. Cho, K., et al.: Learning phrase representations using RNN encoder-decoder for statistical machine translation. In: EMNLP, pp. 1724–1734 (2014)
7. Chorowski, J., Bahdanau, D., Cho, K., Bengio, Y.: End-to-end continuous speech recognition using attention-based recurrent NN: first results (2014)
8. Deng, Y., Kanervisto, A., Ling, J., Rush, A.M.: Image-to-markup generation with coarse-to-fine attention. In: ICML, pp. 980–989 (2017)
9. Endong, Z., Licheng, L.: Design of online handwritten mathematical expression recognition system based on gated recurrent unit recurrent neural network. In: PRAI, pp. 446–451 (2021)
10. Graves, A., Fernández, S., Gomez, F., Schmidhuber, J.: Connectionist temporal classification: labelling unsegmented sequence data with recurrent neural networks. In: ICML, pp. 369–376 (2006)
11. Hou, Y., Kong, Q., Li, S.: Audio tagging with connectionist temporal classification model using sequentially labelled data. In: Liang, Q., Liu, X., Na, Z., Wang, W., Mu, J., Zhang, B. (eds.) CSPS 2018. LNEE, vol. 516, pp. 955–964. Springer, Singapore (2020). https://doi.org/10.1007/978-981-13-6504-1_114
12. Jaderberg, M., Simonyan, K., Vedaldi, A., Zisserman, A.: Deep structured output learning for unconstrained text recognition. ArXiv abs/1412.5903 (2014)
13. Kang, L., Toledo, J.I., Riba, P., Villegas, M., Fornés, A., Rusiñol, M.: Convolve, attend and spell: an attention-based sequence-to-sequence model for handwritten word recognition. In: Brox, T., Bruhn, A., Fritz, M. (eds.) GCPR 2018. LNCS, vol. 11269, pp. 459–472. Springer, Cham (2019). https://doi.org/10.1007/978-3-030-12939-2_32
14. Loshchilov, I., Hutter, F.: Decoupled weight decay regularization (2017). https://arxiv.org/abs/1711.05101

15. Nguyen, C.T., Nguyen, H.T., Morizumi, K., Nakagawa, M.: Temporal classification constraint for improving handwritten mathematical expression recognition. In: ICDAR Workshops, pp. 113–125 (2021)
16. Noya, E., Benedí, J., Sánchez, J., Anitei, D.: Discriminative learning of two-dimensional probabilistic context-free grammars for mathematical expression recognition and retrieval. In: Pinho, A.J., Georgieva, P., Teixeira, L.F., Sánchez, J.A. (eds.) IbPRIA 2022. LNCS, vol. 13256, pp. 333–347. Springer, Cham (2022). https://doi.org/10.1007/978-3-031-04881-4_27
17. Noya, E., Sánchez, J., Benedí, J.: Generation of hypergraphs from the N-best parsing of 2D-probabilistic context-free grammars for mathematical expression recognition. In: ICPR, pp. 5696–5703 (2021)
18. Pang, N., Yang, C., Zhu, X., Li, J., Yin, X.C.: Global Context-Based Network with Transformer for Image2latex. In: ICPR, pp. 4650–4656 (2021)
19. Papineni, K., Roukos, S., Ward, T., Zhu, W.: BLEU: a method for automatic evaluation of machine translation. In: ACL, pp. 311–318 (2002)
20. Peng, S., Gao, L., Yuan, K., Tang, Z.: Image to LaTeX with graph neural network for mathematical formula recognition. In: Lladós, J., Lopresti, D., Uchida, S. (eds.) ICDAR 2021. LNCS, vol. 12822, pp. 648–663. Springer, Cham (2021). https://doi.org/10.1007/978-3-030-86331-9_42
21. Puigcerver, J.: Are multidimensional recurrent layers really necessary for handwritten text recognition? In: ICDAR, vol. 01, pp. 67–72 (2017)
22. Suzuki, M., Tamari, F., Fukuda, R., Uchida, S., Kanahori, T.: Infty - an integrated OCR system for mathematical documents. In: ACM, pp. 95–104 (2003)
23. Wang, J., Sun, Y., Wang, S.: Image to Latex with Densenet encoder and joint attention. Procedia Comput. Sci. **147**, 374–380 (2019)
24. Wang, Z., Liu, J.C.: Translating math formula images to LaTeX sequences using deep neural networks with sequence-level training. IJDAR **24**, 63–75 (2021)
25. Xu, K., et al.: Show, attend and tell: neural image caption generation with visual attention. In: ICML, pp. 2048–2057 (2015)
26. Yan, Z., Zhang, X., Gao, L., Yuan, K., Tang, Z.: ConvMath: a convolutional sequence network for mathematical expression recognition. In: ICPR, pp. 4566–4572 (2021)
27. Yuan, Y., et al.: Syntax-aware network for handwritten mathematical expression recognition. In: CVPR, pp. 4543–4552 (2022)
28. Zanibbi, R., Blostein, D., Cordy, J.: Recognizing mathematical expressions using tree transformation. PAMI **24**(11), 1–13 (2002)
29. Zhang, J., Du, J., Dai, L.: Multi-scale attention with dense encoder for handwritten mathematical expression recognition. In: ICPR, pp. 2245–2250 (2018)
30. Zhang, J., et al.: Watch, attend and parse: an end-to-end neural network based approach to handwritten mathematical expression recognition. Pattern Recogn. **71**, 196–206 (2017)
31. Zhang, T., Mouchère, H., Viard-Gaudin, C.: Using BLSTM for interpretation of 2-D languages: case of handwritten mathematical expressions. In: CORIA, pp. 217–232 (2016)
32. Zhang, W., Bai, Z., Zhu, Y.: An improved approach based on CNN-RNNs for mathematical expression recognition. In: ICMSSP, pp. 57–61 (2019)
33. Zhelezniakov, D., Zaytsev, V., Radyvonenko, O.: Online handwritten mathematical expression recognition and applications: a survey. IEEE Access **9**, 38352–38373 (2021)

34. Zhu, Z., Dai, W., Hu, Y., Wang, J., Li, J.: Speech emotion recognition model based on CRNN-CTC. In: Abawajy, J.H., Choo, K.-K.R., Xu, Z., Atiquzzaman, M. (eds.) ATCI 2020. AISC, vol. 1244, pp. 771–778. Springer, Cham (2021). https://doi.org/10.1007/978-3-030-53980-1_113
35. Álvaro, F., Sánchez, J., Benedí, J.: An integrated grammar-based approach for mathematical expression recognition. Pattern Recogn. **51**, 135–147 (2016)
36. Álvaro, F., Sánchez, J.A., Benedí, J.M.: Recognition of on-line handwritten mathematical expressions using 2D stochastic context-free grammars and hidden Markov models. PRL **35**, 58–67 (2014)

Hierarchical Line Extremity Segmentation U-Net for the SoccerNet 2022 Calibration Challenge - Pitch Localization

Miguel Santos Marques[1(✉)], Ricardo Gomes Faria[3], and José Henrique Brito[1,2]

[1] 2Ai - School of Technology, IPCA, Barcelos, Portugal
{msmarques,jbrito}@ipca.pt
[2] LASI - Associate Laboratory of Intelligent Systems, Guimarães, Portugal
[3] Mobileum, Braga, Portugal
Ricardo.Faria@mobileum.com

Abstract. This paper describes the methodology our team followed for our submission to the SoccerNet Calibration Challenge - Soccer Pitch Markings and Goal Posts Localization. The goal of the challenge is to detect the extremities of soccer pitch lines present in the image. Our method directly infers the line extremities' localization in the image, using a Deep Learning Convolutional Neural Network based on U-Net with a hierarchical output and unbalanced loss weights. The hierarchical output contains three outputs composed of different segmentation masks. Our team's name is 2Ai-IPCA and our method achieved third place in the Pitch Localization task of the SoccerNet Calibration Challenge, with 71.01%, 76.18%, and 77.60% accuracies for the 5 pixel, 10 pixel, and 20 pixel thresholds respectively, and a global average accuracy score of 73.81% on the challenge set.

Keywords: SoccerNet challenge · line segmentation · hierarchical U-Net

1 Introduction

The SoccerNet Calibration challenge comprises two tasks: (1) Soccer Pitch Markings and Goal Posts Localization, and (2) Automatic camera calibration. This paper describes the approach our team used for the first task, Pitch Localization. The objective of this task is to estimate the extremities of each field line type present in a given image. The associated dataset is composed of image-annotation pairs, and the annotation files contain the 2D coordinates of the extremities of the 26 different possible line classes that a soccer pitch is composed of. Some lines may be defined in the annotations with more than two 2D points. Further information about the challenge, context, motivations and data is available on the challenge website [2] and the associated github repository [1].

Our method segments line extremities using a Deep Learning segmentation network based on U-Net to directly infer the line extremities' localization in

A. Pertusa et al. (Eds.): IbPRIA 2023, LNCS 14062, pp. 442–453, 2023.
https://doi.org/10.1007/978-3-031-36616-1_35

the image. The main modification made to the classical U-Net network is the hierarchical classification output, comprised of 3 different outputs: (1) a softmax-activated output with background, floor lines, and goal lines; (2) a sigmoid-activated output with segmentation masks for each line type, and (3) a sigmoid-activated output with line extremity heat maps for each line type.

We did experiment with first segmenting just the lines, and then detect and use their extremities as the coordinate predictions, but that approach yielded results with a much lower accuracy, by a significant margin.

Our method will be used as a building block for a target application, in which the field line segmentation will help guide player detection and tracking in a soccer video analysis tool.

2 Related Work

Our work relates to methods for line detection and its application to sports analysis, more specifically for soccer pitch markings. While line detection and segmentation is one of the classic tasks in computer vision, its application to soccer images is not always straightforward. Several previous works have aimed for the detection of soccer field lines, usually within a system with higher level functionalities. Authors have traditionally used standard methods to detect field lines such as the Hough transform and the Canny Edge Detector, with more or less sophisticated implementations given the application setting.

The paper in [5] describes a system to track basketball players within the field lines, but the lines are manually annotated in the images. In [4] the authors use the Hough transformation applied to line color features to detect field lines. In [12] the authors use colour segmentation and fit an ellipse to the image to find the central circle in the soccer pitch, and fit vertical lines to find the goal posts and the goal bar. In [8] the field lines are used to calculate the homography between successive video frames and the known dimensions of the soccer pitch, and subsequently localize players' and ball's positions in 3D. The pitch region is segmented in the image through the use of a colour histogram and a Gaussian mixture model, to cluster pixels using the histogram peaks, and the homography is computed from line intersections in the field. In [10] the field lines are detected through the Canny Edge Detector and filtered using morphological operations. In [7] the authors detect field lines in successive soccer video frames using the Hough Transform and track them throughout the video by predicting their positions with a Kalman Filter. In [6] the K-means clustering algorithm is applied to cluster lines, and for each cluster only the line with the largest value of Hough intensity was selected as the real field line. In [9] the authors present a system for tactical analysis of soccer teams, by analyzing the trajectory of video objects, namely players, ball, and field lines. To the detect lines the authors use a two component Gaussian Mixture Model to model the Hough parameters for each field line and its orientation angle.

In [3] the authors foresee the SoccerNet calibration challenge and propose to use a U-Net architecture [11] to perform a zone segmentation of the field, where a zone is a field area enclosed by field lines. The SoccerNet calibration challenge's baseline implementation for line segmentation [1] is a DeepLabV3+ network trained on the SoccerNet calibration dataset. The author of [14] also proposes U-Net and DeepLabV3+ to segment field lines in images. U-Net presents better results on situations with low visibility of the field lines, for example in the presence of snow, and DeepLabV3+ show better results on cases with good visibility. The SoccerNet Challenge, to the best of our knowledge, is the first proposal of a comprehensive dataset of soccer images that includes detailed field line annotations, enabling the comparison of results between different algorithms and strategies for field line segmentation. For this reason, we are unable to compare our approach to implementations proposed by other authors, as no published papers of those approaches exist at the present time. Following the work in [3] from the SoccerNet challenge creators, we base our approach on a U-Net network for field line and field line extremity segmentation.

3 Pitch Line and Line Extremity Segmentation

The dataset annotations for the SoccerNet Pitch Localization Challenge describe each pitch line as a set of points. The objective of the task is to infer the extremities of the lines present in the image, according to the class of line they belong to, for all line classes except for the central circle of the pitch, for which there are no extremities, since it is a circle.

The overall strategy of our method is to directly infer the extremity coordinates of soccer pitch lines captured in the image using a Deep Convolutional Neural Network for Semantic Segmentation based on U-Net. Hence, we train the network to produce line extremity masks, in which each extremity is represented by a 2D gaussian centered around the extremity coordinates. When we use the dataset annotations to train the network, we take the first and last point of each line as the extremities the method is meant to detect. We however do also use line segmentation masks to help the network learn more effectively to detect and segment line extremities. The network is therefore trained to segment lines and to segment line extremities. Given the interdependence between class lines and their extremities, we model this dependence by structuring the network outputs in such a manner that the extremity predictions depend on the line predictions. We also make the line predictions for each line class depend on the prediction of floor lines and/or goal lines (independently of their class) against the background.

Finally, since the focus of the task is to predict line class extremities, we train that network with uneven loss weights, so that the global loss is more influenced by the loss associated with the segmentation of line extremities.

3.1 Modified U-Net Network Structure

Our CNN is based on U-Net and it segments soccer pitch elements. Most of the network is based on the re-implementation of U-Net available at [13] which is itself a re-implementation of the original U-Net [11].

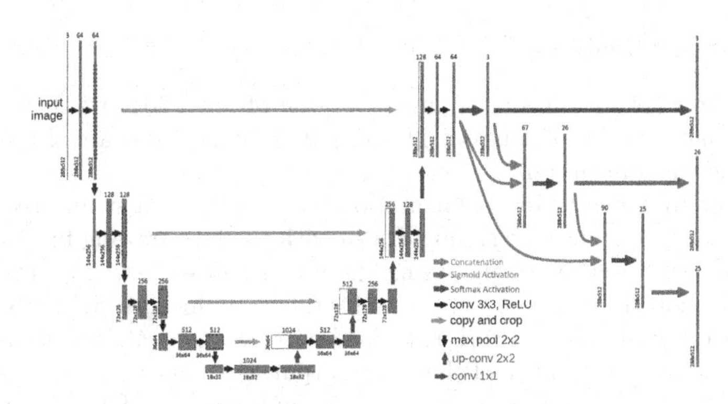

Fig. 1. The architecture of our U-Net

The network has an input dimension of 288×512, and the output masks also have a dimension of 288×512. We chose these dimensions as they allow us to keep the 16:9 aspect ratio when we downsample the dataset images, while also allowing to have integer-sized feature maps at every level of the network after the successive max pooling layers.

A significant difference to the original U-Net is that our network produces three outputs after the usual 5-level backbone, instead of the classic single pixel-wise classification layer. The first output is a convolutional layer with 1×1 kernels with three channels, which are the segmentation masks for Background, Floor Lines and Goal Lines. The second output produces 26 channels corresponding to the segmentation masks of the different line classes (Side line top, Side line left, etc.) including goal line classes (Goal left crossbar, Goal left post right, etc.). The third output has 25 channels, corresponding to the extremities of each Floor Line/Goal Line (excluding Circle Central). The activation of the first output is softmax, and second and the third outputs both use a sigmoid activation.

Furthermore, the outputs are arranged in a hierarchical structure, as shown in Fig. 1. The first output is simply the result of a convolutional layer with 1×1 kernels applied to the 64 feature maps of the last layer of the network backbone, followed by a softmax activation over the 3 channels of the output for each pixel. The second output takes as input the concatenation of the 64 feature maps of the last layer of the backbone network and the 3 feature maps of the first output (before the softmax activation), applies a convolutional layer with 1×1 kernels followed by the sigmoid activation, producing 26 channels, one for each line class. The third output takes as input the concatenation of the 64 feature maps of the last layer of the backbone, the 26 feature maps of the second output (before the activation), applies a convolutional layer with 1×1 kernels, again followed by the sigmoid activation, producing 25 channels with the heatmaps of the line extremities, one for each line class excluding "Circle center". The network therefore produces an total of 54 segmentation masks: 3 for background, floor lines and goal lines, 26 for lines by class, and 25 for line extremities by class.

3.2 Network Training

As the neural network has an input dimension of 288×512, we downscale the training images to this dimension. The output training masks are also generated considering this dimension.

In order to generate the segmentation masks for the first and second outputs, the respective lines are simply drawn with the line drawing function of the function in scikit-image package using the unnormalized extremity coordinates from the annotations. These lines therefore have a thickness of 1 pixel in the segmentation masks the network is trained with. To generate the segmentation masks of the line extremities of the third output, the unnormalized line extremities' coordinates are used to generate gaussian heatmaps. These 2D gaussians are centered in the extremity coordinates and have a standard deviation of 1 pixel. The network is trained to minimize the following loss:

$$L_t = w_{BLG} \cdot CCE_{BLG} + w_{LC} \cdot BCE_{LC} + w_E \cdot BCE_E \quad (1)$$

where L_t is the total loss, CCE_{BLG} is the Categorical Cross-Entropy loss computed over the 3 channels of the first output (background, floor lines and goal lines), BCE_{LC} is the Binary Cross-Entropy loss computed over the 26 channels of the second output (lines by class), BCE_E is the Binary Cross-Entropy loss computed over the 25 channels of the third output (extremities), and w_{BLG}, w_{LC}, w_E are their respective weights.

The weights can therefore be set so that each individual output loss may have a different influence over the total loss. After a few experiments we settled on a weight distribution so that the loss of the first output CCE_{BLG} would be approximately 20% of the total loss, the loss of the second output BCE_{LC} would be approximately 30% of the total loss, and the loss of the third output BCE_E would be approximately 50% of the total loss, since we would prefer to skew the results in favour of having a better accuracy for the extremities' coordinates estimation. The actual weight values used for training the model were $w_{BLG} = 1 * (3/54)$, $w_{LC} = 4 * (26/54)$, $w_E = 80 * (25/54)$.

The network was trained from scratch with Tensorflow. For the final challenge submission we trained the network with the union of the training and validation sets, and we used the test set for validation. We used a batch size of 2, trained for 25 epochs, saving a checkpoint with the model weights for the best validation loss value at the end of each epoch. The rest of the training hyper-parameters were the default values used by TensorFlow, namely the Adam optimizer with a learning rate of 1e−4. The relatively small batch size of 2 is motivated by the memory limitations of the available graphics card. Nevertheless, in most experiments the network would converge after less than 10 epochs.

3.3 Extremity Coordinates Prediction

Our best results were achieved by using the 25 channels of the third output to generate the extremity prediction files. To generate the extremity prediction

Table 1. Experiment results for the test set, for different input image sizes, loss weights and output structures. Trained with 10 Epochs and used the Train Set to train the CNN and used the Validation Set to validate.

I. Size	Loss W	Output	Prec.(5/10/20px)	Rec.(5/10/20px)	Acc.(5/10/20px)
144 × 256	–	1+26	43.3/60.8/68.6%	78.3/80.9/82.4%	42.2/59.0/66.3%
144 × 256	–	3+26	41.3/59.6/67.9%	78.1/80.9/82.9%	40.4/58.1/66.0%
144 × 256	–	3+26 H	42.8/60.4/68.0%	77.9/80.5/82.4%	41.8/58.6/65.9%
144 × 256	–	25	70.6/78.7/81.0%	85.1/86.4/86.9%	66.8/74.1/76.2%
144 × 256	–	3+26+25	68.7/78.8/81.3%	83.8/85.4/85.8%	64.5/73.6/75.7%
144 × 256	–	3+26+25 H	68.0/78.4/81.0%	83.3/85.0/85.5%	64.0/73.3/75.5%
144 × 256	1/4/40	3+26+25	71.9/80.5/82.9%	86.7/87.9/88.3%	68.3/**76.0**/**78.1**%
144 × 256	1/4/40	3+26+25 H	71.8/80.3/82.6%	86.5/87.8/88.2%	**68.4**/**76.0**/78.0%
288 × 512	–	1+26	40.0/52.0/58.3%	66.7/68.5/69.6%	36.8/47.5/52.7%
288 × 512	–	3+26	35.1/49.2/57.4%	64.7/66.8/68.1%	31.5/43.8/50.2%
288 × 512	–	3+26 H	39.2/51.6/58.5%	65.9/67.6/68.8%	35.3/46.2/51.9%
288 × 512	–	25	74.5/80.2/81.8%	82.1/83.1/83.4%	68.7/73.5/74.8%
288 × 512	–	3+26+25	71.9/78.1/79.9%	78.9/79.9/80.2%	65.1/70.3/71.7%
288 × 512	–	3+26+25 H	73.5/79.5/81.3%	78.5/79.4/79.7%	66.0/71.0/72.4%
288 × 512	1/4/80	3+26+25	75.9/81.7/83.5%	82.6/83.5/83.8%	69.9/74.6/76.0%
288 × 512	1/4/80	3+26+25 H	76.3/82.1/83.6%	83.2/84.1/84.3%	**70.2**/**75.1**/**76.4**%
432 × 768	–	1+26	19.8/32.9/45.2%	38.5/44.7/48.6%	13.5/22.4/30.2%
432 × 768	–	3+26	19.2/31.7/42.5%	43.5/48.5/51.5%	13.9/22.6/29.8%
432 × 768	–	3+26 H	16.8/29.4/41.2%	35.6/41.9/46.0%	11.2/19.5/26.8%
432 × 768	–	25	75.3/80.1/81.4%	74.7/75.5/75.7%	64.9/68.7/69.7%
432 × 768	–	3+26+25	76.5/81.4/82.7%	74.6/75.4/75.6%	65.2/69.1/70.0%
432 × 768	–	3+26+25 H	74.1/79.3/80.6%	68.9/69.7/69.9%	60.4/64.3/65.2%
432 × 768	1/4/120	3+26+25	77.1/82.2/83.7%	78.0/78.8/79.0%	**67.8**/**71.9**/**73.0**%
432 × 768	1/4/120	3+26+25 H	77.0/81.9/83.5%	77.2/78.0/78.3%	67.5/71.5/72.6%

files, we first feed the image to the CNN to produce the 25 extremity heat maps of the third output, one for each line type. For each heatmap, we first check if the heatmap has at least one pixel with a prediction confidence of a minimum of 0.1. If it does, we threshold the heatmap with the value of 0.1 and find the blobs in the resulting thresholded heatmap. If the thresholded heat map has at least 2 blobs, we take the 2 blobs with the highest confidence peaks, and compute the normalized coordinates of those peaks to generate the predicted extremities for the corresponding class.

4 Experiments

The training and prediction strategies and other decisions outlined in Sect. 3 produced the best results in our experiments. We have however experimented with several different variants and hyper-parameter values, such as different input and

Table 2. Experiment results for the test set, for different input image sizes and output structures. Trained with 25 Epochs and was used the Train Set and Validation Set to train the CNN and used the Test Set to validate.

I. Size	Loss W	Output	Prec. (5/10/20px)	Rec. (5/10/20px)	Acc. (5/10/20px)
144 × 256	1/4/40	3+26+25 H	73.1/81.6/83.9%	88.0/89.2/89.5%	69.7/**77.4**/**79.3**%
288 × 512	1/4/80	3+26+25 H	77.2/82.9/84.6%	84.8/85.7/86.0%	**71.5**/76.4/77.7%
432 × 768	1/4/120	3+26+25	78.1/83.2/84.6%	78.9/79.6/79.8%	68.8/72.8/73.9%

output sizes, different output structures, different loss weights, with and without hierarchical outputs, and different threshold levels for heatmap predictions.

First we experimented with simply segmenting the lines from the background, with a sigmoid activation and use the line-end pixels to estimate the line extremity coordinates. In this setting the network produces a single 27 channel output (background + 26 line classes). The sigmoid activation is justified by the fact that the pixels on the line intersections belong to both classes of the intersecting lines.

The second experiment was to still produce segmentation masks for lines, but with two independent outputs, the first one with the 3-channel background/floor line/goal line segmentation masks, and the second one with the 26-channel class lines, without any hierarchical connection between them. This produces a total of 29 channels considering both outputs.

For the third experiment we kept the structure from the second experiment but introduced a hierarchical connection between the two outputs, so that the 26 class line segmentation masks also depend on the higher level 3-channel feature maps of the first output. For the fourth experiment we tried to simply segment the extremities with a 25 channel output with only the extremity gaussians.

The rest of the experiments used the network with 3 outputs comprising 54 channels described in Sect. 3, with and without the hierarchical output arrangement, and with or without the unbalanced loss weights. For the fifth experiment we chose a Non-Hierarchical Output with balanced loss weights, the sixth experiment had a Hierarchical Output with balanced loss weights, the seventh experiment consisted of Non-Hierarchical Output and unbalanced loss weights, and, finally, the eighth experiment had a Hierarchical Output and unbalanced loss weights.

The results for all these variants for three image sizes are summarized in Table 1 and were achieved on the test set by models trained for 10 Epochs with the Training set, and using the Validation set for validation.

For the final submission we used our three most promising models (one for each image size) for 25 epochs using the Train and Validation sets for training and using the Test set for validation, and the results are presented in Table 2.

5 Results

Starting with the first experiment, with the 27 channel single output, since the output activation is sigmoid we used a Binary Cross-Entropy loss computed over

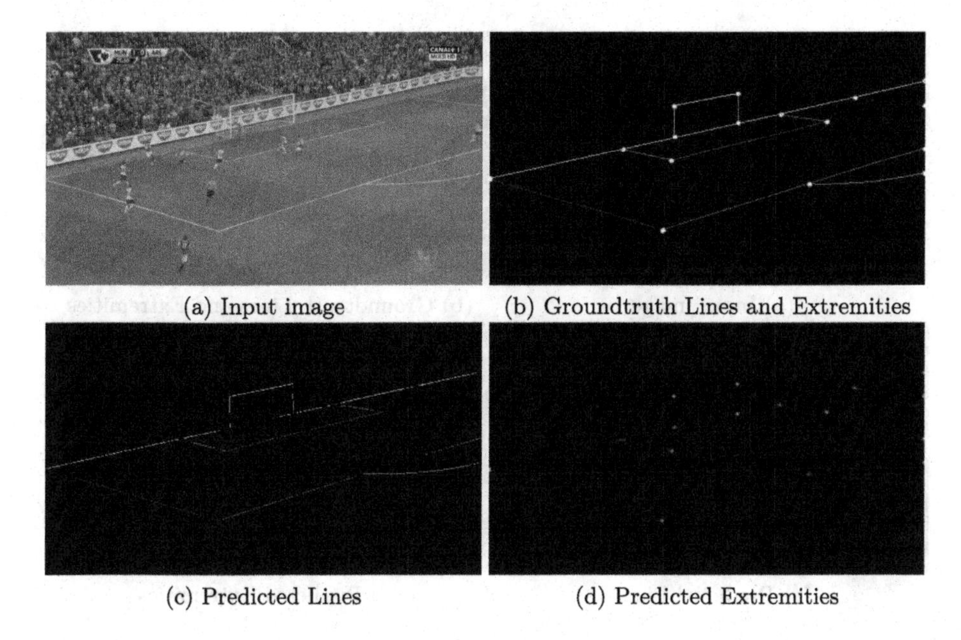

(a) Input image (b) Groundtruth Lines and Extremities

(c) Predicted Lines (d) Predicted Extremities

Fig. 2. Segmentation results for a typical input image from a common view point

the 27 channels of the output. Starting with an input dimension of 144×256, the results were not very encouraging (42.2%, 59.0% and 66.3% for an accuracy of 5, 10 and 20 pixels respectively). Moving to the next dimension (288×512) the results were 36.8%, 47.5% and 52.7% for an accuracy of 5, 10 and 20 pixels respectively and finally we have tried 432×768 where the results were 13.5%, 22.4% and 30.2% for an accuracy of 5, 10 and 20 pixels respectively. We came to the conclusion that this was not the right way for our problem.

We evaluated then the second experiment with 29 channels, where the first output had 3 channels (Background + Floor Lines + Goal Lines) with a softmax activation and the second output had 26 channels (Line Classes) with a sigmoid activation. The total loss value was balanced weighted average between the two output losses, as we used the Categorical Cross-Entropy loss computed over the 3 channels of the first output and the Binary Cross-Entropy for the 26 channels of the second output. We were expecting results would be better than the results of the first model with the 27 classes, but instead the results were lower. This was probably because the Categorical Cross-Entropy loss of the first output is much higher than the Binary Cross-Entropy loss of the second output, as the number of activated pixels in the individual line class segmentation masks is much lower than the pixels activated in the 3-channel segmentation masks with the background pixels and the pixels belonging to floor line classes and goal line classes, all gathered in the same segmentation masks.

In order to improve the metrics, we tried using the hierarchical output arrangement (3rd experiment) and the results were noticeably better for the

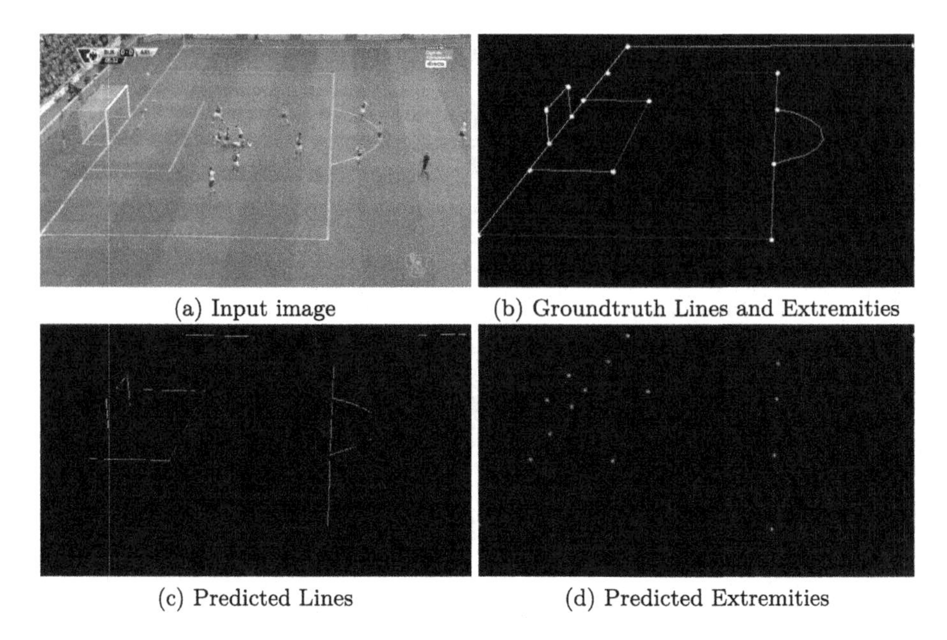

Fig. 3. Segmentation results for an input image from a less common view point

144×256 (41.8%, 58.6% and 65.9% for an accuracy of 5, 10 and 20 pixels respectively) and especially the 288×512 (35.3%, 46.2% and 51.9% for an accuracy of 5, 10 and 20 pixels respectively) dimensions, however the metrics were lower than the first experience. For 432×768 images the metrics were lower (11.2%, 19.5% and 26.8% for an accuracy of 5, 10 and 20 pixels respectively).

The next step was to train with only 25 classes in one output (4th experiment), corresponding to Line Extremities (excluding the Central Circle) with a sigmoid activation and we used the Binary Cross-Entropy loss computed over the 25 channels of the output. This experiment showed that predicting line extremities yields far better results than predicting lines and then find their extremities. For the input dimension of 144×256, the results were 66.8%, 74.1% and 76.2% for an accuracy of 5, 10 and 20 pixels respectively, and for an input dimension of 288×512 the results were 68.7%, 73.5% and 74.8% for an accuracy of 5, 10 and 20 pixels respectively. Finally, we tested with 432×768 images but the results were a bit lower than those with smaller images (64.9%, 68.7% and 69.7% for an accuracy of 5, 10 and 20 pixels respectively), mainly because of lower Recall, which means the network trained for 432×768 images produces many more false negatives, as the network fails to segment many extremities.

Since the experiments for line segmentation with multiple and hierarchical outputs yielded encouraging results, at least for the smaller image sizes, we tried the same approach with the extremity segmentation networks. As explained above, all models with 54 output channels have 3 outputs (3 channels + 26 channels + 25 channels), to force the network to learn the different tasks. We

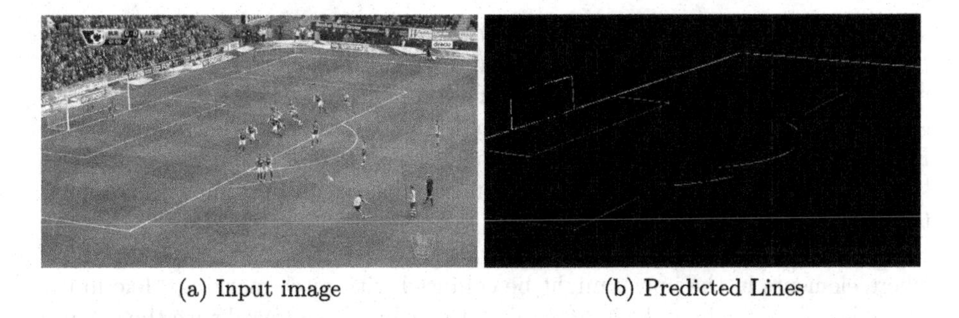

(a) Input image (b) Predicted Lines

Fig. 4. Segmentation results for an input image with misleading content

expected the network to be able to better learn to segment the extremities if it also learns to segment the corresponding lines at the same time.

The first experiment with 54 channels (5th experiment) did not use the hierarchical output and used balanced weights on the output losses, taking into account the number of channels of each output. Comparing with the model with 25 channels the results were lower for small (about 1.2%) and medium (about 3%) image sizes but did improve for large (about 0.5%) input images.

Keeping balanced weights but using hierarchical outputs (6th experiment) did not improve results for small (accuracy has not changed significantly) and large (a 5% loss of accuracy) image sizes but did improve them slightly for 288×512 (about 1%) images.

The biggest gains in segmentation performance were achieved using unbalanced loss weights, whereby we were able to encourage the network to prioritize learning the segmentation of line extremities over the segmentation of lines. We aimed to fix the weights so that the losses of the three outputs would represent 20%, 30% and 50% of the total loss respectively. The metrics were improved relative to the results achieved by the weight-balanced versions of the network, both the non-hierarchical and the hierarchical versions, for all image sizes. Comparing between non-hierarchical (7 h experiment) and hierarchical (8th experiment) versions with unbalanced loss weights, only the networks for larger images didn't take advantage of the hierarchical output, although the 144×256 network produces almost indistinguishable results with or without a hierarchical output. The best overall results were therefore achieved for 288×512 images with a hierarchical output and unbalanced loss weights.

Following all these experiments, we selected the model with the best metrics for each image dimension for our challenge submissions. These models were trained for 25 epochs on the Train + Validation sets, and the Test set was used for validation. The model with the best results was again the model trained with 288×512 images with a hierarchical output and unbalanced loss weights, producing an accuracy of 71.5%, 76.4% and 77.7% for a threshold of 5, 10 and 20 pixels respectively. In terms of qualitative results, a few examples of segmentation outputs are shown in Figs. 2 through 4.

In the first example in Fig. 2 both lines and extremities are quite well segmented, with visual results very similar to the ground truth data. This is a relatively well behaved image from a very common camera viewpoint.

For images taken from more uncommon or challenging viewpoints like in Fig. 3 it is noticeable that the network might struggle a bit more to segment the lines, however the line extremity segmentation seems to be somewhat resilient to changes in the camera pose.

Besides uncommon viewpoints, other potential failure cases include situations where elements in the image might be confused with pitch markings, like in the example of Fig. 4, where the foam used by the referee is confused with the penalty arc in front of the penalty area. In this case the network wrongly detected the foam as a pitch marking.

6 Conclusion

The different experiments and evaluations demonstrate that the approach of segmenting extremities is much more effective than segmenting lines and then detecting their extremities. In fact, from all the variants we experimented with, segmenting line extremities produced the largest differential in segmentation performance.

The results also show that providing supplementary information to the network during training increases the performance. Instead of training the network with just the extremity coordinates, we also trained it with the line segmentations with two levels of class aggregation, and this yielded significant performance gains for all image sizes, even if other supplementary techniques are required.

One of those supplementary techniques is using unbalanced loss weights for the computation of the total loss, so that the network prioritizes the true segmentation objective, in our case the line extremities. Using unbalanced loss weights was, in our estimation, the second most important contributor for the performance gains achieved. Our approach was to set the weights so that the binary-cross entropy loss of the line extremities would be slightly more than 50% of the global loss, and this yielded performance gains for all image sizes.

The other supplementary technique we experimented with was using a hierarchical structure relating the outputs, so that the inference of specific classes would be influenced by the inference of higher level classes for a given pixel, i.e., the classification of a pixel as belonging to a line extremity would be influenced by the classification of the same pixel as belonging to a particular line class and as belonging to a line in general (independently of the line class). This produced mixed results, as the hierarchical output worked better with small (144×256) and medium (288×512) images, but not with large (432×768) images.

Regarding other different but related segmentation tasks, in the near future we intend to develop a network capable of segmenting field areas (left goal area, left penalty area, center circle, etc.). This segmentation of the image will be useful to guide the detection and tracking of soccer player in our target application.

Acknowledgments. This work was partially funded by the project "POCI-01-0247-FEDER-046964", supported by Operational Program for Competitiveness and Internationalization (COMPETE 2020), under the PORTUGAL 2020 Partnership Agreement, through the European Regional Development Fund (ERDF 1). This work was also partially funded by national funds (PIDDAC), through the FCT - Fundação para a Ciência e Tecnologia and FCT/MCTES under the scope of the projects UIDB/05549/2020 and UIDP/05549/2020. This paper was also partially funded by national funds, through the FCT - Fundação para a Ciência e a Tecnologia and FCT/MCTES under the scope of the project LASI-LA/P/0104/2020.

References

1. Soccernet camera calibration challenge development kit. https://github.com/SoccerNet/sn-calibration. Accessed 30 Jan 2023
2. Soccernet field localization challenge. https://www.soccer-net.org/tasks/field-localization. Accessed 30 Jan 2023
3. Cioppa, A., et al.: Camera calibration and player localization in SoccerNet-v2 and investigation of their representations for action spotting. In: Proceedings of the IEEE/CVF Conference on Computer Vision and Pattern Recognition (CVPR) Workshops, pp. 4537–4546 (2021)
4. Ekin, A., Tekalp, A., Mehrotra, R.: Automatic soccer video analysis and summarization. IEEE Trans. Image Process. **12**(7), 796–807 (2003)
5. Fu, X., Zhang, K., Wang, C., Fan, C.: Multiple player tracking in basketball court videos. J. Real-Time Image Proc. **17**(6), 1811–1828 (2020). https://doi.org/10.1007/s11554-020-00968-x
6. Gao, X., Niu, Z., Tao, D., Li, X.: Non-goal scene analysis for soccer video. Neurocomputing **74**(4), 540–548 (2011)
7. Khatoonabadi, S.H., Rahmati, M.: Automatic soccer players tracking in goal scenes by camera motion elimination. Image Vis. Comput. **27**(4), 469–479 (2009)
8. Liu, Y., Liang, D., Huang, Q., Gao, W.: Extracting 3D information from broadcast soccer video. Image Vis. Comput. **24**(10), 1146–1162 (2006)
9. Niu, Z., Gao, X., Tian, Q.: Tactic analysis based on real-world ball trajectory in soccer video. Pattern Recogn. **45**(5), 1937–1947 (2012)
10. Ren, J., Orwell, J., Jones, G.A., Xu, M.: Tracking the soccer ball using multiple fixed cameras. Comput. Vis. Image Underst. **113**(5), 633–642 (2009)
11. Ronneberger, O., Fischer, P., Brox, T.: U-Net: convolutional networks for biomedical image segmentation. In: Navab, N., Hornegger, J., Wells, W.M., Frangi, A.F. (eds.) MICCAI 2015. LNCS, vol. 9351, pp. 234–241. Springer, Cham (2015). https://doi.org/10.1007/978-3-319-24574-4_28
12. Yu, X., Xu, C., Leong, H.W., Tian, Q., Tang, Q., Wan, K.W.: Trajectory-based ball detection and tracking with applications to semantic analysis of broadcast soccer video. In: Proceedings of the Eleventh ACM International Conference on Multimedia, pp. 11–20. Association for Computing Machinery (2003)
13. zhixuhao: Unet. https://github.com/zhixuhao/unet (2017). Accessed 30 Jan 2023
14. Åkeborg, E.: Automating feature-extraction for camera calibration through machine learning and computer vision (2021). Master thesis. https://lup.lub.lu.se/student-papers/search/publication/9054961. Accessed 31 Jan 2023

Object Localization with Multiplanar Fiducial Markers: Accurate Pose Estimation

Pablo García-Ruiz[1(✉)], Rafael Muñoz-Salinas[1,2], Rafael Medina-Carnicer[1,2], and Manuel J. Marín-Jiménez[1,2]

[1] Department of Computing and Numerical Analysis, University of Córdoba, Córdoba, Spain
{pgruiz,rmsalinas,rmedina,mjmarin}@uco.es
[2] Maimonides Institute for Biomedical Research of Córdoba (IMIBIC), Córdoba, Spain

Abstract. Accurate object tracking is an important task in applications such as augmented/virtual reality and tracking of surgical instruments. This paper presents a novel approach for object pose estimation using a dodecahedron with pentagonal fiducial markers. Unlike traditional ArUco markers, which are squared, we propose a pentagonal marker that fits better in the dodecahedron figure. Our proposal improves marker detectability and enhances pose estimation with a novel pose refinement algorithm. Our experiments show the system's performance and the refinement algorithm's efficacy under different configurations.

Keywords: Object localization · Multiplanar fiducial markers · Pose estimation · Augmented reality · Computer vision

1 Introduction

Camera pose estimation consists in estimating the position of a camera with respect to a known reference system and it is a common task in applications such as augmented/virtual reality [16] and medical instrument tracking [22,28].

Fiducial markers have become a popular solution for pose estimation, due to their simplicity, and high performance, consisting in simple black-and-white patterns containing a unique identity and it is possible to estimate the camera pose by a single view of them. Several approaches can be found in the literature, such as [4,17,26]. The ArUco markers [9] is one of these solutions that has been integrated into other projects such as [3,21,24], expanding their use and making it one of the most used. Other authors have proposed customized fiducial markers

Supported by the MCIN Project TED2021-129151B-I00/AEI/10.13039/501100011033/ European Union NextGenerationEU/PRTR, project PID2019-103871GB-I00 of the Spanish Ministry of Economy, Industry and Competitiveness, and project PAIDI P20_00430 of the Junta de Andalucía, FEDER.

A. Pertusa et al. (Eds.): IbPRIA 2023, LNCS 14062, pp. 454–465, 2023.
https://doi.org/10.1007/978-3-031-36616-1_36

[11], allowing markers of various shapes to be generated and providing better adaptability to commercial applications.

Although it is possible to estimate the camera pose from a single marker, it is preferred to use multiple markers in a non-coplanar configuration [25,27]. Sarmadi et al. [18] propose a system that integrates fiducial markers with a multi-faced figure for pose estimation with a system composed of several cameras. The authors proposed a method to obtain the extrinsic parameters of the cameras and the relative poses between the markers and the cameras at each frame. In addition, their method allows them to automatically obtain the three-dimensional configuration of an arbitrary set of markers.

Another system composed of multiplanar fiducial markers is proposed in the work of Wu et al. [23], named DodecaPen, with the aim to calibrate and track a 3D dodecahedron using ArUco markers in its faces. The system allows for real-time camera pose estimation using a single camera, which is applied to develop a digital wand. The authors also develop a set of methods to estimate the 3D relative position of the markers in the dodecahedron. To deal with motion blur, a corner tracking algorithm is used. Nevertheless, their approach has some limitations. First, it does not make effective use of the available area, i.e., square fiducial markers while the faces of the object are pentagonal. Secondly, neither the source code nor the binaries are available, making it impossible for other researchers to use and test their work.

This work presents DoducoPose, an improvement of the DodecaPen system. First, instead of using squared markers, we propose pentagonal markers to fit the object's faces properly. Second, we propose a method to obtain the three-dimensional configuration of the markers placed on a dodecaedron. Third, we propose a camera pose refinement method that can find extra matches from undetected markers. Finally, our solution is publicly available for other researchers to use[1].

The rest of this paper is structured as follows. Section 2 explains the proposed method, including the design and construction of the markers, as well as the algorithms used for detection and mapping, and pose estimation. In Sect. 3, the experimental methodology and results are presented in detail, providing insight into the system's performance. Finally, we conclude by discussing the implications of our findings and suggesting future research directions to further improve the accuracy and efficiency of the proposed system.

2 Proposed Method: DoducoPose

This section explains the basis of the proposed method, DoducoPose. Firstly, the marker design is described, aiming to maximise the available surface's use for efficient tracking. Then, the marker detection process is explained in detail, followed by a description of the mapping approach employed to estimate the marker set configuration. Finally, the proposed pose estimation and refinement algorithm is presented.

[1] https://sourceforge.net/projects/doducopose/.

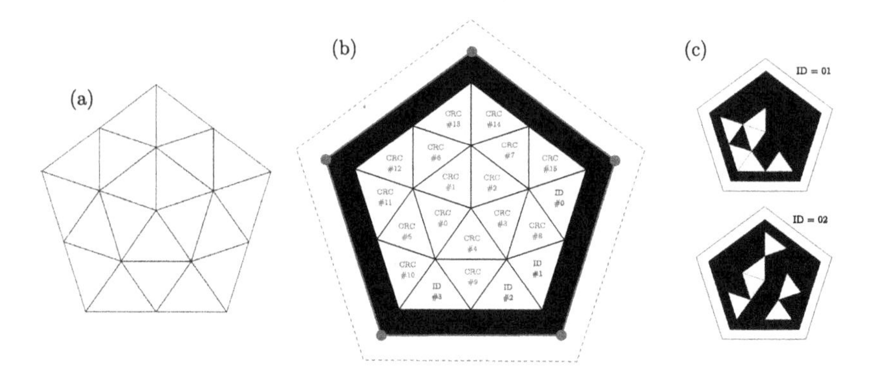

Fig. 1. (a) Division of pentagonal area. (b) Marker design consisting of a total of 20 elements: 16 for cyclic redundancy check (CRC) and 4 for identification (ID). The color scheme is black and white. In the figure, ID bits are indicated in purple, while CRC bits are highlighted in green. The marker contour and corners are highlighted in red. (c) Marker samples. (Color figure online)

2.1 Markers Design

Our marker design is based on the work of Jurado-Rodriguez et al. [11], which allows using customized markers. To do so, we need to indicate the elements as identification bits, the contour, and the layout of the marker.

The first requirement is its contour. It must be a regular pentagon to use the faces of the dodecahedron effectively. Additionally, it must enable the generation of at least 12 different markers. To do this, there must be a structure where at least 4 elements act as identification bits (ID), allowing for 16 different markers. In addition, a minimum of 16 elements is established to act as cyclic redundancy check bits (CRC), to ensure correct detection at all times of the different markers generated, as having a larger number of CRC bits ensures a lower false positive rate. Therefore, there must be 20 elements of ideally the same size, which also fit best within a pentagon, as shown in the Fig. 1.

Once the identification of each bit is arranged, and an initial contour is defined, it is necessary to specify the color scheme. To facilitate their use, it is decided to use only two colors, black and white. In this way, the different identification and redundancy check elements are color-modified for each marker in a unique combination.

After the design is defined, the contour must be rescaled to have a minimum thickness when printed for the size of the pentagon. For the project's development, two different dodecahedrons are used, measuring 3.5 and 2.8 cm per side, respectively.

Knowing these measurements, that the desired thickness should be around 4 mm, and that an additional external contour of 4 mm thickness must be ensured, the basic geometry operations necessary to calculate the desired scaling can be applied, obtaining the final desired scaling for the border. Finally, 12 fiducial markers are generated for each obtained design.

2.2 Marker Detection

For the detection of the custom-designed markers, the algorithm described in the original work of Jurado-Rodriguez et al. [11] is used.

Pentagonal markers are designed with $m_v = 5$ polygonal vertices and $m_B = 20$ bits. Marker bits are partitioned into 16 for cyclic redundancy checking and 4 for identification.

Let $M = V, B$ be a model for marker detection, with V representing the set of vertices of the marker:

$$V = \{\{\boldsymbol{v}_i\}, i \in \{1, 2, 3, 4, 5\}, \boldsymbol{v}_i = (x_i, y_i) \in \mathbb{R}^2\} \tag{1}$$

And B is the set of bits of the marker:

$$B = \{\{\boldsymbol{b}_j\}, j \in \{1, 2, \ldots, 20\}, \boldsymbol{b}_j = (x_j, y_j) \in \mathbb{R}^2\} \tag{2}$$

It should be noted that the coordinates (x, y) represent the coordinates of the elements within the file containing the marker's design.

Given a marker and an image as inputs, the process of detecting the marker in that image is described in the following steps:

1. **Image segmentation**. Since it is ensured that a white contour surrounds the markers, a local adaptive thresholding method is applied to find those contours. The average brightness is calculated for each pixel given a small neighbourhood and compared to a global threshold value.
2. **Contour detection and filtering**. The contours are extracted using the Suzuki and Abe algorithm [19]. Then, a polygonal approximation is obtained using the Douglas-Peucker algorithm [1]. Polygons with a different number of faces than $m_v = 5$ are discarded.
3. **Initial detection of a valid marker**. Here it is determined if a polygon in P is a valid marker. For each of the polygons collected in the previous step V_p, the $m_v = 5$ possible rotations of its vertices are taken into account, and the rotated polygons are defined as:

$$\{\{V_p^r\}, r \in \{1, 2, 3, 4, 5\}\} \tag{3}$$

   For each rotated polygon, the homography that transforms its vertices into those of the original model is calculated. This is necessary to estimate where the different elements of the markers that act as bits should appear in the image.

   To define whether a marker has been detected, one only needs to check the color of these elements and start searching for matches with the dictionary of possible markers. The color of each element will be either white or black. Discrepancies may arise between the colors depicted in the original vector file and the printed colors, due to factors such as printing issues or suboptimal quality. To address this, we propose a method for extracting the color of each individual bit in the image. First, we calculate the mean value of all bit values within the image. Subsequently, if a bit's value falls below the mean, it

is designated as black. If the value equals or exceeds the mean, it is assigned the color white.
The result is the exact position, with coordinates of its vertices, of the markers in the image and, therefore, their detection.

2.3 Marker Mapping

To obtain the 3D coordinates that relate the different markers that make up the figure in a standard reference system, a system that computes these coordinates automatically is necessary if a more remarkable precision is desired, compared to manually creating 3D coordinates.

Simultaneous Localisation and Mapping, or SLAM, is the computational problem of creating a map of the environment while navigating it. There is a multitude of systems that propose a solution to this problem [5,6,15]. The UcoSLAM system [14] incorporates the joint use of fiducial markers with the keypoint system. Although placing fiducial markers in the scene alone would yield a new set of robust keypoints, active and effective use is made of the keypoints in the scene, utilizing additional information provided by the detector. The original system implemented the use of ArUco fiducial markers [9].

To achieve the mapping of our markers, a customized version of the UcoSlam system is employed. Starting from a video where the figure is static and is rotated around it, showing all the markers in the same video, the mapping process is described as follows.

First, a map initialisation process is done. The method used allows the initialization of the map using a minimum of two frames, where the first 3D points are mapped by detecting either keypoints or the markers themselves.

Secondly, further updates of the map are addressed. Once initialized, an update is performed for each reading frame. Using information obtained from previous frames, new points are mapped using both keypoints and the markers in question. A tracking algorithm updates the map to search for correspondences between frames. If it fails, a relocalization process is triggered. This process involves estimating a new pose without using the information given by the tracking algorithm. If a marker has been detected correctly, the new pose is estimated using it. Otherwise, relevant frames similar to the current one are selected for a search and relation of keypoints, ending with an estimation of a new pose using the related information.

The use of environment keypoints is necessary to develop the map with greater accuracy, even though only the 3D coordinates of the marker vertices are exported.

It should be noted that thanks to the use of marker detection in the mapping process, the map is already at the correct scale, allowing it to be used in future applications. This scaling is possible because the marker size is known.

As a result, a mapped scene is obtained where the information attached to our polyhedron is extracted and prepared for use.

2.4 Pose Estimation

The pose estimation of one or a set of markers is defined by the elements $r = (r_x, r_y, r_z)$ and $t = (t_x, t_y, t_z)$. With r, we know the system's rotation with three degrees of freedom. With t, we can determine the information related to the translation with the same number of degrees of freedom. Thus, the pose is defined with 6 degrees of freedom.

It is performed by solving the PnP (Perspective-n-Point) problem, whose objective is to minimize the reprojection errors of the marker vertices, provided their 2D coordinates in the image and 3D coordinates are known, which is our case.

Given the pinhole camera model, we can define the intrinsic parameters as the focal lengths $[f_x, f_y]$, the optical centre (c_x, c_y), and the distortion parameters $[l_1, ..., l_n]$. These parameters characterize the camera and must be considered at all times. The extrinsic parameters are the rotation and translation value sets, r and t, that allow us to relate the real-world and camera coordinate systems.

This transformation is defined as:

$$p = A[R|t]P_w \tag{4}$$

where p is the image pixel that corresponds to the real-world point P_w, A is the set of camera intrinsic parameters, and $[R|t]$ is the calculated set of extrinsic parameters. R is the rotation matrix obtained using the Rodrigues rotation formula from the set of values in r [8].

In more detail, this operation is expressed as:

$$\begin{bmatrix} u_0 \\ v_0 \\ 1 \end{bmatrix} = \begin{bmatrix} f_x & 0 & c_x \\ 0 & f_y & c_y \\ 0 & 0 & 1 \end{bmatrix} \begin{bmatrix} R_{11} & R_{12} & R_{13} & t_x \\ R_{21} & R_{22} & R_{23} & t_y \\ R_{31} & R_{32} & R_{33} & t_z \end{bmatrix} \begin{bmatrix} X_w \\ Y_w \\ Z_w \\ 1 \end{bmatrix} \tag{5}$$

Different algorithms are used for the pose estimation calculation, allowing only one to be used at a time. These algorithms are the Levenberg-Marquardt algorithm (Iterative) [13], the EPnP algorithm [12], and the SQPnP algorithm [20].

Let V_p be the set of detected marker vertices (2D coordinates), V_P the set of mapped vertices (3D coordinates), and d the number of detected markers. We have the sets of values necessary to estimate the pose:

$$\{\{V_p^i, V_P^i\}, i \in \{1, ..., d\}\} \tag{6}$$

Both sets and the camera intrinsic parameters are the inputs to the PnP problem-solving algorithm, and the extrinsic parameters are the output.

If the input to the system is a video sequence, an estimated pose is obtained for each frame where at least one marker is detected.

(a) First detections and pose estimation.

(b) Search for new markers and corners refinement.

(c) Improved pose estimation.

Fig. 2. Refinement pipeline. The main steps of the proposed pose estimation refinement algorithm are depicted. In the final step, the pose estimation is performed using information from both detected and located markers. In images (b) and (c), a portion of the contour of a representative marker has been cropped and its resolution increased, effectively zooming in to facilitate visual comparison and highlight the differences between the two images.

2.5 Pose Refinement

To improve the estimated pose accuracy, a refinement process is developed where the main objective is to make even greater use of the information available from the figure in the image.

Since the marker detector is not perfect, there are occasions where markers are visible in the image but have not been successfully detected. If the markers are still visible despite not being detected, it would be an improvement to add them to the input of the algorithm that solves the PnP problem in order to obtain an even more robust pose against problems associated with the image and the planes where the markers are located. This process is achieved by identifying the remaining markers whose normal lines intersect the camera plane at a positive angle. Upon selecting the presumed markers, it is essential to reproject their corners onto the image, refine the projected corners, and incorporate them into the pose estimation algorithm. The algorithm utilized for refining the corners is the one proposed by Förstner et al. [7]. The refinement process can be observed in Fig. 2.

3 Experiments and Results

This section outlines the experimental methodology used to evaluate the accuracy of the proposed method. The experiments were conducted by recording two sets of videos of the proposed dodecahedron from multiple angles, while moving the camera around it at a distance ranging from 40 to 89.5 cm for the group of experiments without motion defects, and at 50 cm for the group of experiments with motion.

To ensure a precise evaluation, an OptiTrack [2] system, comprised of six cameras strategically placed in the lab ceiling, was employed to estimate the position of the recording camera while it was moving, pointing to the dodecahedron. By placing a set of reflective markers on the recording camera, the Optitrack system obtains the camera trajectory with high precision.

The error of our method is estimated as the difference between the trajectory of the Optitrack system and ours. To compare both trajectories, one has to be transformed into the coordinate system of the other, using the Horn algorithm [10].

First, the Average Translation Error (ATE) is used, which is the mean translation error between two paths and indicates how much one path is, on average, displaced from the other. It is measured to determine how precise the DoducoPose system is. The desirable value for this metric is as close to 0 as possible.

Secondarily we employ the percentage of used frames. It should be noted that the paths generated by DoducoPose and Optitrack are not always complete, and there are frames in both paths where the pose cannot be estimated, and thus that frame does not have an associated pose in one or both paths. The desirable value for this metric is 100% or as close to 100% as possible.

Several paths of DoducoPose are generated to test different possibilities of configurations. Two subsets of DoducoPose trajectories are generated for each video, the first without applying pose refinement and the second with it.

Different execution configurations are tested for each of these subsets. First, we variate the angle threshold. This is the minimum average value of the intersection angle that the detected markers must present is established. The pose is not saved if this value is not satisfied. The following values are tested: {0, 5, ..., 60}.

Secondly, we test different values for a threshold of the number of detected markers. The minimum number of detected markers required is established. The pose is not saved if this value is not satisfied. The following values are tested: {1, ..., 6}.

Therefore, 78 trajectories without applying pose refinement and 78 trajectories with it are obtained for each video using the DoducoPose system. In comparison, only one Optitrack trajectory is obtained. Therefore, two trajectories are obtained for each angle threshold, and several detected markers threshold configurations to compare with the OptiTrack trajectory. For each configuration, we track the values of ATE, with and without refinement applied, plus the percentage of Used frames.

Finally, for the experimentation, a 3.5 cm-sided dodecahedron was the figure to be detected by our method. Taken experiments are distributed into two major application groups to contemplate different application scenarios.

Static Experiments Group. The first group utilized a total of 425 frames, with a minimum and maximum distance of 40 cm and 89.5 cm from the dodecahedron. During the path, the camera made stops, and the frames in which the camera remained stationary were used exclusively for pose estimation, while the others were discarded. Additionally, the camera remained centered throughout the experiment, with the visual center of the image marked as the reference point. This approach allowed us to verify the system's robustness even when the object was not positioned at the center of the frame.

Dynamic Experiments Group. The video dataset consists of 756 frames, each captured at an average distance of 50 cm from the dodecahedron. The camera continuously moved along a route with no stops, and the dodecahedron remained in view at all times due to the camera being centered.

Discussion of the Results. The choice of the solver method is designated as the main configuration variable to keep in mind for these systems. Having a static system, where the object's movement relative to the camera is low, the ITERATIVE solver returns the best results, obtaining the lowest ATE values of the three solvers while maintaining the same percentage of valid poses used. However, when addressing a more dynamic system, the SQPNP solver is the best for moving systems like this one, where no other system is applied to deal with problems associated with motion, such as blurring or defocusing.

Overall, the best configurations for each group coincide, showing their effectiveness. If the percentage of used frames greater than 50% is sufficient, the best configurations require a minimum of 3 detected markers and a minimum average angle of approximately 25°.

The refinement algorithm operates as intended, although some improvements can be made, such as checking the quality of the markers found with good intersection angles. Considering the camera movement as the main factor makes it more difficult for the refinement algorithm. The reason is that this algorithm requires a considerably good initial pose so that the projections of the undetected markers are decent enough to add valuable information to the PnP problem resolution.

Prominently, in our dynamic experimentation shown in Fig. 3, we achieved an error of $ATE \approx 0.82$ cm and %Used frames $\approx 68.12\%$ using the solver SQPNP, our refinement algorithm, and marker and intersection angle thresholds of 3 and 10, respectively. These values were chosen to ensure a usage percentage of at least 60% for frames. A more detailed comparison between configurations is provided in Table 1.

Table 1. Comparison of configurations varying the chosen solver and applying the refinement algorithm. The values shown correspond to the approximated ATE error (cm). The % of frames used is approximately 68.12% for all configurations. In bold is highlighted the best configuration for each solver.

	Iterative	EPnP	SQPnP
w/o Refinement	**1.613**	1.015	0.912
w Refinement	1.624	**0.964**	**0.829**

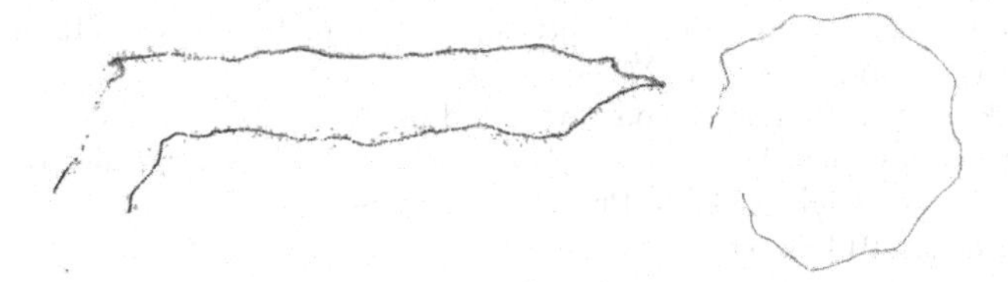

Fig. 3. Drawn paths representation in the dynamic experimentation with the featured configuration. Both sets of points represent the same path from a different point of view. The blue dots are defined by OptiTrack. The red dots are defined by the proposed method. (Color figure online)

4 Conclusions and Future Work

This work presents four main contributions. First, we propose a novel set of pentagonal markers that conform better to the faces of a dodecahedron. Second, we introduce a method to obtain the three-dimensional configuration of the markers placed on a dodecahedron. Third, we present a novel camera pose refinement technique that can identify additional matches from previously undetected markers. Finally, our solution is publicly available to other researchers.

Various experiments have been conducted in real-world scenarios with varying configurations of the proposed system and using various algorithms to solve the PnP problem. The experimental results show that our solution obtains errors below one centimetre from diverse perspectives, even under camera movement. Additionally, the choice of algorithm for solving the PnP problem is of great importance, and our proposed refinement algorithm functions effectively.

Although the system demonstrates positive outcomes, its main drawback lies in the limited dictionary, as it can only use a single dodecahedron at a time. Additional polyhedral designs must be developed to enable multiple markers simultaneously, like with Aruco markers. As an alternative to this approach, we suggest generating a new extended set of pentagonal markers with a greater number of identification bits. This would allow for the creation of distinct dodecahedrons with unique marker configurations, which could be employed simultaneously.

As a future improvement, the application of tracking algorithms could address issues arising from a system in motion, enabling effective use in more diverse environments and conditions and with more complex figures. Additionally, further experimentation should be conducted with different figures and markers to compare the results obtained with those offered by the DodecaPen system.

Moreover, it is recommended to explore various experimental conditions, such as altering distances between the dodecahedron and the camera, introducing different levels of blur, Gaussian noise, and occlusion, in order to evaluate the robustness and adaptability of the system under a wide range of scenarios.

References

1. Douglas-peucker algorithm. https://en.wikipedia.org/wiki/Ramer-Douglas-Peucker_algorithm. Accessed 27 Apr 2023
2. Optitrack. https://optitrack.com. Accessed 07 Sep 2022
3. Babinec, A., Jurišica, L., Hubinský, P., Duchoň, F.: Visual localization of mobile robot using artificial markers. Procedia Eng. **96**, 1–9 (2014). https://doi.org/10.1016/j.proeng.2014.12.091
4. DeGol, J., Bretl, T., Hoiem, D.: ChromaTag: a colored marker and fast detection algorithm. In: IEEE International Conference on Computer Vision, pp. 1481–1490 (2017). https://doi.org/10.1109/ICCV.2017.164
5. Engel, J., Schöps, T., Cremers, D.: LSD-SLAM: large-scale direct monocular SLAM. In: Fleet, D., Pajdla, T., Schiele, B., Tuytelaars, T. (eds.) ECCV 2014. LNCS, vol. 8690, pp. 834–849. Springer, Cham (2014). https://doi.org/10.1007/978-3-319-10605-2_54
6. Forster, C., Pizzoli, M., Scaramuzza, D.: Svo: Fast semi-direct monocular visual odometry. pp. 15–22. IEEE International Conference on Robotics and Automation ICRA (2014)
7. Förstner, W., Gülch, E.: A fast operator for detection and precise location of distinct points, corners and centres of circular features. In: Proceedings of the ISPRS Intercommission Conference on Fast Processing of Photogrammetric Data, vol. 6, pp. 281–305 (1987)
8. Gallego, G., Yezzi, A.J.: A compact formula for the derivative of a 3-D rotation in exponential coordinates. CoRR abs/1312.0788 (2013). http://arxiv.org/abs/1312.0788
9. Garrido-Jurado, S., Munoz-Salinas, R., Madrid-Cuevas, F.J., Marin-Jimenez, M.J.: Automatic generation and detection of highly reliable fiducial markers under occlusion. Pattern Recogn. **47**(6), 2280–2292 (2014). https://doi.org/10.1016/j.patcog.2014.01.005
10. Horn, B.: Closed-form solution of absolute orientation using unit quaternions. J. Opt. Soc. Am. Opt. Image Sci. Vis. **4**(4), 629–642 (1987). https://doi.org/10.1364/JOSAA.4.000629
11. Jurado-Rodriguez, D., Munoz-Salinas, R., Garrido-Jurado, S., Medina-Carnicer, R.: Design, detection, and tracking of customized fiducial markers. IEEE Access **9**, 140066–140078 (2021). https://doi.org/10.1109/ACCESS.2021.3118049
12. Lepetit, V., Moreno-Noguer, F., Fua, P.: EPnP: an accurate o(n) solution to the PnP problem. Int. J. Comput. Vis. **81**(2), 155–166 (2009). https://doi.org/10.1007/s11263-008-0152-6
13. Madsen, K., Nielsen, H., Tingleff, O.: Methods for non-linear least squares problems (2nd ed.) (2004)
14. Munoz-Salinas, R., Medina-Carnicer, R.: UcoSLAM: simultaneous localization and mapping by fusion of keypoints and squared planar markers. Pattern Recogn. **101**, 107193 (2020). https://doi.org/10.1016/j.patcog.2019.107193

15. Mur-Artal, R., Montiel, J.M.M., Tardos, J.D.: ORB-SLAM: a versatile and accurate monocular SLAM system. IEEE Trans. Robot. **31**(5), 1147–1163 (2015). https://doi.org/10.1109/TRO.2015.2463671
16. Nee, A., Ong, S., Chryssolouris, G., Mourtzis, D.: Augmented reality applications in design and manufacturing. CIRP Ann. **61**(2), 657–679 (2012). https://doi.org/10.1016/j.cirp.2012.05.010
17. Olson, E.: AprilTag: a robust and flexible visual fiducial system. In: Proceedings of the IEEE International Conference on Robotics and Automation (ICRA), pp. 3400–3407. IEEE (2011)
18. Sarmadi, H., Muñoz-Salinas, R., Berbís, M.A., Medina-Carnicer, R.: Simultaneous multi-view camera pose estimation and object tracking with squared planar markers. IEEE Access **7**, 22927–22940 (2019). https://doi.org/10.1109/ACCESS.2019.2896648
19. Suzuki, S., Abe, K.: Topological structural-analysis of digitized binary images by border following. Comput. Vis. Graph. Image Process. **30**(1), 32–46 (1985). https://doi.org/10.1016/0734-189X(85)90016-7
20. Terzakis, G., Lourakis, M.: A consistently fast and globally optimal solution to the perspective-n-point problem. In: Vedaldi, A., Bischof, H., Brox, T., Frahm, J.-M. (eds.) ECCV 2020. LNCS, vol. 12346, pp. 478–494. Springer, Cham (2020). https://doi.org/10.1007/978-3-030-58452-8_28
21. Tørdal, S.S., Hovland, G.: Relative vessel motion tracking using sensor fusion, Aruco markers, and MRU sensors. MIC—Model. Identif. Control **38**(2), 79–93 (2017). https://doi.org/10.4173/mic.2017.2.3
22. Wang, J., Meng, M.Q., Ren, H.: Towards occlusion-free surgical instrument tracking: a modular monocular approach and an agile calibration method. IEEE Trans. Autom. Sci. Eng. **12**, 588–595 (2015)
23. Wu, P.C., et al.: Dodecapen: accurate 6D of tracking of a passive stylus. In: 30th Annual ACM Symposium on User Interface Software and Technology, pp. 365–374. (2017). https://doi.org/10.1145/3126594.3126664
24. Xu, Z., Haroutunian, M., Murphy, A.J., Neasham, J., Norman, R.: An underwater visual navigation method based on multiple Aruco markers. J. Mar. Sci. Eng. **9**(12), 1432 (2021). https://doi.org/10.3390/jmse9121432
25. Yoon, J.-H., Park, J.-S., Kim, C.: Increasing camera pose estimation accuracy using multiple markers. In: Pan, Z., Cheok, A., Haller, M., Lau, R.W.H., Saito, H., Liang, R. (eds.) ICAT 2006. LNCS, vol. 4282, pp. 239–248. Springer, Heidelberg (2006). https://doi.org/10.1007/11941354_25
26. Yu, G., Hu, Y., Dai, J.: TopoTag: a robust and scalable topological fiducial marker system. IEEE Trans. Vis. Comput. Graph. **27**(9), 3769–3780 (2021). https://doi.org/10.1109/TVCG.2020.2988466
27. Yu, R., Yang, T., Zheng, J., Zhang, X.: Real-time camera pose estimation based on multiple planar markers, pp. 640–645 (2009). https://doi.org/10.1109/ICIG.2009.93
28. Zhang, L., Ye, M., Chan, P.-L., Yang, G.-Z.: Real-time surgical tool tracking and pose estimation using a hybrid cylindrical marker. Int. J. Comput. Assist. Radiol. Surg. **12**(6), 921–930 (2017). https://doi.org/10.1007/s11548-017-1558-9

Real-Time Unsupervised Object Localization on the Edge for Airport Video Surveillance

Paula Ruiz-Barroso(✉), Francisco M. Castro, and Nicolás Guil

University of Málaga, Málaga, Spain
{pruizb,fcastro,nguil}@uma.es

Abstract. Object localization is vital in computer vision to solve object detection or classification problems. Typically, this task is performed on expensive GPU devices, but edge computing is gaining importance in real-time applications. In this work, we propose a real-time implementation for unsupervised object localization using a low-power device for airport video surveillance. We automatically find regions of objects in video using a region proposal network (RPN) together with an optical flow region proposal (OFRP) based on optical flow maps between frames. In addition, we study the deployment of our solution on an embedded architecture, *i.e.* a Jetson AGX Xavier, using simultaneously CPU, GPU and specific hardware accelerators. Also, three different data representations (FP32, FP16 and INT8) are employed for the RPN. Obtained results show that optimizations can improve up to 4.1× energy consumption and 2.2× execution time while maintaining good accuracy with respect to the baseline model.

Keywords: Edge Computing · Object Localization · Deep Learning

1 Introduction

The goal of *object localization* is to define the spatial extent of objects present in an image or video sequence. In this task, it is not necessary to know the class of the object since the result is usually used for object tracking where it is not important to know the kind of object. Object localization is typically carried out as the first step of the object detection problem in most object detection algorithms. Thus, an object detection algorithm can be used for object localization after discarding the class prediction step. Many different approaches have been proposed for this task [2,9,10,24,30]. Traditional approaches are based on hand-crafted descriptors that are computed and classified along the image in a sliding window setup [9,30]. With the advent of deep learning models, the features are self-learnt by the model, which greatly boosted the performance of the proposed approaches [17,32,34].

Supported by the Junta de Andalucía of Spain (P18-FR-3130 and UMA20-FEDERJA-059), the Ministry of Education of Spain (PID2019-105396RB-I00) and the University of Málaga (B1-2022_04).

A. Pertusa et al. (Eds.): IbPRIA 2023, LNCS 14062, pp. 466–478, 2023.
https://doi.org/10.1007/978-3-031-36616-1_37

However, all those approaches must be trained in a supervised way using labelled data. This requirement, together with the huge amount of data needed for training a deep learning model, limits the use of deep learning models with new classes that are not available in public datasets. Although this problem is less important for object localization since region proposal usually works fine with unknown objects, it is still a limitation for video surveillance tasks where cameras are placed at a certain distance and small objects appear in the scene. This is due to models are trained with public datasets where objects appear in the foreground thus, they are not able to localize small objects since they are not trained for that task. To solve this limitation, since we are working on video surveillance, we compute the optical flow maps between every 15 frames (since it is the frame step in the dataset) to discover areas with moving objects, allowing us to discover small objects in the scene.

Typically, the deployment of these deep learning models needs the use of high-performance architectures with discrete Graphics Processing Units (GPUs) to fulfil both inference time and energy consumption requirements. However, when these applications have to work in embedded platforms, energy consumption is a paramount issue that must also be taken into account. In these platforms, energy-efficient CPUs and accelerators are integrated on the same chip achieving a good trade-off between performance and consumption [23,26]. Different techniques can be applied to reduce both the computational and energy consumption requirements. Thus, using pruning techniques [16] can reduce the number of arithmetic operations by removing filters and even layers of the model. Moreover, more simple data representation can be used using quantization methods [20]. Thus, instead of using 32-bit floating-point data representation, the model could employ 16-bit floating-point or even 8-bit integer data.

In this work, we propose an unsupervised approach for object localization in videos. In Fig. 1 we show a sketch of our pipeline. Firstly, we automatically find regions of objects in a sequence of frames by using the Region Proposal Network (RPN) of a pre-trained object detection model. In addition, we compute the optical flow maps between every 15 frames (it is the frame step in the dataset) to discover areas with small moving objects. Then, a non-maxima suppression step is carried out to filter and combine detections. Finally, the pipeline is optimized and deployed in an NVIDIA Jetson AGX Xavier embedded system.

Therefore, the main contributions of this work are: *(i)* a combination of two complementary region proposals specially designed for video sequences; *(ii)* an optimized pipeline for embedded systems; and, *(iii)* a thorough experimental study to validate the proposed framework and different optimizations.

In our experiments, we use videos obtained from a live RGB camera continuously recording the *apron area* (area where aeroplanes park to load passengers and luggage) of the Gdansk Airport. According to the results, our region proposal approach is robust, especially with small regions or classes with changes in shape such as persons, and can run in real-time in an embedded system.

The rest of the paper is organised as follows. We start by reviewing related work in Sect. 2. Then, Sect. 3 explains the proposed approach. Sect. 4 contains the experiments and results. Finally, we present the conclusions in Sect. 5.

2 Related Work

2.1 Objects Localization Solutions

Object localization is an essential task in the field of object tracking and object classification. Gudovskiy *et al.* [12] present CFLOW-AD model, that is based on a conditional normalizing flow framework adopted for anomaly detection and localization using an encoder and generative decoders. Zimmerer *et al.* [36] complement localization with variational auto-encoders. Gong *et al.* [11] propose a two-step "clustering + localization" procedure. The clustering step provides noisy pseudo-labels for the localization step, and the localization step provides temporal co-attention models that in turn improve the clustering performance.

Moreover, some approaches are related to low-cost devices. In this area, Li *et al.* [15] propose a deep-reinforcement-learning-based unsupervised wireless-localization method. Chen *et al.* [4] present a robust WIFI-Based indoor localization, they use variational autoencoders and train a classifier that adjusts itself to newly collected unlabeled data using a joint classification-reconstruction structure. Chen *et al.* [3] show an unsupervised learning strategy to train the fingerprint roaming model using unlabeled onboard collected data which does not incur any labour costs.

2.2 Implementations on Embedded Systems

Nowadays, the use of embedded platforms to deploy deep learning applications has become more important. These systems are typically employed just for inference due to their limited computational capabilities [14,25]. However, some authors have used them for collaborative training [21,29]. Apply techniques such as pruning, quantization or matrix factorization, among others, that increase inference throughput on embedded platforms have led to multiple contributions. Model compression techniques can be found in surveys of the state-of-the-art [5,7]. Following, we show some contributions in this field related to our work.

Quantization reduces the number of computations using simpler data representation. Some authors apply post-training quantization [18,25]. Another option consists of the application of quantization during the training process [22,35]. Others include quantization with other optimization techniques. Thus, some papers use quantization and pruning techniques separately [25]. Other works jointly use sparsity training, channel and layer pruning and quantization [19].

Furthermore, software development kits are available for optimizing deep learning models such as TensorRT or TensorFlow Lite for GPU or CPU, respectively. Focusing on TensorRT, parallelization methodology is used to maximize performance using GPU and deep learning accelerators (DLAs) [14]. Segmentation approaches and object detection networks can also be optimized using TensorRT [27,33]. Turning our attention to TensorFlow Lite. It has two different targets: microcontrollers [6] and smartphones [1].

3 Methodology

In this section, we present our proposed method to localize small and big objects in real-time using an embedded device. Then, we describe hardware optimizations applied in this work.

3.1 Pipeline Description

Our pipeline is shown in Fig. 1. As input, we use one frame every 15 frames from a video since it is the frame step in the dataset. Then, the region proposal methodology is composed of two steps: (1) object localization using YOLOv4 [31] with ResNet50 [13] as the backbone and (2) object localization using region proposal based on Farneback optical flow (OF). Finally, non-maxima suppression is applied to combine objects from YOLOv4 and OF.

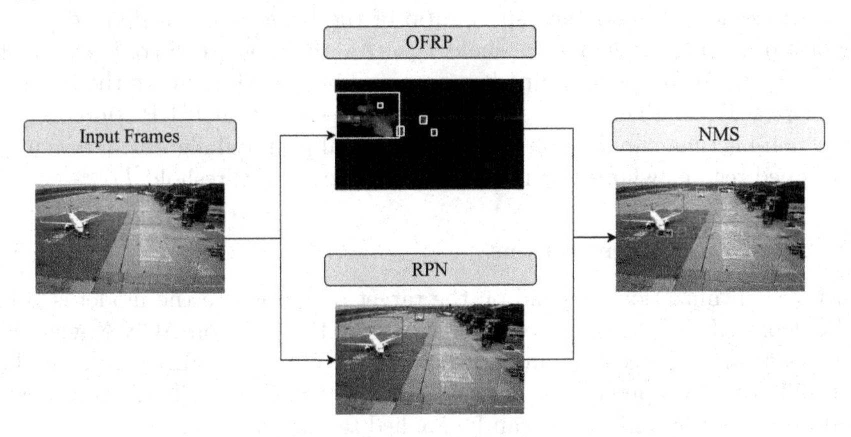

Fig. 1. Object localization method. The input is a subset of RGB video frames. Localizations are jointly obtained from both optical flow region proposal (OFRP) and region proposal network (RPN). Finally, non-maxima suppression (NMS) is applied to combine both localizations.

Input Data. Taking into account the input shape required by our RPN, the original video frames have been resized to 640×384 pixels. On the other hand, optical flow is calculated with a resolution of 720×405 pixels.

Region Proposal Network (RPN). In our approach, we use the Region Proposal Network (RPN) of a pretrained YOLOv4. It has been chosen because is one of the most widespread object detection networks nowadays and it obtains good accuracy for COCO dataset. To improve the performance of our embedded system, we have replaced the original CSPDarknet53 backbone with ResNet50 since it is better optimized for NVIDIA Jetson devices in terms of energy and time. YOLOv4 allows us to obtain the bounding box coordinates, mostly the largest objects which are common detections used for training this kind of network.

Optical Flow Region Proposal (OFRP). On the other hand, region proposal based on OF is obtained from optical flow maps using the FarneBack Dense Optical Flow technique [8]. OF maps F_t are computed every 15 frames (since it is the frame step in the dataset). To remove the optical flow noise produced by changes in the conditions of the scenario, all positions whose optical flow components (x and y) are smaller than a threshold T_F are set to 0. Then, contours containing the objects of the scene are found using the well-known algorithm proposed by Suzuki *et al.* [28]. Finally, to prevent excessively small regions, we remove those proposed regions whose area is smaller than a threshold T_A.

Non-maxima Suppression. To avoid duplicated localizations of a moving object that is localized by the RPN and the OFRP at the same time, we need to combine both localizations. To do that, we compute the Intersection over Union (IoU) metric between these two proposed regions. If the IoU is bigger than a threshold T_I and the aspect ratio of the biggest region divided by the smallest one is bigger than a threshold T_{AR} (*i.e.* the aspect ratio of both regions is similar), we keep the bounding box proposed by the RPN due to the fact that it is more accurate than the bounding box proposed by the OFRP. Moreover, we apply non-maxima suppression to each individual proposal algorithm to remove overlapped regions whose IoU is greater than the same threshold T_I.

3.2 Hardware Optimizations

Hardware optimizations depend on the target device where the model is going to be deployed. In this paper, we employ an NVIDIA Jetson AGX Xavier that includes different processing units such as CPU, GPU, and DLA which can deal with different data precisions, i.e. FP32, FP16, and INT8. Thus, two different hardware-based optimizations can be applied to our platform:

Quantization: We explore, in addition to FP32 representation, FP16 and INT8 precision formats, which provide a more compact numerical representation that can reduce both the inference time and memory consumption. However, lower data precision can degrade model accuracy due to the fact that the numerical precision of the weights and activations is lower. In our case, FP16 optimization is performed with the TensorRT framework, but INT8 optimization requires an additional calibration to reduce the accuracy loss when the network is converted from FP32 to INT8. We have fine-tuned the model using PyTorch *quantization aware training*. During fine-tuning, all calculations are done in floating point, using fake quantization modules, which model the effects of quantization by clamping and rounding to simulate the effects of INT8. Then, the calibrated model is converted using TensorRT to complete the optimization process.

Concurrent Computation: To accelerate our pipeline and take advantage of the different hardware resources available, we perform an efficient mapping of application modules on the available hardware. This way, our RPN and OFRP modules can be executed concurrently since the first one is mapped to the GPU and the second one to the CPU.

4 Experiments and Results

In this section, we first present the apron area dataset used in our experiments. Then, some implementation details are shown. Next, we define our pipeline evaluation, carry out hardware optimizations in our RPN, and show their effect on energy consumption and inference time. Finally, we examine the impact of our two region proposal algorithms according to their individual CorLoc metrics.

4.1 Dataset

In our experiments, we use a video dataset obtained from a live RGB camera continuously recording the apron area (the area where aeroplanes are parked, unloaded, loaded, boarded, refuelled or maintained) of the Gdansk Airport. This camera is publicly available online in a more modern version[1]. The dataset contains 96 video clips of one-minute length recorded by a FULLHD camera which provides a video stream with a resolution of 1920×1080 pixels and a frame rate of approximately 15 fps. In order to deal with different illumination conditions, 60% of the videos were recorded during the morning and the other 40% were recorded during the afternoon/evening. In our experiments, we consider the following categories: car ('car'), fire-truck ('ft'), fuel-truck ('fuel'), luggagetrain-manual ('lgm'), luggagetrain ('lg'), mobile-belt ('mb'), person ('pe'), plane ('pl'), pushback-truck ('pb'), stairs ('st') and van ('van'). Note that the abbreviated name of each class used in the tables is included in parentheses.

4.2 Implementation Details

As it has been indicated in the previous section, we ran our experiments on NVIDIA Jetson AGX Xavier. This platform is designed to be used as a low-power consumption system for embedded computer vision applications providing good performance. It includes an octa-core NVIDIA Carmel ARM V8.2 and a Volta GPU with 512 cores and 64 Tensor Cores accompanied by 32 GB of main memory shared by the ARM processors and the GPU. It also includes two deep learning accelerators (DLAs) and one vision accelerator (VA). Note that DLA is slower and more limited in functionality than GPU so it is expected that execution time increases using DLA and GPU with respect to performing the computation on just GPU. However, DLA consumes less energy than GPU.

Therefore, inference processes have been deployed on GPU or DLA+GPU using TensorRT 8.4.1 and cuDNN 8.4.1. Original YOLOv4 with ResNet50 as backbone has been trained based on Sacaled-YOLOv4 [31] code[2] on a computer with a discrete GPU using PyTorch 1.11. On the other hand, OF has been calculated using an OpenCV library implemented on CPU. Moreover, non-maxima suppression and preprocessing before introducing frames into OFRP or RPN have been calculated on CPU too.

[1] http://www.airport.gdansk.pl/airport/kamery-internetowe.

[2] https://github.com/WongKinYiu/ScaledYOLOv4.

Regarding parameters commented in Sect. 3, after a cross-validation process on a subset of the training data, we have established the following values: $T_F = 0.7$, $T_A = 100$, $T_I = 0.4$, $T_{AR} = 0.5$.

4.3 Performance Evaluation

On the one side, we report energy consumption and execution time for each implementation. Energy consumption is provided by internal sensors incorporated on Jetson boards. We employ 500 Hz sampling frequency which is the maximum value supported by the sensors. Also, the average values for 800 forward steps after 5 warm-up predictions are provided. We use the Energy Delay Product (EDP) metric, which is the product of energy and execution time, to measure hardware performance. Thus, the lower the EDP value, the better.

On the other side, we use the *Correct Localization* metric (CorLoc) to evaluate the precision for region localization. This metric is defined as the percentage of objects correctly localized according to the Pascal criterion: the IoU between the predicted region and the ground-truth region is bigger than a given threshold for the RPN and bigger than another given threshold for the OFRP.

4.4 Experimental Results

Baseline Experiment. In this section, we show our baseline experiment without any optimization, *i.e.* the pipeline has been deployed with PyTorch on GPU using FP32 data representation. The obtained results in terms of energy and time can be observed in Table 1. The pipeline is divided into four stages: frames preprocessing ('Prep.') before introducing them into the next step, region proposal network ('RPN'), optical flow region proposal ('OFRP') and finally, non-maxima suppression ('NMS'). As we can see, the most time and energy consuming part is the RPN. The RPN using PyTorch is a bottleneck for the parallel OFRP+RPN

Table 1. Baseline study: Time and energy. Each row represents a different stage: preprocessing (Prep.), parallel region proposal algorithms: region proposal network (RPN) and optical flow region proposal (OFRP) and non-maxima suppression (NMS). Columns represent inference time and energy consumption together with EDP (Energy Delay Product) value. The last row represents the total time, energy and EDP of our baseline method. Note that for the total time of the baseline, we only consider the highest time of the parallel region proposal part (i.e. 87.49 ms from RPN).

Stage		Time (ms)	Energy (mJ)		EDP
			GPU	CPU	
Prep.		3.32	-	0.05	0.16
Parallel	RPN	87.49	12.68	-	1109.23
	OFRP	73.01	-	3.98	290.89
NMS		3.44	-	0.01	0.03
Total		94.25	16.72		1400.32

process. On the other hand, energy consumption in preprocessing and NMS is almost negligible while inference time represents a 7.17% of the final time. Finally, our complete pipeline takes 94.25 ms with an energy consumption of 16.72 mJ without considering any hardware optimization.

Data Quantization on GPU. In Table 2, we observe energy consumption, inference time and EDP for the RPN after the optimization carries out by TensorRT for different data precisions. The last column shows the EDP reduction ratio (EDP_{rr}) for FP16 and INT8 with respect to FP32 precision. As preprocessing and NMS are executed on the CPU, they have been omitted.

First of all, we compare baseline inference time and energy consumption (third row in Table 1) with FP32 optimization using TensorRT. It can be observed that there is a decrease in inference time of 19.35 ms and a reduction of 4.29 mJ in energy consumption.

Focusing on FP16 data precision, we can see a significative reduction (around 2×) of the computation time for FP16 w.r.t FP32, as it was expected. Energy consumption has also improved due to both the shorter inference time and the use of more compact computing units. Nevertheless, INT8 precision has an unexpected behaviour as it obtains a longer execution time than that achieved for FP16. We analysed the executed kernels with NVIDIA Profiler[3] for INT8 quantization and we discovered that TensorRT employs not only INT8 kernels but also FP32, probably because there is not INT8 implementation for some RPN layers. Thus, the execution improvement is lower than expected since mixed precisions are used together.

Comparing EDP_{rr} among different data precision, there is a great improvement when using FP16. However, the enhancement is not as significant when using INT8 due to the fact that also FP32 is being used as previously discussed.

Finally, we consider the RPN with FP16 data precision the best option due to lower energy consumption and inference time. Thus, the total energy consumption for our approach using FP16 quantization is 6.07 mJ and the total time is 79.77 ms taking into account preprocessing, OFRP+RPN and NMS values. In this case, as the OFRP takes longer than the RPN, it is a bottleneck for the

Table 2. Hardware optimizations: Data quantization. Each row shows different data precision. Columns represent inference time and energy consumption together with EDP (Energy Delay Product) value for GPU. Finally, EDP_{rr} (Energy Delay Product reduction ratio) with respect to the FP32 model is included in the last column.

Quantization	Time (ms)	Energy (mJ)	EDP	EDP_{rr}
FP32	68.14	8.39	571.38	-
FP16	30.90	2.03	62.79	**9.10**×
INT8	52.91	5.52	292.06	1.96×

[3] NVIDIA Nsight Compute documentation can be consulted: https://developer.nvidia.com/nsight-compute.

parallel OFRP+RPN processing. In terms of frames per second (FPS), taking into account that during 79.77 ms we are processing 15 video frames, we are able to process 188 FPS.

Deep Learning Accelerator. Another hardware optimization available on Jetson Xavier is the use of both DLA and GPU to perform model inference using TensorRT. However, it has limitations in the amount and kind of layers supported. Our RPN is too large to be completely mapped on DLA. In this case, TensorRT implementation reaches an internal state which indicates that DLA has exceeded the number of layers in the network that it can handle, so the remaining layers are mapped in the GPU. Moreover, other models cannot be entirely mapped on DLA due to layer incompatibilities[4].

On the other hand, DLA data precision is currently limited to networks running in FP16 and INT8 modes. In our case, INT8 is not available due to previously commented layers incompatibilities with this data precision. Therefore we have only deployed our model on DLA using FP16. The results using DLA and GPU can be observed in the fourth row of Table 3. If we compare inference time and energy consumption with its GPU analogues in FP16 (third row in Table 3), DLA makes the inference process much slower, which increases the energy consumption with respect to GPU only. Thus, due to its lower computational capacity and data transfer rate between GPU and DLA, this combination also consumes more energy and time. However, it allows freeing GPU resources which could be employed for other computations if required.

Comparing EDP reduction ratios (EDP_{rr}) among baseline models without TensorRT optimizations and FP16 models using TensorRT, we can observe significant reduction ratios when using TensorRT and FP16 data precision. For the FP16 model on GPU, it achieves a value of 17.67× with respect to the baseline model and a value of 2.18× for GPU and DLA implementation.

Table 3. Hardware optimizations: DLA study. Columns represent inference time and energy consumption for DLA and GPU together with EDP (Energy Delay Product) for RPN. Finally, EDP_{rr} (Energy Delay Product reduction ratio) with respect to the baseline model is included in the last column. Each row represents a different model: baseline model without optimizations, FP16 model on GPU and FP16 model on DLA and GPU.

Model	Time (ms)	Energy (mJ)			EDP	EDP_{rr}
		GPU	DLA	Total		
Baseline	87.49	12.68	-	12.68	1109.23	-
FP16 on GPU	30.90	2.03	-	2.03	62.79	**17.67**×
FP16 on DLA - GPU	71.41	4.35	2.78	7.13	509.46	2.18×

[4] NVIDIA TensorRT documentation can be consulted to find mapping incompatibilities: https://docs.nvidia.com/deeplearning/tensorrt/archives/tensorrt-841/developer-guide/index.html.

Region Proposal Comparative. In this experiment, we evaluate the performance of each region proposal algorithm (RPN and OFRP) using the CorLoc metric, comparing the annotated ground-truth with the proposals obtained from each approach. Furthermore, we evaluate the performance of each one based on the category object to be detected.

In Table 4, we show the CorLoc results (where higher is better) for our two proposal algorithms. Concerning the RPN algorithm, we have four combinations: FP32, FP16, INT8 using GPU and FP16 using GPU with DLA. Finally, the combination of both has been added. The standard value of 0.5 is used as IoU threshold for the RPN algorithm and 0.1 is used for OFRP. OFRP requires a smaller threshold because Farneback optical flow is computed every 15 frames (since it is the frame step in the dataset), which generates large flow vectors which make the bounding boxes too big in comparison to the ground-truth bounding boxes. With regard to the results, RPN produces more accurate regions than OFRP due to OF is only able to detect moving objects. The RPN that uses INT8 data precision achieves the best results, thus this RPN together with OFRP is the best combination. Note that this improvement in the results for INT8 is because we have fine-tuned the model using PyTorch *quantization aware training*, as commented previously, during four epochs improving the accuracy of the model. The other combinations obtain almost the same CorLoc value.

Table 4. CorLoc results for different data quantization. Each row shows the CorLoc results using only RPN (second column), only OFRP (third column) or both RPN+OFRP (last column) for different data quantization.

Quantization	RPN	OFRP	RPN + OFRP
FP32 GPU	19.60	3.92	23.52
FP16 GPU	19.58	3.87	23.45
INT8 GPU	21.46	3.87	**25.33**
FP16 GPU + DLA	19.58	3.93	23.51

In addition, we measure the CorLoc metric over the true positive set of each class. We obtain the CorLoc metric separately (second and third rows) in Table 5) for each algorithm and each class. Finally, the last row in Table 5 shows the results using both region proposal algorithms together. For brevity, we only show the results per class for the best RPN algorithm, i.e. using INT8 data precision and GPU. It can be observed that RPN proposes better regions for all classes because most of the objects remain stationary, thus, there is no optical flow in this situation. The major contribution of the optical flow is for the person class as very small bounding boxes are generated and sometimes may not be detected by the RPN algorithm. Finally, the OFRP algorithm increases the CorLoc metric of 6 classes and allows us to obtain the best results in our pipeline.

Table 5. CorLoc results per class using true positives. Each row represents a different proposal algorithm and each column represents a different class.

Precision	Car	ft	Fuel	lgm	lg	mb	pe	pl	pb	st	Van
OFRP	8.5	8.0	5.5	2.3	12.0	9.2	36.3	1.4	7.4	37.8	86.7
RPN INT8	**67.6**	97.8	**97.8**	31.2	**15.2**	56.3	37.9	**98.1**	13.3	66.5	**93.3**
OFRP+RPN INT8	**67.6**	**97.9**	**97.8**	**31.3**	**15.2**	**56.7**	**39.7**	**98.1**	**14.0**	**67.1**	**93.3**

5 Conclusions

We have proposed a real-time unsupervised object localization approach using a low-power device for airport video surveillance. Our method has two principal components which produce the bounding boxes: a region proposal network (RPN) and an optical flow region proposal (OFRP). We have evaluated our pipeline on video sequences of the *apron area* of an airport showing that our approach is able to localize objects that appear on videos. Moreover, hardware optimizations allow us to improve energy consumption and inference time in order to obtain real-time localization in addition to low consumption. Regarding the CorLoc metric, it can be observed that the combination of OFRP and RPN improves the results obtained using only one of them. Furthermore, with regard to hardware optimizations, we have demonstrated that the use of different data precisions i.e. FP16 or INT8 or the employment of GPU together with DLA can achieve less energy consumption and inference time than the baseline pipeline.

As future work, we plan to use these localizations to obtain descriptors and group them using a clustering algorithm in order to find similar objects. Finally, we want to label these clusters and a human may optionally revise some samples from each cluster.

References

1. Ahmed, S., Bons, M.: Edge computed NILM: a phone-based implementation using mobilenet compressed by tensorflow lite. In: NILM, pp. 44–48 (2020)
2. Cai, Z., Vasconcelos, N.: Cascade R-CNN: delving into high quality object detection. In: CVPR, pp. 6154–6162 (2018)
3. Chen, M., et al.: MoLoc: unsupervised fingerprint roaming for device-free indoor localization in a mobile ship environment. IEEE Internet Things J. **7**(12), 11851–11862 (2020)
4. Chen, X., Li, H., Zhou, C., Liu, X., Wu, D., Dudek, G.: Fidora: robust wifi-based indoor localization via unsupervised domain adaptation. IEEE Internet Things J. **9**(12), 9872–9888 (2022)
5. Cheng, Y., Wang, D., Zhou, P., Zhang, T.: A survey of model compression and acceleration for deep neural networks. arXiv preprint:1710.09282 (2017)
6. David, R., et al.: Tensorflow lite micro: embedded machine learning for tinyml systems. PMLR **3**, 800–811 (2021)

7. Deng, L., Li, G., Han, S., Shi, L., Xie, Y.: Model compression and hardware acceleration for neural networks: a comprehensive survey. Proc. IEEE **108**(4), 485–532 (2020)
8. Farnebäck, G.: Two-frame motion estimation based on polynomial expansion. In: Bigun, J., Gustavsson, T. (eds.) SCIA 2003. LNCS, vol. 2749, pp. 363–370. Springer, Heidelberg (2003). https://doi.org/10.1007/3-540-45103-X_50
9. Felzenszwalb, P.F., Girshick, R.B., McAllester, D., Ramanan, D.: Object detection with discriminatively trained part-based models. IEEE PAMI **32**(9), 1627–1645 (2010)
10. Girshick, R., Donahue, J., Darrell, T., Malik, J.: Rich feature hierarchies for accurate object detection and semantic segmentation. In: CVPR, pp. 580–587 (2014)
11. Gong, G., Wang, X., Mu, Y., Tian, Q.: Learning temporal co-attention models for unsupervised video action localization. In: CVPR, pp. 9819–9828 (2020)
12. Gudovskiy, D., Ishizaka, S., Kozuka, K.: CFLOW-AD: real-time unsupervised anomaly detection with localization via conditional normalizing flows. In: WACV 2022, pp. 98–107
13. He, K., Zhang, X., Ren, S., Sun, J.: Deep residual learning for image recognition. In: CVPR, pp. 770–778 (2016)
14. Jeong, E., Kim, J., Tan, S., Lee, J., Ha, S.: Deep learning inference parallelization on heterogeneous processors with TensorRT. IEEE Embed. Syst. Lett. **14**(1), 15–18 (2021)
15. Li, Y., Hu, X., Zhuang, Y., Gao, Z., Zhang, P., El-Sheimy, N.: Deep reinforcement learning (DRL): another perspective for unsupervised wireless localization. IEEE Internet Things J. **7**(7), 6279–6287 (2019)
16. Liang, T., Glossner, J., Wang, L., Shi, S., Zhang, X.: Pruning and quantization for deep neural network acceleration: a survey. Neurocomputing **461**, 370–403 (2021)
17. Liu, Z., Lin, Y., Cao, Y., Hu, H., Wei, Y., Zhang, Z., Lin, S., Guo, B.: Swin transformer: hierarchical vision transformer using shifted windows. In: ICCV, pp. 10012–10022 (2021)
18. Liu, Z., Wang, Y., Han, K., Zhang, W., Ma, S., Gao, W.: Post-training quantization for vision transformer. NeurIPS **34**, 28092–28103 (2021)
19. Ma, X., Ji, K., Xiong, B., Zhang, L., Feng, S., Kuang, G.: Light-yolov4: an edge-device oriented target detection method for remote sensing images. IEEE J. Sel. Top. Appl. Earth Obs. Remote Sens. **14**, 10808–10820 (2021)
20. Mathew, M., Desappan, K., Kumar Swami, P., Nagori, S.: Sparse, quantized, full frame cnn for low power embedded devices. In: CVPR (2017)
21. McMahan, B., Moore, E., Ramage, D., Hampson, S., y Arcas, B.A.: Communication-efficient learning of deep networks from decentralized data. In: Artificial Intelligence and Statistics, pp. 1273–1282. PMLR (2017)
22. Park, E., Yoo, S., Vajda, P.: Value-aware quantization for training and inference of neural networks. In: ECCV, pp. 580–595 (2018)
23. Qasaimeh, M., et al.: Benchmarking vision kernels and neural network inference accelerators on embedded platforms. J. Syst. Architect. **113**, 101896 (2021)
24. Ren, S., He, K., Girshick, R., Sun, J.: Faster R-CNN: towards real-time object detection with region proposal networks. In: NIPS, pp. 91–99 (2015)
25. Ruiz-Barroso, P., Castro, F.M., Delgado-Escaño, R., Ramos-Cózar, J., Guil, N.: High performance inference of gait recognition models on embedded systems. Sustain. Comput. Inf. Syst. **36**, 100814 (2022)
26. Saddik, A., Latif, R., Elhoseny, M., Elouardi, A.: Real-time evaluation of different indexes in precision agriculture using a heterogeneous embedded system. Sustain. Comput. Inf. Syst. **30**, 100506 (2021)

27. Seichter, D., Köhler, M., Lewandowski, B., Wengefeld, T., Gross, H.M.: Efficient RGB-D semantic segmentation for indoor scene analysis. In: ICRA, pp. 13525–13531. IEEE (2021)
28. Suzuki, S., et al.: Topological structural analysis of digitized binary images by border following. CVGIP **30**(1), 32–46 (1985)
29. Tao, Z., Li, Q.: eSGD: commutation efficient distributed deep learning on the edge. HotEdge, 6 (2018)
30. Viola, P., Jones, M., et al.: Rapid object detection using a boosted cascade of simple features. In: CVPR, vol. 1, pp. 511–518 (2001)
31. Wang, C.Y., Bochkovskiy, A., Liao, H.Y.M.: Scaled-yolov4: scaling cross stage partial network. In: CVPR, pp. 13029–13038 (2021)
32. Wang, C.Y., Bochkovskiy, A., Liao, H.Y.M.: Yolov7: Trainable bag-of-freebies sets new state-of-the-art for real-time object detectors. arXiv: 2207.02696 (2022)
33. Xia, X., et al.: TRT-ViT: tensorrt-oriented vision transformer. arXiv preprint:2205.09579 (2022)
34. Zhai, X., Kolesnikov, A., Houlsby, N., Beyer, L.: Scaling vision transformers. In: CVPR, pp. 12104–12113 (2022)
35. Zhao, K., et al.: Distribution adaptive int8 quantization for training CNNs. In: Proceedings of the AAAI Conference on Artificial Intelligence, vol. 35, pp. 3483–3491 (2021)
36. Zimmerer, D., Isensee, F., Petersen, J., Kohl, S., Maier-Hein, K.: Unsupervised anomaly localization using variational auto-encoders. In: Shen, D., et al. (eds.) MICCAI 2019. LNCS, vol. 11767, pp. 289–297. Springer, Cham (2019). https://doi.org/10.1007/978-3-030-32251-9_32

Identifying Thermokarst Lakes Using Discrete Wavelet Transform–Based Deep Learning Framework

Andrew Li(✉), Jiahe Liu, Olivia Liu, and Xiaodi Wang

Western Connecticut State University, Danbury, CT 06810, USA
andrew.li.application@gmail.com

Abstract. Thermokarst lakes serve as key signs of permafrost thaw, and as point sources of CH_4 in the present and near future [17]. However, detailed information on the distribution of thermokarst lakes remains sparse across the entire permafrost region on the Qinghai-Tibet Plateau (QTP). In this research, we developed the first discrete wavelet transform (DWT) based dual input deep learning (DL) model using a convolutional neural network (CNN) framework to automatically classify and accurately predict thermokarst lakes. We created a new 3-way tensor dataset based on raw image data from more than 500 Sentinel-2 satellite lake images and decomposed those images using state-of-the-art M-band DWTs. We also incorporated non-image feature data for various climate variables. The special data treatment adds additional features and improves validation accuracy by up to 17%. As our data pre-processing does not require any manual polygon tracing, our method is more robust and can be upscaled easily without having to collect field data. (The code and confusion matrices not present in this paper can be found in this GitHub repository: https://github.com/jliu2006/pingo)

Keywords: climate change · permafrost thaw · thermokarst lake identification · discrete wavelet transform · deep learning · convolutional neural network

1 Introduction

1.1 Background

One of climate change's less-visible consequences is rapid permafrost thaw [6]. Permafrost serves as Earth's largest terrestrial carbon sink, underlying 24% of northern land and containing 1,600 billion tonnes of carbon, twice the amount present in the atmosphere [2,5]. Not only does thermokarst terrain serve as a

A. Li, J. Liu and O. Liu—These authors made equal contribution to this research project and share first authorship.

Supplementary Information The online version contains supplementary material available at https://doi.org/10.1007/978-3-031-36616-1_38.

A. Pertusa et al. (Eds.): IbPRIA 2023, LNCS 14062, pp. 479–489, 2023.
https://doi.org/10.1007/978-3-031-36616-1_38

prominent indicator of permafrost thaw, but thermokarst lakes emit significant amounts of greenhouse gases (GHGs) [17]. However, due to the diversity of their characteristics and the difficulty associated with collecting widespread field data, thermokarst terrain can be difficult to classify.

Other studies have explored the classification of thermokarst landforms. However, these models maintain a manual component in either the post-processing or training stages. Most image-based classification models only use RGB channels and a few infrared light channels. The most frequently used infrared-channel-based indices are the Normalized Difference Water Index (NDWI) and the Modified Normalized Difference Water Index (MNDWI) [19].

More extensive pre-processing has yet to be applied to classifying thermokarst landforms. In fact, some studies have omitted additional data pre-processing altogether [1,12,16]. The few studies using Sentinel data that had performed pre-processing only applied the Sentinel Application Platform (SNAP) [15,18]. These studies do not extract the full extent of pre-processing benefits.

Over the past few years, DL has found applications in permafrost research, from predicting thermokarst landslide susceptibility [16], to mapping retrogressive thaw slumps (RTS) [4], to mapping lake ice [13]. However, we found only two studies that took advantage of CNNs [3,12]. The first study investigated the applicability of DL to classifying RTS. The second study utilized a temporal CNN (TempCNN) to map floating lake ice, but they chose only to employ the temporal dimension, so their model remained one-dimensional.

Despite the use of advanced ML to extract thermokarst lake boundaries, investigations into smaller thermokarst lakes have been poorly resolved. However, in the Arctic circle, small thermokarst lakes were found to be the most active CH_4 generators [10]. It can be reasonably assumed that surface area is not strongly correlated with GHG emissions, and the omission of smaller lakes from inventories could lead to large underestimates of GHG emissions.

2 Methods

2.1 Data Collection

Lake Selection and Distribution. We began our methodology workflow by constructing our own datasets of lakes in the QTP region bounded by longitudes 67–104° E and latitudes 27–46° N. We determined our ground truth labels using a preexisting 2020 thermokarst lake inventory from Wei et al. [15] and a preexisting 2018 glacial lake inventory from Wang et al. [14]. We selected glacial lakes as our non-thermokarst lake type because unlike thermokarst lakes, glacial lakes do not release greenhouse gases. The two lake types are also mutually exclusive, and a glacial lake cannot also be a thermokarst lake. Furthermore, glacial lakes are of similar size as thermokarst lakes, which means that they may be more difficult to distinguish using other methods [15]. These inventories provided coordinates for 114,420 thermokarst lakes and 30,121 glacial lakes, respectively.

We removed glacial lakes from the Altay and Sayan region from our dataset due to excessive distance from the study area and selected lakes in our inventories with surface areas between $0.2\,\text{km}^2$ and $0.5\,\text{km}^2$. These steps left us with coordinates for 1,030 thermokarst lakes and 1,015 glacial lakes, which we used to acquire image data for each lake.

Image Data. We chose to use Sentinel-2 satellite imagery because it has the highest resolution available and captures images at high frequency. As both lake inventories contained coordinates of the centroids of every lake, we were able to extract centered satellite data for a certain bounding box around each lake.

For each lake in our lake collection, we obtained image data for the 1,440 m $\times$ 1,440 m area around it. Because the spatial resolution of our image data was 10 m $\times$ 10 m, our extracted images of each lake contained 144 $\times$ 144 pixels. Due to the range of lake sizes, certain entries in our dataset contained multiple lakes within the given 144 $\times$ 144 pixels.

We used the red, green, blue, near-infrared (NIR), short-wave infrared 1 (SWIR 1) and short-wave infrared 2 (SWIR 2) channels of the Level-2A (Bottom of Atmosphere reflectance) product using high-resolution satellite imagery from the Sentinel-2 MSI. This data was downloaded using the Google Earth Engine API, using Google Earth Engine's `COPERNICUS/S2_SR` product.

Along with these 6 channels, we added 3 spectral indices: the NDWI, the NDVI, and the Brightness Index (BI). These indices were calculated using the formulas in [15].

These three indices can help highlight characteristics of water, vegetation, and soil, respectively. The NDWI and the NDVI can assist with mapping surface water bodies [11].

Because some satellite images were obscured by clouds, we manually selected non-cloudy images from each lake's collection of images between August 1 and October 1, 2020. To avoid biases and ensure a random, representative sample, we first shuffled the lake inventories before manually selecting from them. Each satellite image from the time range was displayed, and we manually selected images where the lake was not blocked by clouds. If we could not find a good picture for a lake, we skipped it entirely. This could skew our data to contain fewer lakes in more cloudy areas, as we removed 47 thermokarst lakes and 45 glacial lakes, approximately 18.3% of the lakes we looked at in total. We performed multiple two independent means two-tailed t-tests on each of our population and sample means of our non-image climate feature data and additional features and found that their distribution to be the same or very similar. Therefore, we believe our sample is representative of the all the lakes that met our area criteria.

In total, we selected 252 thermokarst lakes and 251 glacial lakes to represent our training and validation datasets, about a quarter of the lakes we had originally identified in our study area.

Non-image Climate Data. We incorporated the monthly averaged ERA5-Land dataset from the Copernicus Climate Data Store to supplement our image inputs [9]. This dataset is composed of monthly averages of the hourly ERA5-Land dataset.

The spatial resolution of this data is $0.1^\circ \times 0.1^\circ$, or approximately $81\,\text{km}^2$. Because of the limited resolution of the ERA5-Land dataset, we assign a single data point of variable values to each lake, i.e., the values for the pixel that each lake center falls in. Due to monthly weather data variability between August and September of 2020, our final input values for each lake represent the mean of the value between the two months.

From this dataset, we chose 17 variables, which fell into these main categories:

- Temperature of air, lake, soil, and ice
- Water movement/phase change (precipitation, runoff, snowmelt, and evaporation)
- Snow and ice properties (albedo, depth, coverage, etc.)
- Leaf area index

While no single variable is enough to distinguish between classes of lakes, a lake possessing an outlier value for one or more variables can strongly suggest which class it belongs to. Thus, these variables can help provide additional insight for the model. Variables such as lake bottom temperature, snowmelt, and soil temperature have intuitive reasons why they should help the most in distinguishing between lake classes the most because their distributions may differ between the two classes of lakes.

2.2 Data Pre-processing

M-Band DWT. A family of orthogonal M-band DWT were used to decompose each channel of our images into M^2 different frequency components, or sub-images. DWTs can be useful to obtain some hidden information of an image by separating low-frequency from high-frequency parts. An M-band DWT is determined by a filter bank consisting of M filters ($M \geq 2$), including a low-pass filter α and $M-1$ high-pass filters $\beta^{(j)}$ for $j = 1, \ldots, M-1$.

An M-band wavelet filter bank that has N vanishing moments (or is N-regular) must have its filters satisfy the conditions given in [7]. Intuitively, wavelets with more vanishing moments tend to be smoother and have longer filters.

Since we were dueling with multi-channel images, we applied DWT to each channel separately. Due to higher-band DWTs decompose data into more frequency components and thus may capture more extra detail information, we expect that the highest-band DWTs perform the best. Thus, along with these 2-band Daubechies-6 and -8 wavelets we added a 3-band 2-regular, a 4-band 2-regular, and a 4-band 4-regular wavelet.

Figure 1a is a 144×144 pixel image of a thermokarst lake. Using the 3-band 2-regular wavelet, it is decomposed into approximation and detail components in Figs. 1b and 1c. Figure 1b shows the full result of the wavelet decomposition, with the approximation component in the top-left corner. Because we applied a 3-band DWT to the image, the image is decomposed into 9 48×48-pixel sub-images (components). Because the pixel values of detail components are very small compared to that of the approximation component, the Fig. 1c is the same as the Fig. 1b but with the RGB values multiplied by 64. The detail components directly to the right of or below the approximation component show horizontal or vertical detail, respectively. The other detail components are various diagonal detail components.

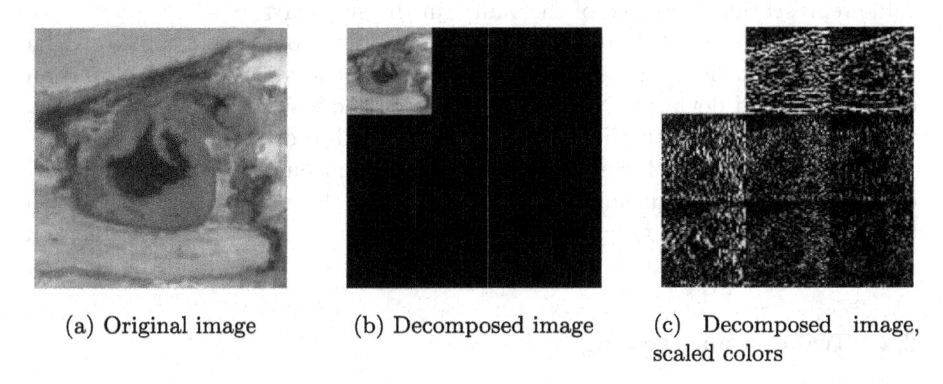

(a) Original image (b) Decomposed image (c) Decomposed image, scaled colors

Fig. 1. An image of a thermokarst lake and the results of applying a DWT to it.

Feature Normalization. After decomposing our image dataset, we also individually normalized each channel of each decomposed image so that the mean of the feature values becomes 0 and the standard deviation becomes 1.

2.3 Model Framework

Convolutional Neural Network. A Convolutional Neural Network (CNN) is a type of Artificial Neural Network (ANN) that is generally used when features need to be learned from input data [8]. In addition to the fully connected layers of an ANN, a CNN makes use of a convolutional layer and a pooling layer. Each channel of our input data is filtered through convolutional layers which each use a $Q \times Q$ convolution window. Starting from the top-left corner, the convolutional layer takes a $Q \times Q$ array of input values from the channel and multiplies it with the corresponding weighted values of the convolution window. This filtering process is repeated for the entire spatial channel as the window is shifted.

The first step in our model framework was to extract features from our image dataset using two such convolutional layers. We chose to use 9 convolutional windows (filters) to match the 9 channels in our input dimension (batch size) $\times$ $144 \times 144 \times 9$, and we specified a large convolutional window shape of $(9, 9)$ to allow our model to prioritize the detection of large land features (i.e., bodies of water) over smaller details. We set the stride value of each convolutional layer to $(1, 1)$ to ensure that each output was the same shape as its input tensor. We selected the Rectified Linear Unit (ReLU) function as the activation function of each convolutional layer, which outputs zeroes for all negative input values and does not change nonnegative input values.

After each convolutional layer, we downsampled our spatial data using average pooling layers with a pool size of $(3, 3)$ and a stride length of 3. Each 3×3 subarray in each channel of our spatial data was therefore condensed into one value representing the mean of the values in the subarray.

Following each average pooling layer, we normalized our downsampled spatial data with batch normalization layers to have a mean value close to 0 and an output standard deviation close to 1, similar to the feature normalization that was performed on our spatial data before it was fed into the model.

Finally, we flattened the extracted features from our CNN layers and concatenated them with our non-image feature data to form a single input. We fed the data through three fully connected layers and a sigmoid activation function to reach a binary output: 1 for a thermokarst lake and 0 for a glacial lake.

2.4 Training and Testing

We split our dataset into training and validation datasets using sklearn's train_test_split so that 70% of our data was used to train our model while the remaining 30% was used to evaluate performance. Then, we developed seven models to be trained on seven modifications of our datasets. This included a true control, which only used non-decomposed image data, a dual input control which included our non-image feature dataset, and five experimental models in which we tested the effects of five distinct DWTs on our image data: Daubechies-6 (db3), Daubechies-8 (db4), a 3-band 2-regular wavelet (wv32), a 4-band 2-regular wavelet (wv42), and a 4-band 4-regular wavelet (wv44).

3 Results and Model Evaluation

After training our model, we tested its classification performance on our validation dataset and recorded each model's confusion matrix and receiver operating characteristics (ROC) curve.

3.1 Confusion Matrices

We produced confusion matrices for each model and collected the results into Table 1. In our study, thermokarst lakes had a positive label and glacial lakes had a negative label. For example, a true positive would be a thermokarst lake that our model correctly classified as a thermokarst lake.

Table 1. A table summary of the confusion matrices gathered for each of our seven models. The more true positives and negatives produced by a model, the better. The fewer false positives and negatives produced, the better. The best results for each classification across all seven models are highlighted in bold.

	True Positives	False Positives	True Negatives	False Negatives
True control	64	31	45	11
Dual input control	**67**	26	50	**8**
Daubechies-6 (db3)	63	15	61	12
Daubechies-8 (db4)	**67**	20	56	**8**
3-band 2-regular (wv32)	63	15	61	12
4-band 2-regular (wv42)	63	**4**	**72**	12
4-band 4-regular (wv44)	62	12	64	13

3.2 Receiver Operating Characteristic (ROC) Curves

Figures 2, 3, 4, and 5 show the ROC curves of our models. The area under an ROC curve is a measurement of how well a model is able to distinguish between two classes; generally, ROC curves that resemble right angles in the top left indicate better performance. Our 4-band 2-regular DWT model performed the best under this metric, with the highest area under the ROC curve (Table 2).

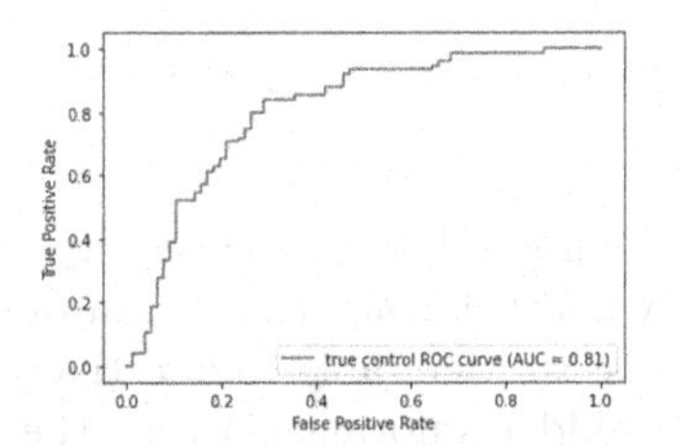

(a) The ROC curve of the true control model.

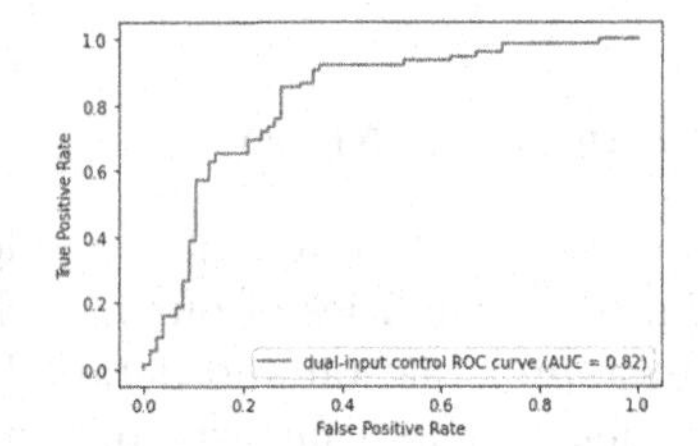

(b) The ROC curve of the dual input control model.

Fig. 2. The ROC curves for models not using DWTs.

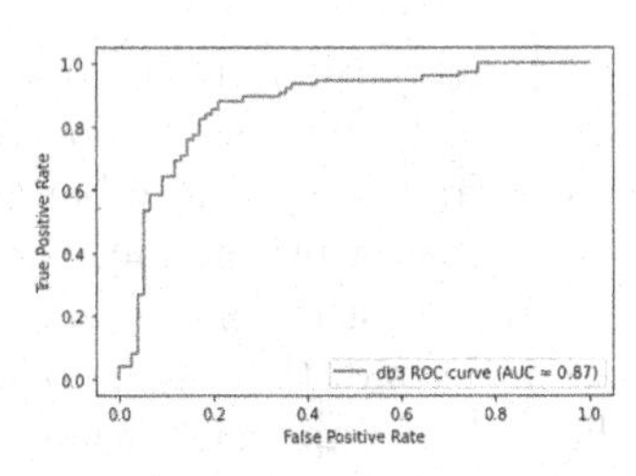

(a) The ROC curve of the model using the db3 DWT

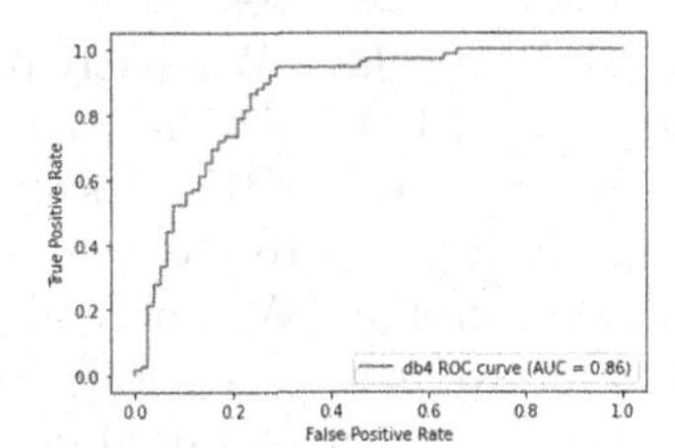

(b) The ROC curve of the model using the db4 DWT

Fig. 3. The ROC curves for the models using the Daubechies DWTs

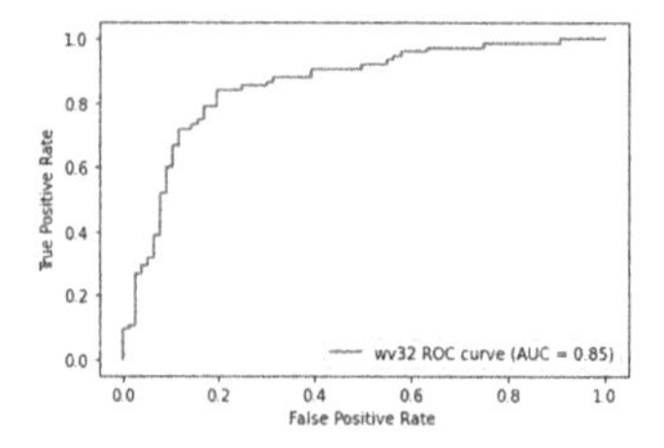

Fig. 4. The ROC curve of the model using the wv32 DWT

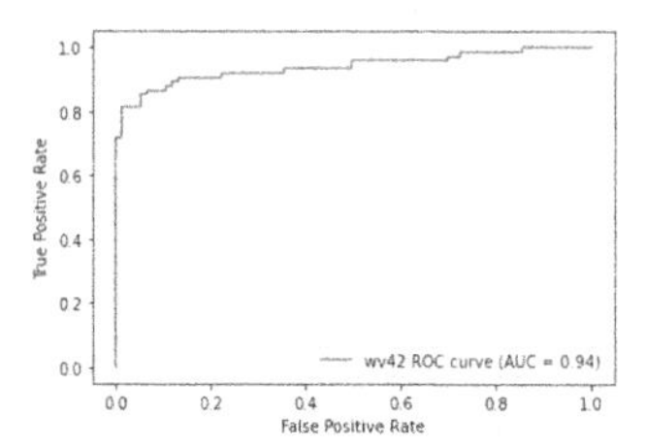

(a) The ROC curve of the model using the wv42 DWT. The model effectively distinguished between glacial and thermokarst lakes.

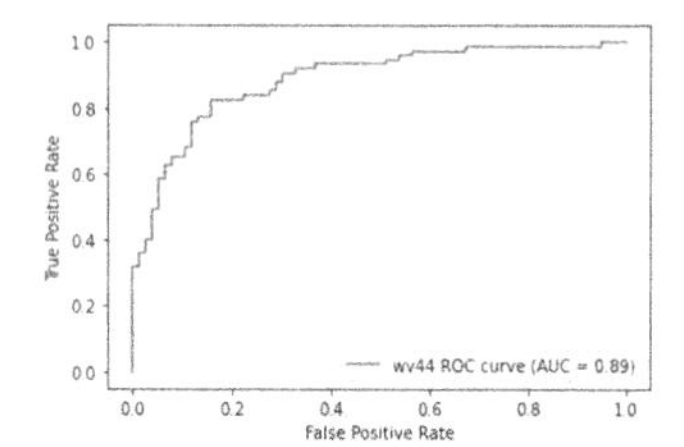

(b) The ROC curve of the model using the wv44 DWT

Fig. 5. The ROC curves for the models using the 3-band and 4-band DWTs

3.3 Summary of Metrics

When compared with our control models, the use of DWTs greatly improved the classification capabilities of our models. By splitting each channel in our image data into approximation and detail channels, we were able to effectively generate more features and uncover potentially hidden information that the models could use to distinguish between thermokarst and glacial lakes. The 4-band 2-regular DWT had the greatest positive impact on the performance of the model, producing an accuracy of almost 90% and an area under the ROC curve of 0.94, demonstrating that the model was able to effectively distinguish between thermokarst and glacial lakes.

The 4-band 2-regular DWT likely produced such positive results for two reasons. First, 4-band DWT divides each image into the highest number of detail channels, giving our model the most features to train from. More bands should generally produce better results, and this is suggested by the fact that as the number of bands in the DWT method increased, its corresponding model's performance in all five metrics generally increased. Second, the 2-regular aspect of this DWT entails that it produced detail channels (components) that were less smooth than the 4-band 4-regular counterpart. This is significant because the images in our dataset were relatively lower-resolution (144×144); each channel was rough, in that the variation in values from pixel to pixel was greater.

Table 2. A table summary of five metrics of evaluation calculated based on each model's testing performance. The highest values in each category are in bold.

	Accuracy	Precision	Recall	F1 Score	Area Under ROC Curve
True control	0.7219	0.6739	0.8533	0.7529	0.81
Dual input control	0.7748	0.7204	**0.8933**	0.7976	0.82
Daubechies-6 (db3)	0.8212	0.8077	0.8400	0.8235	0.87
Daubechies-8 (db4)	0.8146	0.7701	**0.8933**	0.8272	0.86
3-band 2-regular (wv32)	0.8212	0.8077	0.8400	0.8235	0.85
4-band 2-regular (wv42)	**0.8940**	**0.9403**	0.8400	**0.8873**	**0.94**
4-band 4-regular (wv44)	0.8344	0.8378	0.8267	0.8322	0.89

The smooth 4-regular DWT would have negatively impacted the quality of the resulting channels.

To test which parts of the 4-band 2-regular DWT decomposed data were the most important for our model, we extracted the weights of our model's first convolutional layer and calculated the average weight value of each layer. We found that the nine layers with the lowest average weights corresponded to the nine approximation channels that were produced from the DWT on each lake's spatial features. To test this observation, we randomly shuffled the approximation channels of our testing dataset to create noise that would be effectively useless in the model's classification. Then, we tested our model with a dataset that used those noisy channels and found that our model experienced a 5% decrease in accuracy, but its corresponding area under ROC curve only decreased by 0.02, indicating that the model trained on 4-band 2-regular DWT decomposed data relied more heavily on detail channels than approximation channels when classifying between thermokarst and glacial lakes. This result is consistent with our previous observations.

4 Conclusion and Discussion

In addition to successfully producing a relatively high accuracy classification model of thermokarst lakes based on a CNN, the hypotheses that a) incorporating wavelet transforms to decompose our image data into 3-way tensor image data and b) combining image data and non-image climate data would increase accuracy was supported. The use of non-image data and 2D DWTs greatly increased the accuracy of the model by up to 17.21%. When evaluating model accuracy, it is important to consider whether the greatest increase in accuracy was caused by the additional non-image feature data input or the use of DWTs. Although both amendments improved model accuracy compared to the true control, the wavelet decomposition impacted the model performance the most, increasing accuracy by up to 11.9% and F1 score by up to 0.09. Comparing the dual input control to the true control, the addition of the non-image climate feature data improved the model's performance less; it increased accuracy by up to 5.29% and increased the F1 score by up to 0.045.

We believe that the results obtained are significant enough for our model to be used in accurately creating annual inventories of thermokarst lakes, without the need for field data. Given the fact that our image and non-image data sources have global coverage, our model should be applicable to any region with a meaningful amount of thermokarst lakes. This would greatly reduce the cost of tracking thermokarst lakes, as the need for field equipment and research centers would be nullified.

A broader application of annual thermokarst lake inventories would be to increase precision in both global climate models and GHG budgets. Identification of thermokarst landforms and their locations is a prerequisite to understanding their GHG outputs in more detail, thus informing more accurate simulations of the Earth's climate.

Building on our classification of lakes, our next step would be forecasting certain surface dynamics of thermokarst lakes, such as change in surface area or drainage prediction. These predictions would be based off our dual input time-series image data and non-image feature data. Our current limitation is the resolution of our non-image climate data. The highest resolution dataset that includes the features we hope to include is of 9 km horizontal resolution. This would grossly generalize the feature data of thermokarst regions with that of surrounding land, which could potentially include other thermokarst or non-thermokarst water bodies that would not be accounted for.

References

1. Chen, H., Liqiang, T., Zhaocheng, G., Jienan, T., Hua, W., Peng, H.: A dynamics trend analysis method of thermokarst lakes based on the machine learning algorithm, pp. 6484–6487 (2021). https://doi.org/10.1109/IGARSS47720.2021.9554435
2. Grosse, G., Jones, B., Arp, C.: 8.21 thermokarst lakes, drainage, and drained basins, pp. 325–353 (2013). https://doi.org/10.1016/B978-0-12-374739-6.00216-5
3. Huang, L., Liu, L., Jiang, L., Zhang, T.: Automatic mapping of thermokarst landforms from remote sensing images using deep learning: a case study in the northeastern Tibetan plateau. Remote Sens. **10**(12), 2067 (2018). https://doi.org/10.3390/rs10122067
4. Huang, L., Luo, J., Lin, Z., Niu, F., Liu, L.: Using deep learning to map retrogressive thaw slumps in the Beiluhe region (Tibetan Plateau) from CubeSat images. Remote Sens. Environ. **237**, 111534 (2020). https://doi.org/10.1016/j.rse.2019.111534
5. Jin, H., Ma, Q.: Impacts of permafrost degradation on carbon stocks and emissions under a warming climate: a review. Atmosphere **12**(11), 1425 (2021). https://doi.org/10.3390/atmos12111425
6. Lara, M.J.: Driven by climate change, thawing permafrost is radically changing the arctic landscape (2022). https://www.pbs.org/newshour/science/driven-by-climate-change-thawing-permafrost-is-radically-changing-the-arctic-landscape
7. Liu, Z., Liu, T., Sun, W., Zhao, Y., Wang, X.: M-band wavelet-based imputation of scRNA-seq matrix and multi-view clustering of cells. FASEB J. **36**(S1), R5102 (2022). https://doi.org/10.1096/fasebj.2022.36.S1.R5102

8. Martinez, J.C.: Introduction to convolutional neural networks CNNs (2020). https://aigents.co/data-science-blog/publication/introduction-to-convolutional-neural-networks-cnns
9. Muñoz Sabater, J.: Era5-land monthly averaged data from 1981 to present (2019). https://doi.org/10.24381/cds.68d2bb30
10. Polishchuk, V., Polischchuk, Y.: The system of geo-simulation modeling of thermokarst lakes fields based on the log-normal distribution of their sizes. Adv. Intell. Syst. Res. **174**, 195–199 (2020). https://doi.org/10.2991/aisr.k.201029.037
11. Qiao, B., Zhu, L., Yang, R.: Temporal-spatial differences in lake water storage changes and their links to climate change throughout the Tibetan plateau. Remote Sens. Environ. **222**, 232–243 (2019). https://doi.org/10.1016/j.rse.2018.12.037
12. Shaposhnikova, M., Duguay, C.R.: Roy-Léveillée: bedfast and floating ice dynamics of thermokarst lakes using a temporal deep learning mapping approach: case study of the old crow flats, Yukon, Canada. EGUsphere **2022**, 1–36 (2022). https://doi.org/10.5194/egusphere-2022-388
13. Shaposhnikova, M.: Temporal Deep Learning Approach to Bedfast and Floating Thermokarst Lake Ice Mapping using SAR imagery: Old Crow Flats, Yukon, Canada. Master's thesis, University of Waterloo (2021). http://hdl.handle.net/10012/17414
14. Wang, X., et al.: Glacial lake inventory of high-mountain Asia in 1990 and 2018 derived from Landsat images. Earth Syst. Sci. Data **12**(3), 2169–2182 (2020). https://doi.org/10.5194/essd-12-2169-2020
15. Wei, Z., et al.: Sentinel-based inventory of thermokarst lakes and ponds across permafrost landscapes on the Qinghai-Tibet plateau. Earth Space Sci. **8**(11), 154761 (2021). https://doi.org/10.1029/2021EA001950
16. Yin, G., Luo, J., Niu, F., Lin, Z., Liu, M.: Machine learning-based thermokarst landslide susceptibility modeling across the permafrost region on the Qinghai-Tibet Plateau. Landslides **18**(7), 2639–2649 (2021). https://doi.org/10.1007/s10346-021-01669-7
17. Zandt, M.H.i., Liebner, S., Welte, C.U.: Roles of thermokarst lakes in a warming world. Cell Press **28**(10), 769–779 (2020). https://doi.org/10.1016/j.tim.2020.04.002
18. Serban, R.D., Jin, H., Serban, M., Luo, D.: Shrinking thermokarst lakes and ponds on the northeastern Qinghai-Tibet plateau over the past three decades. Permafrost Periglac. Process. **32**(4), 601–617 (2021). https://doi.org/10.1002/ppp.2127
19. Serban, R.D., et al.: Mapping thermokarst lakes and ponds across permafrost landscapes in the headwater area of yellow river on northeastern Qinghai-Tibet plateau. Int. J. Remote Sens. **41**(18), 7042–7067 (2020). https://doi.org/10.1080/01431161.2020.1752954

Object Detection for Rescue Operations by High-Altitude Infrared Thermal Imaging Collected by Unmanned Aerial Vehicles

Andrii Polukhin[1(✉)], Yuri Gordienko[1], Gert Jervan[2], and Sergii Stirenko[1]

[1] National Technical University of Ukraine "Igor Sikorsky Kyiv Polytechnic Institute", Kyiv, Ukraine
pandrii000@gmail.com
[2] Tallinn University of Technology, Tallinn, Estonia
gert.jervan@taltech.ee

Abstract. The analysis of the object detection deep learning model YOLOv5, which was trained on High-altitude Infrared Thermal (HIT) imaging, captured by Unmanned Aerial Vehicles (UAV) is presented. The performance of the several architectures of the YOLOv5 model, specifically 'n', 's', 'm', 'l', and 'x', that were trained with the same hyperparameters and data is analyzed. The dependence of some characteristics, like average precision, inference time, and latency time, on different sizes of deep learning models, is investigated and compared for infrared HIT-UAV and standard COCO datasets. The results show that degradation of average precision with the model size is much lower for the HIT-UAV dataset than for the COCO dataset which can be explained that a significant amount of unnecessary information is removed from infrared thermal pictures ("pseudo segmentation"), facilitating better object detection. According to the findings, the significance and value of the research consist in comparing the performance of the various models on the datasets COCO and HIT-UAV, infrared photos are more effective at capturing the real-world characteristics needed to conduct better object detection.

Keywords: Deep Learning · Object Detection · You Only Look Once · YOLO · Average Precision · AP · Unmanned Aerial Vehicles · UAV · Infrared Thermal Imaging

1 Introduction

Unmanned aerial vehicles (UAVs) are frequently used in many different industries, such as emergency management [1], mapping [2], traffic surveying [3], and environment monitoring [4]. Since UAVs can now load artificial intelligence (AI) algorithms as edge computing devices [5], the utility of the aforementioned applications has increased with the development of deep learning and edge computing

A. Pertusa et al. (Eds.): IbPRIA 2023, LNCS 14062, pp. 490–504, 2023.
https://doi.org/10.1007/978-3-031-36616-1_39

[6]. The rapid expansion of object detection applications has prompted numerous broad datasets to be proposed to boost algorithm training and evaluation [7–10].

This paper proposes the use of unmanned aerial vehicles (UAVs) equipped with thermal infrared (IR) cameras to locate missing individuals in wilderness settings. Detecting abandoned individuals using standard UAV technology, which lacks IR capabilities, is challenging due to a variety of factors, such as terrain, temperature, and obstacles. Infrared thermal imaging has been identified as a promising method for enhancing object detection in adverse weather conditions and challenging environments [11]. Moreover, the use of IR-equipped UAVs has the potential to facilitate rescue operations in conditions such as complete darkness, fog, and heavy rain [1,12]. However, outdated neural network models, which can result in reduced performance, inaccurate positive and false negative predictions, and difficulties in running the software on current and continuously evolving hardware, are the primary limitations of previous studies.

The significance and value of our research are to provide insights into the potential use of infrared thermal imaging on UAVs for object detection in challenging weather conditions, particularly in rescue operations. Special attention will be paid to the investigation of the dependence of some characteristics (average precision, inference time, and latency time) on different sizes of deep learning models, comparing infrared and standard datasets. This is especially important for understanding the feasibility of the usage of relatively small models for Edge Computing devices with regard to the deterioration of their performance with a decrease of model sizes in various applications described in our previous publications [13–15].

2 Background and Related Work

The deep learning methodology has sped up the development of the object detection field in recent years. The development of object detection apps has been facilitated by large datasets for object detection [16–18]. Over time, we've seen advancements in the accuracy and overall performance of object detection systems that have allowed apps to identify and classify objects more accurately.

Many datasets of aerial perspective were presented for the AI job with the UAV platform with the development of AI and the deployment of UAVs for many domains such as forest fire prevention [19], traffic monitoring [3], disaster assistance [20], and package delivery [21]. We can effectively employ object detection to save more people by using infrared imaging.

2.1 Neural Network Object Detection Methods

Convolutional neural networks (CNNs) and large-scale GPU processing have enabled deep learning to achieve remarkable success in modern computer vision [22–25]. CNNs, which concentrate on processing spatially local input to learn the visual representation, are now the de facto method for a variety of vision-related tasks. An object detection model typically consists of two parts: a backbone that extracts features from the image, and a head that predicts object classes and bounding boxes. The choice of backbone architecture for object detectors often depends on the complexity of the model and the platform on which it is intended to be run. For instance, architectures such as VGG [23] or ResNet [24] are typically employed as the backbone for detectors that run on GPU platforms due to their higher computational demands. On the other hand, models such as MobileNet [25] may serve as suitable backbones for detectors that run on CPU platforms, as they have lower computational complexity and are thus better suited for resource-constrained settings.

There are two main architectures of the head: one-stage and two-stage options. The R-CNN series, which includes the Fast R-CNN [26], and Faster R-CNN [27], is the most typical two-stage object detector. The most typical models for one-stage object detectors are YOLO [28] and SSD [29].

The YOLO model [28] family architecture is one of the best object detection algorithms known for its speed and accuracy, which can be pre-trained using the COCO dataset [30]. YOLO applies one neural network to divide the picture into areas and forecast probability and bounding boxes. The architecture of a single-stage object detector like YOLO consists of three parts: backbone, neck, and head. YOLO v5 uses CSPNet [31] as its backbone to extract important features, PANet [32] as its neck to produce feature pyramids, and a similar head to that of YOLO v4 [31] to carry out the final detection step.

2.2 Infrared Object Detection

Apart from neural network approaches [33], there are several traditional methods available for identifying objects in infrared images [34–37]. These methods primarily focus on distinguishing between three elements in infrared images: the object, the background, and the image noise. The main objective is to suppress the background and noise to enhance the object and identify it using various techniques. One such algorithm [34] employs a spatial filtering-based technique for infrared object detection, searching for various background and object gray values. The background is then selected and suppressed to enable the identification of the object. Another technique [35] incorporates shearlet-based histogram thresholding and is based on a practical image denoising approach, offering significant improvement but with a high computational cost. Traditional infrared object identification techniques often use artificially created feature extractors such as Haar [36] or HOG [37], which are effective but not robust to shifts in object diversity.

2.3 UAV Infrared Thermal Datasets

The use of UAVs equipped with infrared thermal cameras can significantly enhance mission accuracy while reducing costs and resource requirements, particularly when dealing with large volumes of data. In this context, several existing datasets have been developed for Infrared Thermal UAV object detection tasks.

For instance, the HIT-UAV dataset [7] comprises 2898 infrared thermal images extracted from 43470 frames captured by a UAV in various scenes, including information such as flight altitude, camera perspective, and daylight intensity. The FLAME dataset [8], on the other hand, includes raw heatmap footage and aerial movies captured by drone cameras, and is used for defining two well-known studies: fire classification and fire segmentation. The PTB-TIR dataset [9], containing 60 annotated sequences with nine attribute labels for pedestrian tracking, is commonly used for evaluating thermal infrared pedestrian trackers. Finally, the BIRDSAI dataset [10] is a long-wave thermal infrared dataset for Surveillance with Aerial Intelligence, which contains images of people and animals at night in Southern Africa, along with actual and fake footage to enable testing of algorithms for the autonomous detection and tracking of people and animals. Although not used in the present study, the availability of these datasets contributes to the development and evaluation of new infrared thermal UAV object detection methods.

2.4 Object Detection Metrics

Supervised object detection methods have recently produced outstanding results, leading to a demand for annotated datasets for their evaluation. A good object detector should locate all ground truth objects with high recall and recognize only relevant objects with high precision, and an ideal model would have high precision with increasing recall. Average precision (AP) summarizes the precision-recall trade-off based on expected bounding box confidence levels. Formally, the equation of AP is defined as follows:

$$AP = \int_0^1 p(r)dr,$$

where p and r are precision and recall values for the same threshold correspondingly.

3 Methodology

We conducted an analysis of the HIT-UAV dataset, which includes five categories of annotated objects: Person, Car, Bicycle, Other Vehicle, and DontCare. We used the YOLOv5 object detection algorithm and split the dataset into three sets for training, validation, and testing. Our analysis showed that the Car, Bicycle, and Person categories make up the majority of the dataset. We utilized all five categories for training but only evaluated the top three due to the shortage of training data for the OtherVehicle and DontCare categories. We also analyzed the distribution of instances across annotated object categories per image and found that most images contain fewer than 10 instances, with some outliers having more than 30 instances per image. We used all available YOLOv5 architectures and evaluated their performance with two GPUs Tesla T4. We also used various augmentation techniques, such as hue, saturation, and value modifications, random translations and scaling, horizontal flipping, and mosaic image generation, to improve the model's robustness and prevent overfitting. We provide more details on the analysis in this section.

3.1 Exploratory Data Analysis (EDA)

We trained and evaluated object detection algorithms on the HIT-UAV dataset. The dataset was split into three sets: 2008 photos for training, 287 images for validation, and 571 images for testing. In total, the training set consisted of 17,628 instances, the validation set had 2,460 instances, and the testing set contained 4,811 instances, introducing this statistics in Table 1.

Table 1. Dataset size for train, test, and validation subsets.

Subset	Images	Instances
Train	2008	17628
Validation	287	2560
Test	571	4811

The HIT-UAV dataset includes five categories of annotated objects: Person, Car, Bicycle, and Other Vehicle, which are frequently observed in rescue and search operations. Additionally, there is an "unidentifiable" category called DontCare, which is used for objects that cannot be assigned to specific classes by an annotator, particularly in cases where the objects appear in high-altitude aerial photos. It can be challenging to determine whether these objects contain something of importance, but there may be an important object present.

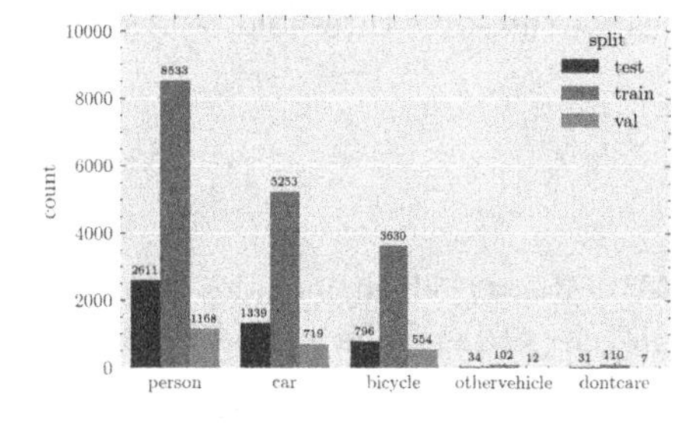

(a) Distribution of instances per category.

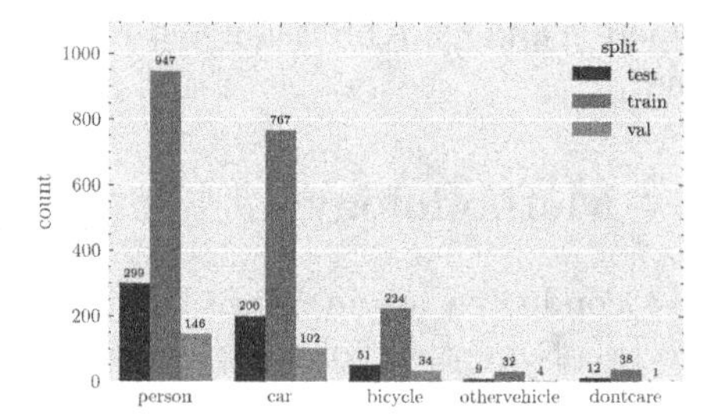

(b) Distribution of images per category.

Fig. 1. Distribution of the categories across the instances and images.

We conducted an analysis of the distribution of the five annotated object categories across instances and images, as depicted in Fig. 1a and Fig. 1b, respectively. Our analysis revealed that the Car, Bicycle, and Person categories make up the majority of the dataset. For subsequent experiments, all five categories were used for training, but only the top three categories were used for evaluation. This is because there is a shortage of training data for the OtherVehicle and DontCare categories, which limited the model's ability to learn them effectively. Consequently, the model performed poorly for these categories, resulting in an underestimation of the average AP metric across all categories.

We analyzed the distribution of instances across all annotated object categories per image, as shown in Fig. 2. Our analysis indicates that the majority of images contain fewer than 10 instances. However, there are some outliers with more than 30 instances per image, which likely correspond to crowded locations with numerous people.

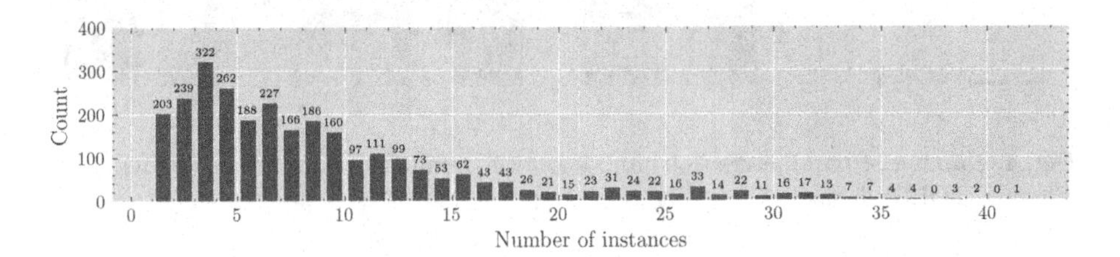

Fig. 2. Distribution of the instances per image.

Also, we have visualized the annotations for the most labeled classes Car in Fig. 3a, Person in Fig. 3b, and Bicycle in Fig. 3c.

(a) Class "Car". (b) Class "Person". (c) Class "Bicycle".

Fig. 3. Example of the annotated classes "Car", "Person", and "Bicycle".

3.2 Model Selection

In order to ensure reliable and reproducible training results, we utilized the standard and freely available YOLOv5 single-stage object detector in conjunction with the open HIT-UAV dataset. YOLOv5 offers five different sizes of architecture, ranging from the extra small (nano) size model denoted by 'n', to the small ('s'), medium ('m'), large ('l'), and extra large ('x') models.

As shown in Fig. 4, the number of parameters for each model size increases nearly exponentially.

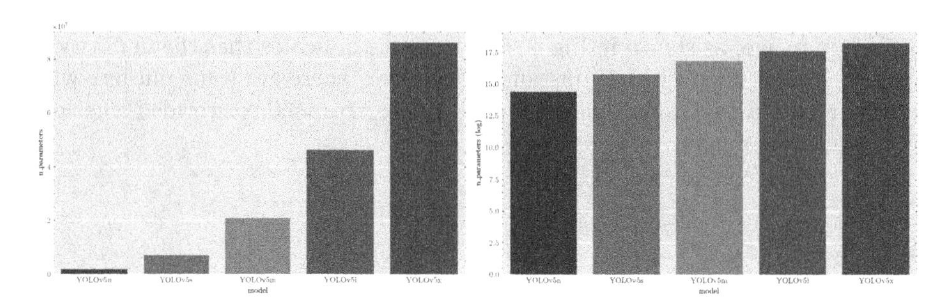

Fig. 4. Number of parameters (left) and log of the number of parameters (right) of the different YOLO v5 sizes.

3.3 Experimental Workflow

The training time required for each variant varies. We conducted an analysis of all the variants using the same dataset, augmentation hyperparameters, training configuration, and hardware. To evaluate performance, we measured the average precision, inference time (ms/image), latency time (ms/image), and frames per second for each category, as well as an average across "Person", "Car", and "Bicycle" categories.

Using the free computational resources, offered by the Kaggle Platform, we trained and evaluated the performance of all YOLO v5 architectures with two GPUs Tesla T4, which have 15 GB of memory, or 30 GB in total, CUDA Version 11.4. All models were trained to similar hyperparameters, which provided 300 training iterations and a batch size of 28. SGD is the optimizer, with a weight decay of 0.0005, a learning rate of 0.01, and a momentum of 0.937. Priorities are set at 0.05 for the box loss function, 0.5 for classification loss, and 0.1 for objectness loss. The thresholding value for IOUs is 0.2.

The YOLOv5 model employs a range of augmentation techniques to improve the performance of object detection. These techniques include the modification of hue, saturation, and value channels, random translations and scaling, horizontal flipping, and generating mosaic images. The hue channel is modified with a portion of 1.5%, while the saturation and value channels are adjusted up to 70% and 40% relative to the original value, respectively. The image is randomly translated up to 10% of its height and width, and the image size is scaled up to 50% of its original size. Additionally, the model applies horizontal flipping with a 50% chance to the image. Finally, the model generates mosaic images by combining multiple images, which is always used during training. These augmentation techniques enhance the model's robustness to different scenarios and prevent overfitting by increasing the diversity of the training dataset.

The hyperparameter values and training setup is aggregated and introduced in Table 2.

Table 2. Training setup and hyperparameters for YOLOv5 models

Parameter	Value	Augmentation
GPUs	2 × Tesla T4	–
Memory	30 GB	–
CUDA Version	11.4	–
Optimizer	SGD	–
Weight decay	0.0005	–
Learning rate	0.01	–
Momentum	0.937	–
Loss function	–	
Box	0.05	–
Classification	0.5	–
Objectness	0.1	–
IOU threshold	0.2	–
Image modifications	–	
Hue	1.5%	–
Saturation	70%	–
Value	40%	–
Random translation	–	Up to 10% of height and width
Image scaling	–	Up to 50% of original size
Horizontal flipping	–	50% chance
Mosaic images	–	Always

In addition, we compare our results with YOLOv5 models trained on the COCO 2017 dataset [38]. The models were trained for 300 epochs using A100 GPUs at an image size of 640. The training process utilized an SGD optimizer with a learning rate of 0.01 and a weight decay of 0.00005. Our training setup closely follows this configuration. These models are used to evaluate the performance of real and infrared thermal imaging.

4 Results and Their Discussion

Our evaluation of the HIT-UAV dataset reveals that YOLOv5 achieves an impressive average precision, as demonstrated by the results presented in Table 3.

We have compared the results of our model trained on the HIT-UAV dataset with the one trained on the COCO dataset. YOLOv5 (x) outperforms other variants in terms of average precision (AP), as evident from Fig. 5a and Fig. 5b. However, it is observed that YOLOv5 (x) has the longest inference and latency times. YOLOv5 (n), on the other hand, exhibits the fastest inference and latency times but performs poorly in terms of both AP and FPS. YOLOv5 (l) and YOLOv5 (m) demonstrate an optimal balance between accuracy and speed, making them a preferable choice for practical applications. Alternatively, YOLOv5 (s) can be used when speed is of higher priority over accuracy.

Table 3. The evaluation for the YOLO v5.

Model	Dataset	Person	Car	Bicycle	FPS	AP
YOLO v5 (n)	HIT-UAV	49.4%	75.4%	50.8%	90	58.5%
YOLO v5 (s)	HIT-UAV	50.8%	73.6%	52.5%	71	58.9%
YOLO v5 (m)	HIT-UAV	**51.0%**	75.1%	55.4%	44	60.5%
YOLO v5 (l)	HIT-UAV	50.9%	**75.7%**	55.4%	25	60.6%
YOLO v5 (x)	HIT-UAV	50.0%	75.4%	**57.0%**	21	**60.8%**
YOLO v5 (n)	COCO	–	–	–	90	28.0%
YOLO v5 (s)	COCO	–	–	–	71	37.4%
YOLO v5 (m)	COCO	–	–	–	44	45.4%
YOLO v5 (l)	COCO	–	–	–	25	49.0%
YOLO v5 (x)	COCO	–	–	–	21	50.7%

Across all YOLOv5 size ranges, the results obtained from the HIT-UAV dataset suggest that the Car category's AP value is significantly higher than those of other categories (Fig. 6). This could be attributed to YOLOv5's superior detection capabilities for large objects such as automobiles, as opposed to relatively smaller objects like bicycles or persons. In highly crowded images, the "Person" category's AP value is not as high as that of "Car". This is because the YOLOv5 algorithm encounters significant difficulties in such scenarios, where the model starts to miss some of the smaller features. Specifically, the model struggles to perform well in very cluttered images.

The COCO results indicate that the original YOLOv5 (n) model achieves an AP of 28%. However, when trained on the HIT-UAV dataset, the YOLOv5 (n) model reaches an AP of 58%. This suggests that aerial image information is more advantageous for detection tasks than natural photos. It is worth noting that while infrared photos are more effective in capturing certain real-world characteristics required for better object recognition, their advantage is limited to certain classes, such as detecting persons who emit heat or vehicles with metal signatures highlighted in infrared images. For other classes, RGB images could be more beneficial.

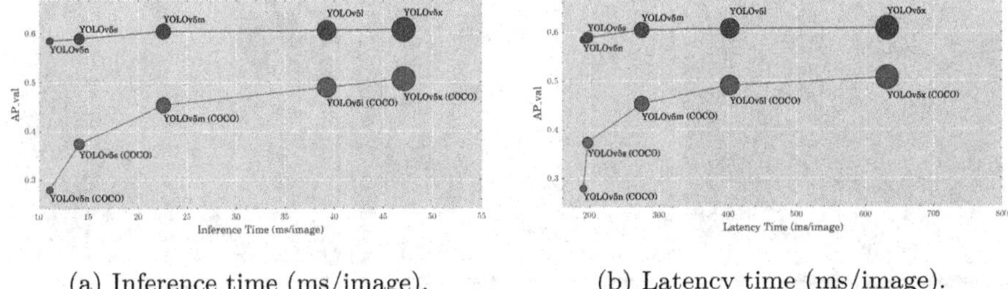

(a) Inference time (ms/image). (b) Latency time (ms/image).

Fig. 5. Comparison of the inference and latency time (ms/image) vs average precision of YOLOv5 models trained on the HIT-UAV and COCO datasets for different model sizes on GPU Tesla T4.

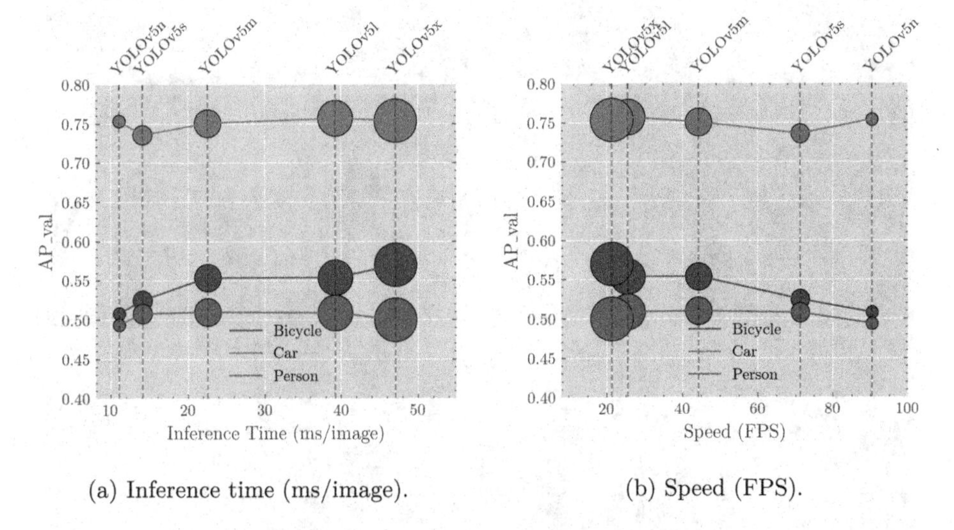

(a) Inference time (ms/image). (b) Speed (FPS).

Fig. 6. Comparison of the inference time (ms/image) and speed (FPS) vs average precision on different object categories of different model sizes on GPU Tesla T4 for HIT-UAV.

The above results indicate the following observations:

1. A significant amount of unnecessary information is removed from infrared thermal pictures, providing a "pseudo segmentation" effect and facilitating object detection. Due to the clearly visible properties of the objects in infrared thermal pictures, the typical detection model may achieve great recognition performance with few images.
2. In the infrared thermal aerial photos, we can see that the model does a good job of capturing large objects. Small objects, including persons, might easily be mistaken for infrared sensor noise, which raises the false positive rate.

The first observation is supported by our previous results on the impact of ground truth annotation (GT) quality on the performance of semantic image segmentation of traffic conditions, where the mean accuracy values of semantic

Fig. 7. Sample results of the YOLO v5. Left is the original annotations, the middle is the smallest YOLO v5 (n), and the right is the largest YOLO v5 (x).

image segmentation for coarse GT annotations are higher than for the fine GT ones [39,40]. The infrared images give similar coarse representations of objects in comparison to their actual visual appearance.

A visual comparison of the bounding box predictions for the trained models is also provided in Fig. 7.

The research shows that YOLOv5 performs better on the HIT-UAV dataset than on COCO, indicating the advantages of using IR aerial imagery over RGB imagery for object detection tasks for certain classes. However, it should be noted that HIT-UAV has a much smaller number of images and classes than COCO, making it an easier benchmark. Additionally, the HIT-UAV dataset only includes a limited number of classes, which may have influenced the reported performance difference between the two datasets. Also, the comparison between the two datasets is not entirely fair due to differences in IoU thresholds and image perspective. Therefore, it is suggested that in the future, the HIT-UAV dataset should be expanded with a larger number of classes to better represent real-world scenarios.

5 Conclusion

This paper presents an analysis of object detection training using the YOLOv5 architecture with an infrared thermal UAV dataset. The dataset comprises five categories of objects with varying sizes, locations, and numbers per image. In this study, YOLOv5 models of sizes (n), (s), (m), (l), and (x) were trained and tested using the HIT-UAV dataset. Our findings reveal that the HIT-UAV dataset shows a lower degradation of average precision with model size compared to the COCO dataset. This can be attributed to the removal of extraneous information from infrared thermal images, resulting in better object detection performance for certain classes, such as persons who emit heat or vehicles with metal signatures highlighted in infrared images. These findings have significant implications for using small deep learning models on Edge Computing devices for rescue operations that utilize HIT-UAV-like imagery, as their performance deterioration decreases with a decrease in model size. Moreover, our study shows that infrared thermal images significantly enhance object detection capabilities by filtering out unnecessary information and improving the recognition of certain classes compared to visual light images. However, for other classes, RGB images may be more effective. These benefits increase the feasibility of autonomous object detection using UAVs in crucial nighttime activities, such as city surveillance, traffic control, and person search and rescue. Further research is necessary to validate these results and expand the dataset with more images and categories to better represent real-world scenarios and facilitate a fair comparison of the model performance with HIT-UAV and COCO 2017 datasets. Overall, this algorithm's evaluation can contribute to the development of efficient object detection techniques for nighttime applications.

Acknowledgements. This research was in part sponsored by the NATO Science for Peace and Security Programme under grant id. G6032.

References

1. Boccardo, P., Chiabrando, F., Dutto, F., Tonolo, F., Lingua, A.: UAV deployment exercise for mapping purposes: evaluation of emergency response applications. Sensors **15**(7), 15717–15737 (2015)
2. de Castro, A., Torres-Sánchez, J., Peña, J., Jiménez-Brenes, F., Csillik, O., López-Granados, F.: An automatic random forest-OBIA algorithm for early weed mapping between and within crop rows using UAV imagery. Remote Sens. **10**(3), 285 (2018)
3. Kanistras, K., Martins, G., Rutherford, M.J., Valavanis, K.P.: A survey of unmanned aerial vehicles (UAVs) for traffic monitoring. In: 2013 International Conference on Unmanned Aircraft Systems (ICUAS), pp. 221–234 (2013)
4. Avola, D., Foresti, G.L., Martinel, N., Micheloni, C., Pannone, D., Piciarelli, C.: Aerial video surveillance system for small-scale UAV environment monitoring. In: 2017 14th IEEE International Conference on Advanced Video and Signal Based Surveillance (AVSS), pp. 1–6 (2017)
5. Liu, Q., Shi, L., Sun, L., Li, J., Ding, M., Shu, F.S.: Path planning for UAV-mounted mobile edge computing with deep reinforcement learning. IEEE Trans. Veh. Technol. **69**(5), 5723–5728 (2020)
6. Wang, F., Zhang, M., Wang, X., Ma, X., Liu, J.: Deep learning for edge computing applications: a state-of-the-art survey. IEEE Access **8**, 58322–58336 (2020)
7. Suo, J., Wang, T., Zhang, X., Chen, H., Zhou, W., Shi, W.: HIT-UAV: a high-altitude infrared thermal dataset for unmanned aerial vehicles (2022)
8. Shamsoshoara, A.: The FLAME dataset: aerial Imagery Pile burn detection using drones (UAVs) (2020)
9. Liu, Q., He, Z., Li, X., Zheng, Y.: PTB-TIR: a thermal infrared pedestrian tracking benchmark. IEEE Trans. Multimedia **22**(3), 666–675 (2020)
10. Bondi, E., et al.: BIRDSAI: a dataset for detection and tracking in aerial thermal infrared videos. In: 2020 IEEE Winter Conference on Applications of Computer Vision (WACV), pp. 1736–1745 (2020)
11. Beyerer, J., Ruf, M., Herrmann, C.: CNN-based thermal infrared person detection by domain adaptation. In: Dudzik, M.C., Ricklin, J.C. (eds.) Autonomous Systems: Sensors, Vehicles, Security, and the Internet of Everything,Orlando, USA, p. 8. SPIE (2018)
12. Levin, E., Zarnowski, A., McCarty, J.L., Bialas, J., Banaszek, A., Banaszek, S.: Feasibility study of inexpensive thermal sensors and small UAS deployment for living human detection in rescue missions application scenarios. Int. Arch. Photogram. Remote Sens. Spat. Inf. Sci. **XLI-B8**, 99–103 (2016)
13. Gordienko, Y., et al.: Scaling analysis of specialized tensor processing architectures for deep learning models. Deep Learn. Concepts Archit. 65–99 (2020)
14. Gordienko, Y., et al.: "Last mile" optimization of edge computing ecosystem with deep learning models and specialized tensor processing architectures. In: Advances in computers, vol. 122, pp. 303–341. Elsevier (2021)
15. Taran, V., Gordienko, Y., Rokovyi, O., Alienin, O., Kochura, Y., Stirenko, S.: Edge intelligence for medical applications under field conditions. In: Hu, Z., Zhang, Q., Petoukhov, S., He, M. (eds.) ICAILE 2022. LNDECT, vol. 135, pp. 71–80. Springer, Cham (2022). https://doi.org/10.1007/978-3-031-04809-8_6
16. Lin, T.-Y., et al.: Microsoft COCO: common objects in context. In: Fleet, D., Pajdla, T., Schiele, B., Tuytelaars, T. (eds.) ECCV 2014. LNCS, vol. 8693, pp. 740–755. Springer, Cham (2014). https://doi.org/10.1007/978-3-319-10602-1_48

17. Geiger, A., Lenz, P., Urtasun, R.: Are we ready for autonomous driving? The KITTI vision benchmark suite. In: 2012 IEEE Conference on Computer Vision and Pattern Recognition, pp. 3354–3361 (2012)
18. Everingham, M., Van Gool, L., Williams, C.K.I., Winn, J., Zisserman, A.: The pascal visual object classes (VOC) challenge. Int. J. Comput. Vision **88**(2), 303–338 (2010)
19. Sudhakar, S., Vijayakumar, V., Kumar, C.S., Priya, V., Ravi, L., Subramaniyaswamy, V.: Unmanned aerial vehicle (UAV) based forest fire detection and monitoring for reducing false alarms in forest-fires. Comput. Commun. **149**, 1–16 (2020)
20. Bendea, H., Boccardo, P., Dequal, S., Tonolo, F.G., Marenchino, D., Piras, M.: Low cost UAV for post-disaster assessment. Int. Arch. Photogram. Remote Sens. Spat. Inf. Sci. **37**, 1373-1379 (2008)
21. John Gunnar Carlsson and Siyuan Song: Coordinated logistics with a truck and a drone. Manage. Sci. **64**(9), 4052–4069 (2018)
22. Krizhevsky, A., Sutskever, I., Hinton, G.E.: ImageNet classification with deep convolutional neural networks. In: Advances in Neural Information Processing Systems, vol. 25. Curran Associates Inc. (2012)
23. Simonyan, K., Zisserman, A.: Very deep convolutional networks for large-scale image recognition (2015)
24. He, K., Zhang, X., Ren, S., Sun, J.: Deep residual learning for image recognition. arXiv:1512.03385 [cs], p. 12 (2015)
25. Howard, A.G., et al.: MobileNets: efficient convolutional neural networks for mobile vision applications. arXiv:1704.04861 [cs], p. 9 (2017)
26. Girshick, R.: Fast R-CNN. In: 2015 IEEE International Conference on Computer Vision (ICCV), pp. 1440–1448 (2015)
27. Ren, S., He, K., Girshick, R., Sun, J.: Faster R-CNN: towards real-time object detection with region proposal networks. arXiv:1506.01497 [cs], p. 14 (2016)
28. Redmon, J., Divvala, S., Girshick, R., Farhadi, A.: You only look once: unified, real-time object detection. In: 2016 IEEE Conference on Computer Vision and Pattern Recognition (CVPR), Las Vegas, NV, USA, p. 10. IEEE (2016)
29. Liu, W., et al.: SSD: single shot multibox detector. arXiv:1512.02325 [cs], 9905:17 (2016)
30. Lin, T.-Y., et al.: Microsoft COCO: common objects in context. In: Fleet, D., Pajdla, T., Schiele, B., Tuytelaars, T. (eds.) ECCV 2014. LNCS, vol. 8693, pp. 740–755. Springer, Cham (2014). https://doi.org/10.1007/978-3-319-10602-1_48
31. Bochkovskiy, A., Wang, C.-Y., Liao, H.-Y.M.: YOLOv4: optimal speed and accuracy of object detection (2020)
32. Liu, S., Qi, L., Qin, H., Shi, J., Jia, J.: Path aggregation network for instance segmentation (2018)
33. Li, S., Li, Y., Li, Y., Li, M., Xiaorong, X.: YOLO-FIRI: improved YOLOv5 for infrared image object detection. IEEE Access **9**, 141861–141875 (2021)
34. Kun, Z.: Background noise suppression in small targets infrared images and its method discussion. Opt. Optoelectron. Technol. **2**, 9–12 (2004)
35. Anju, T.S., Nelwin Raj, N.R.: Shearlet transform based image denoising using histogram thresholding. In: 2016 International Conference on Communication Systems and Networks (ComNet), pp. 162–166 (2016)
36. Viola, P., Jones, M.J.: Robust real-time face detection. Int. J. Comput. Vision **57**(2), 137–154 (2004)

37. Dalal, N., Triggs, B.: Histograms of oriented gradients for human detection. In: 2005 IEEE Computer Society Conference on Computer Vision and Pattern Recognition (CVPR 2005), vol. 1, pp. 886–893 (2005)
38. Jocher, G., et al.: Ultralytics/YOLOv5: V7.0 - YOLOv5 SOTA realtime instance segmentation (2022)
39. Taran, V., et al.: Performance evaluation of deep learning networks for semantic segmentation of traffic stereo-pair images. In: Proceedings of the 19th International Conference on Computer Systems and Technologies, pp. 73–80 (2018)
40. Taran, V., Gordienko, Y., Rokovyi, A., Alienin, O., Stirenko, S.: Impact of ground truth annotation quality on performance of semantic image segmentation of traffic conditions. In: Hu, Z., Petoukhov, S., Dychka, I., He, M. (eds.) ICCSEEA 2019. AISC, vol. 938, pp. 183–193. Springer, Cham (2020). https://doi.org/10.1007/978-3-030-16621-2_17

Medical Imaging and Applications

Inter vs. Intra Domain Study of COVID Chest X-Ray Classification with Imbalanced Datasets

Alejandro Galán-Cuenca, Miguel Mirón, Antonio Javier Gallego(✉), Marcelo Saval-Calvo, and Antonio Pertusa

University Institute for Computer Research, University of Alicante, Alicante, Spain
{a.galan,miguel.ma,m.saval,pertusa}@ua.es, jgallego@dlsi.ua.es

Abstract. Medical image classification datasets usually have a limited availability of annotated data, and pathological samples are usually much scarcer than healthy cases. Furthermore, data is often collected from different sources with different acquisition devices and population characteristics, making the trained models highly dependent on the data domain and thus preventing generalization. In this work, we propose to address these issues by combining transfer learning, data augmentation, a weighted loss function to balance the data, and domain adaptation. We evaluate the proposed approach on different chest X-Ray datasets labeled with COVID positive and negative diagnoses, yielding an average improvement of 15.3% in F_1 compared to the base case of training the model without considering these techniques. A 19.1% improvement is obtained in the intra-domain evaluation and a 7.7% for the inter-domain case.

Keywords: Medical imaging · Imbalanced classification · Transfer learning · Domain adaptation · Convolutional Neural Networks

1 Introduction

Medical imaging is a critical tool in pathology diagnosis and prognosis. With the increasing availability of large medical imaging datasets and the rapid development of deep learning algorithms, machine learning models have shown great potential in automating the diagnosis from medical images in a variety of applications, including brain tumor segmentation [6], chest X-ray classification [16], and skin lesion analysis [13], among others.

However, the performance of these models depends, to a large extent, on a multitude of factors that can considerably deteriorate their results, such as the limited availability of annotated data, the high cost of labeling by experts, the large imbalance between cases of negative and positive samples of a given

This work was supported by the I+D+i project TED2021-132103A-I00 (DOREMI), funded by MCIN/AEI/10.13039/501100011033.

A. Pertusa et al. (Eds.): IbPRIA 2023, LNCS 14062, pp. 507–519, 2023.
https://doi.org/10.1007/978-3-031-36616-1_40

disease—especially in rare diseases—, and the high dependence of the generated models with respect to the domain of the data used for their training, among other challenges that hinder the development of effective, robust and general methods for the processing of this type of images [10].

Concerning the limited availability of annotated medical imaging data and the high cost of obtaining such, one standard solution considered in the machine-learning literature is data augmentation [4], which involves generating synthetic samples from existing images to increase the size of the training dataset. However, the unique properties of medical images, such as their high dimensionality, small structures, and significant inter- and intra-class variability, make applying traditional data augmentation techniques difficult.

Another common solution to the limited availability of annotated data is the use of transfer learning, which consists of taking advantage of the knowledge learned for a domain with sufficient labeled data and applying it to another through a process of fine-tuning—i.e., using the weights of the trained model and use it as initialization for a new model. Due to its good results, there has been a growing interest in this technique in the context of medical imaging [5,12], since in general, it helps to reduce the amount of labeled data needed for training, to make training converge faster, and to obtain models that generalize better, that is, models that can be transferred and applied in other domains (called inter-domain use).

Another related problem is the high dependency on the models concerning the training data domain. In practice, medical imaging data is often collected from different sources, each with its own characteristics, leading to significant differences in image appearance, distribution, and annotation. This leads to a performance drop when models trained on one domain are applied to a different domain (inter-domain use). To address this challenge, domain adaptation techniques have been proposed to leverage the knowledge learned from one domain and apply it to another in an unsupervised way—i.e., without requiring the target domain to be labelled [2,11]. These techniques have been successfully applied to various medical imaging tasks, including classification, segmentation, and registration [15].

Finally, another prevalent problem in medical imaging is highly imbalanced data since fewer positive than negative samples are usually available. Machine learning-based models trained on imbalanced data tend to be biased toward the majority class, ignoring the samples from the minority class [7]. This results in poorly performing methods for the minority class, and, in the context of detecting certain diseases or pathologies, it could drive to erroneous diagnoses that could endanger lives.

This work aims to study the accuracy of learning-based models for medical imaging when faced with these challenges, analyzing their results for intra- and inter-domain use cases and the improvements obtained by applying different types of solutions, such as the use of data augmentation, transfer learning, domain adaptation, or different methods to deal with imbalanced datasets. These techniques will be evaluated considering four public chest x-ray image datasets, with which three corpus pairs will be prepared with positive and negative samples of COVID-19 patients. As it will be shown, the proposed approaches have the potential to address these challenges and improve the performance of medical imaging classification models in the different considered evaluation scenarios.

The rest of the paper is structured as follows: Sect. 2 presents the proposed approach to overcome the previously raised challenges; Sect. 3 describes the experimental setting, as a part of which Sect. 3.1 describes the considered datasets; Sect. 4 reports and discusses the evaluation results; and finally Sect. 5 addresses the conclusions and future work.

2 Methodology

This section presents the methodological proposal to address the problems that learning-based methods usually face when dealing with medical imaging datasets. Figure 1 shows a diagram with the pipeline of the process, starting with the use of different techniques to deal with imbalanced datasets, followed by the application of data augmentation, the inclusion of a domain adaptation technique, and the use of a weighted loss function. The following sections describe each of these processes in detail.

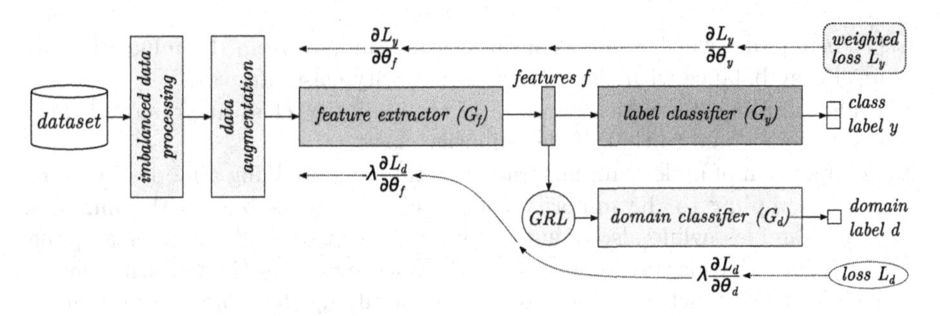

Fig. 1. Diagram with the pipeline of the process. The proposed techniques to be studied are highlighted in yellow. (Color figure online)

2.1 Initialization

The initialization of the neural network weights is a factor that can significantly influence the final result obtained and condition the number of samples needed to obtain an acceptable success rate. For this reason, we will study the application of three strategies: 1) Training from scratch, which usually requires having a larger set of labeled data; 2) Initializing the network with ImageNet [1] pre-trained weights to reduce the training time and the amount of data required; and 3) Initializing the network with a similar X-Ray dataset and fine-tuning it with the target dataset.

2.2 Data Augmentation

Data augmentation has become a standard to improve the training process in learning-based methods. This is a technique focused on generating a set of synthetic samples from the training set. It intends to increase the variability of the training data and thus create models that generalize better. There are different ways to apply this process in the context of image processing, such as rotations, skew, zoom, and contrast or color adjustments, among many others. However, the goodness of each one depends to a large extent on the task in question. In the case of medical imaging, the unique properties of these images force the application of these techniques in a limited manner and are guided by experts. For this reason, a limited range of transformations has been considered since previous tests with other configurations reported poor results. Specifically, the transformations applied were randomly selected from the following set of possible transformations: horizontal and vertical shifts ($[-1, 1]\%$ of the image size), zoom ($[-1, 1]\%$ of the original image size), and rotations (in the range $[-1^\circ, 1^\circ]$).

2.3 Imbalanced Data

To alleviate the bias generated during training when using imbalanced datasets, we compare the four following techniques:

1. Oversampling, which consists of duplicating samples from the minority class until they balance with the number of majority class samples.
2. Undersampling, for which samples of the majority class are removed until their size is equal to that of the minority class.
3. Combination of undersampling and oversampling, applying a modest amount of oversampling to the minority class, which improves bias in the minority class examples, while also doing a modest amount of random undersampling in the majority class to reduce bias in the samples from the majority class.
4. Weighted loss function, which consists of modifying the loss function used to optimize the network weights to increase the importance of the errors made for the minority class. In this way, their contribution to the error committed is balanced and forces during training to consider both classes equally so that a bias towards the majority class is not created.

Although these techniques are well known, they will be studied in the context of medical imaging and when combined.

2.4 Domain Adaptation

Another strategy to overcome the challenges mentioned above is to leverage labeled information from other domains (referred to as *source domains*) to a new *target domain* using domain adaptation techniques. These techniques effectively improve the performance of machine learning models by taking advantage of prior knowledge from related domains for which labeled data is available.

Due to its good results in a multitude of disparate tasks [2,11], we resort to the unsupervised domain adaptation method proposed in [3], the so-called *Domain-Adversarial Neural Networks (DANN)*.

As shown in the right part of Fig. 1, the DANN architecture comprises two branches that perform different classification tasks on the input image: label classification (top branch) and domain classification (bottom branch), which estimates the domain of the sample, either source or target. Note that while depicting different goals, both parts share a common feature extraction G_f block that maps the input data into an f-dimensional space.

This architecture relies on the use of the *Gradient Reversal Layer* (GRL) during the backpropagation phase of the training process, which reverses the gradient of the domain classifier loss L_d and scales it with a $\lambda \in \mathbb{R}$ coefficient to weight its contribution in the overall learning process. This forces the G_f feature extractor to derive a set of domain-invariant descriptors capable of adequately addressing the label classification task disregarding the domain of the input data.

3 Experimental Setup

This section describes the configuration used during the experimentation process, including the considered datasets, the network architecture used in the proposed methodology, the details of the training process, and the evaluation metrics.

3.1 Datasets

The methodology has been evaluated with four different datasets[1], which can be seen in detail in Table 1, including the type of samples they contain—i.e., whether they are negative (−) or positive (+) COVID-19 samples—in addition to the original sizes of the training and evaluation partitions provided. Figure 2 shows some example images from these datasets.

As seen in Table 1, two of these datasets only include negative samples, and the other two, although they include samples of both classes, have imbalance among their classes. Several combinations were generated from these sets to create datasets with positive and negative samples to evaluate the proposed methodology, as shown in Table 2. This table includes an acronym for the combination (to be used in the experimentation section) and details the number of positive and negative samples in each new set. Note that, in order not to overbalance the generated sets and to be able to perform more experiments, the number of samples to be added from the original sets was limited to 10 000. Besides, the *mean imbalance ratio* (MeanIR) index is also provided to report the imbalance level of the corpus [14]. This descriptor ranges in $[1, \infty)$ and denotes a higher imbalance as the value increases.

[1] All the datasets considered are publicly available: ChestX-ray is available at https://nihcc.app.box.com/v/ChestXray-NIHCC, GitHub-COVID at https://github.com/ieee8023/covid-chestxray-dataset, PadChest can be found at https://bimcv.cipf.es/bimcv-projects/padchest, and BIMCV-COVID repositories are available at https://bimcv.cipf.es/bimcv-projects/bimcv-covid19.

Table 1. Original configuration of the considered datasets, including the type of samples (positive + and negative −), the number of samples per class and their total ($\sum$), and the size of the training and test set with its percentage.

Dataset	Classes	Train size		Test size		Total
ChestX-ray	−	86 524	(77%)	25 596	(23%)	112 120
PadChest	−	91 508	(95%)	4 762	(5%)	96 270
BIMCV-COVID	−	3 014		159		3 173
	+	1 610		82		1 692
	$\sum$	4 624	(95%)	241	(5%)	4 865
Github-COVID	−	81		29		110
	+	283		11		294
	$\sum$	364	(90%)	40	(10%)	404

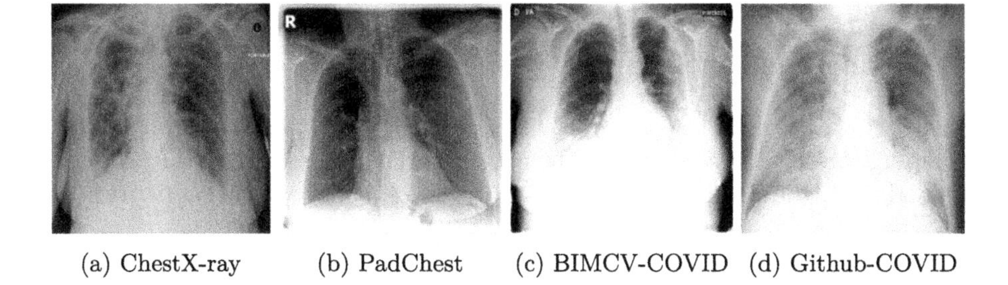

(a) ChestX-ray (b) PadChest (c) BIMCV-COVID (d) Github-COVID

Fig. 2. An image example from each dataset used in the experimentation.

3.2 Network Architecture

To evaluate the proposed methodology, a ResNet-50 v2 [8] is considered since it is a standard architecture for image classification tasks, with state-of-the-art results in several benchmarks, and is usually employed as a backbone to perform other tasks, such as evaluating domain adaptation techniques [2]. This is an updated version of the original ResNet-50 architecture, with a few essential modifications—such as the identity shortcuts and the pre-activation units—that improve its performance and reduce overfitting.

For all the evaluations performed, this architecture was trained under the same conditions, during 200 epochs with a batch size of 32 images and Stochastic Gradient Descent optimizer [9] with a Nesterov momentum parameter of 0.9, a learning rate of 10^{-2}, and a decay factor of 10^{-6}. The images were scaled to a spatial resolution of 224×224 pixels and normalized in the $[0, 1]$ range to favor the model convergence.

3.3 Metrics

In terms of evaluation, we resorted to the F-measure (F_1) figure of merit to avoid possible biases towards any particular class, given the considerable label imbal-

Table 2. Description of the datasets generated for the experimentation from the combination of the original datasets of the Table 1. The acronym of the new dataset, the sizes of each partition, including the number of positive (+) and negative (−) samples, and the percentage they represent are shown. In addition, the MeanIR is included as an indicator of the unbalance of the dataset.

Acronym	Combined data	Train size	Test size	Total	MeanIR
ChestX-Git	ChestX-ray ∪ Github-COVID	−: 10 081 +: 283 $\sum$: 10 364 (51%)	−: 10 029 +: 11 $\sum$: 10 040 (49%)	−: 20 110 (99%) +: 294 (1%) $\sum$: 20 404	34.7
Pad-BIM	PadChest ∪ BIMCV-COVID+	−: 10 000 +: 1 610 $\sum$: 11 610 (71%)	−: 4 762 +: 82 $\sum$: 4 844 (29%)	−: 14 762 (90%) +: 1 692 (10%) $\sum$: 16 454	4.9
BIMCV-COVID	BIMCV-COVID- ∪ BIMCV-COVID+	−: 3 014 +: 1 610 $\sum$: 4 624 (95%)	−: 159 +: 82 $\sum$: 241 (5%)	−: 3 173 (65%) +: 1 692 (35%) $\sum$: 4 865	1.4

ance of the different collections contemplated. Assuming a binary classification scenario, F_1 is computed as the harmonic mean of the Precision (P) and Recall (R) indicators. These figures of merit are defined as:

$$\mathrm{P} = \frac{\mathrm{TP}}{\mathrm{TP} + \mathrm{FP}} \tag{1}$$

$$\mathrm{R} = \frac{\mathrm{TP}}{\mathrm{TP} + \mathrm{FN}} \tag{2}$$

$$\mathrm{F}_1 = \frac{2 \cdot \mathrm{P} \cdot \mathrm{R}}{\mathrm{P} + \mathrm{R}} = \frac{2 \cdot \mathrm{TP}}{2 \cdot \mathrm{TP} + \mathrm{FP} + \mathrm{FN}} \tag{3}$$

where TP, FP, and FN denote the True Positives, False Positives, and False Negatives. Since the experiments were conducted as a multi-class problem, we report the results in terms of macro-F_1 for a global evaluation, which is calculated as the average of the F_1 obtained for each class.

4 Results

This section evaluates the proposed method using the experimental setup previously described. To comprehensively assess the proposal, the results of each stage will be studied separately, starting by analyzing the influence of the initialization process, continuing with a comparison of different techniques for dealing with imbalanced data and ending with the results obtained by applying the adaptation technique to the domain under consideration. Finally, a discussion section that summarizes the results obtained is included.

4.1 Initialization and Data Augmentation

First, we analyze the influence of the initialization process of the considered network architecture (i.e., ResNet-50 v2) and the improvement obtained by applying

Table 3. Comparison of the results obtained by training the model from scratch, starting from a generic initialization with the weights obtained for ImageNet, and applying further data augmentation. The intra-domain case for each scenario is underlined.

From	To	From scratch			ImageNet			ImageNet + Data aug.		
		P	R	F_1	P	R	F_1	P	R	F_1
Chest-Git	<u>Chest-Git</u>	89.0	50.0	64.1	91.0	52.0	66.2	91.0	50.0	64.6
	Pad-BIM	40.9	33.3	36.7	45.9	33.9	38.9	46.9	34.9	40.0
	BIMCV-COVID	43.0	41.4	42.2	43.0	45.4	44.2	45.0	45.4	45.2
Pad-BIM	Chest-Git	50.0	26.2	34.4	55.0	28.2	37.3	54.0	29.2	37.9
	<u>Pad-BIM</u>	87.0	58.5	70.0	91.0	59.5	71.9	90.0	61.5	73.1
	BIMCV-COVID	48.0	36.5	41.4	51.0	39.5	44.5	51.0	40.5	45.1
BIMCV-COVID	Chest-Git	50.2	22.3	30.9	50.9	23.3	32.0	52.9	23.3	32.4
	Pad-BIM	54.2	35.2	42.7	56.2	35.8	43.7	58.2	35.8	44.3
	<u>BIMCV-COVID</u>	73.2	72.3	72.8	74.2	75.3	74.8	76.2	73.3	74.7
Average		59.5	41.7	48.3	62.0	43.6	50.4	62.8	43.8	50.8

data augmentation techniques. For this purpose, we compare the results of training this architecture from scratch and using ImageNet [1] pre-trained weights. Table 3 shows the results of this first experiment, including the detailed result of each possible combination of the datasets considered, taking one of them for training (column "From") and evaluating the generated model on all datasets (column "To"). Of all these possible combinations, the intra-domain cases (when evaluating on the same domain as the one trained on) are underlined, compared to the rest of the inter-domain evaluation cases, where the model is evaluated on domains other than the one considered during training.

As can be seen, in general, the high imbalance of the datasets leads to the model having high precision but low recall. Moreover, this effect is much more noticeable when we analyze the inter-domain results compared to the intra-domain results. On the other hand, initializing the model weights with ImageNet results in an improvement of 2.1% on average over training the network from scratch. Data augmentation only reports an improvement of 0.4% on average, in some cases leading to a worse result (intra-domain Chest-Git case). Different types of data augmentation and combinations were tested without noticeable improvement.

Continuing with the model initialization process, we now analyze the effect of transfer learning pre-training with another X-Ray dataset. Table 4 reports these results, indicating in the first column the model with which each row is pre-trained (a different dataset than the one considered in each *from–to* combination was selected). By pre-training the model with a similar dataset and then applying a fine-tuning process to the target dataset, an average additional improvement of 3.1% was achieved relative to the best average result obtained in Table 3 (initializing with ImageNet and applying data augmentation), and of 5.6% if we

Table 4. Results obtained by applying a transfer learning process. The first column indicates the dataset with which the model is pre-trained, the column "From" indicates the dataset with which fine-tuning is performed, and the column "To" is the set with which it is evaluated. The intra-domain case for each scenario is underlined.

Pre-trained	From	To	P	R	F_1
Pad-BIM	Chest-Git	Chest-Git	92.0	56.0	69.6
BIMCV-COVID		Pad-BIM	55.9	35.3	43.3
Pad-BIM		BIMCV-COVID	51.0	47.4	49.1
BIMCV-COVID	Pad-BIM	Chest-Git	60.0	29.2	39.3
Chest-Git		Pad-BIM	94.0	62.5	75.1
Chest-Git		BIMCV-COVID	58.0	40.5	47.6
Pad-BIM	BIMCV-COVID	Chest-Git	58.2	24.0	34.0
Chest-Git		Pad-BIM	62.2	39.2	48.1
Chest-Git		BIMCV-COVID	81.2	76.3	78.7
Average			68.0	45.6	53.9

compare with the model without initializing. In some cases, it even improves by 5.1% (intra-domain case for Chest-Git pre-trained with Pad-BIM).

Therefore, in the remaining experiments, the network model will be trained by applying transfer learning with another X-Ray dataset and considering the proposed data augmentation process.

4.2 Dealing with Imbalanced Data

Next, we analyze the results using the proposed solutions to alleviate the data imbalance problem. These are undersampling, oversampling, combined under and oversampling, and applying a weighted loss function. For the *combined* case, a 20% random oversampling was applied to the minority class and a 20% of random undersampling to the majority class (during the initial experimentation, other percentages were studied, but this was the one with the best average results). For weighted loss, the weight of each class i is calculated as: *total training samples*/(# *classes* · # *samples of class* i).

Table 5 reports the results of this experiment, in which, in general, it can be seen how these techniques balance the precision and recall results obtained, in this case lowering the precision and slightly increasing the recall. Undersampling is the worst performing case in general, 4.7% worse on average compared to Table 4, getting noticeably worse results for the intra-domain case (8.6%). This is mainly because the number of training samples is considerably reduced, given that a large imbalance in the database means that many samples from the negative class must be discarded to balance them with those from the positive class.

The oversampling technique does achieve a modest improvement of 1.7% over previous results. In this case, it improves inter-domain results. Still, its use is not

Table 5. Comparison of the results obtained by applying the techniques proposed to mitigate the data imbalance problem. The intra-domain case for each scenario is underlined.

From	To	Undersampling			Oversampling			Combined			Weighted loss		
		P	R	F_1	P	R	F_1	P	R	F_1	P	R	F_1
Chest-Git	<u>Chest-Git</u>	65.7	57.2	61.1	88.7	58.2	70.3	91.0	60.3	72.5	94.0	60.3	73.5
	Pad-BIM	34.9	35.3	35.1	52.9	37.3	43.7	53.9	38.3	44.8	58.9	39.3	47.1
	BIMCV-COVID	39.0	51.4	44.3	42.0	53.4	47.0	48.0	51.4	49.6	54.0	56.4	55.1
Pad-BIM	Chest-Git	45.0	35.2	39.5	50.0	36.2	42.0	57.0	34.2	42.7	63.0	35.2	45.2
	<u>Pad-BIM</u>	68.0	60.5	64.0	90.0	63.5	74.5	90.9	64.9	75.7	95.9	65.9	78.1
	BIMCV-COVID	42.0	43.5	42.7	54.9	45.5	49.7	58.0	45.7	51.1	58.0	45.7	51.1
BIMCV-COVID	Chest-Git	46.2	32.3	38.0	57.2	37.3	45.2	57.2	26.0	35.7	61.2	30.0	40.3
	Pad-BIM	50.2	41.2	45.3	59.2	42.2	49.3	62.2	42.2	50.3	64.2	42.2	50.9
	<u>BIMCV-COVID</u>	68.2	78.3	72.9	79.2	78.3	78.8	81.2	77.3	79.2	82.2	81.3	81.8
Average		51.0	48.3	49.2	63.8	50.2	55.6	66.6	48.9	55.7	70.1	50.7	58.1

recommended for intra-domain, given that the likelihood of overfitting increases when positive class samples are replicated numerous times to balance them with those of the negative class. In the case of the combined approach, a similar result to oversampling is obtained, but a more balanced average improvement is obtained for both the inter- and intra-domain cases.

Finally, it should be noted that applying a weighted loss function gives the best results (4.2% improvement compared to previous results). Therefore, in the following experiments, we will consider this last option and evaluate its application with the combined proposal for balancing the number of samples.

4.3 Domain Adaptation

This section explores the proposed domain adaptation approach using the DANN technique to address the intra-domain dependency of the generated models. As discussed in Sect. 2.4, this method requires the computation of a gradient score—performed at the GRL layer—scaled by $\lambda \in \mathbb{R}$ coefficient that must be experimentally tuned. In our case, preliminary testing studied values in the range $\lambda \in \left[10^{-4}, 1\right]$ provide the best performance when $\lambda = 10^{-4}$.

Table 6 compares the results obtained by including domain adaptation and also incorporating the proposal for dealing with unbalanced data, considering both the technique for combining oversampling and undersampling and the use of the weighted loss function. The first to be noted is that the application of DANN obtains a similar average result to that previously obtained using the weighted loss. However, in this case, a notable improvement is observed for the inter-domain cases (7.2%) and a worsening of the result in the intra-domain cases (−2.2%). This worsening is due to the DANN training process itself, in which the network is forced to learn domain-invariant features common to the domains considered.

Table 6. Comparison of the results obtained by applying the DANN domain adaptation technique and combining this approach with previous solutions for dealing with unbalanced datasets. The intra-domain case for each scenario is underlined.

From	To	DANN			DANN + comb. + weighted loss		
		P	R	F_1	P	R	F_1
Chest-Git	Chest-Git	84.7	56.9	68.1	87.7	62.2	72.8
	Pad-BIM	65.4	40.1	49.7	71.4	44.1	54.5
	BIMCV-COVID	69.8	50.4	58.5	75.8	61.4	67.8
Pad-BIM	Chest-Git	63.4	35.7	45.7	69.4	41.7	52.1
	Pad-BIM	85.2	62.6	72.2	90.2	65.0	75.6
	BIMCV-COVID	71.6	46.4	56.3	74.6	52.7	61.7
BIMCV-COVID	Chest-Git	60.9	30.2	40.4	63.9	37.2	47.0
	Pad-BIM	70.2	43.8	53.9	75.2	49.8	59.9
	BIMCV-COVID	79.8	74.4	77.0	83.8	79.4	81.5
Average		72.3	49.0	58.0	76.9	54.8	63.7

Combining this technique with those previously studied to deal with unbalanced data solves the worst result for the intra-domain cases, and the average result obtained is further improved. This better result is mainly because by balancing the number of samples and weighting the loss function so that the errors of the minority class are weighted more heavily, the bias of the model towards the majority class is reduced, notably increasing the recall obtained in this case.

4.4 Discussion

As a final summary, Table 7 reports the improvement ratios of the different proposals considering F_1. Concerning initialization, it is observed that the best solution is to apply the transfer learning technique, being a quite adequate approach for both intra- and inter-domain cases. Data augmentation only reports a rather modest improvement, at least with the dataset configuration considered. Therefore, it is a technique that usually helps the training process but, especially in the case of medical imaging, requires much more fine-tuning to work.

The techniques to solve the imbalance problem also report promising improvement rates, except in the case of undersampling due to the drastic reduction in the size of the training set. Oversampling, on the other hand, produces overfitting by replicating the same samples, reducing the intra-domain improvement ratio. In this case, the combined solution and, above all, the application of the weighted loss function are more recommendable.

Finally, the domain adaptation process achieves a notable improvement for the inter-domain case, worsening for the intra-domain case when generating a model adapted to the target data. However, it is observed that its use combined with the solutions studied for data unbalancing leads to a higher improvement

Table 7. Comparison of the improvement ratios obtained by the different proposals considered the F_1 metric. For each solution, the base result against which it is compared (c/w, compared with) is given.

	Initialization			Imbalanced solutions				Domain adaptation			
	c/w scratch			c/w tr. learning				c/w w. loss		c/w tr. learning	
	ImageNet	ImageNet + data aug.	Tr. learn. + data aug.	Undersamp.	Oversamp.	Combined	W. loss	DANN	DANN + imb. sol.	DANN	DANN + imb. sol.
Average	2.1%	2.5%	5.6%	−4.7%	1.7%	1.8%	4.2%	−0.1%	5.5%	4.1%	9.8%
Inter-domain	2.1%	2.8%	5.5%	−2.8%	2.6%	2.1%	4.7%	2.5%	8.9%	7.2%	13.6%
Intra-domain	2.0%	1.9%	5.7%	−8.6%	−0.1%	1.2%	3.2%	−5.3%	−1.1%	−2.2%	2.0%

ratio, also for the intra-domain case, perhaps because by balancing the importance of the classes, better domain-invariant features are also learned for the minority class.

5 Conclusions

This work explores pre-training strategies, data augmentation, imbalance correction methods, and domain adaptation techniques to overcome the main limitations of chest X-ray image datasets. These datasets have particular features such as highly imbalanced classes, a relatively low number of samples, and different data distribution and acquisition methods depending on the hospitals that collect them.

Due to the characteristics of these images, established methods such as data augmentation, which tends to improve the results in most scenarios, only have a minor relevance. However, transfer learning by pre-training the network with similar images instead of ImageNet is beneficial, even considering the latter has many more samples.

For dealing with unbalanced data, a weighted loss function to balance the data improves the accuracy by increasing the importance of the errors made for the minority class. Moreover, domain adaptation techniques have proven to be an appropriate choice given the different distributions of the evaluated datasets.

The best results were achieved with a combination of all the techniques mentioned above: transfer learning, data augmentation, a weighted loss function, and domain adaptation, obtaining an average improvement of 15.4% in F_1 compared to the base case of training the model without any of these techniques.

However, further research is needed to fully understand the strengths and limitations of these approaches in the context of medical imaging. Future work contemplates the implementation in a real environment with medical images from different sources, which can give a better picture of the improvements done in this analysis. Techniques, such as the use of autoencoders or generative networks to obtain new training images, or the application of SMOTE or Tomek links, can help overcome the limitations of data augmentation in medical images, leading to improvements in performance.

References

1. Deng, J., Dong, W., Socher, R., Li, L.J., Li, K., Fei-Fei, L.: ImageNet: a large-scale hierarchical image database. In: Proceedings of the IEEE/CVF Conference on Computer Vision and Pattern Recognition, pp. 248–255 (2009)
2. Gallego, A.J., Calvo-Zaragoza, J., Fisher, R.B.: Incremental unsupervised domain-adversarial training of neural networks. IEEE Trans. Neural Netw. Learn. Syst. **32**(11), 4864–4878 (2021). https://doi.org/10.1109/TNNLS.2020.3025954
3. Ganin, Y., et al.: Domain-adversarial training of neural networks. J. Mach. Learn. Res. **17**(1), 2096–2030 (2016)
4. Garay-Maestre, U., Gallego, A.-J., Calvo-Zaragoza, J.: Data augmentation via variational auto-encoders. In: Vera-Rodriguez, R., Fierrez, J., Morales, A. (eds.) CIARP 2018. LNCS, vol. 11401, pp. 29–37. Springer, Cham (2019). https://doi.org/10.1007/978-3-030-13469-3_4
5. Ghafoorian, M., et al.: Transfer learning for domain adaptation in MRI: application in brain lesion segmentation. In: Descoteaux, M., Maier-Hein, L., Franz, A., Jannin, P., Collins, D.L., Duchesne, S. (eds.) MICCAI 2017. LNCS, vol. 10435, pp. 516–524. Springer, Cham (2017). https://doi.org/10.1007/978-3-319-66179-7_59
6. Havaei, M., et al.: Brain tumor segmentation with deep neural networks. Med. Image Anal. **35**, 18–31 (2017). https://doi.org/10.1016/j.media.2016.05.004
7. He, H., Garcia, E.A.: Learning from imbalanced data. IEEE Trans. Knowl. Data Eng. **21**(9), 1263–1284 (2009). https://doi.org/10.1109/TKDE.2008.239
8. He, K., Zhang, X., Ren, S., Sun, J.: Identity mappings in deep residual networks. In: Leibe, B., Matas, J., Sebe, N., Welling, M. (eds.) ECCV 2016. LNCS, vol. 9908, pp. 630–645. Springer, Cham (2016). https://doi.org/10.1007/978-3-319-46493-0_38
9. Mitchell, T.M.: Machine Learning, vol. 1. McGraw-Hill, New York (1997)
10. Razzak, M.I., Naz, S., Zaib, A.: Deep learning for medical image processing: overview, challenges and the future. In: Dey, N., Ashour, A., Borra, S. (eds.) Classification in BioApps. LNCVB, vol. 26, pp. 323–350. Springer, Cham (2018). https://doi.org/10.1007/978-3-319-65981-7_12
11. Rosello, A., Valero-Mas, J.J., Gallego, A.J., Sáez-Pérez, J., Calvo-Zaragoza, J.: Kurcuma: a kitchen utensil recognition collection for unsupervised domain adaptation. Pattern Anal. Appl. (2023). https://doi.org/10.1007/s10044-023-01147-x
12. Shin, H.C., et al.: Deep convolutional neural networks for computer-aided detection: CNN architectures, dataset characteristics and transfer learning. IEEE Trans. Med. Imaging **35**(5), 1285–1298 (2016). https://doi.org/10.1109/TMI.2016.2528162
13. Tschandl, P., Rosendahl, C., Kittler, H.: The HAM10000 dataset, a large collection of multi-source dermatoscopic images of common pigmented skin lesions. Sci. Data **5**, 180161 (2018). https://doi.org/10.1038/sdata.2018.161
14. Valero-Mas, J.J., Gallego, A.J., Alonso-Jiménez, P., Serra, X.: Multilabel prototype generation for data reduction in k-nearest neighbour classification. Pattern Recogn. **135**, 109190 (2023). https://doi.org/10.1016/j.patcog.2022.109190
15. Wang, M., Deng, W.: Deep visual domain adaptation: a survey. Neurocomputing **312**, 135–153 (2018). https://doi.org/10.1016/j.neucom.2018.05.083
16. Wang, X., Peng, Y., Lu, L., Lu, Z., Bagheri, M., Summers, R.M.: ChestX-ray8: hospital-scale chest x-ray database and benchmarks on weakly-supervised classification and localization of common thorax diseases. In: 2017 IEEE Conference on Computer Vision and Pattern Recognition (CVPR), pp. 3462–3471 (2017). https://doi.org/10.1109/CVPR.2017.369

Automatic Eye-Tracking-Assisted Chest Radiography Pathology Screening

Rui Santos[1,2(✉)], João Pedrosa[1,2], Ana Maria Mendonça[1,2], and Aurélio Campilho[1,2]

[1] Faculty of Engineering of the University of Porto (FEUP), Porto, Portugal
[2] Institute for Systems and Computer Engineering, Technology and Science (INESC TEC), Porto, Portugal
rui.m.santos@inesctec.pt

Abstract. Chest radiography is increasingly used worldwide to diagnose a series of illnesses targeting the lungs and heart. The high amount of examinations leads to a severe burden on radiologists, which benefit from the introduction of artificial intelligence tools in clinical practice, such as deep learning classification models. Nevertheless, these models are undergoing limited implementation due to the lack of trustworthy explanations that provide insights about their reasoning. In an attempt to increase the level of explainability, the deep learning approaches developed in this work incorporate in their decision process eye-tracking data collected from experts. More specifically, eye-tracking data is used in the form of heatmaps to change the input to the selected classifier, an EfficientNet-b0, and to guide its focus towards relevant parts of the images. Prior to the classification task, UNet-based models are used to perform heatmap reconstruction, making this framework independent of eye-tracking data during inference. The two proposed approaches are applied to all existing public eye-tracking datasets, to our knowledge, regarding chest X-ray screening, namely EGD, REFLACX and CXR-P. For these datasets, the reconstructed heatmaps highlight important anatomical/pathological regions and the area under the curve results are comparable to the state-of-the-art and to the considered baseline. Furthermore, the quality of the explanations derived from the classifier is superior for one of the approaches, which can be attributed to the use of eye-tracking data.

Keywords: Chest X-Ray · Deep Learning · Eye-Tracking Data

This work was funded by the ERDF - European Regional Development Fund, through the Programa Operacional Regional do Norte (NORTE 2020) and by National Funds through the FCT - Portuguese Foundation for Science and Technology, I.P. within the scope of the CMU Portugal Program (NORTE-01-0247-FEDER-045905) and LA/P/0063/2020.

A. Pertusa et al. (Eds.): IbPRIA 2023, LNCS 14062, pp. 520–532, 2023.
https://doi.org/10.1007/978-3-031-36616-1_41

1 Introduction

Chest radiography is one of the most commonly prescribed medical examinations. It is relatively inexpensive, noninvasive, quick to perform and uses only a small dose of radiation. It allows the diagnosis of a myriad of conditions targeting structures like the lungs or the mediastinum. However, the high number of Chest X-Ray (CXR) scans creates a huge burden on radiologists, which paves the way for the introduction of new Artificial Intelligence (AI) methods capable of aiding clinical practice, including Deep Learning (DL) classification models.

A significant limitation regarding the implementation of these models is the overall lack of annotations for most CXR datasets, which decreases the amount of available information during training and leads to a non-existing control over which features are being incorporated in the decision process. One possible way to solve this problem consists in using AI in conjunction with Eye-Tracking Data (ETD). This type of data, if inexpensively collected during CXR screening sessions, could provide inputs to DL models for the automatic annotation of images or for the guided classification of pathologies. The reason behind the broad range of applications is the fact that, if collected from experts, ETD will contain important encoded information, such as pathology locations. Nevertheless, Eye-Tracking (ET) systems are not available in radiology departments, which limits the amount of ETD and halts the production of real-time DL models.

Despite the short supply of ETD, a number of datasets exist, which have been used in the development of different DL approaches. In Saab et al. [12], features were extracted from the ETD and used as a second source of supervision to assist the detection of pneumothoraces in CXR images. In Karargyris et al. [8], ETD was used once more as a second source of supervision in a multitasking framework, this time in the form of heatmaps (a common ETD representation), where heatmap reconstruction and pathology classification were performed simultaneously. In [3], ETD was incorporated in a student-teacher transformer framework that integrates radiologist visual attention to aid CXR classification. The results displayed in these studies showed a positive influence in terms of classification performance with the use of ETD, which attests the utility of such data. Nonetheless, previous work did not evaluate the ETD impact on the relevance of features used in the classification process through the quantitative analysis of explanations extracted from DL models.

In this work, a framework capable of performing ETD heatmap reconstruction and CXR pathology classification was developed. By using only CXR images as input, the proposed framework becomes independent of ETD during the inference stage. The reconstructed heatmaps are processed in two separate ways and incorporated in the input to the classifier through the element-wise product with the CXR images. In both strategies, the goal is to direct the focus of the classifier towards key regions of a chest radiograph. The success of both approaches was quantitatively assessed by computing metrics that measured the overlap between explanation maps and pathology masks, revealing an improvement for one of the strategies. This work extends its scope to all available public ET datasets associated to CXR pathology screening. Furthermore, the developed framework offers new ways of including ETD directly in the classification process, if increasingly available, by bypassing the heatmap reconstruction step.

2 Materials and Methods

2.1 Datasets

In this work, three datasets were used - EGD [8] (named this way for simplicity), REFLACX [9] and CXR-P [12]. The ETD contained in the datasets consists of fixation coordinates and associated time and complementary data regarding the data collection process and images. The datasets differ in their characteristics, such as number of labels, number of radiologists participating in the study and presence of pathology masks, as shown in Table 1. REFLACX is the most complete dataset of the three, with the existence of thorax bounding boxes and pathology masks being important features of this dataset, since they allow a more complete analysis of the reconstructed heatmaps and model explainability. To our knowledge, these three datasets constitute the sole publicly available sources of ETD collected from a CXR screening setting.

Table 1. Description of the datasets used.

Dataset	Images	Classes	Radiologists	Masks	Thorax segmentation/bounding box
EGD	1,083	3	1	No	Yes
REFLACX	2,616	14	5	Yes	Yes
CXR-P	1,250	1	3	Yes	Yes*

*generated with YOLOv5 [7]

2.2 Heatmap Generation

To train a DL model for heatmap reconstruction, first the original heatmaps had to be generated from the ETD. To do so, Gaussian kernels were applied to 2D arrays with the same sizes as the images, which contained the corresponding time values in the coordinates of each fixation. A number of steps was taken prior to this point: i) the angular resolutions in the screen space were computed by taking into account the distance from the radiologist to the screen and the screen size and resolution; ii) the angular resolutions were converted to the image space, which varied depending on the width and height of each image; iii) fixation diameters were computed for every image, through the product between the angular resolution and the selected angle of 5° - equivalent to the foveal vision [2]; iv) the standard deviations of the Gaussian kernels were defined as being half of the fixation diameters.

This process differed for the REFLACX dataset, since angular resolutions in the image space were already provided. Furthermore, zooming was allowed during the CXR screening sessions, contrarily to EGD, which implied the existence of different diameters for each fixation.

For the CXR-P dataset, there was a lack of complementary data regarding the distance to the screen and its characteristics, which also demanded an adaptation of the proposed method. To circumvent this issue, it was assumed that the data collection conditions for this dataset were similar to EGD, i.e., same screen size and distances.

2.3 Eye-Tracking-Assisted Pathology Screening

The proposed framework for the incorporation of ETD in the CXR screening process is depicted in Fig. 1. It receives only a CXR image as input that is firstly used to reconstruct a heatmap highlighting relevant areas for the decision process, with ETD-derived heatmaps, generated as explained in Sect. 2.2, used as ground truth. The reconstructed heatmaps are then processed in two different ways before the element-wise product with the CXR images, ensuring that the input to the classifier is altered according to the information present on the heatmaps. This step aims at conditioning the classification procedure by shifting the focus of the classifier towards more relevant features, in order to produce more reliable models that can be trusted by medical practitioners.

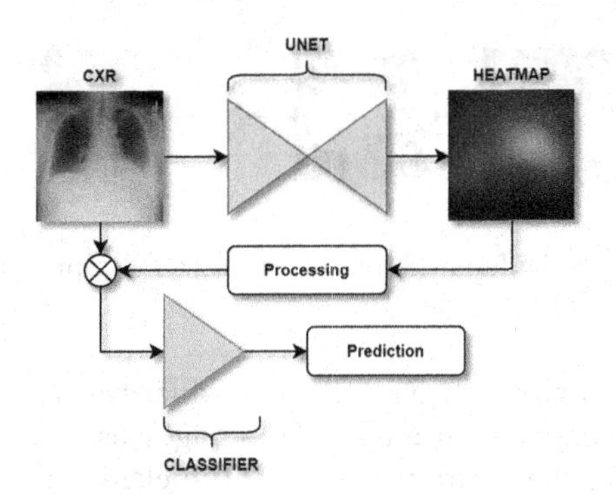

Fig. 1. Schematic of the proposed framework.

Heatmap Reconstruction. The heatmap reconstruction is performed with a UNet [11] architecture, available in [5], containing a DenseNet121 [4] encoder. All models were trained with a batch size of 32, learning rate of 10^{-3} and with a binary cross-entropy loss function. An early stopping strategy was used with a patience of 10. Prior to each batch, the CXR images were submitted to multiple random transforms (rotation, cropping and changes in brightness, contrast, saturation and hue) to augment the data and compensate for limited dataset size.

Reconstructed Heatmap Processing. Before being used for assisting pathology classification, the reconstructed heatmaps undergo a Gaussian filter to increase their smoothness and to avoid the introduction of noise in subsequent steps. Then, two different approaches are followed. The first one consists in binarizing the heatmap using its mean as a threshold and multiplying the resulting mask by the CXR image. In the second approach, the heatmap is simply scaled into the [0,1] range prior to the product with the CXR image. The former was entitled Thorax Segmentation (TS) approach since the binarized heatmaps resemble thorax segmentations. The second approach was named Differential Preservation (DP) since the goal is to preserve areas highlighted by the heatmaps and fade the remaining parts. Figures 2 and 3 represent examples of both strategies.

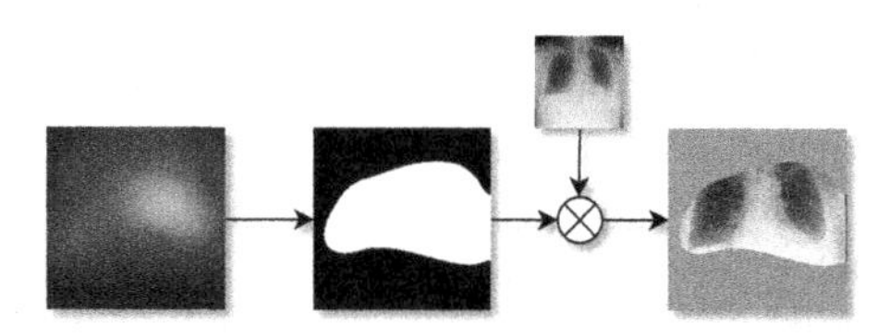

Fig. 2. Thorax segmentation (TS) approach.

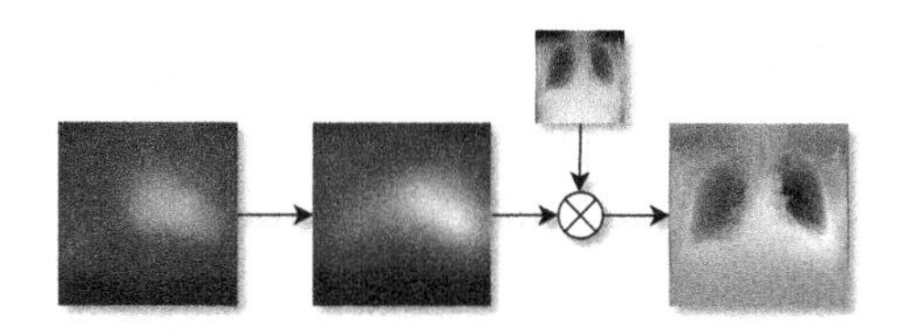

Fig. 3. Differential preservation (DP) approach.

Pathology Classification. The selected classifier was an EfficientNet-b0 [15], responsible for receiving the altered CXR images as input and performing a prediction. Regarding the entire framework, the classifier was either trained in isolation for the TS approach or in conjunction with the UNet for the DP approach, since the gradients could be propagated backwards. In both cases, the UNet for heatmap reconstruction was previously trained until convergence. For the DP approach, this first training stage is essential for a more stable second phase in which the UNet and the classifier are trained together. The same dataset splits, image transformations, hyperparameters and loss function were used to train the UNet and the classifier, except in the second stage of the DP approach where the learning rate of the UNet was 10 times smaller for increased stability.

2.4 Experiments

Two of the datasets were used to separately train the models, namely EGD and REFLACX. A five fold cross-validation approach was used and for each split three folds were used as a training set, one fold as a validation set and the other as a test set, with images from the same patient kept in the same fold. The EGD dataset was created by Karagyris *et al.* and used to develop their model mentioned in Sect. 1, named here EGDmodel and included throughout this work as a term of comparison alongside its respective baseline (EGDbaseline), which contains only a classification block and not the decoder responsible for the heatmap reconstruction. The REFLACX dataset presents an independent and more complex scenario and thus it could be used for an unbiased comparison between the proposed approaches and the state-of-the-art. The models trained on REFLACX were also tested on EGD, since the datasets share two similar classes – Congestive Heart Failure (CHF) and Pneumonia in EGD and Enlarged Cardiac Silhouette and Consolidation in REFLACX –, and on CXR-P, since both contain the Pneumothorax class. This dataset was not used to train any model since only one class was present, which biased the gaze patterns of radiologists towards areas with a higher likelihood for the occurrence of pneumothoraces.

The models and their respective outputs were evaluated using different metrics, such as mean intensities in different locations of the reconstructed heatmaps, Dice Similarity Coefficient (DSC) values between binarized heatmaps and thorax segmentations/bounding boxes and area under the curve (AUC) values. Furthermore, an explainability analysis was performed to assess the impact of the proposed approaches on the quality of the explanations derived from the models.

3 Results

3.1 Heatmap Reconstruction

We first evaluated the mean intensities of heatmaps in different areas, namely inside the pathology masks, inside the thorax bounding box (excluding pathology masks) and outside the thorax, as displayed in Table 2. The goal was to assess the information content of the reconstructed heatmaps in comparison to the original ones. For that purpose, reconstructed heatmaps were first scaled into the same range as the original ones ([0,1]).

The mean intensities for the reconstructed heatmaps derived from both approaches are higher than the original values, most likely because the highlighted areas are wider as shown in Fig. 4. More importantly, it can be seen that there is a preserved contrast in terms of intensity between pathology masks and surrounding regions, albeit smaller for the reconstructed heatmaps. This contrast is slightly higher for the DP approach, probably because of the second UNet training stage in which the classification loss is also considered. Regarding the contrast between the thorax and the outer parts of CXR images, the values are even higher when compared to the original heatmaps. The contrasts obtained for both approaches are comparable to the state-of-the-art. It is worth noting that the heatmaps used in [8] to train the EGDmodel were generated using a Gaussian with a fixed and smaller standard deviation, which might explain lower mean intensities for the reconstructed heatmaps.

Table 2. Mean intensities in original and reconstructed heatmaps from REFLACX.

Heatmaps	Mask	Bounding box	Outside
Original	0.44 ± 0.01	0.26 ± 0.01	0.06 ± 0.01
TS	0.60 ± 0.01	0.50 ± 0.03	0.13 ± 0.01
DP	0.60 ± 0.03	0.47 ± 0.02	0.12 ± 0.01
EGDmodel	0.45 ± 0.01	0.35 ± 0.01	0.06 ± 0.01

As previously mentioned, the reconstructed heatmaps are wider and also more central, and mostly highlight the thorax. In some cases they even seem to more closely match the pathology masks, but in more complex cases, where numerous labels are present, the heatmaps often fail to correctly highlight every pathological location. As expected, these errors are correlated to the number of instances of each label and to the limited dataset size, which advocates for the need to collect more ETD. Regarding the comparison between TS and DP heatmaps it is possible to see that, even though the differences in some cases are insignificant, there are instances where a more considerable overlap occurs between heatmap and mask for the DP approach (Fig. 4, bottom row).

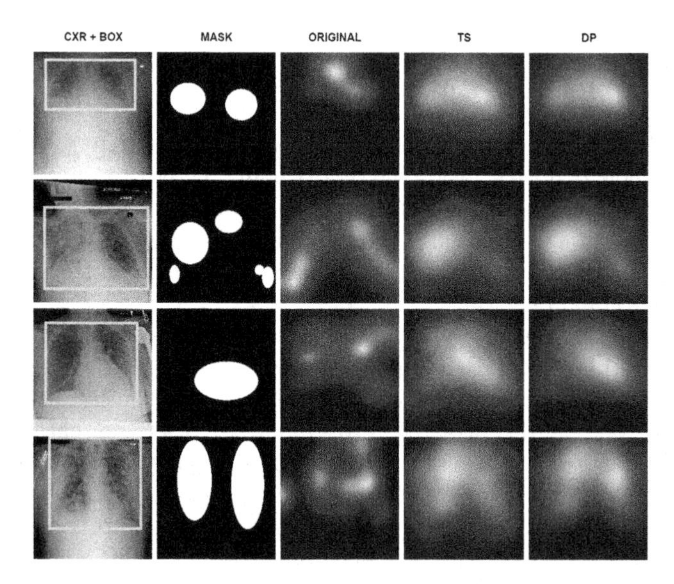

Fig. 4. Examples of original and reconstructed heatmaps for the REFLACX dataset alongside CXR images, bounding boxes and pathology masks.

Since the TS approach consists in binarizing the reconstructed heatmaps into thorax-like segmentations, DSC values were computed for all the datasets and models, as shown in Table 3, with regard to the thorax segmentations/bounding boxes. Even though the main purpose of this strategy is to discard irrelevant parts of the CXR images and not necessarily to segment the thorax, comparing the obtained masks to the ground truth provided in the datasets is a suitable way to assess what areas are being segmented. The obtained values indicate a relatively high similarity between the binarized heatmaps and the thorax segmentations/bounding boxes for all datasets and models. They also indicate that reconstructed heatmaps after binarization resemble the thorax more in comparison to the original heatmaps, which indicates that the models are less discriminative of specific regions and more focused on the whole thorax (Fig. 4). In Fig. 5, examples of segmentations obtained for the TS approach are shown.

Table 3. Mean DSC values between thorax segmentations/bounding boxes and heatmaps binarized using the mean as a threshold.

Train Dataset	Test Dataset	Original	TS	DP	EGDmodel
EGD	EGD	0.77 ± 0.01	0.82 ± 0.01	0.80 ± 0.02	0.86 ± 0.01
REFLACX	REFLACX	0.76 ± 0.01	0.81 ± 0.01	0.81 ± 0.01	0.82 ± 0.01
REFLACX	EGD	0.77*	0.85 ± 0.02	0.84 ± 0.01	0.89 ± 0.01
REFLACX	CXR-P	0.66*	0.82 ± 0.02	0.82 ± 0.01	0.84 ± 0.01

*no standard deviation exists since there are no model/test splits

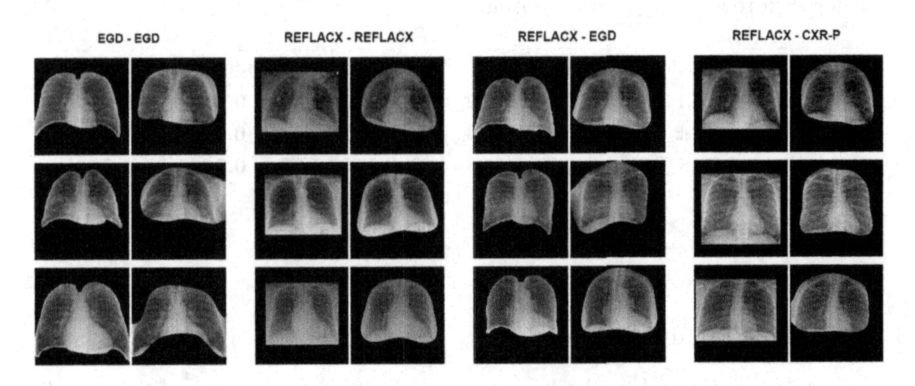

Fig. 5. Examples for the different datasets of thorax segmentations/bounding boxes paired with heatmap-derived segmentations (left and right, respectively).

3.2 Pathology Classification

To assess the pathology classification capacity of the developed approaches, AUC values were computed for models trained and tested on EGD, and for models trained on REFLACX and tested on REFLACX, EGD and CXR-P. AUC values were also obtained for the considered baseline, an EfficientNet-b0 classifier, and for the EGDmodel and EGDbaseline. Since this state-of-the-art model accounts for an extensive optimization and no data augmentation, the models had to be uniformized before an accurate comparison could be made. Therefore, AUC values were also computed for models stripped of their optimizations and with the same data augmentation transforms applied, as shown in Tables 4, 5 and 6.

Mean AUC values for the EGD dataset are higher for the EGDmodel, something not verified in the context of models without their respective optimizations. We can also observe that the Pneumonia class is harder to diagnose, since AUC values are lower. For REFLACX, the optimized models, as expected, perform slightly better on a general basis. Furthermore, no significant differences are verified between the chosen baseline and the proposed approaches, and in comparison to the state-of-the-art. For this dataset, besides the overall mean AUC, an exclusive mean was also computed which excluded classes with fewer examples (Acute fracture, Enlarged hilum, Hiatal hernia and Interstitial lung disease) and that present a high AUC variability between splits. This significantly improves the mean AUC. For the inter-dataset experiments with models trained on REFLACX, it is possible to see that values obtained for CHF are similar to the ones obtained with models trained on EGD. However, that is not the case for the Pneumonia class, since some differences may exist in comparison to the REFLACX class Consolidation. Regarding the model comparison, once more there are no significant differences in terms of classification results (Table 6).

Table 4. Mean AUC values obtained for every class in EGD (bold indicates the best result for each row; w.o. - without optimization).

Class	EfficientNet-b0	TS	DP	EGDmodel	EGDbaseline
Normal	0.87 ± 0.03	0.88 ± 0.02	0.88 ± 0.03	**0.89 ± 0.01**	0.85 ± 0.02
CHF	0.89 ± 0.03	0.87 ± 0.03	0.87 ± 0.05	**0.91 ± 0.03**	0.86 ± 0.03
Pneumonia	0.71 ± 0.03	0.70 ± 0.03	0.69 ± 0.03	**0.73 ± 0.06**	0.68 ± 0.04
Mean	0.83	0.82	0.81	**0.84**	0.80
Mean (w.o.)	**0.82**	0.81	0.81	0.81	**0.82**

3.3 Model Explainability

In order to assess if the described approaches are shifting the focus of the classifier towards relevant parts of the CXR images, LayerCAM [6] maps were produced using layers from the fourth block onwards of the EfficientNet-b0 classifier. LayerCAM differs from the commonly used approach, GradCAM [14], mainly because it incorporates information from several layers and creates more fine-grained class activation maps, which can be especially important for CXR images due to the occurrence of small lesions. A methodology similar to the one performed in [13] was used to evaluate the quality of explanations, where mean Intersection over Union (IoU) and hit rate values were computed from the comparison between explainability maps and pathology masks. Unlike in [13], here LayerCAM maps were binarized using the mask area of the correspondent class to select a given number of highlighted pixels. This means that large masks will be

Table 5. Mean AUC values obtained for every class in REFLACX.

Class	EfficientNet-b0	TS	DP	EGDmodel	EGDbaseline
Abnormal mediastinal contour	0.63 ± 0.05	0.63 ± 0.05	**0.69 ± 0.05**	0.67 ± 0.04	0.63 ± 0.07
Acute fracture	0.64 ± 0.22	0.65 ± 0.05	**0.72 ± 0.07**	0.71 ± 0.09	0.72 ± 0.09
Atelectasis	**0.78 ± 0.01**	0.77 ± 0.03	**0.78 ± 0.03**	**0.78 ± 0.02**	0.76 ± 0.02
Consolidation	0.80 ± 0.02	0.80 ± 0.01	**0.81 ± 0.02**	0.80 ± 0.03	0.75 ± 0.03
Enlarged cardiac silhouette	**0.84 ± 0.02**	**0.84 ± 0.01**	0.83 ± 0.02	0.83 ± 0.02	0.79 ± 0.02
Enlarged hilum	0.57 ± 0.06	0.58 ± 0.07	0.57 ± 0.07	**0.65 ± 0.07**	0.55 ± 0.07
Groundglass opacity	**0.70 ± 0.03**	0.67 ± 0.03	0.66 ± 0.02	**0.70 ± 0.02**	0.62 ± 0.02
Hiatal hernia	0.51 ± 0.19	0.55 ± 0.14	**0.62 ± 0.07**	0.56 ± 0.10	0.58 ± 0.14
High lung volume/Emphysema	**0.87 ± 0.08**	0.85 ± 0.07	0.84 ± 0.09	**0.87 ± 0.06**	0.84 ± 0.06
Interstitial lung disease	0.79 ± 0.11	0.74 ± 0.15	0.78 ± 0.09	**0.83 ± 0.09**	0.69 ± 0.11
Lung nodule or mass	0.63 ± 0.05	0.64 ± 0.05	**0.66 ± 0.05**	0.61 ± 0.05	0.60 ± 0.07
Pleural abnormality	**0.85 ± 0.02**	0.84 ± 0.01	0.84 ± 0.03	**0.85 ± 0.01**	0.82 ± 0.02
Pneumothorax	**0.74 ± 0.08**	0.71 ± 0.04	0.72 ± 0.08	0.71 ± 0.05	0.66 ± 0.09
Pulmonary edema	0.86 ± 0.02	0.84 ± 0.02	0.84 ± 0.03	**0.87 ± 0.02**	0.81 ± 0.03
Overall mean	0.73	0.72	0.74	**0.75**	0.70
Overall mean (w.o.)	0.72	**0.73**	**0.73**	0.71	0.72
Exclusive mean	**0.77**	0.76	**0.77**	**0.77**	0.73
Exclusive mean (w.o.)	**0.76**	**0.76**	**0.76**	0.75	**0.76**

Table 6. Mean AUC values for REFLACX models tested on EGD and CXR-P.

Class	EfficientNet-b0	TS	DP	EGDmodel	EGDbaseline
CHF	0.87 ± 0.01	**0.88 ± 0.01**	0.86 ± 0.02	**0.88 ± 0.01**	0.84 ± 0.02
CHF (w.o.)	0.86 ± 0.01	0.86 ± 0.01	0.86 ± 0.02	0.86 ± 0.01	0.86 ± 0.01
Pneumonia	**0.62 ± 0.01**	0.61 ± 0.02	0.61 ± 0.02	0.58 ± 0.02	0.60 ± 0.04
Pneumonia (w.o.)	**0.61 ± 0.02**	0.59 ± 0.04	0.60 ± 0.01	**0.61 ± 0.02**	**0.61 ± 0.02**
Pneumothorax	0.70 ± 0.01	0.73 ± 0.02	0.74 ± 0.02	**0.76 ± 0.02**	0.64 ± 0.04
Pneumothorax (w.o.)	0.71 ± 0.03	**0.72 ± 0.02**	**0.72 ± 0.01**	0.68 ± 0.05	0.71 ± 0.03

compared to binarized saliency maps with a bigger area and that small masks will be compared to binarized saliency maps of smaller size. It is worth noting that all models considered in this work use an EfficientNet-b0 as a classifier, including the state-of-the-art ones, which allows an unbiased application of the LayerCAM method to perform this explainability analysis.

Overall, the LayerCAM maps for the TS approach are less disperse and more correlated to the pathology masks in comparison with the baseline and the DP approach (Fig. 6), which is in concordance with the mean IoU and hit rate values obtained for the different models across all five REFLACX splits and presented in Fig. 7. It is possible to see that a significant difference exists between both metrics, since the hit rate constitutes a less strict evaluation method. The attained values are similar throughout the models, except for the TS approach which slightly outperforms the remaining ones. This shows that excluding irrelevant parts of the CXR images may benefit the model by shifting its focus towards pathological areas. No improvement in explanation quality was observed for the

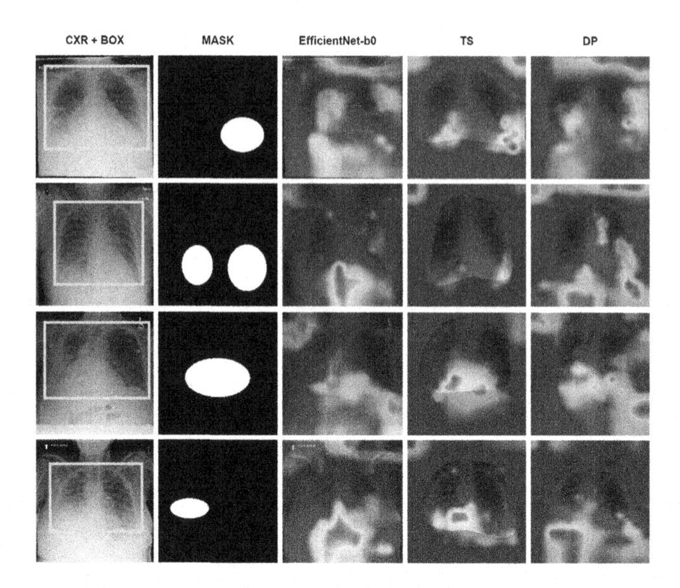

Fig. 6. LayerCAM examples for different models trained on REFLACX alongside CXR images, bounding boxes and pathology masks. Labels are Atelectasis, Consolidation, Enlarged cardiac silhouette and Pleural abnormality, from top to bottom.

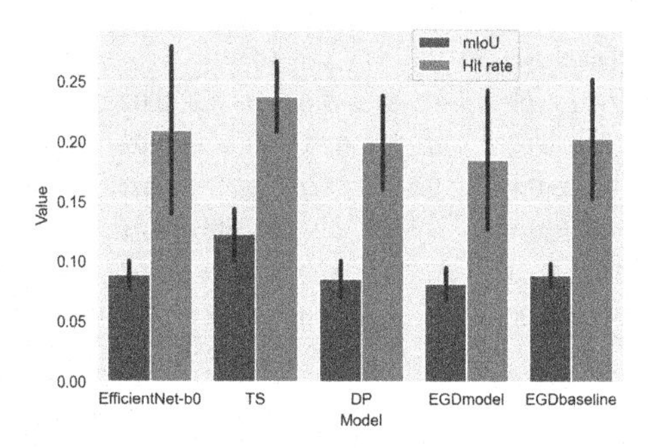

Fig. 7. Mean IoU and hit rate values for LayerCAM explanations from different models trained on REFLACX. Error bars indicate standard deviations across the five splits.

DP approach, probably because irrelevant information is still preserved to some extent although with a lower intensity than the remaining regions. This may lead the model towards less meaningful sets of features, which is also the case for the EGDmodel since no input conditioning occurs.

4 Conclusions

In this work, ETD was used in the form of reconstructed heatmaps to aid pathology classification. The main objective was to guide the classifiers towards important areas of CXR images, in order to perform correct predictions while focusing on meaningful features. This relates to the concept of explainability, which has been increasingly explored in recent times, since a good model is not just capable of classifying an image, but is also capable of outputting trustworthy explanations. This acquires an increased importance for DL applications in the medical domain, since health-related decisions need to be based on strong evidence.

The results obtained here show that the classification performance of the proposed approaches is similar to the considered baseline and to the state-of-the-art. More importantly, the results reveal that for the TS approach there is a higher resemblance between LayerCAM-derived saliency maps and pathology masks. This shows that masking medical images might be a good way of shifting the focus of classifiers towards relevant features of a CXR image, producing higher quality explanations and thus more reliable models. Furthermore, by masking out parts of the images outside of the thorax, common markings on the top corners, which have been shown to influence model decisions [1], are discarded. For the DP approach and for the EGDmodel, however, no positive link was found between the use of heatmaps and model attention.

Altogether, the work developed proposes new ways of incorporating ETD in DL models for CXR screening. The proposed approaches are also applicable in scenarios where a constant source of ETD is available, in case ET systems are included in radiology departments. However, current systems are unpractical and hinder the clinical workflow, which explains the limited number of datasets up to date. Developing new ET systems will therefore be essential for acquiring more data and training more accurate models (e.g., systems that use virtual reality [10]). Regarding the proposed models, future work will focus on performing a more extensive optimization and on testing new explainability methods, accompanied by a strong evaluation of their validity.

References

1. Allen, P.G., Degrave, A.J., Janizek, J.D. Janizek, J.D., Lee, S.I.: AI for radiographic COVID-19 detection selects shortcuts over signal. Nat. Mach. Intell. **3**(7), 610–619 (2021)
2. Aresta, G., et al.: Automatic lung nodule detection combined with gaze information improves radiologists' screening performance. IEEE J. Biomed. Health Inform. **24**, 2894–2901 (2020)
3. Bhattacharya, M., Jain, S., Prasanna, P.: RadioTransformer: a cascaded global-focal transformer for visual attention–guided disease classification. In: Avidan, S., Brostow, G., Cisse, M., Farinella, G.M., Hassner, T. (eds.) Computer Vision – ECCV 2022. ECCV 2022. LNCS, vol. 13681, pp. 679–698. Springer, Cham (2022). https://doi.org/10.1007/978-3-031-19803-8_40
4. Huang, G., Liu, Z., Maaten, L.V.D., Weinberger, K.Q.: Densely connected convolutional networks. In: Proceedings - 30th IEEE Conference on Computer Vision and Pattern Recognition, CVPR 2017 2017-January, 2261–2269, November 2017
5. Iakubovskii, P.: Segmentation Models Pytorch (2019)
6. Jiang, P.T., Zhang, C.B., Hou, Q., Cheng, M.M., Wei, Y.: LayerCAM: exploring hierarchical class activation maps for localization. IEEE Trans. Image Process. **30**, 5875–5888 (2021)

7. Jocher, G., et al.: ultralytics/yolov5: v7.0 - YOLOv5 SOTA Realtime Instance Segmentation, November 2022
8. Karargyris, A., et al.: Creation and validation of a chest X-ray dataset with eye-tracking and report dictation for AI development. Sci. Data **8**(1), 1–18 (2021)
9. Lanfredi, R.B., et al.: Reflacx, a dataset of reports and eye-tracking data for localization of abnormalities in chest x-rays. Sci. Data **9**(1), 1–15 (2022)
10. Moreira, C., Nobre, I.B., Sousa, S.C., Pereira, J.M., Jorge, J.: Improving X-ray diagnostics through eye-tracking and XR. In: Proceedings - 2022 IEEE Conference on Virtual Reality and 3D User Interfaces Abstracts and Workshops, VRW 2022, pp. 450–453 (2022)
11. Ronneberger, O., Fischer, P., Brox, T.: U-Net: convolutional networks for biomedical image segmentation. In: Navab, N., Hornegger, J., Wells, W.M., Frangi, A.F. (eds.) MICCAI 2015. LNCS, vol. 9351, pp. 234–241. Springer, Cham (2015). https://doi.org/10.1007/978-3-319-24574-4_28
12. Saab, K., et al.: Observational supervision for medical image classification using gaze data. In: de Bruijne, M., et al. (eds.) MICCAI 2021. LNCS, vol. 12902, pp. 603–614. Springer, Cham (2021). https://doi.org/10.1007/978-3-030-87196-3_56
13. Saporta, A., et al.: Benchmarking saliency methods for chest x-ray interpretation. Nat. Mach. Intell. **4**(10), 867–878 (2022)
14. Selvaraju, R.R., Cogswell, M., Das, A., Vedantam, R., Parikh, D., Batra, D.: Grad-CAM: visual explanations from deep networks via gradient-based localization. Proceedings of the IEEE International Conference on Computer Vision 2017-October, pp. 618–626, December 2017
15. Tan, M., Le, Q.: EfficientNet: rethinking model scaling for convolutional neural networks. In: Chaudhuri, K., Salakhutdinov, R. (eds.) Proceedings of the 36th International Conference on Machine Learning. Proceedings of Machine Learning Research, vol. 97, pp. 6105–6114. PMLR, 09–15 June 2019

Deep Neural Networks to Distinguish Between Crohn's Disease and Ulcerative Colitis

José Maurício[1] and Inês Domingues[1,2(✉)]

[1] Instituto Politécnico de Coimbra, Instituto Superior de Engenharia, Rua Pedro Nunes - Quinta da Nora, 3030-199 Coimbra, Portugal
{a2018056151,ines.domingues}@isec.pt

[2] Centro de Investigação do Instituto Português de Oncologia do Porto (CI-IPOP): Grupo de Física Médica, Radiobiologia e Protecção Radiológica, Porto, Portugal

Abstract. The number of patients with inflammatory bowel disease (IBD) has been increasing. The diagnosis is a difficult task for the gastroenterologist performing the endoscopic examination. However, in order to prescribe medical treatment and provide quality of life to the patient, the diagnosis must be quick and accurate. Therefore, it is important to develop tools that, based on the characteristics of the inflammation present in the mucosa, automatically recognise the type of inflammatory bowel disease. This paper presents a study where the objective was to collect and analyse endoscopic images referring to Crohn's disease and Ulcerative colitis using six deep learning architectures: AlexNet, ResNet50, InceptionV3, VGG-16, ResNet50+MobileNetV2, and a hybrid model. The hybrid model consists of the combination of two architectures, a CNN and a LSTM. This work also presents techniques that can be used to pre-process the images before the training to remove accessory elements from the image. The obtained results demonstrate that it is possible to automate the process of diagnosing patients with IBD using convolutional networks for processing images collected during an endoscopic examination, and thus develop tools to help the medical specialist diagnose the disease.

Keywords: Inflammatory Bowel Disease · CNN · image classification · Crohn's disease · Ulcerative colitis

1 Introduction

Inflammatory bowel disease (IBD) is characterized as a disease of unknown cause that results from the interaction between genetic and environmental factors, triggering an immune response that causes digestive disorders and inflammation in the gastrointestinal tract. These types of diseases are divided into Ulcerative colitis and Crohn's disease [17]. However, some studies have shown that some of the diagnoses of inflammatory bowel disease refer to a different category of disease, called indeterminate colitis. This pathology refers to colitis with an

A. Pertusa et al. (Eds.): IbPRIA 2023, LNCS 14062, pp. 533–544, 2023.
https://doi.org/10.1007/978-3-031-36616-1_42

extreme degree of severity that has features of both diseases (Ulcerative colitis and Crohn's disease), making it difficult to distinguish between them [16].

It is estimated that in Portugal it already affects 7,000 to 15,000 people, with 2.9 cases per 100,000 inhabitants per year with Ulcerative colitis and 2.4 cases per 100,000 inhabitants per year with Crohn's disease [7]. In Europe, the prevalence of the disease is more than 1 in 198 people for Ulcerative colitis and 1 in 310 people for Crohn's disease [26]. And if the comparison is made worldwide, the prevalence of Ulcerative colitis is 7.6 in 246 cases per 100,000 people per year, preceded by Crohn's disease with 3.6 in 214 cases per 100,000 people per year. This disease has a higher incidence in North America and Europe. However, in recent decades, it has seen an increase in countries such as China and India [11].

As the number of patients worldwide with IBD has been increasing, its diagnosis has become an arduous task for medical specialists and may have some limitations in recognising the disease. To automate the diagnosis of this chronic disease and thus reduce the patient's waiting time in prescribing medical treatment. This work seeks, through the selection of six deep learning architectures, to classify images related to the type of disease (i.e., Crohn's disease and Ulcerative colitis) that were previously collected using tests such as endoscopy and colonoscopy and from the characteristics present in the mucosa according to the type of inflammation, to predict the type of inflammatory bowel disease.

The proposed methodology allows the pre-processing of images to treat the observations made by gastroenterologists during the endoscopic examination, allowing the CNN architectures to focus on the part of the image where the intestinal mucosa is located. In addition, this study presents the development of two hybrid architectures: (i) the junction of a CNN with an LSTM; (ii) the combination of convolutional layers from the ResNet50 network with fully connected layers from the MobileNetV2 network. Which, as demonstrated by the obtained results, are accurate in predicting the type of inflammatory bowel disease.

In summary, the contributions of this work include:

- The derivation of two versions of a new database for the distinction of Crohn's disease from Ulcerative colitis
- The proposal of three pre-processing techniques to minimize the impact of artifacts in the training data
- The development of two new Hybrid Deep Learning Architectures
- Extensive testing and validation (two versions of the database, combined with three pre-processing methods, combined with six architectures, totalling 36 experimental settings)

This paper is divided into four sections: Sect. 2 compiles some works of interest; Sect. 3 presents the methodology developed during this work and the datasets that were used for the classification; Sect. 4 collects the quantitative results; and Sect. 5 summarises the conclusions and directions for future work.

2 Related Work

In this section, a summary of the findings made by some related work within the scope of this study are presented. Along the years, the use of computer vision tools in the diagnosis of inflammatory bowel disease has become essential. Since the evaluation of images obtained through imaging examinations is done by professional gastroenterologists, the criteria used to determine the patient's diagnosis is subjective and may differ among doctors [19]. This process becomes a consequence for the patient as it results in a delay in prescribing an effective treatment to combat the disease.

The authors in [13] developed a study using the CrohnIPI dataset. It consisted of 2124 images without pathology, 1360 images with pathology, and 14 inconclusive images. The goal was to develop a recurrent attention neural network for diagnosing mucosal lesions caused by inflammation. In addition, they also tested networks such as ResNet34, VGGNet16, and VGGNet19 under the same conditions. The experiment was conducted in three rounds: in the first round, the images were labelled by a medical expert, and the networks were trained based on those images; in the second round, the images identified in the first round were validated by three experts; and in the third round, the four medical experts met in several sessions to reduce the number of inconclusive images. They concluded that the proposed network achieved its best performance in the last round with 93.70% accuracy.

Other authors developed a study in [22], where they propose a recurrent attention neural network to classify images with lesions. In this study, the authors used the CrohnIPI dataset to train the proposed network and then used images of several lesions from the CAD-CAP dataset presented during the Giana competition to test [9]. Based on the results presented, the proposed network obtained 90.85% accuracy with the CrohnIPI dataset, and when it was tested with the images from the CAD-CAP dataset, it obtained an accuracy of 99.67%.

In the study developed in [14], the authors sought to develop a system that, through deep learning, could reduce subjectivity and improve confidence in the diagnosis of Ulcerative colitis (UC) disease. Therefore, the authors used the LIMUC dataset to train a proposed deep learning regression-based approach to assess the severity of UC according to the Mayo endoscopic score (MEC). They conclude that this proposed approach after training on $k = 10$ folds and compared to state-of-the-art convolutional neural networks, was more robust in classifying images from the test dataset. This is based on the results obtained on macro F1-score metrics and quadratic weighted Kappa.

Based on the findings made by other authors, this work seeks to stand out by proposing to pool the images by the type of inflammatory bowel disease to distinguish between Crohn's disease and Ulcerative colitis. A methodology that includes some image pre-processing techniques for removing redundant features from endoscopic exam images is also suggested. Moreover, a set of six classification metrics allow the performance of the deep learning architectures proposed in this work to be critically evaluated.

3 Methodology

In order to improve the diagnosis of inflammatory bowel disease, a methodology based on six phases is proposed. In the first phase, the authors collected images of the two inflammatory bowel diseases (i.e. Ulcerative colitis and Crohn's disease). In the second phase, frames were extracted from the videos collected during an endoscopic examination performed on the patients. In the third phase, was applied a pre-processing to the images. In the fourth phase, data augmentation was applied to the training dataset. In the fifth phase, the convolutional networks were implemented and configured. Finally, the performance of the CNNs was evaluated using the classification metrics: Accuracy, Loss, Precision, Recall, F1-Score, Area Under Curve and Mathew's Correlation Coefficient. This methodology is illustrated in Fig. 1 and will be further explained in the following sections.

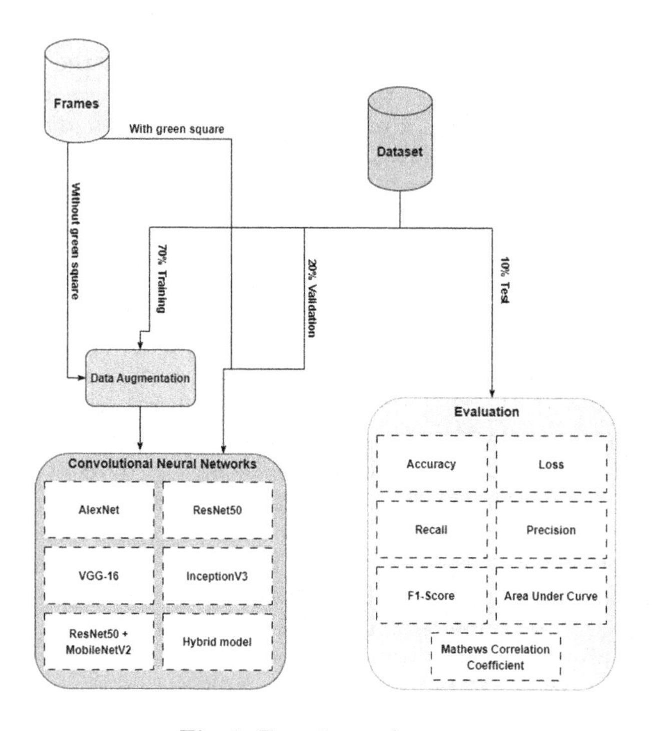

Fig. 1. Experimental setup

3.1 Dataset

Since we were not able to find a dataset with examples from both Crohn's disease and Ulcerative colitis, for the development of this study, three datasets of images referring to Crohn's disease and Ulcerative colitis were used, with the goal of creating a single dataset involving images of both diseases. As illustrated in Fig. 2, HyperKvasir [1,2] and LIMUC [15] datasets were used to get images of the Ulcerative colitis disease, while the CrohnIPI [6] dataset was selected to

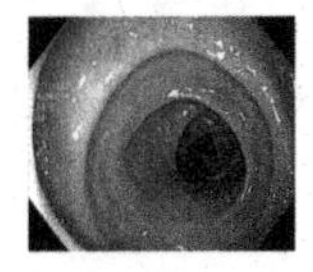
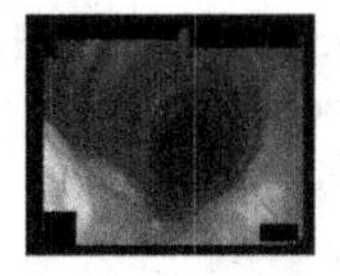
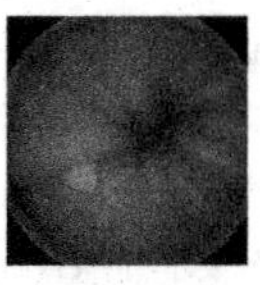

Fig. 2. Ulcerative colitis image from the HyperKvasir dataset (left); Ulcerative colitis image from the LIMUC dataset (center); Crohn's disease image from the CrohnIPI dataset (right).

collect images of Crohn's disease. This last dataset has already been used in other works to develop tools for diagnosing the disease [23,24].

We started by gathering the images present in the CrohnIPI dataset that were divided by lesion type and assigning them to Crohn-disease class. Then we also collected the ulcerative colitis disease images from the HyperKvasir dataset, which were divided by severity according to the Mayo endoscopic score (MAE) and assign them to ulcerative-colitis class. Figure 3 (left) demonstrates the initial distribution of the classes in the created dataset. This dataset does not contain the class "Normal" included in the CrohnIPI dataset, which was excluded. Fourteen images were also ignored due to the lack of labelling.

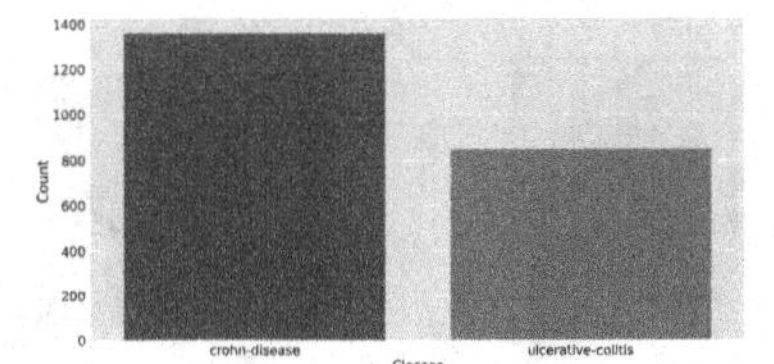

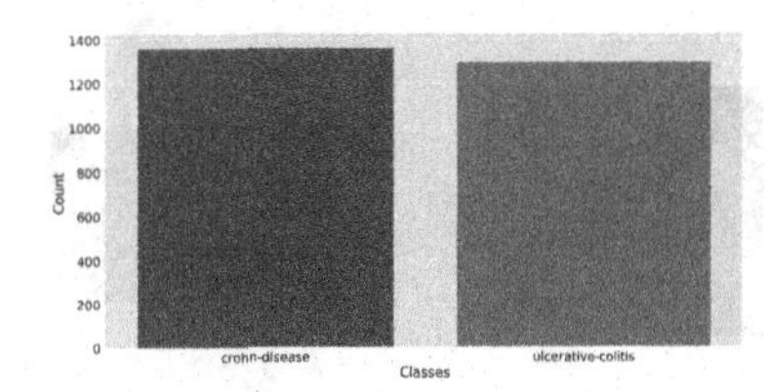

Fig. 3. Class distribution of the imbalanced dataset (left) and the balanced dataset (right). The blue bar represents Crohn's disease, while the orange bar refers to Ulcerative colitis. (Color figure online)

After a first inspection of the distribution of the dataset, it was concluded that the classes are not balanced. The ulcerative-colitis class is in the minority when compared to the Crohn's disease class. To mitigate this problem, an additional dataset was created that maintained the basis of the first dataset with the addition of 444 more images from the LIMUC dataset. Since the LIMUC dataset was divided by the degree of severity according to the Mayo score, the images were collected, taking into account 25% of each severity grade. 111 images from Mayo 0, 111 images from Mayo 1, 111 images from Mayo 2, and 111 images from Mayo 3.

The HyperKvasir dataset, besides the above-mentioned images, contains videos identified by gastroenterologists. With these videos, the goal is to extract more images related to ulcerative colitis disease to balance the classes of the created dataset. We proceeded to extract 20 frames per second with the same dimension as the dataset images (572 × 531). This procedure resulted in the extraction of a total of 64 frames. Figure 3 (right) demonstrates the distribution of the classes of this more balanced dataset.

3.2 Pre-Processing

In the images and frames extracted from a video of the HyperKvasir dataset, there was a green square in the lower left corner, which represents the position of the endoscope during the diagnostic examination. This may become a problem for classifying the type of inflammatory bowel disease present in the images provided to CNN. Therefore, experiments were performed where some transformations were applied when this type of situation was verified.

One of the applied transformations was a horizontal crop across the entire width of the image to remove the green square in the lower left corner. Then a few more experiments were performed, where a vertical crop was applied over the entire height of the image to remove that part of the image. However, during these transformations, some information regarding the mucosa of the intestine is lost. Some more experiments were performed in which a Gaussian blur of 2/16 was applied only to the area where the green square was. The goal was to preserve as much of the image information as possible but to disguise that feature in some images. An illustration of these pre-processing methods is available in Fig. 4.

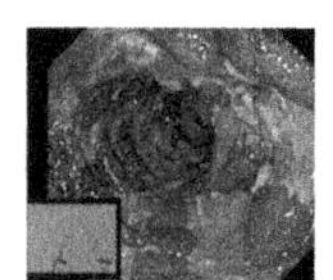
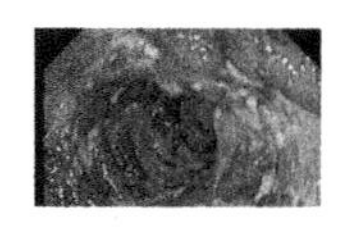
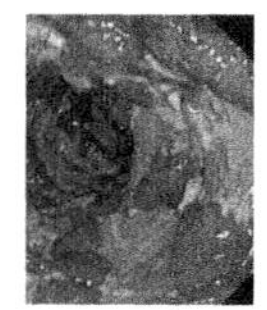
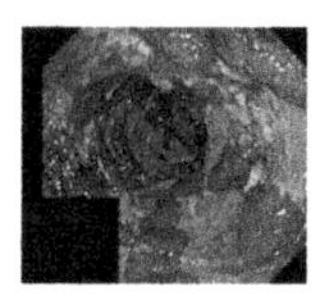

Fig. 4. Pre-processing examples: Original image (left); Horizontal crop (center-left); Vertical crop (center-right); Gaussian blur (right).

3.3 Experimental Setup

In conducting this study, Tensorflow, version 2.11.0, was used. The programming environment for importing the libraries was Google Colab with the NVIDIA T4 Tensor Core GPU.

Data augmentation was performed on the train set only with horizontal and vertical Random Flip, Random Contrast with a factor of 0.15, Random Rotation with a factor of 0.2 and a Random Zoom with a portion of −0.2 for height and −0.3 for width.

The presence of the green square in the lower-left corner of the images obtained from the HyperKavsir dataset, rose the concern that this could lead the models to identify the green square as a disease feature. Therefore, six experiments were performed to test these hypotheses, by combining the three pre-processing techniques presented in Sect. 3.2 with the two dataset versions described in Sect. 3.1. We refer to Table 1 for the configurations of each experiment.

Table 1. Configurations of the six performed experiments

Pre-processing	Dataset	
	Not balanced	Balanced
Horizontal crop	Exp 1	Exp 2
Vertical crop	Exp 3	Exp 4
Gaussian blur	Exp 5	Exp 6

3.4 Convolutional Neural Networks

The following architectures were used to classify images of inflammatory bowel diseases: AlexNet, ResNet50, VGG-16, and InceptionV3. These architectures have been pre-trained with the ImageNet dataset. Besides these architectures, another architecture was built that combines a ResNet50 with a MobileNetV2 network. Finally, a hybrid model was built where a CNN is combined with an LSTM.

The architecture that combines the ResNet50 network and the MobileNetV2 network consists of freezing the weights of all layers of the ResNet50 network and extracting the convolutional layers from it [10]. Then the classification layers of the MobileNetV2 network are extracted to classify the features extracted by the convolutional layers of the ResNet50 network. This architecture is illustrated in Fig. 5.

The hybrid model built in this work was based on a similar architecture developed by other authors [18,21,25]. This architecture is based on: 8 Conv2D layers, 2 BatchNormalization layers, 4 MaxPooling2D layers, 1 Flatten layer, 7 Dense layers, 3 Dropout layers and 3 LSTM layers. The hybrid model is illustrated in Fig. 6. LSTM networks contain a forget gate that learns which information should be forgotten and which should be remembered, sending this information to the next layers [21].

The architectures used to classify the images were trained for 200 epochs, using SparseCategoricalCrossentropy as the loss function, and Adam with a learning rate of 1E-05 was set as the optimizer [12]. During training, an EarlyStopping callback was used with patience of 5 to monitor the validation accuracy.

3.5 Evaluation

In all experiments, the dataset was split into 70% for training, 20% for validation, and 10% for testing [5]. In total, of the 64 frames extracted, 30 frames belonged to the same video in which the green square in the lower left corner was found. These frames were attached to the images in the validation dataset. On the other hand, the remaining frames that belonged to another video and did not have the green square were introduced in the training dataset. Both of the gathered frames have visible ulcerative colitis lesions.

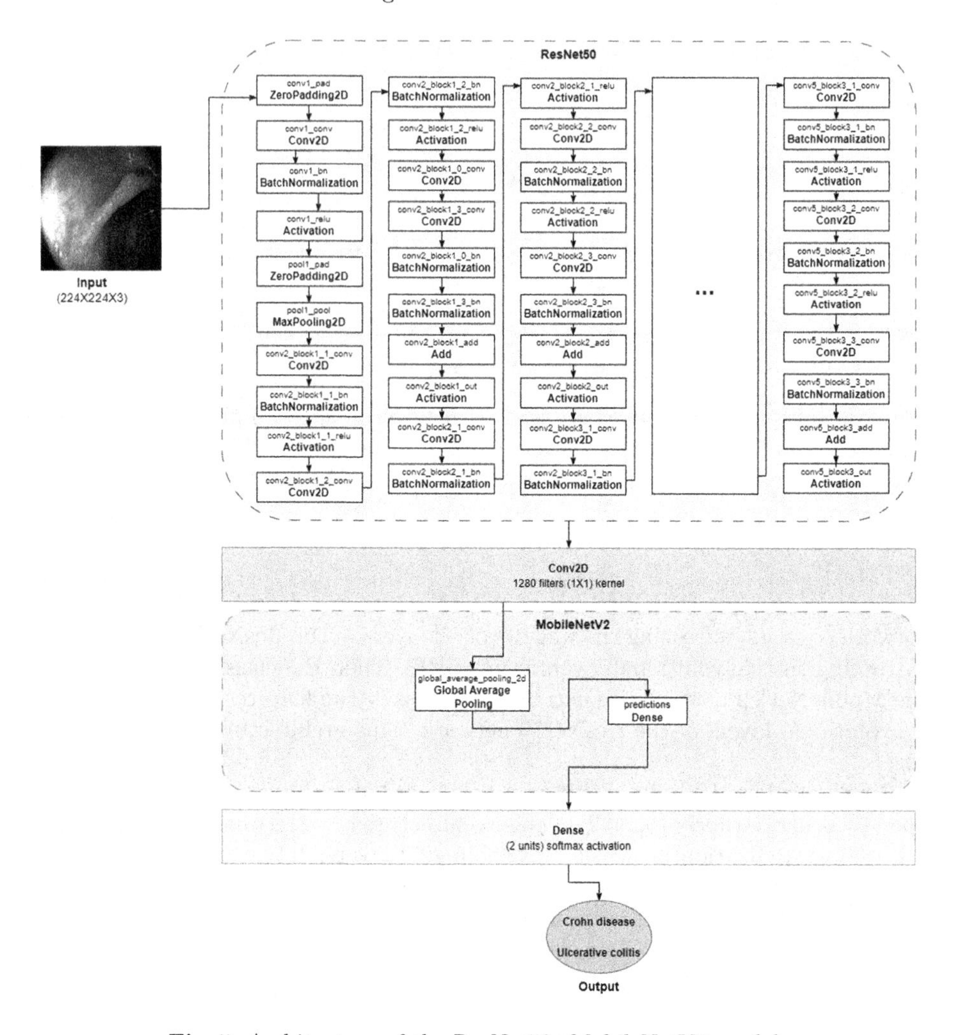

Fig. 5. Architecture of the ResNet50+MobileNetV2 model

Classification metrics were selected to evaluate the network's performance on the test dataset. These include Accuracy, Loss, Precision, Recall, F1-Score, Area Under Curve (AUC) and Mathew's Correlation Coefficient (MCC) [4,20].

$$Loss = -\sum_{i=1}^{n} y_i \cdot \log \hat{y}_i$$

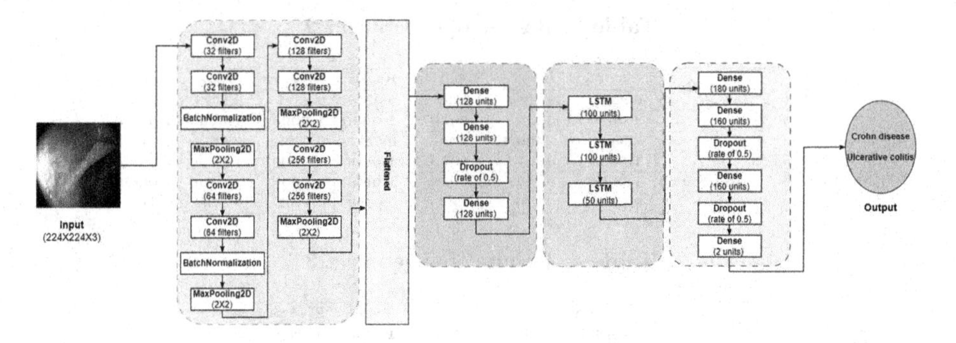

Fig. 6. Architecture of the hybrid model

4 Results

Tables 2 and 3 show the results obtained in experiments 1 and 2, respectively (we refer back to Table 1 for the configurations of each experiment). In Table 3, it is possible to observe the impact that the balanced dataset causes on the results.

Tables 4 and 5 show the results obtained in experiments 3 and 4, respectively. The analysis of the results shows a drop in the results of some architectures when compared to the results of Tables 2 and 3. However, in experiment 4, the hybrid model shows an improvement of 0.70% in Accuracy, 4.46% in Recall, 2.23% in AUC and 4.84% in MCC when comparing the results to experiment 2.

Tables 6 and 7 show the results obtained by the six architectures in experiments 5 and 6, respectively. As can be seen from the results, the chosen architectures perform better in experiment 6 where the classes were balanced.

The experiments performed in this paper serve as a reference for gastroenterologist doctors to improve the process of diagnosing the type of inflammatory bowel disease in patients. Based on the results obtained in the classification metrics, the ResNet50+MobileNetV2 architecture stands out from the others in most of the performed experiments. It is also verified from the values obtained in the AUC metric that the selected architectures demonstrate a good ability to distinguish the classes present in the created datasets, and through the values present in the MCC metric, it is observed that the number of false positives and false negatives is very close to zero or even manages to be zero. Since the MCC metric in most of the experiments is equal to 1.0 and in the remaining ones is very close to 1.0.

Table 2. Results of experiment 1

	Accuracy	Loss	Precision	Recall	F1-Score	AUC	MCC
AlexNet	0.9674	0.1186	0.9733	0.9733	0.9733	0.9795	0.9590
ResNet50	**1.0000**	0.6359	**1.0000**	**1.0000**	**1.0000**	**1.0000**	**1.0000**
VGG-16	0.9767	0.0849	**1.0000**	0.9333	0.9655	0.9667	0.9493
InceptionV3	0.9907	0.6280	**1.0000**	0.9733	0.9865	0.9867	0.9796
ResNet50+MobileNetV2	0.9488	0.6428	**1.0000**	0.8133	0.8971	0.9067	0.8599
Hybrid model	0.9907	**0.0770**	0.9868	**1.0000**	0.9934	0.9964	0.9898

Table 3. Results of experiment 2

	Accuracy	Loss	Precision	Recall	F1-Score	AUC	MCC
AlexNet	0.9648	0.0990	0.9932	0.9236	0.9571	0.9578	0.9109
ResNet50	**1.0000**	0.6326	**1.0000**	**1.0000**	**1.0000**	**1.0000**	**1.0000**
VGG-16	0.9894	**0.0855**	1.0000	0.9427	0.9705	0.9713	0.9382
InceptionV3	**1.0000**	0.6278	**1.0000**	**1.0000**	**1.0000**	**1.0000**	**1.0000**
ResNet50+MobileNetV2	**1.0000**	0.6427	**1.0000**	**1.0000**	**1.0000**	**1.0000**	**1.0000**
Hybrid model	0.9824	0.1000	0.9933	0.9490	0.9707	0.9706	0.9374

Table 4. Results of experiment 3

	Accuracy	Loss	Precision	Recall	F1-Score	AUC	MCC
AlexNet	0.9907	0.0963	0.9762	**1.0000**	0.9880	0.9925	0.9806
ResNet50	**1.0000**	0.6324	**1.0000**	0.9867	0.9933	0.9933	0.9898
VGG-16	0.9628	0.1576	**1.0000**	0.8800	0.9362	0.9400	0.9093
InceptionV3	0.9953	0.6303	**1.0000**	**1.0000**	**1.0000**	**1.0000**	**1.0000**
ResNet50+MobileNetV2	0.9116	0.6553	**1.0000**	0.8000	0.8889	0.9000	0.8500
Hybrid model	0.9953	**0.0814**	0.9868	**1.0000**	0.9934	0.9964	0.9898

Table 5. Results of experiment 4

	Accuracy	Loss	Precision	Recall	F1-Score	AUC	MCC
AlexNet	0.9824	**0.0661**	0.9805	0.9618	0.9711	0.9691	0.9363
ResNet50	0.9930	0.6395	0.9936	0.9873	0.9904	0.9897	0.9787
VGG-16	0.9824	0.1038	0.9935	0.9745	0.9839	0.9833	0.9647
InceptionV3	0.9930	0.6342	**1.0000**	**1.0000**	**1.0000**	**1.0000**	**1.0000**
ResNet50+MobileNetV2	**1.0000**	0.6457	**1.0000**	**1.0000**	**1.0000**	**1.0000**	**1.0000**
Hybrid model	0.9894	0.0715	0.9936	0.9936	0.9936	0.9929	0.9858

Table 6. Results of experiment 5

	Accuracy	Loss	Precision	Recall	F1-Score	AUC	MCC
AlexNet	0.9767	0.0711	**1.0000**	0.9467	0.9726	0.9733	0.9594
ResNet50	**1.0000**	0.6261	**1.0000**	**1.0000**	**1.0000**	**1.0000**	**1.0000**
VGG-16	0.9814	**0.0553**	**1.0000**	0.9600	0.9796	0.9800	0.9695
InceptionV3	0.9953	0.6287	**1.0000**	0.9733	0.9865	0.9867	0.9796
ResNet50+MobileNetV2	0.9953	0.6386	**1.0000**	**1.0000**	**1.0000**	**1.0000**	**1.0000**
Hybrid model	0.9488	0.2109	0.9857	0.9200	0.9517	0.9564	0.9285

Table 7. Results of experiment 6

	Accuracy	Loss	Precision	Recall	F1-Score	AUC	MCC
AlexNet	0.9859	**0.0503**	**1.0000**	0.9618	0.9805	0.9809	0.9583
ResNet50	**1.0000**	0.6312	**1.0000**	**1.0000**	**1.0000**	**1.0000**	**1.0000**
VGG-16	0.9542	0.1613	**1.0000**	0.8662	0.9283	0.9331	0.8622
InceptionV3	**1.0000**	0.6219	**1.0000**	**1.0000**	**1.0000**	**1.0000**	**1.0000**
ResNet50+MobileNetV2	**1.0000**	0.6372	**1.0000**	**1.0000**	**1.0000**	**1.0000**	**1.0000**
Hybrid model	0.9754	0.0830	0.9935	0.9745	0.9839	0.9833	0.9647

5 Conclusion

This study concludes that the use of deep learning architectures is useful in assisting gastroenterologists in the early diagnosis of inflammatory bowel disease. Based on the results of the classification metrics, it can be concluded that the architectures used achieved good accuracy in classifying the disease.

Experiments in which transformations were applied to the images succeeded in maintaining the performance of the architectures. Moreover, this work demonstrates, through the experiments carried out, that applying transformations to the images adds advantages to the classification made by CNNs. It has been discovered that horizontal cropping or the use of a Gaussian blur improves the performance of the architectures.

In future work, it would be important to perform a medical validation with new images of both types of the disease to understand which of the architectures used will be the most accurate in recognising the type of inflammatory bowel disease based on the results obtained and which of the experiments performed will result in better processing.

Also, it is important to use in future work multi-task models to not only predict the type of inflammatory bowel disease but what is the degree of severity according to the Mayo score of ulcerative colitis and what type of lesion a patient with Crohn's disease has in the intestinal mucosa. For this, ordinal techniques can be leveraged [3,8].

References

1. Borgli, H., Riegler, M., Thambawita, V., Jha, D., Hicks, S., Halvorsen, P.: The HyperKvasir Dataset. OSF (2019)
2. Borgli, H., et al.: HyperKvasir, a comprehensive multi-class image and video dataset for gastrointestinal endoscopy. Sci. Data **7**, 283 (2020)
3. Cardoso, J.S., Sousa, R., Domingues, I.: Ordinal data classification using kernel discriminant analysis: a comparison of three approaches. In: 11th International Conference on Machine Learning and Applications. vol. 1, pp. 473–477 (2012)
4. Chicco, D., Jurman, G.: The advantages of the Matthews correlation coefficient (MCC) over F1 score and accuracy in binary classification evaluation. BMC Genomics **21**, 6 (2020)
5. Chierici, M., et al.: Automatically detecting Crohn's disease and ulcerative colitis from endoscopic imaging. BMC Med. Inf. Decis. Making **22**, 300 (2022)
6. CrohnIPI. https://crohnipi.ls2n.fr/en/crohn-ipi-project/. Accessed 21 Feb 2023
7. Doença inflamatória do intestino CUF. https://www.cuf.pt/saude-a-z/doenca-inflamatoria-do-intestino. Accessed 21 Feb 2023
8. Domingues, I., Cardoso, J.S.: Max-ordinal learning. IEEE Trans. Neural Netw. Learn. Syst. **25**(7), 1384–1389 (2014)
9. Dray, X., et al.: CAD-CAP: une base de données française à vocation internationale, pour le développement et la validation d'outils de diagnostic assisté par ordinateur en vidéocapsule endoscopique du grêle. In: Endoscopy, pp. s-0038-1623358 (2018)
10. Fatema, K., Montaha, S., Rony, M.A.H., Azam, S., Hasan, M.Z., Jonkman, M.: A robust framework combining image processing and deep learning hybrid model to classify cardiovascular diseases using a limited number of paper-based complex ECG images. Biomedicines **10**, 2835 (2022)
11. Ghouri, Y.A., Tahan, V., Shen, B.: Secondary causes of inflammatory bowel diseases. World J. Gastroenterol. **26**, 3998–4017 (2020)
12. Khan, M.N., Hasan, M.A., Anwar, S.: Improving the robustness of object detection through a multi-camera-based fusion algorithm using fuzzy logic. Front. Artif. Intell. **4**, 638951 (2021)
13. Maissin, A., et al.: Multi-expert annotation of Crohn's disease images of the small bowel for automatic detection using a convolutional recurrent attention neural network. Endoscopy Int. Open **09**, E1136–E1144 (2021)

14. Polat, G., Kani, H.T., Ergenc, I., Alahdab, Y.O., Temizel, A., Atug, O.: Improving the computer-aided estimation of ulcerative colitis severity according to mayo endoscopic score by using regression-based deep learning. Inflamm. Bowel Dis., izac226 (2022)
15. Polat, G., Kani, H.T., Ergenc, I., Alahdab, Y.O., Temizel, A., Atug, O.: Labeled Images for Ulcerative Colitis (LIMUC) Dataset (2022)
16. Sairenji, T., Collins, K.L., Evans, D.V.: An update on inflammatory bowel disease. Prim. Care: Clin. Off. Pract. **44**, 673–692 (2017)
17. Seyedian, S.S., Nokhostin, F., Malamir, M.D.: A review of the diagnosis, prevention, and treatment methods of inflammatory bowel disease. J. Med. Life **12**, 113–122 (2019)
18. Shahzadi, I., Tang, T.B., Meriadeau, F., Quyyum, A.: CNN-LSTM: cascaded framework for brain tumour classification. In: IEEE-EMBS Conference on Biomedical Engineering and Sciences (IECBES), pp. 633–637 (2018)
19. Stidham, R.W., et al.: Performance of a deep learning model vs human reviewers in grading endoscopic disease severity of patients with ulcerative colitis. JAMA Netw. Open **2**(5), e193963 (2019)
20. Turan, M., Durmus, F.: UC-NfNet: deep learning-enabled assessment of ulcerative colitis from colonoscopy images. Med. Image Anal. **82**, 102587 (2022)
21. Udristoiu, A.L., et al.: Deep learning algorithm for the confirmation of mucosal healing in Crohn's disease, based on confocal laser endomicroscopy images. J. Gastroint. Liver Dis. **30**, 59–65 (2021)
22. Vallée, R., Coutrot, A., Normand, N., Mouchère, H.: Accurate small bowel lesions detection in wireless capsule endoscopy images using deep recurrent attention neural network. In: IEEE 21st International WS on Multimedia Signal Proceedings (MMSP) (2019)
23. Vallée, R., Coutrot, A., Normand, N., Mouchère, H.: Influence of expertise on human and machine visual attention in a medical image classification task. In: European Conference on Visual Perception (2021)
24. Vallée, R., Maissin, A., Coutrot, A., Mouchère, H., Bourreille, A., Normand, N.: CrohnIPI: an endoscopic image database for the evaluation of automatic Crohn's disease lesions recognition algorithms. In: Medical Imaging: Biomedical Applications in Molecular, Structural, and Functional Imaging, p. 61. SPIE (2020)
25. Vankdothu, R., Hameed, M.A., Fatima, H.: A brain tumor identification and classification using deep learning based on CNN-LSTM method. Comput. Electr. Eng. **101**, 107960 (2022)
26. Wehkamp, J., Götz, M., Herrlinger, K., Steurer, W., Stange, E.F.: Inflammatory bowel disease: Crohn's disease and ulcerative colitis. Deutsches Ärzteblatt Int. **113**, 72 (2016)

Few-Shot Image Classification for Automatic COVID-19 Diagnosis

Daniel Cores[1], Nicolás Vila-Blanco[1,2,3](✉), Manuel Mucientes[1,2], and María J. Carreira[1,2,3]

[1] Centro Singular de Investigación en Tecnoloxías Intelixentes (CiTIUS), Universidade de Santiago de Compostela, Santiago de Compostela, Spain
{daniel.cores,nicolas.vila,manuel.mucientes,mariajose.carreira}@usc.es
[2] Departamento de Electrónica e Computación, Escola Técnica Superior de Enxeñaría, Universidade de Santiago de Compostela, Santiago de Compostela, Spain
[3] Instituto de Investigación Sanitaria de Santiago de Compostela (IDIS), Santiago de Compostela, Spain

Abstract. Developing robust and performant methods for diagnosing COVID-19, particularly for triaging processes, is crucial. This study introduces a completely automated system to detect COVID-19 by means of the analysis of Chest X-Ray scans (CXR). The proposed methodology is based on few-shot techniques, enabling to work on small image datasets. Moreover, a set of additions have been done to enhance the diagnostic capabilities. First, a network to extract the lung region to rely only on the most relevant image area. Second, a new cost function to penalize each misclassification according to the clinical consequences. Third, a system to combine different predictions from the same image to increase the robustness of the diagnoses. The proposed approach was validated on the public dataset COVIDGR-1.0, yielding a classification accuracy of 79.10% ± 3.41% and, thus, outperforming other state-of-the-art methods. In conclusion, the proposed methodology has proven to be suitable for the diagnosis of COVID-19.

Keywords: Chest X-Ray · COVID-19 · deep neural networks · few-shot classification

1 Introduction

The assessment of radiological images such as Computerized Tomography (CT) or Chest X-Ray (CXR) scans has demonstrated to be a reliable method for the screening and diagnosis of COVID-19 [9]. Different studies pointed out a set of visual indicators that can be used in this regard, including bilateral and/or interstitial abnormalities, among others [6]. In this regard, automated methods powered by machine learning techniques are a useful tool to improve the diagnosis workflow by reducing the time needed to analyze each scan [5]. In particular, convolutional neural networks (CNN) stand out as the most used technique due

A. Pertusa et al. (Eds.): IbPRIA 2023, LNCS 14062, pp. 545–556, 2023.
https://doi.org/10.1007/978-3-031-36616-1_43

to their ability to extract high-level image characteristics that are useful in the diagnosis.

The particularity of CNN-based systems is that very large datasets are needed for the training process. This led the research groups to collect their own sets of images [3], which turned out to be mostly unbalanced due to the high number of patients with severe condition. This, in addition to the heterogeneous sources used to acquire the images, caused the developed systems to report suspiciously high levels of sensitivity [27]. In the end, this conditioned the applicability of these models, as the detection of patients with low or moderate severity showed very poor performance.

Although more than three years have passed since the start of the pandemic, there are still few high-quality CXR datasets that can be used to build COVID-19 detection systems [17]. Furthermore, the limited size of the available datasets makes it difficult to carry out large-scale studies, and forces the researchers to apply techniques to artificially increase the number of images or to use specific learning algorithms that take into account the lack of data. Among the latter, the use of few-shot techniques is particularly relevant [7].

In a traditional setup, a CNN is fed with a large number of images of each output category during training. The model parameters are modified iteratively to better detect the relevant features in the images and ultimately improve the classification performance. If few images are available during the training process, the model is prone to overfitting the data, that is, learning specific features of each training image instead of general enough features associated with each category. In this regard, the few-shot frameworks address this issue through a meta-learning and fine-tuning approach. The main task in this case is no longer the extraction of relevant characteristics that identify each category, but those that allow to know whether two images belong to the same or to a different category. The strategy consists in comparing template images of each category (which will be referred to as the support set) with each input image and choosing the category that yields the higher affinity. An advantage of this methodology is that the categories used to train the model do not have to be the same as those used during the test phase.

In this paper, we introduce a new approach for the screening of patients with COVID-19 from CXR images. The proposal relies on few-shot techniques, so it is prepared to be used in reduced sets of images. The main contributions of this work are: 1) a system that focuses only on lung regions to classify the images, discarding other meaningless structures that may hinder the results; 2) a novel cost function that penalizes each misclassification according to the specific clinical cost; 3) an ensemble technique that combines different support sets to increase the robustness of the classification; and 4) a validation setup in a public dataset [22] that proves the suitability of the proposed approach regarding the classification performance.

2 Related Work

Automatic COVID-19 diagnosis from CXR images quickly emerged as a very active research area, addressing the automatic triage problem applying image classification techniques. Thus, the initial trend was to rely on well-known traditional architectures for general image classification such as VGG [13], Xception [11], or CapsNets [1]. Due to the promising results achieved with these simple methods, more complex models specifically designed to detect the COVID-19 disease were proposed. COVID-Net [25] seeks execution efficiency while maintaining a high classification performance with the definition of lightweight residual blocks. CVDNet [15] focused on extracting local and global features, including two interconnected paths with different kernel sizes. Also, a lung segmentation network was introduced as a preprocessing stage to force the network to focus only on the lung region [22] or to guide the learning algorithm [14].

Many datasets with CXR COVID-19 images have been released since the pandemic outbreak. Due to the lack of available data, the initial approach followed by early work on automatic COVID-19 diagnosis was to aggregate multiple datasets [2,3]. Unfortunately, most of these datasets are biased towards patients with severe conditions, and generally do not provide a balanced set of positive and negative samples. The severity level bias might explain the abnormal accuracy level reported by some works by not considering the most challenging cases, i.e., patients with mild and moderate conditions [17]. Also, the variability in the data acquisition process caused by integrating different datasets from different sources may cause the model to learn features specific to each device [23].

COVIDGR [22] was presented as a more complete dataset, including high-quality annotated CXR images for both COVID and non-COVID patients extracted under the same conditions. Positive patients are also classified into four severity levels according to the radiological findings. Therefore, this dataset defines a realistic environment to evaluate solutions for automatic COVID-19 triage systems. However, the number of training samples per severity level remains low, which hinders the training of traditional models.

To address the problem of data scarcity, simple data augmentation transformations have been successfully applied to increase the variability of the training set and prevent the overfitting of the model [13]. This overfitting problem was also addressed by including handcrafted features [10]. As a step forward, the generation of synthetic images with generative adversary networks (GAN) was explored in [12]. Alternatively, transfer learning techniques have also proven to be highly effective in adapting a model pretrained in a large dataset to a new domain with limited training samples [12,26].

Few-shot techniques have recently emerged as a popular solution to solve the general image classification problem, dealing with very limited training sets [24]. One of the main lines of research in few-shot learning is the definition of meta-learners. Meta-learning algorithms redefine the classification problem into a similarity metric learning that can identify if two images belong to the same category. This property makes meta-learners more generalizable to novel classes with limited or even no training samples—typically up to 30 labeled samples per category. These techniques have been effectively applied to general medical image processing [4], and to solve the COVID-19 classification problem [7,21].

As data availability in current COVID-19 datasets such as COVID-GR exceeds the usual few-shot setting but does not suffice to properly train a traditional classifier, an intermediate solution is desirable. Therefore, we propose a meta-learning architecture that fully exploits the available labeled information by performing a model fine-tuning in the complete training set and implement a support set ensemble at test time.

3 Methodology

The illustration in Fig. 1 depicts a novel classification framework for diagnosing COVID-19 using CXR scans. Our approach is ideal for cases where a limited number of images are available for training, as it is the situation with many new diseases. Here, few-shot classification methods stand out, as they are specifically developed to handle scenarios where conventional image classification methods would not perform well enough due to the scarcity of data. In the few-shot approach, every input image (query) is categorized based on its similarity to the images of the support set. To accomplish this, the prototypes for each category are computed and compared with the query prototype. The prototypes are derived from the deep feature maps generated by a CNN backbone.

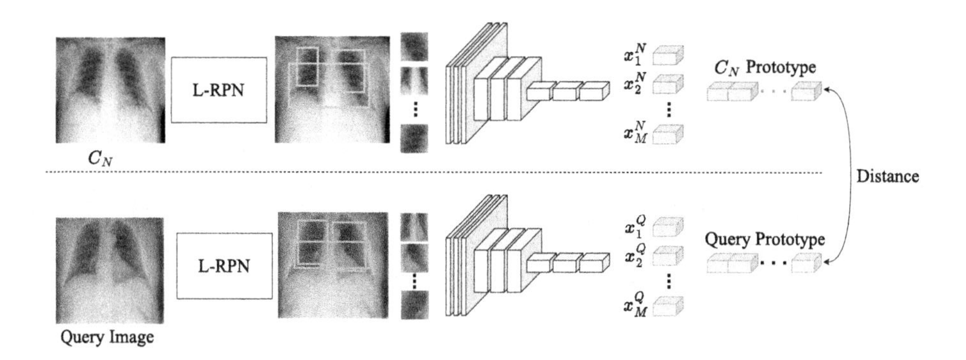

Fig. 1. The proposed workflow for categorizing images using one support set.

We introduce a lung-focused network as the first step of our pipeline. This system aims at proposing regions of interest located in the lung area, both for the query and support images (as seen in Fig. 1 as output of L-RPN). This network design allows the subsequent classification system to process only the most relevant regions for the diagnosis of COVID-19, leading to improved performance. Additional information regarding this component is provided in Subsect. 3.1.

Next, a CNN network is used as the backbone to generate rich representations of the regions proposed in both the query and the support images. The obtained per-region feature maps are combined to create a per-image feature representation. Then, the feature maps corresponding to the support images of each category are averaged to obtain the per-category feature map. In a further step, these maps are refined through gradient descent to obtain robust prototypes which can be utilized to evaluate the similarity of each query image to each category. This evaluation of affinity is performed using Earth Mover's Distance (EMD) [19], which determines the pairwise distance between the prototypes of the query and those belonging to support regions. Therefore, the pair of regions that show the minimum EMD value is deemed to be the most similar.

In typical image classification using few-show methods, a minimal amount of support images is utilized, with the most common cases raging from 1-shot to 10-shot, meaning between one and ten images per category. Nevertheless, these conditions are not applicable in this specific scenario, as more images per category are available in the dataset. Hence, we suggest a different approach employing a group of S sets of k support images each, rather than merely selecting a single set randomly, to mitigate the impact of each support image. This concept is outlined in Subsect. 3.2.

For training purposes, four categories of COVID-19 affection are considered, namely negative, mild, moderate, and severe. Since the magnitude of classification errors should vary depending on the proximity between the correct and the predicted categories, we propose a novel cost function that incorporates expert knowledge in this regard. This technique is explained in more detail in Subsect. 3.3.

3.1 L-RPN: Lung-Aware Region Proposal Network

It is commonly agreed that the lung area is the most affected by COVID-19, and so it has been widely used as the main radiological indicator for diagnosis. However, the position of the patient during a CXR scan is not homogeneous, as it depends on his/her severity and the presence of ventilation or monitoring devices that may not be removable. This situation can lead to a high degree of variability in the CXR scans in terms of the position and scale of the lungs, which inevitably makes diagnosis even more difficult. In this work, a network to propose lung-aware regions of interest is utilized to make the image classification task focus only on the lung area and discard other meaningless regions in the image.

The process of generating region proposals, depicted in Fig. 1, is performed at the beginning of the proposed pipeline. It is presented in Fig. 2, and it has three main steps. First, a U-Net segmentation network [18] is applied to the input CXR image to extract the lung mask. Second, the minimum rectangle which encloses the lung masks is obtained and enlarged by 5% in all dimensions to avoid losing any pixel on the lung border. Given that the lower part of the lungs can often be difficult to see, especially in severe cases, the box is further enlarged on the lower side to a maximum height/width ratio of 1. Third, a set of Q rectangular regions of interest is randomly generated within the limits of the box.

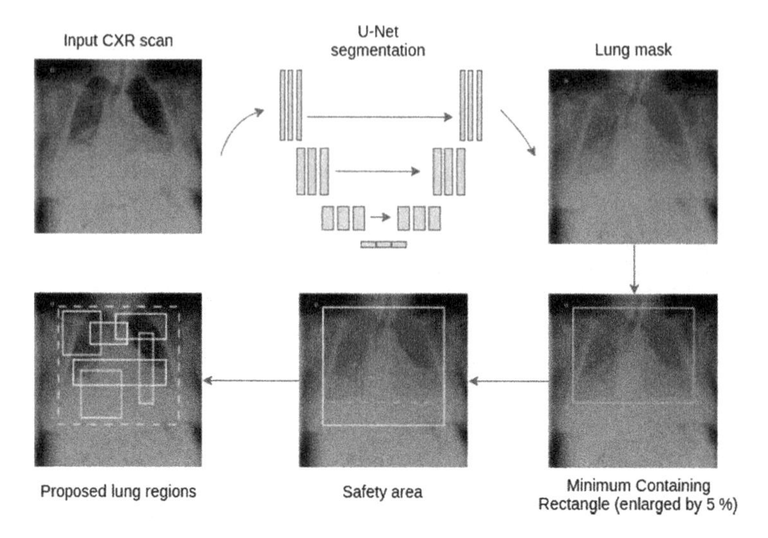

Fig. 2. L-RPN pipeline to focus on the pulmonary zone only.

3.2 Support Set Ensemble

When the data available to train a machine learning model is scarce, an effective way to improve performance is by generating new synthetic training data by applying random transformations. This image augmentation can be used during both training and inference. In the first case, the variability of training data is increased, and so the model is able to learn richer image features. In the second case, different versions of a given image can be used to obtain different predictions, which can be combined for a more robust prediction.

In this work, an inference-time augmentation is performed, so S support sets are generated and kept static throughout the test step. In this regard, a given query image can be compared with all the support sets to obtain the similarity vectors a_i^s, being $i = \{1, \cdots, N\}$ the category index, and $s = \{1, \cdots, S\}$ the index of the corresponding support set. Then, the similarity vectors are aggregated so that the chances that an image has to be assigned to a specific category can be obtained by a voting formula:

$$p_i = \prod_{s=1}^{S} \frac{e^{a_i^s}}{\sum_{j=1}^{N} e^{a_j^s}} \tag{1}$$

where p_i is the probability of belonging to the i-th class, and a_i^s is the similarity value obtained for the i-th class through the s-th support set. Values of p_i are normalized to ensure that $\sum_{i=1}^{|C|} p_i = 1$, being C the set of categories.

3.3 Misdiagnosis-Sensitive Learning

The common clinical approach for COVID-19 triaging is the same as in other diseases. First, a cheap test is used in a large sample of the population. As these tests are characterized by a high number of false positives —negative patients tested positive—, a more accurate and expensive test is used to refine the diagnostic of the patients tested positive.

The aim of this work is to develop a binary classifier to differentiate between COVID-19 positive and COVID-19 negative CXR images. To make the model robust enough, we used the extra categories provided by the COVIDGR-1.0 database. In particular, the images in this dataset are categorized into five classes: negative patients without visible lung abnormalities and positive patients with no, mild, moderate, and severe visual affection. In the following, we will refer to these classes as N_NORMAL, P_NORMAL, P_MILD, P_MODERATE and P_SEVERE.

With the main goal of enhancing the current triaging methods, the followed approach in this work is targeted at reducing the number of positive patients tested negative. To do so, the cost function proposed to train the model is provided with a novel penalty term which is based not only on the distance between classes in terms of visual affection, but also on the clinical implication of each misclassification. Thus, the penalty term follows the principle that the penalty is supposed to be maximum when a severe patient is classified as negative, as the clinical cost is very high. In contrast, a positive patient with moderate affection should lead to a low penalty when he/she is classified as severe, as the clinical cost is negligible and the visual affection can be very similar in both classes.

The addition of this cross-penalties is carried out through a cost matrix M which indicates the penalty of every possible misclassification scenario. This matrix, developed by an expert radiologist, is shown in Table 1, and is included in the modified cross-entropy Eq. (2). Each matrix cell M_{ij} denotes the penalty of a patient of the i-th class when is predicted to belong to the j-th class. As mentioned above, the highest penalties are associated with severe and moderate patients which are assigned to the negative class. It should be noted that the P_NORMAL class is not included in the cost matrix M because, as will be explained in Sect. 4.1, this class is not used in the training process.

Table 1. Cost matrix M included in the new loss function in Eq. (2).

		Predicted			
		Negative	Mild	Moderate	Severe
Real	Negative	0	0.1	0.2	0.3
	Mild	0.3	0	0.1	0.2
	Moderate	0.4	0.1	0	0.1
	Severe	0.5	0.075	0.025	0

The cost of making a prediction on a patient belonging to the i-th class is calculated via

$$\mathcal{L}_i = -log(p_i) + \sum_{j=1}^{|C'|} \left[-log(1-p_j)M_{ij} \right] \quad , \quad C' = C \setminus \{C_i\} \tag{2}$$

where C' is the set consisting of all classes except the class C_i.

4 Experiments

We performed all experiments in the COVIDGR-1.0 dataset [22] to compare the results of our method with previous state-of-the-art approaches for automatic COVID-19 diagnosis with CXR images. This dataset provides a balanced set with 426 positive samples and 426 negative samples. It includes the level of severity for positive patients, including RT-PCR positive cases identified as negatives by expert radiologists.

We report both the overall classification performance and the detailed results for each level of severity. We follow the same experimental setting established by [22], calculating the mean and standard deviation for five 5-fold cross-validation. For a fairer comparison, the same data splits were used for all the methods.

4.1 Implementation Details

The L-RPN module described in Subsect. 3.1 follows a U-net model trained on the Montgomery and SCR datasets [8,20]. The number of regions extracted from the lung area Q is set to 9. All input images are resized to 128 × 128 pixels.

The backbone of the few-shot classifier is implemented as ResNet-12 pre-trained on the miniImageNet dataset [24]. The support set contains 10 randomly selected images for each training episode—each training iteration of the meta-learner—, while we select $S = 5$ static support sets with 10 images each for validation and test. These support images for validation and test are extracted from the training set, performing inference on the complete test set. Positive patients with no radiological findings—category P_NORMAL in COVIDGR-1.0—are not considered for training, as their strong similarity to negative samples could hinder the training process. However, the test set must include images from this category to maintain a more realistic evaluation setting.

The initial learning rate is 0.5×10^{-3}, applying a reduction factor of 0.5 after each 500 iterations. The maximum number of training iterations is set to 5,000 with a validation stage every 50 iterations. Then, the best performing model on the validation set is selected as the final classifier. As the goal is to implement a triage system, both the global classification accuracy and the sensitivity are taken into account to establish the validation iteration. Therefore, the validation performance VP is defined as:

$$VP = \alpha * sensitivity + \beta * accuracy \tag{3}$$

where α and β are hyperparameters heuristically set to 0.3 and 0.7 respectively.

4.2 Results

We conducted a series of experiments and compared our model with the state-of-the-art using binary classification metrics. These include accuracy, specificity, sensitivity, precision, and F1 (the last two are reported for both positive and negative classes). Accuracy measures the global classification performance by means of the percentage of images correctly classified, and so it is a useful metric to compare different approaches. Sensitivity is essential for detecting positive cases in a triage system, but a maximum sensitivity could be obtained if every patient is classified as positive, which in the end would make no sense. Therefore, a good balance between sensitivity and specificity is needed to minimize false positives. The F1 considers this balance, utilizing recall and precision metrics for each class.

Table 2 shows a detailed comparison of our method with state-of-the-art approaches assessed on the COVIDGR-1.0 dataset [22]. The reported results comprise a set of methods developed and/or tested by other authors, including a general-purpose classifier—ResNet-50—and methodologies specifically designed for COVID-19 CXR images. The results in this table do not include the P_NORMAL class for a better comparability, as no previous study uses this category in their experiments. Overall, our method achieves the best results in all metric except for specificity and precision in the positive class. However, the best-performing method in these metrics—COVIDNet-CXR—achieves very poor results in the other metrics, lacking more than 37% behind our method in terms of sensitivity. In fact, our approach outperforms the second-best result by around 11 points in this key metric for any triage system. Also, regarding the overall accuracy, we set a new state-of-the-art result on this dataset, improving previous methods by 2.9% accuracy.

Table 2. Results on COVIDGR-1.0.

	Specificity	N_Precision	N_F1	Sensitivity	P_Precision	P_F1	Accuracy
COVIDNet [25]	**88.8 ± 0.9**	3.4 ± 6.2	73.3 ± 3.8	46.8 ± 17.6	**81.7 ± 6.0**	56.9 ± 15.1	67.8 ± 6.1
CAPS [1]	65.7 ± 9.9	65.6 ± 4.0	65.2 ± 5.0	64.9 ± 9.7	66.1 ± 4.5	64.9 ± 4.9	65.3 ± 3.3
ResNet-50 [22]	79.9 ± 8.9	71.9 ± 3.1	75.4 ± 4.9	68.6 ± 6.1	78.8 ± 6.3	72.7 ± 3.5	74.3 ± 3.6
FuCiTNet [16]	80.8 ± 7.0	72.0 ± 4.5	75.8 ± 3.2	67.9 ± 8.6	78.5 ± 5.0	72.4 ± 4.8	74.4 ± 3.3
SDNet [22]	79.8 ± 6.2	74.7 ± 3.9	76.9 ± 2.8	72.6 ± 6.8	78.7 ± 4.7	75.7 ± 3.4	76.2 ± 2.7
ours	75.1 ± 6.5	**85.3 ± 3.7**	**79.7 ± 3.9**	**84.0 ± 4.9**	73.8 ± 4.9	**78.4 ± 3.1**	**79.1 ± 3.4**

Moreover, we compare our method with the best previous approach—COVID-SDNet—for all severity levels in COVIDGR-1.0, including the P_NORMAL category. As seen in Table 3, our method achieves the best results in every subset, achieving the largest margins in the most demanding subsets. Thus, COVID-SDNet underperforms a random classifier in the P_MILD subset, tending to classify these patients as positive cases. Our approach is capable of increasing the classification accuracy in this demanding subset by more than 16%. This proves that our method is more robust against images with minimal radiological findings.

Table 3. Obtained accuracy by severity level in COVIDGR-1.0.

	COVID-SDNet [22]	ours
P_NORMAL	-	41.6 ± 13.1
P_MILD	46.0 ± 7.1	62.4 ± 9.9
P_MODERATE	85.4 ± 1.9	89.4 ± 6.2
P_SEVERE	97.2 ± 1.9	99.5 ± 1.7

5 Conclusions

We have defined a new few-shot classifier for CXR images, able to correctly identify positive COVID-19 cases. Although the proposed method follows a few-shot architecture, the combination of support sets leverages all the information available, boosting the performance. Furthermore, the L-RPN module ensures that the diagnosis is made only from lung information, which makes the model robust against poorly framed CXR images due to different patient positions or the use of external monitoring devices. Experimental results show that our method scores the best result in the COVIDGR data set, improving the binary classification of COVID-19 cases by around 3 accuracy points. Regarding the four severity levels defined in COVIDGR, our method outperforms the second best approach in all cases, especially in the challenging subset that contains patients with mild condition with an improvement of more than 16%.

Acknowledgements. This work has received financial support from the Spanish Ministry of Science and Innovation under grant PID2020-112623GB-I00, Consellería de Cultura, Educación e Ordenación Universitaria under grants ED431C 2021/48, ED431G-2019/04, ED481A-2018 and ED431C 2018/29 and the European Regional Development Fund (ERDF), which acknowledges the CiTIUS-Research Center on Intelligent Technologies of the University of Santiago de Compostela as a Research Center of the Galician University System.

References

1. Afshar, P., Heidarian, S., Naderkhani, F., Oikonomou, A., Plataniotis, K.N., Mohammadi, A.: COVID-CAPS: a capsule network-based framework for identification of COVID-19 cases from X-ray images. Pattern Recogn. Lett. **138**, 638–43 (2020)
2. Chowdhury, M.E., et al.: Can AI help in screening viral and COVID-19 pneumonia? IEEE Access **8**, 132665–76 (2020)
3. Cohen, J.P., Morrison, P., Dao, L., Roth, K., Duong, T.Q., Ghassemi, M.: COVID-19 image data collection: prospective predictions are the future. arXiv 2006.11988 (2020). https://github.com/ieee8023/covid-chestxray-dataset
4. Cui, H., Wei, D., Ma, K., Gu, S., Zheng, Y.: A unified framework for generalized low-shot medical image segmentation with scarce data. IEEE Trans. Med. Imaging **40**(10), 2656–2671 (2020)

5. Fernández-Miranda, P.M., et al.: Developing a training web application for improving the COVID-19 diagnostic accuracy on chest X-ray. J. Digit. Imaging **34**(2), 242–256 (2021)
6. Huang, C., et al.: Clinical features of patients infected with 2019 novel coronavirus in Wuhan. China Lancet **395**(10223), 497–506 (2020)
7. Jadon, S.: COVID-19 detection from scarce chest X-ray image data using few-shot deep learning approach. In: Proceedings of the SPIE, vol. 11601, p. 116010X (2021)
8. Jaeger, S., Candemir, S., Antani, S., Wáng, Y.X.J., Lu, P.X., Thoma, G.: Two public chest X-ray datasets for computer-aided screening of pulmonary diseases. Quant. Imaging Med. Surg. **4**(6), 475 (2014)
9. Jin, K.N., Do, K.H., Da Nam, B., Hwang, S.H., Choi, M., Yong, H.S.: Korean clinical imaging guidelines for justification of diagnostic imaging study for COVID-19. J. Korean Soc. Radiol. **83**(2), 265–283 (2022)
10. Kang, H., et al.: Diagnosis of coronavirus disease 2019 (COVID-19) with structured latent multi-view representation learning. IEEE Trans. Med. Imaging **39**(8), 2606–2614 (2020)
11. Khan, A.I., Shah, J.L., Bhat, M.M.: CoroNet: a deep neural network for detection and diagnosis of COVID-19 from chest X-ray images. Comput. Meth. Prog. Bio. **196**, 105581 (2020)
12. Loey, M., Smarandache, F., M Khalifa, N.E.: Within the lack of chest COVID-19 X-ray dataset: a novel detection model based on GAN and deep transfer learning. Symmetry **12**(4), 651 (2020)
13. Nishio, M., Noguchi, S., Matsuo, H., Murakami, T.: Automatic classification between COVID-19 pneumonia, non-COVID-19 pneumonia, and the healthy on chest X-ray image: combination of data augmentation methods. Sci. Rep. **10**(1), 1–6 (2020)
14. Oh, Y., Park, S., Ye, J.C.: Deep learning COVID-19 features on CXR using limited training data sets. IEEE Trans. Med. Imaging **39**(8), 2688–700 (2020)
15. Ouchicha, C., Ammor, O., Meknassi, M.: CVDNet: a novel deep learning architecture for detection of coronavirus (COVID-19) from chest X-ray images. Chaos, Solitons Fractals **140**, 110245 (2020)
16. Rey-Area, M., Guirado, E., Tabik, S., Ruiz-Hidalgo, J.: FuCiTNet: improving the generalization of deep learning networks by the fusion of learned class-inherent transformations. Inf. Fusion **63**, 188–95 (2020)
17. Roberts, M., et al.: Common pitfalls and recommendations for using machine learning to detect and prognosticate for COVID-19 using chest radiographs and CT scans. Nat. Mach. Intell. **3**(3), 199–217 (2021)
18. Ronneberger, O., Fischer, P., Brox, T.: U-net: convolutional networks for biomedical image segmentation. In: Proceedings of the MICCAI, pp. 234–41 (2015)
19. Rubner, Y., Tomasi, C., Guibas, L.J.: The earth mover's distance as a metric for image retrieval. Int. J. Comput. Vision **40**(2), 99–121 (2000)
20. Shiraishi, J., et al.: Development of a digital image database for chest radiographs with and without a lung nodule: receiver operating characteristic analysis of radiologists' detection of pulmonary nodules. Am. J. Roentgenol. **174**(1), 71–74 (2000)
21. Shorfuzzaman, M., Hossain, M.S.: MetaCOVID: a siamese neural network framework with contrastive loss for n-shot diagnosis of COVID-19 patients. Pattern Recogn. **113**, 107700 (2021)
22. Tabik, S., et al.: COVIDGR dataset and COVID-SDNet methodology for predicting COVID-19 based on chest X-ray images. IEEE J. Biomed. Health **24**(12), 3595–605 (2020)

23. Teixeira, L.O., et al.: Impact of lung segmentation on the diagnosis and explanation of COVID-19 in chest X-ray images. Sensors **21**(21), 7116 (2021)
24. Vinyals, O., Blundell, C., Lillicrap, T., Wierstra, D., et al.: Matching networks for one shot learning. In: Advances in Neural Information Processing Systems, vol. 29, pp. 3630–8 (2016)
25. Wang, L., Lin, Z.Q., Wong, A.: COVID-Net: a tailored deep convolutional neural network design for detection of COVID-19 cases from chest X-ray images. Sci. Rep. **10**(1), 1–12 (2020)
26. Wang, N., Liu, H., Xu, C.: Deep learning for the detection of COVID-19 using transfer learning and model integration. In: IEEE International Conference on Electronics Information and Emergency Communication, pp. 281–284. IEEE (2020)
27. Wong, H.Y.F., et al.: Frequency and distribution of chest radiographic findings in COVID-19 positive patients. Radiology **296**, 201160 (2020)

An Ensemble-Based Phenotype Classifier to Diagnose Crohn's Disease from 16s rRNA Gene Sequences

Lara Vázquez-González[1,4](✉), Carlos Peña-Reyes[5,6], Carlos Balsa-Castro[1,3,4], Inmaculada Tomás[1,3,4], and María J. Carreira[1,2,4]

[1] Centro Singular de Investigación en Tecnoloxías Intelixentes (CiTIUS), Universidade de Santiago de Compostela, Santiago de Compostela, Spain
{laram.vazquez,inmaculada.tomas,mariajose.carreira}@usc.es

[2] Departamento de Electrónica e Computación, Escola Técnica Superior de Enxeñaría, Universidade de Santiago de Compostela, Santiago de Compostela, Spain

[3] Oral Sciences Research Group, Special Needs Unit, Department of Surgery and Medical Surgical Specialties, School of Medicine and Dentistry, Universidade de Santiago de Compostela, Santiago de Compostela, Spain
cbalsa@coitt.es

[4] Instituto de Investigación Sanitaria de Santiago de Compostela (IDIS), Santiago de Compostela, Spain

[5] School of Management and Engineering Vaud (HES-SO), University of Applied Sciences and Arts Western Switzerland Vaud, Yverdon-les-Bains, Switzerland
carlos.pena@heig-vd.ch

[6] CI4CB—Computational Intelligence for Computational Biology, SIB—Swiss Institute of Bioinformatics, Lausanne, Switzerland

Abstract. In the past few years, one area of bioinformatics that has sparked special interest is the classification of diseases using machine learning. This is especially challenging in solving the classification of dysbiosis-based diseases, i.e., diseases caused by an imbalance in the composition of the microbial community. In this work, a curated pipeline is followed for classifying phenotypes using 16S rRNA gene amplicons, focusing on Crohn's disease. It aims to reduce the dimensionality of data through a feature selection step, decreasing the computational cost, and maintaining an acceptably high f1-score. From this study, an ensemble model is proposed to contain the best-performing techniques from several representative machine learning algorithms. High f1-scores of up to 0.81 were reached thanks to this ensemble joining multilayer perceptron,

This work has received financial support from Instituto de Salud Carlos III (Spain) (PI21/00588), the Xunta de Galicia - Consellería de Cultura, Educación e Universidade (Centro de investigación de Galicia accreditation 2019–2022 ED431G-2019/04, Reference Competitive Group accreditation 2021–2024, GRC2021/48, Group with Growth Potential accreditation 2020–2022 GPC2020/27 and L Vázquez-González support ED481A-2021) and the European Union (European Regional Development Fund-ERDF).

A. Pertusa et al. (Eds.): IbPRIA 2023, LNCS 14062, pp. 557–568, 2023.
https://doi.org/10.1007/978-3-031-36616-1_44

extreme gradient boosting, and support vector machines, with as low as 300 target number of features. The results achieved were similar to or even better than other works studying the same data, so we demonstrated the goodness of our method.

Keywords: Phenotype classification · 16S rRNA gene · ASVs · ensemble-based classification · k-nearest neighbours · support vector machines · extreme gradient boosting · multilayer perceptron

1 Introduction

In recent years, one area of bioinformatics that has sparked special interest is the classification of diseases using machine learning [1,13]. Particularly, of illnesses caused by an imbalance in the composition of the microbial community (dysbiosis), which are harder to predict because there is not a specific bacteria to blame [10].

In this type of analysis, the marker most often used is the 16S ribosomal RNA (rRNA) gene [9], which is present in all bacteria and contains both conserved and hypervariable regions. The former are regions identical or similar in nucleic acids across bacterial species, and the latter have considerable sequence diversity among different bacteria. The conserved regions are usually employed to find the gene inside the bacterial genome, and the hypervariable regions can be used to identify different bacterial species.

In the literature, the 16S rRNA gene sequences are often processed into Operational Taxonomic Units (OTUs), which are collections of 16S rRNA sequences that have a certain percentage of sequence divergence and are standard units in marker gene analysis. However, they have been proven to have some limitations, such as the lack of reusability and comprehensiveness [2], that can affect the quality of the results. An alternative is the use of k-mers, which are length k substrings contained within a biological sequence. These also have limitations of their own, like the lack of interpretability of the results due to the substrings (mers) having no meaning of their own. These constraints can be avoided using Amplicon Sequence Variants (ASVs), which are any one of the inferred single DNA sequences recovered from a high-throughput analysis.

Machine learning algorithms, such as random forest (RF) or support vector machines (SVM), and neural network algorithms, such as multilayer perceptron (MLP), convolutional neural networks (CNN), or recurrent neural networks (RNN), have been applied in different studies to classify patients as either healthy or diseased based on their microbiome composition [11,12]. A recent study proposed MicroPheno, a tool that uses MLP to classify disease based on k-mer frequency tables [1]. They achieved better results with k-mers than with OTU frequency tables and, although they avoided the limitations of OTUs, the results were not interpretable, as k-mers do not have meaning on their own. Another recent study proposed Read2Pheno [13], a tool that uses a combination of CNN, ResNet, Bi-LSTM, and attention mechanisms to predict disease, classifying each read in a sample individually, instead of classifying the samples as a whole. For dysbiosis-caused diseases this may not be the best option, as we are uncertain that a single read has enough information to properly diagnose a sample.

In this work, we reflect on the importance of following a curated pipeline for classifying phenotypes using 16S rRNA gene amplicons. Furthermore, we aim to achieve the best possible results ensuring reproducibility and reducing the computational cost. We selected a wide variety of popular machine learning algorithms, ranging from classical to newer and more complex: k-nearest neighbours (kNN), RF, SVM, extreme gradient boosting (XGBoost), and MLP, and created an ensemble model with the best performing techniques.

2 Materials and Methods

2.1 Dataset

In this study, the dataset from the microbiota of the gut compiled in the Gevers et al. study [5] available at the NCBI Bioproject repository[1] was used. It includes 731 samples belonging to pediatric patients with Crohn's disease (CD) and 628 samples from healthy or diagnosed with other diseases patients (Not-CD), conforming to a total of 1,359 samples. This fairly balanced dataset contains 16S rRNA gene sequences from the V4 hypervariable region, sequenced on the Illumina MiSeq platform (version 2) with 175 base pairs (bp) paired-end reads. However, only the forward read of the paired-end reads is available, therefore this study was performed with only half of the information of the region.

2.2 EPheClass Pipeline for Phenotype Classification

To classify a phenotype, such as a disease, we followed the so called EPheClass pipeline shown in Fig. 1, which starts with samples straight from the sequencing platform, and classifies them into the desired categories. In our proposal, the 16S rRNA gene samples will be classified into CD or Not-CD. This pipeline is composed of two main modules: data processing, and training and evaluation. The data processing step prepares the 16S rRNA gene sequences (reads) for the classification step, selecting the most relevant features from the generated ASV abundance tables to reduce the dimensionality and discard no significant ASVs. The training and evaluation step focuses on the training of the tuned models and subsequent evaluation to propose the best ensemble model as the output.

EPheClass Module 1: Data Processing. In this first module of the pipeline, detailed in Fig. 2, the sequences are prepared for the classification module through four steps: quality filtering, ASV abundance table generation, train/test data split, and feature selection.

In the first step, the reads undergo a process of quality control using DADA2 [3], a software package for modeling and correcting Illumina-sequenced amplicon errors. This is vital due to the commonness of sequencing errors with next-generation sequencers, like Illumina, that degrade the quality of the sequences.

[1] https://www.ncbi.nlm.nih.gov/bioproject/PRJEB13679.

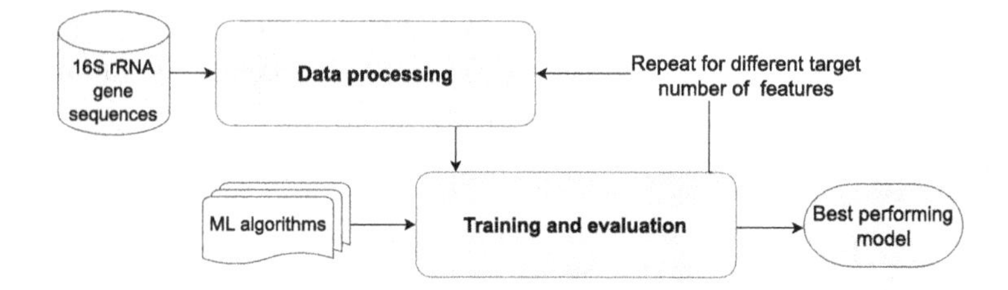

Fig. 1. EPheClass pipeline for ensemble-based phenotype classification from 16s rRNA sequences.

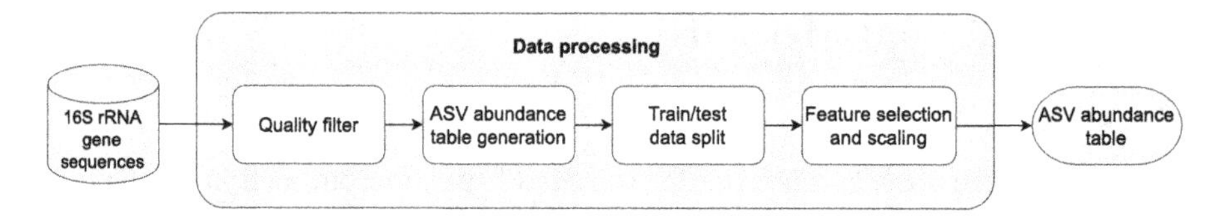

Fig. 2. EPheClass pipeline module 1: data processing.

As each read has a quality or error value associated, it is possible to filter out those that do not meet the quality criteria, which in our case follow the value recommended by Edgar and Flyvbjerg (maximum accumulated error = 1) [4].

After this process, only one sample of the dataset was discarded due to containing very few reads that did not meet the quality criteria.

Then, using DADA2 again, the Amplicon Sequence Variants (ASVs) and their abundance table, which shows the frequency of each ASV is found in each sample, were obtained. For the whole Crohn's disease dataset, we obtained a total of 14,216 ASVs.

Once the reads were properly processed and the ASVs abundance tables were obtained, the dataset is split into training and testing sets in a stratified fashion, so to preserve the same proportions of examples in each class as observed in the original dataset. For the training set, we used 1,154 samples corresponding to 85% of the data. For the testing set, the remaining 204 samples were used, corresponding to 15% of the data. These proportions were selected keeping in mind that 10% of the training subset was to be used as the validation set during cross-validation in models' tuning process.

Finally, a stage of feature selection methods was applied to reduce the number of features (ASVs in our case), decrease the computational cost, and improve the performance of the predictive models. In this stage, three selection methods were applied to prevent any bias in the classification algorithm.

The first and second selection methods were the recursive feature selection (RFE) with RF and SVM as estimators, respectively. This technique iteratively removes features from the feature set and evaluates the performance of the selected model (estimator) on the reduced feature set, discarding the features with the smallest impact on the model's performance.

The third selection method was DESeq2 (differential gene expression analysis using RNA-Seq) [7], a software package for the analysis of RNA-seq data that is used to identify differentially expressed genes between two or more sample groups. The package is based on a statistical model that accounts for the inherent variability in RNA-seq data and allows for the identification of truly differentially expressed genes. Because of that, DEseq2 helps reduce the number of genes tested by removing those unlikely to be significantly differentially expressed and can be applied to select the best ASVs. As these feature selection methods can extract different features depending on their criteria, we picked out ASVs selected by all or at least two methods, to ensure the extraction of the most selected and, thus, best features.

As it was shown in Fig. 2, the last step also includes a scaling process to ensure all features are on the same scale. In our case, the data were scaled by their maximum absolute value so as to preserve sparsity, which is very common and relevant in ASV abundance tables due to most ASVs being present in only a few samples, leading to zero-based tables [8].

At the end of the EPheClass module 1, we obtained a reduced ASV abundance table comprising only the most representative ASVs, depending on the target number of features. Given that we will evaluate n different target number of features, n different ASV abundance tables will be obtained, which we will use to decide on the best classification model and the best target number of features.

EPheClass Module 2: Training and Evaluation. Module 2 of the pipeline, detailed in Fig. 3, performs the classification and evaluation through four steps: hyperparameter tuning, individual models' validation, ensemble model creation, and final training and testing of the models.

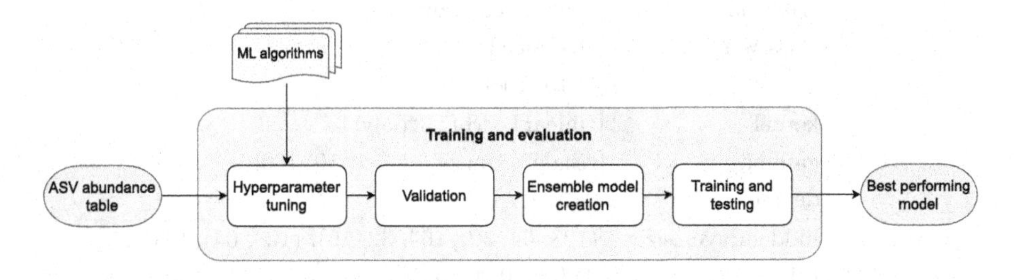

Fig. 3. EPheClass pipeline module 2: training and evaluation.

We must first perform a process of training and evaluation of different models to tackle our classification problem, ranging from more traditional techniques, like kNN, RF or SVM to more complex ones such as XGBoost or MLP. This is an heterogeneous selection of techniques, due to their fundamental differences, with the exception of XGBoost and RF, which are both decision-tree based. By using the best performing techniques we strived to create a diverse ensemble model that surpassed the performance of the individual models and overcame their particular limitations.

The first step is the selection of parameters for each technique, i.e., hyperparameter tuning. We tried different values for each parameter of each algorithm, as shown in Table 1, exhaustively considering all parameter combinations, and then selecting those with the best behaviour using 10-fold stratified cross validation on the training set defined in the previous step. Despite using a fairly balanced dataset, we used the f1-score as the decisive metric to select the best combination of hyperparameters instead of accuracy. In our case, false negatives are more damaging due to the disease being ignored. As the f1-score penalises models that have too many false negatives more than accuracy does, we considered it to be more suitable for this study.

During the hyperparameter tuning process, the parameter combination supposed to be the best often leads to overfitting, causing considerably higher training scores that testing scores. To solve this, we chose a combination that ensured a lower overfitting "ratio", given by the difference of the training score minus the validation score.

Table 1. Hyperparameter tuning step: parameters evaluated for each model.

kNN	Parameter	Evaluated values
	n_neighbors	[1...31]
RF	n_estimators	[50, 100, 300]
	max_features	["sqrt"]
	max_depth	[3, 5, 10]
	min_samples_split	[2, 5]
	min_samples_leaf	[1, 2]
	criterion	["gini", "entropy"]
	class_weight	["balanced"]
SVM	C	[1, 10, 100]
	kernel	["linear", "rbf", "poly"]
	gamma	["scale", "auto", 0.1, 1, 10, 100]
	degree	[2, 3, 4]
MLP	hidden_layer_sizes	[(128, 64, 32), (64, 32, 16), (128, 64), (64, 32)]
	tol	[0.0001, 0.001, 0.01, 0.1,1]
	alpha	[0.0001, 0.001, 0.01, 0.1,1]
	n_iter_no_change	[10, 20, 30, 40, 50]
	max_iter	[50, 100, 500, 1000]
	solver	["adam"]
XGBoost	n_estimators	[60, 100, 140, 180, 220]
	max_depth	[2..10]
	learning_rate	[0.01, 0.05, 0.1]
	min_child_weight	[1, 5, 10]

As shown in Fig. 3, once the hyperparameters for each model were tuned, each model was evaluated individually using the stratified 10-fold cross validation on the training set. Then, in the next step, the three best-performing models were selected to form an ensemble model with a hard voting strategy, where every individual classifier votes for a class, and the majority wins.

Individual models and ensemble models were finally evaluated again through training and testing processes.

3 Results

As described in detail in Sect. 2, we analysed the Crohn's disease dataset following a pipeline to ensure the quality, robustness, and reproducibility of the results. First, we filtered the 16S rRNA gene reads by quality criteria, then obtained the ASVs and generated a table of frequencies. Subsequently, we applied feature selection to find the most representative ASVs, testing several target number of features. In the second module, for each target number of features, we tuned the hyperparameters for RF, SVM, kNN, MLP, and XGBoost models, as to find the combination that achieved the best results. We used 10-fold stratified cross-validation to evaluate the performance of each model individually and selected the three best to compose a promising ensemble model using a voting strategy. All models were finally trained and evaluated with the test set separated from the training and validation sets in the beginning of the experiment.

In EPheClass module 1, we began trying different target number of features, as shown in the first column of Table 2, to be selected by the different methods in order to observe which one performs best. In this table, we can see the number of selected ASVs in common for each of these values, as well as the contribution of each method individually. We can observe that RFE with RF and DESeq2 tend to contribute significantly more than RFE with SVM. We can also see that we have managed to reduce the 14,216 original features to a minimum size of 137, and a maximum size of 891, depending on the target number of features.

Table 2. Target number of features compared with those finally selected and the contribution of each method.

Target # of features	# of selected features	RFE-RF contribution	RFE-SVM contribution	DESeq2 contribution
200	137	127	20	136
300	232	211	75	206
400	331	304	139	290
500	428	391	192	377
800	708	630	417	604
1,000	891	781	566	770

Once obtained the ASV abundance table, we begin EpheClass module 2 evaluating the different algorithms for all target number of features. In Fig. 4, the f1-scores evaluating all five techniques (RF, SVM, XGBoost, kNN, and MLP) with 10-fold stratified cross-validation on the training set (85% of the original data) are plotted. It can be seen that the best-performing model for all features is XGBoost, followed closely by MLP. Tied for the third spot are by RF and SVM, with RF performing better for a lower target number of features (200 and 300), but worse for a larger target number of features (400, 500, 800, and 1,000). Thus, we have two options to form an ensemble model using a hard voting strategy: one being XGBoost, MLP, and RF (XMR ensemble) and the other being XGBoost, MLP, and SVM (XMS ensemble).

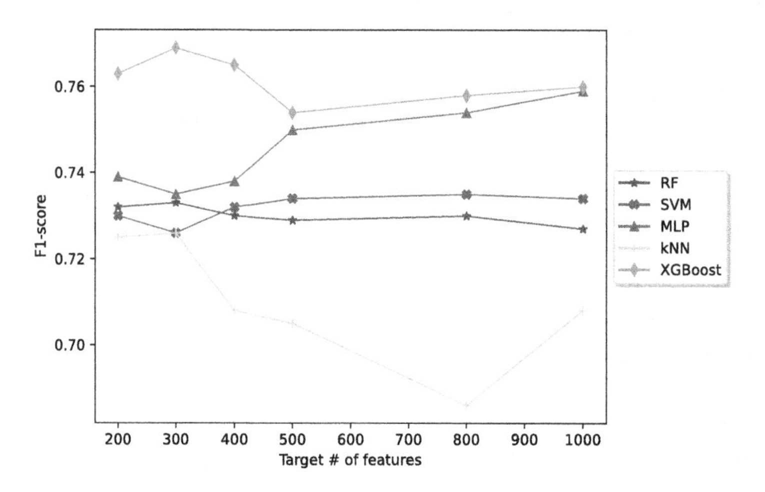

Fig. 4. F1-score evolution with the target number of features for XGBoost, MLP, RF, kNN, and SVM, with 10-fold stratified cross-validation.

After training and testing both ensemble options, in Fig. 5 we plotted the f1-scores for the test set (15% of the original data) for all base models, as well as the two ensemble models. The XMS ensemble model achieved the best results for every number of features tested, with a peak at a target of 300 and an f1-score over 0.8. RF and kNN were the worst performing models for all target number of features tested, although their f1-scores almost always surpassed 0.7. We can also observe that the higher values for all models tend to appear for a lower target number of features, typically under 500. In the confusion matrix obtained by the XMS ensemble with a target of 300 features, which is shown in Table 3, we can see more detected false positives (36) than false negatives (11). That is, if the disease is wrongly predicted it tends to be less ignored and slightly more overdiagnosed.

In Table 4 we present all the metrics used to evaluate the models: f1-score (f1), precision (p), recall (r), accuracy (acc), and area under the curve (roc_auc).

These were calculated using the test set for each model and each target number of features. The best-performing model, as we saw previously in Fig. 5, is the XGS ensemble with an f1-score of 0.808 for 300 features. This is closely followed by XGBoost alone with 0.797 for 400 features, and XMR ensemble with 0.793 for 400 features.

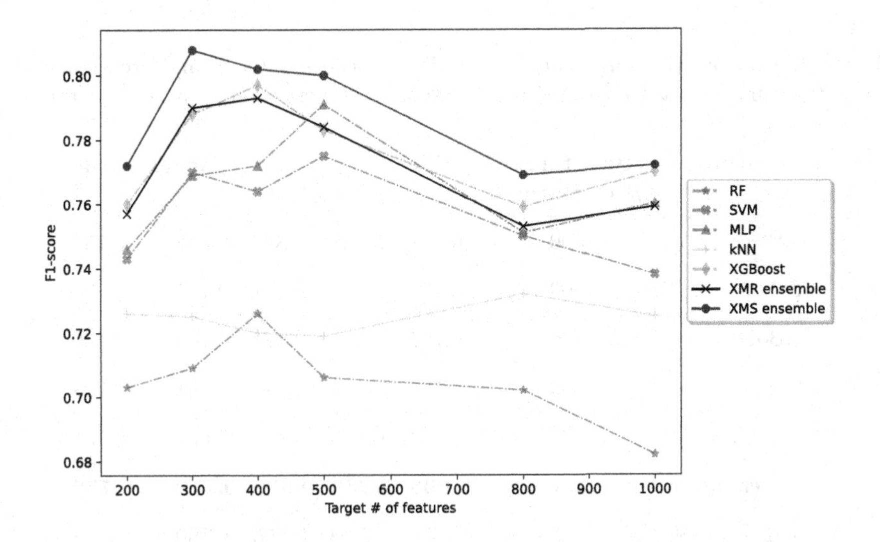

Fig. 5. F1-score evolution with the target number of features, using the test set, for each model individually and for XMR and XMS ensembles.

Table 3. Confusion matrix for best-performing model (XMS ensemble) with a target of 300 features.

		True label	
		CD	Not-CD
Predicted label	CD	99	36
	Not-CD	11	58

Finally, Table 5 compares EPheClass with those works from other classifiers of Crohn's disease, MicroPheno [1] and Read2Pheno [13], in terms of best performing techniques, inputs, and metrics provided by the authors. MicroPheno achieved for this dataset an f1-score of 0.76 in its best performance, using RF with k-mers. These k-mers were obtained using shallow sub-samples, discarding the less representative sequences and, thus, decreasing the number of reads per sample. Read2Pheno achieved for this dataset an accuracy of 0.83 in its best performance. At the beginning of the experiment, they selected 442 samples to use for training and testing with cross-validation. The best results were achieved with a rather complex model, combining different techniques like CNN, Bi-LSTM and ResNet to obtain the prediction of a read and, with an accuracy of 0.73, followed by the generation of pseudo-OTUs to classify the samples from the reads using RF to boost the accuracy.

Table 4. F1-score (f1), precision (p), recall (r), accuracy (acc), and area under the curve (roc_auc) for each evaluated algorithm and the best target number of features.

Algorithm	**Best target # of features**	**f1**	**p**	**r**	**acc**	**roc_auc**
MLP	500	0.791	0.758	0.827	0.765	0.759
SVM	500	0.775	0.685	0.891	0.721	0.706
XGBoost	400	0.797	0.733	0.873	0.760	0.750
RF	400	0.726	0.717	0.736	0.701	0.698
kNN	800	0.732	0.646	0.845	0.667	0.651
XMS ensemble	**300**	**0.808**	**0.733**	**0.900**	**0.770**	**0.759**
XMR ensemble	400	0.793	0.740	0.855	0.760	0.752

Table 5. Comparison of works approaching the phenotype classification of Crohn's disease. AT = abundance table.

Works	**Techniques**	**Input**	**f1**	**acc**	**p**	**r**
EPheClass (ours)	XGBoost, MLP, SVM	ASV AT	0.81	0.77	0.73	0.90
MicroPheno[1]	RF	k-mer AT	0.76	0.76	0.76	0.76
Read2Pheno[13]	ResNet, Bi-LSTM, CNN	read	-	0.73	-	-
	+ RF	pseudo-OTU AT	-	0.83	-	-

4 Discussion and Conclusion

In this work, a curated pipeline is followed for classifying phenotypes on 16S rRNA gene-sequenced samples. It aims to achieve the best possible results ensuring reproducibility and reducing the computational cost of the classification.

We demonstrate that the best results were achieved by the XMS ensemble for all target number of features, with the highest value at 300. The second ensemble model, the XMR ensemble, had a good performance as well, but not as good as the former, and even in some cases worse than XGBoost or MLP alone. The diversity of the first ensemble, where all algorithms are significantly different from one another, may positively influence the results, as it is well known in the literature [6]. As opposed, in the XMR ensemble, XGBoost and RF techniques are quite similar as both use decision trees and are ensemble models themselves, which means that they most likely classify samples similarly.

Even more, thanks to the XMS ensemble model, we managed to achieve the highest f1-score overall with a low target number of features (300, of which 232 are consensus), reducing the dimensionality but keeping the representativeness. In contrast to this ensemble's results, most of the individual models produced their respective best f1-scores with a higher target number of features, like 400 (XGBoost and RF) and 500 (MLP and SVM), and even with 800 (kNN). On top of that the confusion matrix shows that fewer false negatives were detected, which means that a fewer number of false negative classifications was obtained. We can also observe in Fig. 5 how the f1-score decreases as the target number of features increases, highlighting the importance of a proper feature selection. Thanks to removing features that add noise to the data, we can achieve better results, as well as reduce the final classification cost due to the dimensionality reduction.

Results detailed in Table 5 were similar or even better than other studies using the same Crohn's disease dataset aiming to predict the phenotype. On the one hand, the previously mentioned tool MicroPheno [1] achieved for this dataset an f1-score of 0.76 in its best performance, using RF with k-mers. Although they did reduce the dimensionality of the data by generating shallow sub-samples, the best results were achieved with a resample size of 5,000 reads/sample, whereas we managed to reduce the target number of features to 300 achieving better results (f1-score of 0.81). On the other hand, the aforementioned tool Read2Pheno [13] obtained for this dataset an accuracy of 0.83 in its best performance. In this case, the accuracy is the only performance metric provided. In this aspect, we consider that other metrics, like f1-score, precision and recall, an even the confusion matrix, should be shown to properly assess the performance of the model. Besides this, we worked with 1,358 samples, almost the full dataset, whereas Read2Pheno used only 442 samples selected at the beginning. To compare the behaviour of each system, in Read2Pheno they provide an accuracy of 0.82 using RF with an ASV abundance table as previous experimentation. In our work, using RF with tuned hyperparameters and an ASV abundance table containing all the samples, we achieved an accuracy of 0.70. This difference in accuracy using the same methods but a reduced number of samples may be due to the use of the whole dataset. Moreover, the best results were achieved with a rather complex model, combining different techniques like CNN, Bi-LSTM and ResNet to obtain the prediction of a read, achieving an accuracy of 0.73, worst than ours. In order to get higher values, a subsequent generation of pseudo-OTUs to classify the samples from the reads using RF was applied, thus making the system much more complex. In conclusion, the proposed pipeline allowed us to achieve a compromise between acceptable high scores and lower target number of features. Moreover, it ensures the reproducibility of the results and the reduction of computational cost by highly decreasing the dimensionality of the data. We consider it is adequate to correctly classify the Crohn's disease dataset and prepared to work with another diseases.

References

1. Asgari, E., Garakani, K., McHardy, A.C., Mofrad, M.R.K.: MicroPheno: predicting environments and host phenotypes from 16S rRNA gene sequencing using a k-mer based representation of shallow sub-samples. Bioinformatics **34**(13), i32–i42 (2018)
2. Callahan, B.J., McMurdie, P.J., Holmes, S.P.: Exact sequence variants should replace operational taxonomic units in marker-gene data analysis. ISME J. **11**(12), 2639–2643 (2017)
3. Callahan, B.J., McMurdie, P.J., Rosen, Michael Jand Han, A.W., Johnson, A.J.A., Holmes, S.P.: DADA2: high-resolution sample inference from illumina amplicon data. Nat. Meth. **13**(7), 581–583 (2016)
4. Edgar, R.C., Flyvbjerg, H.: Error filtering, pair assembly and error correction for next-generation sequencing reads. Bioinformatics **31**(21), 3476–3482 (2015)
5. Gevers, D., et al.: The treatment-Naive microbiome in new-onset Crohn's disease. Cell Host Microbe **15**(3), 382–392 (2014)
6. Kuncheva, L.I., Whitaker, C.J.: Measures of diversity in classifier ensembles and their relationship with the ensemble accuracy. Mach. Learn. **51**(2), 181–207 (2003)
7. Love, M.I., Huber, W., Anders, S.: Moderated estimation of fold change and dispersion for RNA-seq data with DESeq2. Genome Biol. **15**(12), 550 (2014)
8. Paulson, J.N., Stine, O.C., Bravo, H.C., Pop, M.: Differential abundance analysis for microbial marker-gene surveys. Nat. Methods **10**(12), 1200–1202 (2013)
9. Rajendhran, J., Gunasekaran, P.: Microbial phylogeny and diversity: small subunit ribosomal RNA sequence analysis and beyond. Microbiol. Res. **166**(2), 99–110 (2011)
10. Relvas, M.: Relationship between dental and periodontal health status and the salivary microbiome: bacterial diversity, co-occurrence networks and predictive models. Sci. Rep. **11**(1), 929 (2021)
11. Uddin, S., Khan, A., Hossain, M.E., Moni, M.A.: Comparing different supervised machine learning algorithms for disease prediction. BMC Med. Inform. Decis. Mak. **19**(1), 281 (2019)
12. Yu, Z., Wang, K., Wan, Z., Xie, S., Lv, Z.: Popular deep learning algorithms for disease prediction: a review. Cluster Comput. **26**, 1231–1251 (2022)
13. Zhao, Z., Woloszynek, S., Agbavor, F., Mell, J.C., Sokhansanj, B.A., Rosen, G.L.: Learning, visualizing and exploring 16S rRNA structure using an attention-based deep neural network. PLoS Comput. Biol. **17**(9), 1–36 (2021)

Synthetic Spermatozoa Video Sequences Generation Using Adversarial Imitation Learning

Sergio Hernández-García(✉), Alfredo Cuesta-Infante, and Antonio S. Montemayor

Universidad Rey Juan Carlos, Móstoles, Spain
{sergio.hernandez,alfredo.cuesta,antonio.sanz}@urjc.es

Abstract. Automated sperm sample analysis using computer vision techniques has gained increasing interest due to the tedious and time-consuming nature of manual evaluation. Deep learning models have been applied for sperm detection, tracking, motility analysis, and morphology recognition. However, the lack of labeled data hinders their adoption in laboratories. In this work, we propose a method to generate synthetic spermatozoa video sequences using Generative Adversarial Imitation Learning (GAIL). Our approach uses a parametric model based on Bezier splines to generate frames of a single spermatozoon. We evaluate our method against U-net and GAN-based approaches, and demonstrate its superior performance.

Keywords: Synthetic data · Imitation Learning · Sperm analysis

1 Introduction

Sperm health is a key on fertility issues and their treatments for both humans [7] and the livestock industry [14]. Leaving aside the genetics, the quality of a sperm sample is determined by 4 features: spermatozoa count, morphology, velocity and linearity; the last two are also referred to as motility. To this end, short videos, between 1 and 5 s, are taken of the sample under the microscope and evaluated by an expert according to these features.

Typically, the order of magnitude of the sperm count in the video sequence is 10^2. Hence, measuring the other features for a single video is tedious and time-consuming, which increases the cost of analysis due to the expense per expert hour. For this reason, there is a growing interest in automated sperm sample analysis using computer vision techniques. Specifically, deep learning has been

This research work has been supported by project TED2021-129162B-C22, funded by the Recovery and Resilience Facility program from the NextGenerationEU and the Spanish Research Agency (Agencia Estatal de Investigación); and PID2021-128362OB-I00, funded by the Spanish Plan for Scientific and Technical Research and Innovation of the Spanish Research Agency.

A. Pertusa et al. (Eds.): IbPRIA 2023, LNCS 14062, pp. 569–580, 2023.
https://doi.org/10.1007/978-3-031-36616-1_45

applied to sperm detection, tracking and motility analysis, as well as morphology recognition [11,13,14,21,28]. So far, previous works have followed a supervised approach; but the lack of labeled data sets for training and testing continues to prevent the adoption of such techniques by laboratories.

In this work we present a method to create synthetic video sequences of sperm sample. Synthetic data presents 3 challenges for deep learning based computer vision. First, it should present a realistic aspect. Second, it should be truthful. In the context of this work, that means mimicking the behavior of real sperm. And third, it should be useful, in the sense that models trained and tested on synthetic data must perform similarly on real data. We use a subset of 10 spermatozoa video sequences extracted from the SCASA dataset [11], all of them classified as *normal*, accounting for a total of 260 frames. In each video sequence there is a single spermatozoon with its head centered on the frame moving the tail and changing the bearing. Due to the scarcity of real frames, we discard the use of both generative adversarial networks [2,9] and supervised methods [12] for generating synthetic ones. Instead, our approach consists of using Generative Adversarial Imitation Learning (GAIL). We implicitly assume that there is an expert who produces spermatozoa video sequences. The generating process is unknown, and all we have is a set of demonstrations. GAIL enables training a reinforcement learning (RL) agent to imitate the expert. Specifically, the agent will learn how to modify the parameters of a simplified model of the spermatozoon in a frame conditioned to the previous frames.

The contributions of this work are:

- We present a spermatozoon parametric model which takes advantage of the previous knowledge of the kinematics of spermatozoa as well as incorporate constraints.
- We use GAIL to generate new spermatozoa video sequences based on the SCASA dataset [11].
- We extensively test our approach against U-net and GAN based approaches and show that GAIL outperforms both.

2 Related Works

Limited data availability is a relevant problem for deep learning techniques which require large amounts of data to achieve generalization and invariance or equivariance depending on the task. While the last two are partially mitigated with data augmentation, by applying geometric and pixel-wise transformations, creating new data samples from the same distribution than the real ones is a field of intense research. A popular approach are Generative adversarial networks (GAN). Controlled generation conditioned to a data sample has been proposed in [8], mode collapse issues have been dealt in [10], and GANs with interpretable latent space have been proposed in [4]. GANs have been applied in diverse fields, including medical and biological domains. Cancer detection [23,26] and citrus disease detection [29] are just two examples. In the context of automated sperm analysis GANs have been used in [2,19].

In this work we follow an adversarial imitation learning approach [15,25,27]. These techniques have been applied to solve decision making problems that generate trajectories on robotics [15,20,32], on autonomous vehicles [3,5,6], taxi route prediction [30,31] or modeling animal behavior [16,17]. A comprehensive survey can be found in [1].

Given this background, this paper hypothesizes that it is possible to successfully train an agent to generate spermatozoa trajectories from only a few expert demonstrations using GAIL.

3 Adversarial Imitation Learning on Spermatozoa Sequence Generation

In this paper we train an adversarial imitation learning agent to generate synthetic spermatozoon images. To this end we present a parametric model of the spermatozoon, such that the agent acts on the parameters according to a policy learned from expert demonstrations. In this section we formalize the methodology proposed. Firstly, some background and notation is introduced. Next, the parametric model is presented. Finally, the generating process is described.

3.1 Background

The synthetic video sequence generation by adversarial imitation learning can be expressed as a Markov Decision Process (MDP) formed by the tuple $(\mathcal{S}, \mathcal{A}, \mathcal{P}, r)$ $\mathcal{S}$ and $\mathcal{A}$ are the set of states and actions respectively. $\mathcal{P}(s_{t+1}|s_t, a_t)$ is the state transition probability distribution, where $s_t, s_{t+1} \in \mathcal{S}$, $a_t \in \mathcal{A}$ and subscript t indicates the timestamp. Finally, $r(s_t, s_{t+1}, a_t)$ is the reward function.

Adversarial imitation learning, and specifically in this paper GAIL, avoids the reward shaping task when training a RL agent by introducing a data set of expert trajectories, $\mathcal{E}$. Thus, GAIL produces a reward function that is plugged into the MDP tuple to be solved by RL methods. In this scenario, the resulting agent learns a *policy*, i.e. a parametric distribution over the action space given the current state, denoted by $\pi_\theta(a_t|s_t)$, where θ is the parameter vector.

In this paper, the states are the successive frames of a video sequence and the timestamp indicates their correlative order. The action space is related to the parametric model of the spermatozoon, detailed below, and allows a limited modification of its shape and bearing. The data set of expert trajectories $\mathcal{E} = \{(s_0, a_0), (s_1, a_1), ..., (s_T, a_T)\}$; where s_t is the frame at timestamp t, a_t are the parameters extracted from the spermatozoon in s_t, and T is the index of the last frame in the sequence.

GAIL, as referred before, consists of a RL agent and, as well as GANs, a discriminator $\mathcal{D}_\phi$ that is trained to estimate the probability of a state-action pair belonging to $\mathcal{E}$. This probability is the outcome of the discriminator, and the resulting reward function is

$$r_t = -\ln\bigl(1 - \mathcal{D}_\phi(s_t, a_t, s_{t+1})\bigr). \tag{1}$$

The higher the reward, the more similar the actions of the agent and the expert.

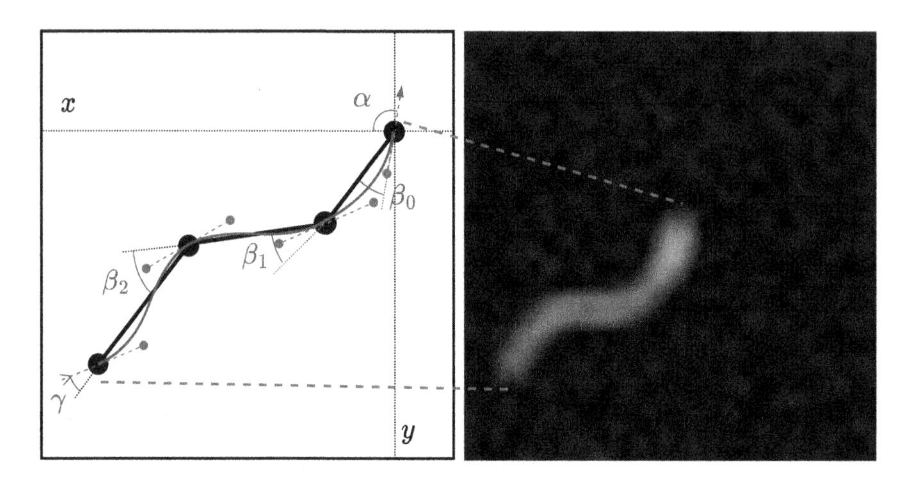

Fig. 1. Left: Spermatozoa parametric model using Bezier splines. **Right:** Image generated given the model on left. (x, y) are the center of the frame, the parameters of the model are in blue, anchor points and junction segments in thick black, spline control points in green with dashed line and the resulting Bezier curve in red. (Color figure online)

3.2 Spermatozoa Parametric Model

The parametric model of spermatozoa has two key purposes: 1) it endows the agent with an action space that consists of the parameter vector that fully determines how the spermatozoa is rendered, and 2) it allows us to constrain the boundaries of the parameters, thus preventing the generation of infeasible spermatozoa morphology.

Our model assumes that the head of the spermatozoon is always located at the center of a gray scale image with resolution 28×28, and the tail is simplified to three segments. Finally, the shape is smoothed using Bezier splines, as shown in Fig. 1. The shape and bearing are determined by the parameter vector $[\alpha, \beta_1, \beta_2, \beta_3, \gamma]$. Parameter $\alpha \in [0, 2\pi]$ controls the direction in which the head is pointing. Parameters $\beta_{1,2,3} \in [-\frac{\pi}{4}, \frac{\pi}{4}]$ control the angle between each segment of the spermatozoon body. Finally, parameter $\gamma \in [-\frac{\pi}{4}, \frac{\pi}{4}]$ determines the direction in which the end of the tail is pointing. We remark that α denotes the global direction of the head, whereas $\beta_1, \beta_2, \beta_3$ and γ correspond to relative values with respect to α.

To ensure a smooth and continuous body shape, we use a Bezier spline to connect each segment junction, known as anchors. Each anchor point is associated with a set of control points that define the smoothness of the curve that connects two adjacent anchor points. The control points are placed at a fixed distance of $\frac{1}{3}$ the distance between anchor points.

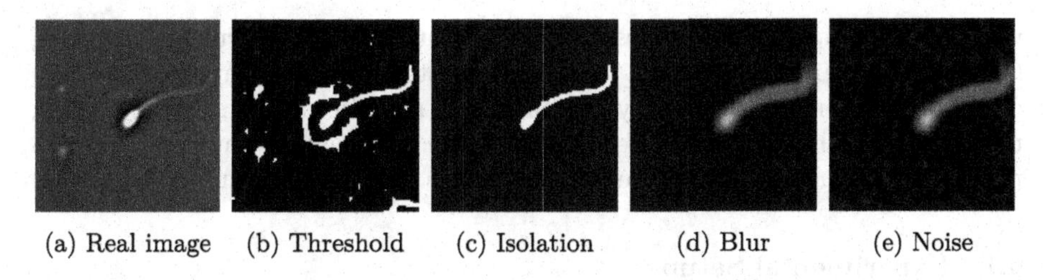

(a) Real image (b) Threshold (c) Isolation (d) Blur (e) Noise

Fig. 2. Sequence of transformations applied to a real image (most left) to obtain the preprocessed one (most right).

3.3 Image Generation

We start from a completely black 28×28 image. The head of the spermatozoon consists of a white ellipse centered on the image with its major axis pointing at the direction given by α. The tail follows the head and consists of the spline obtained given $[\beta_1, \beta_2, \beta_3, \gamma]$, painted in white with 3 pixel thick.

At this point we remark that real frames are preprocessed (Fig. 2) in order to enhance the head and tail, and to clean the background as follows. First an adaptive threshold is applied to extract the spermatozoon silhouette. A global threshold is discarded because the background is not uniform due to the microscope. Next, we apply contour finding to isolate from spurious white pixels. Then, the borders are smoothed with a Gaussian blur. And finally, Gaussian noise is added to the background. The last two transformations (Fig. 2d and 2e) are necessary only for benchmarking our proposal against U-Net and WGAN, which produce blurry and noise images.

3.4 Sequence Generation

Under the GAIL setup, given a state s_t, the agent proposes an action $a_t = (a_t^{(0)}, a_t^{(1)}, a_t^{(2)}, a_t^{(3)} a_t^{(4)})$. The parameters of the spermatozoa model are then calculated using the action proposed by the agent, as follows:

$$\alpha_{t+1} = \alpha_t + a_t^{(0)} \cdot \pi/10 \tag{2}$$

$$\beta_{t+1}^{(M)} = a_t^{(M)} \cdot \pi/4, \quad M \in [1, 3] \tag{3}$$

$$\gamma_{t+1} = a_t^{(4)} \cdot \pi/4 \tag{4}$$

Then, the next image of the sequence is generated according to these parameters. Finally, we get a reward function using Eq. (1) for every time step.

4 Experiments

In this section we carry out a series of experiments to address several questions related to the generation of realistic sperm images and sequences. Firstly, we investigate whether it is possible to generate realistic sperm images using GAIL and compare the results against other state-of-the-art approaches. Secondly, we explore the feasibility of generating realistic sperm sequences using the approaches mentioned above and assess the potential advantages of using a parametric sperm model to generate real-like sequences through GAIL.

4.1 Experimental Setup

We use GAIL in combination with a Proximal Policy Optimization (PPO) [24] agent, whose goal is to learn the *policy* π_θ. The *policy* is represented by a convolutional neural network (CNN). Its input tensor has a shape of $5 \times 28 \times 28$, and consisting of images from the five preceding frames.

The discriminator, $\mathcal{D}_\phi$, also consists of a CNN. Its input tensor has a shape of $6 \times 28 \times 28$, comprising the five preceding frames and the subsequent frame generated by the agent for s_{t+1}.

The SCASA dataset contains sequences of *normal* spermatozoa moving in only a few orientations. Therefore, we apply data augmentation techniques such as flips along the horizontal, vertical, and both axes, as well as rotations. This process yields 960 training frames and 160 evaluation frames.

We provide further details regarding the experimental setup in the Appendix.

4.2 Image Generation Analysis

In this experiment, we compare our GAIL-based approach with 3 generative models: Wasserstein GAN (WGAN) with gradient penalty [10], U-Net [22], and conditional WGAN (CWGAN) [18]. All of them consist of CNNs or convolutional autoencoders that generate a 28×28 image but differ on the expected inputs. Specifically, WGAN was trained to generate frames given a random noise vector. U-Net receives a set of the five preceding frames. The generator model of CWGAN follows the same architecture and workflow as the U-Net model. This model also includes the reconstruction loss from U-Net. We train three CWGAN models with varying levels of importance assigned to the reconstruction. These models include CWGAN with a zero coefficient multiplying the reconstruction loss, CWGAN10 with a coefficient of ten, and CWGAN100 with a coefficient of 100. All GAN-based approaches use a CNN as discriminator.

Figure 3 presents a comparison of the results obtained by attempting to predict a single still sample. The rows depict the results for each algorithm, while the columns display independent examples from the evaluation set. The first row displays frame for s_t, and the second row displays the expected frame for s_{t+1}.

It is worth mentioning that WGAN is used as a baseline for the rest of the algorithms. It is not conditioned to s_t, therefore, it is not expected that the model will generate s_{t+1}. As shown in Fig. 3, its results are very poor.

The U-Net model generates better images than WGAN, which is reasonable because it receives additional information in the form of s_t. However, it fails to generate the tails, which are blurred.

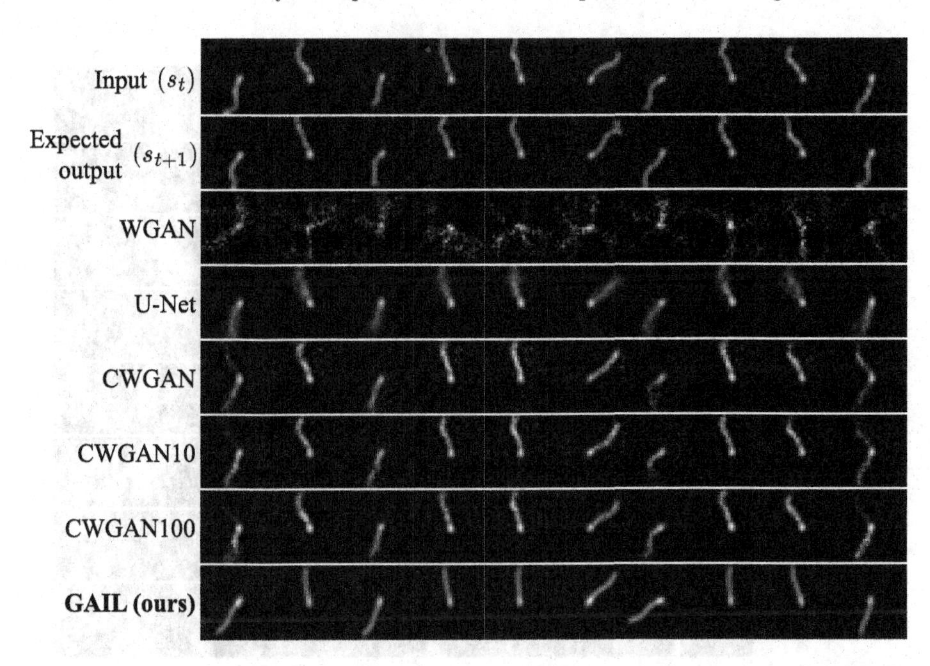

Fig. 3. Single still image generation results. First row shows independent input frames. Second row shows the expected output frames. The remaining rows show predicted frames by each algorithm.

Conditioned WGAN (CWGAN) generates highly realistic images that are not blurred, unlike the ones generated by U-Net. The discriminator in CWGAN enforces the model to generate images with similar biases as the real ones, resulting in sharper images. However, these images appear to resemble images closer to s_t rather than s_{t+1}.

Unlike other methods, our GAIL-based proposal generates spermatozoa in novel positions that may not correspond to the ground truth. This behavior is desirable as our focus is on generating new images. In contrast to other approaches, however, GAIL tends to guide the spermatozoa to a neutral position, resulting in a straight line shape.

We also quantitatively evaluate the quality of our results by comparing the histograms and structural composition of the generated images with those of the evaluation set. Table 1 shows that the histogram of the images generated by GAIL closely resembles that of the validation images, as evidenced by the minimal Kullback-Leibler divergence achieved, which is one order of magnitude lower than the other methods.

Futhermore, while the evaluated algorithms perform similarly in terms of structural similarity, the results demonstrate that our method achieves the best performance in this metric.

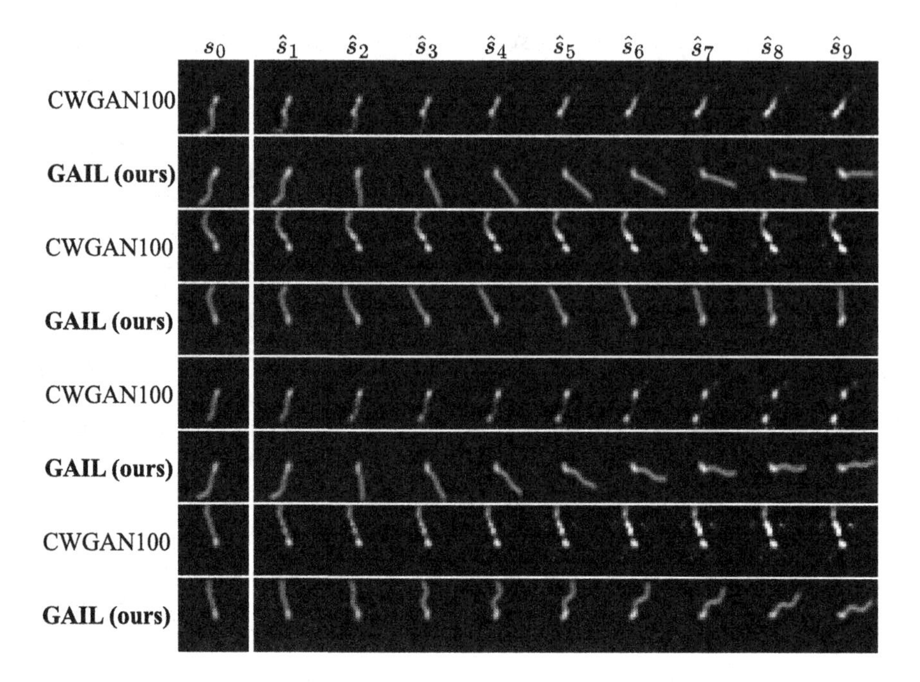

Fig. 4. Image sequences generation results. Left images are the initial state s_0 sampled from the evaluation set. On the right side, respective sequences generated by CWGAN100 and GAIL (ours).

4.3 Sequence Generation Analysis

To generate a sequence of frames, we feed the initial state s_0 to the generator, which returns the subsequent state s_1. This generated state s_1 is then fed back to the generator to obtain the next state s_2, and we repeat this process until we obtain the final state s_T. Figure 4 shows a series of sequences generated running CWGAN100 and GAIL (ours).

The resulting sequences of images generated by CWGAN100 present a continuous degradation. This issue arises because the model is processing an image previously generated by itself, which causes the generated images to deviate from the real distribution of data. Another problem related to this approach, as mentioned before, is that CWGAN100 tends to generate a spermatozoon in the same position as the input one, indicating that it is not replicating a moving spermatozoon as seen in the training sequences.

Our proposal creates new trajectories as shown in Fig. 4 with continuity between frames. Obviously, this approach avoids the degradation from GANs since the image is generated given the parametric model. In this experiment, we appreciate that, rather than leading the spermatozoon to a neutral position, GAIL tends to guide it to a different position depending on the input.

Table 1. Comparison between validation and generated images using the following metrics: Kullback-Leibler divergence (D_{KL}) between histogram distributions and Information Theoretic-based Statistic Similarity Measure (ISSM). Lower values are better for Kullback-Leibler divergence while higher values are better for structural similarity.

Metric	U-Net	WGAN	CWGAN	CWGAN10	CWGAN100	**Ours**
D_{KL}	1.28	1.14	0.37	0.38	0.37	**0.04**
ISSM	0.19	0.12	0.19	0.19	0.19	**0.21**

5 Conclusion

In this paper we show how GAIL can be used as a generative approach for synthetic sperm video sequence. Results show that our solution compromises between the realistic aspect and a truthful imitation of the movement. While CWGAN attains better still images, GAIL outperforms it in the overall sequence. However, with GAIL the image generation strongly depends on the model, which can be replaced by a better one with minimal changes in the rest of the procedure. Thus, the proposal gains explainability, which the other generative models lack.

Appendix

In this section we detail the network architectures

GAIL. We use a PPO agent as a generator and a CCN as a discriminator.

- **PPO** is an actor-critic RL agent that comprises two neural networks referred to as the *actor* and the *critic*. Both networks shared the same backbone, consisting of two convolutional layers alternated with max pooling layers, followed by flatten and passed through a hidden dense layer with ReLU activation. The actor network estimates the *policy* π_θ so its output layer comprises five dense neurons with tanh activation. The critic network evaluates the actor's decisions, so its output layer just has one neuron with linear activation.
- **Discriminator** is similar to the critic but with a sigmoid activation at the output. To prevent overfitting we incorporate dropout after each convolutional and dense layer.

WGAN. It has the conventional architecture.

- **Generator** has an input layer of size 100×1 followed by three transposed convolution layers. We alternate these layers with batch normalization and ReLU activation.
- **Discriminator** has an input layer of size 28×28. Then, it has the same architecture as GAIL's discriminator but with three convolutional layers with a single dense neuron with linear activation at the output.

U-Net. It receives an input tensor of size $5 \times 28 \times 28$ corresponding to the five preceding frames, $[s_{t-4}, \ldots, s_t]$. Then, it compresses the input tensor using four convolutions to a size of $64 \times 4 \times 4$. The image reconstruction is made by three successive transposed convolutions, resulting in an output tensor of size $1 \times 28 \times 28$, corresponding to the next frame s_{t+1}. All hidden activations are ReLU.

CWGAN. We use the same U-Net architecture for the generator and a CNN-based discriminator.

- **Generator** produces an image for s_{t+1} conditioned on the input in s_t.
- **Discriminator** has an input layer of $6 \times 28 \times 28$, corresponding to the five preceding frames concatenated with the one predicted by the generator. The architecture is the same than WGAN's Discriminator.

References

1. Arora, S., Doshi, P.: A survey of inverse reinforcement learning: challenges, methods and progress. Artif. Intell. **297**, 103500 (2021). https://doi.org/10.1016/j.artint.2021.103500
2. Balayev, K., Guluzade, N., Aygün, S., İlhan, H.O.: The implementation of DCGAN in the data augmentation for the sperm morphology datasets. Avrupa Bilim ve Teknoloji Dergisi **26**, 307–314 (2021)
3. Bhattacharyya, R., et al.: Modeling human driving behavior through generative adversarial imitation learning. IEEE Trans. Intell. Transp. Syst. **24**(3), 2874–2887 (2023). https://doi.org/10.1109/TITS.2022.3227738
4. Chen, X., Duan, Y., Houthooft, R., Schulman, J., Sutskever, I., Abbeel, P.: Infogan: interpretable representation learning by information maximizing generative adversarial nets. In: Advances in Neural Information Processing Systems, vol. 29 (2016)
5. Choi, S., Kim, J., Yeo, H.: Trajgail: generating urban vehicle trajectories using generative adversarial imitation learning. Transp. Res. Part C: Emerg. Technol. **128**, 103091 (2021). https://doi.org/10.1016/j.trc.2021.103091
6. Coates, A., Abbeel, P., Ng, A.Y.: Apprenticeship learning for helicopter control. Commun. ACM **52**(7), 97–105 (2009)
7. Dai, C., et al.: Advances in sperm analysis: techniques, discoveries and applications. Nat. Rev. Urology **18**(8), 447–467 (2021)
8. Gauthier, J.: Conditional generative adversarial nets for convolutional face generation. Class Project Stanford CS231N: Convolutional Neural Netw. Vis. Recogn. Winter Semester **2014**(5), 2 (2014)
9. Goodfellow, I., et al.: Generative adversarial nets. In: Ghahramani, Z., Welling, M., Cortes, C., Lawrence, N., Weinberger, K. (eds.) Advances in Neural Information Processing Systems, vol. 27 (2014)
10. Gulrajani, I., Ahmed, F., Arjovsky, M., Dumoulin, V., Courville, A.C.: Improved training of Wasserstein GANs. In: Advances in Neural Information Processing Systems, vol. 30 (2017)
11. Hernández-Ferrándiz, D., Pantrigo, J.J., Cabido, R.: SCASA: from synthetic to real computer-aided sperm analysis. In: Ferrández Vicente, J.M., Álvarez-Sánchez, J.R., de la Paz López, F., Adeli, H. (eds.) Bio-inspired Systems and Applications: from

Robotics to Ambient Intelligence, pp. 233–242. Springer International Publishing, Cham (2022). https://doi.org/10.1007/978-3-031-06527-9_23
12. Hidayatullah, P., Mengko, T., Munir, R., Barlian, A.: A semiautomatic dataset generator for convolutional neural network. In: Proceedings of the International Conference on Electrical Engineering & Computer Science (ICEECS 2018), pp. 17–21 (2018)
13. Hidayatullah, P., Mengko, T.L.E.R., Munir, R., Barlian, A.: Bull sperm tracking and machine learning-based motility classification. IEEE Access **9**, 61159–61170 (2021). https://doi.org/10.1109/ACCESS.2021.3074127
14. Hidayatullah, P., et al.: Deepsperm: a robust and real-time bull sperm-cell detection in densely populated semen videos. Comput. Methods Programs Biomed. **209**, 106302 (2021)
15. Ho, J., Ermon, S.: Generative adversarial imitation learning. In: Advances in Neural Information Processing Systems, vol. 29 (2016)
16. Kim, B., Pineau, J.: Socially adaptive path planning in human environments using inverse reinforcement learning. Int. J. Soc. Rob. **8**, 51–66 (2016)
17. Kretzschmar, H., Spies, M., Sprunk, C., Burgard, W.: Socially compliant mobile robot navigation via inverse reinforcement learning. Int. J. Rob. Res. **35**(11), 1289–1307 (2016). https://doi.org/10.1177/0278364915619772
18. Mirza, M., Osindero, S.: Conditional generative adversarial nets (2014). https://doi.org/10.48550/ARXIV.1411.1784
19. Paul, D., Tewari, A., Jeong, J., Banerjee, I.: Boosting classification accuracy of fertile sperm cell images leveraging cdcgan. In: ICLR, 2021 (2021)
20. Rafailov, R., Yu, T., Rajeswaran, A., Finn, C.: Visual adversarial imitation learning using variational models. In: Ranzato, M., Beygelzimer, A., Dauphin, Y., Liang, P., Vaughan, J.W. (eds.) Advances in Neural Information Processing Systems, vol. 34, pp. 3016–3028. Curran Associates, Inc. (2021)
21. Riordon, J., McCallum, C., Sinton, D.: Deep learning for the classification of human sperm. Comput. Biol. Med. **111**, 103342 (2019). https://doi.org/10.1016/j.compbiomed.2019.103342
22. Ronneberger, O., Fischer, P., Brox, T.: U-Net: convolutional networks for biomedical image segmentation. In: Navab, N., Hornegger, J., Wells, W.M., Frangi, A.F. (eds.) MICCAI 2015. LNCS, vol. 9351, pp. 234–241. Springer, Cham (2015). https://doi.org/10.1007/978-3-319-24574-4_28
23. Rubin, M., et al.: TOP-GAN: stain-free cancer cell classification using deep learning with a small training set. Med. Image Anal. **57**, 176–185 (2019). https://doi.org/10.1016/j.media.2019.06.014
24. Schulman, J., Wolski, F., Dhariwal, P., Radford, A., Klimov, O.: Proximal policy optimization algorithms. arXiv preprint arXiv:1707.06347 (2017)
25. Sun, M., Devlin, S., Hofmann, K., Whiteson, S.: Deterministic and discriminative imitation (d2-imitation): revisiting adversarial imitation for sample efficiency. In: Proceedings of the AAAI Conference on Artificial Intelligence, vol. 36, pp. 8378–8385 (2022)
26. Wu, E., Wu, K., Cox, D., Lotter, W.: Conditional infilling GANs for data augmentation in mammogram classification. In: Stoyanov, D., et al. (eds.) RAMBO/BIA/TIA -2018. LNCS, vol. 11040, pp. 98–106. Springer, Cham (2018). https://doi.org/10.1007/978-3-030-00946-5_11
27. Xiao, H., Herman, M., Wagner, J., Ziesche, S., Etesami, J., Linh, T.H.: Wasserstein adversarial imitation learning. arXiv preprint arXiv:1906.08113 (2019)

28. Yüzkat, M., Ilhan, H.O., Aydin, N.: Multi-model CNN fusion for sperm morphology analysis. Comput. Biol. Med. **137**, 104790 (2021). https://doi.org/10.1016/j.compbiomed.2021.104790
29. Zeng, Q., Ma, X., Cheng, B., Zhou, E., Pang, W.: GANs-based data augmentation for citrus disease severity detection using deep learning. IEEE Access **8**, 172882–172891 (2020). https://doi.org/10.1109/ACCESS.2020.3025196
30. Ziebart, B.D., Maas, A.L., Bagnell, J.A., Dey, A.K., et al.: Maximum entropy inverse reinforcement learning. In: AAAI, Chicago, IL, USA, vol. 8, pp. 1433–1438 (2008)
31. Ziebart, B.D., Maas, A.L., Dey, A.K., Bagnell, J.A.: Navigate like a cabbie: probabilistic reasoning from observed context-aware behavior. In: Proceedings of the 10th International Conference on Ubiquitous Computing, pp. 322–331 (2008)
32. Zuo, G., Chen, K., Lu, J., Huang, X.: Deterministic generative adversarial imitation learning. Neurocomputing **388**, 60–69 (2020). https://doi.org/10.1016/j.neucom.2020.01.016

A Deep Approach for Volumetric Tractography Segmentation

Pablo Rocamora-García, Marcelo Saval-Calvo(✉), Victor Villena-Martinez, and Antonio Javier Gallego

University Institute for Computer Research, University of Alicante, Alicante, Spain
prg66@alu.ua.es, m.saval@ua.es, vvillena@dtic.ua.es, jgallego@dlsi.ua.es

Abstract. The study of tracts—bundles of nerve fibers that are organized together and have a similar function—is of major interest in neurology and related areas of science. Tractography is the medical imaging technique that provides the information to estimate these tracts, which is crucial for clinical applications and scientific research. This is a complex task due to the nature of the nerve fibers, also known as streamlines, and requires human interaction with prior knowledge. In this paper, we propose an automatic volumetric segmentation architecture based on the 3D U-Net architecture to segment each tract individually from streamlines data. We evaluate the impact of different data pre-processing techniques namely Rescaled Density Map (RDM), Gaussian Filter Mask (GFM), and Closing Opening Mask (COM) on the final segmentation results using the Tractoinferno dataset. In our experiments, the average DICE and IoU average performance was 62.2% and 72.2% respectively. Our results show that proper data pre-processing can significantly enhance segmentation performance. Moreover, we achieve similar levels of accuracy for all segmented tracts, despite shape disparity and an unequal number of occurrences in the tract dataset. Overall, this work contributes to the field of neuroimaging by providing a reliable approach for accurately segmenting individual tracts.

Keywords: Tractography · 3D-UNet · 3D segmentation · Volumetric segmentation · Neuroimaging

1 Introduction

The study of neurological pathologies is of major interest in current society. Drugs, such as alcohol, affect brain connectivity [1,2]. To analyze the possible evolution or degradation in the connectivity, it is necessary to individually differentiate the fibers and tracts that compose the brain. Tractography is a medical imaging technique that allows the visualization of brain connectivity from the white matter [3,4]. The data is originally acquired using diffusion-weighted magnetic resonance imaging (dWI), which provides an estimate of water diffusion intensity at a specific location in volumetric space. Next, a fiber orientation distribution function (fODF) is used to describe the directionality and location of

A. Pertusa et al. (Eds.): IbPRIA 2023, LNCS 14062, pp. 581–592, 2023.
https://doi.org/10.1007/978-3-031-36616-1_46

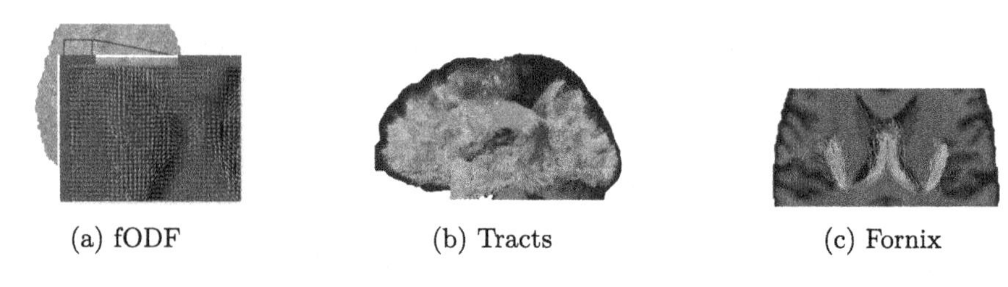

(a) fODF (b) Tracts (c) Fornix

Fig. 1. Tractography segmentation stream. From a dWI, the fODF is estimated (a). A process of tracking the fODFs provides a representation of all the nerve fibers (b), that are segmented in tracts, such as the Formix-fimbria tract (c).

water diffusion (see Fig. 1a). To obtain the tracts from a dWI, a certain number of peaks (principal directions) are extracted from each fODF element. The next step is to track peak by peak in different directions to obtain the nerve fibers, also known as streamlines, Fig. 1b. Finally, the groups of fibers that compose each tract are determined. This process is carried out mainly with probabilistic algorithms requiring human interaction.

The study of the machine learning-based methods working with tractographies has had a growth in recent years [5,6], from fODF automatic generation [7] to tracts segmentation [8]. Automatically segmenting tracts from dWI is a complicated task that has been studied [9,10] but still remains a challenging problem [5]. Recent approaches use Convolutional Neural Network architectures (CNN) in order to tackle this problem [8,9]. Since a dWI, as any other variant of the MRI, is acquired in slices that can be afterward packed in a voxel grid, most current methods use these slices as input of the network. This is the case of TragSeg [9], which proposes a 2D U-Net encoder-decoder to estimate a probability map of the tracts. There are further proposals that use or extend TracSeg [10,11]. Liu et al. [12] use TracSeg to obtain a set of embeddings from each input image which is then projected to the predicted output image of the tracts using a convolutional layer. Lu et al. presented in [13] a transfer learning solution for unseen tracts segmentation based on TracSeg. They apply a fine-tuning process to improve the performance of tract segmentation when a new tract is included in the dataset without the need to re-adapting the entire network. DeepWMA [10] extracts a 2D multi-channel descriptor—coined as FiberMap—that is fed to a classical CNN to predict the labels of the different fibers, they particularly focus on *arcuate fasciculus* (AF) and *corticospinal tracts* (CST). Neuro4Neuro [8] adds additional skip connections to the U-Net to improve the segmentation, with parametric rectified linear units (PReLU) and a voxelwise softmax last layer to create the probability map.

All of the above-mentioned methods use 2D deep Convolutional Neural Networks, mainly based on the U-Net architecture [14]. However, there is still room for improvement due to the complexity of some specific tracts, such as the *formix-fimbria*, which has much fewer fibers compared to other tracts. An interesting way of improvement is the use of the actual 3D volumetric data instead of each slice individually, leveraging the structural information of the whole set of fibers and much more context for their identification.

Volumetric segmentation has been applied to medical imaging for different purposes, for instance, Zhang et al. [15] presented a comparison of 2.5D segmentation approaches. Rather than using the complete 3D volume, they segment simultaneously 2D slices from different planes, including sagittal, coronal, and axial planes, merging the result to obtain the final segmentation. These approaches are more efficient than the 3D methods are require fewer data for model training, but still rely on per-plane geometry for segmentation, when tracts can follow multiple angles and directions.

Traditional 3D volumetric segmentation [16] uses similar principles as 2D segmentation methods, such as global and adaptive thresholding, edge detection, neighborhood homogeneity, etc. New approaches based on Deep Learning architectures have proven better results because the models learn better the data distribution and characteristics. V-Net [17] was presented as a volumetric CNN architecture for medical image segmentation. An extension of the U-Net was proposed for 3D segmentation, coined 3D-UNet [18]. This proposal replaces 2D with 3D convolutional layers, achieving better performance in volumetric segmentation. Variants of 3D-UNet have also been presented. For instance, the proposal by Qamar et al. [19] adds context information to better segment similar tissues, by combining dense and residual connections and an inception module.

Different methods have been proposed in the literature for the automatic segmentation of tractography information using dWI or processed data, such as fODF and peaks, however, they are mainly based on U-Net architectures or other 2D CNN networks [20]. These previous methods based on 2D convolutions exploit the neighborhood relationship for each pixel per slice, however, it does not take into account inter-slice information. Leveraging the geometrical relationship in 3D will provide better performance in the segmentation problem since the tracts are oriented in different 3D directions. To the best of our knowledge, there are no specific proposals using volumetric approaches for tract segmentation using fODF. Hence, this paper presents a novel 3D U-Net-based tractography segmentation architecture that uses as input the fODF data. Furthermore, three different techniques are proposed to pre-process the raw fODF data in order to generate segmentation masks. These techniques are compared on the Tractoinferno dataset with 284 subjects to analyze their impact on the final segmentation results.

The rest of the paper is structured as follows: in Sect. 2 the architecture and the training process are explained; Sect. 3 develops the experimentation and the dataset. Finally, in Sect. 4 we discuss some findings of this work.

2 3D Tractography Segmentation

This section first details the proposed architecture, followed by the loss function used and the training process, and finally, the different proposals for preprocessing the input data are explained.

2.1 Network Architecture

Following the success of "U" shaped networks in medical semantic segmentation tasks, we decided to resort to a U-Net architecture [18] but changing the 2D convolutions layers for 3D convolutions, as shown in Fig. 2. This model therefore allows the volumetric segmentation of 3D images, being a variation of the original U-Net architecture, which was introduced for the segmentation of 2D medical images. The architecture consists of an encoder-decoder structure with skip or residual connections [21] between corresponding encoder and decoder layers.

The encoder part of the 3D U-Net consists of a series of 3D convolutional layers, max-pooling, and batch normalization layers [22], which reduce the spatial dimensionality of the input while increasing the number of feature maps. The input of the network is a voxel grid with the shape of $160 \times 160 \times 160 \times 9$, where each voxel contains a tensor with the 9 main directions extracted from the peaks of the fiber orientation distribution function data (see Fig. 2). In this way, the encoder part of the network increasingly learns complex features from the input data.

The decoder part consists of a series of 3D convolutional and upsampling layers [23], which restore the spatial dimensions of the data while decreasing the number of feature maps. Connections between the encoder and decoder beyond the context vector—or latent representation—are carried out by concatenating feature maps at specific layers. This structure provides additional context information to restore the spatial dimensions of the feature maps. Finally, the decoder outputs a $160 \times 160 \times 160 \times 1$ tensor representing a segmentation map of the estimated region that contains all the streamlines which compose the tract.

2.2 Loss Function

The loss function used to optimize the weights of the proposed architecture is based on a combination of the binary cross-entropy (BCE) loss and the jaccard index. Let $x_i \in \mathbb{R}^{d \times h \times w \times c}$ be the input containing the fODF peaks data from one subject i, and $y_i \in \mathbb{R}^{d \times h \times w \times c}$ the ground truth mask for one tract, where d, h, w, and c are the depth, height, width, and the number of channels of the voxel grid, respectively. To minimize the difference between the ground truth y_i and predicted masks $\hat{y_i}$, the following joint optimization loss function is proposed:

$$\mathcal{L}_{total} = \mathcal{L}_{bce} + \mathcal{L}_{jaccard} \tag{1}$$

The $\mathcal{L}_{bce}$ (see Eq. 2) measures the difference between the predicted mask probabilities and the ground truth. The $\mathcal{L}_{jaccard}$ (see Eq. 3) focuses on the dissimilarity between the predicted and ground truth masks.

$$\mathcal{L}_{bce} = -\frac{1}{N}\sum_{i=1}^{N}[y_i \log(\hat{y}_i) + (1 - y_i)\log(1 - \hat{y}_i)] \tag{2}$$

$$\mathcal{L}_{jaccard} = 1 - \frac{\sum_{i=1}^{N} y_i \hat{y}_i}{\sum_{i=1}^{N} y_i + \sum_{i=1}^{N} \hat{y}_i - \sum_{i=1}^{N} y_i \hat{y}_i} \tag{3}$$

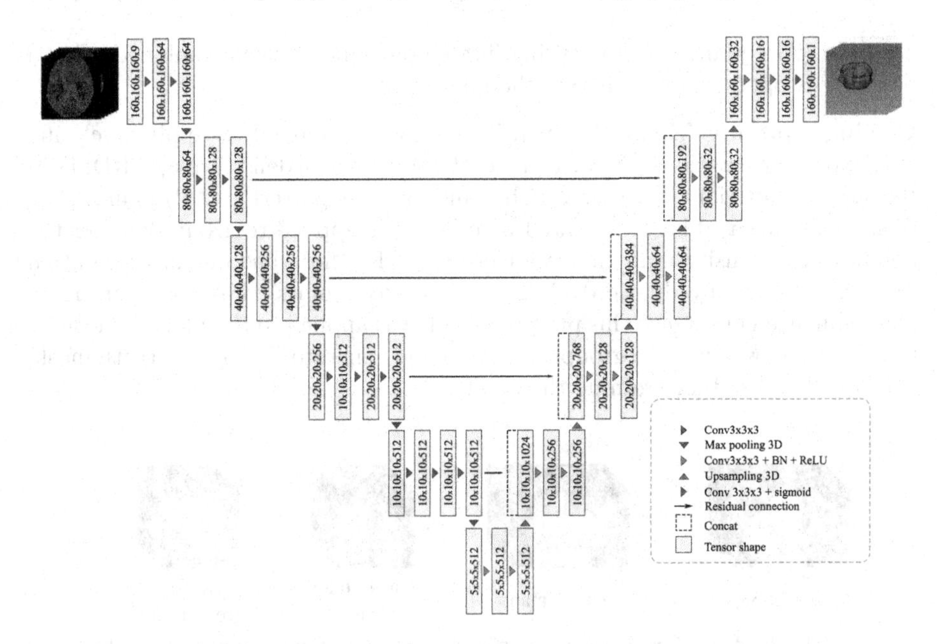

Fig. 2. 3D U-Net architecture for individual tract segmentation in a voxelgrid.

2.3 Pre-processing Proposals

Usually, the datasets for tractography segmentation provide information per fiber, stating which tract each fiber belongs to. Therefore, it becomes necessary to convert the data to a format that can be fed into the network. The purpose is to calculate the mask of a specific tract from an fODF peaks input. Each mask indicates the voxels where the fibers of a specific tract are spatially located.

The process to generate these ground truth masks is detailed in Fig. 3. From an array of streamlines, a 3D density map is obtained by stating the number of fibers crossing each voxel. Once the density map is calculated per tract, a binarization technique is applied to define which voxels are considered *occupied.* This technique of pre-processing is rather simple and provides an initial ground truth. The only parameter to tune is the threshold that defines the minimum number of fibers to consider the voxel as *occupied.* The obtained mask is then rescaled to fit the input of the network.

Two variants of the previous process are also proposed in order to smooth the areas limited by the mask. A good threshold that allows the edges of the tracts to be taken into account is hard to find. Thus, a gaussian filter and morphological operators over the density map are also proposed here to generate the ground truth mask.

Gaussian Filter Mask (GFM). A Gaussian filter is applied to the rescaled density map—the raw density map described above, which we will refer to as RDM from here on–in the lateral (LR), superior-inferior (SI), and anterior-posterior (AP) planes. The resulting filtered outputs are then combined using a bitwise OR operation to generate the final mask.

Closing Opening Mask (COM). This mask is obtained by applying closing and opening operations to each plane of the rescaled density map (RDM) in the superior-inferior (IS), lateral (LR), and anterior-posterior (AP) planes. The closing operation (dilation followed by erosion) is applied to group together the points in the density map that are close to each other. The opening operation (erosion followed by dilation) eliminates isolated points, thus contributing to mask enhancement. Once this pre-processing was applied to the planes, a bitwise OR operation was used to group the planes and generate the ground truth mask, which will be used for training the model.

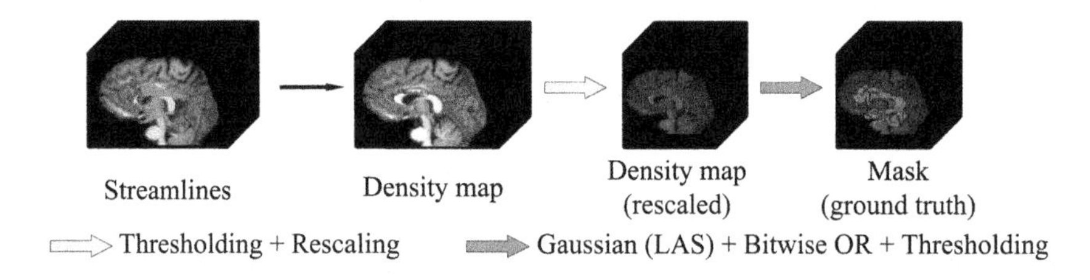

Fig. 3. Scheme of the GFM pre-processing proposal.

3 Experimentation

Setup. The TensorFlow framework [24] and the Dipy library [25] was used for the implementation of the proposal and a 12 GB Nvidia Titan X GPU card for its training and evaluation. The network was trained for 30 epochs with a mini-batch size of 1—due to GPU memory constraints—and *early stopping* with a patience of 7 epochs in order to avoid overfitting. The adaptive Adam optimizer [26] was used with a learning rate of 10^{-4}, which was decreased whenever the validation error increased over 2 epochs (i.e., with a patience value of 2). The input size for the experiments was $160 \times 160 \times 160$. One model was generated for each tract using the training setup described above and using random weights initialization, thus generating specialized models for the segmentation of each type of bundle.

Metrics. The evaluation of the segmentation was carried out using two commonly used metrics: intersection over union (IoU) with a threshold of 0.5 and Dice score. The IoU metric (Eq. 5) calculates the ratio of the intersection of the predicted segmentation and the ground truth segmentation to the union of these two sets of pixels. The threshold of 0.5 is used to define the voxels that are correctly classified as the positive class—the bundle to be segmented. The Dice score (Eq. 4), on the other hand, measures the similarity between the predicted and ground truth segmentations by calculating the ratio of their overlapping area to the average of their total areas.

$$Dice = \frac{2TP}{2TP + FP + FN} \quad (4)$$

$$IoU = \frac{|X \cap Y|}{|X \cup Y|} = \frac{TP}{TP + FP + FN} \quad (5)$$

where TP, FP, and FN respectively denote the True Positives, False Positives, and False Negatives.

3.1 Experimental Dataset

A large-scale tractography dataset called Tractoinferno [27] was used for the training and evaluation of the proposal. This dataset is made of a combination of six dWI datasets and has data from 284 healthy subjects, 256 of whom were used for training and 28 for testing.

Table 1. Distribution of labeled tracts by sets in Tractoinferno.

Bundles	Train samples	Validation samples	Test samples
AF	149	36	25
CC_Fr_1	163	49	23
CC_Fr_2	160	44	22
CC_Oc	169	42	26
CC_Pa	150	41	21
CC_Pr_Po	166	45	22
CG	130	40	20
FAT	148	44	17
FPT	167	48	24
FX	29	7	3
IFOF	117	30	14
ILF	175	54	23
MCP	162	49	23
MdLF	186	54	26
OR_ML	150	42	25
POPT	169	47	23
PYT	167	50	25
SLF	181	52	25
UF	172	44	24

The data—which was originally collected for research and clinical purposes—shows great variability as it was captured from several MRI equipment, each with different characteristics, such as resolution, and from 284 patients of different ages and sex. All the dataset was pre-processed following the same steps in a semi-manual pipeline.

For each subject, the dataset provides T1-weighted images, single-shell diffusion MRI acquisitions, spherical harmonics fitted to the dWI signal, fiber ODFs, and reference streamlines for 30 delineated bundles generated using different tractography algorithms (deterministic tracking [28], probabilistic tracking [29], particle-filtered tractography [30], and surface-enhanced tracking [31]) and masks needed to run tractography algorithms. The number of labeled tracts in each set is detailed in Table 1.

3.2 Results

This section evaluates the performance of the 3D U-Net architecture considered for segmenting tractography fibers and compares the three pre-processing proposals, namely: raw density map (RDM), Gaussian filter mask (GFM), and closing opening mask (COM). These proposals are trained and evaluated following the experimental setup and the previously described dataset.

The detailed results of each pre-processing method evaluated with the IoU and Dice score are shown in Table 2. Although both pre-processing methods—GFM and COM—improve the results of the raw density map, GFM reaches better results with average dice of 73.7 % and an average IoU of 79.5 % (with threshold 0.5). This means that using smoother masks for training the model has a significant impact on the results.

Figure 4 depicts a comparison of the IoU metric with a threshold of 0.5. As expected, tracts with higher occurrences in the dataset (Table 1) achieve better segmentation results, but it is remarkable that the pre-processing methods allow tracts with fewer training samples, like IFOF, to obtain better results than others with more samples, such as SLF. Besides, acceptable results have been obtained for tracts with few training samples, as the FX with only 29 samples. Also, the density of fibers per tract is not equal in all of them (see line chart in Fig. 4). All the pre-processing methods proposed here manage this problem by using density maps, which is an advantage over models performing fiber-level classification.

Finally, Fig. 5 shows a qualitative comparison of the results obtained. It is possible to observe that using a Gaussian filter as pre-processing technique (GFM) leads to a smoother result, while the other two (RDM and COM) provide a more detailed mask.

Table 2. Comparison of the results obtained with the different pre-processing proposals. Elements highlighted in bold represent the best-performing cases for each bundle.

Tract	GFM		RDM		COM	
	IoU	Dice	IoU	Dice	IoU	Dice
AF	**0.744**	**0.656**	0.647	0.495	0.643	0.485
CC_Fr_1	**0.833**	**0.792**	0.733	0.654	0.750	0.682
CC_Fr_2	**0.819**	**0.774**	0.683	0.580	0.704	0.618
CC_Oc	**0.832**	**0.797**	0.730	0.647	0.744	0.670
CC_Pa	**0.802**	**0.753**	0.726	0.642	0.705	0.599
CC_Pr_Po	**0.807**	**0.763**	0.717	0.627	0.720	0.633
CG	**0.793**	**0.734**	0.643	0.469	0.643	0.468
FAT	**0.818**	**0.778**	0.728	0.646	0.723	0.637
FPT	**0.797**	**0.750**	0.669	0.555	0.668	0.556
FX	**0.686**	**0.525**	0.578	0.300	0.577	0.301
IFOF	**0.800**	**0.750**	0.671	0.562	0.672	0.562
ILF	**0.833**	**0.800**	0.707	0.623	0.717	0.636
MCP	**0.814**	**0.768**	0.685	0.573	0.698	0.584
MdLF	**0.781**	**0.706**	0.655	0.501	0.668	0.526
OR_ML	**0.787**	**0.728**	0.704	0.592	0.690	0.569
POPT	**0.800**	**0.755**	0.669	0.544	0.686	0.579
PYT	**0.855**	**0.832**	0.735	0.662	0.734	0.665
SLF	**0.746**	**0.665**	0.661	0.520	0.661	0.528
UF	**0.763**	**0.681**	0.637	0.462	0.639	0.469
Mean	**0.795**	**0.737**	0.683	0.561	0.686	0.567

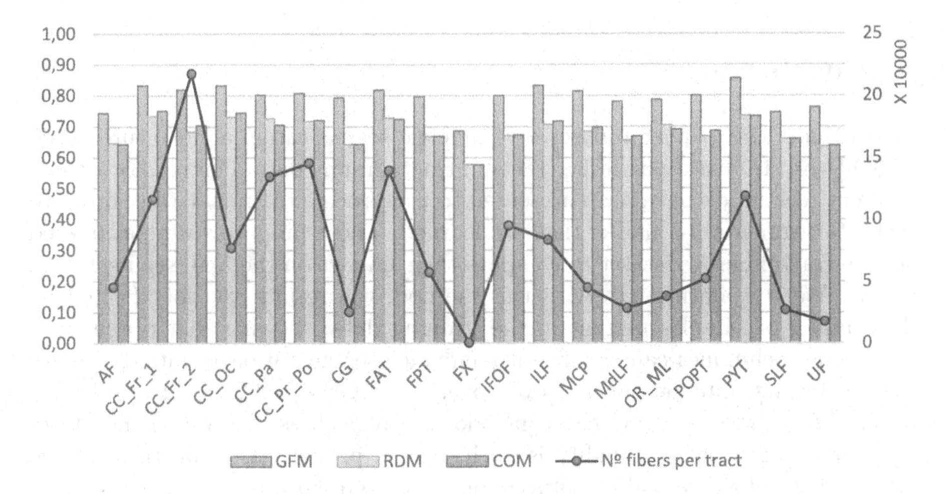

Fig. 4. Results of the IoU metric evaluation for the segmentation of each tract individually with three different pre-processing techniques. Also, the number of fibers per tract is included.

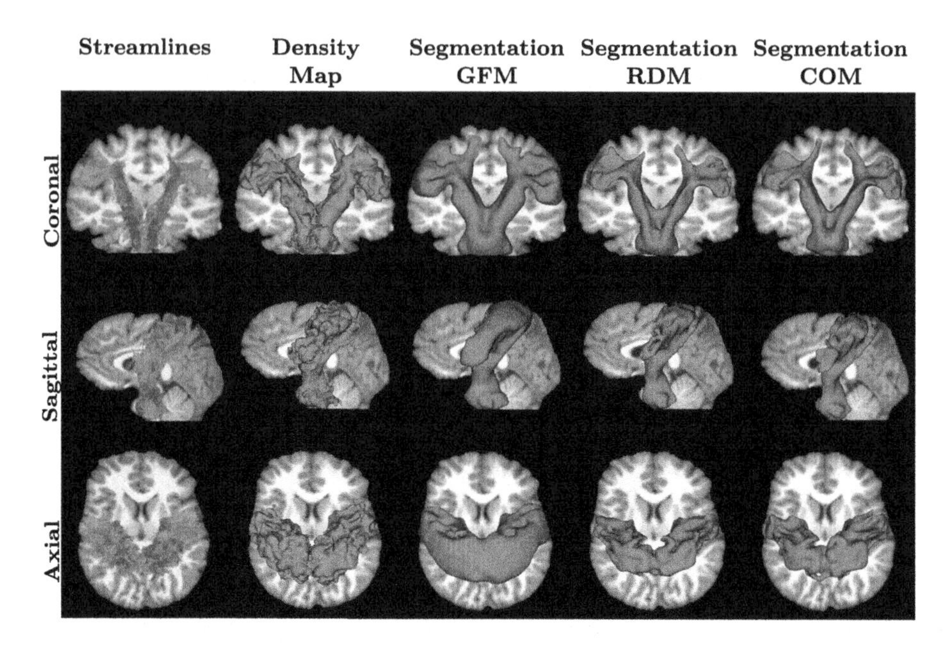

Fig. 5. Visual comparison of the results using different pre-processing techniques with the *Pyramidal tract (PYT)*. The first two columns show the streamlines and the density map (obtained from a binarization of the prior streamlines) of that tract. The three last columns compare the segmentation results provided by the model with different pre-processing techniques: GFM, RDM, and COM.

4 Conclusions

To conclude, this paper proposes a volumetric segmentation architecture based on 3D U-Net to mask individual tracts in tractography data. Furthermore, three different techniques have been presented to pre-process the raw data in order to generate segmentation masks that are learned. These proposals were evaluated on the Tractoinferno dataset to compare their impact on the final segmentation results. We can affirm that, with this method, the results are not affected by the number of samples present in the training data, showing that proper pre-processing techniques can significantly enhance segmentation results. As quantitative results, our method achieved average DICE of 62.2% and an average IoU of 72.2%. Moreover, we can conclude that regardless of the fiber density of the tracts, segmentation quality is fairly stable, proving the robustness of the proposed architecture and pre-processing to uneven data.

In future works, we aim to compare with other state-of-the-art solutions and other datasets. Moreover, we intend to develop a method to quantify physical or cognitive brain deterioration.

References

1. De Santis, S., Sommer, W.H., Canals, S.: Detecting alcohol-induced brain damage noninvasively using diffusion tensor imaging. ACS Chem. Neurosci. **10**(10), 4187–4189 (2019)
2. De Santis, S., et al.: Chronic alcohol consumption alters extracellular space geometry and transmitter diffusion in the brain. Sci. Adv. **6**(26), eaba0154 (2020)
3. Assaf, Y., Pasternak, O.: Diffusion tensor imaging (DTI)-based white matter mapping in brain research: a review. J. Mol. Neurosci. **34**(1), 51–61 (2008)
4. Jeurissen, B., et al.: Diffusion MRI fiber tractography of the brain. NMR Biomed. **32**(4), e3785 (2019)
5. Poulin, P., et al.: Tractography and machine learning: current state and open challenges. Magn. Reson. Imaging **64**, 37–48 (2019)
6. Zhang, F., et al.: Quantitative mapping of the brain's structural connectivity using diffusion MRI tractography: a review. Neuroimage **249**, 118870 (2022)
7. Hosseini, S., et al.: CTtrack: a CNN+transformer-based framework for fiber orientation estimation & tractography. Neurosci. Inform. **2**(4), 100099 (2022)
8. Li, B., et al.: Neuro4Neuro: a neural network approach for neural tract segmentation using large-scale population-based diffusion imaging. Neuroimage **218**, 116993 (2020)
9. Wasserthal, J., Neher, P., Maier-Hein, K.H.: TractSeg - Fast and accurate white matter tract segmentation. Neuroimage **183**, 239–253 (2018)
10. Zhang, F., et al.: Deep white matter analysis (DeepWMA): fast and consistent tractography segmentation. Med. Image Anal. **65**, 101761 (2020)
11. Lu, Q., Li, Y., Ye, C.: Volumetric white matter tract segmentation with nested self-supervised learning using sequential pretext tasks. Med. Image Anal. **72**, 102094 (2021)
12. Liu, W., et al.: Volumetric segmentation of white matter tracts with label embedding. Neuroimage **250**, 118934 (2022)
13. Lu, Q., et al.: A transfer learning approach to few-shot segmentation of novel white matter tracts. Med. Image Anal. **79**, 102454 (2022)
14. Ronneberger, O., Fischer, P., Brox, T.: U-Net: convolutional networks for biomedical image segmentation. In: International Conference on Medical Image Computing and Computer-Assisted Intervention - MICCAI 2015, pp. 234–241 (2015)
15. Zhang, Y., et al.: Bridging 2D and 3D segmentation networks for computation-efficient volumetric medical image segmentation: an empirical study of 2.5D solutions. Comput. Med. Imaging Graph. **99**, 102088 (2022)
16. Burdescu, D.D., et al.: Efficient volumetric segmentation method. In: 2014 Federated Conference on Computer Science and Information Systems, pp. 659–668 (2014)
17. Milletari, F., Navab, N., Ahmadi, S.-A.: V-Net: fully convolutional neural networks for volumetric medical image segmentation. In: 2016 Fourth International Conference on 3D Vision (3DV), pp. 565–571. IEEE (2016)
18. Çiçek, Ö., Abdulkadir, A., Lienkamp, S.S., Brox, T., Ronneberger, O.: 3D U-Net: learning dense volumetric segmentation from sparse annotation. In: Ourselin, S., Joskowicz, L., Sabuncu, M.R., Unal, G., Wells, W. (eds.) MICCAI 2016. LNCS, vol. 9901, pp. 424–432. Springer, Cham (2016). https://doi.org/10.1007/978-3-319-46723-8_49
19. Qamar, S., et al.: A variant form of 3D-UNet for infant brain segmentation. Futur. Gener. Comput. Syst. **108**, 613–623 (2020)

20. Mukherjee, P., et al.: Diffusion tensor MR imaging and fiber tractography: technical considerations. Am. J. Neuroradiol. **29**(5), 843–852 (2008)
21. Xu, K., et al.: Optimization of graph neural networks: Implicit acceleration by skip connections and more depth. In: Meila, M., Zhang, T. (eds.) Proceedings of the 38th International Conference on Machine Learning, vol. 139 of Proceedings of Machine Learning Research, pp. 11592–11602, PMLR (2021)
22. Ioffe, S., Szegedy, C.: Batch normalization: accelerating deep network training by reducing internal covariate shift (2015)
23. Im, D., et al.: DT-CNN: an energy-efficient dilated and transposed convolutional neural network processor for region of interest based image segmentation. IEEE Trans. Circuits Syst. I Regul. Pap. **67**(10), 3471–3483 (2020)
24. Abadi, M., et al.: TensorFlow: large-scale machine learning on heterogeneous systems (2015). software available from tensorflow.org, https://www.tensorflow.org/
25. Garyfallidis, E., et al.: Dipy, a library for the analysis of diffusion MRI data. Front. Neuroinform. **8**(FEB), 8 (2014)
26. Kingma, D.P., Ba, J.L.: Adam: a method for stochastic optimization. In: 3rd International Conference on Learning Representations, ICLR 2015 - Conference Track Proceedings (2014)
27. Poulin, P., et al.: TractoInferno: a large-scale, open-source, multi-site database for machine learning dMRI tractography. bioRxiv, 2021.11.29.470422 (2021)
28. Basser, P.J., et al.: In Vivo Fiber Tractography Using DT-MRI Data, Technical report (2000)
29. Tournier, J.-D., Calamante, F., Connelly, A.: MRtrix: diffusion tractography in crossing fiber regions. Int. J. Imaging Syst. Technol. **22**(1), 53–66 (2012)
30. Girard, G., et al.: Towards quantitative connectivity analysis: reducing tractography biases. Neuroimage **98**, 266–278 (2014)
31. St-Onge, E., et al.: Surface-enhanced tractography (set). Neuroimage **169**, 524–539 (2018)

MicrogliaJ: An Automatic Tool for Microglial Cell Detection and Segmentation

Ángela Casado-García[1(✉)], Estefanía Carlos[2], César Domínguez[1], Jónathan Heras[1], María Izco[2], Eloy Mata[1], Vico Pascual[1], and Lydia Álvarez-Erviti[2]

[1] Department of Mathematics and Computer Science, University of La Rioja, 26004 Logroño, Spain
{angela.casado,cesar.dominguez,jonathan.heras,eloy.mata, mvico}@unirioja.es

[2] Laboratory of Molecular Neurobiology, Center for Biomedical Research of La Rioja (CIBIR), Piqueras 98, 3th floor, 26006 Logroño, Spain
{ecarlos,mizco,laerviti}@riojasalud.es

Abstract. Microglial cells are now recognized as crucial players in the development of neurodegenerative diseases. The analysis and quantification of microglia changes is necesary to better understand the contribution of microglia in neurodegenerative processes or drug treatments. However, the manual quantification of microglial cells is a time-consuming and subjective tasks; and, therefore, reliable tools that automate this process are desirable. In this paper, we present MicrogliaJ, an ImageJ macro, that can measure both the number and area of microglial cells. The automatic procedure implemented in MicrogliaJ is based on classical image processing techniques, and the results can be manually validated by experts with a simple-to-use interface. MicrogliaJ has been tested by experts and it obtains analogous results to those manually produced, but considerably reducing the time required for such analysis. Thanks to this work, the analysis of microglia images will be faster and more reliable, and this will help us to advance our understanding of the behaviour of microglial cells.

Keywords: Microglia · ImageJ · Detection · and Segmentation

1 Introduction

Microglia are cells that play an important role in the central nervous system; in particular, they are the primary immune cells in the brain [1]. Microglia mediate responses such as inflammation and phagocytosis associated with neurodegeneration, and are pivotal players in exacerbating or relieving disease progression.

This work was partially supported by Grant PID2020-115225RB-I00 funded by MCIN/AEI/ 10.13039/501100011033.

A. Pertusa et al. (Eds.): IbPRIA 2023, LNCS 14062, pp. 593–602, 2023.
https://doi.org/10.1007/978-3-031-36616-1_47

Numerous studies have described microglial changes in patients, as well as in genetic and neurotoxic-based animal models, of Parkinson's disease, the second most prevalent neurodegenerative disorder worldwide [6].

Microglial cells monitor the brain for damage and invading pathogens, and upon changes in their surrounding environment, they activate and change their morphology [1]. Hence, it is necessary to measure those changes in order to better understand the role of microglia in both health and disease and to test the efficacy of treatments; however, most researchers still manually count and measure microglial cells. This measurement process is a tedious, error-prone, time-consuming and subjective task since experts have to analyse hundreds of images that contain dozens of cells. Therefore, it is instrumental to provide reliable tools that not only automate this process, but also provide the necessary features to modify the automatic detections.

In this paper, we present a pipeline to analyse microglial cell images using image processing techniques. Using this pipeline we can produce an automatic annotation of images that can be easily validated and corrected by experts; hence, the tedious task of annotating images is simplified. The pipeline has been included in an open-source tool called MicrogliaJ that is implemented on top of ImageJ [7]—a Java platform for image processing that can be easily extended by means of macros and plugins and that is widely used in the bioimaging community. MicrogliaJ is freely available at the project webpage https://github.com/ancasag/microglia.

The rest of the paper is organised as follows. In the next section, an overview of the literature related to microglial cell detection and segmentation is provided. Subsequently, in Sect. 3, we detail the biological and image-acquisition methods employed in this work to obtain the microglial cell images (from now on, microglia images); moreover, we describe the classical method employed to manually analyse those images. MicrogliaJ and the pipeline designed in this work to analyse microglia images are explained in Sect. 4. Subsequently, we compare, in Sect. 5, the results obtained with MicrogliaJ against those manually obtained. The paper ends with the conclusions and further work.

2 Related Work

Interest in computer-assisted analysis of microglia images has increased recently, which is reflected in the number of recent papers published on different aspects related to microglia cell detection and segmentation [3–5,8].

In [4], an end-to-end system for automatic characterisation of Iba-1 immunopositive microglia cells in whole slide images was presented. This system, based on MATLAB, had a two phased process. They worked with large images that are divided into patches. In a first step, the patches are automatically classified into white matter or gray matter sections using an SVM model. In those classified as white matter an algorithm to identify Iba-1 immunopositive microglia cells is proposed in a second step. In particular, those patches are converted to grayscale and normalised to have the same mean intensity. Patches with more than 15% white pixels are removed. Then, imaging artifacts and noise are removed using the Munford-Shah total variation smooting algorithm. Then, a peak detection algorithm is used. That algorithm selects components of constant intensity with fronts of less intensity. Those components with area smaller than a threshold are removed, and those with a size larger than 5000 pixels are scrutinised separately looking for multiple components. These thresholds had been determined empirically. A second screening was used to eliminate false positives using supervised learning techniques. The method achieved 84% accuracy classifying white matter, and detected microglia with a performance of 0.70 F1-score [4].

In [8] the commercial deep learning tool Aiforia was used to develop and validate a microglia activation model. They developed, in a first step, a model to distinguish between Iba-1+ tissue and background. Then, in a second step, other model was trained to identify microglia and its overall cell diameter. The microglia detected (manually reviewed and adjusted if necessary) was included for a subsequent round of training of the model. The detection model was compared with the results from human annotators. No differences were obtained between the two methods; with human annotations containing 1.1% false positive and 1.3% false negative, and model annotations containing 0.77% false positive and 1.1% false negative.

Another commercial tool named DeePathology STUDIO was included in [5]. Using deep learning models obtained a precision of 0.91 and a recall of 0.84 identifying only somas; and a precision of 0.61 and a recall of 0.99 identifying ramified microglia.

The previous commented works used either commercial software for their development [4] or included commercial tools [5,8] which could limit the availability for their use by some research groups. The implementation of open-source tools which enable the democratisation of their use is a different approach. This approach was followed in the recent work by Khakpour, et al. [3]. They developed an ImageJ script to perform microglial density analysis by automatic detection of microglial cells from brightfield microscopy images. In particular, three parameters were studied: the density of cells, the average distance of each cell to its nearest neighbour, and the spacing index (obtained from the two previous parameters). The image processing analysis consisted in a series of steps. Namely, (1) image scaling (duplication, conversion to 8-bit format, and scaling the image), (2) ROI cropping, (3) ROI saving, (4) threshold adjustment validated by an expert for each specific data set (enhances contrast, despeckles the image, removes outliers, and thresholds the image using max entropy), (5) detecting and analyzing cells, using Analyze Particles function (particle size fixed to 40–100 μm^2), and (6) measuring the average distance of each cell to its nearest neighbour. The work suggested that similar results are obtained from manual versus automatic methods. Although the results included in this work are quite interesting, the objective of our work (also by developing open-source tools) is different. It consists in building an automatic tool which obtains the number of cells and the area covered by them, which allows the validation and correction by an expert.

3 Materials and Methods

In this section, the biological and image-acquisition methods employed in this work to obtain the microglia images are detailed. We also describe the classical method employed to manually analyse those images using ImageJ.

3.1 Animals

All housing and experimental procedures involving animals were conducted in accordance with the European Communities Council Directive (2010/63/UE) and Spanish legislation (RD53/2013) on animal experiments and with approval from the ethical committee on animal welfare for our institution (*Órgano Encargado del Bienestar Animal del Centro de Investigación Biomédica de La Rioja, OEBA-CIBIR*). Animals were housed under environmentally controlled standard conditions with a 12-hour light/dark cycle and provided with food and water ad libitum.

In compliance with the 3Rs for refining, reducing, and replacing animals for research purposes, we obtained the brain sections of the present study from a previous published study [2]. We used a progressive mouse model of Parkinson's disease based on the injection of murine alpha-synuclein preformed fibrils (PFF) into the striatum of normal C57BL6/C3H F1 mice. Mice received two 2.5-μl injections of sonicated alpha-synuclein PFF (5 μg) into the dorsal striatum by stereotaxic injection. Control animals were injected with an equal volume of sterile mouse alpha-synuclein monomer or PBS. Animals were sacrificed at various time points 15, 30 and 90 days post injection.

3.2 Tissue Preparation

Mice were anesthetized with isoflurane and transcardially perfused with cold PBS followed by 4% paraformaldehyde (PFA). Whole brains were carefully removed from the skull, post-fixed for 24 h in 4% PFA, transferred into 30% sucrose for cryoprotection, rapidly frozen and stored at −80 °C until use. Frozen brains were cut into 30μm thick coronal sections using a cryostat and stored frozen in a cryoprotectant solution until immunohistochemical staining.

3.3 Iba-1 Immunohistochemistry

To identify microglial cells, free-floating brain sections were washed with TBS and incubated with 3% hydrogen peroxidase to inactivate the endogenous peroxidase. Afterwards, sections were blocked in blocking solution containing 5% normal goat serum and 0.04% Triton X-100 followed by incubation with the primary antibody Iba-1 (Wako, dilution 1:500) overnight at 4 °C. The next day, brain sections were rinsed and incubated with the biotinylated secondary antibody of the appropriate species. After wash steps, sections were incubated with ABC Peroxidase Staining Kit (Thermo Scientific), washed again and revealed with a solution of 3.3′-diaminobenzidine substrate (Dako). Finally, brain sections were mounted onto microscopic slides, dehydrated and covered with DPX mounting medium and coverslips.

3.4 Image Acquisition

Digital images of microglial cells from brain sections including the substantia nigra pars compacta (SNpc), striatum and cerebral cortex were acquired at 20x magnification using a Leica microscope equipped with a camera. The acquiring conditions and microscope settings were maintained throughout all the imaging sessions. Equivalent areas of the brain regions were consistently selected and photographed. For the analyses of the SNpc, we included 4 coronal sections (240 μm intervals between sections) for each animal, whereas for the analyses of microglial cells in the striatum and cortex, we included 3 selected representative rostro-caudal sections (bregma coordinates AP: +1.3, +0.5 and −0.4) for each animal. Two images were taken at the same brain area of interest (1 image from each hemisphere) for each section at 1372 × 1024 resolution. A total of 846 images were captured using this procedure.

3.5 ImageJ Manual Threshold Method

Images were processed and analysed using directly the ImageJ software (NIH) [7]. The analysis requires loading each image one by one. All the images were processed with background subtraction and automatic thresholding to generate binary masks of Iba-1 microglia. Total number of Iba-1 positive cells was estimated by using the "Analyse Particles" function of ImageJ with particle size and circularity parameters for microglial cell differentiation and to exclude small pixel noise. The parameters used to determine microglial cells were circularity = 0.05–1.00 and particle size = 0.001–infinity inch^2.

Quantification of the percentage of Iba-1 positive immunostained area was carried out using the threshold method of ImageJ. The pictures were converted to grayscale and a threshold was manually set to convert the grayscale images to binary images. Relative staining intensities were semi-quantified in binary converted images and the results were transformed into percentage area values.

It is worth noting that all this process is manually conducted for each image and has several subjective steps. Therefore, such a process is a time-consuming and non-reproducible task, and its automation will reduce the burden of experts.

4 MicrogliaJ

MicrogliaJ has been implemented as an ImageJ macro to automatically analyse microglia images. The procedure implemented in MicrogliaJ consists of three steps: soma detection, soma segmentation, and cell segmentation. Those three steps are automatically conducted; but, after each step, users have the option to manually modify the detection and segmentation results using the ImageJ functionality. The analysis can be applied to a single images or to a batch of images (by simply providing the folder containing that set of images). The rest of this section is devoted to detail the automatic methods implemented in MicrogliaJ.

The first functionality implemented in MicrogliaJ is in charge of soma detection, and it is diagrammatically described in Fig. 1. First, the image is converted to grayscale since colour does not provide relevant information. Subsequently, noise is removed by applying a median filter with radius 5. Finally, since the soma regions are the darker areas of the resulting image, the local minima of the image are detected using a noise tolerance value of 50.

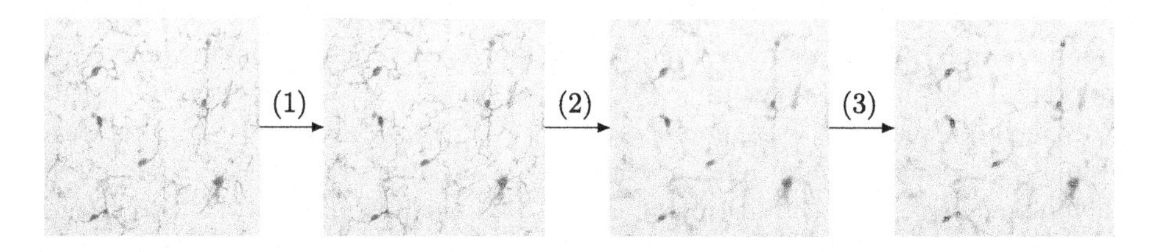

Fig. 1. Steps for soma detection. (1) Conversion to grayscale. (2) Noise removal. (3) Local minima detection.

After the detected soma are validated by the experts, MicrogliaJ automatically segments soma—this process is summarised in Fig. 2. As for the soma detection functionality, the first step is a conversion of the image to grayscale followed by an inversion of colours. Subsequently, the background of the image is removed by obtaining two versions of the image (one where a Gaussian blur filter with radius 2 is applied, and the other where a Gaussian blur filter with radius 30 is applied), and then substracting them. After that, the IsoData threshold algorithm is applied to obtain the masks of soma segmentations. In order to remove noise, two erosion steps, followed by two dilation steps, are applied. Finally, the segmented regions that are smaller than a given size or that were not detected in the soma detection step are removed.

As previously explained, after the soma segmentation step, the users validate the results, and then, cells are automatically segmented using the process diagramatically described in Fig. 3. The first step for cell segmentation is the conversion to grayscale; then, the contrast of the image is enhanced, and a variance filter with radius 2 is applied to detect the contours of the cells. Afterwards, the image is binarized using the IsoData threshold algorithm, and holes of the image are filled. The previous steps produce several regions but only those that have a non-empty intersection with the soma segmentation are kept.

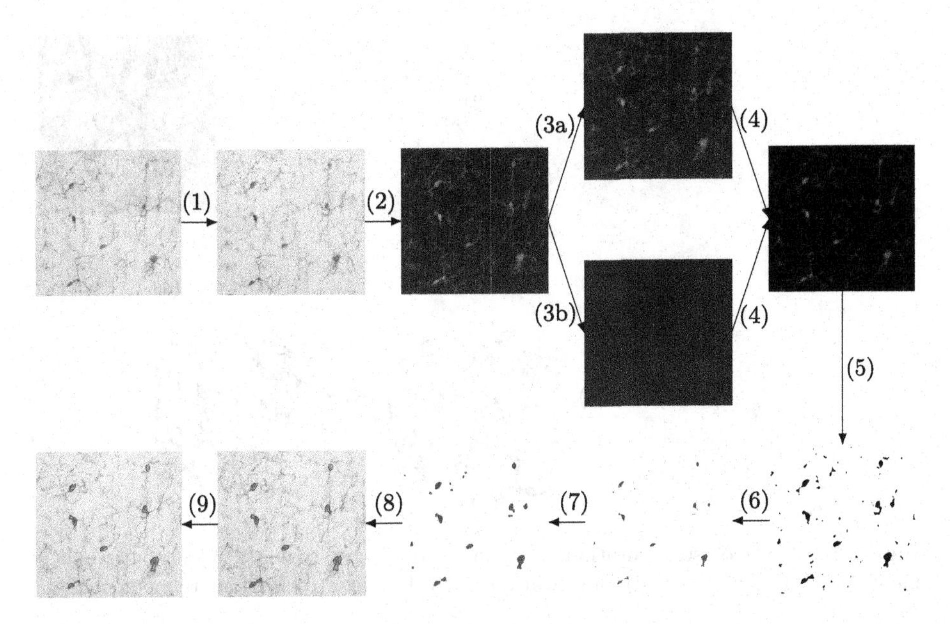

Fig. 2. Steps for soma segmentation. (1) Conversion to grayscale. (2) Colour inversion. (3–4) Background removal. (5) Binarization. (6) Mask erosion. (7) Mask dilation. (8) Removal of small regions. (9) Removal of regions that were not detected in the soma detection step.

5 Results and Discussion

In order to optimise and validate MicrogliaJ, we analysed a dataset of 830 brain microglia images using the ImageJ threshold microglia detection method (described in Subsect. 3.5), MicrogliaJ automatic detection method (without the validation of an expert) and MicrogliaJ detection method corrected by an expert. These images were obtained using the materials and methods described in Sect. 3. We included images from 3 brain areas, cerebral cortex, striatum and SNpc, from control mice and mice injected into the striatum with alpha-synuclein monomer or alpha-synuclein PFF. This dataset contained different subtypes of microglia and microglia activation state (Fig. 4a) [2] in order to test the methods with different types of images. The results of the validation study are summarized in Fig. 4b. The ImageJ threshold analysis counted an average of 41.25 neurons per image, the MicrogliaJ analysis 49.92 neurons per image, MicrogliaJ corrected by expert-1 37.21 neurons per image and MicrogliaJ corrected by expert-2 39.34 neurons per image (Fig. 4b).

These results indicated that MicrogliaJ automatic count over-estimated the number of microglia cells; however, the analysis corrected by an expert allowed a quick and easy correction of the false positive signals. ImageJ threshold also over-estimated the number of microglia cells to a lesser extent, but it takes more time to produce and its correction is more complex.

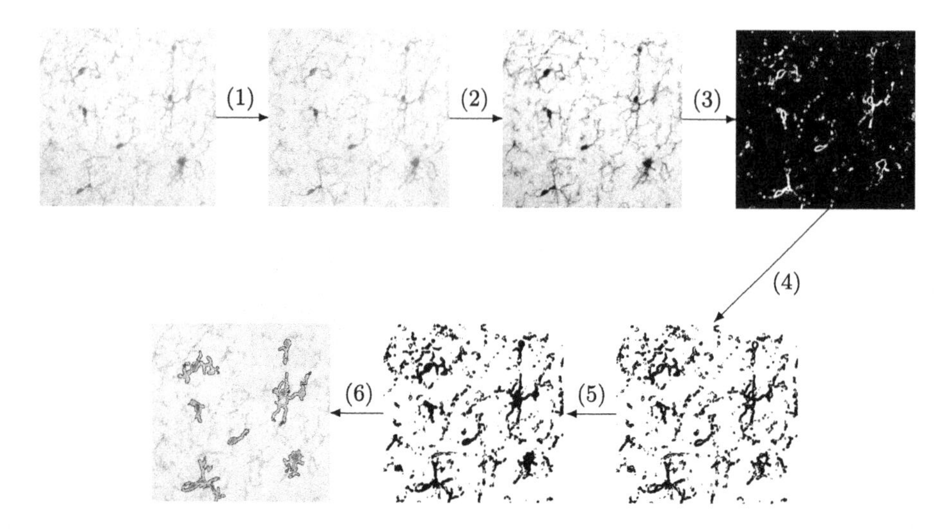

Fig. 3. Steps for cell segmentation. (1) Conversion to grayscale. (2) Contrast enhancement. (3) Variance filter. (4) Binarization. (5) Fill holes. (6) Regions refinement.

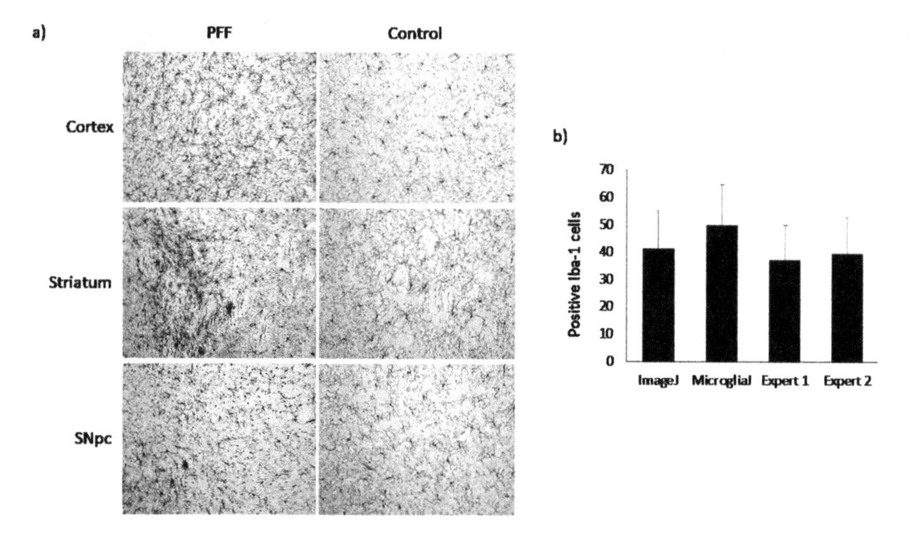

Fig. 4. a) Representative images of the dataset analysed in the study showing different microglial activation state in cortex, striatum and SNpc. b) Average number of Iba-1 positive cells per image quantified by ImageJ threshold method, MicrogliaJ, and MicrogliaJ corrected by expert-1 or expert-2.

We performed a second study to compare the ImageJ threshold method for soma segmentation with the MicrogliaJ method after the validation for soma detection made by an expert. We analysed a set of images from a previous study comprised by 3 group of animals showing morphological changes and different numbers of Iba-1 positive cells [2]. The results of the analysis including false positive and false negative cells were assessed by a different researcher. The analysis demonstrated that ImageJ threshold method over-estimated the number of positive microglia cells with an average 32 false positive cells per image; however, the quantitation using MicrogliaJ was significantly more precise (Fig. 5 and Table 1).

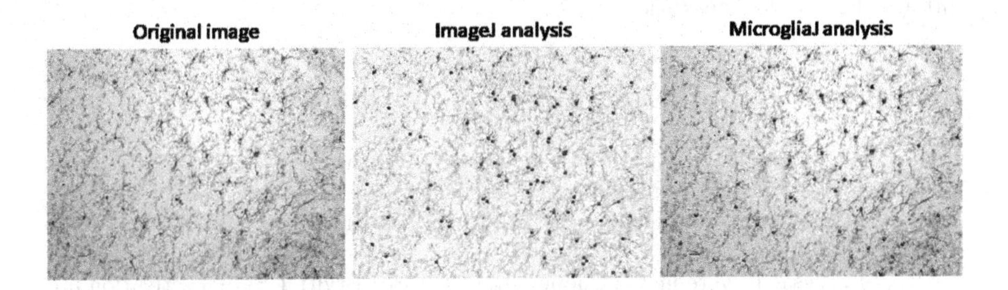

Fig. 5. Soma segmentation images of the ImageJ and MicrogliaJ interfaces

Table 1. Mean (standard derivation) of the detected microglia number, false positives and false negatives by the ImageJ and MicrogliaJ methods.

	ImageJ	MicrogliaJ	Statistic[a]	Cohen's d
Number	72.65 (21.16)	40.65 (8.6)	t= −11.131***	−1.61
False positive	32.02 (19.99)	0.15 (0.36)	t= −11.070***	−1.60
False negative	0 (0)	0.44 (0.63)	Z= −3.464***	0.59

[a]Paired t-test or Wilcoxon test; ***ρ <.001

Finally, the results for cell segmentation were validated by the experts, but it was not possible to compare them against a gold standard. This is due to the fact that, up to now, the experts did not measure cell area since it was infeasible to manually obtain such information. This opens the door to advances in the study of microglial cells, and requires the improvement and validation of our tool—a task that remains as further work.

6 Conclusions and Further Work

In this paper, we have presented MicrogliaJ, an ImageJ macro, that allows users to automatically measure the number and the area of microglial cells. The underlying algorithms of MicrogliaJ are based on classical image processing techniques that generate a detection and segmentation of microglial cells comparable to those produced manually by experts; but, considerably reducing the effort required to obtain them. In addition, MicrogliaJ provides a simple-to-use interface for editing the results.

The developed tool will help to advance our understanding of the behaviour of microglial cells. To reach such a goal, the pipeline included in MicrogliaJ must generalise to cells from other animals and to images captured under different camera conditions. Preliminary experiments have shown promising results in this direction; but, a further study is necessary to address the domain shift problem in this context. Furthermore, we plan to build a dataset with MicrogliaJ that can be used for training deep learning models for soma detection and segmentation, and also for cell segmentation.

References

1. Ho, M.S.: Microglia in Parkinson's disease. In: Verkhratsky, A., Ho, M.S., Zorec, R., Parpura, V. (eds.) Neuroglia in Neurodegenerative Diseases. AEMB, vol. 1175, pp. 335–353. Springer, Singapore (2019). https://doi.org/10.1007/978-981-13-9913-8_13
2. Izco, M., Blesa, J., Verona, G., Cooper, J.M., Alvarez-Erviti, L.: Glial activation precedes alpha-synuclein pathology in a mouse model of Parkinson's disease. Neurosci. Res. **170**, 330–340 (2021). https://doi.org/10.1016/j.neures.2020.11.004
3. Khakpour, M., et al.: Manual versus automatic analysis of microglial density and distribution: a comparison in the hippocampus of healthy and lipopolysaccharide-challenged mature male mice. Micron **161**, 103334 (2022). https://doi.org/10.1016/j.micron.2022.103334
4. Kyriazis, A.D., et al.: An end-to-end system for automatic characterization of Iba1 immunopositive microglia in whole slide imaging. Neuroinformatics **17**(3), 373–389 (2018). https://doi.org/10.1007/s12021-018-9405-x
5. Möhle, L., Bascuñana, P., Brackhan, M., Pahnke, J.: Development of deep learning models for microglia analyses in brain tissue using DeePathology™STUDIO. J. Neurosci. Methods **364**, 109371 (2021). https://doi.org/10.1016/j.jneumeth.2021.109371
6. Olanow, C., Kieburtz, K., Katz, R.: Clinical approaches to the development of a neuroprotective therapy for PD. Exp. Neurol. **298**, 246–251 (2017). https://doi.org/10.1016/j.expneurol.2017.06.018
7. Rueden, C.T., et al.: Image J2: ImageJ for the next generation of scientific image data. BMC Bioinf. **18**, 1–26 (2017). https://doi.org/10.1186/s12859-017-1934-z
8. Stetzik, L., et al.: A novel automated morphological analysis of microglia activation using a deep learning assisted model. Front. Cell. Neurosci. **16** (2022). https://doi.org/10.3389/fncel.2022.944875

Automated Orientation Detection of 3D Head Reconstructions from sMRI Using Multiview Orthographic Projections: An Image Classification-Based Approach

Álvaro Heredia-Lidón[1(✉)], Alejandro González[1], Carlos Guerrero-Mosquera[1], Rubèn Gonzàlez-Colom[2], Luis M. Echeverry[2], Noemí Hostalet[3,4], Raymond Salvador[3,4], Edith Pomarol-Clotet[3,4], Juan Fortea[5], Neus Martínez-Abadías[2], Mar Fatjó-Vilas[3,4], and Xavier Sevillano[1]

[1] HER - Human-Environment Research Group, La Salle - Universitat Ramon Llull, Barcelona, Spain
alvaro.heredia@salle.url.edu

[2] Departament de Biologia Evolutiva, Ecologia i Ciències Ambientals (BEECA), Facultat de Biologia, Universitat de Barcelona (UB), Barcelona, Spain

[3] FIDMAG, Sisters Hospitallers Research Foundation, Barcelona, Spain

[4] CIBERSAM (Biomedical Research Network in Mental Health, Instituto de Salud Carlos III), Madrid, Spain

[5] Sant Pau Memory Unit, Hospital de Sant Pau i la Santa Creu, Barcelona, Spain

Abstract. Recent studies in neuropsychiatry have highlighted the correlation between facial and brain dysmorphologies. One way of simultaneously analysing the brain and the face of a subject is by reconstructing a whole-head 3D model from structural magnetic resonance imaging (sMRI). However, the use of different reconstruction protocols generates undesired orthogonal rotations of the 3D models. This is a likely situation in multicentric studies that hampers the combination of data from different centers. Although the original sMRI files contain the subject orientation, it is not always possible to access this data. To solve this issue, in this work we propose a novel method to estimate the orientation of 3D heads with rotations of 90° or multiples thereof around any of the three Cartesian axes as a required step for generating a normalised dataset in terms of orientation. Our proposal creates 2D images from orthogonal projections of the 3D object, transforming orientation estimation into an image classification problem. Experimental results show that our method, using three orthographic views of the 3D head to create the projection image and ResNet50 for classification, achieves an accuracy of 99.7%, which corresponds to 0.15 mean absolute error in rotation, outperforming state-of-the-art point cloud registration methods like DeepBBS and PRNet.

Keywords: Structural magnetic resonance imaging · 3D head orientation · Multiview orthographic projections · Image classification · Point cloud registration

A. Pertusa et al. (Eds.): IbPRIA 2023, LNCS 14062, pp. 603–614, 2023.
https://doi.org/10.1007/978-3-031-36616-1_48

1 Introduction

Recent studies have highlighted the integrated development and correlation between facial and brain dysmorphologies in schizophrenia and bipolar disorder [1], and therefore the need to simultaneously analyse both structures to further understand the etiology of these neuropsychiatric disorders [2]. In this field, the prevalent technique for studying brain anatomy is structural Magnetic Resonance Imaging (sMRI) of the head. However, by modifying existing scanning protocols, it is also possible to capture facial tissues [3]. This enables the reconstruction of whole-head 3D models of the subjects, thus allowing both brain and face anatomical studies (Fig. 1a).

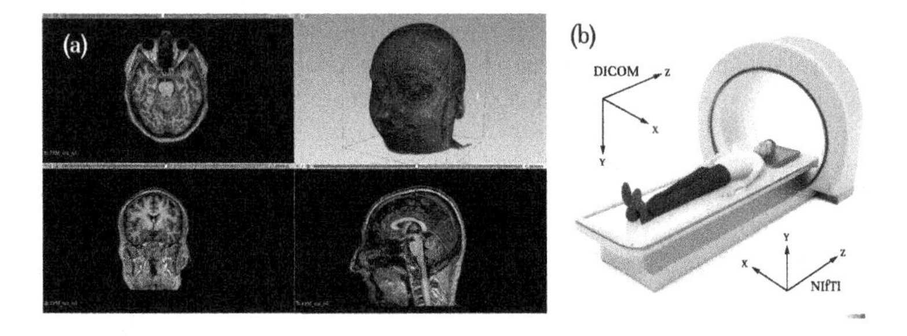

Fig. 1. (a) Visualisation of 3D head reconstruction from sMRI with 3D Slicer software. **(b)** DICOM Patient Coordinate System (PCS) and NIfTI coordinate system related to the Cartesian axes. Image created with the assistance of DALL·E 2.

To ensure the significance of the conclusions of these studies, it is crucial to analyse and process large samples of sMRI scans of both healthy subjects and patients. This often requires collecting data from different centres and integrating them into a single dataset.

However, this can be challenging in a multicentric scenario due to the diversity of scanners, protocols, software tools and people involved. This makes it highly likely that the data is heterogeneous. In this context, one of the main problems is that the reconstructed 3D head models can present arbitrary rotations towards one of the Cartesian axes due to differences in the process of exporting the 3D model from the sMRI file. This constitutes a hurdle because positioning and orienting each head model into a common reference space is necessary to simplify subsequent processing and analysis steps [4], like automatic 3D facial landmarking and morphometric analyses [5].

It is important to remark that determining the orientation of a 3D head extracted from sMRI should not pose a problem if we have access to the original data obtained from the scanner. In fact, standards such as DICOM or NIfTI have their own internal coordinate system, in which the subjects' head orientation is known (Fig. 1b). However, it is not always possible to access the original sMRI scanner metadata, either because they have not been correctly stored during the acquisition process, have been altered during file processing or conversion, or have been deleted for privacy or legality concerns [6].

In this case, the alternative is to develop an automatic method to estimate the orientation using the reconstructed whole-head 3D model alone. This could be addressed as a computer vision problem such as Head Pose Estimation (HPE) [7] or Point Cloud Registration (PCR) [8]. However, these techniques have certain limitations in solving the severe rotations (90° or multiples thereof around any of the Cartesian axes) that can be found in 3D head models exported from sMRI.

For these reasons, in this paper we present a novel method to determine the orientation of a 3D head extracted from an sMRI, as a prior step required for producing a normalised dataset. Assuming that 3D head models are orientated towards an unknown coordinate axis, our proposal creates 2D images from projections orthogonal to the coordinate planes of the 3D object. This allows transforming the orientation detection into an image classification problem.

In the experimental section of this work, we present a performance study of different variants of our method, and a comparison against two state-of-the-art deep learning-based PCR methods. The obtained results show that *i)* the best configuration of our method uses three orthographic views of the 3D head model to create the projection image and ResNet50 [9] for classification, achieving an accuracy of 99.7%, which corresponds to 0.15 mean absolute error in rotation, and *ii)* outperforms current state-of-the-art PCR methods by a wide margin.

2 Related Work

The alignment of 3D medical image data to a canonical coordinate space has been treated in the literature before. Some approaches are based on processing the original files from either ultrasound or MRI scanners, predicting the 3D alignment transformation from the 2D images slices [10,11]. However, our approach differs in the sense that we operate on a reconstructed 3D model obtained from the medical image data.

The automatic orientation and alignment of a 3D object in space is a topic addressed by various methods within the field of computer vision. Their aim is to find the rigid transformation that best aligns the data with respect to a reference in terms of rotation (3 degrees of freedom or DoF) or rotation and translation (6 DoF).

This problem could be addressed using Point Cloud Registration, defined as the 6 DoF process of aligning an input point cloud to a reference in a common coordinate system. The accuracy of these methods largely depends on the level of similarity between both point clouds [12].

Classic PCR techniques, like the Iterative Closest Point (ICP) algorithm [13], are able to perform slight alignments to register 3D objects [5]. However, these methods are sensitive to their initialisation parameters and prone to fall into local minima. Recently, deep learning-based registration methods [8] have overcome the limitations of ICP-based techniques, either extracting global features from the point cloud [14] or local features from each point and its neighbourhood [12].

Typically, these methods are evaluated on standard datasets of point clouds downsampled to a few thousand of points, and with transformations involving random rotations of up to 45° and random translations in the range of [-0.5,0.5] in each axis, like in [12,15].

Considering that the 3D objects analysed in this work are human heads, another related technique that could be used is Head Pose Estimation (HPE) [7]. These 3 DoF techniques primarily focus on determining the orientation of the natural movements of the head from three angles: pitch (±30°), yaw (±90°), and roll (±45°). However, very recent techniques ensure HPE for the full range of angles [16]. Although HPE methods typically work on 2D images, depth images and videos, recent contributions have started to work on 3D point clouds generated from depth images [17].

Nevertheless, the nature of the 3D head models reconstructed from sMRI poses a significant challenge for HPE or PCR, since these techniques cannot efficiently handle the severe rotations presented by the 3D models. One approach to working with these 3D structures is to make 2D projections of them as in [18], which enables the 3D orientation problem to be tackled as a 2D classification problem without the need for a reference. This is the main concept of our solution, a 3 DoF method for determining the orientation of 3D heads reconstructed from sMRI. Once the orientation is detected, a rotation matrix $R \in SO_3$ is applied to rotate all heads towards a common direction.

3 Proposed Method

The key characteristic of our method is that we are addressing 3D orientation detection by converting it into a 2D image classification problem. This is possible due to the discrete nature of the possible rotations of our data, which implies a set of 24 possible orientations. These are all the possible orientations that a 3D head model can present considering that the face is pointing towards one of the Cartesian axes (positive or negative) and the right-hand rule.

The block diagram of the proposed method is shown in Fig. 2 and can be summarised in four blocks: *i)* mesh pre-processing, *ii)* projection image generation, *iii)* classification and *iv)* rotation.

Mesh Pre-processing. Segmented 3D models from sMRI are noisy and often contain artefacts and translations. To clean the data, a filtering of the 3D object is performed to remove isolated small elements with a diameter below a threshold value of 150. Moreover, to normalise the data in terms of translation, we compute the centroid of the bounding box of the 3D head and use this value to translate the whole mesh to the origin of the Cartesian coordinate system.

Projection Image Generation. One of the challenges of our method is to find a unique representation of 3D heads for their orientation classification. To address this issue, we propose generating a projection image (PI) from 2D views

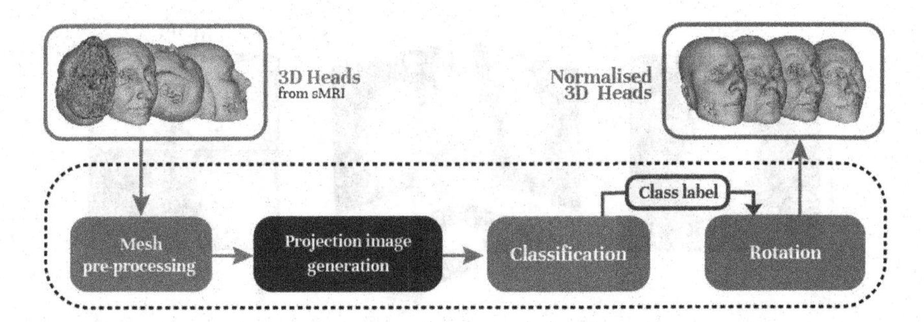

Fig. 2. Block diagram of the method. Starting from a 3D mesh of the head, a pre-processing of the mesh is performed. Then, 2D projections are computed to generate the projection image. Next, the projection image is the input for the classification module, which outputs the class label corresponding to the orientation of the 3D head. Finally, the 3D head is rotated to a reference direction.

of the 3D object. To that end, we generate three projections following the multiview orthographic projection concept used in computer graphics [19], which represents a 3D object from three primary views: front, top and side. Each of these projections represents the 3D object viewed in a direction parallel to each Cartesian axes. From a practical standpoint, a projection consists of a 2D RGB capture of the 3D rendered model.

Herewith, we define two types of PI: single projection image (SPI) and multiview orthographic projection image (MOPI) (Fig. 3).

In the case of SPI, we use only the top view, which is the projection perpendicular to the XY plane. As for the MOPI, we use the three views arranged according to the first dihedral European projection method, where the front is placed at the top left, the top at the bottom left and the side at the top right, applying the rotations defined by the standard. To adapt the views to the MOPI, it is necessary to segment the head of the 2D projections using a binary mask and perform a resize of the image.

Classification. As mentioned before, the use of PI allows solving orientation detection as a multi-class image classification problem of 24 categories.

Each of these categories is defined according to the anatomical coordinate system used in the clinical field. For this reason, each class has a label consisting of three letters relating the anatomical coordinate system to the Cartesian coordinate axes: the first letter indicates the direction to which the x-axis points, the second letter the y-axis and the third letter the z-axis, from the patient's point of view.

Each letter of the class label can have six possible orientations: [A] anterior, [P] posterior, [L] left, [R] right, [S] superior and [I] inferior. Within a class label there can be no repeated letters and the coordinates must always be orthogonal following the right-hand rule, giving rise to the 24 possible orientations or classes to be classified (Fig. 4).

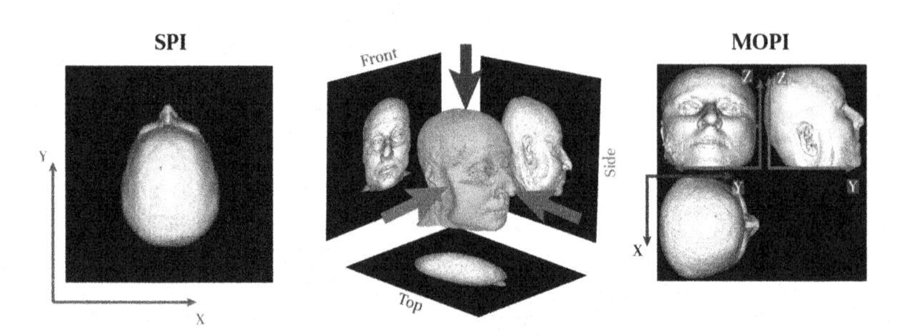

Fig. 3. Generation of the projection images from the primary views: front, top and side based on the concept of multiview orthographic projections [19]. Example of an SPI (left) and a MOPI (right), where the three views are arranged according to the first dihedral European projection method. Projection image size is 1024×1024 px.

Determining the orientation of a given 3D head involves training a classifier capable to determine which of the 24 possible classes corresponds to the projection image of that head.

Rotation. Once the orientation has been detected, the final step consists of applying the rotation matrix R that passes from the detected to the target orientation. For this purpose, it was necessary to create a dictionary of rotations between all possible orientations. This is the concluding stage to obtain a normalised dataset in terms of orientation.

4 Experiments and Results

In this section, we first describe the data used, the experimental setup and the evaluation metrics employed in this work. Then, we present two experiments: the first one aims at finding the variant of our method (type of PI and classifier) that provides the best performance. The second experiment compares our proposal against two state-of-the-art PCR methods in the task of finding the orientation of 3D head models in our dataset.

4.1 Data

The data used in this work comprised 693 3D heads manually segmented from high resolution structural T1 MRI scans provided by Hospital Sant Joan de Déu (1.5 GE Sigma scanner, 185 scans), Fundació Pasqual Maragall (Siemens Magnetom PRISMA 3T, 259 scans), and Hospital Sant Pau Memory Unit (Philips 3 T X Series Achieva, 249 scans), all from Barcelona, Spain. It is important to note that each subject has been scanned only in one center.

The 3D head models were reconstructed from sMRI adjusting by hand the grey-scale threshold that optimises the segmentation of skin. The resulting 3D

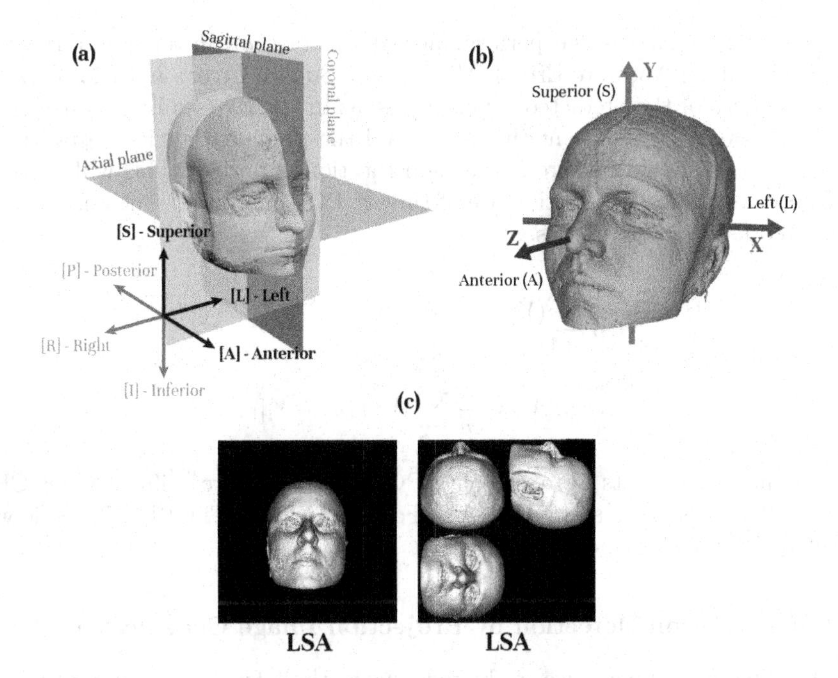

Fig. 4. (a) Definition of the anatomical coordinate system (b) Example of a head belonging to 'LSA' class (x-axis pointing to left, y-axis pointing to superior and z-axis pointing to anterior). (c) Examples of SPI (left) and MOPI (right) of the 'LSA' class. Please visit https://bit.ly/42fBxKF to view to SPI and MOPI corresponding to the 24 possible classes.

meshes were exported as PLY files without following a standard protocol, producing a diversity of orientations in the original data.

To test our method under all possible conditions, we rotated each 3D head model in every possible orientation, obtaining a total of 16632 3D heads (693 heads x 24 possible orientations). For the ground truth, each head was labelled with its corresponding orientation label.

4.2 Experimental Setup and Evaluation Metrics

In these experiments, we have split the available data into training and test subsets by centres of origin: 444×24 3D heads (64%) (Hospital Sant Joan de Déu and Pasqual Maragall Foundation) for training, and 249×24 (36%) (Sant Pau Memory Unit) for test.

To evaluate the orientation detection performance of our method, the first evaluation metric is the accuracy of the classifier averaged over all 24 classes and on the class with minimum accuracy (worst class). This latter metric is important because, since we are working with severe rotations, even small misclassification rates result in large angular estimation errors.

Moreover, to assess the performance of the method in angular terms, we use MAE (Mean Absolute Error), MSE (Mean Square Error) and RMSE (Root MSE) metrics of the predicted Euler angles, as in [12,15] (see Equations (1, 2)) where Y and Y_{pred} represent the target and the predicted angles, respectively. These metrics should be zero if the rigid rotation is perfect. This will allow us to compare our proposal with state-of-the-art PCR methods, which are usually evaluated using these metrics.

$$MSE = \frac{1}{n}\sum_{i=1}^{n}(Ypred_i - Y_i)^2 \qquad RMSE = \sqrt{MSE} \tag{1}$$

$$MAE = \frac{1}{n}\sum_{i=1}^{n}|Ypred_i - Y_i| \tag{2}$$

All the experiments were run on a PC with Intel Core™i9-10980XE CPU @ 3.00 GHz × 36 cores and NVIDIA GeForce RTX 2080 Ti GPU. The code was developed in Matlab and Python.

4.3 Orientation Detection by Projection Image Classification

In this first experiment, we evaluated our method from two perspectives: *i)* depending on the type of PI used, either SPI or MOPI, and *ii)* based on the algorithm used for image classification.

In this sense, we implemented the image classifier using the following approaches. First, we used a ResNet50 [9] model, pre-trained on ImageNet [20] and fine-tuned on our training set. This fine-tuning consisted in replacing the last three layers of ResNet50 with a fully connected layer, a softmax layer and a classification output layer with the 24 possible categories. The initial learning rate was set to 0.001, the batch size to 10 and the maximum number of epochs was set to 6, as longer training did not provide better learning. And second, we applied a transfer learning approach using ResNet50 as a feature extractor to train a linear Support Vector Machine classifier, which we refer to as SVM^{RN50}.

Table 1. Comparison of different variants of our method using SPI, MOPI, and SVM^{RN50} and ResNet50 classifiers.

Metrics	SPI		MOPI	
	SVM^{RN50}	ResNet50	SVM^{RN50}	ResNet50
Accuracy (average %)	99.12	99.30	99.32	**99.77**
Accuracy (worst class %)	90.40	96.40	98.80	**98.00**
MSE	554.37	376.80	178.91	**105.72**
RMSE	23.54	19.41	13.38	**10.28**
MAE	0.93	0.65	0.33	**0.15**

Table 1 presents the performance results of four variants of our method: SPI+SVM^{RN50}, SPI+ResNet50, MOPI+SVM^{RN50} and MOPI+ResNet50.

We detected that all variants provide high average accuracy (over 99%), which validates the use of PI and image classification to detect the orientation of the 3D head models.

However, if we consider the accuracy of the worst class in each case, there are variants such as SPI+SVM^{RN50} where the accuracy is around 90%. Taking into account that any misclassification is equivalent to detecting the orientation of the head with an error of at least 90°, this boosts the angular error metrics (see the three bottom rows of Table 1). This justifies the need to achieve high classification accuracy for all 24 classes.

A detailed analysis of the different configurations of our method reveals that the use of MOPI outperforms SPI. This suggests that the use of multiple views of the 3D object for generating the PI allows to obtain a more discriminant representation of each orientation. Also, for a given type of PI, ResNet50 obtains better results than SVM^{RN50} in terms of average accuracy.

In this sense, the MOPI+ResNet50 implementation of our method yields the best performance (highlighted in boldface in Table 1), with an accuracy higher than 98% for all 24 classes, and angular error metrics significantly lower than any other variant.

4.4 Comparison Against State-of-the-Art PCR Methods

In this experiment, we selected the best variant of our method (MOPI+ResNet50) and compared it with two state-of-the-art PCR end-to-end methods: DeepBBS [12] and PRNet [15]. The choice of these methods is motivated by the fact that they are among the most accurate PCR techniques reported in recent literature [12].

DeepBBS is an accurate registration method based on training a deep learning model using a representation that takes into account the distance between pairs of points nearest to each other. Whereas, PRNet is self-supervised deep learning approach that learns a geometric representation, a keypoint detector, and keypoint-to-keypoint correspondences to perform registration.

Due to the high computational complexity of these methods, it was necessary to empty the internal region of the 3D head meshes, randomly downsampling them to 1024 points and normalising the point coordinates into [-0.5,0.5] to have zero means, as previously done [17].

To validate the function of these methods on our downsampled 3D head models, first we ran a preliminary test: we trained both models with slightly rotated heads (up to 45° on each axis), obtaining state-of-the-art results with a MAE of 0.0005. Figure 5a shows the training loss curve of DeepBBS and PRNet, which are able to learn from our data.

Finally, to evaluate if these methods were able to solve our problem, we trained[1] them with our data, which presents severe rotations as mentioned

[1] In these experiments, the batch size was 16 for DeepBBS and 4 for PRNet and the rest of training parameters was the same as in [12,15].

before. Table 2 presents the results of the experiment expressed in terms of angular error metrics.

Both PCR methods show fairly similar angular errors, while our proposal (MOPI+ResNet50) has a much lower error. Furthermore, during the training of the PCR methods, we observed that PRNet was not able to learn and the loss of DeepBBS stagnated at a high value. This indicates that, with the network hyperparameters proposed and the data available, these methods are able to work with our data and slight rotations (<45°). However, they are not able to give a reliable outcome when dealing with severe rotations (Fig. 5b).

Table 2. Comparison of our method against DeepBBS and PRNet.

Method	MSE	RMSE	MAE
DeepBBS	20703.79	143.88	99.16
PRNet	20751.44	144.05	100.75
Ours (MOPI+ResNet50)	**105.72**	**10.28**	**0.15**

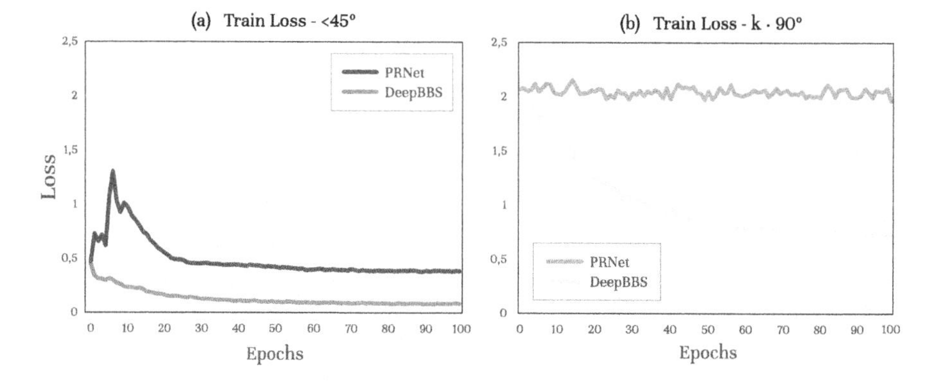

Fig. 5. Training loss curve comparison between DeepBBS and PRNet with the rotations proposed in the methods (<45°) and our severe rotations (multiples of 90°, noted as $k \cdot 90°$).

5 Conclusions

In this paper, we have focused on a specific orientation problem that arises when exporting 3D head models from sMRI files. Depending on the export parameters, the 3D head models will be randomly orientated towards any of the Cartesian axes. If different protocols are used in this step –which is likely to happen in multicentric scenarios– it will result in an unnormalised dataset.

For this reason, this work proposes a novel method for the orientation of 3D heads segmented from sMRI. In our proposal, we considered that transforming the 3D rotation problem into a 2D image classification problem through head PI allows determining with high accuracy the orientation of 3D heads with rotations of 90° or multiples thereof around any of the Cartesian axes.

Our approach allows to determine the orientation of 3D heads under these severe rotations, which PCR methods cannot solve. Additionally, in contrast to PCR methods, *i)* the computational complexity of our proposal is invariant to point cloud size –as we work with projection images–, making it highly scalable, and *ii)* our method is intrinsic, i.e. it does not need a reference point cloud for alignment.

From the obtained results, we have observed the importance of achieving very high classification accuracy for all 24 orientation classes. Otherwise, the angular error metrics become very high. Results have shown that the most accurate variant of our method is MOPI+ResNet50, and the angular errors of our method are much lower than those of state-of-the-art PCR methods.

As future work, we foresee the following avenues of research. First, we will explore the extension of our method to detect any possible type of angulation, not just multiples of 90°. Second, we will investigate different strategies for creating the projection images. Third, we will compare out proposal against HPE methods. And finally, we will study the use of our method on other anatomical structures extracted from sMRI.

Acknowledgements. The research was supported by the Joan Oró grant (FI 2022) from the DRU of the Generalitat de Catalunya and the European Social Fund (2023 FI-2 00160). The authors would also like to thank the Agència de Gestió d'Ajuts Universitaris i de Recerca (AGAUR) of the Generalitat de Catalunya (2021 SGR01396, 2021 SGR00706, 2021 SGR1475), the Spanish Ministry of Science, Innovation, and Universities under grant PID2020-113609RB-C21, and Fondation Jerome Lejeune under grant 2020b cycle-Project No.2001.

References

1. Martínez-Abadías, N., et al.: Understanding brain/face integration from neuropsychiatric disorders. FASEB J. **34**, 1–1 (2020). https://doi.org/10.1096/fasebj.2020.34.s1.05095
2. Myers, L., et al.: Minor physical anomalies in neurodevelopmental disorders: a twin study. Child Adolesc. Psychiatry Ment. Health **11**, 57 (2017). https://doi.org/10.1186/s13034-017-0195-y
3. Hammond, P., Suttie, M.: Large-scale objective phenotyping of 3D facial morphology. Hum. Mutat. **33**, 817–825 (2012). https://doi.org/10.1002/humu.22054
4. Li, M., et al.: Rapid automated landmarking for morphometric analysis of three-dimensional facial scans. J. Anat. **230**, 607–618 (2017). https://doi.org/10.1111/joa.12576
5. Hallgrímsson, B., et al.: Automated syndrome diagnosis by three-dimensional facial imaging. Genet. Med. **22**, 1682–1693 (2020). https://doi.org/10.1038/s41436-020-0845-y

6. Poldrack, R.A., Gorgolewski, K.J.: Making big data open: data sharing in neuroimaging. Nat. Neurosci. **17**, 1510–1517 (2014). https://doi.org/10.1038/nn.3818
7. Abate, A., Bisogni, C., Castiglione, A., Nappi, M.: Head pose estimation: an extensive survey on recent techniques and applications. Pattern Recognit. **127**, 108591 (2022). https://doi.org/10.1016/j.patcog.2022.108591 https://doi.org/10.1016j.patcog.2022.108591
8. Zhang, Z., Dai, Y., Sun, J.: Deep learning based point cloud registration: an overview. Virtual Reality Intell. Hardw. **2**, 222–246 (2020). https://doi.org/10.1016/j.vrih.2020.05.002
9. He, K., Zhang, X., Ren, S., Sun, J.: Deep residual learning for image recognition. In: IEEE Conference on Computer Vision and Pattern Recognition, pp. 770–778 (2016). https://doi.org/10.1109/CVPR.2016.90
10. Hou, B., et al.: 3-D reconstruction in canonical co-ordinate space from arbitrarily oriented 2-D images. IEEE Trans. Med. Imaging **37**(8), 1737–1750 (2018). https://doi.org/10.1109/TMI.2018.2798801
11. Namburete, A.I.L. et al.: Fully-automated alignment of 3D fetal brain ultrasound to a canonical reference space using multi-task learning. Med. Image Anal. **46**, 1–14 (2018). https://doi.org/10.1016/j.media.2018.02.006
12. Hezroni I., Drory, A., Giryes, R., Avidan S.: DeepBBS: deep best buddies for point cloud registration. In: 2021 International Conference on 3D Vision (3DV), London, United Kingdom, pp. 342–351 (2021). https://doi.org/10.1109/3DV53792.2021.00044
13. Besl, P.J., McKay, N.D.: A method for registration of 3-D shapes. IEEE Trans. Pattern Anal. Mach. Intell. **14**, 239–256 (1992). https://doi.org/10.1109/34.121791
14. Aoki, Y., Goforth, H., Srivatsan, R.A., Lucey, S.: PointNetLK: robust & efficient point cloud registration using PointNet. In: 2019 IEEE/CVF Conference on Computer Vision and Pattern Recognition (CVPR), pp. 7156–7165 (2019). https://doi.org/10.1109/CVPR.2019.00733
15. Wang, Y., Solomon, J.M.: PRNet: self-supervised learning for partial-to-partial registration. In: Advances in Neural Information Processing Systems. Curran Associates, Inc. (2019)
16. Hempel, T., Abdelrahman, A.A., Al-Hamadi, A.: 6D rotation representation for unconstrained head pose estimation. In: 2022 IEEE International Conference on Image Processing (ICIP), pp. 2496–2500 (2022). https://doi.org/10.1109/ICIP46576.2022.9897219
17. Xu, Y., Jung, C., Chang, Y.: Head pose estimation using deep neural networks and 3D point clouds. Pattern Recogn. **121**, 108210 (2022). https://doi.org/10.1016/j.patcog.2021.108210
18. Gomez-Donoso, F., Garcia-Garcia, A., Garcia-Rodriguez, J., Orts-Escolano, S., Cazorla, M.: LonchaNet: a sliced-based CNN architecture for real-time 3D object recognition. In: 2017 International Joint Conference on Neural Networks (IJCNN), pp. 412–418 (2017). https://doi.org/10.1109/IJCNN.2017.7965883
19. Carlbom, I., Paciorek, J.: Planar geometric projections and viewing transformations. ACM Comput. Surv. **10**, 465–502 (1978). https://doi.org/10.1145/356744.356750
20. Krizhevsky, A., Sutskever, I., Hinton, G.E.: ImageNet classification with deep convolutional neural networks. Commun. ACM **60**, 84–90 (2017). https://doi.org/10.1145/3065386

Machine Learning Applications

Enhancing Transferability of Adversarial Audio in Speaker Recognition Systems

Umang Patel[1(✉)], Shruti Bhilare[1], and Avik Hati[2]

[1] Dhirubhai Ambani Institute of Information and Communication Technology, Gandhinagar, Gujarat, India
{202021006,shruti_bhilare}@daiict.ac.in
[2] NIT Tiruchirappalli, Tiruchirappalli, India
avikhati@nitt.edu
https://www.daiict.ac.in/, https://www.nitt.edu/

Abstract. Although deep neural networks have demonstrated state-of-the-art performance in several tasks such as speaker recognition among others, they are highly vulnerable to adversarial attacks. These attacks involve the transformation of the original speech signal in order to fool the trained model with minimal alteration in the auditory perception. These attacks have been shown to succeed in white-box settings, however, they are less likely to succeed in a realistic black-box setting. However, it is imperative to investigate the extent of the threat posed by transferability of such attacks to target models to strengthen the defense against them. Therefore, in this work, to enhance the transferability of adversarial examples in black-box setting, the source model's architecture has been minimally modified. Particularly, by skipping selected ReLU activation functions during backpropagation. Experiments on the VoxCeleb dataset resulted in average transferability of 18.7% and 20.5% on two target models.

Keywords: transferability · black-box attack · speaker recognition · adversarial machine learning

1 Introduction

Owing to the rapid progress in the field of machine learning, nearly perfect accuracy has been achieved in several image, speech, and text-based applications such as automated driving and speaker recognition (SR). SR systems have been largely adopted for person authentication [1] in various critical security applications such as banking and border security. However, in the year 2014, Goodfellow et al. [2] detected a vulnerability in the deep learning models in which the trained model stumbled even when a small imperceptible perturbation was introduced into the original image resulting in an adversarial example (AE). This form of attack is called an adversarial attack and has been successfully carried out in image [2,3], speech [4–7] and, text [8,9] domains in recent years.

A. Pertusa et al. (Eds.): IbPRIA 2023, LNCS 14062, pp. 617–628, 2023.
https://doi.org/10.1007/978-3-031-36616-1_49

An AE's ability to continue to adversely impact the models other than the one employed to create it is a common property called transferability [10]. For instance, an AE generated for one model can mislead another model trained on the same or a different dataset [7]. In the domain of computer vision, authors [11,12] have shown highly transferable AEs which pose a serious threat to the credibility of the deployed model. Though significant research has been conducted on transferability in the image domain, not much attention has been given to addressing this issue in speech-based applications [13]. Therefore, it is crucial to investigate the extent to which a modern SR system can be fooled by transferable AEs.

Assume a trained SR system $\phi(\cdot)$ that takes an original audio signal x as input and predicts the speaker ID. In order to perform an untargeted adversarial attack on this system, a perturbation δ is added to the signal x, resulting in the adversarial audio x' that leads to misclassification by the model. Perturbation δ is minimized to obtain δ_{min} as shown in Eq. (1) resulting in the adversarial audio x' that leads to misclassification by the model.

$$\delta_{min} = \arg\min_{\delta} ||x' - x||_p \qquad \text{s.t. } \phi(x') \neq \phi(x) \tag{1}$$

where $x' = x + \delta$

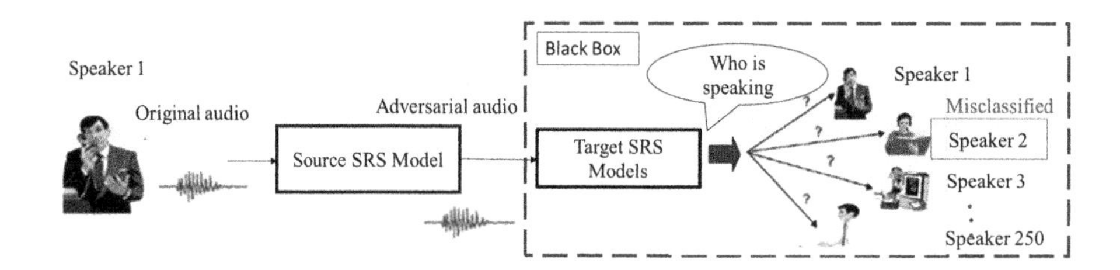

Fig. 1. Overview of the proposed methodology

It has been observed that achieving transferability is far more challenging in the speaker identification systems as opposed to speaker verification systems [5,7,14,15]. Therefore, in this work, we have considered speaker identification systems to evaluate transferability. Figure 1 shows an overview of the proposed work. Adversarial audio generated from the source SR model is fed to the target SR model in order to evaluate its transferability in the black-box setting. For instance, adversarial audio of *Speaker 1* when fed to the target model gets misclassified as being from *Speaker 2*.

The rest of the paper is structured as: Sect. 2 looks into prior studies on transferability, adversarial attacks on SR systems, and DNN models used for SR systems. Section 3 describes the proposed methodology for the generation of transferable AEs. Section 4 presents the results and experimental setup we have used to test the transferability of our proposed method against other state-of-the-art models. Finally, we conclude in Sect. 5.

2 Related Work

Adversarial attacks can be categorized into targeted and untargeted attacks. Targeted attacks are designed to cause a trained model to produce a desired incorrect output, whereas untargeted attacks are intended to generate any incorrect output, not necessarily a particular target output. These attacks can be carried out in a white-box setting where the adversary has complete knowledge of the target model's architecture or in a black-box setting in which the model's details are unknown. It has been shown that the adversarial examples generated in the white-box setting are highly transferable to other target models for various tasks [5,7,14,15]. This may be attributed to the congruence of decision boundaries across models [16]. However, it is more challenging to generate transferable adversarial examples in black-box setting [14,15]. Moreover, untargeted attacks are more easily transferable as compared to targeted ones. Transferability of AEs can be analyzed in various settings such as cross-architecture, cross-dataset, cross-parameter, or any combination thereof [7]. Spoofing countermeasures are effective in dealing with presentation attacks such as replay attacks, speech synthesis attacks, and voice conversion attacks. However, they are susceptible to black-box adversarial attacks.

Wang et al. [6] proposed a way of generating inaudible adversarial perturbations based on the psycho-acoustic principle of frequency masking - instead of relying on the traditional l_p norm, this technique involves limiting the perturbation to be within the masking threshold of the original audio. Authors achieved up to 98.5% adversarial attack success rate (AASR) on the Aishell-1 corpus for speaker identification system while keeping the adversarial audio indistinguishable from the original audio for listeners. In addition to speech data, authors tested the approach on non-speech data as well, achieving 91.5% AASR for attacks on target speakers. Xie et al. [17] proposed a real-time targeted adversarial attack against SR systems. They generate a universal perturbation that can be added to any enrolled speaker's voice input to make the SR system predict the adversary-desired speaker label. Evaluation on a public dataset of 109 speakers shows the effectiveness and robustness of the attack with a high AASR of over 90%. The attack launching time is only 0.015s, a 100x speedup over contemporary non-universal attacks.

However, very few researchers have focused on evaluating the transferability of AEs generated for the source model to the target model [7,14,15,18,19]. Kreuk et al. [5] conducted two experiments to demonstrate the transferability of AEs in the speaker verification system. A combination of cross-features and cross-datasets has been tested by the authors for the transferability of the AEs. First, they considered a source model trained on the YOHO dataset with MFCC features and a target model trained on the NTIMIT dataset with Mel spectrum features resulting in a drastic increase in false positive rate (FPR) from 12% to 46%. Conversely, they trained the model on the NTIMIT dataset with Mel spectrum features and tested the attack on a model trained on the YOHO dataset with MFCC features, and the FPR worsened from 16% to 46%.

Li et al. [19] evaluated the transferability in a speaker verification system in the cross-feature and cross-architecture settings and a combination of both. They employed fast gradient sign method (FGSM) for generating AEs in the following settings: (i) cross-feature setting considering a GMM i-vector model with Mel-frequency cepstral coefficient (MFCC) and log power magnitude spectrum (LPMS) features; (ii) cross-model setting using MFCC feature with i-vector as the source model and x-vector as the target model; and (iii) a cross-feature-model assuming LPMS features in i-vector as the source model and MFCC along with x-vector as the target model. They achieved the highest transferability in the cross-feature setting. Zhang et al. [18] presented an iterative ensemble method (IEM) fused with Momentum based Iterative Fast Gradient Sign Method (MI-FGSM) which could generate highly transferable AEs and boost AASR by 4–30% relative to the baseline logit ensemble model, tested on four distinct anti-spoofing models indicating a requirement for more robust defensive methods.

Tan et al. [15] proposed temporal and spatial momentum-based iterative FGSM (TSMI-FGSM) to increase the robustness and transferability of targeted AEs in SR systems. Unlike the gradient-based approaches such as FGSM, PGD, and MI-FGSM which accumulate the gradients in the temporal domain alone, authors also consider gradients in the spatial domain. Particularly, they consider the gradients generated in the sample neighborhood space as well as internal space, and employ mean and variance tuning to steady the gradient's path during iterations, thus increasing their capacity to be transferred across different architectures. They experimented with nine distinct architectures for target models, three of which were used as source models to generate AEs. Their approach achieves considerable transferability in the white-box setting. In the black-box setting, the ECAPA model [20] exhibited the greatest transferability rate of 41.2% on the TDy-H model and the least transferability rate of 21.5% on the x-vec-C. Similarly, the Res34-V model showed a notable difference in transferability rate, with 79.2% on the TDy-H and 13.5% on the i-vec. Finally, the x-vec-P model had poor performance on the Res34-V with only 9% transferability, while it fared best on the i-vec with 50.4%.

Chan et al. [7] achieved transferability of 5% and 62% in cross-architecture setting under black-box environment for speaker verification and open speaker identification (OSI), but 100% in cross-parameter settings for untargeted OSI tasks. However, cross-parameter experiments were performed in a white-box setting, which may not be useful in a realistic situation. Zhang et al. [14] improved the transferability rate up to 89.5% using an ensemble technique for generating AEs. However, this had two significant limitations. First, a relatively small dataset consisting of only 10 speakers was used. Second, the computational cost is excessively high due to the employment of two models for AE generation.

3 Methodology

In this paper, we have adopted the cross-architecture approach to evaluate the transferability of AEs from a source model to two target models. Attacks have been carried out in the black-box setting assuming that the adversary has complete knowledge of the source model. The foundation of our approach is linearization of the source model's architecture during backpropagation in such a way that the model's classification accuracy is not compromised. The proposed methodology enhances the transferability of the speaker identification system by modifying its time delay neural network (TDNN) blocks. To boost its transferability, we have reduced non-linear activation functions during the backpropagation phase. The idea is originated by Guo et al. [11], and further advanced by Shah et al. [21] in the image domain.

Assume a source SR model $\phi : \mathbb{R}^D \rightarrow \mathbb{R}^C$ which takes $x \in \mathbb{R}^D$ as input and is trained for C subjects. It consists of d trainable layers pertaining to linear and nonlinear functions with weight matrices W_i corresponding to the ith layer as shown in Eq. (2)

$$\phi(x) = W_d^T\ \sigma_{d-1}(W_{d-1}^T \cdots\ \sigma_{i+1}(W_{i+1}^T\ \sigma_i(W_i^T\ \sigma_{i-1}(W_{i-1}^T \cdots\ \sigma_1(W_1^T(x)))))) \quad (2)$$

where σ_i is a non-linear activation function of ith layer, which can be tanh, sigmoid or ReLU.

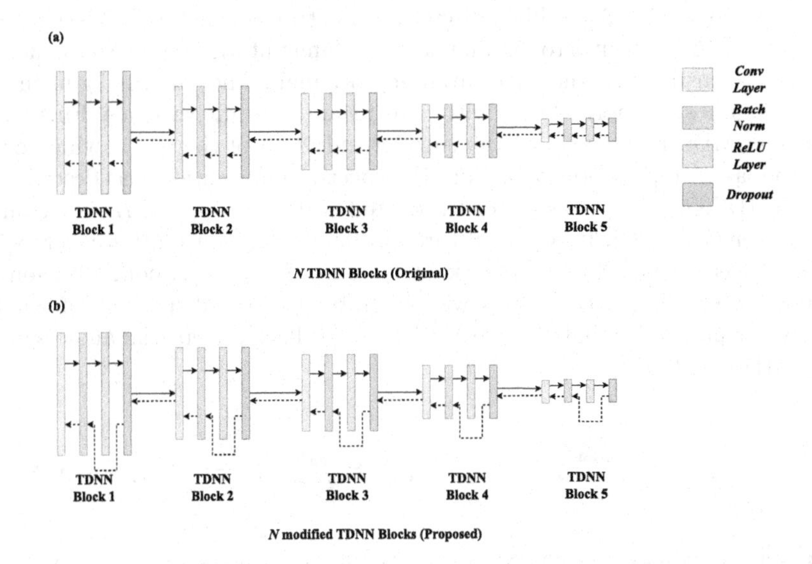

Fig. 2. (a) TDNN block used in original model [4] (b) Modified TDNN (mTDNN) block used in our work

We employed FoolHD [4] to generate AEs since it results in high AASR and exploits the advantage of encoder-decoder architecture. Specifically, AEs were generated using a minimally modified source model which consists of a gated convolution auto-encoder (GCA) and a time-delayed neural network (TDNN) based classifier. This classifier consists of N number of TDNN blocks, a statistical pooling layer, three fully connected layers, and a softmax layer, as illustrated in Fig. 3. The original TDNN architecture of classifier has five blocks as shown in Fig. 2(a) with each block comprising convolution, batch normalization, ReLU, and dropout layers. Inspired by the success of Guo et al. [11] in transferring adversarial images, we proposed a modified TDNN architecture (henceforth referred to as mTDNN) featuring skip connections as depicted in Fig. 2(b). During forward propagation, the model architecture remains unchanged, while during backpropagation we selectively removed non-linear activation functions (ReLU) from the architecture to enhance transferability. Our proposed classifier model (mTDNN) $\phi' : \mathbb{R}^D \rightarrow \mathbb{R}^C$. This source model has many linear and nonlinear functions during forward pass but non-linearity of ith layer (σ_i) is removed during backpropagation as shown in Eq. (3). The value of i can range from 1 to d.

$$\phi'(x) = W_d^T \ \sigma_{d-1}(W_{d-1}^T ... \ \sigma_{i+1}(W_{i+1}^T \ W_i^T \ \sigma_{i-1}(W_{i-1}^T ... \ \sigma_1(W_1^T(x))))) \quad (3)$$

Shamsabadi et al. proposed an adversarial attack, FoolHD employing a model architecture of an encoder-decoder [4], particularly, a gated convolution autoencoder model as depicted in Fig. 4. It leverages speech steganography techniques to generate imperceptible audio perturbations against a white-box speaker identification model [22]. A modified discrete cosine transform (MDCT) is applied to the original audio input x to obtain real-valued magnitude and phase coefficients. In order to ensure that the adversarial and original audios are in alignment, the original audios are normalized, with zero mean and unit variance. Finally, the output of MDCT (s) is fed to the encoder unit $E(.)$ of the GCA, which generates the latent representation h from the spectral information s as described by $h = E(\text{MDCT}(x))$. h and s are concatenated as $H' = [h; s]$ and H' is forwarded to decoder layer $D(.)$. It is passed through four gated convolution layers where each layer is composed of 64 3×3 kernels, followed by batch normalization and dropout layers. The decoder unit will generate adversarial spectral information s' which is passed through inverse MDCT to get back adversarial audio samples $x' = \text{iMDCT}(D(H'))$

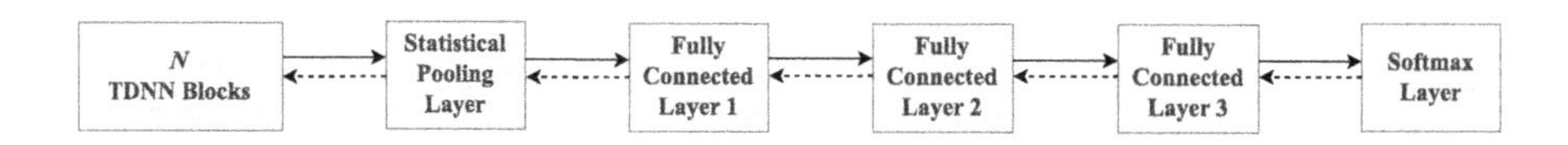

Fig. 3. Block diagram of TDNN-based classifier of source model, where $N = 5$

The classifier used to generate AEs in Fig. 4 is based on modified N-TDNN blocks as shown in Fig. 2(b). We have skipped certain ReLU non-linear activation functions during backpropagation to enhance the linearity. The GCA model is used to generate AEs, with mTDNN being part of the classifier. We have removed one or more ReLU layers during backpropagation from different TDNN blocks. $L^{(n)}$ refers to the scenario in which the ReLU layer has been removed from the nth TDNN block. AEs generated from each scenario from source model were considered for evaluating the transferability on the target models. It is important

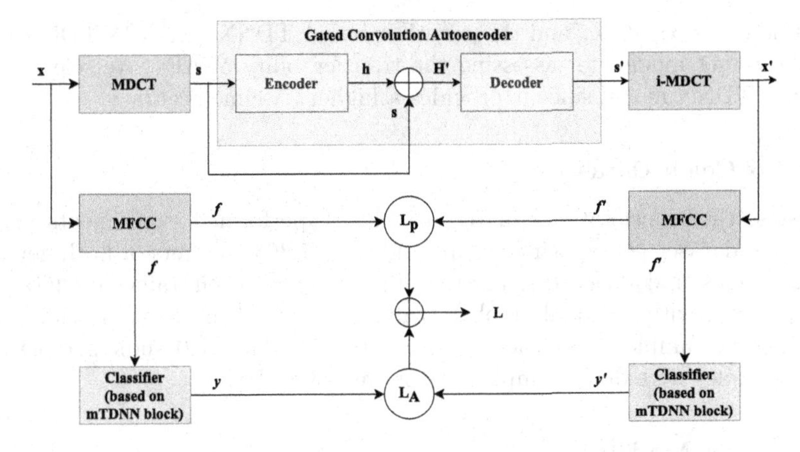

Fig. 4. Overview of the FoolHD attack considered in our work with modified TDNN (mTDNN) block in classifier

to note that, as the attacks have taken place in a black-box setting, the target SR model architectures remain unchanged. The loss function comprises perceptual loss $L_p(f, f')$ and adversarial loss $L_a(y, y')$. $L_p(f, f')$ is calculated by computing the cosine distance between the MFCC features f and f' of the original and AEs respectively, as represented in the Eq. (4). While $L_a(y, y')$ is calculated using the logits of the original Z and adversarial audio Z' of the softmax layer as shown in Eq. (5), where y and y' are output of the classifier for original audio x and adversarial audio x'. T is the number of frames in an audio sample and F is the number of MFCC features of each frame. We have extracted 30 MFCC features per frame for input x.

$$L_p(f, f') = \sum_{t=1}^{T} (1 - \frac{\sum_{i=1}^{F} f_{it} f'_{it}}{\sqrt{\sum_{i=1}^{F} f_{it}^2 \sum_{i=1}^{F} {f'_{it}}^2}}) \tag{4}$$

$$L_a(y, y') = Z_y - \max_{\substack{i=1,....C \\ i \neq y}} Z'_i \tag{5}$$

$$L(x, x') = L_p(f, f') + L_a(y, y') \tag{6}$$

Overall loss function $L(x, x')$ is optimized using Adam optimizer with various hyperparameters such as learning rate, decay learning rate, and dropout.

4 Experiments and Results

For generation of AEs, we have considered the GCA-mTDNN as the source model. Densely connected TDNN (DTDNN-SS) [23] and Emphasized Channel

Attention, Propagation, and Aggregation based TDNN (ECAPA-TDNN) [20] are the target models for assessing the transferability of AEs. We selected the x-vector TDNN model since it provides a higher baseline accuracy.

4.1 VoxCeleb Dataset

To assess the transferability of the attack, we performed experiments on the VoxCeleb dataset [24] which contains audios of 7,363 speakers of both genders, all age groups, and all accents. The target models have been trained on data from 250 speakers, with an equal number of male and female speakers. Each speaker has 10 audio samples of 4 s each, resulting in a total of 2500 audio samples. All audio samples were down-sampled at the rate of 8 kHz.

4.2 Target Models

Table 1. Architecture of DTDNN-SS

(# - Layer number, C - number of classes, t - time frame)

#	Component Type	Context	Size
1	TDNN-ReLU	t − 2 : t : t + 2	128
2–7	DTDNN	t − 1, t, t + 1	64
8	ReLU-FNN	t	256
9–20	DTDNN	t − 3, t, t + 3	64
21	ReLU-FNN	t	512
22	Pooling (mean+stddev)	full-seq	
23	FNN-BN (Emb.)		512
	Softmax		C

DTDNN-SS: DTDNN utilizes bottleneck layers and dense connectivity to compress the number of parameters and still achieve a high level of accuracy. It has been further improved in its variant, DTDNN-SS (Statistics-and-Selection), by utilizing multiple TDNN branches with short-term and long-term contexts [23]. To integrate the information from these sources, a channel-wise selection mechanism is employed, allowing the selection of the most pertinent information from the TDNN branches. This architecture has five components as outlined in Table 1: the first component is for initializing the number of channels; the second component contains layers 2–8, each with a frame offset of one to capture local features in short-term contexts $t-1$ to $t+1$, activated via ReLU and feed-forward neural networks (FNN); the third component, containing layers 9–21, similarly contains a frame offset of three ($t-3$ to $t+3$) for learning long-term contexts, also using ReLU and FNN. The last two components consist of a statistical pooling layer with a full sequence of context, FNN, and softmax layer which combines the frame-level features to create the output utterance-level features.

ECAPA-TDNN: Desplanques et al. [20] put forward a TDNN-based speaker embedding extractor named Emphasized Channel Attention, Propagation, and Aggregation (ECAPA-TDNN) for SR systems (Table 2). They enhanced the original X-vector architecture by incorporating Squeeze-Excitation blocks, multiscale Res2Net features, additional skip connections, and channel-dependent attentive statistics pooling. Dilation spacing helps expand kernel size by skipping consecutive elements. These enhancements led to a significant improvement of 19% in EER relative to baseline systems on the VoxCeleb dataset.

Table 2. Architecture of ECAPA+TDNN

(# - Layer number, C - number of classes)

#	Component Type	kernel size(k)	dilation spacing(d)
0	Input (80 × T)		
1	Conv1D+ReLU+BN	5	1
2	SE-Res2Block	3	2
3	SE-Res2Block	3	3
4	SE-Res2Block	3	4
5	Conv1D+ReLU	1	1
6	Attentive Statistical Pooling + BN		
7	FC+BN		
	AAM-Softmax		C

4.3 Experiments

In our study, we have used a pre-trained model of GCA and TDNN that was trained on 2500 audio files from 250 speakers. Both target models are trained by us with a batch size of 20 and 100 epochs, using an Adam optimizer with a learning rate of 0.001. To analyze the transferability of adversarial attacks, we have performed the following experiments. In the first set of experiments, the baseline performance of the source model as well as the target models are evaluated. In the second set of experiments, we have computed the AASR on the source model and target models through AEs generated with the help of FoolHD attack and original source model. This experiment used 2500 audio samples. An AE is generated using a single batch size with 1000 epochs and a learning rate of 0.001 in the Adam optimizer. In the next set of experiments, we removed the ReLU activation function during backpropagation from one or more TDNN blocks. $L^{(n)}$ refers to the scenario in which the ReLU layer in the nth TDNN block is removed. We considered $n = 1, 3, 5, (1, 2), (2, 3, 4), (4, 5)$ for our experiments. The AEs generated in each scenario are fed to the target models to assess their vulnerability. For each scenario, we report the AASR which is the percentage of adversarial examples misclassified by the classifier to the total AEs. All experiments have been performed on a system with the following specifications: Intel (R) Xeon(R) Gold 6145 CPU @ 2.00 GHz 20 Cores, RAM of 96 GB, two NVidia Quadro P400 GPUs of each 2 GB, and NVidia Quadro GP100 of 16 GB.

4.4 Results

Table 3 shows that the source as well as target models exhibit decent baseline accuracy for speaker recognition. Table 4 shows the AASR of FoolHD attack on source and target models without making any modifications to the original model architecture. Comparing target models with similar AEs, 12.53% and 8.34% AASR values were observed, demonstrating an 87.02% and 91.21% decrease in AASR compared to the source model. To enhance the transferability, experiments were conducted in six different scenarios $L^{(n)}$, as explained earlier.

Table 3. Baseline accuracy for source Model (GCA-mTDNN) and two target models (DTDNN-SS and ECAPA-TDNN) tested over 2500 audio samples.

Baseline Accuracy (%)		
GCA-mTDNN	DTDNN-SS	ECAPA-TDNN
98.39	100	99.8

Table 4. 500 audio samples were used to assess the AASR of six scenarios for the source model (GCA-mTDNN) and two target models (DTDNN-SS and ECAPA-TDNN), while 2500 audio samples were used to evaluate the original model. **Bold** indicates highest AASR among six scenarios.

Adversarial Attack Success Rate (%)			
Segment	GCA-mTDNN	DTDNN-SS	ECAPA-TDNN
original model	99.55	12.53	8.34
$L^{(1)}$ – *first*	**96.96**	**20.25**	21.42
$L^{(3)}$ – *middle*	96.88	19.6	**22.78**
$L^{(5)}$ – *last*	96.44	19.23	20.93
$L^{(1,2)}$ – *first two*	95.77	18.23	19.78
$L^{(2,3,4)}$ – *middle three*	93.87	16.85	17.95
$L^{(4,5)}$ – *last two*	96.27	18.27	20.04

In Table 4, second row onwards results of the six scenarios are presented, which show a significant improvement in the transferability rate of the AEs. It was observed that $L^{(1)}$ and $L^{(3)}$ yielded the greatest AASR on target model 1 and target model 2, respectively. Therefore, omitting the ReLU layer of 1st and 3rd blocks evidently increased the transferability rate. Both DTDNN-SS and ECAPA-TDNN target models saw an improvement in attack transferability of up to 7.73% and 14.4%, respectively. Interestingly, $L^{(2,3,4)}$ performed the worst, signifying that skipping more non-linear layers during backpropagation could decrease the performance and transferability of the model. The AASR on the original source model did not deteriorate after the ReLU layers were removed. As highlighted in Sect. 2, speaker verification performed better than speaker identification for transferring AEs. However, the only two state-of-the-art methods outlined in [14,15] adopted different attacks and experimental setups with limited and distinct datasets having just 10 speakers. Therefore, it is difficult to properly compare the results of these models with our results.

5 Conclusion

Studying transferability has two main goals: an assessment of threat posed by the attack and to create a more robust model. High transferability rate requires the model to be reinforced. Our findings indicated a lower rate of transferability than in the image classification task. This could be due to SR models being more resilient than image-based classifiers. However, not many conclusions can be drawn yet as SR system research is still in its infancy. In our experiments, we observed that cross-architecture transferability was enhanced by 7.7% and 14.4% for target model 1 and target model 2, respectively. We have assessed the effects of taking out the ReLU activation function from different blocks of TDNN during backpropagation, and it has been shown to improve transferability without detrimentally affecting the accuracy of the original model. Our experiments showed that skipping the ReLU layer in the first TDNN block yields the best result. Moving forward, we will explore targeted attack strategies and assess whether similar concepts can be applied for defence mechanisms against AEs. Moreover, our current study focuses on a single type of attack. Future research could investigate the impact of multiple attacks on the system.

References

1. Singh, N., Agrawal, A., Khan, R.: Voice biometric: a technology for voice based authentication. Adv. Sci. Eng. Med. **10**(7–8), 754–759 (2018)
2. Goodfellow, I.J., Shlens, J., Szegedy, C.: Explaining and harnessing adversarial examples, ArXiv Preprint ArXiv:1412.6572 (2014)
3. Moosavi-Dezfooli, S.-M., Fawzi, A., Frossard, P.: DeepFool: a simple and accurate method to fool deep neural networks. In: Proceedings of the IEEE Conference on Computer Vision and Pattern Recognition, pp. 2574–2582 (2016)
4. Shamsabadi, A.S., Teixeira, F.S., Abad, A., Raj, B., Cavallaro, A., Trancoso, I.: FoolHD: fooling speaker identification by highly imperceptible adversarial disturbances. In: IEEE International Conference on Acoustics, Speech and Signal Processing (ICASSP), pp. 6159–6163 (2021)
5. Kreuk, F., Adi, Y., Cisse, M., Keshet, J.: Fooling end-to-end speaker verification with adversarial examples. In: IEEE International Conference on Acoustics, Speech and Signal Processing (ICASSP), pp. 1962–1966 (2018)

6. Wang, Q., Guo, P., Xie, L.: Inaudible adversarial perturbations for targeted attack in speaker recognition. ArXiv Preprint ArXiv:2005.10637 (2020)
7. Chen, G., et al.: Who is real bob? Adversarial attacks on speaker recognition systems. In: IEEE Symposium on Security and Privacy (SP), pp. 694–711 (2021)
8. Miyato, T., Dai, A.M., Goodfellow, I.: Adversarial training methods for semi-supervised text classification. ArXiv Preprint ArXiv:1605.07725 (2016)
9. Liang, B., Li, H., Su, M., Bian, P., Li, X., Shi, W.: Deep text classification can be fooled. ArXiv Preprint ArXiv:1704.08006 (2017)
10. Vakhshiteh, F., Nickabadi, A., Ramachandra, R.: Adversarial attacks against face recognition: a comprehensive study. IEEE Access **9**, 92735–92756 (2021)
11. Guo, Y., Li, Q., Chen, H.: Backpropagating linearly improves transferability of adversarial examples. Adv. Neural. Inf. Process. Syst. **33**, 85–95 (2020)
12. Papernot, N., McDaniel, P., Goodfellow, I.: Transferability in machine learning: from phenomena to black-box attacks using adversarial samples. ArXiv Preprint ArXiv:1605.07277 (2016)
13. Jiang, W., He, Z., Zhan, J., Pan, W., Adhikari, D.: Research progress and challenges on application-driven adversarial examples: a survey. ACM Trans. Cyber-Phys. Syst. (TCPS) **5**(4), 1–25 (2021)
14. Zhang, J., et al.: NMI-FGSM-Tri: an efficient and targeted method for generating adversarial examples for speaker recognition. In: 7th IEEE International Conference on Data Science in Cyberspace (DSC), pp. 167–174 (2022)
15. Tan, H., Gu, Z., Wang, L., Zhang, H., Gupta, B.B., Tian, Z.: Improving adversarial transferability by temporal and spatial momentum in urban speaker recognition systems. Comput. Electr. Eng. **104**, 108446 (2022)
16. Abdullah, H., Karlekar, A., Bindschaedler, V., Traynor, P.: Demystifying limited adversarial transferability in automatic speech recognition systems. In: International Conference on Learning Representations (ICLR) (2021)
17. Xie, Y., Li, Z., Shi, C., Liu, J., Chen, Y., Yuan, B.: Real-time, robust and adaptive universal adversarial attacks against speaker recognition systems. J. Sign. Proc. Syst. **93**, 1–14 (2021). https://doi.org/10.1007/s11265-020-01629-9
18. Zhang, Y., Jiang, Z., Villalba, J., Dehak, N.: Black-box attacks on spoofing countermeasures using transferability of adversarial examples. In: Interspeech, pp. 4238–4242 (2020)
19. Li, X., Zhong, J., Wu, X., Yu, J., Liu, X., Meng, H.: Adversarial attacks on GMM i-vector based speaker verification systems. In: IEEE International Conference on Acoustics, Speech and Signal Processing (ICASSP), pp. 6579–6583 (2020)
20. Desplanques, B., Thienpondt, J., Demuynck, K.: ECAPA-TDNN: emphasized channel attention, propagation and aggregation in TDNN based speaker verification. ArXiv Preprint ArXiv:2005.07143 (2020)
21. Shah, M., Mandal, S., Bhilare, S., Dhirubhai, A.H.: Increasing transferability by imposing linearity and perturbation in intermediate layer with diverse input patterns. In: IEEE International Conference on Signal Processing and Communications (SPCOM), pp. 1–5 (2022)
22. Kreuk, F., Adi, Y., Raj, B., Singh, R., Keshet, J.: Hide and speak: towards deep neural networks for speech steganography. ArXiv Preprint ArXiv:1902.03083 (2019)
23. Yu, Y.-Q., Li, W.-J.: Densely connected time delay neural network for speaker verification. In: Interspeech, pp. 921–925 (2020)
24. Nagrani, A., Chung, J.S., Zisserman, A.: VoxCeleb: a large-scale speaker identification dataset. ArXiv Preprint ArXiv:1706.08612 (2017)

Fishing Gear Classification from Vessel Trajectories and Velocity Profiles: Database and Benchmark

Pietro Melzi(✉), Juan Manuel Rodriguez-Albala, Aythami Morales, Ruben Tolosana, Julian Fierrez, and Ruben Vera-Rodriguez

BiDA-Lab, Universidad Autonoma de Madrid, Madrid, Spain
pietro.melzi@uam.es, juanmanuel.rodrigueza@alumni.uam.es

Abstract. International Organizations demand to take care of our oceans and their ecosystems since they are of incalculable value to humanity. The illegal fishing activity does irreparable damage to these ecosystems and these organism are pushing to detect and combat illegal fishing activities. Fishing vessels are equipped with a radio frequency beacon that emits their GPS position and other information relevant to the Automatic Identification System (AIS). The GPS positions can be used to infer the vessel trajectories and detect illegal fishing activities. In this study we present a new database (https://github.com/BiDAlab/TrFGdb) including trajectories representing 5 different fishing gears, and analyze them as in a problem of time sequence analysis. We extract global and local features from the trajectories of vessels, and propose several supervised learning algorithms to classify the kinematics of vessels according to different fishing gears. Compared to previous works, we highlight the importance of considering trajectories with sampling period in the order of minutes instead of hours, to detect activities carried out in a short time that could help to distinguish fishing gears. A considerable effort has been dedicated to pre-processing the real data at our disposal, to generate a quality dataset with highly reliable labels. The best classification accuracy obtained in this study is 90%. We expect to improve it if more trajectories describing the different fishing gears were available.

Keywords: Fishing Gear Classification · Database · Illegal Fishing

1 Introduction

According to the Food and Agriculture Organization of the United Nations (FAO), illegal, unreported and unregulated (IUU) fishing is a broad term that encompasses a wide variety of fishing activities. IUU fishing exists in all types and extents of fishing, occurs both on the high seas and in areas under national

P. Melzi and J. M. Rodriguez-Albala—These authors contributed equally to this work.

A. Pertusa et al. (Eds.): IbPRIA 2023, LNCS 14062, pp. 629–638, 2023.
https://doi.org/10.1007/978-3-031-36616-1_50

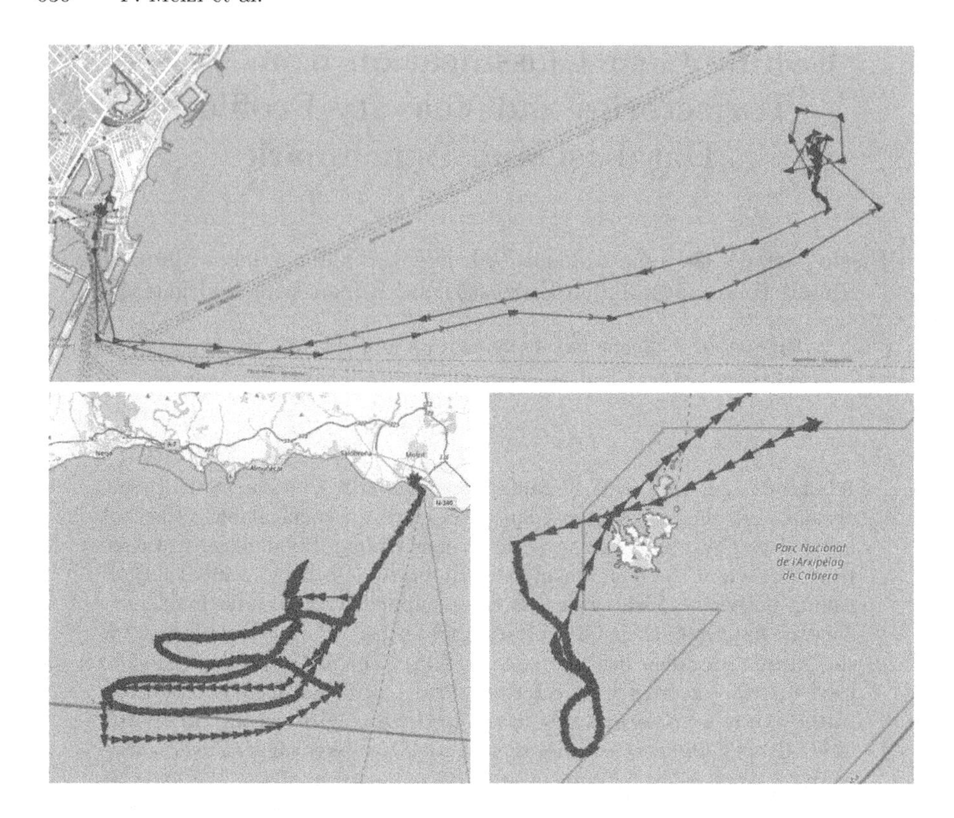

Fig. 1. Example of trajectories for different fishing gears: surroundings (above), longlines (left), and trawls (right).

jurisdiction, affects all aspects and stages of the capture and use of fish, and can sometimes be associated with the organized crime [6].

IUU fishing represents a great disadvantage and discrimination for fishermen who act responsibly, honestly, and in accordance with the conditions of their fishing authorizations. If IUU fishers operate with vulnerable populations subject to strict moratoriums, healthy levels of those populations cannot be restored, threatening marine biodiversity, the food supply of communities that depend on fishing resources for protein intake, and the subsistence of people related to the sector. According to estimates, IUU fishing involves around 26 million tons of fish per year in the whole world, equivalent to more than 15% of the total annual amount of fish products [5].

With this study, we aim to contribute to the monitoring and surveillance of IUU fishing for social good. We process the records provided by Tragsatec's Management of Agricultural and Fisheries Information Systems, that detail the fishing activities carried out by 357 vessels leaving the ports of Spain. Such records contain information about the geographical position, speed, and direction of vessels over time, together with the description of the fishing gear transported.

The skippers of fishing vessels declare fishing gears at the exit of port, sail in search of schools of fish and begin their fishing operations describing a trajectory with speed usually below 5 knots. As shown in Fig. 1, different fishing gears present trajectories with peculiar characteristics, that we can classify by means of supervised learning algorithms [2,9].

The number of public dataset available to model the trajectories of vessels is limited. In [9] authors presented a database with 1,227 fishing vessels operating in the Indonesian regulatory area, one of the countries with the highest rate of IUU fishing in the world. Each fishing vessel is provided with time series representing its GPS positions over one year, with an approximate resolution of one hour. In such work, features of interest are obtained with unsupervised analysis and combined in classification models, namely Random Forests (RFs) and Support Vector Machines (SVMs). With the sole use of behavioural features, accuracy of 93% is obtained for the following fishing gears: *i)* trawls, *ii)* longlines, *iii)* pole-and-lines, and *iv)* purse seines. The additional use of mean GPS positions increases accuracy to 97%, but this information discriminates region-specific fisheries.

In another work [1], combinations of local and global features are extracted from trajectories of Thai vessels. An accuracy above 90% is obtained for the classification of the following fishing gears: *i)* trawls, *ii)* purse seines, *iii)* longlines, and *iv)* reefer. However, the sampling period of the employed time series is two hours. Such time interval provides insufficient information for purse seine because the speed of fishing vessels suddenly changes, causing lower classification performance for the specific class. The study suggests to reduce the sampling period to 15–30 min, as for the time series considered in our study. A comparison of our database with the ones considered in [9] and [1] is provided in Table 1.

The contributions of this study are the following:

- A new database containing more than one thousand trajectories recorded from 357 fishing vessels with a sampling period of 5 min, to overcome some limitations of previous study based on hourly sampling periods. This database reduce by more than 10 the Nyquist bandlimit of existing databases.
- A novel method based on the fusion of global and local features to classify the trajectories of vessels according to their fishing gear with high reliability.

The remaining of the paper is organized as follows: Sect. 2 describes the database and methods proposed, Sect. 3 presents the results obtained, and Sect. 4 draws the conclusions of this study.

2 Database and Methods

2.1 Database

The data provided by Tragsatec's Management of Agricultural and Fisheries Information Systems, with the authorization of the General Secretariat of Fisheries of the Spanish Ministry of Agriculture, Fisheries and Food, are not originally captured and processed for the purpose of this study. As a consequence,

Table 1. Comparison of databases describing the trajectories of fishing vessels equipped with different fishing gears.

	Our	[9]	[1]
Fishing vessels	357	1,227	32
VMS positions	184,489	5,263,158	184,577
Trajectories	1,045	–	771
Trajectories/classes	209	–	–
Observed days	66	365	–
Sampling period (minutes)	**5 ± 0.83**	**60 ± 15**	**120**
Nyquist Bandlimit (Hz)	**1/600**	**1/7200**	**1/14400**
Classes (Fishing Gear)	Trawls, Purse seines, Trammels, Longlines, Gillnets	Trawls, Longlines, Pole-and-liners, Purse seines	Trawls, Purse seines, Longlines, Reefer

detailed operations of cleaning and data preparation were required. We describe the data curation task that generates a quality dataset with highly reliable labels [10]. The original raw data contain the information described in Table 2. In addition, we consider the expert knowledge provided by Tragsatec's Management and the General Secretariat of Fisheries, about data format and properties of the different fishing gears.

Given the high detail of "Fishing gears" table, we group fishing gears according to the Annex III of Regulation (EU) n° 1379/2013 [3]. The resulting classes of fishing gear that we consider in our study are[1]:

- Trawls: is a fishing practice that herds and captures the target species by towing a net along the ocean floor.
- Purse seines, or surrounding: is a large wall of netting deployed around an entire area or school of fish.
- Trammels: are similar to a gill net but are made up of three layers of netting.
- Longlines: consist of a mainline, gangions, and baited hooks.
- Gillnets: is a wall of netting hanging in the water column, typically made of monofilament or multifilament nylon.

The Tragsatec database presents a Nyquist bandlimit $B = 1/600$. The Nyquist theorem stablish that *"If a function $x(t)$ contains no frequencies higher than B hertz, then it can be completely determined from its ordinates at a sequence of points spaced less than $1/(2B)$ seconds apart."* Thus, the Tragsatec database outperform by 12 and 24 the bandlimit of existing databases [1,9] (see Table 2). This bandlimit is critical when implementing frequency analysis used in time-based feature extraction methods (e.g., Recurrent Neural Networks, Hidden Markov Models, etc.).

[1] https://www.seafish.org/.

Table 2. Description of the data provided by Tragsatec's Management of Agricultural and Fisheries Information Systems and used in this study. "Samples" column refers to the period between 2021-12-15 and 2022-02-19.

Table	Description	Samples	Fields
AIS messages	Messages issued by vessel's AIS beacon	5,007,208	Geom. position, date, hour, speed, course, vessel id
Vessels	Basic data of a vessel	1,647	Vessel id, usual fishing gears
Diary statements	Declarations of when and where vessels start and end their navigation	31,794	Diary id, vessel id, departure date, return date, departure port id, return port id
Fishing gears carried	Declarations of fishing gears carried on board by vessels before going to sea	33,129	Record id, diary id, fishing gear id
Fishing gears	Information about fishing gears	157	Fishing gear id, name, details, code
Ports	Information about ports	31,794	Port id, geometric outline of the port, name

2.2 Data Curation

We rule out diary statements with more than one fishing gear, as we do not know which gear is used at what time. We identify vessel's departures and returns to port by combining two consecutive vessel's GPS positions with the port outline. Due to the variability of the AIS beacon, in some trajectories there is no intersection between vessel's positions and port outline. It may be confused with loss of coverage. Hence, we decided to consider only the trajectories that intersect the outline of a port at their beginning and end, given that the correct use of the AIS beacon provides more reliability. We use the starting and ending time and place of a trajectory, determined from the vessel's GPS positions, to obtain the fishing gear declared in the diary statement.

The messages issued by the AIS beacon do not always have a fixed period of 300 s. We fix the threshold to 350 s, to cover 95.45% (2σ) of AIS messages and detect outliers, according to the empirical rule of 68-95-99.7 (three-sigmas rule) [7]. This threshold represents the maximum time that can elapse between two consecutive messages, which guarantees continuous sampling of GPS positions without loss of coverage and outliers. On the other hand, we significantly reduce the number of diary statements because trajectories with at least a message exceeding the threshold are completely discarded (19,633 of 31,794 diary statements). Additionally, we discard trajectories with a low percentage of AIS messages at fishing speed (lower than 5 kn), or with a total duration minor than 180 min. Such trajectories represent other activities, for instance docking at intermediate ports. The number of valid diary statements decreases to 9,399. Finally, we select 209 diary statements and their trajectories (undersampling) to equally represent the five fishing gears described above.

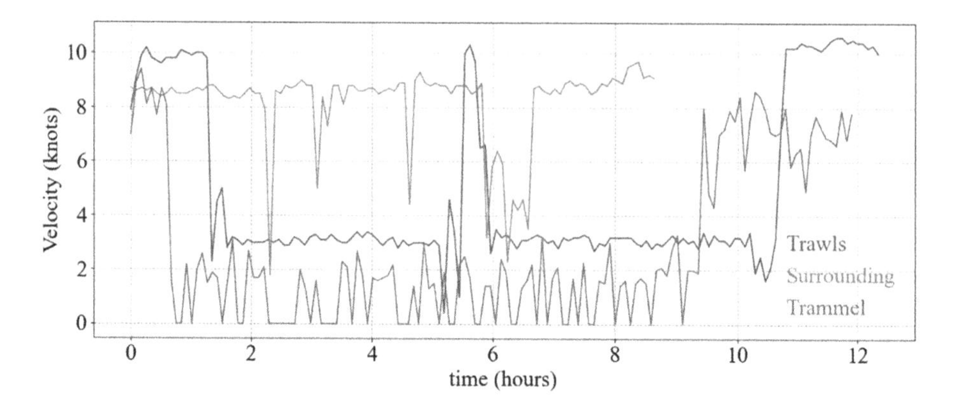

Fig. 2. Velocity profiles of three fishing gears: trawls (blue), surrounding (orange), and trammel (green). (Color figure online)

2.3 Feature Extraction

The course and speed of the vessel are affected by the fishing gear (see Fig. 2). The trajectory over time t of a vessel is described as time sequences of geographical coordinates, longitude(t) and latitude(t). They are analogous to the $x(t)$ and $y(t)$ coordinates of a trajectory over two-dimensional space (x, y) over time t. The literature on modeling trajectories using machine learning approaches is very broad. Among the different applications of these methods, the biometric recognition of dynamic signatures is interesting for the present work because of the high intra-class variability of signers and the low inter-class variability of forgeries. Based on this, we adapt state-of-the-art techniques for dynamic handwritten signature recognition to the kinematics of vessels. Moreover, portions of trajectories representing fishing activities, usually with a speed lower than $5kn$, provide an analogy with the contact of digital pens with electronic tablets. Hence, we establish a relationship of inverse proportionality between the fishing speed and the pressure of the digital pen $p(t)$.

Global Features. A trajectory can be described by an n-dimensional vector, containing features related to its shape and temporal events. In [8], a set of 100 global features is considered, many of them providing high performance in the literature for the task of online signature trajectory recognition. Global features are extracted from discrete time signals of digital pen trajectories: $x(t)$, $y(t)$, and $p(t)$, with $p > 0$ to indicate digital pen down and $p = 0$ to indicate digital pen up. Global features are normalized in $[0, 1]$ with hyperbolic tangent. They can be divided into four categories:

- Time: 25 features related to the duration of the trajectory, events such as raising the digital pen, or local maximums.
- Velocity and acceleration: 25 features obtained from the first and second order temporal derivatives of position-temporal functions.

- Direction: 18 features extracted from the trajectory, for instance the starting and average direction.
- Geometry: 32 features associated with the line or aspect of the dynamic trajectory.

In this study we adapted the extraction of global features from [8] to fishing vessel's trajectories. We considered the following input parameters: $x(t)$ and $y(t)$ representing the GPS vessel's position converted to nautical miles, $p(t)$ analogous to the stylus pressure, with $p > 0$ to indicate vessel at fishing speed, and $p = 0$ to indicate vessel at navigation speed, $TSAMPLE$, *i.e.* the average sampling period, and a time vector indicating the real-time instant of each point, since it is not equally distributed as assumed in the original function. The output was the *100*-dimension vector proposed in [8], with no elimination of any global feature.

Local Features. We adapted the set of local features proposed in [8]. This set of features was an extension of the set described in [4], comprising seven discrete time functions extracted from the trajectory and pressure of the digital pen, their first and second order derivatives, and other for a total of 27 features. A detailed description of the global and local features extracted can be found in [8].

3 Experiments and Results

We evaluated the performance of several supervised learning classifiers according to the mean accuracy obtained with *10*-fold stratified cross-validation (CV). We split our database into training and test sets with a 70:30 ratio. We considered three classifiers for global features, and set optimal parameters with CV: *i)* Support Vector Machine (SVM) with Gaussian kernel, as data are not linearly separable, $C = 100$ and $\gamma = 0.01$, *ii)* Random Forest (RF) with 101 estimators, and *iii)* Multilayer Perceptron (MLP) with hidden layer of size $10,000$, Rectified Linear Unit (ReLU) activation function, max iterations $= 1,000$, and $\alpha = 0.0001$.

For classification with local features, we used a Bidirectional Gated Recurrent Unit (BiGRU) with the following layers: *i)* input layer, *ii)* masking layer, to ignore trajectory positions without information, *iii)* bidirectional layer surrounding a GRU layer, and *iv)* fully connected layer with 5 outputs and softmax activation function. Finally, the possibility of RF and BiGRU fusion at score level was also investigated. It consists in a weighted sum of the scores provided by RF (s_{RF}) and BiGRU (s_{BiGRU}) classifiers ($s_{fusion} = 0.2s_{BiGRU} + 0.8s_{RF}$).

The mean accuracy obtained with *10*-fold CV is reported in Table 3 for the different classifiers. The best accuracy provided by a single classifier was 86.22%, obtained with RF with the Global feature set. The MLP and SVM classifiers showed lower performances with 82.69% and 83.16% respectively. The BiGRU classifier provided 75.6% of accuracy using the Local feature set. However, the best performance is obtained when combining the Global and Local feature set scores. We considered at the same time global and local features by fusing

Table 3. Classification accuracy of the different approaches.

Classifier	Features	Mean Acc. [%]
BiGRU	Local	75.60
MLP	Global	82.69
SVC	Global	83.16
RF	Global	86.22
RF + BiGRU	Fusion	90.13

the scores provided by RF and BiGRU, and obtained an increase of accuracy above 90%. This is a relative error reduction of 28%. We consider it a promising result and expect to increase it with improvements in feature selection and data availability.

In Fig. 3 we provide the confusion matrices obtained for the five fishing gears with the following classifiers: *i)* RF (left), *ii)* BiGRU (center), and *iii)* fusion of

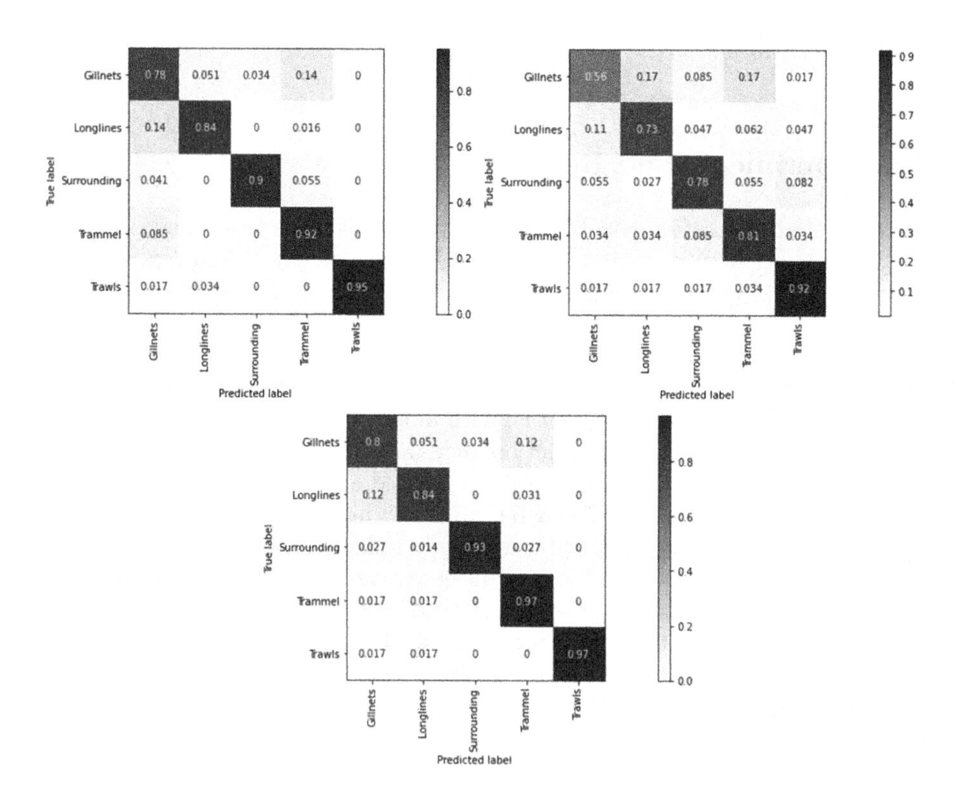

Fig. 3. Confusion matrices obtained for RF based on global features (left), BiGRU based on local features (center), and fusion of RF and BiGRU based on both features (right).

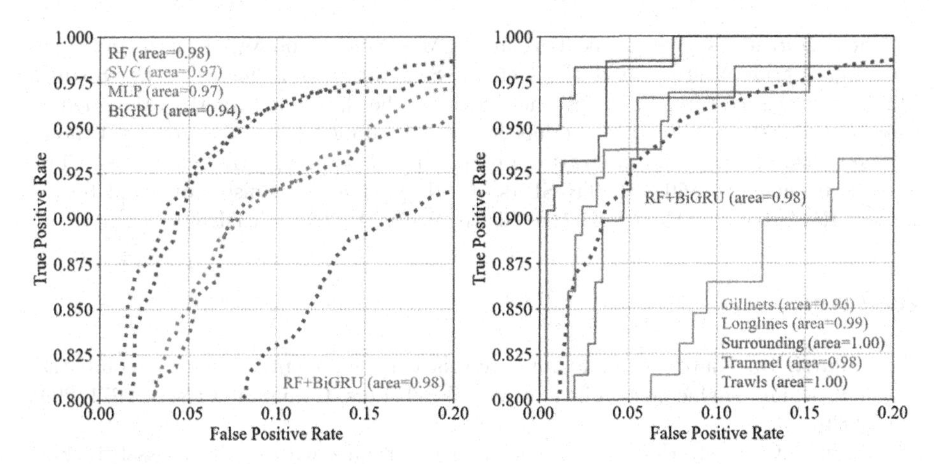

Fig. 4. ROC curves of different classifiers (left) and different fishing gears in the case of fused scheme with RF + BiGRU (right).

RF and BiGRU at score level (right). We observe in Fig. 3 (right) that we obtain accuracy of 93% for the "surrounding" class, greater than the total accuracy of the classifier (90.13%). This is not the case of [1], where time series with two-hours sampling period have been employed. Such sampling interval is too big to classify the "surrounding" gear ("purse seine" in [1]) with an accuracy similar to the other classes.

Finally, in Fig. 4 we report the Receiver operating characteristic (ROC) curves of different classifiers (left) and different fishing gears when RF + BiGRU is considered (right). Trammels consist in a variant of gillnets. This may explain the lower area under the curve obtained for the "gillnets" gear and the error of 12% between "gillnets" and "trammels" provided by the confusion matrix in Fig. 3 (right).

4 Conclusions

This work present a new database to train and evaluate fishing vessel classification from GPS trayectories. The database comprises more than one thousand trajectories recorded from 357 fishing vessels and 5 different fishing gears. This database reduce by more than 10 the Nyquist bandlimit of existing databases and represent a new resource to combat illegal fishing activities.

In this study we carried out the analysis of fishing vessel's dynamic trajectories to classify them according to fishing gear classes. The kinematics of vessels have been modeled according to the global and local features extracted from their trajectory, exploiting the analogy with the problem of dynamic handwritten signature recognition. The results obtained in terms of accuracy with multiple supervised learning classifiers confirm our choice to adapt the feature extraction process proposed in [8] to the analysis and classification of fishing vessel's trajectories.

Acknowledgments. We thank Tragsatec's Management of Agricultural and Fisheries Information Systems and the General Secretariat of Fisheries of the Spanish Ministry of Agriculture, Fisheries and Food for the data and expertise provided to carry out the study. This project has received funding from the European Union's Horizon 2020 research and innovation programme under the Marie Skłodowska-Curie grant agreement No 860813 - TReSPAsS-ETN. This study is also supported by the project INTER-ACTION (PID2021- 126521OB-I00 MICINN/FEDER).

References

1. Chuaysi, B., Kiattisin, S.: Fishing vessels behavior identification for combating IUU fishing: enable traceability at sea. Wirel. Pers. Commun. **115**(4), 2971–2993 (2020)
2. Dunn, D.C., et al.: Empowering high seas governance with satellite vessel tracking data. Fish Fish. **19**(4), 729–739 (2018)
3. European Union: Regulation (EU) no 1379/2013 of the European parliament and of the council of 11 December 2013 on the common organisation of the markets in fishery and aquaculture products, amending council regulations (EC) no 1184/2006 and (EC) no 1224/2009 and repealing council regulation (EC) no 104/2000 (2013). https://eur-lex.europa.eu/legal-content/EN/TXT/?uri=celex%3A32013R1379. Accessed 21 Nov 2022
4. Fierrez, J., Ortega-Garcia, J., Ramos, D., Gonzalez-Rodriguez, J.: Hmm-based on-line signature verification: feature extraction and signature modeling. Pattern Recogn. Lett. **28**(16), 2325–2334 (2007)
5. Food and Agriculture Organization of the United Nations: What is IUU fishing? (2022). https://www.fao.org/iuu-fishing/background/what-is-iuu-fishing/en/. Accessed 17 Nov 2022
6. Jarvis, R.M., Young, T.: Pressing questions for science, policy, and governance in the high seas. Environ. Sci. Policy **139**, 177–184 (2023)
7. Leonard, K.: Schaum's Outline of Business Statistics, 4th edn. (2003)
8. Martinez-Diaz, M., Fierrez, J., Krish, R.P., Galbally, J.: Mobile signature verification: feature robustness and performance comparison. IET Biometrics **3**(4), 267–277 (2014)
9. Marzuki, M.I., Gaspar, P., Garello, R., Kerbaol, V., Fablet, R.: Fishing gear identification from vessel-monitoring-system-based fishing vessel trajectories. IEEE J. Oceanic Eng. **43**(3), 689–699 (2017)
10. Miller, R.J., et al.: Big data curation. In: COMAD, p. 4 (2014)

Multi-view Infant Cry Classification

Yadisbel Martinez-Cañete[1](✉), Hichem Sahli[2,3](✉), and Abel Díaz Berenguer[2](✉)

[1] Universidad de Oriente, Las Americas Ave, 90900 Santiago de Cuba, Cuba
ymartine@etrovub.be
[2] Department of Electronics and Informatics, Vrije Universiteit Brussel, Pleinlaan 2, 1050 Brussels, Belgium
{hsahli,aberengu}@etrovub.be
[3] Interuniversity Microelectronics Centre, Kapeldreef 75, 3001 Heverlee, Belgium

Abstract. This paper addresses infant cry classification in multi-view settings, that is, settings where the typical low-level representations, commonly used for audio recognition tasks, are considered as different views of the target data. We show that through the use of multi-view methods, such as Structured Latent Multi-View Representation Learning, we are able to reliably discriminate between normal and pathological infant cry signals. Extensive experimental results on two benchmark infant cry data sets indicate that the proposed method outperforms state-of-the-art models.

Keywords: Classification · Infant cry · Multi-view · Representation learning

1 Introduction

Newborn cry is an essential means of communication available to babies during early infancy, from birth to three months of age. It allows them to interact with the outside environment. At this early stage of life, the infant's cry constitutes a fundamental physiological action. It might be sparked due to health-related or psychological factors. Indeed, infant cry signals hold essential information, ranging from health problems, sleep issues, hunger, or asphyxia to emotional states of high arousal triggered by the nervous system due to stimulations motivated by unusual situations [1,2]. Therefore, correctly recognizing the cues delivered by infant crying is paramount in ensuring the children's survival, health, well-being, and development.

To meet their needs at this early stage of their development, infants almost entirely depend on parents, doctors, nurses, or, roughly speaking, on their caregivers for interpreting their cry. However, infant cry can be differentiated auditorily by trained adult listeners [3]. Consequently, automatically identifying the

Thanks to VLIRUOS for financial support in the framework of the Institutional University Cooperation project with Universidad de Oriente, Cuba.

A. Pertusa et al. (Eds.): IbPRIA 2023, LNCS 14062, pp. 639–653, 2023.
https://doi.org/10.1007/978-3-031-36616-1_51

hidden patterns in the audio signals of infants cry is of great significance to providing non-experienced caregivers with the tools for better understanding infants and timely acting upon cry triggering factors.

Automatic Infant Cry Recognition (AICR) methods can be roughly categorized into hand-crafted and data-driven [4]. Hand-crafted AICR methods generally consist of the following sequential processing steps: i) pre-processing, ii) cry signal segmentation, iii) hand-crafted feature extraction, iv) feature selection, and v) classification using machine learning. A major issue of these methods is that they typically use different types of features extracted from the input signal, and the integration of this variety of features is often done by a simple concatenation diminishing the capacity to encode complementary information from the different types of low-level features, as to reveal the underlying representations of the cry signals.

Data-driven AICR models also involve i) pre-processing, ii) cry signal segmentation, and iii) classification but are capable of learning useful feature representation from data. These methods mostly rely on deep learning techniques. Two categories of methods are often used, either end-to-end methods or two steps methods that first focus on learning deep feature representations which are subsequently used for classification [5,6]. Data-driven AICR models typically built on the outstanding success of Convolutional Neural Networks (CNN) architectures through the use of Spectrogram-based images for infant cry recognition [7,8]. However, while Spectogram-based AICR deep learning methods have shown promising results, they do not comprehensively leverage different types of predictive features from cry signals proven effective for cry signal recognition [9,10].

Moreover, the recognition results obtained from different low-level features vary in the sense that one feature is good for some target classes while another is good for some other target classes. Intuitively, owing to their complementary, jointly learning from these features should leverage their individual strength and gives rise to performance gain on cry classification. However, in practice, this depends on the fusion method. Naive approaches, such as often-seen concatenation or late fusion, can result in decreased performance rather than improvement, and the multi-view methods can be worse than the best single-view [11,12]. The reasons for degradation might be various; for example, single-view subnetworks can learn at different rates, and one can converge while another overfit, or simply because the multi-view method cannot capture the proper interaction between the views to exploit their strengths.

To alleviate the above problem of integrating different types of features and motivated by the significant success of multi-view learning methods [13,14] on a broad array of application domains [15], in this paper, we propose to adopt a multi-view strategy to encode the information from different type of features into a unique latent representation. Our proposal combines domain knowledge engineering using different hand-crafted feature domains per view and deep learning techniques by utilizing neural networks to encode the information from each feature domain into a discriminate latent space for the cry signal representation.

We build upon the recent work of Zhang *et al.* [14] to map the training and testing samples into a latent space, where the latent representations are expected to encode complementary information from the different types of features with promising structures revealing the underlying class distribution. Our experiments on two benchmark data sets for infant cry classification tasks show that the proposed multi-view embedding learned via the Cross Partial Multi-View Networks (CPM-Nets) approach of Zhang *et al.* [14–16], consistently results in better performance than the obtained by the single-view baselines and previous state-of-the-art methods.

1.1 Multi-view Learning

Multi-view learning (MVL) aims to learn the common feature or shared patterns by combining multiple views from different data modality sources and can learn more revealing and compressed representation, improving predictors' performance [13]. Multi-view Deep Learning can be useful for analyzing the implicit feature correlation and internal dynamic changes in data and solving the incompleteness and uncertainty in sequential data analysis [17]. The multi-view representation learning algorithms have shown promising performance in different applications of audio signals for diverse problems [12,18–20,22,23]. For a comprehensive overview of multi-view methods and models, the interested reader is referred to [13].

In the area of sound classification, Casseber *et al.* [18,19] proposed multiview networks for sound classification with multiple sensors that can handle channel disturbances in a dynamic environment with simulated Room Impulse Responses several times larger than Short Time Fourier Transform (STFT) frames, by using the capability of Recurrent Neural Networks (RNN) to generalize to sequences of any length by unrolling across time and channels. In the same area, Singh *et al.* [20] proposed a feature representation framework for acoustic scene classification using various intermediate levels of the pre-trained CNN SoundNet [21]. The combined features from the intermediate layers provided better discrimination as compared to the features from each of the individual layers. In their strategy, one could consider that the features provide different types of abstraction and exhibit complementary information because they were obtained from different intermediate layers of a pre-trained model.

More focused on the audio scene recognition area, specifically music emotion recognition, Chandrakala [22] proposed two variants of the multi-view representation approach achieving promising experimental results. Their first approach combines the auditory image-based features (spectrogram) and the cepstral features from sound signals. Their second approach combines the statistical features extracted from the auditory images and the cepstral features of the sound signals. The authors also explored other visual image-based features such as constant Q-transform and variable Q-transform. In the work of He and Fergunson [23], the authors introduced a multi-view CNN model to extract feature representations from the raw audio signals. Based on different kernel sizes, their model was defined for two parallel CNN modules, fine-view CNN and coarse-view CNN. The advantage of these models is that they do not require the use of human-engineered audio features. Phan *et al.* [12] proposed a multi-view embedding model for audio and music classification. Their approach considers four low-level features, most generally used for audio and music analysis, Mel-scale spectrogram, Gammatone spectrogram, CQT spectrogram, and raw waveform. The architecture of their model consists of four subnetworks, each of which is designed to process one of the low-level inputs, and the multi-view embedding is formed by concatenating the embeddings learned by the view-specific subnetworks. Their experimental results demonstrated the model outperforms the single-view baselines but also is superior to the multi-view baselines based on concatenation and late fusion.

2 Proposed Method

In general, infant cry is a combination of vocalization, silence, coughing, and interruptions, which includes a diversity of acoustic and prosodic information at different levels. In this work, we adopt the four categories of the audio features that are applied to research related to infant cry analysis. Namely, prosodic features, cepstral domain features, spectral domain features, and image-based features. They are considered as different views of the underlying data distribution of cry signal classification. The proposed latent-representation-based cry signal classification is illustrated in Fig. 1. It is composed of three components in the training stage, as shown in Fig. 1. First, based on the CPM-Nets [14], we learn latent representations. Second, for the consistency of the latent space between [14,16] training and testing, we train a projection model termed as Latent-representation regressor between the four types of features and the latent representations. Third, a latent-representation-based classifier for infant cry classification is trained. Accordingly, in the testing stage, the original features are projected into the latent space with a latent-representation regressor, and then the final classification results are obtained with the latent-representation-based classifier (see Fig. 1. Inference). In the following, we give an overview of the different steps, the reader is referred to [15] for the details.

2.1 Problem Statement

Our aim is a multi-class classification, by subjects, of infant cry into five classes (asphyxia, pain, normal, deaf, and hungry) by subjects or patients. In our study, 4 types of features are extracted from four domains (views), namely cepstral, spectral, image, and prosodic. Given the training set $\{X_n, y_n\}_{n=1}^{N}$, where $X_n = \{{x_n}^{(v)}\}_{v=1}^{V}$ is a multi-view sample and is the corresponding class label $y_n \in \{1, \cdots, C\}$. With, N the number of training samples, V the number of views, and C the number of classes.

2.2 Method

Our method for learning multi-view embedding for cry classifications adopts the CPM-Nets proposed in [14]. Illustrated in Fig. 1, it consists of 3 steps in the training stage:

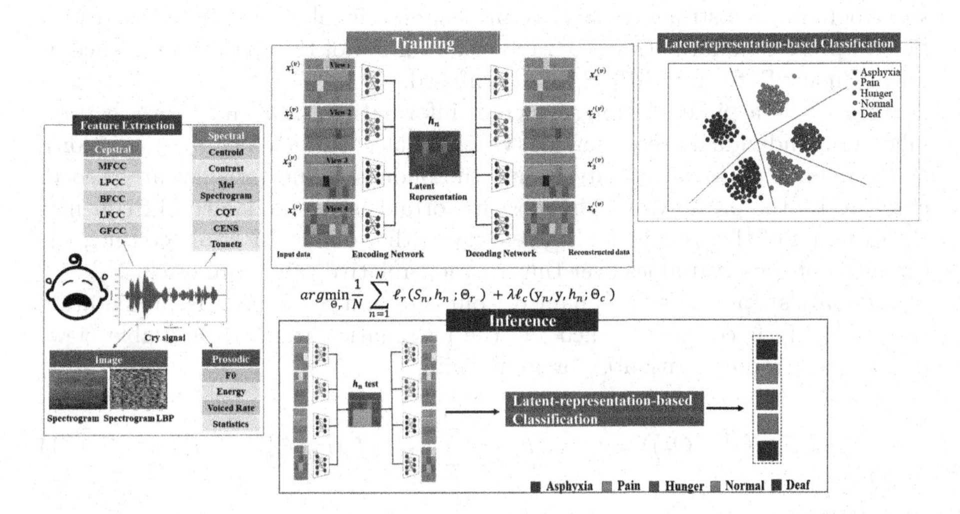

Fig. 1. Illustration of the proposed multi-view cry infant classification framework (adapted from [15]).

First step, is the feature extraction step consisting of extracting the features from the 4 domains and concatenating them into sub-arrays of $M \times S$, with M the dimension of the feature vectors and S the number of subjects or samples.

Cepstral features are calculated by applying a pre-emphasis filter on an audio signal, taking the STFT of that signal, applying the filter banks according to the features, taking a Discrete Cosine Transform (DCT), and normalizing the output. The Mel-Frequency Cepstral Coefficient (MFCC) and Linear Predictive Cepstral Coefficients (LPCC) are widely used in infant cry recognition [24–27]. However, recent researchers have obtained better results by combining these features with other cepstral features, such as Bark-Frequency Cepstral Coefficients (BFCC), Linear-Frequency Cepstral Coefficients (LFCC) [28,29], and Gammatone-Frequency Cepstral Coefficients (GFCC), the latter is mostly used in recognition tasks on speech signals [30]. We decide to include in our experiments the five cepstral features MFCC, LPCC, BFCC, LFCC, and GFCC, using different filters Mel, linear, bark, linear-frequency, and gamma, respectively.

Spectral analysis refers to the process of studying the frequency content of a signal. It is used to determine the relative distribution of energy in a signal across different frequencies. Spectral analysis can be performed using various techniques, including the Fourier transform, the STFT, and the wavelet transform, among others. The result of spectral analysis is typically a plot or graph of the signal's frequency content, known as a spectrum.

The image domain features can be represented differently, and the most popular is the spectrogram image. A spectrogram provides a visual representation of how the frequency content of a signal changes over time, making it a useful tool for analyzing signals with time-varying spectral content, such as speech signals, music signals, and signals in the biomedical domain. It is recognized that a spectrogram has a strong capacity to present the signal and include both acoustic and prosodic information. Extra transformation of the spectrogram, such as Local Binary Pattern (LBP) [8] has been used.

Prosody is a piece of suprasegmental information, over and above features inherent in individual speech sounds (voicing, place, and manner of articulation), and is defined on segments larger than the phones. Some variables are used to characterize the prosody, but the most important is the fundamental frequency, the duration of the sounds, and the energy of the sounds. Most of the time, the evolution of these variables over time or their relative values are used.

Second step, this step is named Complete and Structured Representation Learning [14], it consists in encoding the information from the available views into a latent representation h_n by minimizing:

$$\Delta(S_n, f(h_n; \Theta_r)) = \ell_r(S_n, h_n) = \sum_{v=1}^{V} s_{nv} \| f_v(h_n; \Theta_r^{(v)}) - x_n^{(v)} \|^2 \tag{1}$$

where $\ell_r(S_n, h_n)$ represents the reconstruction loss, s_{nv} is an indicator of the availability for the nth sample in the vth view. $f_v(h_n; \Theta_r^{(v)})$ is the reconstruction network for the vth view parameterized by $\Theta_r^{(v)}$, $f(\cdot)$ parameters governing the mapping from common representation h_n that encodes comprehensive information from different available views, and different samples. To make the learned latent representation consistent with the different classes we define:

$$\ell_c(y_n, y, h_n) = \max_{y \in Y}(0, \Delta(y_n, y)) + E_{h\tau(y)} \phi(h; \Theta_c)^T \phi(h_n; \Theta_c)^T - E_{h\tau(y_n)} \phi(h; \Theta_c)^T \phi(h_n; \Theta_c)^T \tag{2}$$

where $\phi(\cdot; \Theta_c)$ is the mapping function from h to the class prediction output with parameters Θ_c and $\tau(y)$ the set of latent representations from class y. Thus, to jointly consider informativeness and class labels separability, the following objective function is optimized:

$$\arg\min_{\Theta_r} \frac{1}{N} \sum_{n=1}^{N} \ell_r(S_n, h_n; \Theta_r) + \lambda \ell_c(y_n, y, h_n; \Theta_c) \tag{3}$$

where $\lambda > 0$ balances the confidence margin of information from multiple views and class labels.

Third step, denoted as Latent-Representation-Based Classifier dealing with training a classifier, using as inputs the latent representation, to classify the input infant cry to one of the classes asphyxia, deaf, hunger, normal, or pain. In our implementation we investigated several state-of-the-art classifiers, namely, Support Vector Machine (SVM), Ridge Classifier (RC), and Logistic-Regression Classifier (LR).

3 Experimental Results

3.1 Data Sets

We conducted experiments on publicly available data sets Baby Chillanto [31] and donateacry-corpus[1]. Baby Chillanto is a data set from the National Institute of Optical and Electronic Astrophysics (INAOE) - CONACYT, Mexico. It consists of 138 crying signals sampled at 16000 Hz and categorized into 5 classes and distributed as follows: 21 normal, 34 hunger, 52 deaf, 6 asphyxia, and 25 pain. The donateacry-corpus is a data set of infant cries recorded by mobile phone and uploaded by volunteers. It consists of 457 babies crying signals distributed into five classes: 16 belly pain, 8 burping, 27 discomfort, 382 hungry, and 24 tired.

3.2 Feature Extraction

As mentioned in previous sections, we use prosodic, cepstral domain features, spectral domain features, and image-based features. These features are computed as follows:

Cepstral, MFCC, LPCC, LFCC, BFCC, and GFCC were extracted using the Spafe library[2]. For all these features, we extracted 13 coefficients for each frame of 50ms as window size and 10ms as the hop size, with hamming as the window function.

We used a linear layer for each view to obtain the latent representation. The cepstral features were extracted as a matrix with dimensions equal to the number of frames $\times$ 13 coefficients, and the results were embedded into a 1D vector features. The number of frames depends on the signal length. For instance, for 1 s signal length the 1D vector has a dimension 1248 (i.e., 96×13), for 20 s signal length the vector dimension has a dimension 25948 (1996×13) and 7 s signal length 9048 (696×13).

Spectral, 6 spectral features were generated via Librosa library[3]. Namely, spectral centroid, spectral contrast, mel spectrogram, constant-Q chromagram, chroma energy normalized statistics, and tonal centroid. We set the time frames $T \in 32, 219, 626$ for 1 s, 7 s, and 20 s signal length. The obtained 2D features for spectral contrast, mel spectrogram, constant-Q chromagram, chroma energy normalized statistics, and tonal centroid are with size $7 \times$ T, $128 \times$ T, $12 \times$ T, $12 \times$ T, $6 \times$ T respectively and the matrices were embedded into a 1D feature vector.

Image, we used the OpenSoundscape library[4] to extract the spectrogram image and the Local Binary Pattern (LBP) transformation in gray-scale, using a Hamming window with a window size of 50 ms and a hop size of 10 ms. Finally, we reshaped the spectrogram image to (224×224). The images were embedded into a 1D feature vector with size 50176.

[1] https://github.com/gveres/donateacry-corpus.
[2] https://github.com/SuperKogito/spafe.
[3] https://librosa.org/.
[4] https://github.com/kitzeslab/opensoundscape.

Prosodic, we used the Disvoice library[5] to extract fundamental frequency (F0), energy, and voiced rate, and we computed their statistics: average, standard deviation, maximum, minimum, skewness, and kurtosis.

30 values for relative to F0, F0-contour, Tilt of F0 in voiced segment, MSE of F0 in voiced segment, F0 on the first voiced segment, F0 on the last voiced segment and theirs 6 statistical values.

47 for energy, energy-contour, Tilt of energy, MSE of energy, energy in the first segment, energy in the last segment, energy in the voiced and unvoiced segments, and theirs 6 statistical values.

29 for voiced rate and duration of voiced, unvoiced, pauses with their 6 statistical values per each one and duration ratios.

We also considered the following features: (Voiced+Unvoiced), Pause/Unvoiced, Unvoiced/(Voiced+Unvoiced), Voiced/(Voiced+Unvoiced), Voiced/Pause, Unvoiced/Pause. In summary, 103 prosodic features have been used.

3.3 Experimental Setting

For the Baby Chillanto data set, state-of-the-art methods [7,26,32,33] evaluated short-term time signals corresponding to 1 s of recordings. Thereby, we also evaluated the short-term time scenario using 1 s signals. Moreover, in this work, we evaluated the proposed model's effectiveness for recognizing long-term time signals because other relevant patterns, such as the frequency measures, and the melody contour, can be only recognized in a long sequence of the acoustic signal. Thus, based on the signal duration statistics of the Baby Chillanto, we decided to use 20 s as the long-term signal length. Signals shorter than 20 s were zero-padded at the beginning, and those longer than 20 s were truncated. For the Donateacry-corpus, we followed previous studies [9,34–36] who conducted experiments on long-term signals of 7 s, we also considered 1 s signals for the sake of completeness of our experiments.

The data sets were randomly divided into 80% and 20% for training and testing, respectively. We tuned the parameters with a 5-fold cross-validation strategy on the training data. For the CPM-Nets, we empirically set the dimensionality (i.e., k) of the latent representation to 128 for short-term time signals (i.e., 1 s) and to 512 for long-term time signals (i.e., 7 and 20 s signals). The hyper-parameter λ (see Eq. 3) was set to 10 for all the experiments conducted in this work. For the representation-based classification, we evaluated three classifiers: Support Vector Machine (SVM), Ridge Classifier (RC), and Logistic-Regression Classifier (LR). The performance of the evaluated classifiers is reported in terms of Accuracy, in order to compare with the state-of-the-art methods reported in the literature, even when this measure is inadequate for imbalanced data.

[5] https://pypi.org/project/disvoice-prosody/.

3.4 Multi-view Based Classification

We conducted experiments using different feature domain combinations to validate our proposal to harness audio features complementarity for multi-class infant cry classification. In particular, we evaluated: i) integrating two views using cepstral and spectral, ii) integrating three views employing cepstral, spectral, and image, and iii) integrating four views using cepstral, spectral, image, and prosodic.

The results on Baby Chillanto for long-term time and short-term time signals are summarized in Fig. 2 and Fig. 3, respectively. As one can observe, regardless of the classifier used and the input signal length, our method's performance almost steadily increased by exploiting features from more domains to encode their complementary statistics into the latent representation. Furthermore, analyzing Fig. 4 one can also notice that our multi-view method outperformed the previous state-of-the-art methods.

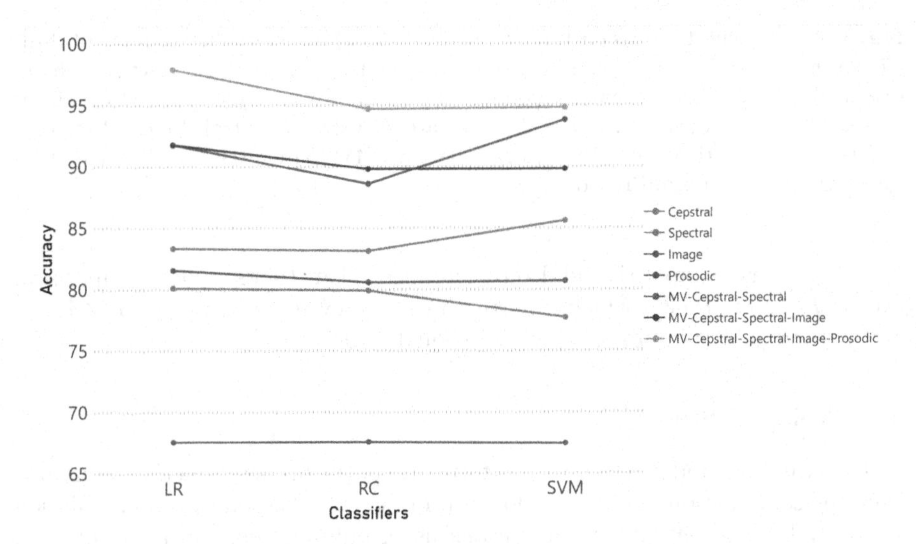

Fig. 2. Experimental results on Baby Chillanto in 20 s. We report the accuracy of three off-the-shell classifiers using four domains independently in a single-view (i.e., baseline single-view) classification setting and adopting multi-view (MV) based classification as proposed in this work. The domains are: Cepstral, Spectral, Image, Prosodic, and their combination are MV-Cepstral-Spectral, MV-Cepstral-Spectral-Image, MV-Cepstral-Spectral-Image-Prosodic.

In Fig. 5 and Fig. 6 we report the results on Donateacry-corpus corresponding to the short-term time and long-term time signals. One can notice that the proposed multi-view method achieved its highest effectiveness when the latent representation encoded information from the four domains. Besides, compared to the state-of-the-art in this data set, one can see in Fig. 7, that the proposed

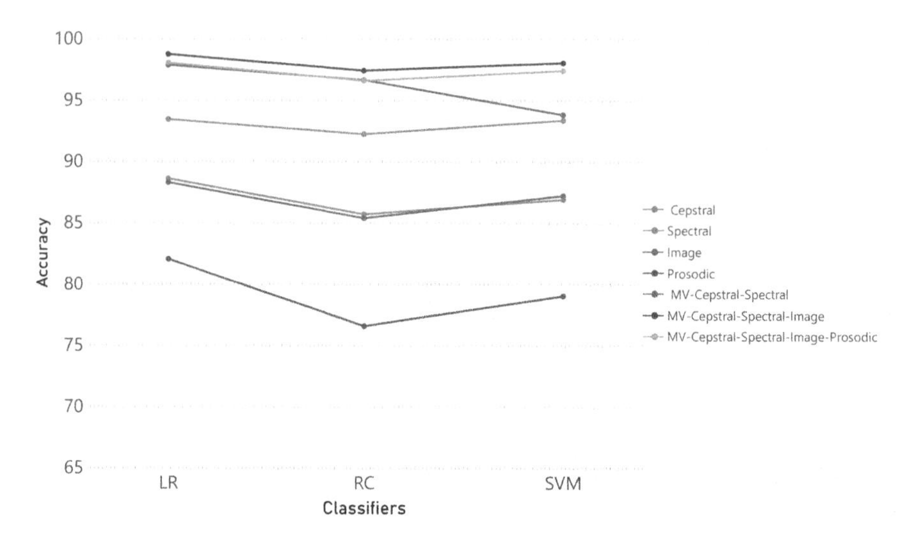

Fig. 3. Experimental results on Baby Chillanto in 1 s. We report the accuracy of three off-the-shell classifiers using four domains independently in a single-view (i.e., baseline single-view) classification setting and adopting multi-view (MV) based classification as proposed in this work. The domains are: Cepstral, Spectral, Image, Prosodic, and their combination are MV-Cepstral-Spectral, MV-Cepstral-Spectral-Image, MV-Cepstral-Spectral-Image-Prosodic.

method is very competitive and outperformed all methods under comparison, except Jiang *et al.* [36], which reported an accuracy score equal to 93.43% and our method achieved an accuracy score equal to 93%.

3.5 Ablation Study

To better understand whether the performance of the proposed method stemmed from the combination of different domains or not, we also conducted an ablation study by learning latent representations using only one domain as input (i.e., v=1). The results on Baby Chillanto and donateacry-corpus for long-term and short-term time signals are shown in Fig. 2, Fig. 3, Fig. 5 and Fig. 6. One can see that the proposed method does not perform adequately using only one domain as input. It suggests that the effectiveness of the proposed method for infant cry classification is derived from its capacity to exploit domains' complementary information demonstrating that the multi-view feature representation adopted in this work is effective for infant cry classification.

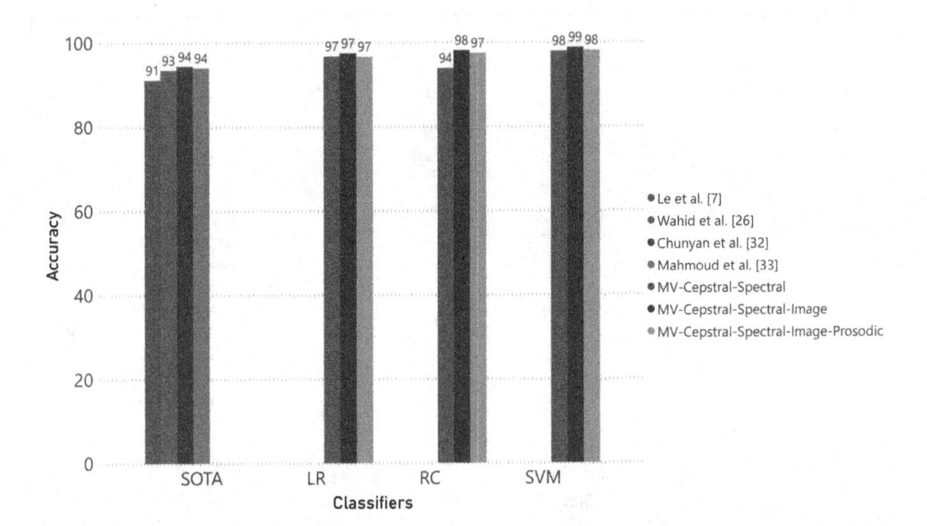

Fig. 4. Comparison with state-of-the-art methods on Baby Chillanto in 1 s with four domains multi-view based classification MV-Cepstral-Spectral, MV-Cepstral-Spectral-Image, MV-Cepstral-Spectral-Image-Prosodic achieved the highest performance and outperformed most previous methods.

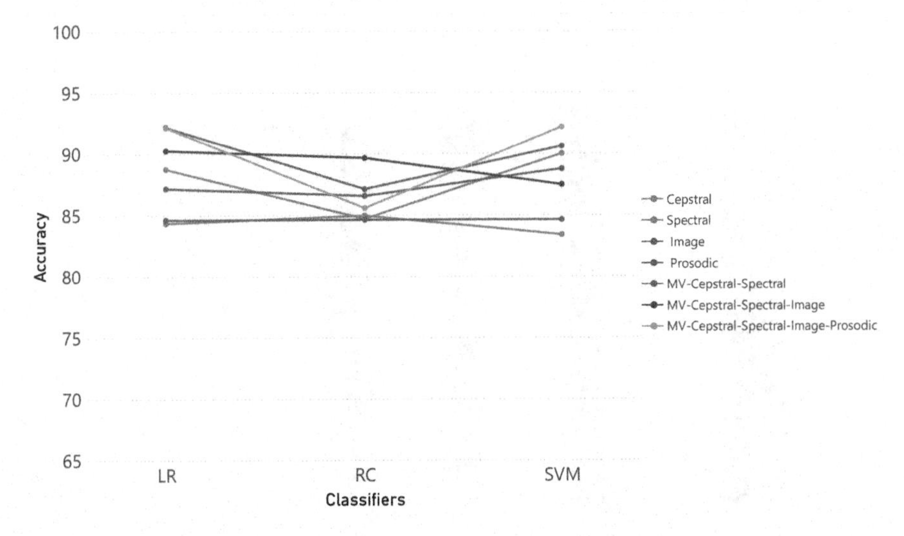

Fig. 5. Experimental results on donateacry-corpus in 7 s. Accuracy scores of three off-the-shell classifiers using four domains independently for single-view classification versus our proposal using multi-view (MV) based classification. The domains are: Cepstral, Spectral, Image, Prosodic, and their combination are MV-Cepstral-Spectral, MV-Cepstral-Spectral-Image, MV-Cepstral-Spectral-Image-Prosodic.

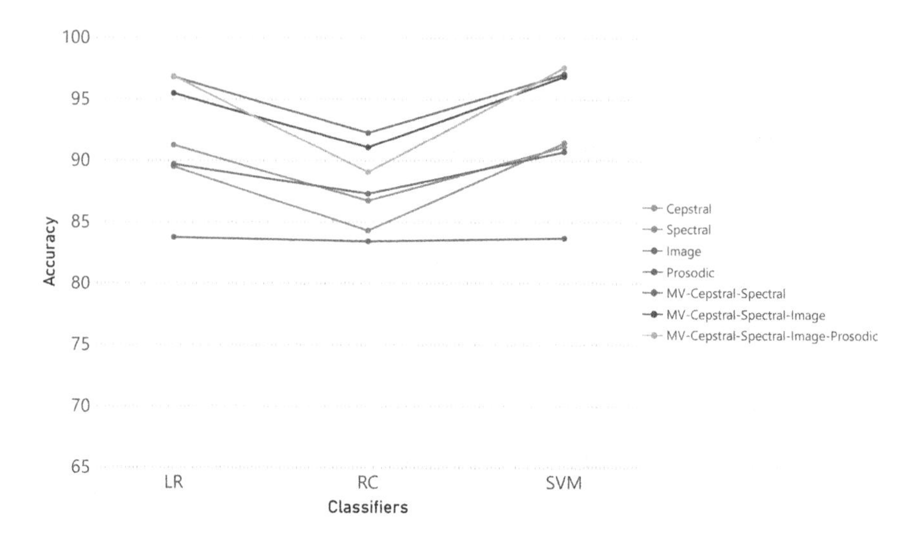

Fig. 6. Experimental results on donateacry-corpus in 1 s. We report the accuracy of three off-the-shell classifiers using four domains independently for single-view classification and our proposal utilizing multi-view (MV) based classification. The domains are: Cepstral, Spectral, Image, Prosodic, and their combination are MV-Cepstral-Spectral, MV-Cepstral-Spectral-Image, MV-Cepstral-Spectral-Image-Prosodic.

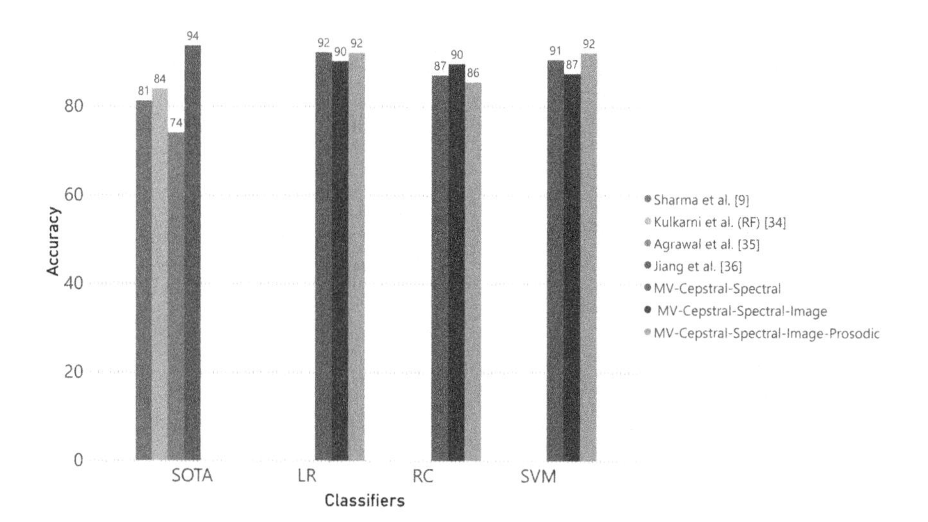

Fig. 7. Comparison with state-of-the-art methods on donateacry-corpus in 7 s with four domains multi-view based classification MV-Cepstral-Spectral, MV-Cepstral-Spectral-Image, MV-Cepstral-Spectral-Image-Prosodic achieved the highest performance and outperformed most previous methods.

4 Conclusions

We presented a multi-view learning method for infant cry classification in this work. The multi-view latent representation was produced using the CPM-Nets [14]. In this manner, the different audio domain features contributed to effective cry classification by jointly considering informativeness and separability between the classes. The efficacy of the proposed method was demonstrated in two infant cry data sets and outperformed most state-of-the-art. Our future work will consider data augmentation to increase the number of patient signals, balance the data and extend the views to other acoustic feature domains.

References

1. LaGasse, L.L., Neal, A.R., Lester, B.M.: Assessment of infant cry: acoustic cry analysis and parental perception. Ment. Retard. Dev. Disabil. Res. Rev. **11**(1), 83–93 (2005)
2. Zeman, J.: Emotional development-early infancy, later infancy months. JRank Psychology Encyclopedia. https://psychology.jrank.org/pages/212/Emotional-Development.html. Accessed 29 Nov 2022
3. Bashiri, A., Hosseinkhani, R.: Infant crying classification by using genetic algorithm and artificial neural network. Acta Medica Iranica 531–539 (2020)
4. Ji, C., Mudiyanselage, T.B., Gao, Y., Pan, Y.: A review of infant cry analysis and classification. EURASIP J. Audio Speech Music Process. **2021**(1), 1–17 (2021). https://doi.org/10.1186/s13636-021-00197-5
5. Chunyan, J., Xueli, X., Sunitha, B., Yi, P.: Deep learning for asphyxiated infant cry classification based on acoustic features and weighted prosodic features. In: 2019 International Conference on Internet of Things and IEEE Green Computing and Communications and IEEE Cyber. Physical and Social Computing and IEEE Smart Data, Atlanta, USA, pp. 1233–1240. IEEE (2019)
6. Maghfira, T.N., Basaruddin, T., Krisnadhiand, A.: Infant cry classification using CNN-RNN. In: Journal of Physics: Conference Series, vol. 1528, pp. 012–019 (2020)
7. Le, L., Kabir, A.N.M., Ji, C., Basodi, S., Pan, Y.: Using transfer learning, SVM, and ensemble classification to classify baby cries based on their spectrogram images. In: 2019 IEEE 16th International Conference on Mobile Ad Hoc and Sensor Systems Workshops (MASSW), Monterey, USA, pp. 106–110. IEEE (2019)
8. Felipe, G.Z., et al.: Identification of infants' cry motivation using spectrograms. In: 2019 International Conference on Systems. Signals and Image Processing (IWSSIP), Osijek, Croatia, pp. 181–186. IEEE (2019)
9. Sharma, S., Mittal, V.K.: A qualitative assessment of different sound types of an infant cry. In: 2017 4th IEEE Uttar Pradesh Section International Conference on Electrical. Computer and Electronics (UPCON), Mathura, India, pp. 532–537. IEEE (2017)
10. Dewi, S.P., Prasasti, A.L., Irawan, B.: The study of baby crying analysis using MFCC and LFCC in different classification methods. In: 2019 IEEE International Conference on Signals and Systems (ICSigSys), Bandung, Indonesia, pp. 18–23. IEEE (2019)
11. Wang, W., Tran, D., Feiszli, M.: What makes training multi-modal classification networks hard?. In: Proceedings of the IEEE/CVF Conference on Computer Vision and Pattern Recognition, Seattle, USA, pp. 12695–12705. IEEE (2020)

12. Phan, H., et al.: Multi-view audio and music classification. In: 2021 IEEE International Conference on Acoustics. Speech and Signal Processing (ICASSP), ICASSP 2021, Toronto, Canada, pp. 611–615. IEEE (2021)
13. Yan, X., Hu, S., Mao, Y., Ye, Y., Yu, H.: Deep multiview learning methods: a review. Neurocomputing **448**, 106–129 (2021)
14. Zhang, C., Han, Z., Fu, H., Zhou, J.T., Hu, Q.: CPM-Nets: cross partial multi-view networks. In: Advances in Neural Information Processing Systems, vol. 32 (2019)
15. Kang, H., et al.: Diagnosis of coronavirus disease 2019 (Covid-19) with structured latent multi-view representation learning. IEEE Trans. Med. Imaging **39**(8), 2606–2614 (2020)
16. Zhang, C., Cui, Y., Zongbo, Z.H., Zhou, J.T., Fu, H., Hu, Q.: Deep partial multi-view learning. IEEE Trans. Pattern Anal. Mach. Intell. **44**(5), 2402–2415 (2022)
17. Xie, Z., Yang, Y., Zhang, Y., Wang, J., Du, S.: Deep learning on multi-view sequential data: a survey. Artif. Intell. Rev. **56**, 6661–6704 (2022)
18. Casebeer, J., Luc, B., Smaragdis, P.: Multi-view networks for denoising of arbitrary numbers of channels. In: 2018 16th International Workshop on Acoustic Signal Enhancement (IWAENC), Tokyo, Japan, pp. 496–500. IEEE (2018)
19. Casebeer, J., Wang, Z., Smaragdis, P.: Multi-view networks for multi-channel audio classification. In: 2019 IEEE International Conference on Acoustics. Speech and Signal Processing (ICASSP), ICASSP 2019, Brighton, UK, pp. 940–944. IEEE (2019)
20. Singh, A., Rajan, P., Bhavsar, A.: Deep multi-view features from raw audio for acoustic scene classification. In: Detection and Classification of Acoustic Scenes and Events 2019, New York, USA. IEEE (2019)
21. Aytar, Y., Vondrick, C., Torralba, A.: SoundNet: learning sound representations from unlabeled video. In: Advances in Neural Information Processing Systems, vol. 29 (2016)
22. Chandrakala, S.: Multi-view representation for sound event recognition. SIViP **15**(6), 1211–1219 (2021). https://doi.org/10.1007/s11760-020-01851-9
23. He, N., Ferguson, S.: Multi-view neural networks for raw audio-based music emotion recognition. In: 2020 IEEE International Symposium on Multimedia (ISM), Naples, Italy, pp. 168–172. IEEE (2020)
24. Badreldine, O.M., Elbeheiry, N.A., Haroon, A.N.M., ElShehaby, S., Marzook, E.M.: Automatic diagnosis of asphyxia infant cry signals using wavelet based Mel frequency cepstrum features. In: 2018 14th International Computer Engineering Conference (ICENCO), Giza, Egypt, pp. 96–100. IEEE (2018)
25. Hariharan, M., et al.: Improved binary dragonfly optimization algorithm and wavelet packet based non-linear features for infant cry classification. Comput. Methods Programs Biomed. **155**, 39–51 (2018)
26. Wahid, N.S.A., Saad, P., Hariharan, M.: Automatic infant cry pattern classification for a multiclass problem. J. Telecommun. Electron. Comput. Eng. (JTEC) **8**(9), 45–52 (2016)
27. Martinez-Cañete, Y., Cano-Ortiz, S.D., Lombardía-Legrá, L., Rodríguez-Fernández, E., Veranes-Vicet, L.: Data mining techniques in normal or pathological infant cry. In: Hernández Heredia, Y., Milián Núñez, V., Ruiz Shulcloper, J. (eds.) IWAIPR 2018. LNCS, vol. 11047, pp. 141–148. Springer, Cham (2018). https://doi.org/10.1007/978-3-030-01132-1_16
28. Liu, L., Li, Y., Kuo, K.: Infant cry signal detection, pattern extraction and recognition. In: 2018 International Conference on Information and Computer Technologies (ICICT), Illinois, USA, pp. 159–163. IEEE (2018)

29. Liu, L., Li, W., Wu, X., Zhou, B.X.: Infant cry language analysis and recognition: an experimental approach. IEEE/CAA J. Automatica Sinica **6**(3), 778–788 (2019)
30. Patni, H., Jagtap, A., Bhoyar, V., Gupta, A.: Speech emotion recognition using MFCC, GFCC, chromagram and RMSE features. In: 2021 8th International Conference on Signal Processing and Integrated Networks (SPIN), Delhi, India, pp. 892–897. IEEE (2021)
31. Reyes-Galaviz, O.F., Cano-Ortiz, S.D., ReyesGarc'ıa, C.A.: Evolutionary-neural system to classify infant cry units for pathologies identification in recently born babies. In: 2008 Seventh Mexican International Conference on Artificial Intelligence, Atizapan de Zaragoza, pp. 330–335. IEEE (2008)
32. Chunyan, J., Chen, M., Bin, L., Pan, Y.: Infant cry classification with graph convolutional networks. In: 2021 IEEE 6th International Conference on Computer and Communication Systems (ICCCS), Chengdu, China, pp. 322–327. IEEE (2021)
33. Mahmoud, A.M., Swilem, S.M., Alqarni, A.S., Haron, F.: Infant cry classification using semisupervised k-nearest neighbor approach. In: 2020 13th International Conference on Developments in eSystems Engineering (DeSE), Wuhan, China, pp. 305–310. IEEE (2020)
34. Kulkarni, P., Umarani, S., Diwan, V., Korde, V., Rege, P.P.: Child cry classification-an analysis of features and models. In: 2021 6th International Conference for Convergence in Technology (I2CT), Pune, India, pp. 1–7. IEEE (2021)
35. Agarwal, P., Kumar, M., Sriramoju, V., Deshpande, K., Shaikh, N.: New-born's cry analysis using machine learning algorithm. Available at SSRN 4091262. https://ssrn.com/abstract=4091262 or https://doi.org/10.2139/ssrn. Accessed 23 Apr 2022
36. Jiang, L., Yi, Y., Chen, D., Tan, P., Liu, X.: A novel infant cry recognition system using auditory model-based robust feature and GMM-UBM. Concurr. Comput.: Pract. Exp. **33**(11), e5405 (2021)

Study and Automatic Translation of Toki Pona

Pablo Baggetto[1], Damián López[2], and Antonio M. Larriba[2](✉)

[1] valgrAI: Valencian Graduate School and Research Network of Artificial Intelligence, Camí de Vera S/N, Edificio 3Q, 46022 Valencia, Spain
[2] vrAIn: Valencian Research Institute for Artificial Intelligence, Universitat Politècnica de València, 46022 Valencia, Spain
anlarflo@dsic.upv.es

Abstract. In this work, we explore the use of Neural Machine Translation models for Toki Pona, a small, low resourced, and minimalistic language. We tackle the problem by employing the transformer model and transfer learning. Despite the challenges posed by the language's limited resources and the scarcity of available data, we demonstrate that a high-accuracy machine translator can be developed for Toki Pona using these advanced techniques. Through transfer learning from existing models, we were able to adapt to the unique characteristics of Toki Pona and overcome the challenges of limited data. Our results show the potential for applying these methods to other under-resourced languages, providing a valuable tool for increasing accessibility and communication in these communities.

Keywords: Neural Machine Translation · Toki Pona · Transformers · Few Resources · Machine Learning

1 Introduction

Machine translation has become an indispensable tool for facilitating communication and increasing accessibility in the digital age. However, the development of neural machine translation systems for under-resourced languages, such as Toki Pona, has been limited due to the scarcity of data and resources. In this paper, we present a machine translator for Toki Pona using transformers and transfer learning.

Toki Pona is a minimalistic, artificial language spoken only by a niche online community. As a consequence, Toki pona has some unique characteristics such as its small vocabulary and high variability, which make it an interesting case study for machine translation as it presents a challenge for the use of transfer learning. In addition, the limited data available presents a challenge for developing accurate models in these circumstances. We here address this task by developing two machine translation models for Toki Pona, one which was trained from scratch, and, a second one trained using transfer learning.

Transformers [12] are a type of neural network architecture that have achieved state-of-the-art results in multiple natural language processing tasks, including

A. Pertusa et al. (Eds.): IbPRIA 2023, LNCS 14062, pp. 654–664, 2023.
https://doi.org/10.1007/978-3-031-36616-1_52

machine translation. They use self-attention mechanisms to capture long-range dependencies and contextual information in the input text. Transfer learning [1] is a technique where pre-trained models on large datasets are fine-tuned on a smaller target dataset, allowing for more efficient and effective training.

We evaluated our models using the BLEU [7] metric, which measures the overlap between the predicted and reference translations. Our results show that using transformers and transfer learning, we were able to achieve high BLEU scores on the Toki Pona dataset, proving the potential for these methods in under-resourced language translation. Additionally, the comparison of the two models allowed us to examine the effects of transfer learning on the peculiarities of Toki Pona and its limited data availability.

We were interested in exploring transfer learning as it is common to other neural machine translation works with few resources [13] and has been proven useful even applying it from unrelated languages [5], which fits the needs for Toki Pona.

This paper is structured as follows: in Sect. 2 the unique characteristics of Toki Pona are explained, in Sect. 3 we describe the way the corpus was obtained, in Sect. 4 we explain our approach for the neural machine translation and in Sect. 5 the results obtained in this work are presented. Finally, some conclusions are summarized in Sect. 6.

2 Toki Pona

Toki Pona is a philosophical and artistic artificial language created by the Canadian linguist and translator Sonja Lang [6]. The first drafts of the language were published on the internet in 2001 and since then, an increasingly growing community of online speakers emerged fascinated by this language. Even though the community provides a great number of resources, they are insufficient to create models with the performance of widely spoken languages.

This language is lexically, phonetically and semantically minimalist. It uses the minimum amount of elements and the simplest ones in order to express as much as possible. That is why the language is constructed with around 120 words and an alphabet of 14 letters.

As a consequence of its minimalist approach, Toki Pona often lacks the ability to distinguish finer shades of meaning, being too general and vague. This is a serious problem for machine translation models, as it might evaluate correct sentences as wrong because the dataset cannot consider every possible translation, making it hard for these language models to train and evaluate.

In addition, Toki Pona can be written using different writing systems. The most frequently used online is the roman alphabet, but it also has two own writing systems: *sitelen pona*, which is an ideographic system where each character or glyph represents a word created by Sonja Lang, and *sitelen sitelen*, created by Jonathan Gabel and introduced in *Toki Pona: The language of good* [6]. An example of these writing systems appear in Fig. 1. We only considered the Latin alphabet for this paper.

(a) Sitelen pona

(b) Sitelen sitelen

Fig. 1. *o olin e jan poka*, which means "love your neighbor", written in 2 different writing systems: *sitelen pona* and in *sitelen sitelen*

Another important aspect of Toki Pona, for the matter of machine translation, is that it is not close to any other language, thus, making it difficult to apply transfer learning techniques which is the most popular option for machine translation models with few resources.

All these features make Toki Pona a great testbed for machine translation techniques in languages with limited resources. As a consequence, this study is valuable not only by providing useful resources for the Toki Pona community, but also supposes a complete revision on the problems that may arise with minor languages.

3 Dataset

The main problem in order to address a Toki Pona translator is the lack of the usually big corpus needed to address a deep learning approach. Fortunately, this language has an active community that uploads translations made by the speakers to https://tatoeba.org, an online collection of sentences and translations. We downloaded from there every pair of sentences that were translated between Toki Pona and English.

All these sentences were divided into three different groups: one for training, another one for validation, and a final set for testing. We decided to apply a series of transformations and preprocessing of these datasets in order to improve the performance of the training and simplify the task that was going to be performed, thus, we carried out the following processes:

1. Removal of the repeated sentences.
2. Removal of every punctuation mark.
3. Transformation of every letter into lowercase.
4. Random shuffle of the pairs of sentences.

Even though it is a common practice, the number of words per sentence in the dataset was not limited as, in our experience, does not affect the performance of the model. After the transformations, we ended up with 15579 pairs for training, 2005 for validation and 1004 for testing. In Table 1, we provide some metrics to better characterize the dataset and its different sets. These metrics represent the (average, median and modal) length of sentences and words in the dataset, and therefore provide a high-level glimpse of the difficulty of the translation task.

Table 1. Statistical information about the Toki Pona dataset

Set	Avg. Sent.	Median Sent.	Mod Sent.	Avg. Words	Median Word.	Mod Word.
Training	8.47	8	7	3.19	3	4
Dev	8.56	8	8	3.20	3	4
Test	8.60	8	7	3.18	3	4

4 Our Implementation

To create a high-accuracy machine translator for Toki Pona, we developed two different models using transformers.

The first model was trained from scratch with random initial weights. We performed an empiric exploration of the hyperparameters to achieve the best performance. The second model used transfer learning, leveraging the knowledge of pre-trained transformer models from the OPUS collection [9]. We fine-tuned several of these models on our Toki Pona dataset, and compared their performance to select the best-performing model for further analysis and comparison. This allowed us to benefit from the knowledge of these models, and adapt them to the unique characteristics of Toki Pona. The training of the models was performed on Linux Ubuntu 20.04 with an NVIDIA GeForce GTX 1060 with 6 GB of VRAM.

4.1 Model from Scratch

To implement the model trained from scratch, we used the OpenNMT-py [4] library. The transformer architecture is known to be sensitive to hyperparameters, and therefore careful tuning is necessary to achieve good performance. The following hyperparameters were explored in the model:

- Word vector size: Number of bits in the vector space used to represent words.
- Number of heads: Number of attention heads in the transformer's attention module.
- Transformer_ff: Size of hidden transformer feed-forward layer.
- Number of layers: Number of identical layers in the encoder and decoder sections.
- Dropout rate [8].

In addition to these transformations, we also applied byte-pair encoding (BPE [2]) preprocessing to study if it improved the obtained BLEU score.

As there were many parameters, in order to explore the best combination, we created a base model with standard parameters and changed one or two parameters each time to test the different performances of the configurations. Moreover, some parameters were fixed and not explored in this study. The base configuration and the values of the unexplored hyperparameters can be seen in Table 2.

Table 2. Configuration of the base model

Hyperparameter	Value
accum_count	8
optim	adam
adam_beta1	0.9
adam_beta2	0.998
decay_method	noam
learning_rate	2.0
max_grad_norm	0.0
batch_size	128
batch_type	sents
normalization	sents
label_smoothing	0.1

(a) Unexplored parameters of the model

Hyperparameter	Value
word_vec_size	128
layers	2
transformer_ff	512
heads	2
dropout	0.1

(b) Base configuration of the model

4.2 Transfer Learning with OPUS

In this Section, we wanted to explore if the Transfer Learning technique could benefit our model if it was trained in larger models in other languages. For this reason, we decided to use HuggingFace datasets and transformer model libraries.

The pretrained models used for the experiment are the OPUS models developed by Jörg Tiedemann [10]. These models had been trained using the transformer model in the MarianNMT framework [3]. The data used to train the model was obtained from the OPUS dataset [11].

The languages chosen for our investigation are Spanish (es), French (fr), German (de), Russian (ru), Arabic (ar), Japanese (ja), Chinese (zh), Esperanto (eo) and Tok Pisin (tpi). These languages were chosen because they were among the languages with the higher number of training examples or, in the case of Esperanto and Tok Pisin, because, even though Toki Pona is an isolated language with few similarities to any other languages, they have been the inspiration for some words in the vocabulary, and we thought it could be interesting to explore them. Table 3 shows the number of sentences and the hyperparameters which have been used for pretraining each model.

Table 3. Information about the pretraining

Language	Sentences
eo	19.4M
tpi	405k
fr	479.1M
de	349.0M
ja	68.1M
ru	213.8M
ar	102.8M
es	553.1M
zh	103.2M

(a) Sentences used to pretrain each translation model

Hyperparameter	Value
word_vec_size	64
layers	6
transformer_ff	2048
heads	8
dropout	0.1

(b) Hyperparameters of the pretrained models

In developing our fine-tuning architecture, we followed the steps here described.

Initially, we obtained each one of the pre-trained models with the OPUS dataset from HuggingFace. Models were trained until the early stop condition was met: no BLEU score improvement over the validation set for 10 consecutive evaluations. Models were assessed every 10000 steps.

Subsequently, the selected models of each language were fine-tuned with our Toki Pona dataset, with a batch size of 8 sentences. To safeguard against the fine-tuning process inadvertently negating the effects of the initial training, and to benefit from cross-language models, the learning rate was reduced from 3^{-4} to 2^{-5}. Once again, the models were trained until the early stop condition was met. Figure 2 depicts this evaluation process against the validation set for each of the chosen languages.

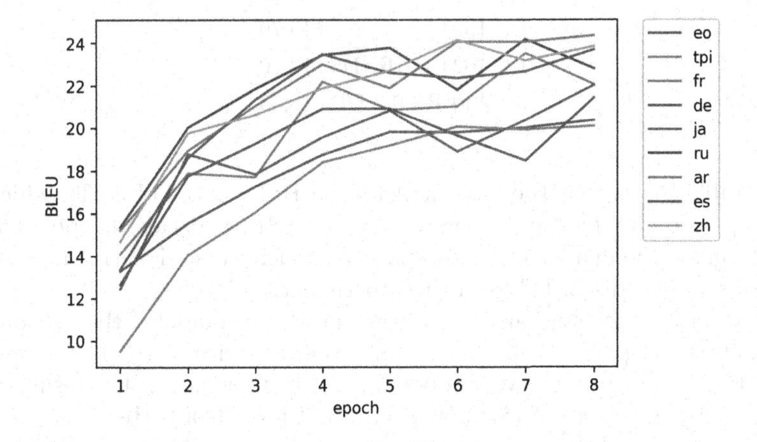

Fig. 2. BLEU in the validation set for each epoch

5 Results

In this section, we present the results obtained by our models and the comparison among them. For the evaluation, we used the BLEU [7] metric, specifically BLEU-4. This metric measures the geometric mean of n-grams between the inference sentence produced by the model and the reference sentence. It also has a penalty for sentence length to ensure that translated sentences have a similar length. Although it is far from reliable human evaluation, BLEU allows us to quickly, deterministically, and efficiently obtain a view of the complexity of the task in terms of translation quality. BLEU has been established as a reference metric for these tasks, and that's why we use it to present our results.

5.1 Model from Scratch

Taking into account the base configuration defined in Table 2, we explored different hyperparameters for our model trained from scratch. The results of this experimentation can be found in Table 4.

Table 4. BLEU of the different configurations of hyperparameters

Configuration	BLEU
BASE	20.33
4 HEADS	15.80
8 HEADS	14.80
BPE	8.49
3 LAYERS	21.07
4 LAYERS	18.44
256 WVS	15.29
0.05 DROPOUT	15.61
0.2 DROPOUT	19.32
FF 1024	15.79
FF 1024 4 HEADS	13.60
3 LAYERS 0.2 DP	18.33

The table shows that the base model has a BLEU score of 20.33, which has been surpassed only by the 3 layers version with 21.07. It can also be seen that the BPE made the quality of the translations much worse. For this reason, this preprocessing was discarded from the experiment.

In order to try to build an even better model, we combined the version with a dropout of 0.2 (with 19.32 BLEU) and the one with the 3 layers, as they are the two features that gave the best result. Unfortunately, the result of this combination had a score of 18.33, which did not provide a better score than the other versions alone. Using the model with 3 layers, we obtained a BLEU score of 21.07, improving the result.

5.2 Transfer Learning with OPUS

Regarding the transfer learning models, we selected the epochs with the bests scores, with regard to the validation dataset, as portrayed in Fig. 3. Those exact checkpoints were later evaluated against the test set.

Results compared with the model trained from scratch in Fig. 3. In the figure, it can be seen that some models (Arabic, Russian, Japanese and German) perform better than the model from scratch, being Arabic the best model with a BLEU score of 23.33. In contrast, five languages performed worse than the model from scratch. This difference between languages is mainly caused by the grammatical similarity of the language to the Toki Pona.

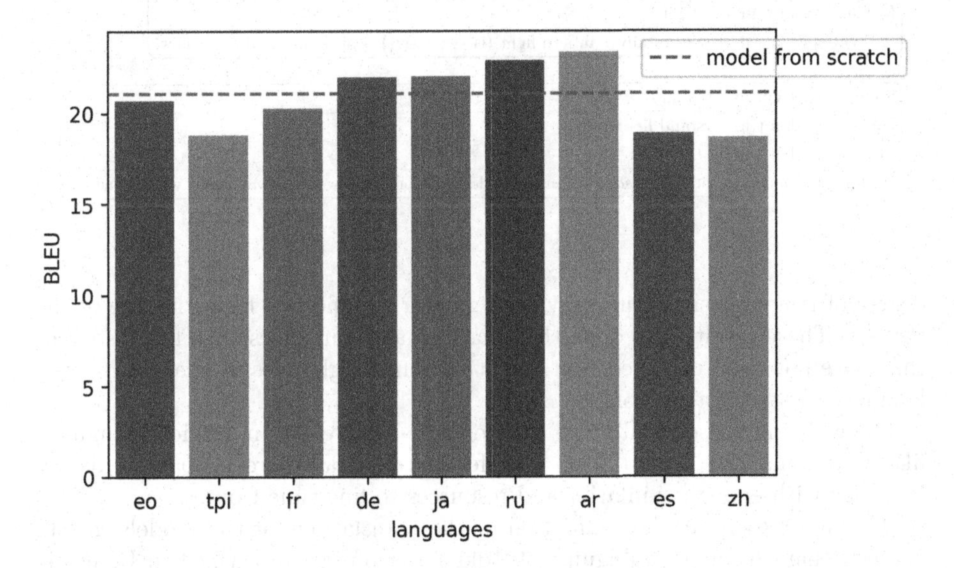

Fig. 3. BLEU in the test set

However, there are more factors that can cause a difference in the score like the amount of data the model was trained with, the writing system, which might have been detrimental for the Chinese as it uses only a logographic system or many other peculiarities of the languages that are unknown to us.

With these results, it can also be seen that our model built from scratch performed very well taking into account that requires a total training time much lower than the models which use transfer learning as their pretraining are very computing intensive, often by a factor of ten, even though this work had already been done.

Interestingly, the Tok Pisin model, pretrained on a smaller dataset, achieved marginally higher scores than the Spanish model, which was trained on a more extensive dataset (405k vs. 553M of training sentences as shown in Table 3). This observation suggests that, in the context of our transfer learning scenario, the

Table 5. Example of incorrectly evaluated sentences

Source:	mi jo ala e telo nasa
Human:	i don t have vodka
Machine:	i don t have any wine
Here the type of alcohol is not specified and depends on the context.	

Source:	sina pilin ala pilin e ni jan sewi li lon
Human:	do you believe that god exists
Machine:	do you believe in god
The two translations mean the same but the structure is different.	

Source:	jan ton li wile utala
Human:	tom is aggressive
Machine:	tom wants to fight
It is not specified if Tom generally wants to fight (is aggressive) or if he wants to fight now.	

Source:	mi sona ala e nimi ona
Human:	i don t understand her words
Machine:	i don t know her name
Both translations are correct as *sona* means to understand or to know and *nimi* means word or name.	

degree of language similarity may hold greater significance than the size of the dataset. These results also open the door to many languages with low resources that have not been explored that might obtain a higher result than any of the languages tested in this work.

After seeing these results, we can conclude that even an artificial language like Toki Pona can benefit from transfer learning, as there probably exists a language with enough similarity and resources to fine-tune from.

Looking closely at the evaluation of the translations of the models, it can be seen that the high ambiguity of Toki Pona often cause the BLEU metric to give low scores to correctly translated sentences because BLEU only counts right n-grams of words, and therefore it is not able to capture semantic valid transformations. Which specially harms translations from ambiguous languages such as Toki Pona, where a word has multiple translation depending on context. This effect can be seen in Table 5.

6 Conclusion

In this paper, we address a machine translation task for a low resources language using transformers and transfer learning. We developed two models, one trained from scratch and one using transfer learning, and evaluated their performance using the BLEU metric. Our results showed that using transformers and transfer learning, it should be possible to achieve high accuracy on the Toki Pona dataset, demonstrating the potential for these methods in under-resourced language translation, even if they are language isolates, following the work on unrelated transfer learning by Tom Kocmi and Ondřej Bojar [5].

Additionally, our comparison of the two models showed that the transfer learning model outperformed the model trained from scratch, highlighting the effectiveness of this technique for limited data scenarios. Despite that, the model from scratch performed better than some of the transfer-learning models, showing that this model can still be useful if we cannot find an appropriate model to fine-tune from.

Overall, our work demonstrates the feasibility and benefits of using transformer-based models and transfer learning for machine translation in under-resourced languages. We propose as a line of work to extend, that this approach can be extended and applied to other languages and tasks, and can contribute to increasing the accessibility and availability of machine translation for a wider range of languages. This can help to facilitate communication and increase the reach and impact of digital content in under-resourced languages, enabling more people to access and benefit from the wealth of knowledge and information available online.

Acknowledgements. Pablo Baggetto is supported by **ValgrAI - Valencian Graduate School and Research Network for Artificial Intelligence**, and **Generalitat Valenciana** under a grant for predoctoral studies.

References

1. Cowan, J., Tesauro, G., Alspector, J.: Advances in Neural Information Processing Systems 6. Advances in Neural Information Processing Systems. Morgan Kaufmann (1994). https://books.google.es/books?id=6tGHlwEACAAJ
2. Gage, P.: A new algorithm for data compression. C Users J. Arch. **12**, 23–38 (1994)
3. Junczys-Dowmunt, M., et al.: Marian: fast neural machine translation in C++. In: Proceedings of ACL 2018, System Demonstrations, pp. 116–121. Association for Computational Linguistics, Melbourne (2018). https://www.aclweb.org/anthology/P18-4020
4. Klein, G., Kim, Y., Deng, Y., Senellart, J., Rush, A.M.: OpenNMT: open-source toolkit for neural machine translation. arXiv preprint arXiv:1701.02810 (2017)
5. Kocmi, T., Bojar, O.: Trivial transfer learning for low-resource neural machine translation. In: Proceedings of the Third Conference on Machine Translation: Research Papers, pp. 244–252. Association for Computational Linguistics, Brussels (2018). https://doi.org/10.18653/v1/W18-6325, https://aclanthology.org/W18-6325
6. Lang, S.: Toki Pona: the language of good. Sonja Lang (2014). https://books.google.es/books?id=5P0ZjwEACAAJ
7. Papineni, K., Roukos, S., Ward, T., Zhu, W.J.: Bleu: a method for automatic evaluation of machine translation. In: Proceedings of the 40th Annual Meeting of the Association for Computational Linguistics, pp. 311–318. Association for Computational Linguistics, Philadelphia (2002). https://doi.org/10.3115/1073083.1073135, https://aclanthology.org/P02-1040
8. Srivastava, N., Hinton, G., Krizhevsky, A., Sutskever, I., Salakhutdinov, R.: Dropout: a simple way to prevent neural networks from overfitting. J. Mach. Learn. Res. **15**(1), 1929–1958 (2014)

9. Tiedemann, J., Nygaard, L.: The OPUS corpus - parallel and free: http://logos.uio.no/opus. In: Proceedings of the Fourth International Conference on Language Resources and Evaluation, LREC 2004, 26–28 May 2004, Lisbon, Portugal. European Language Resources Association (2004). https://www.lrec-conf.org/proceedings/lrec2004/summaries/320.htm
10. Tiedemann, J., Thottingal, S.: OPUS-MT—Building open translation services for the World. In: Proceedings of the 22nd Annual Conference of the European Association for Machine Translation (EAMT), Lisbon, Portugal (2020)
11. Tiedemann, J.: Parallel data, tools and interfaces in opus. In: Chair, N.C.C., et al. (eds.) Proceedings of the Eight International Conference on Language Resources and Evaluation (LREC 2012). European Language Resources Association (ELRA), Istanbul (2012)
12. Vaswani, A., et al.: Attention is all you need (2017). https://arxiv.org/pdf/1706.03762.pdf
13. Zoph, B., Yuret, D., May, J., Knight, K.: Transfer learning for low-resource neural machine translation (2016). https://doi.org/10.48550/ARXIV.1604.02201, https://arxiv.org/abs/1604.02201

Detecting Loose Wheel Bolts of a Vehicle Using Accelerometers in the Chassis

Jonas Schmidt[1,2](✉), Kai-Uwe Kühnberger[1], Dennis Pape[2], and Tobias Pobandt[2]

[1] Institute of Cognitive Science, University of Osnabrück, Osnabrück, Germany
{jonschmidt,kkuehnbe}@uos.de
[2] ZF Friedrichshafen AG, Friedrichshafen, Germany
{dennis.pape,tobias.pobandt}@zf.com

Abstract. Increasing road safety has been a society's goal since the automobile's invention. One safety aspect that has not been the focus of research so far is that of a loose wheel. Potential accidents could be prevented with the help of early detection of loose wheel bolts. This work investigates how acceleration sensors in the chassis can be used to detect loose wheel bolts. Test drives with tightened and loosened wheel bolts were carried out. Several state-of-the-art semi-supervised anomalous sound detection algorithms are trained on the test drive data. Evaluation and optimization of anomalous sound detection algorithms shows that loose wheel bolts can be reliably detected when at least three out of five wheel bolts are loose. Our study indicates that acoustic preprocessing and careful selection of acoustic features is crucial for performance and more important than the choice of a special algorithm for detecting loose wheel bolts.

Keywords: Time Series Analysis · Anomaly Detection · Anomalous Sound Detection · Loose Wheel Bolts

1 Introduction

In 2021, more than 2500 people died in road accidents in Germany. This figure is the lowest in 60 years. At the same time, the amount of traffic has multiplied during this period. Driving a car has thus become significantly safer in the past decades. The government and manufacturers are trying to further reduce traffic accidents with the help of new standards and technological innovations.

One aspect of vehicle driving safety that has been less of a focus is that of a loose wheel or even wheel loss, although this leads to a significant number of accidents each year. The number of annual non-injury accidents in the UK alone due to wheel loss is estimated to be between 50 and 134. The number of personal injuries is estimated to be between 10 and 27 annually, and the number of fatalities is between 3 and 7 [5]. When a wheel breaks loose, there are high risks for occupants of the affected vehicle and all other road users, since the detaching wheel reaches very high speeds and can hit anyone in the vicinity. Traffic law regulations are actually intended to prevent such accidents

Supported by ZF Friedrichshafen AG.

A. Pertusa et al. (Eds.): IbPRIA 2023, LNCS 14062, pp. 665–679, 2023.
https://doi.org/10.1007/978-3-031-36616-1_53

before they happen. §23 Abs. 1 of the Straßenverkehrsordnung (German Road Traffic Act) states, that the driver of a vehicle is responsible for ensuring that the vehicle is in a roadworthy condition when participating in public traffic [9]. For employers, Germany has even stricter regulations regarding the use of company vehicles. If the vehicle is provided by the employer in a professional context, it is considered as work equipment, such that §4 Abs. 1 Betriebssicherheitsverordnung (German Ordinance on Industrial Safety) is applied [14]. This Ordinance states that the employer must ensure that a vehicle is inspected for defects before it is put into service. Unfortunately, however, these inspections are only sporadically carried out, which increases the risk of accidents. Additionally, loose wheel bolts may not be detected easily at first glance.

One solution to this problem is to continuously monitor vehicle dynamics by in car sensors which shows anomalies when a wheel is loose. Since 2018, Audi has been using NIRA Dynamics software in luxury cars such as the Audi A6, A7, A8, and Q8 that can detect a loose wheel by analyzing the wheel speed sensor signal [8]. A loose wheel, in this case, means that all wheel bolts are loose. This is detected by the wheel speed sensor signal having an increased energy level. The software detects a loose wheel within one minute of driving at a speed of 25 km/h or more.

In this work, we examine whether it is possible to detect a loose wheel before all the wheel bolts are loose by using accelerometers, which may react more sensitively to changes in vehicle dynamics. We investigate this for accelerometers mounted in the chassis of the vehicle. The main contributions of this work are an evaluation and optimization of state-of-the-art anomalous sound detection algorithms on our data set. A comparison with the wheel speed sensor based detection of loose wheel bolts is not within the scope of this work. For this comparison, the raw data from the wheel speed sensor would have been required, unfortunately, this was not available to us.

This paper is structured as followed. In the second chapter we describe the measurement techniques and test drives carried out with loosened wheel bolts. The third chapter gives an introduction to time series anomaly detection while the fourth chapter outlines anomalous sound detection as a special time series anomaly detection case. Within this section also the most popular acoustic features for sound data are discussed, various anomalous sound detection algorithms are introduced, and metrics for evaluating them are defined. The fifth chapter discusses the results of training and optimizing the presented anomalous sound detection algorithms. A summary and an outlook concludes the paper.

2 Test Drives and Data Collection

Test drives are carried out with a BMW F10 (5 series) to generate training data for the anomaly detection algorithms. The test vehicle is equipped with tri-axial acceleration sensors that record vibrations in X-, Y- and Z-direction. The vibration directions follow the vehicle coordinate system and are illustrated in Fig. 1. The acceleration sensors are mounted on the front axle's upper right and left control arms. The accelerometer data, the vehicle speed, and the GPS signal are recorded and stored at a sampling rate of 10 kHz. The additional data should provide information if certain driving maneuvers allow for an easier detection of loosened wheel bolts than others. All maneuvers are carried

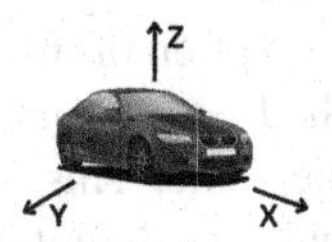

Fig. 1. Vehicle coordinate system used for the acceleration sensors.

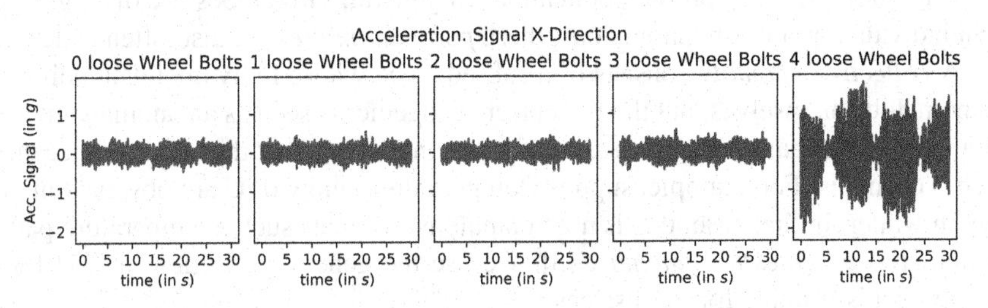

Fig. 2. Acceleration signal in X-direction from 30-s sections of two complete laps on the test track are shown.

out on the same route and at the same speed to ensure comparability. The test drives were performed in a parking lot and only at low speeds for safety reasons. The speed was approximately 30 km/h. The test track had a figure eight shape to accommodate both left and right turns. First, a reference run is carried out in which all wheel bolts are correctly tightened. This test drive lasted about five minutes. The corresponding time series of acceleration data contains approximately three million data points for each direction. On after another wheel bolt is loosened during each of the next 4 test runs, i.e., all wheel bolts except one are loosened by the last test run. We did not loosen the last wheel bolt because we already noticed with four loosened wheel bolts a distinctive change in vehicle dynamics, and we could furthermore perceive acoustically in the interior that there was a problem with the wheels. Since the safety of our road tests had top priority, we wanted to eliminate the risk of losing a wheel as much as possible. The acceleration data of these test runs each had a length of about 72.5 s, i.e., generating 725.000 data points for each acceleration direction. When loosening, the wheel bolt is loosened by precisely two revolutions. Each newly loosened wheel bolt was on the opposite side of the previous one. Due to time constraints, we did not test different orders of loosening the wheel bolts.

It can be seen in Fig. 2 that the amplitudes of the acceleration signal in X-direction for four loosened wheel bolts stand out significantly. The acceleration, in this case, has higher amplitudes in both positive and negative directions. The test drives with no loosened wheel bolts up to three loosened wheel bolts can hardly be distinguished visually from each other by the amplitude of their acceleration sensor signal (this applies to all directions).

3 Time Series Anomaly Detection

A time series $Y = (y_0, \ldots, y_m)$ is an ordered sequence of data points $y_i \in \mathbb{R}^n$. If $n = 1$, a time series is called univariate and for $n > 1$ multivariate. Subsequences

$Y_{i,j} = \{y_i, \ldots, y_j\}$ of a time series Y are often of interest. They are a contiguous section of data points. A time series is called equidistant, if the time interval between two consecutive data points is always the same. This is ensured in our test drives by the measurement system. There is no uniform definition of a time series anomaly in the literature. This is because there are several types of anomalies in time series, which can look different depending on the application. In general, three types are distinguished: When individual points are anomalous, these *point anomalies* are also often called *outliers*. A *collective anomaly* consists of a subsequence whose individual data points are not anomalous themselves, but their occurrence together is seen as an anomaly. In addition, it is possible that a subsequence itself is not anomalous, but the circumstance under which it occurs is. For example, suppose temperatures below 0 °C are observed in Germany in winter. In that case, this is not anomalous, whereas such a temperature pattern in summer is a *contextual anomaly* [3]. We use the general definition of [21] for an anomaly that is suitable for our use case.

Definition 1. *A time series anomaly is a sequence of data points $Y_{i,j}$ of length $j-i+1 \geq 1$ that deviates w. r. t. some characteristic embedding, model, and/or similarity measure from frequent patterns in the time series Y.*

Time series anomaly detection is related to *time series classification*, in which entire time series are assigned to individual classes [1]. Classification is a step that usually takes place after anomaly detection. For this, anomalous subsequences are evaluated by a classifier that assigns them to predefined classes. For example, the acceleration signal could show anomalies due to low tire pressure or worn chassis components, which also would influence the acceleration signal. Therefore, anomalies in the acceleration data need to be processed further to determine the cause of the anomaly. However, this work is limited to investigate if an anomaly detection algorithm can detect anomalies caused by loosened wheel bolts. In a later step, post-processing could be applied for anomaly classification. Nevertheless, this is beyond the scope of this work.

For each anomaly detection algorithm we want to compare, we need a standardized output format that assigns a so-called *anomaly score* to each data point or subsequence that describes degree of anomaly. Following the definition from [21]:

Definition 2. *An anomaly scoring $S = \{s_0, \ldots, s_m\}$ with $s_i \in \mathbb{R}$ is the result of a time series anomaly detection algorithm that assigns each data point $y_i \in Y$ an anomaly score $s_i \in S$. For any two scores s_i and s_j, it must be true that if $s_i > s_j$, then y_i is more anomalous than y_j (in their respective contexts).*

If an algorithm receives a subsequence $Y_{i,j}$ as input and its output is a single anomaly score s, then the anomaly score for all $y_k \in Y_{i,j}$ is equal to s. If we use overlapping subsequences, we average the anomaly scores of overlapping segments. The definition of anomaly score thresholds to classify data points and subsequences into the classes 'anomaly' and 'no anomaly' is a challenge being independent of the applied algorithm. The threshold value influences the ratio of false positives (choice of a low threshold increases number of false positives) and false negatives (choice of a high threshold increases number of false negatives). Here, the application context plays a decisive role since false positives or false negatives can be of varying importance.

4 Anomalous Structure-Borne Sound Detection

The acceleration sensors measure the structure-borne sound at the upper control arm. They record, therefore, a special kind of time series that falls into the category of sound data. *Anomalous sound detection* (ASD) has recently gained increasing attention in the research community through competitions such as DCASE [6,12]. In addition, further development of specialized algorithms is also driven by economic applications in the field of surveillance systems [23], machine condition monitoring [2], or health care monitoring [18]. For this type of time series, ASD algorithms usually aim to learn characteristic properties of the (structure-borne) sound signal, such as changes in amplitudes or the increased occurrence of specific frequencies. Therefore, methods from signal processing are used to preprocess the sound data, extracting certain *acoustic features* from them.

4.1 Acoustic Features

Acoustic features can be categorized as time-based, frequency-based, or time-frequency-based. In this paper, we restrict ourselves to acoustic features that have performed best at the DCASE 2022 competitions in the area of unsupervised ASD for machine condition monitoring [6]. These are Mel-spectrograms, spectrograms and the raw waveform [15]. Due to the similarity to our task and their proven record, we assume that they can also be applied to detect loose wheel bolts.

For feature extraction, a given time series $Y = \{y_0, \dots, y_m\}$ is divided into subsequences of *frame length* $w \in \mathbb{N}$ and *hop length* $h \in \mathbb{N}$, i.e.,

$$Y_{i(w-h),i(w-h)+w-1} = \{y_{i(w-h)}, \dots, y_{i(w-h)+w-1}\}.$$

and $i(w-h)+w-1 \leq m$. The hop length defines the overlap between two subsequences. We call the subsequences extracted by framing Y *windows* and reference them using $W = \{W_0, \dots, W_m\}$, where $W_i = Y_{i(w-h),i(w-h)+w-1} = \{w_{i,0}, \dots, w_{i,w-1}\}$. The simplest time-based acoustic features are the windows themselves. These raw waveforms contain all the available information of the input. However, their greatest advantage is also their greatest disadvantage because the ASD algorithm must first learn the relevant information in the high-dimensional input signal. This requires complex ASD algorithms with low bias and carries the risk of overfitting.

The magnitude spectrum of a frame W_i, is a frequency-based feature, and obtained by calculating the magnitudes of the *Discrete Fourier Transform* (DFT) [22]. If W_i has been extracted from a multi-variate time series, then the DFT and all other frequency-based features are calculated for each uni-variate time series individually, i.e. the acceleration directions in our application. To prevent spectral leakage, we use a *Hanning window* $H = \{h_0, \dots, h_{w-1}\}$ before transferring the signal into the frequency domain. See [16] for details about spectral leakage and the concrete implementation of the Hanning window. We refer to the windows after applying the window function with $\tilde{W} = \{\tilde{W}_0, \dots, \tilde{W}_m\}$, where

$$\begin{aligned}\tilde{W}_i &= Y_{i(w-h),i(w-h)+w-1} \cdot H \\ &= \{w_{i,0}h_0, \dots, w_{i,w-1}h_{w-1}\} \\ &= \{\tilde{w}_{i,0}, \dots, \tilde{w}_{i,w-1}\}.\end{aligned}$$

The kth frequency bin f_k of the DFT is equal to

$$f_k = \sum_{j=0}^{w-1} e^{-2\pi i \frac{jk}{w}} \cdot \tilde{w}_{i,k} \text{ for } k = 0, \dots, w-1.$$

The magnitude spectrum $|F|$ is then $|F| = \{|f_0|, \dots, |f_{w-1}|\}$. A real valued signal has a hermitian spectrum, such that $f_{w-i} = f_i$ for $i = 1, \dots, w-1$ and the second half thus contains redundant information that we can omit. In the following, we assume that w is even, i.e., we use windows with an even number of data points.

Another feature that can be calculated from the DFT, is the *Mel-spectrum M*. For this, the $\frac{w}{2}+1$ frequency bins corresponding to frequencies in *Hz* are converted into *Mel* bins using a triangluar Mel-filter bank B. Let n_{mels} be the number of Mel bins. B is a precomputed matrix with the shape $(n_{mels}, \frac{w}{2}+1)$ and $M = B \cdot F$. As an acoustic feature we use the magnitude of the Mel-spectrum $|M| = \{|m_0|, \dots, |m_{n_{mels}-1}|\}$. The Mel scale is based on the pitch perceived by humans. A value twice as high as another on the Mel scale is also perceived by a person as twice as high. This linear relationship does not apply to frequencies in hertz since the human ear is more sensitive to low frequencies and less discriminative to high frequencies. Mel filter banks transfer this idea to the frequency space, providing higher resolution at low frequencies and lower resolution at high frequencies. Mel-based acoustic features regularly achieve the best results in audio classification competitions. An advantage of them is that the dimension of the frequency range can be reduced considerably, thereby reducing the input dimension of an ASD algorithm.

Other considered time-frequency based features are the *Spectrogram* and the *Mel-Spectrogram*. For this, the magnitude spectra or Mel-spectra of $n_{wind} \in \mathbb{N}$ windows are joined together to form a matrix of the form $(\frac{w}{2}+1, n_{wind})$ or (n_{mels}, n_{wind}). These time-frequency based features have potential advantages for stochastic signals like our acceleration signals. Stochastic signals do not have a constant spectrum. It changes over time. Time-frequency features can make the change over time in the frequency range visible and learn valuable patterns from this. For algorithms that expect a one-dimensional input, i.e., $n_{wind} = 1$, we test the raw waveform, Mel-spectrum and magnitude spectrum as features. For the other algorithms that expect a two-dimensional input, i.e., $n_{wind} > 1$, we test the Mel-spectrogram and spectrogram as features. As it is common in acoustics, we calculate all frequency-based features in decibels. This involves taking a logarithm of the amplitudes in the frequency domain and thus emphasizing especially higher-frequency parts in the signal.

4.2 Anomalous Sound Detection Algorithms

ASD algorithms can be divided into three groups *supervised*, *semi-supervised*, and *unsupervised*. *Supervised* ASD algorithms require a training dataset with labeled normal data and anomalies. They are equivalent to standard supervised machine learning classification algorithms. However, it is often difficult to collect a large number of anomalies, resulting in class imbalances. *Semi-supersived* methods have only the normal class available as training data. They learn the distribution of the normal state during the training process. Deviations from this learned distribution are then detected as anomalies. The most straightforward methods are *unsupervised* ASD algorithms because they do not have a training process at all. No labeled data is needed for them either. They are based on the assumption that anomalies in the test data occur less frequently or are very different from the rest of the distribution of the test data. Nevertheless, if this is not the case, they can have a very high false positive rate. In our use case, however, we have labeled data, so we will no longer consider unsupervised ASD algorithms. Furthermore, we do not classify how many wheel bolts are loose, but rather only whether the acceleration data changes from the normal behavior due to wheel bolts loosening. Therefore, we merely consider semi-supervised ASD algorithms in the following. Within this group of ASD algorithms, there exist algorithms with different methods to learn normal behavior as well as to score the degree of anomaly for individual test data sets. We test various ASD algorithms for our use case, which also differ in complexity. In this way, we determine whether it is necessary to use complex deep-learning-based ASD algorithms or whether classical algorithms already provide good results. For all considered ASD algorithms, suppose a test window $W_i \in \mathbb{R}^{n_{acc} \times w}$, where n_{acc} is the number of used acceleration directions, is given. Let $n_{feat} \in \mathbb{N}$ be the number of features, e.g., the number of extracted frequency or Mel bins. We define the acoustic feature of W_i as $A_i \in \mathbb{R}^{n_{acc} \times n_{feat} \times n_{wind}}$.

As examples of distance-based ASD algorithms, we use the classical machine learning algorithms k-nearest neighbor (KNN) and k-means in a semi-supervised version [10, 13]. This means that both receive only the acoustic features of the normal class as training data. We use the one-dimensional acoustic features for KNN and k-means as training data. If the features of several acceleration directions are used, the feature A_i is flattened. The anomaly score s of W_i is calculated for KNN as follows: First, the k nearest neighbors $N_1, \ldots N_k$ of A_i are determined. The anomaly score s is then the mean Euclidian distance of the neighbors to A_i, i.e., $s = \frac{1}{k} \sum_{j=1}^{k} ||N_j - A_i||_2$. For k-means, the cluster centroids $C_1, \ldots, C_k$ are determined first. The anomaly score s here is analogously calculated to KNN but based on the cluster centroids, i.e., $s = \frac{1}{k} \sum_{j=1}^{k} ||C_j - A_i||_2$. As reconstruction-based ASD algorithms, we use a dense autoencoder and a 1D-CNN autoencoder. The autoencoders AE are trained on the acoustic features of the normal class. They consist of an encoder and a decoder part. The encoder part projects the acoustic features into a low-dimensional latent space. The decoder tries to reconstruct the original acoustic features from the encodings in the latent space, i.e.,

$$\mathrm{AE}(A_i) = \mathrm{DEC}_{\mathrm{AE}}(\mathrm{ENC}_{\mathrm{AE}}(A_i)) \approx A_i.$$

Their training objective is $\min_{\omega \in \Omega} \sum_{i=0}^{m} |\mathrm{AE}(A_i) - A_i|_2$, where Ω is the training parameter set of the respective autoencoder. If the one-dimensional features of several acceleration directions are used as input, they are also concatenated into one feature vector for the dense autoencoder. The encoder part of the dense autoencoder consists of 4 blocks. Each block is a dense layer with ReLU as the activation function and a dropout of $p = 0.5$. The input shape corresponds to the dimension of the acoustic features, and the first dense layer has 128 neurons. The following layers of the encoder each halve the dimension of the input, which means that the output of the last encoding layer is 16-dimensional. The decoder works the other way around. It consists of the same four blocks, but each dense layer doubles the input dimension of the previous one, with the last dense layer of the decoder generating a vector as output with the dimension of the initial input feature.

The encoder part of the 1D-CNN autoencoder consists of three blocks of Conv1D, ReLU, and BatchNorm layers, where for the Conv1D layer $kernel\ size = 3$, $stride = 2$ and $padding = 1$ is chosen, which leads halving the input dimension after each block [19]. The number of filters for the first Conv1D layer corresponds to n_{acc}, the number of acceleration directions from which the acoustic features were extracted. All further Conv1D layers use n_{filter} as the number of filters determined during hyperparameter optimization. The final output of these three blocks is flattened, and a dense layer with 16 neurons follows so that the latent space has the same dimension as the latent space of the dense autoencoder. The decoder receives the latent vector of the encoder as an input and starts with a dense layer with one-eighth of the input dimension of the selected acoustic feature. The three following blocks of ConvTranspose1D, ReLU, and BatchNorm layers with $kernel\ size = 3$, $stride = 2$, and $padding = 1$ each double the input dimension so that the final output again has the shape of the acoustic feature. The last Conv1D layer of the decoder has n_{acc} filters exactly like the first Conv1D layer of the encoder to reconstruct all acceleration directions.

Forecasting-based ASD algorithms learn to predict the next windows features based on learned information from previous windows features. The Euclidian distance between the forecast and the next windows real feature is then displayed as an anomaly score. The idea is that they perform worse at forecasting anomalies because they have only been trained on normal data, leading to higher forecasting errors. We test two ASD algorithms of this type that use time-frequency features, i.e., two-dimensional features as input. One is based on LSTMs called LSTM-AD and Interpolating Deep Neural Network (IDNN) [17,24]. LSTM-AD consists of $n_{lstm} \in \mathbb{N}$ stacked LSTM layers, where the input size is n_{feat}, followed by a dense layer with also n_{feat} as input and output shape. IDNN is not an actual forecasting algorithm because no future frame is predicted but the middle frame of the time-frequency feature. We use IDNN because it achieved the best results in the DCASE Challenge 2020 Task 2 "Unsupervised Detection of Anomalous Sounds for Machine Condition Monitoring" [12]. It works similarly to the dense autoencoder. The difference is that we use time-frequency features as input here and do not want to reconstruct the entire input but only learn to predict the middle window. This approach proved to be more robust in the DCASE Challenge for non-stationary sounds. Details on the architecture can be found in [24].

As a final approach, we test an algorithm based on outlier exposure [7]. The idea is to train first a classifier on an auxiliary classification problem. In our case, we train to predict whether an acoustic feature comes from an acceleration signal from the X-, Y-, or Z-direction and if it comes from the accelerometer on the left or right side. During hyperparameter optimization, we determine whether all or only some of the mentioned classes should be used for training the auxiliary classifier. The anomaly score is calculated in three steps. First, the acoustic feature $A_i \in \mathbb{R}^{n_{acc} \times n_{feat} \times n_{wind}}$ is divided into sub-features of the corresponding n_{acc} used acceleration directions named $class_1, \dots, class_{n_{acc}}$. This results in $A_{class_1}, \dots, A_{class_{n_{acc}}}$ with $A_{class_j} \in \mathbb{R}^{n_{feat} \times n_{wind}}$. Second, the auxiliary classifier is applied for all A_{class_j}. Suppose $\text{Prob}_{class_j}(A_{class_j})$ is the probability output by the auxiliary classifier that A_{class_j} belongs to $class_j$. Since the auxiliary classifier is trained only on normal data, it is assumed that for this data it predicts well to which class respectively which acceleration direction a certain acoustic feature belongs (high values for $\text{Prob}_{class_j}(A_{class_j})$). If the acceleration data of loosened wheel bolts behaves differently and shows discrepancies in its distribution, the hypothesis is that this classifier performs worse (low values for $\text{Prob}_{class_j}(A_{class_j})$). Third, the anomaly score s for the frame W_i of A_i is calculated using the formula

$$s = \frac{1}{n_{acc}} \sum_{j=1}^{n_{acc}} \log\left(\frac{1 - \text{Prob}_{class_j}(A_{class_j})}{\text{Prob}_{class_j}(A_{class_j})}\right).$$

Outlier exposure based ASD algorithms achieved the best results in the DCASE Challenge 2022 for Task 2, covering anomalous sound detection for machine condition monitoring [6]. As an auxiliary classifier, we train MobilenetV2, a CNN model optimized for mobile devices, which provides a good trade-off between the number of parameters and performance [20].

4.3 Metrics

Let the anomaly class (loose wheel bolts) be the positive class and the normal class (no loose wheel bolts) be the negative class. We classify predictions as true positives (TPs), false positives (FPs), true negatives (TNs) and false negatives (FNs). The most popular threshold-independent *Area under the Curve* (AUC) metrics are AUC-PR and AUC-ROC [4]. Their values range between 0 and 1. They take the value 1 when an algorithm produces scorings so that the anomalies can be clearly separated from the normal data. In the case of a balanced data set, both metrics take the value 0.5. AUC-PR is the area under the *precision-recall curve*, which represents the *precision* ($\frac{TP}{TP+FP}$) as a function of the *recall* ($\frac{TP}{TP+FN}$). In contrast, AUC-ROC is the area under the curve representing recall as a function of *false positive rate* ($\frac{FP}{FP+TN}$). From the definition of the two metrics, it follows that AUC-PR rates precise algorithms higher and AUC-ROC sensitive algorithms. It is important to note that AUC-ROC, unlike AUC-PR, is not influenced by class imbalances. However, these often occur in anomaly detection problems. Usually, there is much more data in the normal class and it is more time-consuming or expensive to generate data from the anomaly class.

We choose AUC-PR as the metric because precision is of utmost importance in our use case. We calculate these individually for the test data of the normal class versus the anomaly class of i loosened wheel bolts for $1 \leq i \leq 4$. We also calculate the average AUC-PR to provide a metric to evaluate the overall performance of the ASD algorithm. If the ASD algorithm predicts that a wheel nut is loose, we want this to be correct with high probability. Otherwise, users in the vehicle would not accept the system because they would then keep checking their wheels for loose bolts. We do not use AUC-ROC because even low FP rates can lead to a very high number of FPs if the class imbalance is very high. ASD algorithms with high AUC-ROC values would thus provide deceptive reliability and cause the user to carry out unnecessary checks in real driving situations.

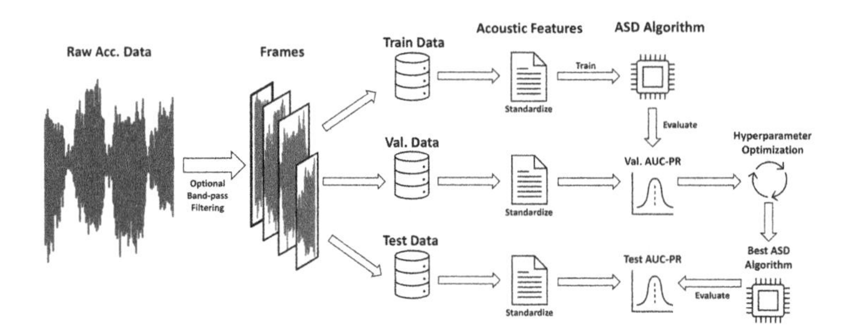

Fig. 3. Data preprocessing steps and training procedure

5 Experiments

We describe our procedure for training and evaluating the ASD algorithms in the following (see an overview in Fig. 3). First, we divide the collected data from the test drives into training, validation, and test data. Since we use semi-supervised ASD algorithms, the training data consists only of the no loosened wheel bolts class. We utilize contiguous sections of the test drives in the splits. The reason for this is that we will perform windowing of the sections and calculate anomaly scores for individual windows. If these windows overlap, this ensures that they were used exclusively in training, validation, or testing. Otherwise, it is possible that, for example, part of the data is used for testing but simultaneously for training. The training data comprises a section of 1.87 million data points. The validation data consists of 500.000 data points for the class with no loosened wheel bolts and 50.000 data points for all others. The test data consists of 600.000 data points from each class. This split makes it possible to optimize individual ASD algorithms on the validation data, even with a small number of anomaly data.

5.1 Feature Extraction

Our preprocessing starts with an optional band-pass filter, which allows us to limit the frequencies in the acceleration signals. The respective frequency ranges are determined for each acceleration direction during hyperparameter optimization. However, information that were present in the raw signal are also lost through filtering. With a huge number of training data, such a step could be omitted because the ASD algorithm can learn the relevant frequencies from the frequency spectrum. Nevertheless, with a smaller amount of training data like ours, we can speed up the learning process and reduce overfitting by filtering. We then divide the training, validation, and test data into windows $W_i \in \mathbb{R}^{n_{acc} \times w}$ of length w using a hop length h. From these windows, we extract the selected acoustic features $A_i \in \mathbb{R}^{n_{acc} \times n_{feat} \times n_{wind}}$. If a frequency-based feature is used and, simultaneously, a band-pass filter limits the frequency range in the output signal, only the frequency bins of the feature that lie in the frequency range of the filter are used, i.e., n_{feat} is reduced. This significantly decreases the input dimension of the ASD algorithm depending on the bandwidth of the filter. We then standardize the features. The standardized features are used as input to train the respective ASD algorithm and determine the AUC-PR value on the validation data. We perform a Bayesian hyperparameter optimization to find the appropriate parameters of the ASD algorithms and preprocessing that maximize the AUC-PR value on the validation set.

5.2 Training and Hyperparameters

The hardware setup for training and evaluating the ASD algorithms consisted of an NVIDIA Tesla P100 GPU (16GB GPU memory). Several hyperparameters related to acoustic feature extraction must be optimized individually for all algorithms. These include the window length $w \in \{512, 1024, 2048, 4096, 8192\}$ and the hop length h. Let $h = w \cdot h_{factor}$, where $h_{factor} \in \{0.125, 0.25, 0.5, 1\}$. It is also necessary to determine which acceleration directions achieve the best results for each algorithm. For this purpose, a subset of $\{LX, LY, LZ, RX, RY, RZ\}$ is selected during hyperparameter optimization, where L and R describe the acceleration sensor on the left (loosened wheel bolts) and right (no loosened wheel bolts) side respectively and X, Y, and Z the direction. Using the acceleration information from the opposite side, the ASD algorithm can possibly learn the relative changes of the accelerations to each other. In addition, the lower bound $f_{min} \in [0, 5000]$ and upper bound $f_{max} \in [0, 5000]$ of the band-pass filter are optimized for each selected acceleration direction. For all algorithms using gradient descent, we choose *Adam* as optimiser, a learning rate $lr \in \{10^{-1}, 10^{-2}, 10^{-3}, 10^{-4}, 10^{-5}\}$, a batch size $b \in \{16, 32, 64, 128, 256\}$ and a number of epochs $e \in [10, 100]$ [11]. To avoid overfitting, we use early stopping with a patience of 5.

5.3 Results and Discussion

We divide the analysis of the results of the Bayesian hyperparameter optimization (see the details in Table 1) into the following different areas: impact of the number of loosened wheel bolts, performance of acoustic features, acceleration direction and side, frequency ranges, and algorithms.

The results of all algorithms show that very high AUC-PR values up to 100% are achieved for four loosened wheel bolts. However, this result must be put into perspective as the differences in the amplitudes of the individual acceleration signals between no loosened wheel bolts and four are large. Also, during the test drive, it was even possible to hear and feel that there was a problem with the wheel. Here it is questionable whether complex machine learning algorithms are needed. Even for three loosened wheel bolts, some algorithms still achieve 100% AUC-PR. During this test drive, the influence of the loosened wheel bolts could neither be heard nor felt in the car. Likewise, no significant differences in the amplitudes and distribution can yet be observed in the raw acceleration signal. The difference becomes visible only in the frequency range, which explains why frequency-based features, in particular, still achieve AUC-PR values far above 90% here. In this case, a loose wheel bolt detection system would add value to the driver's safety. Even with two loosened wheel bolts, the best-performing algorithm LSTM-AD still achieves an AUC-PR of 97.5%. In the case of one loosened wheel bolt, the AUC-PR of almost all algorithms is the worst, which also corresponds to the expectation. No algorithm achieves AUC-PR values of more than 80.5% here. Interestingly,

Table 1. Performance comparison between all algorithms and acoustic features with optimal hyperparameters

	Metrics					Inputs and Bandwith in Hz								
Algorithm and Feature	avg. AUC-PR	AUC-PR 1 loose	AUC-PR 2 loose	AUC-PR 3 loose	AUC-PR 4 loose	LX	LY	LZ	RX	RY	RZ	w	h	Other Hyperparams
KNN + Mel Spectrum	0.853	0.598	0.867	0.947	0.999	[0, 2400]						8192	8192	$n_{mels} = 64,\ k = 2$
KNN + Magn. Spectrum	0.795	0.61	0.69	0.886	0.994	[4800, 5000]	[0, 3300]	[3200, 4200]	[0, 4900]	[800, 1300]	[2400, 2500]	8192	2048	$k = 5$
KNN + Raw Waveform	0.548	0.445	0.485	0.478	0.785	[1400, 5000]						8192	8192	$k = 21$
k-means + Mel Spectrum	0.809	0.553	0.823	0.883	0.997	[0, 2400]						8192	8192	$n_{mels} = 64,\ k = 7$
k-means + Magn. Spectrum	0.756	0.457	0.750	0.819	0.996	[300, 1600]	[500, 1600]	[3200, 3500]				4096	4096	$k = 91$
k-means + Raw Waveform	0.679	0.612	0.533	0.679	0.891	[3600, 5000]						4096	4096	$k = 5$
1D-CNN + Mel Spectrum	0.885	0.624	0.929	0.986	1	[3200, 5000]	[0, 5000]	[100, 1500]				4096	2048	$n_{mels} = 128,\ n_{filters} = 8,\ lr = 0.001,\ b = 64,\ e = 10$
1D-CNN + Magn. Spectrum	0.805	0.449	0.926	0.847	0.999	[500, 4100]	[3700, 5000]	[3200, 4200]				4096	4096	$n_{filters} = 64,\ lr = 0.01,\ b = 32,\ e = 60$
1D-CNN + Raw Waveform	0.685	0.622	0.514	0.706	0.900	[4800, 5000]						1024	128	$n_{filters} = 16,\ lr = 0.01,\ b = 32,\ e = 10$
LSTM-AD + Mel Spectrogram	0.945	0.805	0.975	0.999	1			[100, 2600]				8192	1024	$n_{mels} = 64,\ n_{wind} = 6,\ lr = 0.01,\ n_{lstm} = 3,\ b = 16,\ e = 50$
LSTM-AD + Spectrogram	0.884	0.513	0.884	0.957	0.988	[2700, 5000]	[200, 2500]	[2700, 5000]				4096	2048	$n_{wind} = 4,\ lr = 0.001,\ n_{lstm} = 3,\ b = 64,\ e = 20$
IDNN + Mel Spectrogram	0.915	0.773	0.886	1	1			[200, 1000]				8192	1024	$n_{mels} = 128\ n_{wind} = 5,\ lr = 0.001,\ b = 32,\ e = 30$
IDNN + Spectrogram	0.885	0.650	0.955	0.933	1	[1300, 5000]	[1200, 4900]	[500, 4200]				4096	1024	$n_{wind} = 3,\ lr = 0.0001,\ b = 128,\ e = 30$
MobileNetV2 + Mel Spectrogram	0.855	0.434	0.988	1	1	[300, 1100]	[4200, 5000]	[200, 1000]				2048	1024	$n_{wind} = 8,\ n_{mels} = 128,\ b = 32,\ lr = 0.001,\ e = 50$
MobileNetV2 + Spectrogram	0.778	0.603	0.760	0.747	1	[2200, 4000]	[3200, 5000]	[2700, 4500]				4096	4096	$n_{wind} = 8,\ lr = 0.01,\ b = 64,\ e = 30$
Dense AE + Mel Spectrum	0.876	0.680	0.824	1	1	[2100, 4300]	[3700, 5000]	[3600, 5000]				2048	256	$n_{mels} = 64,\ lr = 0.001,\ b = 16,\ e = 30$
Dense AE + Magn. Spectrum	0.852	0.777	0.756	0.885	0.991	[3100, 3800]			[800, 3300]			8192	1024	$lr = 0.001,\ b = 16,\ e = 30$
Dense AE + Raw Waveform	0.663	0.584	0.524	0.652	0.891	[1400, 5000]						4096	4096	$lr = 0.001,\ b = 16,\ e = 20$

there are even some algorithms, such as MobileNetV2, with spectrograms as a feature, where the performance drops very sharply between two and one loosened wheel bolt, in this case from 98.8% to only 43.4%.

A comparison of the performance of the acoustic features shows that Mel frequency-based features produced the best results for all algorithms. The magnitude spectra features performed slightly worse. This observation is likewise found in results from other audio classification tasks, such as DCASE, where Mel frequency-based features outperform regular magnitude spectra. This behavior is particularly interesting because the former is calculated from the latter using a triangular filter bank. The advantage here is that this operation significantly reduces the number of dimensions while retaining enough information from the frequency domain. The raw waveforms performed significantly worse, suggesting that learning from these meaningful representations that generalize well is more challenging.

In all but two combinations, the best results were obtained with the acceleration data from the upper left control arm, on whose side the wheel bolts were loosened. In the other two cases, the acceleration signals from the right side, where all wheel bolts were tight, were additionally included. This indicates that no comparison signals from the other side are generally needed to detect loose wheel bolts. The most common input was the acceleration in the X-direction. No clear tendencies can be identified when looking at the selected frequency ranges. Noticeably is that no frequency ranges that contain only low frequencies are selected. This means that if low frequencies are not filtered out, higher frequencies are also left in the signal up to at least 1000 Hz. On the other hand, satisfactory results are also achieved with avg. AUC-PR of over 87% when only frequencies above 2000 Hz are left in the signal, such as for the dense AE. Overall, we conclude that higher frequencies in the signal are more crucial for the detection of loose wheel bolts. In further consideration of the use case, interference frequencies of the engine and the road impacts must also be taken into account here, which influence the acceleration signal.

Comparing the performance of classical algorithms (KNN and k-Means) versus deep-learning-based algorithms shows that the latters achieve higher AUC-PR values. However, the differences are often only a few percentage points. LSTM-AD and IDNN achieve an avg. AUC-PR of 94.5% and 91.5%, respectively. Then their AUC-PR values are more than five percentage points above those of KNN. However, these are both the algorithms with the best overall performance. One possible reason they perform so well is that they can learn time-dependent patterns in the frequency spectrum. This is very helpful for our acceleration signal with a non-constant spectrum. MobileNetV2, with the outlier exposure approach, only achieves results similar to those of KNN. After careful hyperparameter optimization, very high AUC-PR values can also be achieved with less complex algorithms. This indicates that huge models or very deep networks are unnecessary to solve the use case. It is more important to select the right acoustic features and to optimize the other hyperparameters, such as the window length. Here, hyperparameter optimization has shown that large differences in the performance of algorithms arise when these are selected suboptimally. Finally, it should be noted that we only collected data from test drives in a limited speed range (approximately 30 km/h) and on a single road surface with a limited number of driving maneuvers. Further test drives on more different road sections and speeds need to be carried out to confirm the results.

6 Summary and Outlook

In this paper, we investigated the detection of loosened wheel bolts with acceleration sensors in the chassis. Test drives with tightened and loosened wheel bolts were carried out. Several state-of-the-art anomalous sound detection algorithms are introduced. Hyperparameter optimization is performed to find the best acoustic features and parameters for the specific algorithms. The best-performing algorithm is the forecasting-based LSTM-AD algorithm, with an avg. AUC-PR of 94.5%. From two loosened wheel bolts onwards, these are already detected very reliably by LSTM-AD. The study shows that classical algorithms such as KNN can also detect loose wheel bolts with an AUC-PR of 94.7% if three out of five are loosened. The most important factor influencing the performance is not the algorithm but the careful selection of the acoustic features and the window length. In future work, the extent to which the anomalies of loosened wheel bolts differ from other possible anomalies occurring, such as low tire pressure or worn components, should be investigated. In addition, future work should investigate how well the algorithms generalize and work on vehicles other than the test vehicle.

References

1. Bagnall, A., Lines, J., Bostrom, A., Large, J., Keogh, E.: The great time series classification bake off: a review and experimental evaluation of recent algorithmic advances. Data Min. Knowl. Disc. **31**(3), 606–660 (2016). https://doi.org/10.1007/s10618-016-0483-9
2. Bernhard, J., Schmidt, J., Schutera, M.: Density based anomaly detection for wind turbine condition monitoring. In: Proceedings of the 1st International Joint Conference on Energy and Environmental Engineering - CoEEE, pp. 87–93. INSTICC, SciTePress (2022). https://doi.org/10.5220/0011358600003355
3. Braei, M., Wagner, S.: Anomaly detection in univariate time-series: a survey on the state-of-the-art. CoRR (2020). https://doi.org/10.48550/arXiv.2004.00433
4. Davis, J., Goadrich, M.: The relationship between precision-recall and ROC curves. In: Proceedings of the 23rd International Conference on Machine Learning, ICML 2006, pp. 233–240. Association for Computing Machinery, New York (2006). https://doi.org/10.1145/1143844.1143874
5. Dodd, M.: Heavy vehicle wheel detachment and possible solutions-phase 2-final report (2010)
6. Dohi, K., et al.: Description and discussion on DCASE 2022 challenge task 2: unsupervised anomalous sound detection for machine condition monitoring applying domain generalization techniques (2022). https://doi.org/10.48550/ARXIV.2206.05876
7. Hendrycks, D., Mazeika, M., Dietterich, T.: Deep anomaly detection with outlier exposure. In: Proceedings of the International Conference on Learning Representations (2019). https://doi.org/10.48550/ARXIV.1812.04606
8. Hägg, J.: Loose wheel indicator selected by Audi for a range of car models (2019). https://www.mynewsdesk.com/se/nira-dynamics/pressreleases/loose-wheel-indicator-selected-by-audi-for-a-range-of-car-models-2966123

9. Straßenverkehrs-ordnung (stvo) §23 sonstige pflichten von fahrzeugführenden (2013). https://www.gesetze-im-internet.de/stvo_2013/_23.html. Accessed 04 Mar 2023
10. Keller, J.M., Gray, M.R., Givens, J.A.: A fuzzy k-nearest neighbor algorithm. IEEE Trans. Syst. Man Cybern. 580–585 (1985)
11. Kingma, D.P., Ba, J.: Adam: a method for stochastic optimization (2014). https://doi.org/10.48550/ARXIV.1412.6980
12. Koizumi, Y., et al.: Description and discussion on DCASE2020 challenge task2: unsupervised anomalous sound detection for machine condition monitoring (2020). https://doi.org/10.48550/ARXIV.2006.05822
13. Krishna, K., Murty, M.N.: Genetic k-means algorithm. IEEE Trans. Syst. Man Cybern. Part B (Cybern.) 433–439 (1999)
14. Verordnung über sicherheit und gesundheitsschutz bei der verwendung von arbeitsmitteln (betriebssicherheitsverordnung - betrsichv) §4 grundpflichten des arbeitgebers (2015). https://www.gesetze-im-internet.de/betrsichv_2015/_4.html. Accessed 04 Mar 2023
15. Logan, B.: Mel frequency cepstral coefficients for music modeling. In: International Society for Music Information Retrieval Conference (2000)
16. Lyon, D.A.: The discrete Fourier transform, part 4: spectral leakage. J. Object Technol. 23–34 (2009). https://doi.org/10.5381/jot.2009.8.7.c2
17. Malhotra, P., Vig, L., Shroff, G., Agarwal, P., et al.: Long short term memory networks for anomaly detection in time series. In: Proceedings of the European Symposium on Artificial Neural Networks, Computational Intelligence and Machine Learning (ESANN) (2015)
18. Nannavecchia, A., Girardi, F., Fina, P.R., Scalera, M., Dimauro, G.: Personal heart health monitoring based on 1D convolutional neural network. J. Imaging (2021)
19. O'Shea, K., Nash, R.: An introduction to convolutional neural networks. CoRR (2015). https://doi.org/10.48550/arXiv.1511.08458
20. Sandler, M., Howard, A., Zhu, M., Zhmoginov, A., Chen, L.C.: MobileNetV2: inverted residuals and linear bottlenecks (2018). https://doi.org/10.48550/ARXIV.1801.04381
21. Schmidl, S., Wenig, P., Papenbrock, T.: Anomaly detection in time series: a comprehensive evaluation. Proc. VLDB Endow. 1779–1797 (2022). https://doi.org/10.14778/3538598.3538602
22. Smith, J.O.: Mathematics of the Discrete Fourier Transform (DFT): With Audio Applications. BookSurge Publishing (2008)
23. Socoró, J.C., Alías, F., Alsina-Pagés, R.M.: An anomalous noise events detector for dynamic road traffic noise mapping in real-life urban and suburban environments. Sensors **17**(10) (2017). https://doi.org/10.3390/s17102323
24. Suefusa, K., Nishida, T., Purohit, H., Tanabe, R., Endo, T., Kawaguchi, Y.: Anomalous sound detection based on interpolation deep neural network. In: 2020 IEEE International Conference on Acoustics, Speech and Signal Processing (ICASSP), ICASSP 2020, pp. 271–275 (2020). https://doi.org/10.1109/ICASSP40776.2020.9054344

Clustering ECG Time Series for the Quantification of Physiological Reactions to Emotional Stimuli

Beatriz Henriques[1(✉)], Susana Brás[1,2], and Sónia Gouveia[1,2]

[1] Institute of Electronics and Informatics Engineering of Aveiro (IEETA) and Department of Electronics, Telecommunications and Informatics (DETI), University of Aveiro, Aveiro, Portugal
{beatriz.henriques,susana.bras,sonia.gouveia}@ua.pt
[2] Intelligent Systems Associate Laboratory (LASI), University of Aveiro, Aveiro, Portugal

Abstract. Emotion recognition systems aim to develop tools that help in the identification of our emotions, which are related to learning, decision-making and treatment and diagnosis in mental health contexts. The research in this area explores different topics ranging from the information contained in different physiological signals and their characteristics, to different methods aiming feature selection and classification tasks. This work implements a dedicated experimental protocol, consisting of sessions to collect physiological data, such as the electrocardiogram (ECG), while the participants watched emotional videos to provoke reactions of fear, happiness, and neutrality. Data analysis was restricted to features extracted directly from the ECG, being possible to verify that the intended stimuli effectively provoked variation in the heart rhythm and other ECG features of the participants. In addition, it was observed that each emotional stimulus presents different degrees of reactions that can be clearly distinguished by a clustering procedure.

Keywords: Emotion classification · affective computing · ECG · cluster analysis · pattern recognition

1 Introduction

Emotions are in the basis of cognition, learning, and decision-making. Therefore, the understanding and description of emotional context allow the interaction

This work was partially funded by FCT - Fundação para a Ciência e a Tecnologia (FCT), I.P., through national funds, within the scope of the UIDB/00127/2020 project (IEETA/UA, http://www.ieeta.pt/). S. Brás acknowledges the support by national funds, European Regional Development Fund, FSE through COMPETE2020, through FCT, in the scope of the framework contract foreseen in the numbers 4, 5, and 6 of the article 23, of the Decree-Law 57/2016, of August 29, changed by Law 57/2017, of July 19.

A. Pertusa et al. (Eds.): IbPRIA 2023, LNCS 14062, pp. 680–692, 2023.
https://doi.org/10.1007/978-3-031-36616-1_54

between different persons most easily and more productively. Good health and well-being are one of the United Nations' sustainable goals, in particular, the promotion of mental health [9]. Mental health problems are one of the major issues in modern society. Depression, anxiety, and stress are usual words nowadays. These mental health problems are usually associated with a magnification of negative feelings (fear, disgust, etc.), and difficulty in enjoying positive moments. Identifying and describing emotions allow us to interpret human psychological mechanisms.

A consensual definition is that an emotion is an event-focused, two-step, fast process consisting of relevance-based emotion elicitation mechanisms that shape a multiple and brief emotional response, for example, action tendency, automatic reaction, expression, and feeling [1]. According to this description, it is evident that the body reacts to emotion by alterations in mechanical, physical, chemical, and electrical activity. This evidence opens the opportunity to use sensors and monitoring to quantify body alterations through emotional stimulation. The research for emotion recognition systems is growing along with technological advancement. There are three important aspects in building this system: protocol; data processing; classification techniques. For the protocol, the norm over the literature is to carry out one session with a group of participants, where the subjects are stimulated by a defined modality while acquiring some physiological signals. Additionally, the subjects usually answer a questionnaire to report their emotions during the session/stimulus [3].

The vast majority of the literature makes use of audio-visual stimuli, such as movie or music video clips, because this is the most affordable setup that stimulates both visual and audio sensors. Other approaches, such as images, audio, and video games, are also implemented in the literature. More methods-oriented works involve the use of databases available online [3]. As an example, Bota et al. used the ITMDER, WESAD, DEAP, MAHNOB-HCI, and ESD databases to classify several emotions, such as anger, joy, stress, fear, amusement, sad, and neutral [2].

Emotion recognition systems are generally based on information from two different approaches. The explicit approach is built on visible information, such as facial expression, posture, speech, etc. On one hand, this approach is easier to capture for analysis. On the other hand, these signals can be influenced by the subject, which turns the information non-reliable. The second methodology, the implicit approach, uses physiological signals, which are generated inside the body by the autonomous nervous system. Physiological signals have been largely used for the detection of various diseases. Besides that, these signals can be also applied to evaluate various mental and psychological conditions, like emotions. Examples of physiological signals are electroencephalogram (EEG), electrocardiogram (ECG), electromyogram (EMG), and electrodermal activity (EDA) [3].

The final goal is to incorporate the most relevant information conveyed in the features extracted from the acquired physiological signals into an algorithm capable of correctly labelling the samples with the maximized performance. Choosing the proper classifier is crucial to ensure good performance. Numerous machine learning algorithms have been used in the existing studies, including Support Vector Machine (SVM), Neural Networks, k-Nearest Neighbours, and others [4].

The recent advances in machine learning and data analysis techniques have opened up new opportunities for exploring the relationship between physiological signals and emotions. However, few studies have focused on developing a comprehensive framework for characterizing degrees of emotions. Considering this information, the main objective of the present work is to address this gap by exploring the feasibility of characterizing the intensity of a given emotion based on physiological signals. Specifically, the study aims to establish a reliable physiological signals database from participants watching emotional videos, develop an exploratory data analysis framework to identify the most informative features and apply clustering techniques to evaluate discriminate degrees of intensity in an emotion. Thus, Sect. 2 presents the methodology used to acquire the experimental data and subsequent analysis. Section 3 presents the results and outlines the discussion. Finally, Sect. 4 presents the conclusions of this study.

2 Materials and Methods

Briefly, ECG, EDA and EMG signals were recorded simultaneously from each subject enrolled in the protocol while visualizing videos with different emotional content (56 participants). The physiological signals were processed and beat-to-beat ECG time series (namely HR) were extracted. The exploratory analysis focused on HR variations which feed a clustering procedure to disclosure classes with different levels of intensity to the emotional stimuli. All algorithms were implemented in Python, mainly using *Neurokit2* [5] (advanced biosignal processing tools), *SciPy* (statistics) and *scikit-learn* [8] (clustering algorithms and supervised learning) modules.

2.1 Experimental Protocol and ECG Acquisition

This study aims at all gender participants with Portuguese nationality or with Portuguese mother tongue, between 18 and 70 years old, with no psychological or psychiatric diagnosed disorders, no physical condition nor medication that could impact the data collection, and with normal or corrected-to-normal vision. This work was built on a previous research work [6] developed at the University of Aveiro (UA), and the protocol was already established and validated. The ethical approval was obtained by the Ethics Committee of the UA (12-CED/2020) and the Commissioner of Data Protection of the UA. The experiment took place in a room of the Institute of Electronic Engineering and Informatics of Aveiro (IEETA) in Aveiro, Portugal, from September 2020 through early January 2022.

The participants were recruited by invitation via e-mail with a sociodemographic questionnaire attached. This questionnaire accessed the social and demographic characteristics of the participants to select the subjects. The participants selected answered the questionnaires STICSA Trait, STICSA State, and TAS-20 in order to assess the traits of anxiety and evaluate the alexithymia construct. Additionally, at the beginning and at the end of the experimental session, the participants answered respectively the Vas-pre and Vas-pos questionnaires, which document the emotional state of the participants via a psychometric response scale regarding anxiety, happiness, fear, stress, arousal, and valence.

The experimental procedure consisted of recording the physiological signals while the participant was induced to experience different emotional states (Neutral (N), Happy (H), and Fear(F)) through the visualization of movie excerpts. All sessions were done in a controlled room, with no exterior light font. The subjects were instructed to put on the headphones, sit in a chair, and try to remain still during data acquisition. The data collection started automatically at the moment the video was initiated, allowing a temporal synchronization between the stimulus and the reaction. The initial 5 min of the presented video focused on a baseline measurement. The three emotional states were provoked by a 10-minute duration video, in which the order was randomized between participants.

To acquire the physiological signals the multi-sensor BioSignalPlux Explorer kit from PLUX®was used (https://www.pluxbiosignals.com/). Although other signals have been acquired, this study focuses on the ECG. Concerning the placement of the electrodes, a three-lead methodology was conducted, with the potential electrodes placed in the abdominal zone and the reference electrode above the left clavicle. The signals were collected with the software OpenSignals version 2.2.0.

2.2 ECG Pre-processing and Feature Extraction

The preprocessing of the ECG signals aimed the computational analysis of the signal to remove the noise, and detect and delineate the different waves of the cardiac cycle. This procedure was implemented with the Neurokit2 functions `nk.ECG_clean`, `nk.ECG_peaks`, `nk.ECG_delineate`, `nk.ECG_rate`, and `nk.hrv`, with the default inner parameters. Firstly, to improve peak-detection accuracy, the signal noise was removed with a 0.5 Hz high-pass Butterworth filter with order 5, followed by powerline filtering 50 Hz [5]. Then, the R-peaks were detected by locating the local maxima in the QRS complexes and removing artefacts using the method by Lipponen and Tarvainen [7]. The samples in which the peak, onset and offset of the waves P, R and T were returned by the delineation of the different waves of the cardiac cycles with prominence for the QRS complex. Also, the calculation of the instantaneous heart rate (HR) was processed with a monotone cubic interpolation.

With the information of activation's onsets, offsets, and peaks, the durations of waves, segments, and intervals of the cardiac cycle heartbeat to heartbeat were calculated, specifically, the duration of the T wave, P wave, ST interval, PR interval, TP interval, and QRS complex duration.

2.3 Detection of Time Instants with a Significant Group Response

The HR series of the participants were used to identify time instants exhibiting a significant physiological response based on a two-step procedure. The first step aims to identify the protocol's time instants with a significant response based on the group average. Then, the second step further inspects if the individual response, at the identified time instants, is significant with respect to the pairwise variations of the individual baseline.

The evaluation of the group response in the first step demands the individual HR evaluation at the same time instants in order to quantify the average response for the group. Since the heartbeats of different participants are not synchronized in the same time instants (and neither are of the same number) a piecewise cubic Hermite interpolating spline was applied for every individual ECG time series. The average response of the group was then evaluated after individual standardization of the HR data by detrending the time series, to diminish the effect of the variability between participants.

An envelop approach was performed to outline the various time series, as well as extract the average response. Through the application of the same strategy to the Baseline signal, a 95% confidence interval (CI) was calculated for each time instant of the group average. During a set of instances, if the CI did not include the value of zero, the set was considered a significant event. This is expected to occur either due to a rapid increase/decrease of the mean value, which indicates that a strong stimulus was produced at that moment, or a very narrow CI interval, which indicates a great synchronization between participants since the CI width is directly associated to the inter-participants variability. The same approach was applied to every time series and to the individual signals. The individual events were extracted in the condition of the presence of an individual significant instance during a group significant event.

The individual response to the stimulus was then evaluated for each significant group event. The intensity of the response was quantified from the individual signals used to compute the group average, considering two approaches: the area under the curve normalized by the duration of the event or the same using the absolute curve. Additionally, for each significant individual event, it was extracted the mean of the duration of waves, segments, and intervals previously mentioned as features of the event.

2.4 Cluster Analysis of Individual Responses

Recalling that in this study the emotional states are provoked by emotional videos with different scenes and consecutive different types of stimuli, it is natural to assume that although it is the same emotion stimulated, it has multiple degrees of emotion throughout the video. With this in mind, the individual responses as evaluated by the area under the curve were then processed into a cluster analysis in order to find groups of responses that exhibited high cohesion and high separability in terms of intensity. This analysis can provide valuable information about the degree of separation between two or more classes, as well as the specific attributes that contribute to optimal class separation.

Firstly, it was fundamental to define the number of divisions to be performed. There is no established answer to the number of optimal clusters since it is a subjective factor dependent on the method used for measuring similarities and the parameters used for partitioning. A state-of-the-art technique is the Elbow method. This method selects the optimal number of clusters by fitting the model with a range of values. If the line chart resembles an arm, then the point of inflexion on the curve, named the elbow, is a good indication that the underlying model fits best at that point. Additionally, a dendrogram was generated as a hierarchical clustering representation. The dendrogram is constructed by drawing a U-shaped link between a non-singleton cluster and its children, where the top of the U-link denotes a cluster merge, and the two legs of the U-link indicate the clusters that were merged. The length of the legs of the U-link represents the distance between the child clusters [8].

The optimal number of clusters was then considered into a K-means clustering algorithm to produce distinct non-overlapping clusters. This algorithm separates the data into k clusters of equal variance, minimizing the within-cluster sum of squares

$$\sum_{i=0}^{n} \min_{\mu_j \in C} \left(\|x_i - \mu_j\|^2 \right). \quad (1)$$

The classes were ordered by intensity, meaning, the higher the area the higher the intensity of the event and consequently the degree of the class. The degree was represented by the first letter of emotion followed by the degree number, e.g. the ith class of Fear is represented as "F_i" for different levels $i = 1, 2, \cdots, k$.

The evaluation of the clustering procedure was based on the idea that members belonging to the same class are more similar (cohesion) than members of different classes (separation). In this work, performance was evaluated from the Silhouette Coefficient (SC), Calinski-Harabasz index (CH) and Davies-Bouldin index (DBI) [8].

The SC is defined by the mean of the relation between the two scores for each sample. This score is bounded between -1 for clustering is not appropriate and 1 for highly dense clustering, where scores around zero indicate overlapping clusters.

$$S = \frac{1}{n} \sum_{i=1}^{n} \frac{b_i - a_i}{\max(a_i, b_i)}, \quad (2)$$

where a is the mean distance between a sample and all other points in the same class, and b is the mean distance between a sample and all other points in the next nearest cluster.

The CH, also known as the Variance Ratio Criterion, measures the similarity of an object to its cluster compared to other clusters. The index is the ratio of the sum of between-clusters dispersion and of within-cluster dispersion for all clusters, where dispersion is defined as the sum of distances squared. A higher CH score is associated with a model with a better-defined cluster.

$$CH = \frac{b}{a}, \tag{3}$$

The DBI has also been successfully used in studies of pattern recognition of physiological signals. The index is defined as the average similarity between each cluster In contrast to the previous metrics, a lower DBI is related to a model with better cluster separation, with 0 being the lowest possible score.

$$DB = \frac{1}{k}\sum_{i=1}^{k} \max_{i \neq j} \frac{\bar{d}_i + \bar{d}_j}{d_{ij}}, \tag{4}$$

where $\bar{d}$ is the average distance between each object and its cluster centroid, and $\bar{d}_i j$ is the Euclidean distance between the centroids of clusters i and j.

There are different techniques to deal with the problem of comparing two clustering approaches (agreement between different approaches) or even comparing the results of a clustering approach where the true classes are known a prior. An example of these evaluation methods is the Adjusted Rand Index, which is based on the Rand Index proposed in *Rand* work in 1971. This coefficient varies between 0 and 1, where a value 0 represents the formation of random clusters and 1 is when the formed clusters are equivalent to true values.

$$ARI = \frac{RI - E[RI]}{max(RI) - E[RI]}, \tag{5}$$

where $E[RI]$ e $max(RI)$ represents, respectively, the expected value and the maximum RI for two random clustering processes.

3 Results and Discussion

The database is constituted of the physiological data collected from 56 participants during the experimental procedure, as well as their answers to all the questionnaires. At a sociodemographic level, the participants were between 18 and 58 years old (31.2 ± 11.5). Regarding the genre, 34 participants were female and 22 were male, which consists of 61% female and 39% male. Although two participants had another language as their mother tongue than Portuguese, all of them had Portuguese or Brazilian nationality, with 95% and 5% respectively. Concerning academic qualifications, 45% had a bachelor's degree, 21% had finished high school, 20% had a master's degree, 12% are doctorate, and 2% had a professional course. Respecting coffee habits, most of the participants, 73%, had the habit of drinking three or more coffee a week, followed by 16% of participants that do not drink coffee, 9% that drink once or twice a week, and 2% of participants that drink once a week. Lastly, 34% of participants reported practising physical exercise three or more times a week, 32% no exercise, 27% once or twice a week, and 7% at least once a week.

The average group response was analysed in the Baseline moment and in the three emotions (F, H, N). For the Baseline, the group profile presents a small deviation from zero, where the CI for the group averaged signal does not contain the zero value in 9% of the protocol time instants. In the initial phase of the Baseline condition, an HR peak was detected as a consequence of a "short-term nervousness" feeling at the beginning of the session. This was followed by a slow decrease in the average HR curve representing the physiological recovery of the "short-term nervousness" feeling. Therefore, this first minute of Baseline recording was disregarded from the analysis as it did not characterize a baseline behaviour. The group profile for the Neutral emotion is similar to that of the Baseline, with 10% of time instances where zero was outside of the CI interval. This is in agreement with the protocol since the Neutral stimulation is intended to provoke an emotion with neutral arousal or valence i.e. no stimulation. The Happy emotion exhibits moments of stimulation with a more evident group response, where 32% of time instances had the CI excluding the zero and, in general, the deviations towards zero were superior with respect to the Baseline. These results clearly indicated a relevant reaction of the group to the sound and comedy moments of the movies. Finally, the mean profile for the Fear emotion had a larger deviation towards zero where 30% of the time instances had the CI excluding the zero and many events of positive area were observed. These peaks over the zero are coincident with the time instants of the jump scare moments of horror movies after which follows a recovery moment characterized by a concave-shaped curve with negative values. Around 30% of CI excluded the zero for Happy and Fear. While Fear induced fast changing increasing HR values (which respect to Baseline) followed by a slowing decay of recovery, Happy produced a reaction prone to deviate from zero more often towards the negative values. As Fear and Happy hold opposite valence dimensions, this result should be further explored to investigate its connections to valence.

For illustrative purposes, Fig. 1 represents a time excerpt of several ECG extracted time series for the Fear emotion, distinguishing reaction (red) and recovery (blue). The group response displayed in Fig. 1(a) shows a reaction characterized by a positive peak. For the same time excerpt, Fig. 1(b) shows three selected individual responses with different (individual) intensities in the reaction to the stimulus. Two participants react similarly to the group (i.e. HR increase with respect to its Baseline) while one participant presented a more neutral response. This result demonstrates that the protocol provides an effective method of stimulation but it is quite dependent on the participant. Finally, Fig. 1(c–d) exhibit the time series of beat-to-beat QRS complex duration and ST interval duration demonstrating that the protocol elicits changes in the duration of the cardiac cycle waves.

The average group response for the 56 participants during the F, H, and N emotions allowed the detection of, respectively, 54, 87 and 26 significant events. However, in the detection of the significant individual responses, respectively, only 489, 1002, and 122 events were detected. Furthermore, some participants showed no significant individual responses to any of the emotions. These results highlight the need for a careful data selection mechanism since there are many time instances where the group shows a significant reaction while a particular individual does not hold a significant reaction to the intended emotion.

The intensity of the reaction was proposed to be quantified from areas of the curve or of its absolute value. Figure 2 shows the boxplots of quantified intensities by these two approaches. Despite the negative areas in the Happy emotion being camouflaged with the normalization, this approach creates a greater variability between emotions.

After this analysis, it was clear the presence of significant events with different degrees of reaction, which lead to the analysis and clustering of the significant events. The dendrogram suggested two or three as the optimal number of clusters based on the intensity of the response for all emotions. Therefore, three clusters were used to differentiate low, medium and high-intensity responses for each emotion. This was performed by running the K-means algorithm with K= 3 for each emotion, where the input feature was the intensity. Table 1 gives the various performance metrics calculated to analyse the clusters generated by the two approaches to quantify intensity. Overall, these metrics demonstrate that this is an effective technique for identifying the different degrees of response to emotional stimulation. The approach based on absolute curves presents the

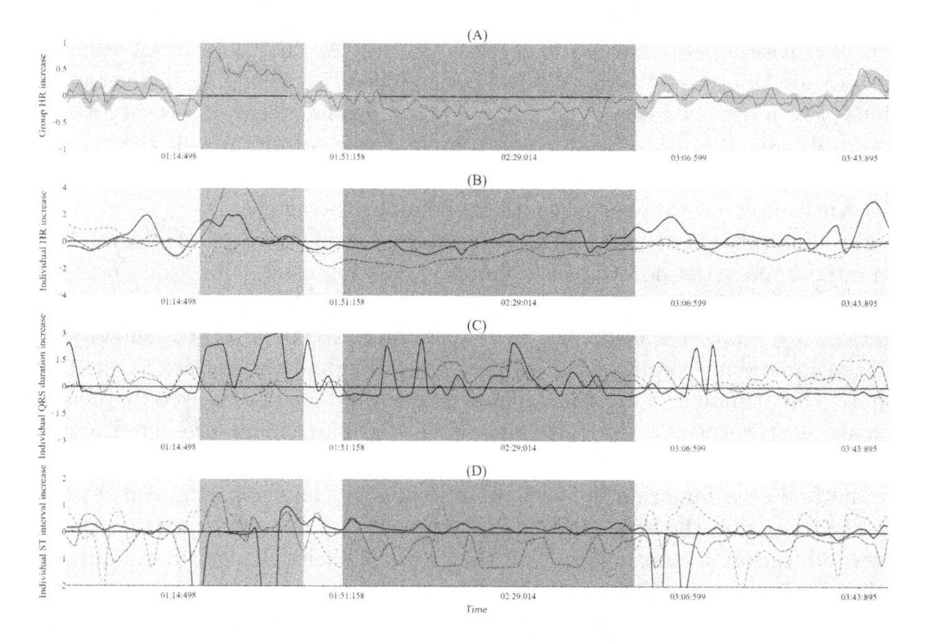

Fig. 1. Time series excerpt of the (A) HR group response, (B) HR individual response, (C) QRS interval individual response, and (D) ST interval individual response for the Fear emotion. The different elicitation moments are represented by the colour zones: The red zone shows a reaction event; the blue zone shows a recovery moment; the grey zone represents the remaining time series with an overall null response. (Color figure online)

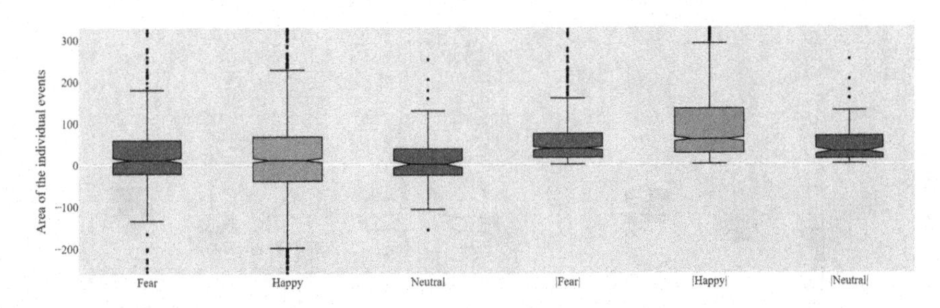

Fig. 2. Boxplot of the distribution of the areas of the significant individual events for the two approaches. The colour scheme relates green to Fear, yellow to Happy, and grey to Neutral, and is applied in all this study. (Color figure online)

highest performance, which supports the use of the absolute curve to quantify the intensity of the reaction.

Table 1. Performance results of the clustering carried out for each emotion where the intensity of the response is quantified from the area under the curve normalized by the duration of the event (A) or the same using the absolute curve (|A|).

	Fear		Happy		Neutral	
Performance metrics	A	|A|	A	|A|	A	|A|
Silhouette Coefficient	0.52	0.63	0.52	0.57	0.55	0.61
Calinski-Harabasz	775.15	1994.89	1587.39	2787.75	200.46	916.00
Davies-Bouldin index	0.59	0.36	0.58	0.49	0.54	0.39

To compare the two approaches, the Adjust Rand Index was computed. The Fear emotion obtained an index of 0.16, which points to random labelling. Moreover, the Happy and Neutral emotions obtained 0.39 and 0.59.

Additionally, in Fig. 3 the boxplot of the created clusters for the three emotions was computed for both approaches. It is clear that the normalization with the Baseline is extremely important to emphasize the variations between clusters, especially in the Fear emotion, since the significant events are related to the deviation of the values. While the heart rate increases with intensity, the ST and TP intervals decrease.

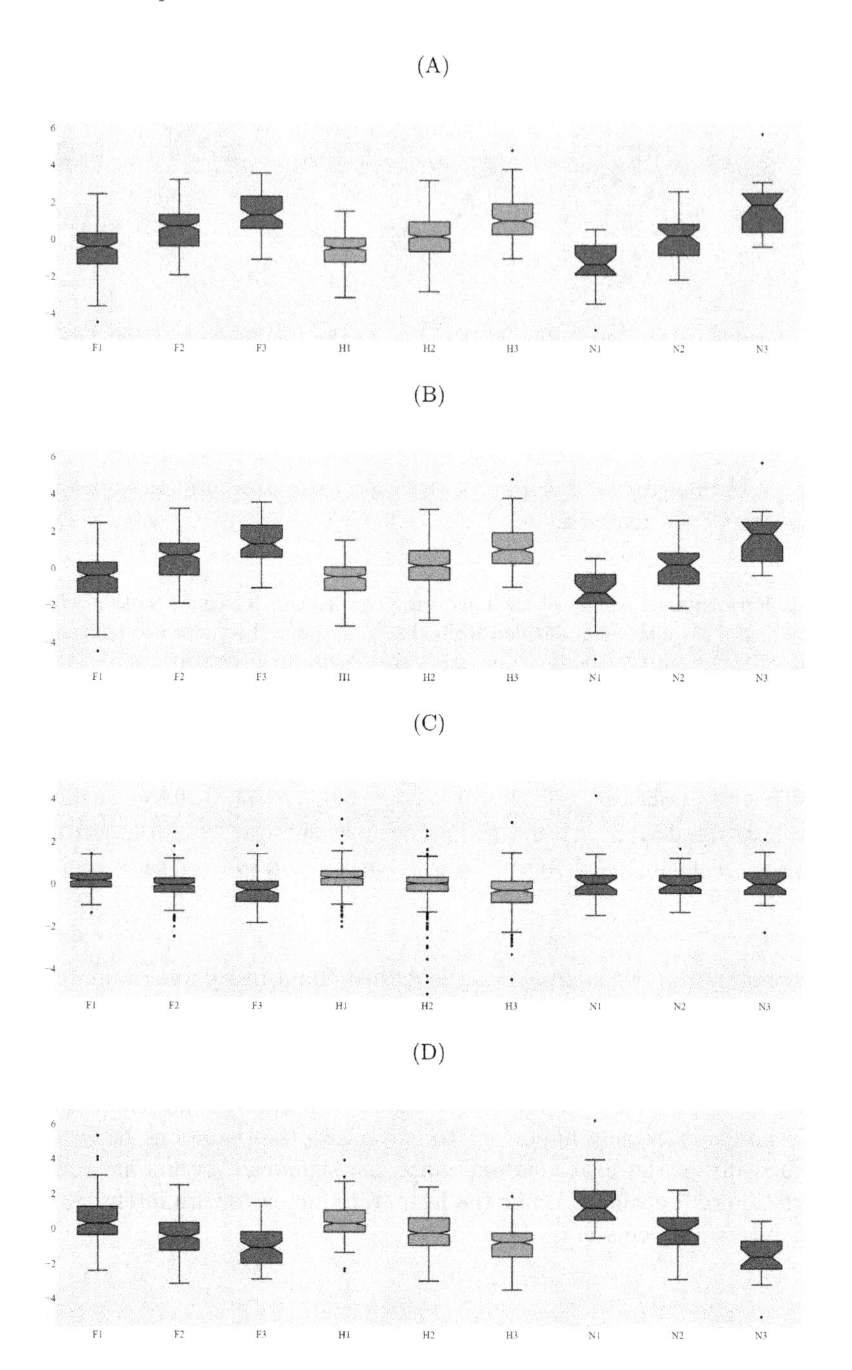

Fig. 3. Boxplots showing the distribution of the averaged feature values (A) Heart rate, (B) Heart rate normalized, (D) ST interval normalized and (D) TP interval normalized. Averages are obtained over all the events of each subject. The horizontal axis identifies the cluster of each emotion.

4 Conclusions

In this study, the implementation of the protocol enabled the collection of physiological signals using movies as a method of stimulation for the emotions of Fear, Happy and Neutral. From the questionnaire's answers, it was concluded that the emotional videos produced an accordance emotional response with the expected, as well as a uniform and general response among all the participants. This protocol demonstrated adequate results for affective data generation, along with the advantage of being easily reproducible.

In the exploratory analysis, the normalization of the HR data by the baseline was carried out to minimize inter-subject variability. In the profile graph, it was visually verified that the Fear emotional videos provoke several rapid increases in the HR, which were associated with the jump scare moments. Contrasting, the Happy profile presented negative variations. Also, both Happy and Fear data demonstrated a higher percentage of significant moments, which supported that the protocol provoked the expected stimuli in the participants: a larger response for Happy and Fear rather than for Neutral and Baseline segments.

With respect to the individual responses, it was clear from the data that different individuals do not share the same intensity in their reactions to the same stimuli. Therefore, the clustering procedure was applied to the individual responses by just taking into consideration the intensity of the response, quantified from the integral of the absolute (individual) curve in time instants of significant group response. The clustering performance metrics demonstrated that this approach constitutes an efficient technique to identify the different degrees of response to an emotional stimulation.

This study presents a novel solution to the classification problem underlying emotional stimulation and physiological responses. In conclusion, the application of this approach in machine learning classifiers, such as SVM, can enable the quantification of response intensity given the physiological condition, ultimately leading to revolutionary advancements in practical applications across various fields, from healthcare and marketing to industrial engineering.

References

1. Meiselman, H.L.: Theoretical Approaches to Emotion and Its Measurement. Woodhead Publishing, Sawston (2021)
2. Bota, P., Wang, C., Fred, A., Silva, H.: Emotion assessment using feature fusion and decision fusion classification based on physiological data: are we there yet? Sensors **20**(17), 1–17 (2020)
3. Shu, L., et al.: A review of emotion recognition using physiological signals. Sensors **18**(7), 2074 (2018)
4. Bulagang, A.F., Weng, N.G., Mountstephens, J., Teo, J.: A review of recent approaches for emotion classification using electrocardiography and electrodermography signals. Inform. Med. Unlocked **20**, 100363 (2020)
5. Makowski, D., et al.: NeuroKit2: a python toolbox for neurophysiological signal processing. Behav. Res. Meth. **53**(4), 1689–1696 (2021). https://doi.org/10.3758/s13428-020-01516-y

6. Henriques, B.: Machine learning in physiological signals for emotion classification (Master's thesis). University of Aveiro, Portugal (2022)
7. Lipponen, J.A., Tarvainen, M.P.: A robust algorithm for heart rate variability time series artefact correction using novel beat classification. J. Med. Eng. Technol. **43**(3), 173–181 (2019)
8. Pedregosa, F., et al.: Scikit-learn: machine learning in python. J. Mach. Learn. Res. **12**, 2825–2830 (2011)
9. United Nations: Goal 3: Good Health and Well-being — The Global Goals. https://www.globalgoals.org/goals/3-good-health-and-well-being/. Accessed 6 May 2023

Predicting the Subjective Responses' Emotion in Dialogues with Multi-Task Learning

Hassan Hayat, Carles Ventura(✉), and Agata Lapedriza(✉)

Universitat Oberta de Catalunya, 08018 Barcelona, Spain
{hhassan0,cventuraroy,alapedriza}@uoc.edu

Abstract. Anticipating the subjective emotional responses of the user is an interesting capacity for automatic dialogue systems. In this work, given a piece of a dialog, we addressed the problem of predicting the subjective emotional response of the upcoming utterances (i.e. the emotion that will be expressed by the next speaker when the speaker talks). For that, we also take into account, as input, the personality trait of the next speaker. We compare two approaches: a Single-Task architecture (ST) and a Multi-Task architecture (MT). Our hypothesis is that the MT architecture can learn a richer representation of the features that are important to predict emotional reactions. We tested both models using the Personality EmotionLines Dataset (PELD), which is the only publicly available dataset in English that provides individual information about the participants. The results show that our proposed MT approach outperforms both the ST and the state-of-the-art approaches in predicting the subjective emotional response of the next utterance.

Keywords: Subjective emotional responses · Dialogue systems · Multi-Task (MT) learning

1 Introduction

Dialogue systems are an important tool to achieve intelligent interaction between humans and machines [1–3]. In the quest for generating more human-like conversations, researchers are designing dialogue systems that are more understandable not only in contextual meaning but also emotionally [4–7]. Emotional understanding is an essential feature for many conversation scenarios such as social interaction and mental health support [8,9].

In this paper, we address the problem of predicting the emotion of the next utterance in a dialogue. Concretely, given $t - 1$ turns of dialogue, the goal is to predict the emotional response of the next utterance t with respect to the speaker (see Fig. 1). This problem was introduced by [10], where the authors state that predicting the emotion of the next turn can be used to automatically select the emotion for response in the conversation. Additionally, we think that

A. Pertusa et al. (Eds.): IbPRIA 2023, LNCS 14062, pp. 693–704, 2023.
https://doi.org/10.1007/978-3-031-36616-1_55

the prediction of the emotion of the next turn can also help to anticipate the user's emotional response, which is also interesting to create dialogue systems, in order to adapt their communication style to the user's preferences. In our study, we followed two emotional models: (*i*) **3 classes**, where the emotion of the next utterance is classified into negative, neutral, and positive emotions, and (*ii*) **7 classes**, where the emotion of the next utterance is classified into anger, disgust, fear, joy, neutral, sadness, and surprise classes.

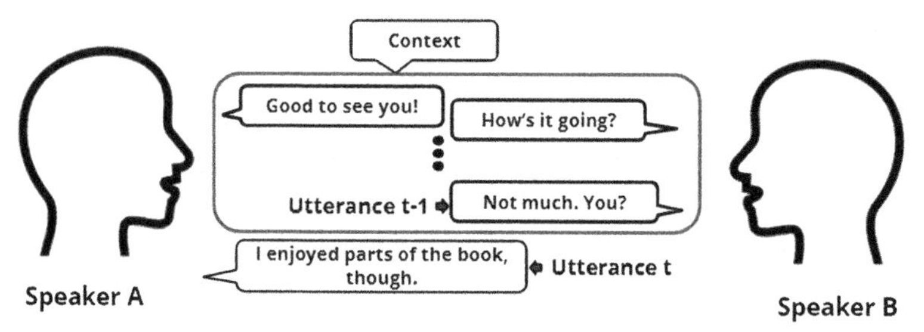

The task is to predict the emotion category of the upcoming utterance *(i.e. I enjoyed parts of the book, though)* with respect to Speaker A. The emotion category of the upcoming utterance will be estimated from the context *(represented in red outline)*, preceding emotions E_i and eventually, the personality traits of the Speaker A.

Fig. 1. This figure replicates the scenario of a typical conversation between two speakers. The first utterance is from Speaker A, the second is from Speaker B, and so on. In this example, our aim is to predict the emotion category of the upcoming utterance (i.e. utterance t) with respect to Speaker A.

Our work differs from previous approaches because it addresses the prediction of the emotion in a Multi-Task manner, with a model that can learn the particularities of each speaker regarding to their emotional reactions in the same dialog context. Furthermore, our model also takes as input the personality trait of the speaker as in [10]. In fact, most existing works on incorporating emotional intelligence into the dialogue systems (see Sect. 2 for a review of related work) ignore the individual differences in emotional reactions. For example, Zhou et al. [4] only highlighted the issue of emotional subjectivity in their research work and left it for future work. On the contrary, Wen et al. [10] explicitly consider emotional subjectivity while predicting the emotional responses of upcoming utterances. More concretely, the authors proposed a Personality-Affected Emotion Transition architecture, which takes the personality information into account while predicting the emotion class of the next utterance. The authors first created the dialogue dataset with emotions and personality labels, called Personality EmotionLines Dataset (PELD). Currently, PELD is the only publicly available dataset for emotional reaction prediction in dialogues that contains individual information (personality traits) about the participants. Later, the authors proposed a deep-learning architecture to predict emotional responses. Their experiments showed that considering personality traits as an input outperforms a generic model that does not take the personality traits of the next speaker into account.

Our work builds on top of [10] to better predict the emotional response of the next utterance with respect to each individual speaker. In particular, we are also taking into account the personality trait of the next speaker to address subjectivity. However, instead of creating a single model where the personality trait is one of the inputs, we propose a Multi-Task architecture with a separate output head for each speaker. That means each head is only responsible to learn the subjective emotional responses with respect to a specific speaker. Our hypothesis is that the Multi-Task (MT) architecture can learn a richer feature representation (with respect to ST) to predict emotional reactions with respect to each individual speaker. The Multi-Task (MT) architecture is presented in Sect. 3.2.

Our experimental results show that the MT architecture, where multiple subjective emotional responses are modeled jointly, outperforms the ST architecture, as well as the architecture proposed by [10] in most of the cases[1]. The interest of the obtained results are the following. First, we empirically show that the subjective emotional responses of different people are related tasks that can take benefit from a Multi-Task modeling approach. Second, our results support the findings of [10] about the interest in considering individual information when modeling emotional responses. We expect that these results motivate the practice of collecting individual information of the participants in future dataset collection efforts.

2 Related Work

In the context of dialogue systems, most of the work done around considering subjective information has focused on the generation of the next utterance, such as [11–13]. In turn, other works considering subjectivity have explored the generation of empathic responses. For example, a virtual robot named "Zara the Supergirl" [14] was developed in 2016. One of the goals of this virtual robot was to empathize with users during the interaction. In order to generate subjective empathizing responses with respect to the user, the authors considered the personality information of users during training. Another study on subjective empathic response was proposed by Zhong et al. [15]. In this work, the authors developed a model called CoBERT, which efficiently generates the empathic response in a conversation based on subjective information about the users. Similarly, Zandie et al. [16] proposed a multi-head Transformer model called EmpTransfo for generating the emotional response. The proposed architecture recognizes the emotions of the user first and then generates empathetic responses.

On the other hand, predicting subjective emotional responses to the next utterance is rarely approached in previous research. The potential reason is that

[1] Our code and the supplementary materials are available at: https://github.com/HassanHayat08/Predicting-the-Subjective-Responses-Emotion-in-Dialogues-with-Multi-Task-Learning.

most state-of-the-art dialogue datasets do not have any subjective information about speakers annotated with dialogues. Li et al. [17] developed a multi-turn dialogue dataset based on daily communications. They manually labeled the emotional keywords in each utterance into act emotion classes. Chen et al. [18] developed EmotionLines dataset in which each utterance of dialogue is annotated with emotional labels. These labels are selected based on the textual content presented in each individual utterance. An enhanced version of EmotionLines dataset is proposed by Poria et al. [19] named MELD. In this dataset, each utterance is annotated into emotion and sentiment labels. Another similar dataset in which the utterances are only annotated with emotion labels based on their contextual representation was developed by Chatterjee et al. [20].

To our best knowledge, Personality EmotionLines Dataset (PELD) [10] is the first dyadic dataset in which dialogues are annotated with emotions and the personality traits of speakers. The authors also proposed a deep learning system that uses personality traits while predicting the emotional responses of the next utterance. To automatically select the subjective emotional responses, the system first simulates the transition of emotions in the conversation. Then this transitioned emotion is triggered by two factors: the preceding dialogue context and the personality traits. Finally, the response emotion is the sum of the preceding emotion and the transitioned emotion. In our work, the emotional responses E_r of the next utterance t with respect to a speaker are also based on contextual understanding of the preceding $t - 1$ utterances, the emotions expressed in the previous utterances E_i, and the personality traits P of a speaker. However, we compare two types of architectures: (*i*) Single-Task (ST) and (*ii*) Multi-Task (MT). In the ST architecture, each time a separate model is trained that only predicts the emotional responses with respect to a single speaker. On the other hand, in the MT architecture, we train a unique model to predict the emotional responses with respect to each individual speaker jointly, exploiting in one single model all the available training data while still learning the subjective emotional responses of each individual.

3 Deep Learning Architectures for Emotional Response Prediction

In this section, we describe the two architectures tested in our work (i.e. Single-Task (ST) and Multi-Task (MT)) as well as the implementation details.

3.1 Single-Task (ST) Architecture

In our Single-Task (ST) learning, each time the emotional response prediction model is trained on dialogues that are associated with one single speaker. This means that the model has the capability to generate subjective emotional responses with respect to that particular speaker. The ST architecture is illustrated in Fig. 2.

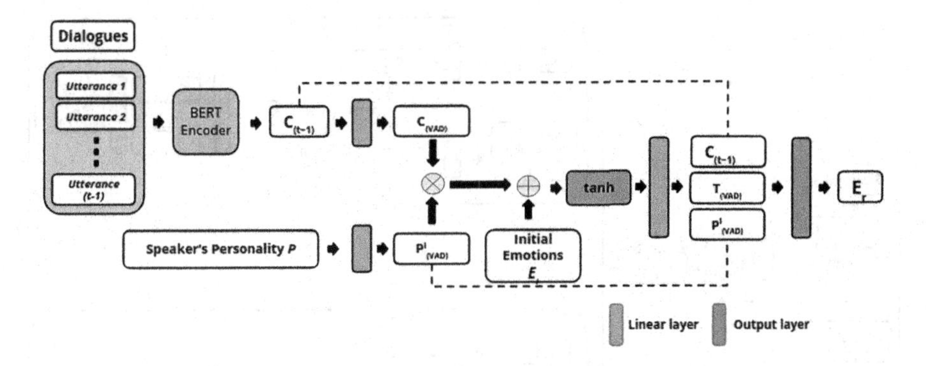

Fig. 2. Architecture of the Single-Task model. At a time, the model only considers a single speaker, and the training is done independently per speaker. Section 3.3 provides the details on the computation of $C_{(VAD)}$, $P_{(VAD)}$, and $T_{(VAD)}$.

3.2 Multi-Task (MT) Architecture

Unlike ST architecture, where ST always predicted single subjective emotional responses, the Multi-Task (MT) has the capability to predict multiple subjective emotional responses. The MT architecture is illustrated in Fig. 3. The MT architecture consists of a single BERT encoder connected with multiple separate branches. Each branch only designates to learning single subjective emotional responses. In a single training loop, the model only considers the dialogues that are associated with the same personality information. After getting the semantics representation of the dialogues, only a single branch that is associated with that personality information is active for further processing. The objective function is calculated using dialogue labels associated with that personality information and at a backpropagation step, only the selected branch and the BERT encoder hyperparameters are updated. Similarly, for a second training loop, dialogues that are associated with another personality information are selected and processed accordingly. This is iteratively done until all the available dialogues having unique personality information available in the dataset are processed.

3.3 Implementation Details

Our ST architecture is inspired by [10], which achieves state-of-the-art results with the PELT dataset. Before feeding the dialog and the personality information into the model, we perform two preliminary steps (which we describe below).

Preliminary Steps: First, convert the categorical emotions into continuous emotional space known as VAD (Valence, Arousal, and Dominance) [21]. Concretely, each utterance in the dialogue is categorized into six basic emotions [22]: Anger, Disgust, Fear, Joy, Sadness, and Surprise. Russell et al. [23] proposed an analysis that is used to convert categorical emotions into the VAD emotional

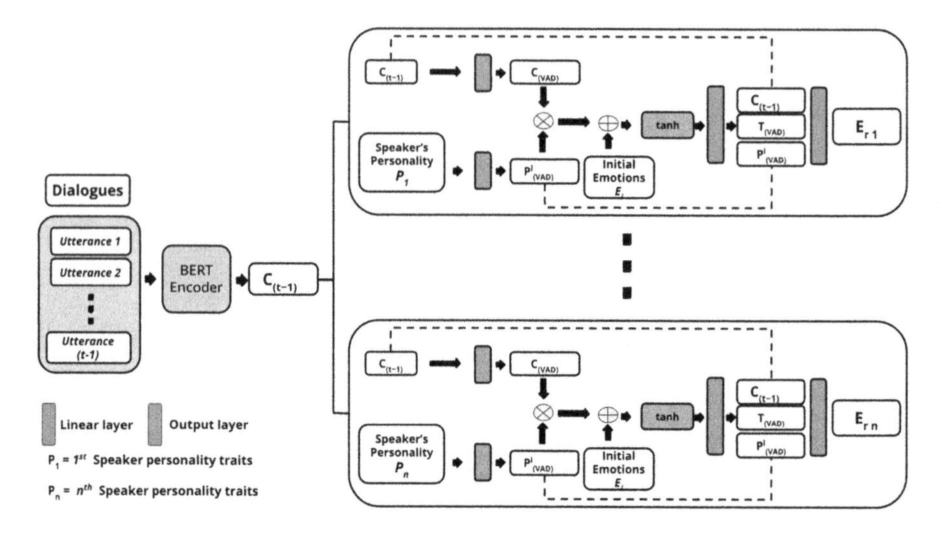

Fig. 3. Architecture of the Multi-Task model. The model has multiple separate branches which are designated for each speaker. Each branch learns speaker-specific emotional responses by using the speaker's personality traits. Section 3.3 provides the details on the computation of $C_{(VAD)}$, $P_{(VAD)}$, and $T_{(VAD)}$.

space (see Table 1). The VAD space indicates emotion intensity in three different dimensions, where Valence measures the positivity/negativity, Arousal the agitation/calmness, and Dominance the control/no-control.

Table 1. Categorical emotions into the VAD emotional space [23].

Emotion Category	Corresponding VAD Vector
Anger	(−0.51, 0.59, 0.25)
Disgust	(−0.60, 0.35, 0.11)
Fear	(−0.62, 0.82, −0.43)
Joy	(0.81, 0.51, 0.46)
Neutral	(0.00, 0.00, 0.00)
Sadness	(−0.63, −0.27, −0.33)
Surprise	(0.40, 0.67, −0.13)

Secondly, we estimate the valence, arousal, and dominance expression of a speaker using personality information, i.e. big five (OCEAN) traits. For this, we use a temperament model that was developed by Mehrabian et al. [24]. The model is derived through a linear regression to show the VAD emotional scale of the personality traits as specified in the following equation:

$$\begin{aligned} P_A &= 0.15(O) + 0.30(A) - 0.57(N) \\ P_V &= 0.21(E) + 0.59(A) + 0.19(N) \\ P_D &= 0.25(O) + 0.17(C) + 0.60(E) - 0.32(A) \end{aligned} \tag{1}$$

where, P_V is the personality-influenced valence, P_A is the personality-influenced arousal, and P_D represents the personality-influenced dominance emotional vectors. O for Openness, C for Conscientiousness, E for Extraversion, A for Agreeableness, and N for Neuroticism.

Contextual Understanding: In order to predict the emotional expression of the upcoming utterance t, it is necessary to understand the context of all the preceding $t-1$ utterances. For this, the BERT Base [25] model is fine-tuned to get the textual embeddings of $t-1$ utterances. BERT is a famous pre-trained language model whose performance is widely validated in many natural language tasks. BERT encodes each utterance into a $768 - \mathit{dimensional}$ vector.

Learning Personality-Based VAD Vector: After the conversion of the big five personality traits vector into the VAD emotional vector, still, the VAD values are not the true representation of emotions with respect to the data. The reason behind this is that the temperament model [24] was based on the analysis of 72 participants and hence represents the weights related to the data generated by these 72 participants. In order to use this VAD emotional vector in our experiments, it is necessary to learn the appropriate weights with respect to the underlying data. This is why a linear layer that transforms $P_{(VAD)}$ to $P^l_{(VAD)}$ is applied.

Contextual-Based Emotional Variations: The dialogue context is one of the main factors in order to generate a certain emotion in the speaker while speaking an utterance [19]. Since the dialogues consist of multiple utterances, the emotional representation also changes with respect to each utterance. It means that the dialogue as a whole represents the transition of emotions from the first utterance to the last utterance. Similarly, in order to generate emotional responses for the next utterance, the preceding variations of emotions should be known. To compute this emotional variation, the dialogue context $C \in \{U_1, U_2, ..., U_{(t-1)}\}$ is encoded into emotional space, i.e. $C_{(VAD)}$ (see Eq. 2).

$$\begin{aligned} C_{(t-1)} &= [B_r(U_1), B_r(U_2), ..., B_r(U_{(t-1)})] \\ C_{(VAD)} &= LinearLayer(C_{(t-1)}) \end{aligned} \tag{2}$$

where B_r is the BERT-Base encoder, $C_{(t-1)}$ is the contextual semantics of preceding utterances, and $C_{(VAD)}$ is the context-based emotional variations presented in the preceding utterances.

Personality-Based Emotional Variations. After obtaining the weighting parameters of the personality $P^l_{(VAD)}$ and the contextual-based emotional variations in preceding utterances, i.e. $C_{(VAD)}$, the personality-influenced emotional variations are generated by the sum of two different VAD vectors: the first VAD vector represents the initial emotions ($E_{i(VAD)}$) and the second VAD vector is the contextual-based emotional variation $C_{(VAD)}$ affected by the personality $P^l_{(VAD)}$ vectors (see Eq. 3).

$$\begin{aligned} T_{(VAD)} &= E_{i(VAD)} + P^l_{(VAD)} * C_{(VAD)} \\ T_{(VAD)} &= Tanh(T_{(VAD)}) \end{aligned} \tag{3}$$

where $T_{(VAD)}$ is the personality-based emotion variations, $E_{i(VAD)}$ is the initial emotions, and $C_{(VAD)}$ is the emotional variations due to the context.

Response Emotions: To generate the subjective emotional responses, the personality-based emotional variations $T_{(VAD)}$ is combined with the personality vector $P^l_{(VAD)}$ of a speaker and the preceding $C_{(t-1)}$ utterances. Lastly, we feed this concatenated vector to a linear layer to transform it into a probability distribution on the discrete emotion category. The output E_r is the response emotion that has the largest probability (see Eq. 4).

$$\begin{aligned} L_1 &= [T_{(VAD)}, P^l_{(VAD)}, C_{(t-1)}] \\ E_r &= OutputLayer(L_1) \end{aligned} \tag{4}$$

where L_1 is the concatenation of personality-based emotional variations, personality-based emotions, and contextual semantics of preceding utterances. E_r is the response emotions.

4 Experiments and Results

4.1 PELD Dataset

Personality EmotionLines Dataset (PELD) consists of dyadic conversations taken from a famous TV series named Friends. The authors only included the conversations with six speakers: Joey, Chandler, Phoebe, Monica, Rachel, and Ross. Dialogues and their emotional labels are taken from MELD [19] and the EmoryNLP [26] datasets. The personality information of each speaker is taken from the FriendsPersona dataset [27]. The number of dialogues per emotion class with respect to each speaker is mapped in Fig. 4.

4.2 Results

In our experiments, we first follow the simplest approach to emotion classification, i.e. predicting emotional responses into 3 valence classes (negative, neutral, and positive). Later, we raised the difficulty level and predict emotional responses in 7 categories (anger, disgust, fear, joy, neutral, sadness, and surprise). Both Single-Task (ST) and Multi-Task (MT) approaches were tested for predicting emotional responses into these 3 and 7 classes. The performance of each ST and MT was evaluated by the F-score for each emotional category and two aggregated evaluation measures: the macro average (m-avg) and the weighted average (w-avg) of the F-score values.

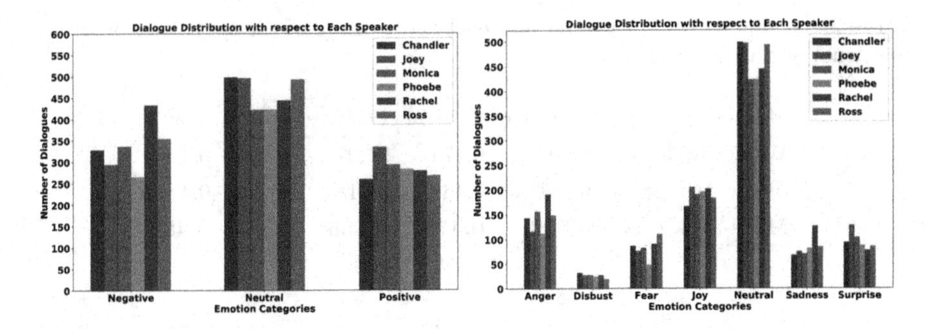

Fig. 4. The Left and Right histograms represent the number of dialogues with respect to each speaker in 3 and 7 emotion categories respectively.

Predicting Response Emotions with 3 Classes. We first group 7 emotions into 3 valence classes: positive, negative, and neutral. Specifically, anger, disgust, fear, and sadness emotions are considered as negative, whereas joy and surprise are considered as positive. The neutral category only considers neutral emotion.

In a Single-Task (ST) learning approach, a single model is trained with respect to each speaker. This means the trained model is only able to predict the emotional responses with respect to the speaker (i.e. subjective) whose personality traits are considered during training. Since the PELD dataset consists of conversations belonging to six different speakers, this is why six ST models were trained independently. Each model only considers a single speaker's personality traits during training. So by using Single-Task (ST) learning approach, when we need emotional responses with respect to a speaker (i.e. subjective), we need to train a separate model.

On the other hand, Multi-Task (MT) learning approach has multiple separate branches, i.e. one for each speaker. Each branch is designated to learn the speaker's specific emotional response. During training, each separate branch is reserved for a single personality speaker, but these separate branches are jointly trained. Therefore, by using Multi-Task (MT) learning approach, a single model needs to be trained for getting multiple subjective emotional responses. For comparison, the mean values of each evaluation measure of all six ST models were calculated. The comparison of ST, MT, and state-of-the-art [10] approaches is presented in Table 2. The results show that the Multi-Task(MT) approach has got considerable improvement with respect to the Single-Task (ST) approach and [10] for predicting the subjective emotional responses in 3 valence classes.

Predicting Response Emotions with 7 Classes. The same procedure previously described for training ST and MT has been followed for predicting the 7-class emotional responses. The only difference is that now the emotional responses of the next utterance are classified into 7 classes: anger, disgust, fear, joy, neutral, sadness, and surprise. Table 3 shows the comparison of ST and MT approaches with state-of-the-art [10]. The MT approach outperforms the

Table 2. Comparison of MT and ST approaches with state-of-the-art [10] for predicting emotional responses in 3 classes.

Method	Negative	Neutral	Positive	m-avg	w-avg
Wen et al. [10]	0.492	0.474	0.327	0.431	0.445
Single-Task (ST)	0.401	0.468	0.418	0.434	0.448
Multi-Task (MT)	**0.509**	**0.475**	**0.436**	**0.473**	**0.482**

majority of emotion categories. In particular, we observe the MT approach has obtained significant gains in the categories disgust, fear, joy, and sadness with respect to the ST and [10]. The m-avg of MT has obtained significant improvement, i.e. 10 points higher as compared to [10] and 9 points higher than ST approach. Furthermore, Multi-Task learning also helps to tackle the class imbalance problem. Figure 4 shows that the emotion classes disgust, fear, sadness, and surprise have a lower number of samples as compared to others (anger, neutral, and joy) and the results show that the MT approach gives benefits in learning emotional patterns for most of these classes.

Table 3. Comparison of MT and ST approaches with state-of-the-art [10] for predicting emotional responses in 7 classes.

Method	Anger	Disgust	Fear	Joy	Neutral	Sadness	Surprise	m-avg	w-avg
Wen et al. [10]	**0.320**	0.070	0.140	0.198	0.528	0.155	0.098	0.203	**0.424**
Single-Task (ST)	0.226	0.017	0.129	0.266	**0.533**	0.108	**0.242**	0.218	0.352
Multi-Task (MT)	0.279	**0.122**	**0.162**	**0.286**	0.531	**0.221**	0.163	**0.308**	0.413

5 Conclusion

This work is inspired by [10] to address the inference problem that considers the information of the speaker in order to predict the emotional responses to upcoming utterances. For the modeling approach, we are motivated by previous studies on MultiTask Learning (MTL), that show the benefit of jointly modeling related tasks. The main contribution of this work is to show that the MultiTask learning approach learns the underlying data better as compared to Single-Task. Concretely, we proposed two different types of approaches: Single-Task (ST) and Multi-Task (MT). In the ST approach, the model is only capable to predict single subjective emotional responses. On the other hand, Multi-Task (MT) approach has the the capability to predict multiple subjective emotional responses. The results show that joint modeling (MT) improves the generalization capacity of the model when compared to the ST approach. Although the results are promising, the proposed ST and MT approaches were tested on a small dataset (PELD). We consider that further research on larger datasets is necessary to be able of making stronger conclusions. Nonetheless, it is also important to notice that the results obtained in this study are aligned with previous works [28,29] that show the capacity of MT approaches to better model subjective perceptions or reactions in the context of affect-related tasks. However, despite these empirical evidences, considering individual annotations or user-specific information such as personality traits is still an uncommon practice in affect-related research. We hope the results provided by this study further encourage both to release individual annotations as well as more efforts on better understanding and modeling subjective expressions and perceptions.

Acknowledgements. This work was partially supported by the Spanish Ministry of Science, Innovation and Universities, RTI2018-095232-B-C22. We thank NVIDIA for their generous hardware donations.

References

1. Caldarini, G., Jaf, S., McGarry, K.: A literature survey of recent advances in chatbots. Information **13**(1), 41 (2022)
2. Dale, R.: The return of the chatbots. Nat. Lang. Eng. **22**(5), 811–817 (2016)
3. Maedche, A., et al.: Ai-based digital assistants. Bus. Inf. Syst. Eng. **61**(4), 535–544 (2019)
4. Zhou, H., Huang, M., Zhang, T., Zhu, X., Liu, B.: Emotional chatting machine: Emotional conversation generation with internal and external memory. In: Proceedings of the AAAI Conference on Artificial Intelligence, vol. 32 (2018)
5. Zhou, X., Wang, W.Y.: MojiTalk: generating emotional responses at scale. arXiv preprint arXiv:1711.04090 (2017)
6. Huber, B., McDuff, D., Brockett, C., Galley, M., Dolan, B.: Emotional dialogue generation using image-grounded language models. In: Proceedings of the 2018 CHI Conference on Human Factors in Computing Systems, pp. 1–12 (2018)
7. Huang, M., Zhu, X., Gao, J.: Challenges in building intelligent open-domain dialog systems. ACM Trans. Inf. Syst. (TOIS) **38**(3), 1–32 (2020)
8. Zhou, L., Gao, J., Li, D., Shum, H.Y.: The design and implementation of XiaoIce, an empathetic social chatbot. Comput. Linguist. **46**(1), 53–93 (2020)
9. Van der Zwaan, J.M., Dignum, V., Jonker, C.M.: A BDI dialogue agent for social support: specification and evaluation method. In: AAMAS 2012: Proceedings of the 11th International Conference on Autonomous Agents and Multiagent Systems, Workshop on Emotional and Empathic Agents, Valencia, Spain, 4–8 June 2012; authors version. International Foundation for Autonomous Agents and Multiagent Systems (IFAAMAS) (2012)
10. Wen, Z., Cao, J., Yang, R., Liu, S., Shen, J.: Automatically select emotion for response via personality-affected emotion transition. In: Findings of the Association for Computational Linguistics: ACL-IJCNLP 2021, pp. 5010–5020 (2021)
11. Zhang, S., Dinan, E., Urbanek, J., Szlam, A., Kiela, D., Weston, J.: Personalizing dialogue agents: i have a dog, do you have pets too? arXiv preprint arXiv:1801.07243 (2018)
12. Mazaré, P.E., Humeau, S., Raison, M., Bordes, A.: Training millions of personalized dialogue agents. arXiv preprint arXiv:1809.01984 (2018)

13. Madotto, A., Lin, Z., Wu, C.S., Fung, P.: Personalizing dialogue agents via meta-learning. In: Proceedings of the 57th Annual Meeting of the Association for Computational Linguistics, pp. 5454–5459 (2019)
14. Fung, P., et al.: Zara the SuperGirl: an empathetic personality recognition system. In: Proceedings of the 2016 Conference of the North American Chapter of the Association for Computational Linguistics: Demonstrations, pp. 87–91. ACL (2016)
15. Zhong, P., et al.: Endowing empathetic conversational models with personas. arXiv preprint arXiv:2004.12316 (2020)
16. Zandie, R., Mahoor, M.H.: EmpTransfo: a multi-head transformer architecture for creating empathetic dialog systems. In: The Thirty-Third International Flairs Conference (2020)
17. Li, Y., Su, H., Shen, X., Li, W., Cao, Z., Niu, S.: DailyDialog: a manually labelled multi-turn dialogue dataset. arXiv preprint arXiv:1710.03957 (2017)
18. Chen, S.Y., Hsu, C.C., Kuo, C.C., Ku, L.W., et al.: EmotionLines: an emotion corpus of multi-party conversations. arXiv preprint arXiv:1802.08379 (2018)
19. Poria, S., Hazarika, D., Majumder, N., Naik, G., Cambria, E., Mihalcea, R.: Meld: a multimodal multi-party dataset for emotion recognition in conversations. arXiv preprint arXiv:1810.02508 (2018)
20. Chatterjee, A., Gupta, U., Chinnakotla, M.K., Srikanth, R., Galley, M., Agrawal, P.: Understanding emotions in text using deep learning and big data. Comput. Hum. Behav. **93**, 309–317 (2019)
21. Mehrabian, A.: Pleasure-arousal-dominance: a general framework for describing and measuring individual differences in temperament. Curr. Psychol. **14**(4), 261–292 (1996)
22. Fox, A.S., Lapate, R.C., Shackman, A.J., Davidson, R.J.: The Nature of Emotion: Fundamental Questions. Oxford University Press, Oxford (2018)
23. Russell, J.A., Mehrabian, A.: Evidence for a three-factor theory of emotions. J. Res. Pers. **11**(3), 273–294 (1977)
24. Mehrabian, A.: Analysis of the big-five personality factors in terms of the pad temperament model. Aust. J. Psychol. **48**(2), 86–92 (1996)
25. Devlin, J., Chang, M.W., Lee, K., Toutanova, K.: BERT: pre-training of deep bidirectional transformers for language understanding. arXiv preprint arXiv:1810.04805 (2018)
26. Zahiri, S.M., Choi, J.D.: Emotion detection on tv show transcripts with sequence-based convolutional neural networks. In: Workshops at the thirty-second AAAI conference on artificial intelligence (2018)
27. Jiang, H., Zhang, X., Choi, J.D.: Automatic text-based personality recognition on monologues and multiparty dialogues using attentive networks and contextual embeddings (student abstract). In: Proceedings of the AAAI Conference on Artificial Intelligence, vol. 34, pp. 13821–13822 (2020)
28. Hayat, H., Ventura, C., Lapedriza, A.: Modeling subjective affect annotations with multi-task learning. Sensors **22**(14), 5245 (2022)
29. Jaques, N., Taylor, T.S., Nosakhare, N.E., Sano, S.A., Picard R,P.R.: Multi-task learning for predicting health, stress, and happiness. In: Neural Information Processing Systems (NeurIPS) Workshop on Machine Learning for Healthcare (2016)

Few-Shot Learning for Prediction of Electricity Consumption Patterns

Javier García-Sigüenza[1(✉)], José F. Vicent[1], Faraón Llorens-Largo[1], and José-Vicente Berná-Martínez[2]

[1] Department of Computer Science and Artificial Intelligence, University of Alicante, Campus de San Vicente, Ap. Correos 99, 03080 Alicante, Spain
{javierg.siguenza,jvicent,faraon.llorens}@ua.es

[2] Department of Computer Science and Technology, University of Alicante, Campus de San Vicente, Ap. Correos 99, 03080 Alicante, Spain
jvberna@ua.es

Abstract. Deep learning models have achieved extensive popularity due to their capability for providing an end-to-end solution. But, these models require training a massive amount of data, which is a challenging issue and not always enough data is available. In order to get around this problem, a few shot learning methods emerged with the aim to achieve a level of prediction based only on a small number of data. This paper proposes a few-shot learning approach that can successfully learn and predict the electricity consumption combining both the use of temporal and spatial data. Furthermore, to use all the available information, both spatial and temporal, models that combine the use of Recurrent Neural Networks and Graph Neural Networks have been used. Finally, with the objective of validate the approach, some experiments using electricity data of consumption of thirty-six buildings of the University of Alicante have been conducted.

Keywords: Few-shot learning · Graph neural networks · Electricity consumption · Pattern recognition

1 Introduction

Currently, the energy consumption of buildings represents more than one third of total energy consumption globally [1]. Therefore, reducing the energy consumption of buildings is important to achieve sustainable goals. Thus, the prediction of the energy consumption of a building helps to achieve better control of the energy system and improve energy utilization [2].

Generally, the predictive methods of building energy consumption can be divided into two: methods based on physical models and methods based on data. Methods based on physical models use the principles of physics to assess the

Supported by CENID (Centro de Inteligencia Digital) in the framework of the Agreement between the Diputación Provincial de Alicante and the University of Alicante.

A. Pertusa et al. (Eds.): IbPRIA 2023, LNCS 14062, pp. 705–715, 2023.
https://doi.org/10.1007/978-3-031-36616-1_56

energy consumption of the building, for instance, the calculation of the energy consumption of the heating or air conditioning system. When simulating energy consumption, these methods require detailed input information, such as physical characteristics of the building, air conditioning systems, or occupants' schedule. This input information helps achieve an accurate prediction of the building's energy consumption, but makes the process cumbersome and time consuming [3,4]. On the other hand, data-driven methods are usually divided into two categories: statistical methods and machine learning methods. Statistical methods are generally predictive models of regression analysis [5] or time series [6]. The regression analysis simulates and predicts energy consumption by establishing mathematical expressions between independent and dependent variables [7,8]. For instance, in [9] the authors evaluated the prediction of the energy consumption of a building by means of several models. The results of their research showed that the developed fuzzy systems and models based on neural networks obtained the best performance and accuracy indicators. Models based on time series are statistical methods for dynamic data processing. In [10] a time series prediction algorithm based on a hybrid system of neurofuzzy inference was proposed to predict the energy consumption of buildings. In this case, the results showed that this method is more robust than traditional adaptive neurofuzzy inference system models but its efficiency is not high.

In general, statistical methods have the advantage of a simple structure and relatively easy modeling. However, due to the complex interaction between the input elements, they are likely to suffer from low computational efficiency and low prediction accuracy.

In order to overcome the shortcomings of the methods based on physical models and statistical methods, machine learning methods have been developed [11]. These prediction methods for energy consumption in buildings have become a focus of research. As such, Artificial Neural Networks (ANNs) are utilized to forecast buildings energy consumption [12]. ANNs have many advantages over other methods in terms of their ability to solve complex nonlinear problems and resistance to failure [13]. Li and Yao [12] developed a framework to predict the energy consumption of residential and non-residential buildings through the generation of an energy database that used to train different machine learning models. In [14] the authors used a simulation-based technique to forecast the life cycle energy of residential buildings. D'Amico et al. [15] developed an ANN model to determine the energy performance of non-residential buildings.

In this paper, we propose the use of fine-tuning for the transfer of knowledge in order to be able to perform few-shot learning with the aim to obtain patterns of electrical consumption of buildings through predictive methods. For this purpose, we use electricity consumption data from 36 buildings of the University of Alicante in a period of six months.

2 Preliminaries

This section describes the data processing performed, as well as the metrics used to validate the accuracy obtained with the models.

2.1 Dataset

The dataset is composed of a time series, which covers 6 months of sensorization, in periods of 1h, of the electricity consumption, measured in kWh, of 36 buildings of the University of Alicante. This dataset has been obtained through the Smart University project, which is a system that integrates and centralizes all the information coming from the different types of sensing devices that the University may have.

First, data cleaning, preprocessing, and transformation have been performed. For this purpose, outliers have been eliminated, since they can have a critical effect on the performance of the network, preventing it from finding relevant characteristics and the effect of the relationships between nodes. Once the data has been cleaned, the dataset has been generated as a graph, encoding the information as a time sequence representing the evolution of the nodes through the different time steps.

In addition, the hour, day of the week and month information has been encoded for each of the time steps. To encode this information, instead of using one-hot encoding, an encoding based on sine and cosine has been used. For this purpose, each of the time variables has been divided into a value based on sine and cosine [16] according to the following equations:

$$h_{sin} = \sin\left(\frac{2\pi h}{24}\right) \quad (1) \qquad h_{cos} = \cos\left(\frac{2\pi h}{24}\right) \quad (2)$$

$$w_{sin} = \sin\left(\frac{2\pi w}{7}\right) \quad (3) \qquad w_{cos} = \cos\left(\frac{2\pi w}{7}\right) \quad (4)$$

$$m_{sin} = \sin\left(\frac{2\pi m}{12}\right) \quad (5) \qquad m_{cos} = \cos\left(\frac{2\pi m}{12}\right) \quad (6)$$

The Eqs. 1 and 2 refer to the hour encoding based on sine and cosine, the Eqs. 3 and 4 encode the day of the week and the Eqs. 5 and 6 encode the month. The variables h, w and m encode the time, day of the week and month respectively, being encoded to start their values at 0 in all the three cases.

Once the data were generated, they were divided into two datasets, the first one, composed of 6 months of data and 30 nodes, and the second one, also composed of 6 months of data but 6 nodes. The first dataset has been used to train a model that is used as a baseline for fine-tuning, while the second one has been used to train two more models, one has been trained using fine-tuning and another has been trained from scratch, using between 1 and 4 weeks for training, one week for validation and the remaining weeks for testing. In this way the second dataset will be used to simulate a scenario where the amount of data is limited in time and to test the improvement that the transfer of information from the baseline model can bring.

2.2 Precision Metrics

The root mean square error (RMSE) was used to compare the results, since it allows the model to give greater importance to larger failures. The RMSE is defined as:

$$\mathrm{RMSE} = \sqrt{\frac{1}{N}\sum_{i=1}^{N}(y_i - \hat{y}_i)^2}, \tag{7}$$

where N is the number of values, y_i represents the true value and $\hat{y}_i$ the predicted value.

3 Definition of the Problem

The prediction of electricity consumption in *smart cities* combines the use of both temporal and spatial data. For its exploitation and prediction, there are both models that only work with the temporal element, and others that try to make use of both the spatial and temporal components of the problem. This aspect is essential when selecting the model to be used to predict electricity consumption, since those models that only use temporal information are losing a large part of the data available to make the prediction. To address the problem, both models belonging to the field of statistical methods and to the field of classical machine learning and deep learning have been explored.

3.1 Models

In the field of statistical methods we can find the *Auto-Regressive Integrated Moving Average* (ARIMA) with Kalman filter [17] and *Vector Auto-regressive (VAR)* models. [18]. Within the field of classical machine learning we have *Linear Support Vector Regression* (SVR) [19], and in the field of deep learning there are a wide variety of architectures for the models. Among all the models currently available, the most accurate and state-of-the-art are those in the field of deep learning.

When exploring deep learning models, we come across architectures based on recurrent neural networks (RNNs), which are capable of detecting patterns in Euclidean spaces, being able to treat and predict temporal sequences. However, they are not capable of processing and representing spatial information and the relationship between them. To solve this problem, it is necessary to use models that are capable of processing and predicting data represented through non-Euclidean spaces, making use of graphs. This change in data representation implies a change in the structure of the information that is problematic for many deep learning models. For example, the possibility of different nodes in the graph having a different number of neighbors is problematic for architectures based on convolutional neural networks (CNNs). To address these challenges, *Graph Neural Networks* (GNNs) [20] were proposed. Therefore, in order to use all the available information, both spatial and temporal, models combining the use of RNNs and GNNs have been developed.

3.2 Graph Neural Networks

Among the different models that combine RNNs and GNNs, *Adaptive Graph Convolutional Recurrent Network* (AGCRN) [21], a deep learning model that uses both temporal and spatial information of the problem, stands out, combining the architecture of *Graph Convolutional Network* (GCN), a subtype of GNN [22], with *Gated recurrent unit* (GRU) [23], a temporal layer used by RNNs. This model proposes an alternative to other architectures also based on GNNs, which detect common patterns between nodes with the help of predefined graphs. Instead, AGCRN seeks to learn node-specific patterns and interdependencies between nodes during the training of the model itself.

To achieve these objectives, two characteristic modules of AGCRN are proposed, which modify the original implementation of GCN. The first one, *Node Adaptive Parameter Learning (NAPL)*, seeks to capture the specific patterns of each node, and the second one, *Data Adaptive Graph Generation (DAGG)*, focuses on inferring the interdependencies between nodes, in order to obtain information about the existing spatial relationships. Finally, using the modified implementation of GCN, for spatial information, and GRU, for temporal information, the *Adaptive Graph Convolutional Recurrent Network (AGCRN)* layer is created, which gives its name to the model itself.

By adapting the AGCRN model to electricity consumption data, we obtain a model that is capable of taking into account both the distribution of consumption in the different nodes of an electrical installation and the evolution of consumption in each of these nodes, using both the spatial and temporal information available.

3.3 Embeddings and Weight Pools

Embeddings are a low-dimensional vector representation that captures the characteristics of an element in a continuous and dense space. In this case the embeddings are applied to define the nodes, allowing to capture characteristics of these during the training of the model, based on the data on which the model is being trained.

These embeddings are unique to each node, so they cannot be used directly to characterize new nodes. However, they are especially relevant for knowledge transfer, since AGCRN in the NAPL module uses a pool of weights for filter generation to detect node-specific patterns in each of its layers. The NAPL filter generation pool, by allowing to generate customized filters for each node from its embeddings, allows that if new embeddings are learned for new nodes but the filter generation pool is maintained, a latent space trained to determine the best filters for each node is obtained, being able to extrapolate this knowledge to new nodes that the model has never seen before.

To obtain this latent space it is necessary to apply NAPL to the convolution performed by GCN, which is defined as:

$$Z = (I_n + D^{-\frac{1}{2}} A D^{-\frac{1}{2}}) X \Theta + b \tag{8}$$

where:

$Z \in R^{n \times f}$ = Output of the GCN.
$I_n \in R^{n \times n}$ = Identity matrix of size $n \times n$.
$D \in R^{n \times n}$ = Degree matrix.
$A \in R^{n \times n}$ = Adjacency matrix of G
$X \in R^{n \times c}$ = Input of the GCN layer.
$\Theta \in R^{c \times f}$ = Weights of the GCN Layer.
$b \in R^{f}$ = Bias of the GCN layer.
n = Number of nodes.
c = Number of input variables.
f = Number of output variables.

The NAPL module is applied to the Eq. 8 so that it redefines Θ and b as:

$$\Theta = E_G \cdot W_G \tag{9}$$

$$b = E_G \cdot b_G \tag{10}$$

where $E_G \in R^{n \times d}$ represents the matrix of the embeddings of the nodes, $W_G \in R^{d \times c \times f}$ is the weight pool containing the latent space for the generation of filters, $b_G \in R^f$ is a weight pool for bias generation and d is the size of the embeddings.

In order to transfer the information between models, W_G is used, so its parameters must be frozen while the rest of the model is trained to learn the embeddings of the nodes and adjust the other parameters of the model. To achieve this goal, the fine-tuning technique has been used.

Fine-tuning is a technique used to adapt the knowledge of pre-trained neural network models on a different dataset to new tasks by reusing part of the pre-trained model weights and adjusting them to the new data. To perform the fine-tuning, a set of layers is selected whose parameters will not be modified, while the rest of the parameters will be updated during the training of the new model.

When fine-tuning is applied to GNNs, usually the aim is to learn a Θ capable to extract information that can be used to make a prediction on a new dataset. However, in this case we have sought to learn W_G instead, thus being able to generate filters for each node based on the value of E_G.

4 Methodology

In this section we present the different steps performed to train both the model that will be used as the baseline for the knowledge transfer and the two models trained on the dataset with limited information, one model trained using fine-tuning and another trained from scratch.

4.1 Training Standard and Fine-Tuned Model

First, one model, based on AGCRN, has been trained on the first dataset, consisting of 6 months of data and 30 nodes. This model is referred to as *baseline model*, since it is the model that has been trained on a larger dataset, being able to learn a latent space containing information about the different electricity consumption patterns that different buildings may have.

Using the baseline model, a new model has been trained using fine-tuning, freezing the parameters of W_G, which represents the latent space for the generation of customized filters for the nodes, and reinitializing the weights of the final output layer, in order to facilitate the last layer to learn how to generate the prediction. The rest of the parameters have been trained keeping the baseline model parameters as the initial state. In this way, the model can learn the embeddings that represent the building characteristics and the rest of the model parameters, but it cannot change the existing information in the latent space of the weight pool, since the model is instructed that when backpropagation is performed, the parameters belonging to W_G should not be updated.

In order to compare the results of applying fine-tuning, a third model has been trained on the dataset of limited information, being this one trained from scratch, which we refer to as *standard model*. Therefore, this third model does not have any restriction when performing backpropagation nor does it use information from the baseline model, allowing to compare whether the use of the latent space of the baseline model with fine-tuning is an improvement when performing few-shot learning for the prediction of electricity consumption patterns. Both models have been trained for 500 epochs, using RAdam optimizer [24], with a learning rate of 0.001.

4.2 Generating Test Splits

To measure the impact of the amount of data used for training, as well as the evolution of model accuracy as the time difference between the dataset used for training and testing increases, the limited information dataset has been generated in 4 different ways. These datasets set 1 to 4 weeks of data for training, 1 week for validation and the following 20 weeks to measure the accuracy on test partitions grouped by weeks, thus allowing to test the accuracy week by week, checking the effect of the temporal distance on the accuracy.

5 Experimental Results

To compare the accuracy of the models when fine-tuning is applied and when the model is trained from scratch, a series of tests have been performed, using between 1 and 4 weeks of data for training and then showing the RMSE over 20 test sets grouping the data in one-week periods.

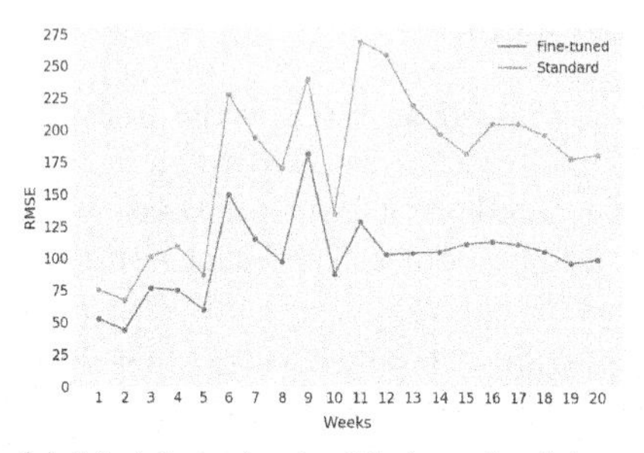

(a) Models trained with 1 week of data.

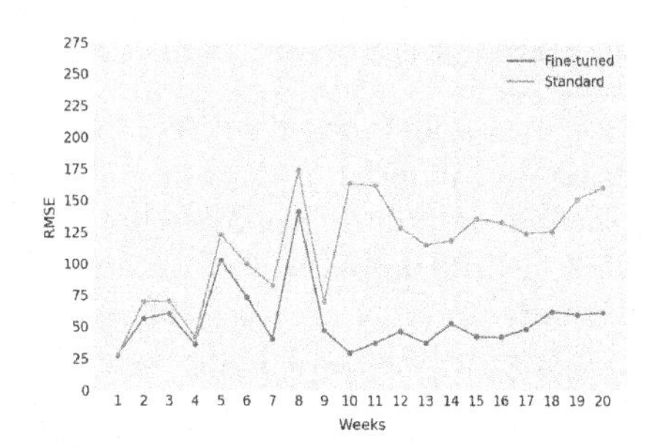

(b) Models trained with 2 weeks of data.

Fig. 1. Evolution of the RMSE of the standard and fine-tuned model with respect to the time difference between the train and val dataset and the test dataset with models trained with 1 and 2 weeks of data.

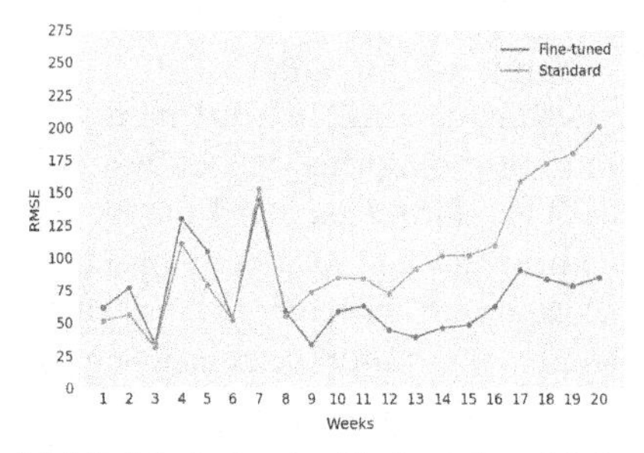

(a) Models trained with 3 weeks of data.

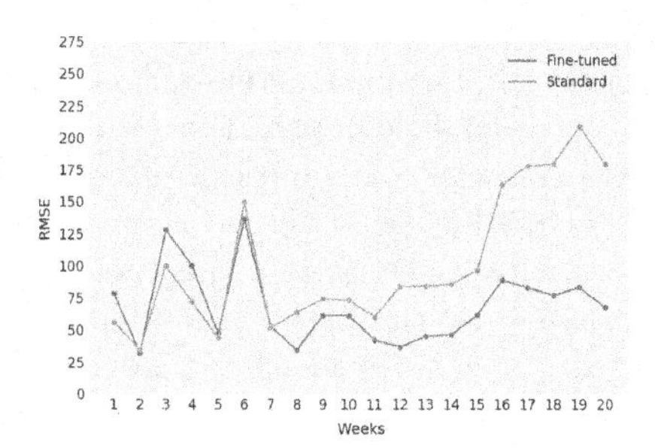

(b) Models trained with 4 weeks of data.

Fig. 2. Evolution of the RMSE of the standard and fine-tuned model with respect to the time difference between the train and val dataset and the test dataset with models trained with 3 and 4 weeks of data.

In Fig. 1a it can be seen how for the case of the two models trained with one week of data, the model with fine-tuning has a higher accuracy in all cases, being the difference greater as the period to be predicted is more distant in time with respect to the date of the training data. In the case of Fig. 1b the difference between the two models is smaller, being the accuracy of both similar until week 9, where the difference between both models increases, being remarkable the improvement of the model with fine-tuning with respect to the standard model from this week on.

In Fig. 2a a similar behavior to that observed in Fig. 1b can be appreciated, but the difference between both models from week 9 onwards is not so pronounced. In the case of the Fig. 2b the same trend as in Fig. 2a can be observed, so the difference between the accuracy of the two models in the long term continues decreasing, although the fine-tuned model continues being better as the date of the data to be predicted becomes more distant, although in the nearby dates the standard model improves slightly in accuracy with respect to the fine-tuned model, since the standard model has been trained in consumption patterns of nearby dates.

Therefore, by checking Figs. 1 and 2 it can be seen that the use of fine-tuning allows the model to generalize better when performing the prediction, and to take advantage of the existing information in the latent space to extract knowledge that cannot be extracted from the dataset due to the limited information it contains.

This difference is greater the less information is available for training, being especially difficult for the standard model to predict trends with 1 week of data. Figure 3 shows the prediction performed by the fine-tuned model (Fig. 3a) and the prediction realized by the model trained from scratch (Fig. 3b), together with the ground truth of the electricity consumption of the second week, with both models having been trained with one week of data. This comparative shows how the fine-tuned model is able to capture the power consumption patterns while the standard model cannot, despite sharing the model architecture and having been trained on the same dataset.

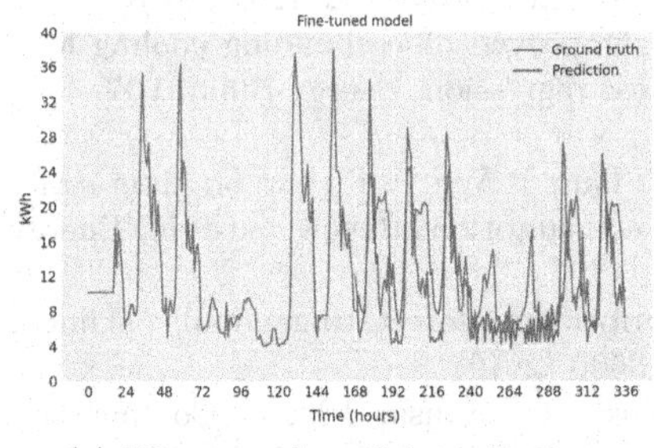

(a) Fine-tuned model prediction.

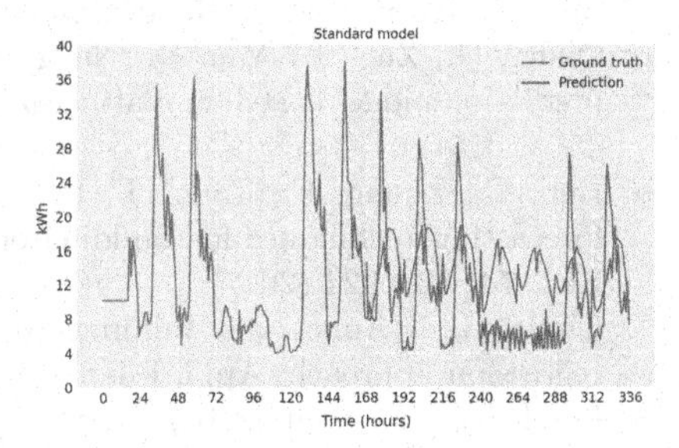

(b) Standard model prediction.

Fig. 3. Comparison between the prediction made by the fine-tuned and standard models trained with one week of data.

6 Conclusions

In this paper we have sought to train a model to detect patterns of electricity consumption with a limited dataset, with which the detection of patterns without previous information would be complex, thus performing a few-shot learning.

In order to perform few-show learning and face this problem, fine-tuning has been used, so that through the use of the information in a latent space of a model previously trained on another dataset, patterns can be detected without this information being available in the dataset itself.

In this work three models have been trained, one used as baseline, and other two used to predict on a dataset of limited information, training one using fine-tuning and the other from scratch, and then comparing the results of both. In this way, it has been experimentally verified that the use of the latent space information of the baseline model helps to generalize the behavior of electricity consumption patterns with only one week of data, allowing to improve the accuracy of the models trained on the limited information dataset.

Therefore, the use of fine-tuning to transfer the knowledge in a latent space can be useful for contexts in which the amount of information is limited, allowing a higher accuracy than could be obtained only by making use of the information available in the training data.

References

1. Zhong, H., Wang, J., Jia, H., Mu, Y., Lv, S.: Vector field-based support vector regression for building energy consumption prediction. Appl. Energy **242**, 403–414 (2019)
2. Li, C., Ding, Z., Zhao, D., Yi, J., Zhang, G.: Building energy consumption prediction: an extreme deep learning approach. Energies **10**(10) (2017)
3. Li, X., Yao, R.: Modelling heating and cooling energy demand for building stock using a hybrid approach. Energy Build. **235**, 110740 (2021)
4. Amasyali, K., El-Gohary, N.M.: A review of data-driven building energy consumption prediction studies. Renew. Sustain. Energy Rev. **81**, 1192–1205 (2018)
5. Qiang, G., Zhe, T., Yan, D., Neng, Z.: An improved office building cooling load prediction model based on multivariable linear regression. Energy Build. **107**, 445–455 (2015)
6. Deb, C., Zhang, F., Yang, J., Lee, S.E., Shah, K.W.: A review on time series forecasting techniques for building energy consumption. Renew. Sustain. Energy Rev. **74**, 902–924 (2017)
7. Ciulla, G., D'Amico, A.: Building energy performance forecasting: a multiple linear regression approach. Appl. Energy **253**, 113500 (2019)
8. Pombeiro, H., Santos, R., Carreira, P., Silva, C., Sousa, J.M.C.: Comparative assessment of low-complexity models to predict electricity consumption in an institutional building: linear regression vs. fuzzy modeling vs. neural networks. Energy Build. **146**, 141–151 (2017)
9. Jallal, M.A., González-Vidal, A., Skarmeta, A.F., Chabaa, S., Zeroual, A.: A hybrid neuro-fuzzy inference system-based algorithm for time series forecasting applied to energy consumption prediction. Appl. Energy **268**, 114977 (2020)
10. Li, Z., Han, Y., Xu, P.: Methods for benchmarking building energy consumption against its past or intended performance: an overview. Appl. Energy **124**, 325–334 (2014)
11. Braun, J.E., Chaturvedi, N.: An inverse gray-box model for transient building load prediction. HVAC&R Res. **8**(1), 73–99 (2002)
12. Chen, H., Li, B.L.Z., Dai, J.: An ANN-based fast building energy consumption prediction method for complex architectural form at the early design stage. Build. Simul. **12**, 665–681 (2019)

13. Zhao, H.X., Magoulès, F.: A review on the prediction of building energy consumption. Renew. Sustain. Energy Rev. **16**(6), 3586–3592 (2012)
14. Zou, Y., Xiang, K., Zhan, Q., Li, Z.: A simulation-based method to predict the life cycle energy performance of residential buildings in different climate zones of China. Build. Environ. **193**, 107663 (2021)
15. D'Amico, A., Ciulla, G., Traverso, M., Lo Brano, V., Palumbo, E.: Artificial neural networks to assess energy and environmental performance of buildings: an Italian case study. J. Clean. Prod. **239**, 117993 (2019)
16. Mahajan, T., Singh, G., Bruns, G., Bruns, G., Mahajan, T., Singh, G.: An experimental assessment of treatments for cyclical data. In: Proceedings of the 2021 Computer Science Conference for CSU Undergraduates, Virtual, vol. 6 (2021)
17. Lippi, M., Bertini, M., Frasconi, P.: Short-term traffic flow forecasting: an experimental comparison of time-series analysis and supervised learning. IEEE Trans. Intell. Transp. Syst. **14**(2), 871–882 (2013)
18. Hamilton, J.D.: Time Series Analysis. Princeton University Press, Princeton (2020)
19. Drucker, H., Burges, C.J., Kaufman, L., Smola, A., Vapnik, V.: Support vector regression machines. In: Advances in Neural Information Processing Systems, vol. 9 (1996)
20. Gori, M., Monfardini, G., Scarselli, F.: A new model for learning in graph domains. In: Proceedings of the 2005 IEEE International Joint Conference on Neural Networks, vol. 2, pp. 729–734 (2005)
21. Bai, L., Yao, L., Li, C., Wang, X., Wang, C.: Adaptive graph convolutional recurrent network for traffic forecasting. CoRR, abs/2007.02842 (2020)
22. Kipf, T.N., Welling, M.: Semi-supervised classification with graph convolutional networks. CoRR, abs/1609.02907 (2016)
23. Cho, K., van Merrienboer, B., Bahdanau, D., Bengio, Y.: On the properties of neural machine translation: encoder-decoder approaches. CoRR, abs/1409.1259 (2014)
24. Liu, L., et al.: On the variance of the adaptive learning rate and beyond. In: Proceedings of the Eighth International Conference on Learning Representations (ICLR 2020) (2020)

Author Index

A
Achaibou, Amina 323
Alfaro-Contreras, María 108, 146
Almeida, Ana 28
Almutairi, Khleef 54
Alonso, Carlos 121
Álvarez-Erviti, Lydia 593
Anitei, Dan 426
Atlas, James 351
Ayllon, Eric 170

B
Baggetto, Pablo 654
Balsa-Castro, Carlos 557
Batchelor, Oliver 351
Becerra, David 121
Benedí, José Miguel 426
Berenguer, Abel Díaz 639
Bermudez-Cameo, Jesus 288
Bermudez-Vargas, James 288
Berná-Martínez, José-Vicente 705
Berral-Soler, Rafael 199
Bhilare, Shruti 617
Bimbo, Alberto Del 238
Brás, Susana 28, 680
Brea, Víctor M. 238
Breznik, Eva 134
Bribiesca, Ernesto 363
Bribiesca-Sánchez, Andrés 363
Brito, José Henrique 442

C
Calpe, Javier 323
Calvo-Zaragoza, Jorge 15, 108, 146, 158, 170
Campilho, Aurélio 520
Cardot, Hubert 40
Carlos, Estefanía 593
Carreira, María J. 545, 557
Carrión, Salvador 94
Casacuberta, Francisco 94
Casado-García, Ángela 593
Castellanos, Francisco J. 158, 170
Castro, Francisco M. 389, 466
Chanda, Sukalpa 182
Cores, Daniel 238, 545
Corkidi, Gabriel 363
Costa, Dinis 82
Costa, Joana 82
Cózar, Julián R. 389
Cruz, Ricardo 276
Cubero, Nicolás 389
Cuesta-Infante, Alfredo 375, 569

D
Dansoko, Makan 54
Darszon, Alberto 363
De Ketelaere, B. 249
Dehaeck, S. 249
Dobbs, Harry 351
Domingues, Inês 533
Domínguez, César 593

E
Echeverry, Luis M. 603

F
Fatjó-Vilas, Mar 603
Fierrez, Julian 629
Fortea, Juan 603

G
Galán-Cuenca, Alejandro 507
Gallego, Antonio Javier 15, 158, 507, 581
García-Ruiz, Pablo 454
García-Sigüenza, Javier 705
Garrido-Munoz, Carlos 108
Gomes Faria, Ricardo 442
Gonçalves, Tiago 276
González, Alejandro 603
González-Barrachina, Pedro 146

A. Pertusa et al. (Eds.): IbPRIA 2023, LNCS 14062, pp. 717–719, 2023.
https://doi.org/10.1007/978-3-031-36616-1

Gonzàlez-Colom, Rubèn 603
Gordienko, Yuri 490
Gouveia, Sónia 680
Green, Richard 351
Guerrero, Jose J. 288
Guerrero-Mosquera, Carlos 603
Guil, Nicolás 389, 466
Gurav, Aniket 182
Guzmán, Adolfo 363

H
Hati, Avik 617
Hayat, Hassan 693
Heil, Raphaela 134
Henriques, Beatriz 680
Heras, Jónathan 593
Heredia-Lidón, Álvaro 603
Hernández-García, Sergio 569
Hostalet, Noemí 603

I
Izco, María 593

J
Jensen, Joakim 182
Jervan, Gert 490
Jiménez-Velasco, Isabel 402

K
Krishnan, Narayanan C. 182
Kühnberger, Kai-Uwe 665

L
Lacharme, Guillaume 40
Lapedriza, Agata 693
Larriba, Antonio M. 654
Latorre-Carmona, Pedro 54
Lenté, Christophe 40
Li, Andrew 479
Liao, W. 249
Lindblad, Joakim 261
Liu, Jiahe 479
Liu, Olivia 479
Llorens-Largo, Faraón 705
López, Damián 654

M
Magalhães, João 67
María-Arribas, David 375
Marín-Jiménez, Manuel J. 199, 389, 402, 454
Marques, Ricardo 3
Martínez-Abadías, Neus 603
Martinez-Cañete, Yadisbel 639
Martinez-Esteso, Juan P. 158
Mata, Eloy 593
Maurício, José 533
Medina-Carnicer, Rafael 199, 454
Melzi, Pietro 629
Mendonça, Ana Maria 520
Mirón, Miguel 507
Monmarché, Nicolas 40
Montemayor, Antonio S. 569
Morales, Aythami 629
Moravec, Jaroslav 336
Moreno, Plinio 223
Morillas, Samuel 54
Mucientes, Manuel 238, 545
Muñoz-Salinas, Rafael 199, 402, 454

N
Nagarajan, Bhalaji 3
Nieto-Hidalgo, Mario 146

O
Oliveira, Hélder P. 211, 312
Oliveira, Hugo S. 312
Oliveira, Lino 211

P
Pantrigo, Juan J. 375
Pape, Dennis 665
Pardo, Xosé M. 300
Paredes, Roberto 414
Parga, César D. 300
Parres, Daniel 414
Pascual, Vico 593
Patel, Umang 617
Paula, Beatriz 223
Pedrosa, João 520
Peña-Reyes, Carlos 557
Penarrubia, Carlos 15
Pertusa, Antonio 507
Pinto, Filipe Cabral 28
Pla, Filiberto 323
Pobandt, Tobias 665
Polukhin, Andrii 490

Pomarol-Clotet, Edith 603
Prieto, Jose Ramón 121

R
Radeva, Petia 3
Regueiro, Carlos V. 300
Ribeiro, Bernardete 82
Ribeiro, Pedro P. 312
Rocamora-García, Pablo 581
Rodriguez-Albala, Juan Manuel 629
Rosello, Adrian 158
Ruiz-Barroso, Paula 466

S
Sahli, Hichem 639
Salvador, Raymond 603
Sánchez, Joan Andreu 426
Sanmartín-Vich, Nofre 323
Santos Marques, Miguel 442
Santos, Rui 520
Šára, Radim 336
Sargento, Susana 28
Saval-Calvo, Marcelo 507, 581
Schmidt, Jonas 665
Seidenari, Lorenzo 238
Serrano e Silva, Pedro 276
Sevillano, Xavier 603
Shihavuddin, A. S. M. 276
Silva, Catarina 82
Sladoje, Nataša 261
Stirenko, Sergii 490

T
Tatjer, Albert 3
Tolosana, Ruben 629
Tomás, Inmaculada 557
Toselli, Alejandro Hector 121

V
Valério, Rodrigo 67
Valero-Mas, Jose J. 15
Van Belleghem, R. 249
Vázquez-González, Lara 557
Ventura, Carles 693
Vera-Rodriguez, Ruben 629
Vicent, José F. 705
Victoriano, Margarida 211
Vidal, Enrique 121
Vila-Blanco, Nicolás 545
Vilariño, Gabriel 300
Villena-Martinez, Victor 581

W
Wang, Xiaodi 479
Wetzer, Elisabeth 261
Wouters, N. 249

Printed in the United States
by Baker & Taylor Publisher Services

Printed in the United States
by Baker & Taylor Publisher Services